Green Energy and Technology

More information about this series at http://www.springer.com/series/8059

Gerasimos Rigatos

Intelligent Renewable Energy Systems

Modelling and Control

Springer

Gerasimos Rigatos
Unit of Industrial Automation
Industrial Systems Institute
Rion Patras
Greece

ISSN 1865-3529 ISSN 1865-3537 (electronic)
Green Energy and Technology
ISBN 978-3-319-39154-0 ISBN 978-3-319-39156-4 (eBook)
DOI 10.1007/978-3-319-39156-4

Library of Congress Control Number: 2016943390

Printed on acid-free paper

This Springer imprint is published by Springer Nature
The registered company is Springer International Publishing AG Switzerland

To Elektra

Foreword

Renewable energy systems become progressively a significant part of the electric power network worldwide. During the last years there has been a raise in the use of renewable energy sources, in the form of hydroelectric systems, wind power systems, photovoltaic units, marine power systems, fuel cells, biomass systems, etc. It is necessary renewable energy sources to become more reliable and more profitable. This will allow countries to reduce their dependence on energy coming from fossil fuels or natural gas while it will also contribute to the elimination of the negative environmental effects of thermal power stations.

To deploy the exploitation of renewable energy sources, technologies of the smart grid have gained particular importance. At a first stage one has to solve in an efficient and intelligent manner problems associated with the connection of renewable energy sources to the electricity grid, as for instance perturbations in the quality of electric power or synchronization and stability issues of distributed power generators. Thanks to information technology, one can better integrate renewable energy sources into the electricity grid.

The monograph focuses on the modelling and control of renewable energy sources and on technologies which allow for the more efficient integration of such sources in the electricity grid. The monograph presents recent results and methods which assure that the production of electric power coming from renewable energy systems, will be free of interruptions, that power losses will be minimized while also the generators functioning will be robust to adverse operating conditions (for example to faults of the electricity grid). The monograph analyses advanced technologies for renewable energy systems, as for example:

(i) intelligent control for power generators (synchronous and asynchronous) finding application in the exploitation of renewable energy sources (control of the rotors speed of power generators, control of the magnetic flux and control of the produced active and reactive power).

(ii) intelligent control of power electronics which are used for the connection of renewable energy sources to the electricity grid (control of AC to DC converters, control of inverters, control of distortions in voltage harmonics with

the use of active filters and control of high voltage transmission lines of the HVDC type).
(iii) optimization methods for the synchronization and stabilization of distributed power generators (interconnected generators of the PMSG or DFIG type, DC electric power units such as photovoltaics, optimal planning and management of electric power production coming from renewable energy sources).
(iv) Computational intelligence techniques for the financial exploitation of the electric power that is produced by renewable energy sources, in the framework of open and interconnected electricity markets.

The monograph contains new results and findings on control and estimation problems for renewable energy sources within a smart grid context. Actually, it presents new nonlinear control and estimation methods, which are based on the exact linearization of the dynamics of renewable power sources. Such methods make use of differential flatness theory. These methods are shown to have specific advantages in terms of accuracy of estimation and computation speed comparing to other control and filtering methods. Moreover, the monograph analyses approximate linearization methods for renewable energy sources based on H-infinity control theory. Such methods succeed solution of control and estimation problems for the electric machines and the power electronics constituting the renewable energy systems, without the need for computing linearizing transformations. The latter methods, yet being computationally and conceptually simple succeed robustness for the control and estimation schemes of renewable energy sources. Finally, the monograph analyses methods for control of renewable energy systems which are based on the definition and minimization of suitable Lyapunov functions. Such methods can be completely model-free, and the lack of knowledge about the dynamics of the renewable energy systems can be compensated by adaptive learning schemes of proven convergence and stability.

The content of the monograph can be used for teaching undergraduate or postgraduate courses in renewable energy systems. Therefore it can be used by both academic tutors and students as a reference book for such a course. The monograph's methods and findings on renewable energy systems can be also of interest for the engineering community. The nonlinear control and estimation methods analysed in the monograph can be a powerful tool and useful companion for engineers working on practical problems of renewable energy systems.

Athens, Greece
June 2016

Gerasimos Rigatos

Preface

The use of renewable energy systems, such as wind power, hydropower, tidal power, solar power, geothermal power and biomass burn is growing. Research in electric power generation from renewable sources is continuously expanding and stands for an area of high technological and financial importance. The implementation of new technologies for the functioning and management of renewable energy systems will help to further develop the renewable energy sector. The application of advanced scientific methods (for the modelling, identification and control of the dynamics of renewable energy systems and for the management of the electric power that is generated by them) can improve the reliability and efficiency of renewable power generation units. To this end, the present monograph analyses newer methods in the design of renewable energy systems based on intelligent information processing.

The monograph gives emphasis to the following key topics and technologies for renewable energy systems: (i) intelligent control of power generators (synchronous or asynchronous) used in renewable energy systems (control of rotational motion, electromagnetic field, as well as control of active and reactive power), (ii) intelligent control of the power electronics that connect renewable power generation units to the grid (control of AC to DC voltage source converters, control of inverters, AC lines control through active power filters, control of HVDC lines), (iii) synchronization and stability of distributed power generation units (interconnected PMSGs or DFIGs), as well as power planning and management in distributed renewable energy sources, (iv) fault diagnosis for power generators and power electronics used for producing power from renewable sources, as well as condition monitoring of the associated transmission and distribution grid.

The monograph comprises ten chapters.

Chapter 1 overviews the basic components of renewable energy systems, such as (i) power generators (PMSG, DFIG, multi-phase machines, and Doubly Fed Reluctance machines), (ii) power electronics for connecting generators to the grid (DC to AC inverters, AC to DC and DC to DC converters, active power filters, power transformers. First, doubly fed induction generators (DFIGs) are analysed.

The concept of field orientation is explained and the complete 6th order model of the asynchronous machine is introduced. Next, the model of synchronous generators is analysed. Additionally, more elaborated schemes for power generation from renewable sources arise in the case of using power generators of multiple phases (windings) at the stator. Thus, the model of multi-phase synchronous generators and doubly fed reluctance generators is introduced. Next, the dynamics of power electronics used for connecting renewable energy sources to the grid is explained. To this end the model of three-phase AC to DC converters, of inverters (DC to AC converters), of DC to DC converters and of HVDC transmission lines is given. Moreover, the dynamic model of electrochemical power generation devices, such as fuel cells, is explained. Additionally, the distributed parameters (PDE) model of power storage devices, such as Li–ion batteries is explained.

Chapter 2 analyses nonlinear control for doubly fed induction generators through different approaches: (i) flatness-based control in cascading loops, (ii) flatness-based control through transformation to the canonical form, (iii) differential geometry-based control, (iv) nonlinear H-infinity control, (v) flatness-based adaptive control (vi) field orientation and backstepping control.

First, the chapter proposes a solution to the problem of sensorless control of doubly fed induction generators, which is based on differential flatness theory and which is implemented in successive loops. Next, the chapter studies differential flatness properties and an input–output linearization procedure for control of doubly fed induction generators. Additionally, the chapter introduces a new method for feedback control of asynchronous electrical machines, which consists of a repetitive solution of an H-infinity control problem for the DFIG. The method makes use of a locally linearized model of the generator and is based on the solution of an algebraic Riccati equation at each iteration of the control algorithm. Finally, the chapter proposes an adaptive control approach, being based on differential flatness theory, that is capable of compensating for model uncertainty and parametric changes of the DFIG, as well as for lack of measurements for the DFIG's state vector elements. The stability of the adaptive control method is proven through Lyapunov analysis.

Chapter 3 analyses different approaches for nonlinear control of synchronous generators: (i) flatness-based control through transformation to the canonical form, (ii) flatness-based control in cascading loops, (iii) differential geometry-based control, (iv) nonlinear H-infinity control, (v) stabilizing control of synchronous power generators in the s-frequency domain. First, the chapter is concerned with proving differential flatness of the synchronous generator. Using differential flatness properties a sensorless control scheme for the electric machine is developed. Next, the chapter proposes a new method for the control of nonlinear dynamics of synchronous generators which is based on demonstrating that differential flatness properties hold for the subsystems which are obtained from the per-row decomposition of the machine's state-space models into subsystems. The controller's design is implemented in successive loops, while asymptotic stability of the control scheme is analytically proven. Finally, the chapter introduces an s-frequency domain method for the stabilization of the synchronous generators connected to the electricity grid. The method is based on interval polynomials theory and on frequency response analysis techniques (root locus diagrams or Nyquist plots) for local transfer functions through which the generator's model is represented.

Chapter 4 analyses different approaches for nonlinear control of multi-phase electric machines: (i) H-infinity control of multi-phase synchronous electric machines, (ii) H-infinity control of doubly fed reluctance machines, (iii) flatness-based adaptive control of doubly fed reluctance machines. First, in this chapter power generation with the use of multi-phase synchronous machines is considered. The dynamic model of the 6-phase synchronous electric machine undergoes an approximate linearization, through Taylor series expansion and an H-infinity feedback controller is applied for its stabilization. The controller requires the solution of a Riccati equation at each iteration of the control algorithm. Next, the chapter develops an H-infinity approach for the problem of nonlinear optimal control of doubly fed reluctance machines. Again the dynamic model of the machines is subjected to linearization round local operating points, through Taylor series expansion and the computation of Jacobian matrices. For the linearized model an H-infinity feedback controller is designed. Finally, the chapter proposes a flatness-based adaptive control approach that is capable of compensating for model uncertainty and parametric changes of the doubly fed reluctance machines (DFRMs), as well as for the lack of measurements about the DFRM's state vector elements. First it is proven that the DFRM's model is a differentially flat one. The unknown dynamics of the system is identified in real time with the use of neuro-fuzzy approximators. The stability of the considered observer-based adaptive control approach is proven using Lyapunov analysis.

Chapter 5 analyses different approaches for nonlinear control of AC to DC converters (rectifiers), such as: (i) flatness-based control through transformation to the canonical form, (ii) differential geometry-based control, (iii) nonlinear H-infinity control. Moreover, the chapter presents flatness-based control for the HVDC-VSC transmission system. First, the chapter proposes a method for nonlinear control of the dynamical system that is formed by a DC–DC converter and a DC motor, making use of differential flatness theory. To compensate for parametric uncertainties and external perturbations a Kalman filter is used as a disturbance observer. Next, the chapter is concerned with proving differential flatness of the three-phase voltage source converter (VSC) model. This enables transformation of the VSC model to the linear canonical form and the design of a stabilizing feedback controller. At the second stage, a Kalman filter is used as disturbance observer for estimating and compensating for additive input disturbances to the VSC model. Additionally, the chapter proposes nonlinear H-infinity control for the model of the three-phase voltage source converter. The converter's model is locally linearized round its current operating point through the computation of the associated Jacobian matrices. An H-infinity feedback control law is computed through the solution of an algebraic Riccati equation at each iteration of the control algorithm. Finally, the chapter develops a control method for the VSC-HVDC system, that is for an AC to DC voltage source converter connected to the electricity grid through a high-voltage DC transmission line terminating at an inverter. By showing that the VSC-HVDC system is a differentially flat one, its transformation to the linear canonical form and the design of a stabilizing feedback controller becomes possible. To compensate for model uncertainties and external perturbations a Kalman filter is used as a disturbance observer.

Chapter 6 analyses different approaches for nonlinear control of DC to AC converters (inverters), such as: (i) flatness-based control through transformation to the canonical form, (ii) differential geometry-based control. Moreover, the chapter presents flatness-based control and nonlinear H-infinity control for DC to DC converters. Finally, the chapter analyses adaptive control of active power filters. First, the chapter proposes a nonlinear feedback control method for three-phase inverters, which is based on differential flatness theory. By exploiting differential flatness properties, it is shown that the inverter's model can be transformed to the linear canonical form. For the latter description the design of a state feedback controller becomes possible. Moreover, to estimate and compensate for model uncertainty and external perturbation inputs that affect the inverter's model, a Kalman filter algorithm is used as a disturbance observer. Next, the chapter proposes a differential flatness theory-based approach for adaptive fuzzy control of active power filters. By proving that the active power filter is a differentially flat system, its transformation to the linear canonical (Brunovsky) form becomes possible. In this new description, the control input of the active power filter comprises unknown nonlinear terms which are identified with the use of neuro-fuzzy networks and through an adaptation/learning procedure. These estimated parts of the system's dynamics are used in an indirect adaptive control scheme, which finally makes the outputs of the active power filter converge to the desirable set points. The stability of the flatness-based control scheme is confirmed through Lyapunov analysis.

Chapter 7 analyses flatness-based control for proton exchange membrane (PEM) fuel cells. Additionally, it gives results on nonlinear H-infinity control of PEM fuel cells. Moreover, it proposes flatness-based control for the PDE model of the Li–ions batteries. First, the chapter presents an approach to nonlinear control of fuel cells using differential flatness theory and Kalman filtering. By exploiting the differential flatness properties of the model its transformation to an equivalent linear canonical form and the design of a state-feedback controller become possible. A Kalman filter-based disturbance observer is applied to the linearized extended model of the fuel cells aiming at the compensation of perturbations effects. Moreover, the chapter applies nonlinear H-infinity control to PEM fuel cells. The dynamic model of the PEM fuel cells undergoes approximate linearization through Taylor series expansion, and an H-infinity feedback controller is designed. The computation of the optimal control input requires the solution of an algebraic Riccati equation at each iteration of the control algorithm. The known robustness properties of H-infinity control enable compensation of model uncertainty and rejection of the perturbation terms that affect the PEM fuel cells system. The stability of the control loop is proven through Lyapunov analysis. Finally, the chapter analyses estimation and control for the PDE dynamics describing particles diffusion in Li–ion batteries. It is proven that this PDE dynamics satisfies differential flatness properties and this enables to solve the associated state estimation problem and to design a stabilizing feedback controller. By applying semi-discretization and the finite differences method the particles diffusion PDE model is decomposed into an equivalent set of nonlinear coupled ODEs and a state-space description is obtained. It is shown that each row of the state-space model is a differentially flat subsystem for which a feedback control law can be found. The feedback controller is implemented in successive loops.

Chapter 8 analyses synchronization and stabilization for distributed power generation units. First, the problem of stabilization and synchronization is solved for a model of distributed interconnected synchronous power generators with the use of flatness-based control. Next, the problem of stabilization and synchronization of the distributed synchronous generators is solved with the use of flatness-based adaptive control. Moreover, the problem of decentralized control and synchronization is solved for a model of distributed interconnected DC power generation units of photovoltaics. Initially, the chapter develops a flatness-based control method for distributed interconnected power generation units. The power system comprises Permanent Magnet Synchronous Generators (PMSGs) which are connected to each other through transformers and tie lines. Differential flatness properties for this model are proven and its transformation to a canonical form is performed. A nonlinear Kalman filter is used as a disturbance observer, thus making possible to estimate at the same time the non-measurable elements of each generator's state vector, the unknown input power (torque) and the disturbance terms induced by interarea oscillations. Next, the chapter proposes a new nonlinear H-infinity control method for stabilization and synchronization of distributed interconnected synchronous generators. Using the linearized description of the distributed generators' dynamics, that is obtained through Taylor series expansion, an H-infinity feedback controller is designed through the solution of a Riccati equation at each step of the control algorithm. The stability of the H-infinity control scheme is proven through Lyapunov analysis. Additionally, the chapter presents an adaptive fuzzy approach to the problem of synchronization and control of distributed power generators, which is based on differential flatness theory and which uses exclusively output feedback. To identify the unknown dynamics of the system, adaptive fuzzy approximators are used in the control loop. This adaptive control scheme is exclusively implemented with the use of output feedback, while the state vector elements which are not directly measured are estimated with the use of a state observer that operates in the control loop. Finally, the chapter develops decentralized control for parallel inverters connected to the power grid using differential flatness theory. By exploiting differential flatness properties it is shown that the multiple inverters model can be transformed into a set of local inverter models which are decoupled and linearized. For each local inverter the design of a state feedback controller becomes possible. To estimate and compensate for disturbance terms that affect each local inverter, the Kalman filter is used as a disturbance observer.

Chapter 9 analyses condition monitoring and fault diagnosis for electric power generators with the following approaches: (i) fault diagnosis in the time domain for distributed interconnected synchronous generators and with the use of the local statistical approach to fault diagnosis, (ii) fault diagnosis for asynchronous generators and with the use of neural networks having activation functions that remain invariant to the Fourier transform, (iii) fault diagnosis in the time domain for synchronous generators and with the use of nonlinear filtering methods. First the chapter analyses a Kalman filtering-based approach for fault diagnosis in distributed and interconnected power generators. The method is based on a new nonlinear filtering scheme under the name derivative-free nonlinear Kalman filter. Statistical processing is performed for the obtained residuals, that is for the differences between the state vector of the monitored power system and the state vector provided by the aforementioned filter when the latter makes use of a fault-free model. It is shown that the suitably weighted square of the residuals vector follows the χ^2 statistical distribution. This property allows to use confidence intervals and to define thresholds that demonstrate the existence of parametric changes (faults). Next, the chapter proposes neural modelling and fault diagnosis methods for the early detection of cascading events in electric power systems. A neural fuzzy network is used to model the dynamics of a distributed power system in fault-free conditions. The output of the neural fuzzy network is compared to measurements from the power system and the obtained residuals undergo statistical processing using the local statistical approach to fault diagnosis. This allows the detection of incipient parametric changes in the model. Finally, the chapter proposes the use of a neural network with Gauss-Hermite polynomial activation functions. Knowing that the Gauss-Hermite basis functions satisfy the orthogonality property and remain unchanged under the Fourier transform, subjected only to a change of scale, one has that the considered neural network provides the spectral analysis of the output of the monitored system. By observing changes in the amplitude of the aforementioned spectral components one can have also an indication about malfunctioning of the monitored system and can detect the existence of failures.

Chapter 10 analyses condition monitoring and fault diagnosis for components of the electric power transmission and distribution system. Initially, neural modelling and the local statistical approach to fault diagnosis is used for incipient fault detection and isolation in power transformers. Next, the chapter proposes distributed nonlinear filtering for condition monitoring of the power distribution system and for the detection of changes in the quality of the provided electric power (voltage sags, change of harmonics, etc). A neural fuzzy network is used to model the thermal condition of the power transformer in fault-free operation its output is compared to measurements from the power transformer and the obtained residuals undergo statistical processing using the local statistical approach to fault diagnosis. If a fault threshold (that is optimally defined according to detection theory) is exceeded, then deviation from normal operation can be detected at its early stages and an alarm can be launched. Moreover, fault isolation can be also performed. Next, the chapter analyses distributed state estimation methods for condition monitoring of electric power transmission and distribution systems. As suitable approaches for distributed state estimation the chapter proposes the Extended

Information Filter (EIF) and the Unscented Information Filter (UIF). With the use of the aforementioned filtering algorithms in processing units located at different parts of the power grid, one can produce local estimates of the system's state vector which in turn can be fused into an aggregate state estimation. The produced global state estimate enables continuous monitoring of the condition of the electric power system.

Regarding the control theoretic part, the monograph introduces systematic methods for compensating the nonlinear dynamics of renewable energy system using global linearization methods (differential flatness theory), approximate linearization methods (H-infinity control) and Lyapunov-based methods. The monograph treats also the problems of (i) uncertainty in the model of the renewable power systems which can be due to incomplete knowledge about the associated dynamics, external perturbations or varying operating conditions, (ii) stochasticity in the model of renewable energy systems which can be due to inability to measure the complete state vector of such systems or due to randomly varying perturbation inputs. To handle (i) the monograph studies extensively adaptive and robust control methods for renewable energy systems, while to handle (ii) the monograph introduces new nonlinear filtering approaches which are proven to be more accurate and computationally more efficient.

The monograph is primarily addressed to the academic community. Its content can be used for teaching undergraduate or postgraduate courses on renewable energy systems. Therefore it can be used by both academic tutors and students as a reference book for such a course. The monograph is also addressed to the engineering community. The nonlinear control and estimation methods analysed in the monograph can be a powerful tool and useful companion for engineers working on practical problems of renewable energy systems. Since the monograph analyses several critical engineering systems associated with the exploitation of renewable energy sources it is likely to be a useful reference for all engineering sectors involved in the development of such systems (electrical and computer engineers, mechanical engineers, physicists and scientists working on relevant computation problems).

Athens, Greece
June 2016

Dr. Gerasimos Rigatos
Electrical and Computer Engineer, Ph.D.

Acknowledgments

The author of this monograph would like to thank researchers in the area of renewable energy systems, as well as in the area of dynamic systems modelling and control for contributing to the development and completion of this research work, through reviews, comments and meaningful remarks.

Contents

Acronyms

AC	Alternate Current
ARE	Algebraic Riccati Equation
DC	Direct Current
DFIG	Doubly Fed Induction Generator
DFKF	Derivative-Free Nonlinear Kalman Filtering
DFRM	Doubly Fed Reluctance Machine
DPS	Distributed Parameter Systems
EIF	Extended Information Filter
EKF	Extended Kalman Filtering
FDI	Fault Detection and Isolation
HST	Hot Spot Temperature
HVDC	High Voltage DC line
KF	Kalman Filtering
MAMM	Multi-Area Multi-Machine Power System
MIMO	Multi-Input Multi-Output
MMSE	Minimum Mean Square Estimator
MSE	Mean Square Error
ODE	Ordinary Differential Equation
PDE	Partial Differential Equation
PMSG	Permanent Magnet Synchronous Generator
PMSM	Permanent Magnet Synchronous Machine
PSS	Power System Stabilizer
RMSE	Root Mean Square Error
SISO	Single Input Single Output
SMIB	Single Machine Infinite Bus model
SNR	Signal-to-Noise Ratio
UIF	Unscented Information Filter
UKF	Unscented Kalman Filtering
VSC	Voltage Source Converter

Chapter 1
Electric Machines and Power Electronics

1.1 Outline

Renewable energy systems produce electricity out of renewable energy sources, such as wind power , solar power , hydro-power and tidal power. The generated electric power in AC or DC form has to be transmitted and distributed through the electricity grid, therefore another major issue in renewable energy systems is their connection to the grid through the use of power electronics (Fig. 1.1). The chapter overviews the basic components of renewable energy systems, such as (i) power generators (PMSG, DFIG, multi-phase machines, and Doubly-Fed Reluctance machines), (ii) power electronics for connecting generators to the grid (DC to AC inverters, AC to DC and DC to DC converters, active power filters, power transformers).

First, doubly-fed induction generators (DFIGs) are analyzed. In DFIGs the rotor's magnetic field is considered to be rotating at an asynchronous speed that is not dependent on grid's reference frequency. DFIGs have been widely used in variable-speed fixed frequency hydro-power generation systems, wind-power generation systems and turbine engine power generation systems. Doubly-fed induction generators have proven to be more efficient than squirrel-cage induction generator systems (SCIG) and the synchronous generator systems in terms of cost and losses of the associated power electronics converters. DFIG systems can operate either in grid-connected mode or in stand-alone mode. Several field-oriented control schemes have been proposed for both operation modes. Additionally, to control electric power generators and the power electronics that enable their connection to the grid, feedback linearization approaches have been developed. In parallel, several results have been published on sensorless control of DFIGs. One can also note nonlinear optimal control approaches for DFIGs. There exist also several results on model-free control of DFIGs implemented through indirect adaptive control schemes.

Next, synchronous generators are analyzed. In Synchronous generators the rotor's magnetic field is considered to be rotating at the synchronous speed which is determined by grid's reference frequency. Comparing to DFIGs the associated dynamic model is of smaller dimension, however its control can be achieved by applying the

G. Rigatos, *Intelligent Renewable Energy Systems*,
Green Energy and Technology, DOI 10.1007/978-3-319-39156-4_1

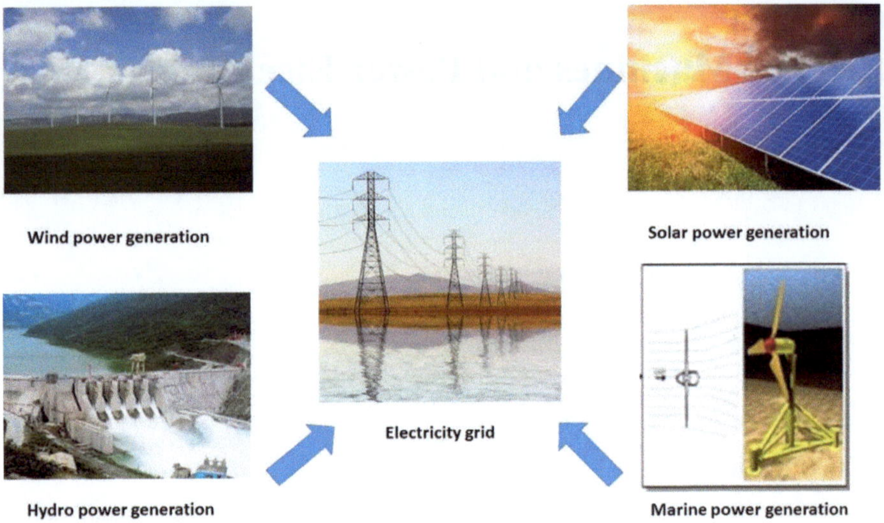

Fig. 1.1 Renewable energy systems

nonlinear control methods that were mentioned of the DFIG case. Thus one may develop control based on global linearization of the machine's model with the use of differential flatness theory or Lie algebra-based methods. Besides one can implement control based on approximate linearization of the machine's model round local operating points, thus arriving at nonlinear H-infinity control methods or methods based on the solution of linear matrix inequalities (LMIs). Besides one can implement feedback control with the use of Lyapunov methods. In case that the dynamic model of the synchronous generators is taken to be unknown and to be identified in real-time through estimation procedures then one arrives at nonlinear indirect adaptive control schemes for such machines.

More elaborated schemes for power generation from renewable sources arise in the case of using power generators of multiple phases (windings) at the stator. Thus one can use multi-phase synchronous generators or doubly-fed reluctance generators. Multi-phase synchronous machines are a generalization of the case of synchronous machines after including additional windings at the stator. On the other side, in Doubly-fed reluctance machines there is no winding and no permanent magnet at the rotor while an additional winding is included in the stator through which the power of the machine is controlled.

The power generated by renewable energy sources has to be transmitted and distributed through the electricity grid and to this end the renewable power generation units have to be connected to the grid with the use of power electronics. These can be three-phase AC to DC converters, inverters (DC to AC converters), DC to DC converters and power transformers.

Voltage source converters, are three-phase filtered rectifiers, that are widely used in the electric power grid (mainly for power flow control). Voltage source converters are the main building blocks of power flow controllers in transmission lines. For example, VSCs are contained in Unified Power Flow Controllers (UPFCs), or distribution-static synchronous compensators (D-STATCOMs). VSCs enable control of the amplitude and phase angle of the AC terminal voltages. Moreover, their bidirectional power flow capabilities allow VSCs to perform real and/or reactive power flow control in AC transmission lines. The dynamic model of voltage source converters is a nonlinear one and requires the development of nonlinear control methods.

Additionally DC-DC converters are used for the exploitation of the power produced by photovoltaic units or fuel cells, since the DC-DC converters control the level of the produced output voltage. In turn the DC output voltage can be fed into DC motors, as for instance in the case of actuators, mechatronic devices and pumps. Pulse Width Modulation (PWM) is applied for such converters' control. The amplitude of the output voltage is determined by the duty cycle of the PWM.

The dynamic model of the inverter is a nonlinear multi-variable one. The control inputs are the voltages generated by an H-bridge circuit while the state vector comprises as elements the voltages at the load's side and the currents at the H-bridge circuit output. Efficient control of inverters is important for the reliable operation of the electric power grid and the connection to it of distributed DC power generation units (such as photovoltaic arrays providing power to the main grid or local photovoltaic cells connected to a microgrid infrastructure). Moreover, inverters find extensive use in HVDCs (High Voltage DC) lines and in their connection with the rest of the AC electric power network.

Fuel cells are electrochemical energy devices that convert the chemical energy of the reaction between hydrogen and oxygen into electricity and heat giving also water as byproduct of the reaction. Fuel cells are a renewable power generation source and their use gets widely deployed in the smart grid. Fuel Cells of the Proton Exchange Membrane type consist of a polymer electrolyte membrane which is placed between the electrodes (anode and cathode). Ions can be diffused through the membrane. If an electrical circuit is established between the anode and the cathode, there will be also a flow of electrons and a potential will appear between the electrodes. Gas supply at the fuel cells' cathode and anode controls the produced electric power. Fuel cells generate DC voltage output, at levels which can be varied with the use of a DC to DC converter.

To assure uninterrupted power supply from renewable sources, the generated power has often to be stored. The need for more efficient power storage has led to the development of new control and estimation methods of the batteries dynamics which are not based on lumped parameter models but which make use of distributed parameter (PDE) models. Specific interest has been shown in Lithium-ion batteries because these exhibit higher power density than other types of batteries, low self-discharge and prolonged life cycle. The distributed parameter models of Li-ion batteries comprise several coupled PDEs which describe the diffusion of Li ions at the solid electrodes as well as their diffusion in the electrolyte. Control of these diffusion phenomena, by applying external voltage or currents can result in making the battery dynamics follow specific charge or discharge patterns as well to keep its functioning within desirable charge levels. Obviously, this can further improve the performance indexes of Li-ion batteries.

As more and more renewable power generation units get connected to the electricity grid, the functioning of renewable energy systems has to be reconsidered within a networked control perspective. A large-scale power system consists of multiple control areas where local power generation units operate, while these areas are connected through tie lines and power transformers. Due to change in active power loads, the frequencies of the areas variate from their nominal values (synchronous speed) and power exchange through the tie lines will also deviate from the associated setpoints. Therefore, it is necessary to supply power generator units with local controllers capable of maintaining the generators' turn speed at its nominal value. To stabilize the functioning of the distributed power generation system and eliminate frequency deviations elaborated nonlinear control approaches have to be implemented. Moreover, the integration of distributed DC power generation units in microgrids, such as photovoltaics and fuel cells requires the interfacing of such devices with the electricity network through inverters. Another problem that arises is the synchronization between multiple parallel inverters. To this end, it is again necessary to develop suitable distributed nonlinear control schemes. By solving the problem of synchronization between DC distributed power generation units which are connected to the microgrid through voltage inverters a more efficient operation of the power grid can be succeeded in terms of power quality and stability.

Apart from control and stabilization another problem that has to be dealt with in the functioning of renewable power generation units is that of condition monitoring and fault diagnosis. There are several fault detection and isolation problems associated with power generators, either in stand alone or in grid-connected functioning. Furthermore, there are several fault detection and isolation problems associated with the power electronics which are used for connecting the renewable energy sources to the grid, such converters, inverters and power transformers. Finally, there are fault detection and isolation issues relation to the transmission and distribution grid. The development of elaborated fault diagnosis tools can result if efficient preventive maintenance of renewable power generation units, early detection of their components' failure, safer operation of the power grid and uninterrupted power supply.

1.2 Main Types of Power Generators

1.2.1 Asynchronous Generators

1.2.1.1 The Complete Sixth-Order Model of the Induction Generator

The doubly-fed induction generator (DFIG) is not only the most widely exploited technology in wind turbines due to its good performance, but it is also used in many other fields such as hydro-power generation, pumped storage plants and flywheel energy storage systems [56, 140, 377, 472, 481, 489, 490]. The DFIG model is derived from the voltage equations of the stator and rotor. It is assumed that the stator and rotor windings are symmetrical and symmetrically fed. The saturation of the inductances, iron losses, skin effect, and bearing friction is neglected. The winding resistance is considered to be constant. A model of the doubly-fed induction generator is as follows:

Dynamic equations:

$$J\dot{\omega} = T_m - K_f\omega - T_e \tag{1.1}$$

where J is the moment of inertia of the rotor, T_m is the externally applied mechanical torque that makes the turbine rotate, T_e is the electrical torque which is associated to the generated active power and finally the term $k_f\omega$ expresses friction, with K_f being the friction coefficient. The wind generated mechanical torque is given by

$$T_m = \frac{1}{2}\rho\pi R^3 C_q(\lambda, \beta)v^2 \tag{1.2}$$

where v is the wind's speed [56]. C_q is a torque coefficient which depends on the blade pitch angle β and the tip-speed ratio which is provided by $\lambda = \frac{\omega_r R}{v}$, with ω_r being the rotor's angular velocity, R is the rotor radius and ρ is the air density.

This type of wound-rotor machine is connected to the grid by both the rotor and stator side. The DFIG stator can be directly connected to the electric power

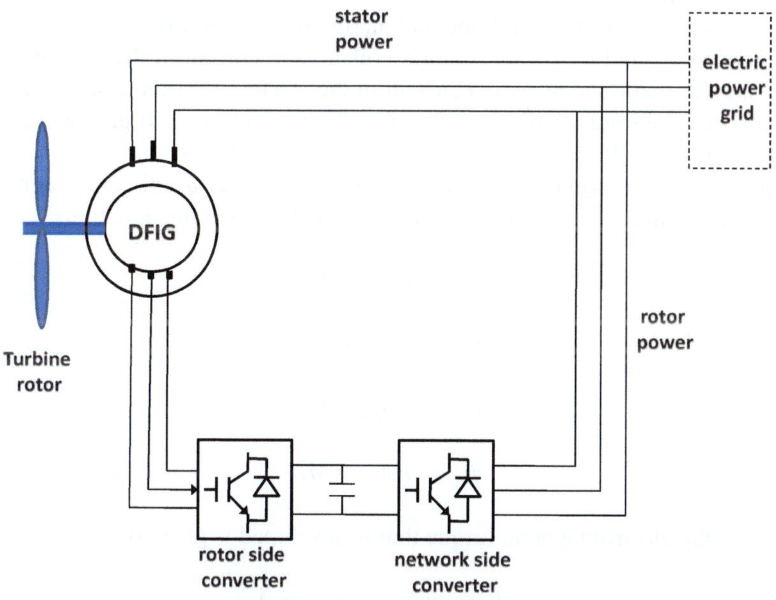

Fig. 1.2 Configuration of a doubly-fed induction generator unit in the power grid

grid while the rotor is interfaced through back-to-back converters (see Fig. 1.2). By decoupling the power system electrical frequency and the rotor mechanical frequency the converter allows a variable speed operation of the wind turbine.

The doubly-fed induction generator is analogous to the induction motor. In an induction motor the stator voltage plays the role of an input variable, while the rotor voltage is a constant (it is usually zero). In case of the doubly-fed induction machine it is very similar but the other way round, with a dual analogy to hold between the stator and rotor parameters of the generator and the motor. This means that the rotor voltage now acts as an input, while the stator voltage depends on the voltage at the bus to which the DFIG is connected and is a constant parameter [352, 353].

Electrical equations: Using the Park transform the DFIG is described in the $d-q$ reference frame by the following set of equations:

$$v_{s_d} = R_s i_{s_d} + \frac{d\psi_{s_d}}{dt} - \omega_{dq}\psi_{s_q} \tag{1.3}$$

$$v_{s_q} = R_s i_{s_q} + \frac{d\psi_{s_q}}{dt} + \omega_{dq}\psi_{s_d} \tag{1.4}$$

$$v_{r_d} = R_r i_{r_d} + \frac{d\psi_{r_d}}{dt} - \omega_r\psi_{r_q} \tag{1.5}$$

$$v_{r_q} = R_r i_{r_q} + \frac{d\psi_{r_q}}{dt} + \omega_r\psi_{r_d} \tag{1.6}$$

where ω_{dq} is the synchronous frequency, ω_r is the rotation frequency of the rotor, ψ_{s_d} is the stator flux component along the d-axis, ψ_{s_q} is the stator flux component along the q-axis and equivalently ψ_{r_d} is the rotor flux component along the d-axis, while ψ_{r_q} is the stator flux component along the q-axis (see Fig. 1.3).

Moreover, v_{s_d} and i_{s_d} are the stator's voltage and current in the d reference, v_{s_q} and i_{s_q} are the stator's voltage and current in the q reference frame and equivalently v_{r_d} and i_{r_d} are the rotor's voltage and current in the d reference frame, while v_{r_q} and i_{r_q} are the rotor's voltage and current in the q reference frame.

Under the assumption of field orientation, the d and q axis are magnetically decoupled the flux components are described by the following equations:

$$\psi_{s_d} = L_s i_{s_d} + M i_{r_d} \tag{1.7}$$

$$\psi_{s_q} = L_s i_{s_q} + M i_{r_q} \tag{1.8}$$

$$\psi_{r_d} = L_r i_{r_d} + M i_{s_d} \tag{1.9}$$

$$\psi_{r_q} = L_r i_{r_q} + M i_{s_q} \tag{1.10}$$

Moreover, the electromagnetic torque that is developed is given by

$$T_e = \eta(i_{s_q}\psi_{s_d} - i_{s_d}\psi_{s_q}) \tag{1.11}$$

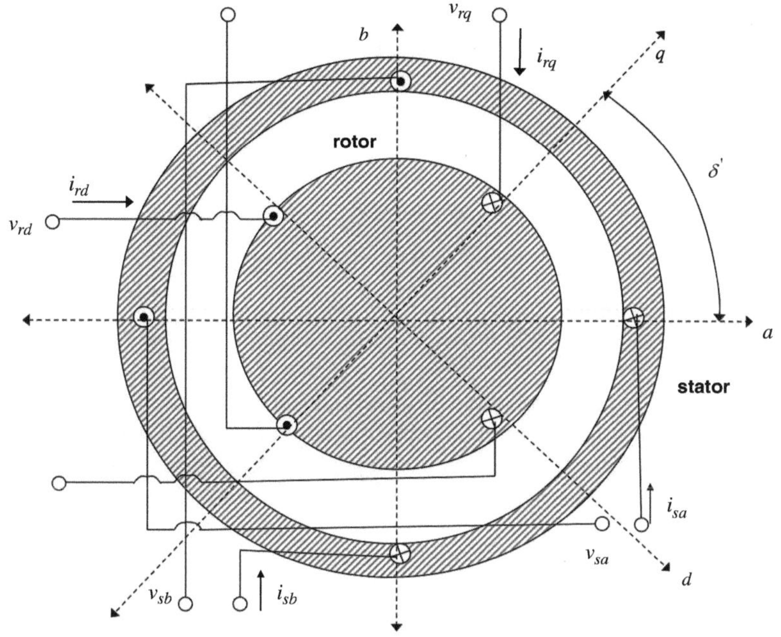

Fig. 1.3 The $a - b$ stator reference frame and the $d - q$ rotor reference frame of the induction generator

where η is a variable that is associated to the number of poles and to the mutual inductance M. Additionally, active and reactive power delivered by the DFIG stator are associated to the real and imaginary part of the apparent power at the stator's terminals, i.e.

$$P_s = Re\{U_s I_s^*\} = v_{s_d} i_{s_d} + v_{s_q} i_{s_q} \tag{1.12}$$

$$Q_s = Im\{U_s I_s^*\} = v_{s_d} i_{s_q} - v_{s_q} i_{s_d} \tag{1.13}$$

The angle of the vectors that describe the magnetic flux $\psi_{s\alpha}$ and ψ_{sb} is first defined for the stator, i.e.

$$\rho = tan^{-1}(\frac{\psi_{s_b}}{\psi_{s_a}}) \tag{1.14}$$

The angle between the inertial reference frame and the rotating reference frame is taken to be equal to ρ.

Moreover, it holds that $cos(\rho) = \frac{\psi_{sa}}{||\psi||}$, $sin(\rho) = \frac{\psi_{s_b}}{||\psi||}$, and $||\psi|| = \sqrt{\psi_{s_\alpha}^2 + \psi_{s_b}^2}$. Therefore, in the rotating frame $d-q$ of the generator there will be only one non-zero component of the magnetic flux ψ_{s_d}, while the component of the flux along the q axis equals 0.

In a compact form the doubly-fed induction generator can be described by the following set of equations in the $d-q$ reference frame that rotates at an arbitrary speed denoted as ω_{dq} [140]

$$\frac{d\psi_{sq}}{dt} = -\frac{1}{\tau_s}\psi_{s_q} - \omega_{dq}\psi_{s_d} + \frac{M}{\tau_s}i_{r_q} + v_{s_q} \tag{1.15}$$

$$\frac{d\psi_{s_d}}{dt} = \omega_{dq}\psi_{s_q} - \frac{1}{\tau_s}\psi_{s_d} + \frac{M}{\tau_s}i_{r_d} + v_{s_d} \tag{1.16}$$

$$\frac{di_{r_q}}{dt} = \frac{\beta}{\tau_s}\psi_{s_q} + \beta\omega_r\psi_{s_d} - \gamma_2 i_{r_q} - \omega_{dq} - \omega_r i_{r_d} - \beta v_{s_q} + \frac{1}{\sigma L_r}v_{r_q} \tag{1.17}$$

$$\frac{di_{r_d}}{dt} = -\beta\omega_r\psi_{s_q} + \frac{\beta}{\tau_s}\psi_{s_d} + (\omega_{dq} - \omega_r)i_{r_q} - \gamma_2 i_{r_d} - \beta v_{s_d} + \frac{1}{\sigma L_r}v_{r_d} \tag{1.18}$$

where λ_{s_q}, λ_{s_d}, i_{r_q}, i_{r_d} are the stator flux and the rotor currents, v_{s_q}, v_{s_d}, v_{r_q}, v_{r_d} are the stator and rotor voltages, L_s and L_r are the stator and rotor inductances, ω_r is the rotor's angular velocity, M is the magnetizing inductance. Moreover, denoting as R_s and R_r the stator and rotor resistances the following parameters are defined

$$\begin{array}{ccc} \sigma = 1 - \frac{M^2}{L_r L_s} & \beta = \frac{1-\sigma}{M\sigma} & \tau_s = \frac{L_s}{R_s} \\ \tau_r = \frac{L_r}{R_r} & \gamma_2 = (\frac{1-\sigma}{\sigma\tau_s}) & \end{array} \tag{1.19}$$

The dynamic model of the doubly-fed induction generator can be also written in state space equations form by defining the following state variables: $x_1 = \theta$, $x_2 = \omega_r$, $x_3 = \psi_{s_d}$, $x_4 = \psi_{s_q}$, $x_5 = i_{r_d}$ and $x_6 = i_{r_q}$. It holds that

$$\dot{x}_1 = x_2 \tag{1.20}$$

$$\dot{x}_2 = -\frac{K_m}{J}x_2 - \frac{T_m}{J} + \frac{\eta}{J}(i_{s_q}x_3 - i_{s_d}x_4) \tag{1.21}$$

$$\dot{x}_3 = -\frac{1}{\tau_s}x_3 + \omega_{dq}x_4 + \frac{M}{\tau_s}x_5 + v_{s_d} \tag{1.22}$$

$$\dot{x}_4 = -\omega_{dq}x_3 - \frac{1}{\tau_s}x_4 + \frac{M}{\tau_s}x_6 + v_{s_q} \tag{1.23}$$

$$\dot{x}_5 = -\beta x_2 x_4 + \frac{\beta}{\tau_s}x_3 + (\omega_{dq} - x_2)x_6 - \gamma_2 x_5 + \frac{1}{\sigma L_s}v_{r_d} - \beta v_{s_d} \tag{1.24}$$

$$\dot{x}_6 = \frac{\beta}{\tau_s}x_4 + \beta x_2 x_3 + (\omega_{dq} - x_2)x_5 - \gamma_2 x_6 \tag{1.25}$$

which can be written also in the form

$$\dot{x} = f(x) + g_a(x)v_{r_d} + g_b(x)v_{r_q} \tag{1.26}$$

where $x = [x_1, x_2, x_3, x_4, x_5, x_6]^T$ and

$$f(x) = \begin{pmatrix} x_2 \\ -\frac{K_m}{J}x_2 - \frac{T_m}{J} + \frac{n}{J}(i_{s_q}x_3 - i_{s_d}x_4) \\ -\frac{1}{\tau_s}x_3 + \omega_{dq}x_4 + \frac{M}{\tau_s}x_5 + v_{s_d} \\ -\omega_{dq}x_3 - \frac{1}{\tau_s}x_4 + \frac{M}{\tau_s}x_6 + v_{s_q} \\ -\beta x_2x_4 + \frac{\beta}{\tau_s}x_3 + (\omega_{dq} - x_2)x_6 - \gamma_2x_5 - \beta v_{s_d} \\ \frac{\beta}{\tau_s}x_4 + \beta x_2x_3 + (\omega_{dq} - x_2)x_5 - \gamma_2x_6 \end{pmatrix} \tag{1.27}$$

$$g_a(x) = \left(0\ 0\ 0\ 0\ \frac{1}{\sigma L_s}\ 0\right)^T$$

$$g_b(x) = \left(0\ 0\ 0\ 0\ 0\ \frac{1}{\sigma L_s}\right)^T$$

1.2.2 *Synchronous Generators*

1.2.2.1 Dynamic Model of the Permanent Magnet Synchronous Generator

It is considered that the third order model of a Permanent Magnet Synchronous Generator (PMSG) is connected to the power grid as shown in Fig. 1.4 thus forming the model of a Single Machine Infinite Bus (SMIB) system.

The PMSG mechanical dynamics can be represented as follows

$$\begin{aligned} \dot{\delta} &= \omega \\ \dot{\omega} &= -\frac{D}{2J}(\omega - \omega_0) + \frac{\omega_0}{2J}(P_m - P_e) \end{aligned} \tag{1.28}$$

where δ is the turn angle of the generator's rotor, ω is the rotation speed of the rotor with respect to synchronous reference, ω_0 is the synchronous speed of the generator, J is the moment of inertia of the rotor, P_m is the mechanical input torque to the

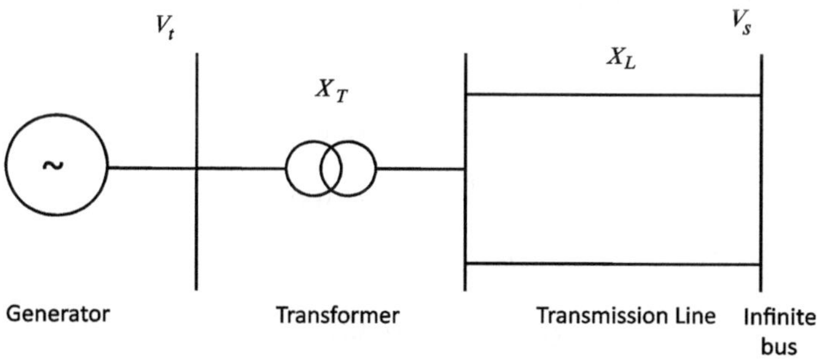

Fig. 1.4 PMSG connected to the power grid according to the single machine infinite bus (SMIB) model

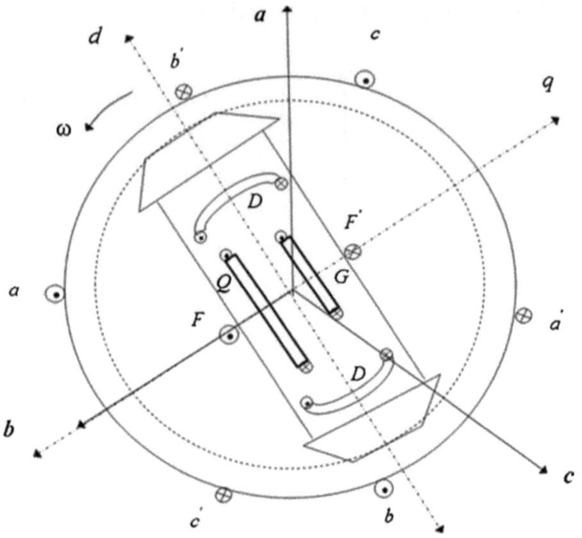

Fig. 1.5 Rotation frame for the magnetic field of the synchronous generator

generator which is associated with the mechanical input power, D is the damping constant of the generator and T_e is the electrical torque which is associated to the generated active power. Moreover, the following variables are defined: $\Delta\delta = \delta - \delta_0$ and $\Delta\omega = \omega - \omega_0$ with ω_0 denoting the synchronous speed. The reference frame for the magnetic field of the synchronous generator turns at the synchronous speed which is defined by the grid's frequency (Fig. 1.5).

The generator's electric dynamics is described by [20, 112, 197, 249, 266, 374]:

$$\dot{E}_q^{'} = \frac{1}{T_{d_o}}(E_f - E_q) \tag{1.29}$$

where $E_q^{'}$ is the quadrature-axis transient voltage of the generator, E_q is the quadrature axis voltage of the generator, T_{d_o} is the direct axis open-circuit transient time constant of the generator and E_f is the equivalent voltage in the excitation coil.

Considering a dq reference frame that rotates at the rotor's turn speed, the algebraic equations of the synchronous generator are given by

$$\begin{aligned}
E_q &= \frac{x_{d\Sigma}}{x_{d\Sigma}^{'}}E_q^{'} - (x_d - x_d^{'})\frac{V_s}{x_{d\Sigma}^{'}}cos(\Delta\delta) \\
I_q &= \frac{V_s}{x_{d\Sigma}^{'}}sin(\Delta\delta) \\
I_d &= \frac{E_q^{'}}{x_{d\Sigma}^{'}} - \frac{V_s}{x_{d\Sigma}^{'}}cos(\Delta\delta) \\
P_e &= \frac{V_s E_q^{'}}{x_{d\Sigma}^{'}}sin(\Delta\delta) \\
Q_e &= \frac{V_s E_q^{'}}{x_{d\Sigma}^{'}}cos(\Delta\delta) - \frac{V_s^2}{x_{d\Sigma}} \\
V_t &= \sqrt{(E_q^{'} - X_d^{'}I_d)^2 + (X_d^{'}I_q)^2}
\end{aligned} \tag{1.30}$$

where $x_{d_\Sigma} = x_d + x_T + x_L$, $x'_{d_\Sigma} = x'_d + x_T + x_L$, x_d is the direct-axis synchronous reactance, x_T is the reactance of the transformer, x_L is the reactance of the transmission line, I_d and I_q are direct and quadrature axis currents of the generator, V_s is the infinite bus voltage, Q_e is the generator reactive power delivered to the infinite bus, and V_t is the terminal voltage of the generator.

Substituting the electrical equations of the PMSG given in Eq. (1.30) into the equations of the electrical and mechanical dynamics of the rotor given in Eqs. (1.28) and (1.29) respectively, the complete description of the Single Machine Infinite Bus model is obtained

$$\begin{aligned} \dot{\delta} &= \omega - \omega_0 \\ \dot{\omega} &= -\frac{D}{2J}(\omega - \omega_0) + \omega_0\frac{P_m}{2J} - \omega_0\frac{1}{2J}\frac{V_s E'_q}{x'_{d_\Sigma}} sin(\Delta\delta) \\ \dot{E}'_q &= -\frac{1}{T'_d}E'_q + \frac{1}{T_{d_o}}\frac{x_d - x'_d}{x'_{d_\Sigma}} V_s cos(\Delta\delta) + \frac{1}{T_{d_o}}E_f \end{aligned} \tag{1.31}$$

where $T'_d = \frac{x'_{d_\Sigma}}{x_{d_\Sigma}} T_{d_o}$ is the time constant of the field winding.

The previously analyzed single-machine infinite-bus model of the PMSG is described by a nonlinear state-space model of the form

$$\dot{x} = f(x) + g(x)u \tag{1.32}$$

where the state vector x is defined as

$$x = \left(\Delta\delta \ \Delta\omega \ E'_q\right)^T \tag{1.33}$$

The vector fields $f(x)$ and $g(x)$ are defined as

$$f(x) = \begin{pmatrix} \omega - \omega_0 \\ -\frac{D}{2J}(\omega - \omega_0) + \omega_0\frac{P_m}{2H} - \omega_0\frac{1}{2J}\frac{V_s E'_q}{x'_{d\sigma}} sin(\Delta\delta) \\ -\frac{1}{T'_d}E'_q + \frac{1}{T_{do}}\frac{x_d - x'_d}{x'_{d\sigma}} V_s cos(\Delta\delta) \end{pmatrix} \tag{1.34}$$

$$g(x) = \left(0 \ 0 \ \frac{1}{T_{do}}\right)^T \tag{1.35}$$

with control input $u = E_f$ the field voltage (equivalent voltage in the excitation coil) and measurable output the rotation speed of the rotor

$$y = h(x) = \delta - \delta_0 \tag{1.36}$$

1.3 Main Types of Multi-phase Machines

1.3.1 The 6-Phase Synchronous Machine

1.3.1.1 Dynamics of the 6-Phase PMSM

The dynamic model of a six-phase Permanent Magnet Synchronous Machine (PMSM) will be analyzed next. There are two three-phase systems, the first one is denoted as *abc* and the second one is denoted as *xyz* and results from the first system by a 30° rotation (Fig. 1.6) [23, 250, 286, 400]. By applying Kirchhoff's law in these phase systems one obtains:

$$v_{abc} = \bar{R}_s i_{abc} + \tfrac{d}{dt}\lambda_{abc} \tag{1.37}$$

$$\lambda_{abc} = \bar{L}_{11} i_{abc} + \bar{L}_{12} i_{xyz} + \lambda'_{mabc} \tag{1.38}$$

$$v_{xyz} = \bar{R}_s i_{xyz} + \tfrac{d}{dt}\lambda_{xyz} \tag{1.39}$$

$$\lambda_{xyz} = \bar{L}_{22} i_{xyz} + \bar{L}_{21} i_{abc} + \lambda_{mxyz} \tag{1.40}$$

where $\bar{R}_s = diag[R_s, R_s, R_s]^T$ is the stator resistance matrix, $v_{abc} = [v_a, v_b, v_c]^T$ is the phase voltage vector of the *abc* winding, $i_{abc} = [i_a, i_b, i_c]^T$ is the current vector of the *abc* winding, $v_{xyz} = [v_x, v_y, v_z]^T$ is the phase voltage vector of the *xyz* winding, $i_{xyz} = [i_x, i_y, i_z]^T$ is the current vector of the *xyz* winding, $\lambda_{abc} = [\lambda_a, \lambda_b, \lambda_c]^T$ is the stator flux linkage vector of the *abc* winding, $\lambda_{xyz} = [\lambda_x, \lambda_y, \lambda_z]^T$ is the stator flux linkage vector of the *xyz* winding, $\bar{L}_{11} = diag[L_{11}, L_{11}, L_{11}]^T$ is the stator inductance vector of the *abc* winding, $\bar{L}_{12} = diag[L_{12}, L_{12}, L_{12}]^T$ is a mutual inductance vector,

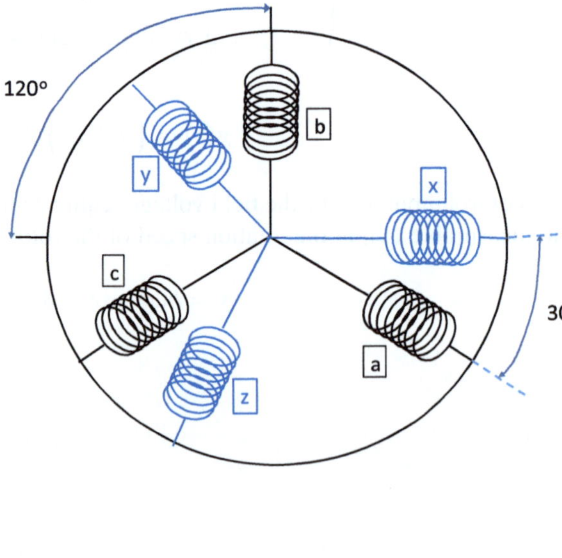

Fig. 1.6 A six-phase electric machine

$\bar{L}_{22} = diag[L_{22}, L_{22}, L_{22}]^T$ is the stator inductance vector of the xyz winding, and $\bar{L}_{21} = diag[L_{21}, L_{21}, L_{21}]^T$ is also a mutual inductance vector. Moreover λ'_{mabc} is the permanent-magnet flux linkage vector of the abc winding and λ'_{mxyz} is the permanent-magnet flux linkage vector of the xyz winding.

In aggregate one has the following equations

$$\begin{aligned} v_a &= R_s i_a + \tfrac{d}{dt}\lambda_a \\ v_b &= R_s i_b + \tfrac{d}{dt}\lambda_b \\ v_c &= R_s i_c + \tfrac{d}{dt}\lambda_c \end{aligned} \tag{1.41}$$

$$\begin{aligned} \lambda_a &= L_{11} i_a + L_{12} i_x + \lambda'_{ma} \\ \lambda_b &= L_{11} i_b + L_{12} i_y + \lambda'_{mb} \\ \lambda_c &= L_{11} i_c + L_{12} i_z + \lambda'_{mc} \end{aligned} \tag{1.42}$$

$$\begin{aligned} v_x &= R_s i_x + \tfrac{d}{dt}\lambda_x \\ v_y &= R_s i_y + \tfrac{d}{dt}\lambda_y \\ v_z &= R_s i_z + \tfrac{d}{dt}\lambda_z \end{aligned} \tag{1.43}$$

$$\begin{aligned} \lambda_x &= L_{22} i_x + L_{21} i_a + \lambda'_{mx} \\ \lambda_y &= L_{22} i_y + L_{21} i_b + \lambda'_{my} \\ \lambda_z &= L_{22} i_z + L_{21} i_c + \lambda'_{mz} \end{aligned} \tag{1.44}$$

To simplify the dynamic model of the multi-phase PMSM the following Park transformations are applied:

$$T_{qd1} = \tfrac{2}{3} \begin{pmatrix} cos(\theta_e) & cos(\theta_e - 120°) & cos(\theta_e + 120°) \\ sin(\theta_e) & sin(\theta_e - 120°) & sin(\theta_e + 120°) \\ \frac{1}{2} & \frac{1}{2} & \frac{1}{2} \end{pmatrix} \tag{1.45}$$

$$T_{qd2} = \tfrac{2}{3} \begin{pmatrix} cos(\theta_e - 30°) & cos(\theta_e - 150°) & cos(\theta_e + 90°) \\ sin(\theta_e - 30°) & sin(\theta_e - 150°) & sin(\theta_e + 90°) \\ \frac{1}{2} & \frac{1}{2} & \frac{1}{2} \end{pmatrix} \tag{1.46}$$

where T_{dq1} is the transformation leading from the abc reference frame to the $dq1$ reference frame, T_{dq2} is the transformation leading from the xyz reference frame to the $dq2$ reference frame, and θ_e is the rotor flux angle. In the new reference frame, the six-phase PMSM model becomes

$$\begin{aligned} v_{q_1} &= R_s i_{q_1} + L_{q11}\tfrac{d}{dt} i_{q_1} + \omega_e(L_{d11} i_{d1}) + \lambda'_m \\ u_{d_1} &= R_s i_{d_1} + L_{d11}\tfrac{d}{dt} i_{d_1} - \omega_r L_{q11} i_{q_1} \\ v_{q_2} &= R_s i_{q_2} + L_{q22}\tfrac{d}{dt} i_{q_2} + \omega_e(L_{d22} i_{d2}) + \lambda'_m \\ u_{d_2} &= R_s i_{d_2} + L_{d22}\tfrac{d}{dt} i_{d_2} - \omega_r L_{q22} i_{q_2} \\ \omega_e &= \tfrac{p}{2}\omega_r \end{aligned} \tag{1.47}$$

where v_{d_1} and v_{q_1} are the stator abc windings' voltages expressed in the $dq1$ reference frame, v_{d_2} and v_{q_2} are the stator xyz windings' voltages expressed in the $dq2$ reference frame, i_{d_1} and i_{q_1} are the stator *abc* windings' currents expressed in the $dq1$ reference frame, i_{d_2} and i_{q_2} are the stator *xyz* windings' currents expressed in the $dq2$ reference frame, L_{d11} and L_{q11} are the the stator *abc* windings' inductances expressed in the $dq1$ reference frame, L_{d22} and L_{q22} are the the stator *xyz* windings' inductances expressed in the $dq2$ reference frame, ω_r is the rotor's angular velocity, ω_e is the rotation speed of the dq reference frame, $\lambda_m^{'}$ is the permanent magnet flux linkage, and p is the number of poles of the electric machine.

The electric torque of the six-phase PMSM is given by

$$\begin{aligned} T_e = \tfrac{3}{2}\tfrac{p}{2}([(L_{d11}i_{d1} + \lambda_m^{'})i_{q1} + (L_{d11} - L_{q11})i_{d1}i_{q1}] + \\ + [(L_{d22} + \lambda_m^{'})i_{q2} + (L_{d22} - L_{q22})i_{d2}i_{q2}]) \end{aligned} \tag{1.48}$$

Moreover, about the rotational motion of the rotor of this electric machine one has

$$J\tfrac{d}{dt}\omega_r + B\omega_r = T_e - T_L \tag{1.49}$$

where ω_r is the rotor's turn speed, J is the moment of inertia, B is a friction coefficient, T_L is an external load's torque and T_e is the electric torque exerted on the rotor due to magnetic field developed at the stator.

1.3.1.2 State-Space Description of the 6-Phase PMSM

The following state variables are defined for the 6-phase PMSM machine. These are: $x_1 = \omega_r$, $x_2 = i_{d1}$, $x_3 = i_{q1}$, $x_4 = i_{d2}$, $x_5 = i_{q2}$. Moreover, the following control inputs are defined: $u_1 = v_{d1}$, $u_2 = v_{q1}$, $u_3 = v_{d2}$, $u_4 = v_{q2}$. Using this notation, the state-space model of the PMSM is written as:

$$\begin{aligned} \dot{x}_1 = -\tfrac{B}{J}x_1 - \tfrac{3}{2}\tfrac{p}{2J}([(L_{d11}x_2 + \lambda_m^{'})x_3 + (L_{d11} - L_{q11})x_2x_3] + \\ + [(L_{d22} + \lambda_m^{'})x_5 + (L_{d22} - L_{q22})x_4x_5]) + \tfrac{T_L}{J} \end{aligned} \tag{1.50}$$

$$\dot{x}_2 = -\tfrac{R_s}{L_{d11}}x_2 - \tfrac{p}{2}x_1\tfrac{L_{q11}}{L_{d11}}x_3 + \tfrac{1}{L_{d11}}u_1 \tag{1.51}$$

$$\dot{x}_3 = -\tfrac{R_s}{L_{q11}}x_3 - \tfrac{p}{2}x_1(\tfrac{L_{d11}}{L_{q11}}x_2 + \tfrac{1}{L_{q11}}\lambda_m^{'}) + \tfrac{1}{L_{q11}}u_2 \tag{1.52}$$

$$\dot{x}_4 = -\tfrac{R_s}{L_{d22}}x_4 - \tfrac{p}{2}x_1\tfrac{L_{q22}}{L_{d22}}x_5 + \tfrac{1}{L_{d22}}u_3 \tag{1.53}$$

$$\dot{x}_5 = -\tfrac{R_s}{L_{q22}}x_5 - \tfrac{p}{2}x_1(\tfrac{L_{d22}}{L_{q22}}x_4 + \tfrac{1}{L_{q22}}\lambda_m^{'}) + \tfrac{1}{L_{q22}}u_2 \tag{1.54}$$

It can be noticed that the equations that constitute the dynamics of the electric part of the 6-phase machine for the two different frames are decoupled, while coupling appears in the part describing the mechanical part of the machine and the associated rotational motion.

Using the previous notation one can also arrive at a state-space description of the 6-phase machine in matrix form

$$\dot{x} = f(x) + g(x)u \tag{1.55}$$

where $x \in R^{5\times1}$, $u \in R$ and $f(x) \in R^{5\times1}$, $g(x) \in R^{5\times4}$ and

$$f(x) = \begin{pmatrix} -\frac{B}{J}x_1 + \frac{3}{2}\frac{p}{2J}([(L_{d11}x_2 + \lambda_m^{'})x_3 + (L_{d11} - L_{q11})x_2x_3] + \\ +[(L_{d22}x_4 + \lambda_m^{'})x_5 + (L_{d22} - L_{q22})x_4x_5]) + \frac{T_L}{J} \\ -\frac{R_s}{L_{d11}}x_2 - \frac{p}{2}x_1\frac{L_{q11}}{L_{d11}}x_3 \\ -\frac{R_s}{L_{q11}}x_3 - \frac{p}{2}x_1(\frac{L_{d11}}{L_{q11}}x_2 + \frac{1}{L_{q11}}\lambda_m^{'}) \\ -\frac{R_s}{L_{d22}}x_4 - \frac{p}{2}x_1\frac{L_{q22}}{L_{d22}}x_5 \\ -\frac{R_s}{L_{q22}}x_5 - \frac{p}{2}x_1(\frac{L_{d22}}{L_{q22}}x_4 + \frac{1}{L_{q22}}\lambda_m^{'}) \end{pmatrix} \tag{1.56}$$

$$g(x) = \begin{pmatrix} 0 & 0 & 0 & 0 \\ \frac{1}{L_{d11}} & 0 & 0 & 0 \\ 0 & \frac{1}{L_{q11}} & 0 & 0 \\ 0 & 0 & \frac{1}{L_{d22}} & 0 \\ 0 & 0 & 0 & \frac{1}{L_{q22}} \end{pmatrix} \tag{1.57}$$

1.3.2 Doubly-Fed Reluctance Machine

1.3.2.1 Dynamics of the Electrical and Mechanical Part of the Machine

The brushless doubly-fed reluctance machine has two separate stator windings where the first one is noted as power winding while the second one is noted as control winding (Fig. 1.7). The power winding is directly connected to the 3-phase grid and using the dq reference notation one has two voltage components v_d and v_q [6, 179, 232, 318, 409, 449]. The frequency of current at the primary winding is the one of the grid and is denoted as ω_1. The secondary winding is connected to the grid through an AC/DC/AC converter thus it can have current at a frequency different from the one of the grid, which is denoted as ω_2. The power winding has p_1 poles while the secondary (control) winding has p_2 poles. The rotor of the machine has no windings and the number of poles in it is usually chosen to be $p_r = (p_1 + p_2)/2$. About the electrical angular speed of the reluctance rotor one has $\omega_r = \omega_1 - \omega_2$ and that $p_r\omega_r = \omega_m$.

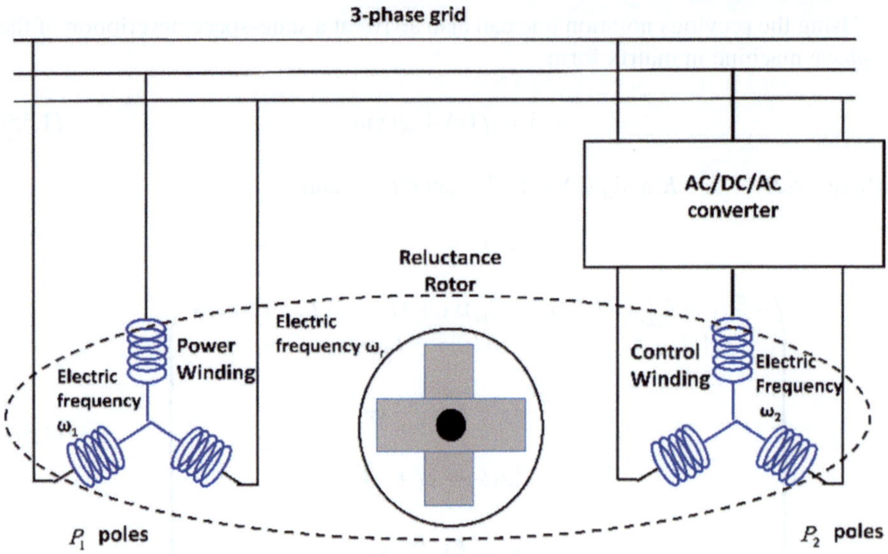

Fig. 1.7 Doubly-fed reluctance machine and its windings

Next, by applying Kirchhoff's laws at the primary and the secondary winding of the reluctance machine one obtains the dynamic model of its electrical part. Thus, at the primary winding one has [449]

$$\dot{\lambda}_{1d} = -R_1 i_{1d} + \omega_1 \lambda_{1q} + v_{1d} \tag{1.58}$$

$$\dot{\lambda}_{1q} = -R_1 i_{1q} - \omega_1 \lambda_{1d} + v_{1q} \tag{1.59}$$

while at the secondary winding it holds

$$\dot{\lambda}_{2d} = -R_2 i_{2d} + \omega_2 \lambda_{2q} + v_{2d} \tag{1.60}$$

$$\dot{\lambda}_{2q} = -R_2 i_{2q} - \omega_2 \lambda_{2d} + v_{2q} \tag{1.61}$$

The magnetic flux at the primary winding is the result of the inductance of this winding and of the mutual inductance (coupling) with the secondary winding

$$\lambda_{1d} = L_1 i_{1d} + L_{12} i_{2d} \tag{1.62}$$

$$\lambda_{1q} = L_1 i_{1q} - L_{12} i_{2q} \tag{1.63}$$

Equivalently, the magnetic flux at the secondary winding is the result of the inductance of this winding and of the mutual inductance (coupling) with the primary winding

$$\lambda_{2d} = L_2 i_{2d} + L_{12} i_{1d} \tag{1.64}$$

$$\lambda_{2q} = L_2 i_{2q} - L_{12} i_{1q} \tag{1.65}$$

In the above equations variables v_{1d}, v_{1q} are the grid voltage components in the dq reference frame which are exerted as inputs to the circuit of the primary winding, while v_{2d}, v_{2q} are the voltage components at the output of the AC/DC/AC converter, expressed in the dq reference frame, which are exerted as inputs to the circuit of the secondary winding.

Equivalently i_{1d}, i_{1q} are the currents of the primary winding expressed in the dq reference frame, while i_{2d}, i_{2q} are the currents of the secondary winding represented again in the dq reference frame.

Moreover, λ_{1d}, λ_{1q} are the components of the magnetic flux of the primary winding written in the dq reference frame, while λ_{2d}, λ_{2q} are components of the magnetic flux of the secondary winding expressed again in the dq reference frame.

Additionally, R_1 and L_1 is the resistance and the inductance of the primary winding, while R_2 and L_2 is the resistance and the inductance of the secondary winding. Moreover, L_{12} is the mutual inductance (coupling) between the primary and the secondary winding.

The electromagnetic torque of the machine is a function of the currents and the magnetic flux expressed in the dq reference frame. It also depends on the number of rotor's poles. Thus, one has

$$T_e = \frac{3}{2}\frac{L_{12}}{L_1} p_r (\lambda_{1d} i_{2q} + \lambda_{1q} i_{2d}) \tag{1.66}$$

Next, the dynamics of the mechanical part of the reluctance machine (generator) can be formulated. Considering that the moment of inertia of the rotor is J_r, and denoting the rotation speed of the rotor as ω one has that

$$\begin{aligned} \dot{\omega} &= \frac{1}{J_r}(-T_e + T_m) \Rightarrow \\ \dot{\omega} &= \frac{1}{J_r}[-\frac{3}{2}\frac{L_{12}}{L_1} p_r (\lambda_{1d} i_{2q} + \lambda_{1q} i_{2d}) + T_m] \end{aligned} \tag{1.67}$$

Moreover, the active power of the reluctance machine is computed as follows [449]

$$P^* = \frac{3}{2}(v_{1d} i_{1d} + v_{1q} i_{1d}) \tag{1.68}$$

or an equivalent relation can be obtained from $P = T_e \omega$ that is

$$P^* = \frac{3}{2}\frac{L_{12}}{L_1} p_r (\lambda_{1d} i_{2q} + \lambda_{1q} i_{2d})\omega \tag{1.69}$$

whereas the reactive power of the machine is given by

$$Q^* = 3/2(v_{1d}i_{1q} - v_{1q}i_{1d}) \tag{1.70}$$

1.3.2.2 State-Space Model of the Reluctance Machine

By substituting the relations about the magnetic flux in the primary and the secondary winding which are given in Eqs. (1.62)–(1.65) into the Eqs. (1.58)–(1.61) and after solving with respect to $\dot{i}_{1d}$, $\dot{i}_{1q}$, $\dot{i}_{2d}$, and $\dot{i}_{2q}$, one arrives at the following description:

$$\begin{pmatrix} L_1 & 0 & L_{12} & 0 \\ 0 & L_1 & 0 & -L_{12} \\ L_{12} & 0 & L_2 & 0 \\ 0 & -L_{12} & 0 & L_2 \end{pmatrix} \begin{pmatrix} d/dt i_{1d} \\ d/dt i_{1q} \\ d/dt i_{2d} \\ d/dt i_{2q} \end{pmatrix} = \begin{pmatrix} -R_1 i_{1d} + \omega_1(L_1 i_{1q} - L_{12} i_{2q}) + v_{1d} \\ -R_1 i_{1q} - \omega_1(L_1 i_{1d} + L_{12} i_{2d}) + v_{1q} \\ -R_2 i_{2d} + \omega_2(L_2 i_{2q} - L_{12} i_{1q}) + v_{2d} \\ -R_2 i_{2q} - \omega_2(L_2 i_{2d} + L_{12} i_{1d}) + v_{1q} \end{pmatrix} \tag{1.71}$$

By denoting $\sigma = 1/(L_1L_2 - L_{12}^2)$, and by using that about the grid voltage holds v_{1d} is constant and v_{1q} is zero, the previous relation provides finally the following equations about the electrical dynamics of the machine:

$$\begin{aligned} \tfrac{d}{dt} i_{1d} = \sigma\{&-R_1L_2i_{1d} + \omega_1L_1L_2i_{1q} - \omega_1L_2L_{12}i_{2q} + L_2V_s + \\ &+ R_2L_{12}i_{2d} - \omega_2L_2L_{12}i_{2q} + \omega_2L_{12}^2i_{1q} - L_{12}v_{2d}\} \end{aligned} \tag{1.72}$$

$$\begin{aligned} \tfrac{d}{dt} i_{1q} = \sigma\{&-R_1L_2i_{1q} + \omega_1L_1L_2i_{1d} + \omega_1L_2L_{12}i_{2d} - \\ &- R_2L_{12}i_{2q} - \omega_2L_2L_{12}i_{2d} + \omega_2L_{12}^2i_{1d} - L_{12}v_{2q}\} \end{aligned} \tag{1.73}$$

$$\begin{aligned} \tfrac{d}{dt} i_{2d} = \sigma\{&-R_1L_{12}i_{2d} - \omega_1L_1L_{12}i_{1q} + \omega_1L_{12}^2i_{2q} - L_{12}V_s - \\ &- R_2L_1i_{2d} + \omega_2L_1L_2i_{2q} - \omega_2L_1L_{12}i_{1q} + L_1v_{2d}\} \end{aligned} \tag{1.74}$$

$$\begin{aligned} \tfrac{d}{dt} i_{2q} = \sigma\{&-R_1L_{12}i_{1q} - \omega_1L_1L_{12}i_{1d} - \omega_1L_{12}^2i_{2d} - \\ &- R_2L_1i_{2q} - \omega_2L_1L_2i_{2d} - \omega_2L_1L_{12}i_{1d} + L_1v_{2q}\} \end{aligned} \tag{1.75}$$

Moreover, from Eq. (1.67) about the rotor's turn speed one has

$$\dot{\omega} = \tfrac{1}{J_r}(\tfrac{3}{2}\tfrac{L_{12}}{L_1}p_r(L_1i_{1d}i_{2q} + L_{12}i_{2d}i_{2q} + L_1i_{1q}i_{2d} + L_{12}i_{2q}i_{2d})) \tag{1.76}$$

Additionally, by considering damping (friction) effects in the rotor's motion one has

$$\dot{\omega} = -\tfrac{b}{J}\omega + \tfrac{1}{J_r}(\tfrac{3}{2}\tfrac{L_{12}}{L_1}p_r(L_1i_{1d}i_{2q} + L_{12}i_{2d}i_{2q} + L_1i_{1q}i_{2d} + L_{12}i_{2q}i_{2d})) \tag{1.77}$$

Next, by defining the state variables of the model as $x_1 = \omega$, $x_2 = i_{1d}$, $x_3 = i_{1q}$, $x_4 = i_{2d}$, $x_5 = i_{2q}$ and the control inputs $u_1 = v_{2d}$ and $u_2 = v_{2q}$ the following state-space description for the reluctance machine is obtained:

$$\dot{x}_1 = -\tfrac{b}{J}x_1 + \tfrac{3}{2J}\tfrac{L_{12}}{L_1}p_r(L_1x_2x_5 + L_{12}x_3x_5 + L_1x_4x_5 + L_{12}x_3x_5) + \tfrac{T_m}{J} \tag{1.78}$$

$$\begin{aligned}\dot{x}_2 = &-\sigma R_1L_2x_2 - \sigma\omega_1L_1L_2x_3 - \sigma\omega_1L_2L_{12}x_5 + \sigma L_2V_s - \\ &-\sigma R_2L_{12}x_4 - \sigma\omega_2L_2L_{12}x_5 + \sigma\omega_2L_{12}^2x_3 - \sigma L_{12}u_1\end{aligned} \tag{1.79}$$

$$\begin{aligned}\dot{x}_3 = &-\sigma R_1L_2x_3 - \sigma\omega_1L_1L_2x_2 + \sigma\omega_1L_2L_{12}x_4 - \\ &-\sigma R_2L_{12}x_5 - \sigma\omega_2L_2L_{12}x_4 + \sigma\omega_2L_{12}^2x_2 + \sigma L_{12}u_2\end{aligned} \tag{1.80}$$

$$\begin{aligned}\dot{x}_4 = &-\sigma R_1L_{12}x_2 - \sigma\omega_1L_1L_{12}x_3 + \sigma\omega_1L_{12}^2x_5 - \sigma L_{12}V_s - \\ &-\sigma R_2L_1x_4 + \sigma\omega_2L_1L_2x_5 - \sigma\omega_2L_1L_{12}x_3 + \sigma L_1u_1\end{aligned} \tag{1.81}$$

$$\begin{aligned}\dot{x}_5 = &-\sigma R_1L_{12}x_3 - \sigma\omega_1L_1L_{12}x_2 - \sigma\omega_1L_{12}^2x_4 - \\ &-\sigma R_2L_1x_5 - \sigma\omega_2L_1L_2x_4 - \sigma\omega_2L_1L_{12}x_2 + \sigma L_1u_2\end{aligned} \tag{1.82}$$

Consequently, the state-space description of the system becomes:

$$\dot{x} = f(x) + g(x)u \tag{1.83}$$

where $x \in R^{5\times1}$, $u \in R^{2\times1}$, $f(x) \in R^{5\times1}$ and $g(x) \in R^{5\times2}$, with the drift matrix to be given by

$$f(x) = \begin{pmatrix} -\tfrac{b}{J}x_1 + \tfrac{3}{2J}\tfrac{L_{12}}{L_1}p_r(L_1x_2x_5 + L_{12}x_3x_5 + L_1x_4x_5 + L_{12}x_3x_5) + \tfrac{T_m}{J} \\ \\ -\sigma R_1L_2x_2 - \sigma\omega_1L_1L_2x_3 - \sigma\omega_1L_2L_{12}x_5 + \sigma L_2V_s - \\ -\sigma R_2L_{12}x_4 - \sigma\omega_2L_2L_{12}x_5 + \sigma\omega_2L_{12}^2x_3 \\ \\ -\sigma R_1L_2x_3 - \sigma\omega_1L_1L_2x_2 + \sigma\omega_1L_2L_{12}x_4 - \\ -\sigma R_2L_{12}x_5 - \sigma\omega_2L_2L_{12}x_4 + \sigma\omega_2L_{12}^2x_2 \\ \\ -\sigma R_1L_{12}x_2 - \sigma\omega_1L_1L_{12}x_3 + \sigma\omega_1L_{12}^2x_5 - \sigma L_{12}V_s - \\ -\sigma R_2L_1x_4 + \sigma\omega_2L_1L_2x_5 - \sigma\omega_2L_1L_{12}x_3 \\ \\ -\sigma R_1L_{12}x_3 - \sigma\omega_1L_1L_{12}x_2 - \sigma\omega_1L_{12}^2x_4 - \\ -\sigma R_2L_1x_5 - \sigma\omega_2L_1L_2x_4 - \sigma\omega_2L_1L_{12}x_2 \end{pmatrix} \tag{1.84}$$

and the inputs gain matrix to be given by

$$g(x) = \begin{pmatrix} 0 & 0 \\ -\sigma L_{12} & 0 \\ 0 & \sigma L_{12} \\ \sigma L_1 & 0 \\ 0 & \sigma L_2 \end{pmatrix} \tag{1.85}$$

1.4 Main Types of Power Electronics

1.4.1 Voltage Source Converters

The voltage source converter turns AC voltage to DC. The voltage source converter model in the rotating dq reference frame is given by [73, 76, 236, 424]:

$$\begin{aligned} L\dot{i}_d &= -Ri_d + L\omega i_q + v_d - \tfrac{V_{dc}}{2}u_1 \\ L\dot{i}_q &= -L\omega i_d - Ri_q + v_q - \tfrac{V_{dc}}{2}u_2 \\ C_{dc}\dot{V}_{dc} &= -\tfrac{1}{R_c}V_{dc} + \tfrac{3}{4}i_d u_1 + \tfrac{3}{4}i_q u_2 \end{aligned} \tag{1.86}$$

where i_d, i_q are the line currents (i_a, i_b, i_c) after transformation in the dq reference frame, and equivalently v_d, v_q are the phase voltages v_a, v_b, v_c after transformation in the dq reference frame. Variable V_{dc} denotes the DC voltage output of the converter, $u_1 = \eta_d$ and $u_2 = \eta_q$ stand for control inputs. The line losses and the transformer conduction losses are modelled by R and the inverter switching losses are modeled by R_c. Moreover, v_q is taken to be 0 (Fig. 1.8).

In Fig. 1.8 the electric circuit of the voltage source converter is depicted. Denoting $x = [i_d, i_q, V_{dc}]^T$ as the state vector, $y = [e_c, i_q]^T$ as the output and $u = [\eta_d, \eta_q]^T$ as the input vector, the MIMO nonlinear model of the VSC is written in the state-space form

$$\begin{aligned} \dot{x} &= f(x) + G(x)u \\ y &= h(x) \end{aligned} \tag{1.87}$$

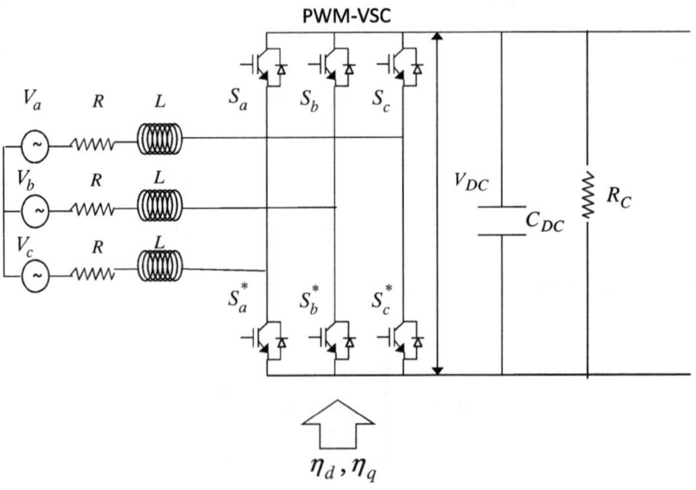

Fig. 1.8 Electrical circuit of the voltage source converter

where

$$f = \begin{pmatrix} -\frac{R}{L}x_1 + \omega x_2 + \frac{v_d}{L} \\ -\omega x_1 - \frac{R}{L}x_2 \\ -\frac{x_3}{C_d c R_c} \end{pmatrix} \quad G = [g_1 \; g_2] = \begin{pmatrix} -\frac{x_3}{2L} & 0 \\ 0 & -\frac{x_3}{2L} \\ \frac{3x_1}{4C_{dc}} & \frac{3x_1}{4C_{dc}} \end{pmatrix} \tag{1.88}$$

$$h = \begin{pmatrix} h_1 \\ h_2 \end{pmatrix} = \begin{pmatrix} e_c \\ i_q \end{pmatrix} = \begin{pmatrix} \frac{3}{4}L(x_1^2 + x_2^2) + \frac{C_{dc}}{2}x_3^2 \\ x_2 \end{pmatrix} \tag{1.89}$$

1.4.2 Inverters

Inverters aim at converting DC electric voltage to AC voltage, thus allowing DC power generation units get connected to the grid. The inverter's (DC to AC converter's) circuit is depicted in Fig. 1.9 [216, 257, 261, 437]. By applying Kirchoff's voltage and current laws one obtains

$$\begin{aligned} \frac{d}{dt}i_I &= \frac{1}{L_f}V_I - \frac{1}{L_f}V_L \\ \frac{d}{dt}V_L &= \frac{1}{C_f}i_I - \frac{1}{C_f}i_L \end{aligned} \tag{1.90}$$

For the representation of the voltage and current variables, denoted as $X = \{I, V\}$ in the ab static reference frame one has

$$X_{ab} = X_a e^{j0} + X_b e^{\frac{j2\pi}{3}} + X_c e^{\frac{j4\pi}{3}} \tag{1.91}$$

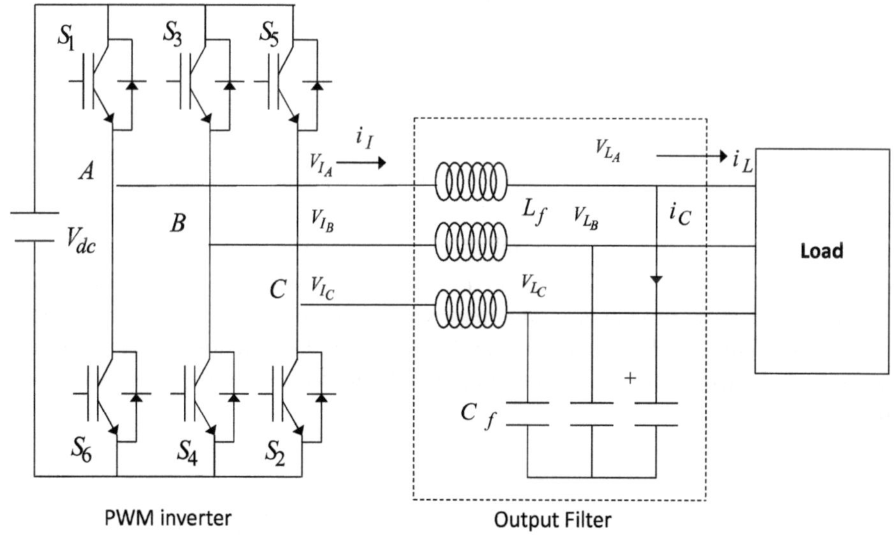

Fig. 1.9 Circuit of the three-phase voltage inverter

which finally gives a complex variable of the form

$$X_{ab} = X_a + jX_b \tag{1.92}$$

Next, the voltage and current variables are represented in the rotating dq reference frame [115]. It holds that $X_{dq} = x_d + jx_q$ and

$$\begin{array}{c} X_{dq} = X_{ab}e^{-j\theta} \Rightarrow X_{ab} = X_{dq}e^{j\theta} \\ \text{where } \theta(t) = \int_0^t \omega(t)dt + \theta_0 \end{array} \tag{1.93}$$

By differentiating with respect to time one obtains the following description of the system's dynamics

$$\dot{X}ab = \frac{d}{dt}X_{dq} + j\omega X_{dq} \tag{1.94}$$

Thus, one has for the current and voltage variables respectively,

$$\begin{array}{c} \dot{i}_{I,ab} = \frac{d}{dt}i_{I,dq} + (j\omega)i_{I,dq} \\ \dot{V}_{L,ab} = \frac{d}{dt}V_{L,dq} + (j\omega)V_{L,dq} \end{array} \tag{1.95}$$

By substituting Eq. (1.95) into Eq. (1.90) one obtains

$$\begin{array}{c} \frac{d}{dt}i_{I,dq} + j\omega i_{I,dq} = \frac{1}{L_f}V_{I,dq} - \frac{1}{L_f}V_{L,dq} \\ \frac{d}{dt}V_{L,dq} + j\omega V_{L,dq} = \frac{1}{C_f}i_{I,dq} - \frac{1}{C_f}i_{L,dq} \end{array} \tag{1.96}$$

Using Eq. (1.96) and by rearranging rows one finally obtains the inverter's dynamic model expressed in the dq reference frame:

$$\begin{array}{c} \frac{d}{dt}V_{L,d} = \omega V_{L,q} + \frac{1}{C_f}i_{I,d} - \frac{1}{C_f}i_{L,d} \\ \frac{d}{dt}V_{L,q} = -\omega V_{L,d} + \frac{1}{C_f}i_{I,q} - \frac{1}{C_f}i_{L,q} \\ \frac{d}{dt}i_{I,d} = \omega i_{I,q} + \frac{1}{L_f}V_{I,d} - \frac{1}{L_f}V_{L,d} \\ \frac{d}{dt}i_{I,q} = -\omega i_{I,d} + \frac{1}{L_f}V_{I,q} - \frac{1}{L_f}V_{L,q} \end{array} \tag{1.97}$$

The state vector of the system is taken to be $\tilde{X} = [V_{L_d}, V_{L_q}, i_{I,d}, i_{I,q}]^T$ while the control input is taken to be the vector $\tilde{U} = [V_{I,d}, V_{I,q}]$. The load currents $i_{L,d}$ and $i_{L,q}$ are taken to be unknown parameters which can be considered as perturbation terms. Alternatively, these currents can be expressed as functions of the inverter's active and reactive power. In the latter approach one has that the active power of the inverter is [216]

$$p_f = V_{L_d}i_{L_d} + V_{L_q}i_{L_q} \tag{1.98}$$

while the reactive power, consisting of reactive power at the load, reactive power at the capacitor and reactive power at the inductance is given by

$$q_f = V_{L_q} i_{L_d} - V_{L_d} i_{L_q} - \omega C_f (V_{L_d}^2 + V_{L_q}^2) + \omega L_f (i_{I,d}^2 + i_{I,q}^2) \tag{1.99}$$

By solving Eqs. (1.98) and (1.99), with respect to the currents i_{L_d} and i_{L_q} one obtains

$$i_{L_d} = \frac{p_f V_{L_d} + q_f V_{L_q}}{V_{L_d}^2 + V_{L_q}^2} + \omega C_f V_{L_q} - \frac{\omega L_f V_{L_q} (i_{I_d}^2 + i_{I_q}^2)}{(V_{L_d}^2 + V_{L_q}^2)} \tag{1.100}$$

$$i_{L_q} = \frac{p_f V_{L_q} - q_f V_{L_d}}{V_{L_d}^2 + V_{L_q}^2} - \omega C_f V_{L_d} + \frac{\omega L_f V_{L_d} (i_{I_d}^2 + i_{I_q}^2)}{(V_{L_d}^2 + V_{L_q}^2)} \tag{1.101}$$

Using Eqs. (1.97), (1.100) and (1.101) one obtains the state-space description of the inverter's dynamics

$$\frac{d}{dt}\begin{pmatrix} V_{L_d} \\ V_{L_q} \\ i_{I_d} \\ i_{I_q} \end{pmatrix} = \begin{pmatrix} \omega V_{L_q} + \frac{1}{C_f} i_{I_d} - \frac{1}{C_f}\frac{p_f V_{L_d} + q_f V_{L_q}}{V_{L_d}^2 + V_{L_q}^2} + \omega C_f V_{L_q} - \frac{\omega L_f V_{L_q}(i_{I_d}^2 + i_{I_q}^2)}{(V_{L_d}^2 + V_{L_q}^2)} \\ -\omega V_{L_d} + \frac{1}{C_f} i_{I_q} - \frac{1}{C_f}\frac{p_f V_{L_q} - q_f V_{L_d}}{V_{L_d}^2 + V_{L_q}^2} - \omega C_f V_{L_d} + \frac{\omega L_f V_{L_d}(i_{I_d}^2 + i_{I_q}^2)}{(V_{L_d}^2 + V_{L_q}^2)} \\ \omega i_{I_q} - \frac{1}{L_f} V_{L_d} \\ -\omega i_{I_d} - \frac{1}{L_f} V_{L_q} \end{pmatrix}$$

$$+ \begin{pmatrix} 0 & 0 \\ 0 & 0 \\ \frac{1}{L_f} & 0 \\ 0 & \frac{1}{L_f} \end{pmatrix} \begin{pmatrix} V_{I_d} \\ V_{I_q} \end{pmatrix} \tag{1.102}$$

while the measurement equation of the inverter's model is

$$\begin{pmatrix} y_1 \\ y_2 \end{pmatrix} = \begin{pmatrix} V_{L_d} \\ V_{L_q} \end{pmatrix} = \begin{pmatrix} 1 & 0 & 0 & 0 \\ 0 & 1 & 0 & 0 \end{pmatrix} \begin{pmatrix} V_{L_d} \\ V_{L_q} \\ i_{L_d} \\ i_{L_q} \end{pmatrix} \tag{1.103}$$

and by using the state variables notation $x_1 = V_{L_d}$, $x_2 = V_{L_q}$, $x_3 = i_{L_d}$ and $x_4 = i_{L_q}$ one has

$$\frac{d}{dt}\begin{pmatrix} x_1 \\ x_2 \\ x_3 \\ x_4 \end{pmatrix} = \begin{pmatrix} \omega x_2 + \frac{1}{C_f} x_3 - \frac{1}{C_f}\frac{p_f x_1 + q_f x_2}{x_1^2 + x_2^2} + \omega C_f x_2 - \frac{\omega L_f x_2 (x_3^2 + x_4^2)}{(x_1^2 + x_2^2)} \\ -\omega x_1 + \frac{1}{C_f} x_4 - \frac{1}{C_f}\frac{p_f x_2 - q_f x_1}{x_1^2 + x_2^2} - \omega C_f x_1 + \frac{\omega L_f x_1 (x_3^2 + x_4^2)}{(x_1^2 + x_2^2)} \\ \omega x_4 - \frac{1}{L_f} x_1 \\ -\omega x_3 - \frac{1}{L_f} x_2 \end{pmatrix} + \tag{1.104}$$

$$+ \begin{pmatrix} 0 & 0 \\ 0 & 0 \\ \frac{1}{L_f} & 0 \\ 0 & \frac{1}{L_f} \end{pmatrix} \begin{pmatrix} u_1 \\ u_2 \end{pmatrix}$$

while the measurement equation of the inverter's model is

$$\begin{pmatrix} y_1 \\ y_2 \end{pmatrix} = \begin{pmatrix} V_{L_d} \\ V_{L_q} \end{pmatrix} = \begin{pmatrix} 1\,0\,0\,0 \\ 0\,1\,0\,0 \end{pmatrix} \begin{pmatrix} x_1 \\ x_2 \\ x_3 \\ x_4 \end{pmatrix} \quad (1.105)$$

thus, the inverter's model is written in the nonlinear state-space form

$$\begin{aligned} \dot{x} &= f(x) + G(x)u \\ y &= h(x) \end{aligned} \quad (1.106)$$

where $f(x) \in R^{4\times 1}$, $G(x) \in R^{4\times 1}$ and $h(x) \in R^{2\times 4}$.

1.4.3 Active Power Filters

Active power filters are inverters which inject to the grid AC voltage of specific frequency, aiming at compensating for harmonics' perturbations. The mathematical model of the active power filter is obtained from the application of Kirchhoff's voltage and currents laws (see Fig. 1.10) [95, 135, 173, 210]:

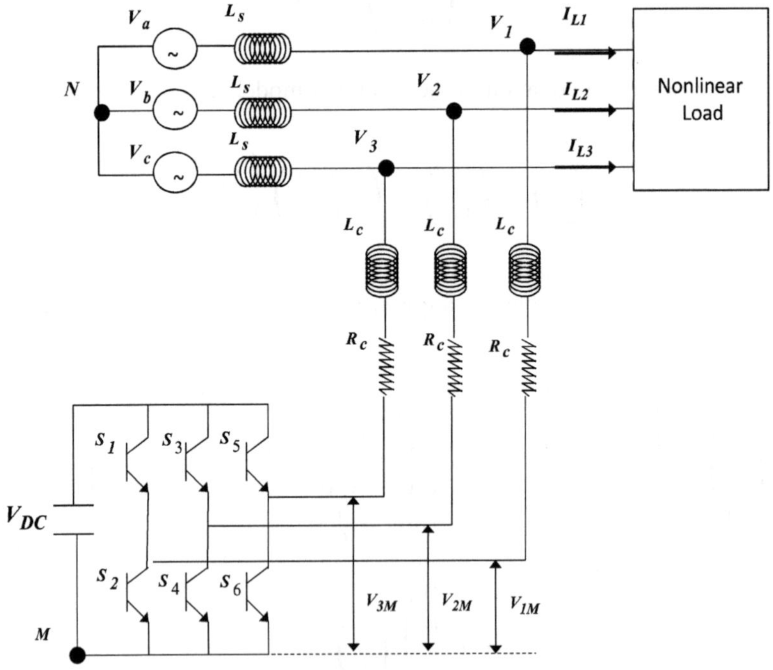

Fig. 1.10 Diagram of the three-phase active power filter

$$\begin{aligned} V_1 &= L_c\frac{dI_1}{dt} + R_cI_1 + V_{1M} + V_{MN} \\ V_2 &= L_c\frac{dI_2}{dt} + R_cI_2 + V_{2M} + V_{MN} \\ V_3 &= L_c\frac{dI_3}{dt} + R_cI_3 + V_{3M} + V_{MN} \end{aligned} \tag{1.107}$$

Considering that the AC supply voltages are balanced one has that

$$V_{MN} = -\tfrac{1}{3}\sum_{m=1}^{3} V_{mM} \tag{1.108}$$

The switching function c_k denotes the ON/OFF status of the IGBT in the two legs of the bridge

$$c_k = \begin{cases} 1 \text{ if } S_K \text{ is ON and } S_{k+1} \text{is OFF} \\ 0 \text{ if } S_K \text{ is OFF and } S_{k+1} \text{is ON} \end{cases} \tag{1.109}$$

where $k = 1, 2, 3$. Using that $V_{kM}, k = 1, 2, 3$ is given by $V_{kM} = C_k V_{DC}$, the previous voltage-current equations become

$$\begin{aligned} \frac{dI_1}{dt} &= -\frac{R_c}{L_c}I_1 + \frac{V_1}{V_c} - \frac{V_{DC}}{L_c}(c_1 - \tfrac{1}{3}\sum_{m=1}^{3} c_m) \\ \frac{dI_2}{dt} &= -\frac{R_c}{L_c}I_2 + \frac{V_2}{V_c} - \frac{V_{DC}}{L_c}(c_2 - \tfrac{1}{3}\sum_{m=1}^{3} c_m) \\ \frac{dI_3}{dt} &= -\frac{R_c}{L_c}I_3 + \frac{V_3}{V_c} - \frac{V_{DC}}{L_c}(c_3 - \tfrac{1}{3}\sum_{m=1}^{3} c_m) \end{aligned} \tag{1.110}$$

The state-vector of the active power filter is taken to be $x = [I_1, I_2, I_3]$. Thus one arrives at a state-space description of the system in the form

$$\dot{x} = f(x,t) + g(x,t)u \tag{1.111}$$

where $f(x,t) = [f_1, f_2, f_3]^T$ and $g(x,t) = diag[g_1, g_2, g_3]$. In Eq. (1.111) the drift function $f(x,t)$, the input gain function $g(x,t)$ and the input vector $u(x,t)$ are defined as follows

$$\begin{aligned} f(x) &= \begin{pmatrix} -\frac{R_c}{L_c}x_1 + \frac{V_1}{V_c} \\ -\frac{R_c}{L_c}x_2 + \frac{V_2}{V_c} \\ -\frac{R_c}{L_c}x_3 + \frac{V_3}{V_c} \end{pmatrix} \qquad g(x) = \begin{pmatrix} -\frac{1}{L_c} \\ -\frac{1}{L_c} \\ -\frac{1}{L_c} \end{pmatrix} \\ u(x,t) &= \begin{pmatrix} V_{DC}[c_1 - \frac{1}{3}\sum_{m=1}^{3} c_m] \\ V_{DC}[c_2 - \frac{1}{3}\sum_{m=1}^{3} c_m] \\ V_{DC}[c_3 - \frac{1}{3}\sum_{m=1}^{3} c_m] \end{pmatrix} \end{aligned} \tag{1.112}$$

The dynamic model of the active power filter can be extended by including in it vector $\tilde{d}$ which denotes additive nonlinear disturbance inputs. Thus the model of Eq. (1.111) becomes

$$\dot{x} = f(x,t) + g(x,t)u + \tilde{d} \tag{1.113}$$

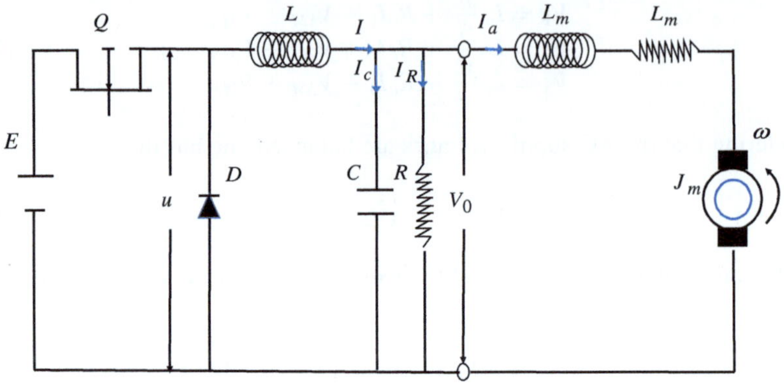

Fig. 1.11 Circuit of the DC-DC converter connected to a DC motor

1.4.4 DC to DC Converters

DC to DC converters are used for generating DC voltage output of variable amplitude, out of DC voltage input of constant amplitude. The equivalent circuit of the system that is formed after connecting a DC motor to a DC-DC (buck) converter is depicted in Fig. 1.11.

Pulse Width Modulation (PWM) is applied for the converter's control. The amplitude of the output voltage V_o is determined by the duty cycle of the PWM. The on/off state of the switch Q sets voltage u to E or to 0 for specific time intervals within the sampling period. The ratio between the time interval in which $u = E$ and the sampling period defines the duty cycle. By varying the duty cycle one can control the voltage output V_o, as if a variable input voltage E was applied to the circuit.

The dynamics of the electrical part of the circuit comes from the application of Kirchhoff's laws. It holds:

$$L\frac{dI}{dt} + V_c = u$$

$$L_m\frac{dI_a}{dt} + R_m I_a + K_e\omega = V_c \tag{1.114}$$

$$I = I_c + I_R + I_a \; or \; I = C\frac{dV_c}{dt} + \frac{V_c}{R} + I_a$$

The dynamics of the mechanical part of the converter and motor system comes from the laws of rotation motion of the rotor. It holds that:

$$\dot{\theta} = \omega$$

$$J\dot{\omega} = -B\omega + K_a I_a + \tau_L \tag{1.115}$$

where τ_L is the load's torque. By defining the state variables $x_1 = \theta, x_2 = \omega, x_3 = I$, $x_4 = V_c$ and $x_5 = I_a$ one obtains the following state-space model:

$$\begin{aligned}
\dot{x}_1 &= x_2 \\
\dot{x}_2 &= -\tfrac{B}{J}x_2 + \tfrac{K_a}{J}x_5 + \tfrac{1}{J}\tau_d \\
\dot{x}_3 &= -\tfrac{1}{L}x_4 + \tfrac{1}{L}u \\
\dot{x}_4 &= \tfrac{1}{C}x_3 - \tfrac{1}{RC}x_4 + \tfrac{1}{C}x_5 \\
\dot{x}_5 &= \tfrac{1}{L_m}x_4 - \tfrac{R_m}{L_m}x_5 + \tfrac{K_e}{L_m}x_2
\end{aligned} \tag{1.116}$$

Without loss of generality it is considered that the load's torque is $\tau_L = mglsin(x_1)$ (the motor is considered to be rotating a rod of length l having a a mass m at its end), the previous state-space description is written in the following matrix form

$$\begin{pmatrix} \dot{x}_1 \\ \dot{x}_2 \\ \dot{x}_3 \\ \dot{x}_4 \\ \dot{x}_5 \end{pmatrix} = \begin{pmatrix} x_2 \\ -\tfrac{B}{J}x_2 + \tfrac{K_a}{J}x_5 + \tfrac{1}{J}mglsin(x_1) \\ -\tfrac{1}{L}x_4 \\ \tfrac{1}{C}x_3 - \tfrac{1}{RC}x_4 + \tfrac{1}{C}x_5 \\ \tfrac{K_e}{L_m}x_2 + \tfrac{1}{L_m}x_4 - \tfrac{R_m}{L_m}x_5 \end{pmatrix} + \begin{pmatrix} 0 \\ 0 \\ \tfrac{1}{L} \\ 0 \\ 0 \end{pmatrix} u \tag{1.117}$$

1.4.5 Fuel Cells

1.4.5.1 Nonlinear Dynamics of PEM Fuel Cells

Fuel cells have emerged as a significant renewable energy source which finds wide application in electric traction systems (e.g. electric vehicles). The Proton Exchange Membrane (PEM) Fuel Cell consists of a polymer electrolyte membrane which is placed between the electrodes (anode and cathode), as shown in Fig. 1.12. Ions can be diffused through the membrane. If an electrical circuit is established between the anode and the cathode, there will be also a flow of electrons and a potential will appear between the electrodes [100, 257, 298, 332].

In the considered PEM fuel cell the anode is supplied with gas that contains hydrogen while the cathode is supplied with gas which contains oxygen. The overall electrochemical dynamics is

$$\begin{array}{lc}
\text{Anode} & 2H_2 \leftrightarrow 4H^+ + 4e^- \\
\text{Cathode} & O_2 + 4H^+ + 4e^- \leftrightarrow 2H_2O \\
\text{Overall} & 2H_2 + O_2 \leftrightarrow 2H_2O + \text{electricity} + \text{heat}
\end{array} \tag{1.118}$$

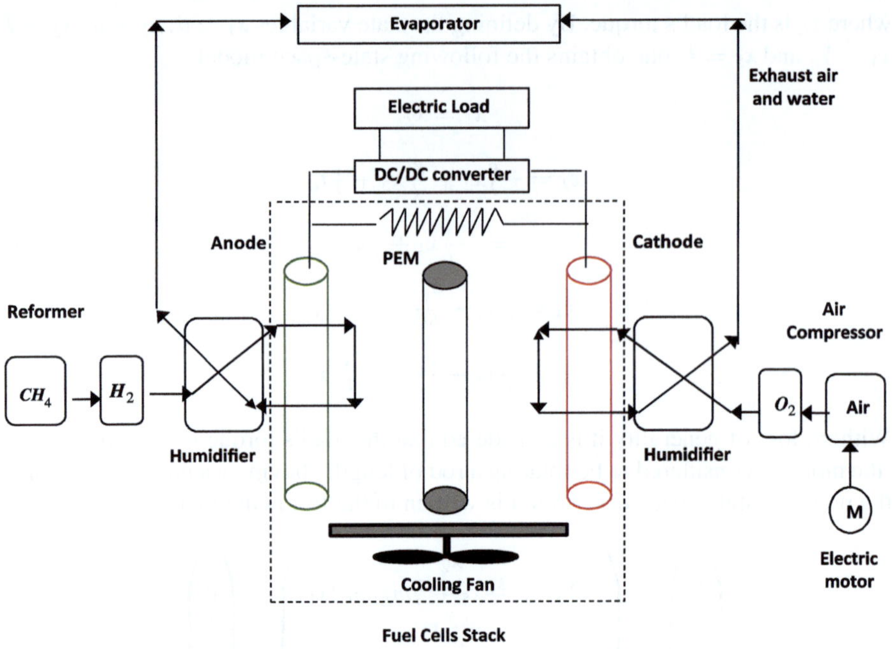

Fig. 1.12 A PEM fuel cells model

The anode can either be supplied with H_2 under pressure or can be supplied with hydrogen by the *reformer* which generates H_2 from methane or other natural gas. The cathode is supplied with oxygen through an air compressor connected to an air filter and finally connected to an air flow controller (valve). On both sides a humidifier is used to prevent drying of the PEM. To produce a higher voltage, multiple cells are connected in series and this forms a stack of fuel cells. A singe cell provides voltage between 0 and 1 V.

Conditions about the PEM fuel cells functioning are outlined as follows: (1) The temperature of the fuel cells both at the anode's and at the cathode's side is assumed to remain constant, (2) The anode and the cathode are sufficiently humidified, (3) it is assumed that the produced water is evaporated (4) the inlet reactants are assumed to be supplied in constant mole fractions. This means that pure hydrogen 100 % is fed to the anode. The air supply to the cathode consists of nitrogen and oxygen at rations 79 % and 21 %, respectively, (5) the gases are assumed to follow the ideal gases low.

The dynamics of the fuel cells system is given through the following two sets of differential equations [299]:

Anode mole conservation:

$$\frac{dP_{H_2}}{dt} = \frac{RT}{V_a}[H_{2in} - H_{2used} - H_{2out}] \tag{1.119}$$

$$\frac{dP_{H_2O_A}}{dt} = \frac{RT}{V_a}[H_2O_{in} - H_2O_{used} - H_2O_{out}] \tag{1.120}$$

Cathode mole conservation:

$$\frac{dP_{O_2}}{dt} = \frac{RT}{V_c}[O_{2in} - O_{2used} - O_{2out}] \tag{1.121}$$

$$\frac{dP_{N_2}}{dt} = \frac{RT}{V_c}[N_{2in} - N_{2out}] \tag{1.122}$$

$$\frac{dP_{H_2O_c}}{dt} = \frac{RT}{V_c}[H_2O_{cin} - H_2O_{cproduced} - H_2O_{cout} + H_2O_{cmbr}] \tag{1.123}$$

In the above state equations H_{2in}, O_{2in}, H_2O_{Ain}, N_{2in}, and H_2O_{cin} are the inlet flow rates of hydrogen, oxygen, nitrogen, anode-side water and cathode-side water. Moreover, H_{2out}, O_{2out}, H_2O_{Aout}, N_{2out} and H_2O_{cout} are the outlet flow rates of each reactant. Additionally, H_{2used}, O_{2used} and $H_2O_{cproduced}$ are the usage and produced concentrations of the reactants. Furthermore, H_2O_{mbr} is the water concentration transferred through the membrane and is a function of the stack current and of the humidity (which is assumed to remain constant). It is also noted that V_a is the anode's volume and V_c is the cathode's volume (multiplied by the reactant's mass concentration in mole).

1.4.5.2 A Nonlinear State Equations Model of the PEM Fuel Cells

In continuation to the previous analysis a nonlinear model of the PEM fuel cells system is presented. Focusing on the cathode, the state vector of the model is defined as $x = [p_{O_2}, p_{N_2}, \omega_{cp}, p_{sin}]^T$, where p_{O_2} is the oxygen pressure at the cathode, p_{N_2} is the nitrogen pressure at the cathode, ω_{cp} is the compressor's rotational speed (r/min), and p_{sm} is the supply manifold pressure [257].

By applying the ideal gas law and by considering that the volume of the cathode is known one has

$$\begin{aligned} \frac{dp_{O_2}}{dt} &= \frac{RT}{M_{O_2}V_{ca}}(W_{O_{2,in}} - W_{O_2,out} - W_{O_{2,react}}) \\ \frac{dp_{N_2}}{dt} &= \frac{RT}{M_{N_2}V_{ca}}(W_{N_{2,in}} - W_{N_2,out}) \end{aligned} \tag{1.124}$$

where V is the volume of the cathode, R is the universal gas constant, and M_{O_2}, M_{N_2} are the mass concentrations (in mole) of oxygen and nitrogen. The incoming flow rates of oxygen and nitrogen are given by

$$\begin{aligned} W_{O_2,in} &= x_{O_2}W_{ca,in} \\ W_{N_2,in} &= (1 - x_{O_2})W_{ca,in} \end{aligned} \tag{1.125}$$

where x_{O_2} is the oxygen mass fraction of the inlet air, $1 - x_{O_2}$ is the nitrogen mass fraction of the inlet air, and $W_{ca,in}$ is the mass flow rate entering the cathode which is given by

$$W_{ca,in} = \frac{1}{1+\omega_{atm}} k_{ca,in}(p_{sm} - p_{in}) \tag{1.126}$$

where ω_{atm} is the humidity ratio

$$\omega_{atm} = \frac{M_v}{M_{a,ca,in}} \frac{\phi_{ca} p_{sat}(T_{atm})}{p_{atm} - \phi_{ca} p_{sat}(T_{atm})} \tag{1.127}$$

M_v is the mass of the vapor in mole, $M_{a,ca,in}$ is the mass of the air in mole, ϕ_{ca} is the relative humidity in ambient conditions, $p_{sat}(T_{atm})$ is the saturation pressure in ambient temperature, p_{atm} is the atmospheric pressure and $k_{ca,in}$ is the cathode inlet orifice constant.

The outlet flow rates of oxygen and nitrogen $W_{O_2,out}$ and $W_{N_2,out}$ are calculated from the mass fraction of oxygen and nitrogen in the stack after reaction

$$\begin{aligned} W_{O_2,out} &= \frac{M_{O_2} p_{O_2}}{M_{O_2} p_{O_2} + M_{N_2} p_{N_2} + M_v p_{sat}} W_{ca,out} \\ W_{N_2,out} &= \frac{M_{N_2} p_{N_2}}{M_{O_2} p_{O_2} + M_{N_2} p_{N_2} + M_v p_{sat}} W_{ca,out} \end{aligned} \tag{1.128}$$

The flow rate at the cathode's exit $W_{ca,out}$ is calculated by the nozzle flow equation

$$\begin{aligned} W_{ca,out} &= \frac{C_D A_T p_{ca}}{\sqrt{RT}} \left(\frac{p_{atm}}{p_{ca}}\right)^{\frac{1}{T}} \\ \text{if } \frac{p_{atm}}{p_{ca}} &> \left(\frac{2}{\gamma+1}\right)^{\frac{\gamma}{\gamma-1}} \left\{\frac{2\gamma}{\gamma-1}\left[1 - \left(\frac{p_{atm}}{p_{ca}}\right)^{\frac{\gamma-1}{\gamma}}\right]\right\} \end{aligned} \tag{1.129}$$

where γ is the ratio of the specific heat capacities of the air, $p_{ca} = p_{O_2} + p_{N_2} + P_{sat}$. The mass flow rate of oxygen is expressed as

$$W_{O_2,react} = \frac{nI_{st}}{4F} M_{O_2} \tag{1.130}$$

where n is the number of cells in the stack, F is the Faraday number and I_{st} is the stack current. The compressor's turn speed is related to the associated mechanical torque

$$\frac{d\omega_{cp}}{dt} = \frac{1}{J_{cp}}(\tau_{cm} - \tau_{cp}) \tag{1.131}$$

where τ_{cm} is the mechanical input torque, τ_{cp} is the load torque [257]

$$\tau_{cm} = \eta_{cm} \frac{K_v}{R_{cm}} (v_{cm}) k_v \omega_{cp} \tag{1.132}$$

$$\tau_{cp} = \frac{C_p}{\omega_{cp}} \frac{T_{atm}}{\eta_{cp}} \left[\left(\frac{p_{atm}}{p_{ca}}\right)^{\frac{\gamma-1}{\gamma}} - 1\right] W_{cp} \tag{1.133}$$

where k_t, R_{cm} and k_v are motor constants, η_{cm} is a coefficient that denotes the motor's mechanical efficiency. C_p is the specific heat capacity of air and W_{cp} is the compressor mass flow rate. The dynamics of the air pressure in the supply manifold depend on the compressor flow into the supply manifold $W_{cp} = A\omega_{cp}$, on the flow out of the supply manifold into the cathode $W_{co,in}$ and on the compressor flow temperature T_{cp} [257]

$$\frac{dp_{sm}}{dt} = \frac{RT_{cp}}{M_a V_{sm}}[W_{cp} - k_{ca,in}(p_{sm} - p_{ca})] \tag{1.134}$$

where V_{sm} is the supply manifold volume and T_{cp} is the temperature of the air leaving the compressor

$$T_{cp} = T_{atm} + \frac{T_{atm}}{\eta_{cp}}[(\frac{p_{sm}}{p_{atm}})^{\frac{\gamma-1}{\gamma}} - 1] \tag{1.135}$$

The nonlinear state-space model of the PEM fuel-cells model is based on Eqs. (1.124), (1.131) and (1.134) [257]

$$\dot{x}_1 = c_1(x_4 - x_1 - x_2 - c_2) - \frac{c_3 x_1 W_{co,out}}{c_4 x_1 + c_5 x_2 + c_6} - c_7\zeta \tag{1.136}$$

$$\dot{x}_2 = c_8(x_4 - x_1 - x_2 - c_2) - \frac{c_3 x_2 W_{co,out}}{c_4 x_1 + c_5 x_2 + c_6} \tag{1.137}$$

$$\dot{x}_3 = -c_9 x_3 - c_{10}[(\frac{x_4}{c_{11}})^{c_{12}} - 1] + c_{13}u \tag{1.138}$$

$$\dot{x}_4 = c_{14}\{1 + c_{15}[(\frac{x_4}{c_{11}})^{c_{12}} - 1]\} \cdot [W_{cp} - c_{16}(x_4 - x_1 - x_2 - c_2)] \tag{1.139}$$

where the coefficients $c_1, c_2, \ldots, c_{16}$ are constants. The control input u depends the motor's current. The control input ζ is the stack current (which can be considered as an external perturbation to the model).

The previous set of state-space equations is also written in the form

$$\dot{x} = f(x) + g(x)u \Rightarrow$$

$$\begin{pmatrix} \dot{x}_1 \\ \dot{x}_2 \\ \dot{x}_3 \\ \dot{x}_4 \end{pmatrix} = \begin{pmatrix} c_1(x_4 - x_1 - x_2 - c_2) - \frac{c_3 x_1 W_{co,out}}{c_4 x_1 + c_5 x_2 + c_6} - c_7\zeta \\ \dot{x}_2 = c_8(x_4 - x_1 - x_2 - c_2) - \frac{c_3 x_2 W_{co,out}}{c_4 x_1 + c_5 x_2 + c_6} \\ \dot{x}_3 = -c_9 x_3 - c_{10}[(\frac{x_4}{c_{11}})^{c_{12}} - 1] \\ c_{14}\{1 + c_{15}[(\frac{x_4}{c_{11}})^{c_{12}} - 1]\} \cdot [W_{cp} - c_{16}(x_4 - x_1 - x_2 - c_2)] \end{pmatrix} + \begin{pmatrix} 0 \\ 0 \\ c_{13} \\ 0 \end{pmatrix} u \tag{1.140}$$

1.4.6 Batteries

Batteries are needed for the storage of electric power. Diffusion PDE dynamics in Li-ion batteries is considered. The conservation of Li in a single spherical active material is described by Fick's law of diffusion [81, 218, 219, 416, 417]

$$\frac{\partial c_s}{\partial t} = \frac{D_s}{r^2}\frac{\partial}{\partial r}\left(r^2\frac{\partial c_s}{\partial r}\right) \tag{1.141}$$

with boundary conditions

$$\frac{\partial c_s}{\partial r}\mid_{r=0} = 0 \tag{1.142}$$

$$D_s\frac{\partial c_s}{\partial r}\mid_{r=R_s} = -\frac{j^{Li}}{\alpha_s F} \tag{1.143}$$

In Eqs. (1.141)–(1.143) c_s denotes the concentration of Li and the subscript s stands for the solid phase. Moreover, D_s is a solid-phase diffusion coefficient, j^{Li} is the volumetric rate of electrochemical reaction at the particle surface with $j^{Li} > 0$ to denote ion discharge, a_s is the specific interfacial surface area, and F is Faraday's constant (96.487 C/mol.) For spherical active material particles of radius R_s occupying electrode volume fraction equal to ε_0, the interfacial surface area is $\alpha_s = 3\varepsilon_0/R_s$.

Equations (1.141)–(1.143) are applied on a continuum basis across each electrode, thus giving to the solid-phase concentration of Li the form of a 2D spatial variation $c_s(x, r, t)$, where x is the particle's position, r is the radial position within a particle and t is time. The electrochemical model depends only on concentration on a particle's surface $c_{s,e} = c_s(x, R_s, t)$. Where the subscript s denotes the solid condition and the subscript e denotes the electrolyte condition (Fig. 1.13).

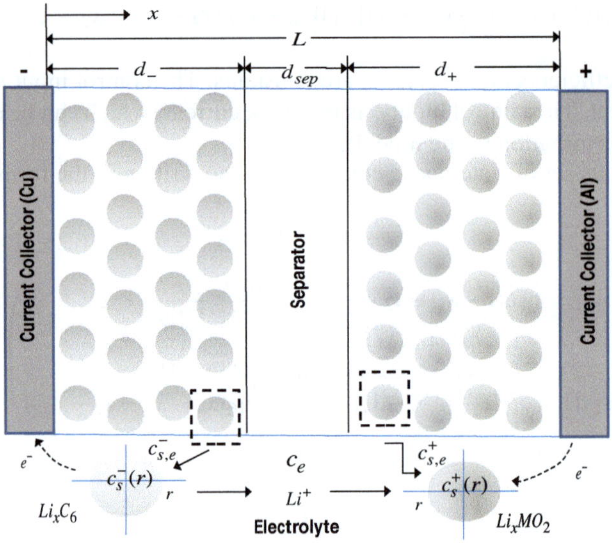

Fig. 1.13 Diffusion of lithium ions in a Li-ion battery [417]

Conservation of Li in the electrolyte phase gives [417]

$$\frac{\partial(\varepsilon_0 c_e)}{\partial t} = \frac{\partial}{\partial x}(D_e^{\text{eff}} \frac{\partial}{\partial x} c_e) + \frac{(1-t_+^0)}{F} j^{Li} \tag{1.144}$$

with boundary conditions

$$\frac{\partial c_e}{\partial x} \mid_{x=0} = 0 \tag{1.145}$$

$$\frac{\partial c_e}{\partial x} \mid_{x=L} = 0 \tag{1.146}$$

where $c_e(x, t)$ is Li concentration in the electrolyte phase, ε_e is the electrolyte phase volume fraction, t_+^0 is the transference number of Li^+ with respect to the velocity of the solvent. The effective diffusion coefficient is calculated from a reference coefficient using the Bruggeman relation $D_e^{\text{eff}} = D_e \varepsilon_e^p$ that accounts for the path that the Li^+ ions follow through the porous media. Moreover, it is assumed that t_+^0 is constant.

Charge conservation in the solid phase of each electrode is described by Ohm's law

$$\frac{\partial}{\partial x}(\sigma^{\text{eff}}) \frac{\partial}{\partial x} \phi_s - j^{Li} = 0 \tag{1.147}$$

with boundary conditions at the current collectors to be proportional to the applied current

$$-\sigma_-^{\text{eff}} \frac{\partial \phi_s}{\partial x} \mid_{x=0} = \sigma_+^{\text{eff}} \frac{\partial \phi_s}{\partial x} \mid_{x=L} = \frac{I}{A} \tag{1.148}$$

while the electronic current at the separator is zero

$$\frac{\partial \phi_s}{\partial x} \mid_{x=\delta_-} = \frac{\partial \phi_s}{\partial x} \mid_{x=\delta_- + \delta_{sep}} = 0 \tag{1.149}$$

In Eqs. (1.147)–(1.149) $\phi_s(x, t)$ is the potential, σ^{eff} is the effective conductivity, and is evaluated from the active material conductivity $\sigma^{\text{eff}} = \sigma \varepsilon_s$, A is the electrode plate area, and $I(t)$ is the applied current, following the sign convention that a positive current discharges the battery.

Electrolyte phase charge conservation gives

$$\frac{\partial}{\partial x}(\kappa^{\text{eff}} \frac{\partial}{\partial x} \phi_e) + \frac{\partial}{\partial x}(\kappa_D^{\text{eff}} \frac{\partial}{\partial x} ln c_e) + j^{Li} = 0 \tag{1.150}$$

with boundary conditions

$$\frac{\partial \phi_e}{\partial x} \mid_{x=0} = \frac{\partial \phi_e}{\partial x} \mid_{x=L} = 0 \tag{1.151}$$

In Eqs. (1.150) and (1.151), $\phi_e(x, t)$ is the electrolyte phase potential and κ^{eff} is the effective ionic conductivity calculated from the Bruggeman relation $\kappa^{\text{eff}} = \kappa \varepsilon_e^D$.

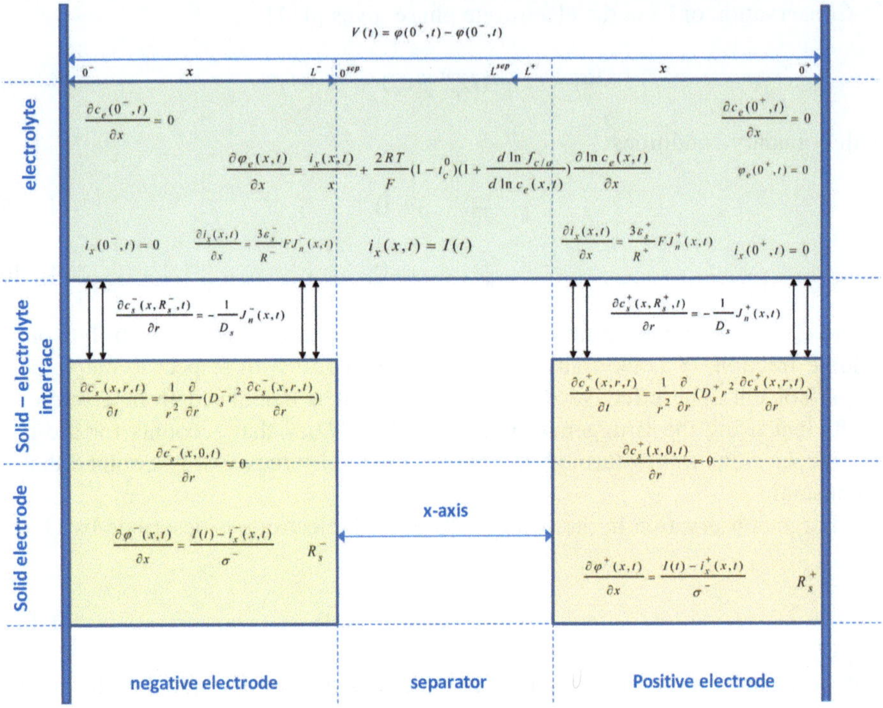

Fig. 1.14 Diagram of the Li-ions diffusion dynamics in the battery [218]

The effective diffusion conductivity is given by

$$\kappa_D^{\text{eff}} = \frac{2RT\kappa^{\text{eff}}}{F}(t_+^0 - 1)(1 + \frac{d \cdot ln(f_\perp)}{d \cdot ln(c_e)}) \tag{1.152}$$

where R is the universal gas constant (8.3143 J/molK), T is the temperature and $f_\perp$ is the activity coefficient which is taken to be constant.

The battery's dynamics consists of four PDEs which describe (i) the lithium concentration at the electrode $c_s(x, t)$ given in Eq. (1.141), (ii) the lithium concentration at the electrolyte $c_e(x, t)$ given in Eq. (1.144), (iii) the potential at the electrode $\phi_s(x, t)$ given in Eq. (1.147) and (iv) the potential at the electrolyte $\phi_e(x, t)$ given in Eq. (1.150). The aggregate PDE dynamics of the Li-ions battery is shown in Fig. 1.14. These PDEs are coupled by the Butler-Volmer electrochemical kinetic expression

$$j^{Li} = \alpha_s i_0\{exp[\frac{\alpha_a F}{RT}\eta] - exp[-\frac{\alpha_c F}{RT}\eta]\} \tag{1.153}$$

where η is the difference between the solid and the electrolytic phase potential minus the thermodynamic equilibrium potential U. Thus one has

$$\eta = \phi_s - \phi_e - U \tag{1.154}$$

The equilibrium potential $U(c_{s,e})$ is evaluated as a function of the solid phase concentration at the particle surface. In Eq. (1.153) the exchange current density i_0 is a function of ion concentrations both at the solid electrode and at the electrolyte

$$i_0 = k(c_e)^{\alpha_0}(c_{s_{max}} - c_{s,0})^{\alpha_0}(c_{s,e})^{\alpha_c} \tag{1.155}$$

where k is a kinetic rate constant, while α_0 and α_c are the anodic and cathodic transfer coefficients. The current $I(t)$ is provided by the boundary conditions of Eq. (1.148). Then the voltage across the cell terminals is given by

$$V(t) = \phi_s(L,t) - \phi_s(0,t) - \frac{R_I}{A}I(t) \tag{1.156}$$

where R_I is an empirical contact resistance.

1.5 Components of the Transmission and Distribution System

1.5.1 Power Transformers

1.5.1.1 Main Types of Power Transformers

Power transformers are the most expensive and strategic components of a power system (Fig. 1.15). One can distinguish between several classes of power transformers using two major classification criteria. The first criterion has to do with the insulating material used in the transformer (e.g. oil-immersed, gas-immersed and dry-type transformers). The second criterion has to do with the incoming and outgoing voltage levels of the power transformers and their role in the electric power grid (e.g. power transformers in generation stations, power transformers in the transmission system, power transformers in the distribution system, distribution substation transformers or distribution network transformers) [3, 27, 196, 289, 430, 452].

1.5.1.2 Classification According to the Insulating Material

Considering the insulation material, power transformers are classified in three main categories: (i) oil-immersed transformers, (ii) gas-insulated transformers and (iii) dry-type transformers [289, 452]. Oil-immersed transformers are used for a wide range of voltage and power ratings from distribution to transmission levels, i.e. from medium voltage to ultra-high voltage operation ranges. They use paper-wrapped windings immersed in mineral oil, which serves as both the insulation and the cooling medium. On the other hand, gas-insulated transformers use polyethylene terephthalate (PET) to insulate the windings, which are placed in a sealed tank filled with

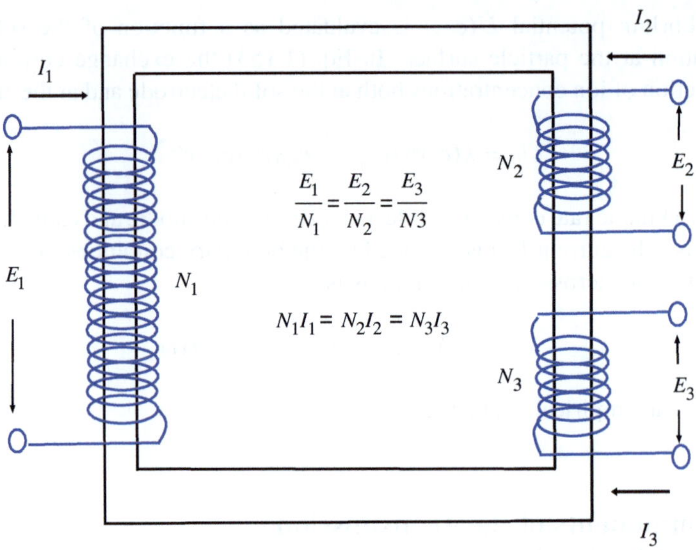

Fig. 1.15 A single-phase power transformer

SF_6 gas that cools the windings and protects them from moisture and dust. There are also dry-type transformers which can have their windings insulated in several ways. A basic insulation method is to preheat the windings and then dip them into varnish at a high temperature. The coils are then baked to cure the varnish. Another version of dry-type transformers are the ones using a cast coil insulation system. This provides additional coil strength and protection. The latter type of transformers is used in harsh environments such as chemical and cement plants and in outdoors installations, where moisture, salt spray, dust, corrosive fumes and metal particles can destroy other types of transformers.

1.5.1.3 Classification According to Voltage Levels

(a) *HV and EHV transformers in the power transmission system.* Such transformers can connect different parts of the power transmission system, e.g. 400–132 kV. Indicative MVA ratings are of several hundred of MVA (e.g. for a part of the transmission system between 400 and 220 kV).
(b) *Distribution substation transformers*: Distribution substation transformers connect the transmission system (400–132 kV) to the distribution system (46–2.2 kV). Distribution substation transformers have MVA ratings (e.g. 75–2.5 MVA) that indicate the continuous load that the transformers carry without exceeding a specified temperature rise of either 55 °C (for older transformers) or 65 °C (for newer transformers) above a specified ambient (typically 40 °C).
(c) *Distribution network transformers*: Distribution network transformers connect the primary system (46–2.2 kV) to the secondary system (480 V and lower).

The kVA ratings of distribution transformers are based again on the continuous load the transformers can carry out without exceeding a specified temperature rise of either 55 °C (for older transformers) or 65 °C (for newer transformers), above a specified ambient temperature (usually 40 °C).

1.5.2 AC Lines

The AC transmission line parameters R, L and C are considered to be distributed over the entire line's length and are given in per-length values, having units Ω/m, H/h and F/m. To take into account the distributed nature of the transmission line's parameters, the following circuit is considered, which represents a line section of length equal to Δx [157].

Parameters $V(x)$ and $I(x)$ denote the voltage and current at position x, which is measured starting from the receiving end of the line. In a similar manner $V(x+\Delta x)$ and $I(x+\Delta x)$ denote the voltage and current at position $(x+\Delta x)$. The circuit constants, such as series impedance and shunt conductance are

$$\begin{aligned} z &= R + j\omega L \ \ \Omega/m \\ y &= G + j\omega C \ \ S/m \end{aligned} \tag{1.157}$$

where the shunt conductance is usually neglected for 60 Hz overhead lines. By applying Kirchhoff's voltage law for the circuit of Fig. 1.16

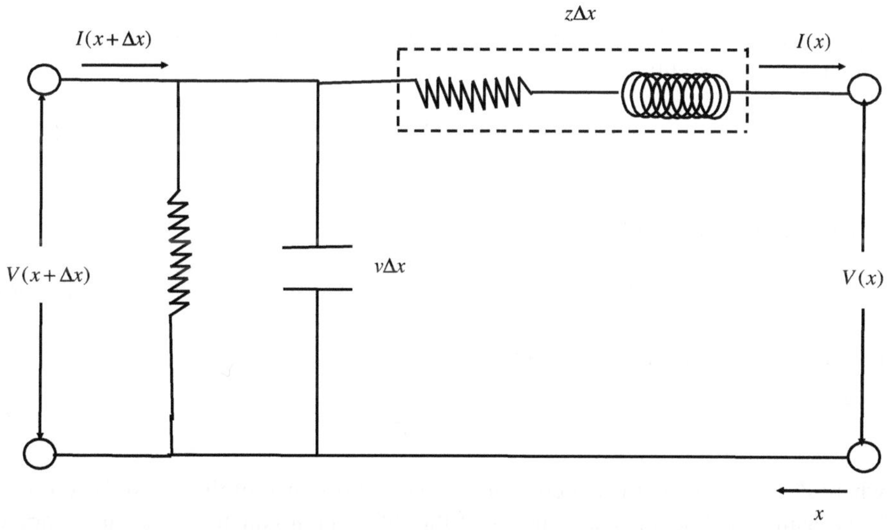

Fig. 1.16 Equivalent circuit of the segment of an AC transmission line

$$V(x+\Delta x) = V(x) + (z\Delta x)I(x) \tag{1.158}$$

Eq. (1.158) is also written as

$$\frac{V(x+\Delta x)-V(x)}{\Delta x} = zI(x) \tag{1.159}$$

and taking the limit as Δx approaches 0 one has

$$\frac{dV(x)}{dx} = zI(x) \tag{1.160}$$

Similarly, Kirchhoff's current law gives

$$I(x+\Delta x) = I(x) + (y\Delta x)V(x+\Delta x) \quad A \tag{1.161}$$

By rearranging terms one obtains

$$\frac{dI(x)}{dx} = yV(x) \tag{1.162}$$

By differentiating Eq. (1.160) with respect to x and by substituting in it Eq. (1.162) one obtains

$$\frac{d^2V(x)}{dx^2} = \frac{zdI(x)}{dx} = zyV(x) \tag{1.163}$$

or equivalently

$$\frac{d^2V(x)}{dx^2} - zyV(x) = 0 \tag{1.164}$$

Equation (1.165) is a linear second-order homogeneous differential equation of $V(x)$, which admits as solution

$$V(x) = A_1e^{\gamma x} + A_2e^{-\gamma x} \tag{1.165}$$

where $\gamma = sqrtzy$, is the propagation constant and can take complex values. By differentiating Eq. (1.165) one gets

$$\frac{dV(x)}{dx} = \gamma A_1e^{\gamma x} - \gamma A_2e^{-\gamma x} = zI(x) \tag{1.166}$$

By solving Eq. 1.166 with respect to $I(x)$ one gets

$$I(x) = \frac{A_1e^{\gamma x} - A_2e^{-\gamma x}}{Z_c} \tag{1.167}$$

where $Z_c = \sqrt{z/y}$. and is called characteristic impedance of the transmission line. Next, using the boundary conditions of the differential equation one can compute coefficients A_1 and A_2. At the receiving end of the transmission line, that is at $x = 0$ one has the boundary conditions

$$V_R = V(0) \quad I_R = I(0) \tag{1.168}$$

Moreover, at $x = 0$, from Eqs. (1.165) and (1.167) one has

$$V_R = A_1 + A_2 \quad I_R = \frac{A_1 - A_2}{Z_c} \tag{1.169}$$

Solving Eq. 1.169 with respect to A_1 and A_2 one gets

$$A_1 = \frac{V_R + Z_c I_R}{2} \quad A_2 = \frac{V_R - Z_c I_R}{2} \tag{1.170}$$

Substituting A_1 and A_2 into Eqs. (1.160) and 1.162 one obtains

$$\begin{aligned} V(x) &= \frac{V_R + Z_c I_R}{2} e^{\gamma x} + \frac{V_R - Z_c I_R}{2} e^{-\gamma x} \\ I(x) &= \frac{V_R + Z_c I_R}{2Z_c} e^{\gamma x} + \frac{V_R - Z_c I_R}{2Z_c} e^{-\gamma x} \end{aligned} \tag{1.171}$$

By rearranging terms Eq. (1.171) can be also written as

$$\begin{aligned} V(x) &= \frac{e^{\gamma x} + e^{-\gamma x}}{2} V_R + Z_c \frac{e^{\gamma x} - e^{-\gamma x}}{2} I_R \\ I(x) &= \frac{1}{Z_c} \frac{e^{\gamma x} - e^{-\gamma x}}{2} V_R + \frac{e^{\gamma x} + e^{-\gamma x}}{2} I_R \end{aligned} \tag{1.172}$$

By substituting in the previous equations the hyperbolic functions *cosh* and *sinh* the relations about voltage and current along the transmission line take the form

$$\begin{aligned} V(x) &= cosh(\gamma x) V_R + Z_c sinh(\gamma x) I_R \\ I(x) &= \frac{1}{Z_c} sinh(\gamma x) V_R + cosh(\gamma x) I_R \end{aligned} \tag{1.173}$$

In matrix form, the equation of the distributed line is written as

$$\begin{pmatrix} V(x) \\ I(x) \end{pmatrix} = \begin{pmatrix} A(x) & B(x) \\ C(x) & D(x) \end{pmatrix} \begin{pmatrix} V_R \\ I_R \end{pmatrix} \tag{1.174}$$

where

$$\begin{aligned} A(x) &= cosh(\gamma x) \quad B(x) = Z_c sinh(\gamma x) \\ C(x) &= \frac{1}{Z_c} sinh(\gamma x) \quad D(x) = cosh(\gamma x) \end{aligned} \tag{1.175}$$

Equation (1.174) gives the voltage and current at any point of the transmission line, provided that the voltage and current at the receiving end are known. In a similar manner one can compute the voltage and current at the sending end

$$\begin{pmatrix} V_s(x) \\ I_s(x) \end{pmatrix} = \begin{pmatrix} A(x) & B(x) \\ C(x) & D(x) \end{pmatrix} \begin{pmatrix} V_R \\ I_R \end{pmatrix} \tag{1.176}$$

where at the sending end $x = l$

$$\begin{array}{ll} A(x) = cosh(\gamma l) & B(x) = Z_c sinh(\gamma l) \\ C(x) = \frac{1}{Z_c} sin(\gamma l) & D(x) = cosh(\gamma l) \end{array} \quad (1.177)$$

1.5.3 HVDC Lines

HVDC lines are used to transmit DC power over long distances and are connected to the rest of the grid through AC-DC converters [25, 26, 89, 90, 104, 303]. By connecting the three-phase voltage source converter, which is found at the generator's side with the three-phase inverter which is found at the load's side through an HVDC line (see Fig. 1.17) the high voltage DC transmission system is formed. Moreover, by applying Kirchhoff's voltage and current laws the dynamic of the HVDC line is obtained. Thus at the converter's side it holds

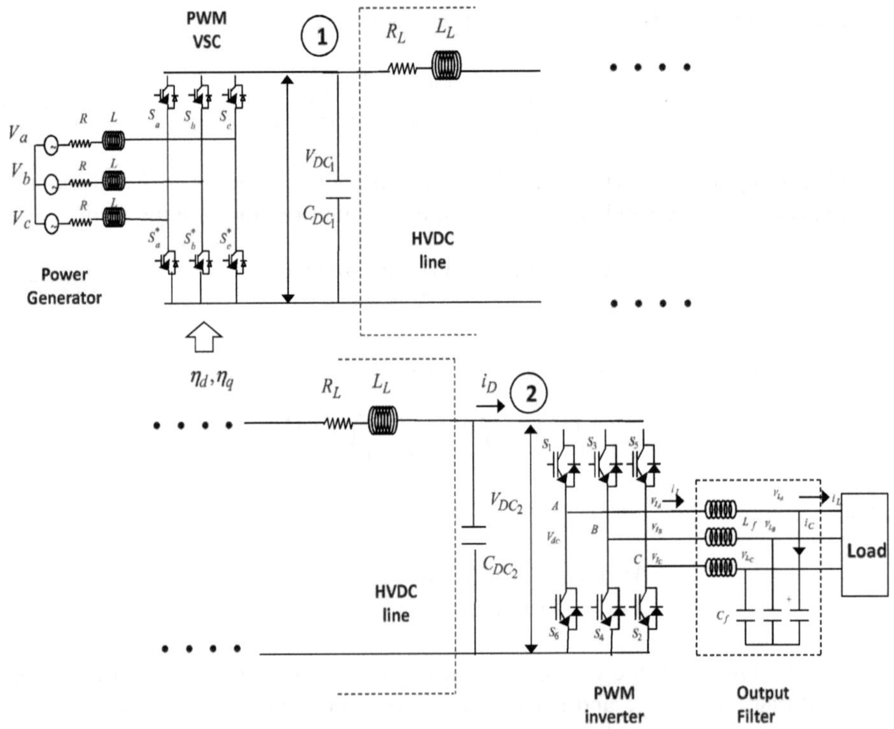

Fig. 1.17 Transmission system comprising a voltage source converter (VSC), a high voltage DC line (HVDC) and an inverter

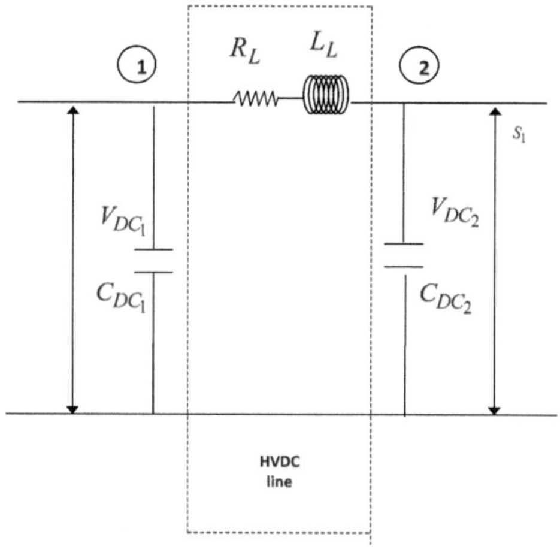

Fig. 1.18 Components of high voltage DC line (HVDC)

$$C_1\dot{V}_{DC_1} = -\frac{1}{Z_L}V_{DC_1} + \frac{1}{Z_L}V_{DC_2} + \frac{3}{4}i_d u_1 + \frac{3}{4}i_q u_2 \tag{1.178}$$

where Z_L is the impedance of the transmission line (Fig. 1.18). Here Z_L is considered to consist of only a resistance part (if inductance part is also included in the transmission line's dynamics then a higher order state-space model for the VSC-HVDC system will be obtained). It is also possible to model the transmission line's dynamics as a distributed parameters system, however this latter assumption will not be followed in this manuscript.

At the inverter's side it holds

$$C_2\dot{V}_{DC_2} = \frac{1}{Z_L}V_{DC_1} - \frac{1}{Z_L}V_{DC_2} \tag{1.179}$$

The state-space model of the VSC-HVDC transmission line, can be obtained using established models about the dynamics of the voltage source converter [370, 424]:

$$\begin{aligned} L\dot{i}_d &= -Ri_d + L\omega i_q + v_d - \frac{V_{DC_1}}{2}u_1 \\ L\dot{i}_q &= -L\omega i_d - Ri_q + v_q - \frac{V_{DC_1}}{2}u_2 \\ C_{DC_2}\dot{V}_{DC_2} &= \frac{1}{Z_L}V_{DC_1} - \frac{1}{Z_L}V_{DC_2} \\ C_{DC_1}\dot{V}_{DC_1} &= -\frac{1}{Z_L}V_{DC_1} + \frac{1}{Z_L}V_{DC_2} + \frac{3}{4}i_d u_1 + \frac{3}{4}i_q u_2 \end{aligned} \tag{1.180}$$

By defining the new state vector of the VSC-HVDC system as $x = [x_1, x_2, x_3, x_4]^T = [i_d, i_q, V_{DC_2}, V_{DC_1}]$ one obtains the state-space description

$$\begin{aligned} L\dot{x}_1 &= -Rx_1 + L\omega x_2 + v_d - \frac{x_4}{2}u_1 \\ L\dot{x}_2 &= -L\omega x_1 - Rx_2 + v_q - \frac{x_4}{2}u_2 \\ C_{DC_2}\dot{x}_3 &= \frac{1}{Z_L}x_4 - \frac{1}{Z_L}x_3 \\ C_{DC_1}\dot{x}_4 &= -\frac{1}{Z_L}x_4 + \frac{1}{Z_L}x_3 + \frac{3}{4}x_1u_1 + \frac{3}{4}x_2u_2 \end{aligned} \tag{1.181}$$

which can be also written in matrix form

$$\begin{pmatrix}\dot{x}_1\\ \dot{x}_2\\ \dot{x}_3\\ \dot{x}_4\end{pmatrix} = \begin{pmatrix} -\frac{R}{L}x_1 + \omega x_2 + \frac{v_d}{L} \\ -\omega x_1 - \frac{R}{L}x_2 + \frac{v_q}{L} \\ \frac{1}{Z_L C_{DC_2}}x_4 - \frac{1}{Z_L C_{DC_2}}x_3 \\ -\frac{1}{Z_L C_{DC_1}}x_4 + \frac{1}{Z_L C_{DC_1}}x_3 \end{pmatrix} + \begin{pmatrix} -\frac{x_4}{2L} & 0 \\ 0 & -\frac{x_4}{2L} \\ 0 & 0 \\ \frac{3}{4C_{DC_1}}x_1 & \frac{3}{4C_{DC_1}}x_2 \end{pmatrix}\begin{pmatrix}u_1\\ u_2\end{pmatrix} \tag{1.182}$$

while, as in the case of the stand-alone VSC the output of the voltage source converter is taken to be

$$h = \begin{pmatrix}h_1\\ h_2\end{pmatrix} = \begin{pmatrix}e_c\\ i_q\end{pmatrix} = \begin{pmatrix}\frac{3}{4}L(x_1^2 + x_2^2) + \frac{C_{DC_2}}{2}x_3^2 \\ x_2\end{pmatrix} \tag{1.183}$$

Consequently, the VSC-HVDC system is written again in the state-space form

$$\begin{aligned} \dot{x} &= f(x) + G(x)u \\ y &= h(x) \end{aligned} \tag{1.184}$$

where $f(x)$, $G(x)$ and $h(x)$ are given by

$$f(x) = \begin{pmatrix} -\frac{R}{L}x_1 + \omega x_2 + \frac{v_d}{L} \\ -\omega x_1 - \frac{R}{L}x_2 + \frac{v_q}{L} \\ \frac{1}{Z_L C_{DC_2}}x_4 - \frac{1}{Z_L C_{DC_2}}x_3 \\ -\frac{1}{Z_L C_{DC_1}}x_4 + \frac{1}{Z_L C_{DC_1}}x_3 \end{pmatrix} \quad G(x) = \begin{pmatrix} -\frac{x_4}{2L} & 0 \\ 0 & -\frac{x_4}{2L} \\ 0 & 0 \\ \frac{3}{4C_{DC_1}}x_1 & \frac{3}{4C_{DC_1}}x_2 \end{pmatrix} \tag{1.185}$$

$$h(x) = \begin{pmatrix}\frac{3}{4}L(x_1^2 + x_2^2) + \frac{C_{DC_2}}{2}x_3^2 \\ x_3\end{pmatrix} \tag{1.186}$$

The impedance of the line can be considered to consist of both a resistance part and of an inductance part. Should one have adopted this description the dynamic model of the HVDC-VSC system would not be described by Eq. (1.181) but would be given a higher-order state-space description. In several research articles, the assumption made is that for short transmission lines (up to tens of Km) the line can be modeled by only a resistance part [225].

Chapter 2
Control of the Functioning of Doubly-Fed Induction Generators

2.1 Outline

The chapter analyzes nonlinear control for Doubly-fed induction generators, through different approaches: (i) flatness-based control in cascading loops, (ii) flatness-based control through transformation to the canonical form, (iii) differential geometry-based control, (iv) nonlinear H-infinity control, (v) flatness-based adaptive control (vi) field orientation and backstepping control.

First, the chapter proposes a solution to the problem of sensorless control of doubly-fed induction generators, which is implemented in successive loops. Using the stator and rotor electrical equations the complete sixth order model of the doubly-fed induction generator is derived. It is proven that the doubly-fed induction generator model is differentially flat since all its state variables and control inputs can be written as functions of the flat output and of the associated derivatives. The differential flatness property shows that the design of a DFIG controller is possible using feed-forward control terms which are complemented by suitable error feedback terms. Next, sensorless control for the doubly-fed induction generator is implemented. The Extended Kalman Filter is used to estimate the non-measurable elements of the DFIG state vector such as the rotation speed and the magnetic flux through the processing of measurements of the rotor's angle and of the rotor'currents. The efficiency of the proposed state estimation-based nonlinear control scheme is evaluated through simulation experiments.

Next, the chapter studies differential flatness properties and an input-output linearization procedure for doubly-fed induction generators (DFIGs). By defining flat outputs which are associated with the rotor's turn angle and the magnetic flux of the stator, an equivalent DFIG description in the Brunovksy (canonical) form is obtained. For the linearized canonical model of the generator a feedback controller is designed. Moreover, a comparison of the differential flatness theory-based control method against Lie algebra-based control is provided. At a second stage, a novel Kalman Filtering method (Derivative-free nonlinear Kalman Filtering) is introduced. The proposed Kalman Filter is redesigned as disturbance observer for estimating

G. Rigatos, *Intelligent Renewable Energy Systems*,
Green Energy and Technology, DOI 10.1007/978-3-319-39156-4_2

additive input disturbances to the DFIG model. These estimated disturbance terms are finally used by a feedback controller that enables the generator's state variables to track desirable setpoints. The efficiency of the proposed state estimation-based control scheme is tested through simulation experiments.

Additionally, the chapter introduces a nonlinear H-infinity control method for feedback control of asynchronous electrical machines, with application example the control problem of doubly-fed induction generators (DFIGs). The control method consists of a repetitive solution of an H-infinity control problem for the DFIG, that makes use of a locally linearized model of the generator and takes place at each iteration of the control algorithm. The asynchronous generator's model is locally linearized round its current operating point through the computation of the associated Jacobian matrices. Using the linearized model of the generator an H-infinity feedback control law is computed. The known robustness features of H-infinity control enable to compensate for the errors of the approximate linearization, as well as to eliminate the effects of external perturbations. The efficiency of the proposed control scheme is shown analytically and is confirmed through simulation experiments.

Finally, the chapter proposes an adaptive control approach that is capable of compensating for model uncertainty and parametric changes of the DFIG, as well as for lack of measurements for the DFIG's state vector elements. First it is proven that the DFIG's model is a differentially flat one. This means that all its state variables and its control inputs can be written as differential functions of key state variables which are the so-called flat outputs. Moreover, this implies that the flat output and its derivatives are not connected to each other through a relation of the form of a differential equation. By exploiting differential flatness properties it is shown that the 6-th order DFIG model can be transformed into the linear canonical form. For the latter description, the new control inputs comprise unknown nonlinear functions which can be identified with the use of neurofuzzy approximators. The estimated dynamics of the generator is used by a feedback controller thus establishing an indirect adaptive control scheme. Moreover, to robustify the control loop a supplementary control term is computed using H-infinity control theory. Another problem that has to be dealt with comes from the inability to measure the complete state vector of the generator. Thus, a state-observer is implemented in the control loop. The stability of the considered observer-based adaptive control approach is proven using Lyapunov analysis. Moreover, the performance of the control scheme is evaluated through simulation experiments.

2.2 Flatness-Based Control of the DFIG in Successive Loops

2.2.1 Overview

As noted in Chap. 1, doubly-fed induction generators (DFIG) have been widely used in variable-speed fixed frequency hydro-power generation systems, wind-power generation systems and turbine engine power generation systems. Doubly-fed induction generators have proven to be more efficient than squirrel-cage induction generator systems (SCIG) and the synchronous generator systems in terms of cost and losses of the associated power electronics converters. DFIG systems can operate either in grid-connected mode or in stand-alone mode [56, 64, 114, 140, 141, 148, 295, 345]. Moreover, several field-oriented control schemes have been proposed for both operation modes. Additionally, several results have been published on sensorless control of DFIG [8, 11, 62, 69, 143, 473, 486]. Taking into account that the installation and maintenance of sensors for measuring several parameters of the generator's state vector can be technically difficult or costly, the need for developing sensorless control schemes for DFIG becomes apparent. In this section a novel sensorless control scheme is developed using flatness-based control theory and a state estimation method that is based on Extended Kalman Filtering.

Using the electric equations of the stator and rotor a dynamic model for the doubly fed induction generator is derived. The doubly-fed induction generator is analogous to the induction motor. In an induction motor the stator voltage plays the role of an input variable, while the rotor voltage is a constant. In case of the doubly-fed induction machine it is quite similar but the other way round, with a dual analogy to hold between the stator and rotor parameters of the generator and the motor. This means that the rotor voltage now acts as an input, while the stator voltage is a constant parameter. The stator's and rotor's voltages, currents and magnetic flux are represented as vectors in a rotating orthogonal axis frame. The complete sixth order model of the DFIG captures efficiently transients at both the stator and the rotor side.

In this section, it is shown that the doubly-fed induction generator model is differentially flat. The theory of differential flatness has significant advantages over other approaches such as backstepping control [280, 137, 388, 369, 456]. In the DFIG case, the proposed flat output is the pair consisting of the rotor's angle θ and of the orientation of the magnetic flux ρ. It is shown that the doubly-fed induction generator model is differentially flat since all its state variables and control inputs can be written as functions of the flat output and of the associated derivatives. The differential flatness property shows that the design of a DFIG controller is possible using feed-forward control terms which are complemented by suitable error feedback terms. The design of the DFIG controller consists of two stages: (i) in the outer-loop the controller enables convergence of the stator's magnetic flux and of the rotor's angular velocity to the associated reference setpoint. The control input consists of the rotor currents expressed in the rotating reference frame (ii) in the inner-loop the controller enables convergence of the rotor's dq currents to the associated reference setpoints, where these setpoints are the control inputs used by the outer control loop. The control inputs for the inner loop are now the rotor voltages also expressed in the rotating reference frames.

Next, sensorless control for the doubly-fed induction generator is implemented. A main difficulty in the machine's control is that some of its state vector elements are not directly measured and their monitoring requires the installation of specific sensors that can be costly, or difficult to install and maintain. The Extended Kalman Filter is used to estimate the non-measurable elements of the DFIG state vector such as the rotation speed and the magnetic flux through the processing of measurements of the rotor's angle and of the rotor'currents. The Kalman Filter is a special kind of observer that provides optimal filtering of process and measurement noises if the covariances of these noises are known. For linear systems subjected to Gaussian measurement or process noise the Kalman Filter is the optimal state estimator, since it results in minimization of the trace of the estimation error's covariance matrix [45, 209, 341]. For nonlinear systems, subjected to Gaussian noise one can use the generalization of the Kalman Filter as formulated in terms of the Extended Kalman Filter (EKF). The Extended Kalman Filter is based on a linearization of the systems' dynamics using a first order Taylor expansion [334, 358]. In this section it is shown that with the use of nonlinear Kalman Filtering it is possible to reduce the number of sensors involved in the control loops of the doubly-fed induction generator and to implement efficient state estimation-based control.

2.2.2 *Field Orientation for Induction Machines*

The classical method for induction machines control was introduced by Blascke (1971) and in the DFIG case is based on a transformation of the rotor's currents ($i_{r\alpha}$) and (i_{rb}) and of the magnetic fluxes of the stator ($\psi_{s\alpha}$ and ψ_{sb}) to the reference frame $d-q$ which rotates together with the rotor. Thus the controller's design uses the currents i_{rd} and i_{rq} and the fluxes ψ_{sd} and ψ_{sq} [353]. The angle of the vectors that describe the magnetic fluxes $\psi_{s\alpha}$ and ψ_{sb} is first defined, i.e.

$$\rho = tan^{-1}(\frac{\psi_{S_b}}{\psi_{S_a}}) \tag{2.1}$$

The angle between the inertial reference frame of the stator and the rotating reference frame of the rotor is taken to be equal to ρ. The transition from $(i_{r\alpha}, i_{rb})$ to (i_{rd}, i_{rq}) is given by

$$\begin{pmatrix} i_{rd} \\ i_{rq} \end{pmatrix} = \begin{pmatrix} cos(\rho) & sin(\rho) \\ -sin(\rho) & cos(\rho) \end{pmatrix} \begin{pmatrix} i_{r\alpha} \\ i_{rb} \end{pmatrix} \tag{2.2}$$

The transition from $(\psi_{s\alpha}, \psi_{rb})$ to (ψ_{sd}, ψ_{sq}) is given by

$$\begin{pmatrix} \psi_{sd} \\ \psi_{sq} \end{pmatrix} = \begin{pmatrix} cos(\rho) & sin(\rho) \\ -sin(\rho) & cos(\rho) \end{pmatrix} \begin{pmatrix} \psi_{s\alpha} \\ \psi_{sb} \end{pmatrix} \tag{2.3}$$

Moreover, it holds that $cos(\rho) = \frac{\psi_{S_a}}{||\psi||}$, $sin(\rho) = \frac{\psi_{S_b}}{||\psi||}$, and $||\psi|| = \sqrt{\psi_{S_a}^2 + \psi_{S_b}^2}$. Using the above transformation ones obtains

$$
\begin{aligned}
i_{rd} &= \tfrac{\psi_{s\alpha}i_{r\alpha}+\psi_{sb}i_{rb}}{||\psi||} \quad \psi_{sd} = ||\psi|| \\
i_{rq} &= \tfrac{\psi_{s\alpha}i_{rb}-\psi_{sb}i_{r\alpha}}{||\psi||} \quad \psi_{sq} = 0
\end{aligned} \tag{2.4}
$$

Therefore, in the rotating frame $d-q$ of the generator there will be only one non-zero component of the magnetic flux ψ_{s_d}, while the component of the flux along the d axis equals 0. The new inputs of the system are considered to be v_{rd}, v_{rq}, which are connected to v_{ra}, v_{rb} according to the relation

$$
\begin{pmatrix} v_{r\alpha} \\ v_{rb} \end{pmatrix} = ||\psi||\cdot \begin{pmatrix} \psi_{sa} & \psi_{sb} \\ \psi_{sb} & \psi_{sa} \end{pmatrix}^{-1} \begin{pmatrix} v_{rd} \\ v_{rq} \end{pmatrix} \tag{2.5}
$$

In the new coordinates, the induction generator model has been described in Eqs. (1.20)–(1.25). The state-space model of the induction generator has been defined in Eqs. (1.26) and (1.27). Using the state variables notation the DFIG model was written in the form

$$
\begin{aligned}
\dot{x}_1 &= x_2 \\
\dot{x}_2 &= -\tfrac{K_m}{J}x_2 - \tfrac{T_m}{J} + \tfrac{\eta}{J}(x_6x_3 - x_5x_4) \\
\dot{x}_3 &= -\tfrac{1}{\tau_s}x_3 + \omega_{dq}x_4 + \tfrac{M}{\tau_s}x_5 + v_{s_d} \\
\dot{x}_4 &= -\omega_{dq}x_3 - \tfrac{1}{\tau_s}x_4 + \tfrac{M}{\tau_s}x_6 + v_{s_q} \\
\dot{x}_5 &= -\beta x_2x_4 + \tfrac{\beta}{\tau_s}x_3 + (\omega_{dq} - x_2)x_6 - \gamma_2x_5 + \tfrac{1}{\sigma L_r}v_{r_d} - \beta v_{s_d} \\
\dot{x}_6 &= -\tfrac{\beta}{\tau_s}x_4 + \beta x_2x_3 - (\omega_{dq} - x_2)x_5 - \gamma_2x_6 + \tfrac{1}{\sigma L_r}v_{r_q} - \beta v_{s_q}
\end{aligned} \tag{2.6}
$$

Next, the following nonlinear feedback control law is defined

$$
\begin{pmatrix} v_{rd} \\ v_{rq} \end{pmatrix} = \sigma L_r \begin{pmatrix} \beta x_2x_4 - \tfrac{\beta}{\tau_s}x_3 - (\omega_{dq} - x_2)x_6 + \beta v_{s_d} + \beta v_1 \\ \tfrac{\beta}{\tau_s}x_4 - \beta x_2x_3 + (\omega_{dq} - x_2)x_5 + \beta v_{s_q} + \beta v_2 \end{pmatrix} \tag{2.7}
$$

The terms in Eq. (2.7) have been selected so as to linearize the fifth and sixth row of the state space model of the induction generator in Eq. (2.6) and to produce first-order linear differential equations. The control signal in the inertial coordinates system $a-b$ is

$$
\begin{aligned}
\begin{pmatrix} v_{r\alpha} \\ v_{rb} \end{pmatrix} &= ||\psi||\sigma L_r \begin{pmatrix} \psi_{s\alpha} & \psi_{sb} \\ -\psi_{sb} & \psi_{s\alpha} \end{pmatrix}^{-1} \cdot \\
&\begin{pmatrix} \beta x_2x_4 - \tfrac{\beta}{\tau_s}x_3 - (\omega_{dq} - x_2)x_6 + \beta v_{s_d} + \beta v_1 \\ \tfrac{\beta}{\tau_s}x_4 - \beta x_2x_3 + (\omega_{dq} - x_2)x_5 + \beta v_{s_q} + \beta v_2 \end{pmatrix}
\end{aligned} \tag{2.8}
$$

Substituting Eq. (2.7) into Eq. (2.6) one obtains:

$$
\dot{x}_1 = x_2 \tag{2.9}
$$

$$
\dot{x}_2 = -\frac{K_m}{J}x_2 - \frac{T_m}{J} + \frac{\eta}{J}(x_6x_3 - x_5x_4) \tag{2.10}
$$

$$\dot{x}_3 = -\frac{1}{\tau_s}x_3 + \omega_{dq}x_4 + \frac{M}{\tau_s}x_5 + v_{s_d} \tag{2.11}$$

$$\dot{x}_4 = -\omega_{dq}x_3 - \frac{1}{\tau_s}x_4 + \frac{M}{\tau_s}x_6 + v_{s_q} \tag{2.12}$$

$$\dot{x}_5 = -\gamma_2 x_5 + \beta v_1 \tag{2.13}$$

$$\dot{x}_6 = -\gamma_2 x_6 + \beta v_2 \tag{2.14}$$

The system of Eqs. (2.9)–(2.14) comprises two linear subsystems, where the first one has as output the magnetic flux $x_3 = \psi_{sd}$ and the second has as output the rotation speed $x_2 = \omega$ [353].

Thus, from Eqs. (2.11) and (2.13) one obtains

$$\dot{x}_3 = -\frac{1}{\tau_s}x_3 + \frac{M}{\tau_s}x_5 + v_{s_d} \tag{2.15}$$

$$\dot{x}_5 = -\gamma_2 x_5 + \beta v_1 \tag{2.16}$$

while from Eqs. (2.10) and (2.14) one obtains

$$\dot{x}_2 = -\frac{K_m}{J}x_2 - \frac{T_m}{J} + \frac{\eta}{J}x_3 x_6 \tag{2.17}$$

$$\dot{x}_6 = -\gamma_2 x_6 + \beta v_2 \tag{2.18}$$

For $x_3 = \psi_{s_d}$, it holds that if $\psi_{sd} \rightarrow \psi_{sd}^{\text{ref}}$, i.e. the transient phenomena for ψ_{sd} have been eliminated and therefore ψ_{sd} has converged to a steady state value, then the two subsystems described by Eqs. (2.15)–(2.16) and Eqs. (2.17)–(2.18) are decoupled.

The subsystem that is described by Eqs. (2.15) and (2.16) is linear with control input v_1, and can be controlled using methods of linear control, such as optimal control, or PID control. For instance the following PI controller has been proposed for the control of the magnetic flux [353]

$$v_1(t) = -k_{d1}(\psi_{sd} - \psi_{sd}^{\text{ref}}) - k_{d2}\int_0^t (\psi_{sd}(\tau) - \psi_{sd}{}^{\text{ref}}(\tau)d\tau \tag{2.19}$$

Thus, if Eq. (2.19) is applied to the subsystem that is described by Eqs. (2.15) and (2.16), one anticipates to succeed $\psi_{sd}(t) \rightarrow \psi_{sd}^{\text{ref}}(t)$.

Now, the subsystem that consists of Eqs. (2.17) and (2.18) is examined. The term $T = \frac{\eta}{J}x_6 x_3$ denotes the torque developed in the rotor. After achieving $\psi_{sd} \rightarrow \psi_{sd}^{\text{ref}}$, one can also control the generator's speed ω, using linear feedback control algorithms. A first approach to the control of the speed ω is to use nested PI loops, i.e.

$$\begin{aligned} v_2 &= -K_{q_1}(T - T_{\text{ref}}) - K_{q_2}\int_0^t (T(t) - T_{\text{ref}}(t))d\tau \\ T_{\text{ref}} &= -K_{q_3}(\omega - \omega_{\text{ref}}) - K_{q_4}\int_0^t (\omega(t) - \omega_{\text{ref}}(t))d\tau \end{aligned} \tag{2.20}$$

From the above analysis it becomes clear that a remaining problem in the implementation of field-oriented control for DFIGs is how to measure efficiently $x_3 = \psi_{sd}(t)$. Therefore one comes against the need for applying a state observer or Kalman Filtering. Besides, the tuning of the multiple PID and PI controllers that constitute the field-oriented control scheme, as described in Eqs. (2.19) and (2.20), remains valid only round local operating points and thus the stability and robustness of the field-oriented control for DFIGs cannot be assured.

2.2.3 *Differential Flatness of the Doubly-Fed Induction Generator*

2.2.3.1 Properties of Differentially Flat Systems

Having pointed out the shortfalls of field oriented control for DFGIs, the need for developing nonlinear control schemes of proven stability for such electrical machines becomes apparent. The main principles of flatness-based control are as follows [137, 280, 388, 456]. Flatness-based control can be applied to finite dimensional system of the form

$$S_i(w, \dot{w}, \ddot{w}, \ldots, w^{(i)}), \; i = 1, 2, \ldots, q \tag{2.21}$$

The term w denotes the system variable while $w^{(i)}, i = 1, 2, \ldots, q$ are its derivatives (these can be for instance the elements of the system's state vector). The system of Eq. (2.21) is said to be differentially flat if there exists a set of m functions $y = (y_1, \ldots, y_m)$ of the system variables and of their time-derivatives, i.e.

$$y_i = \phi(w, \dot{w}, \ddot{w}, \ldots, w^{(\alpha_i)}), \; i = 1, \ldots, m \tag{2.22}$$

such that the following two conditions are satisfied [137–456]:
1. There does not exist any differential relation of the form

$$R(y, \dot{y}, \ldots, y^{\beta}) = 0 \tag{2.23}$$

which implies that the derivatives of the flat output are not coupled in the sense of an ordinary differential equation (ODE) denoted by function R, or equivalently it can be said that the flat output is differentially independent.
2. All system variables, i.e. the components of w (elements of the system's state vectors) can be expressed using only the flat output y and its time derivatives

$$w_i = \psi_i(y, \dot{y}, \ldots, y^{(\gamma_i)}), \; i = 1, \ldots, s \tag{2.24}$$

An equivalent definition of differentially flat systems is as follows:
Definition: The system $\dot{x} = f(x, u)$, $x \in R^n$, $u \in R^m$ is differentially flat if there exist relations

$$\begin{aligned} h &: R^n \times (R^m)^{r+1} \rightarrow R^m \\ \phi &: (R^m)^r \rightarrow R^n \\ \psi &: (R^m)^{r+1} \rightarrow R^m \end{aligned} \tag{2.25}$$

such that

$$\begin{aligned} y &= h(x, u, \dot{u}, \ldots, u^{(r)}) \\ x &= \phi(y, \dot{y}, \ldots, y^{(r-1)}, y^{(r-1)}), \text{ and} \\ u &= \psi(y, \dot{y}, \ldots, y^{(r-1)}, y^{(r)}) \end{aligned} \tag{2.26}$$

This means that all system dynamics can be expressed as a function of the flat output and its derivatives, therefore the state vector and the control input can be written as $x(t) = \phi(y(t), \dot{y}(t), \ldots, y^{(r)}(t))$ and $u(t) = \psi(y(t), \dot{y}(t), \ldots, y^{(r+1)}(t))$.

2.2.3.2 Classes of Differentially Flat Systems

Main classes of nonlinear differentially flat systems are [281]:
1. Affine in-the-input systems: The dynamics of such systems is given by:

$$\dot{x} = f(x) + \sum_{i=1}^{m} g_i(x) u_i \tag{2.27}$$

From Eq. (2.27) one can notice that the above state equation can also describe MIMO dynamical systems. Without out loss of generality it is assumed that $G = [g_1, \ldots, g_m]$ is of rank m. In case that the flat outputs of the aforementioned system are only functions of states x, then this class of dynamical systems is called 0-flat. It has been proven that a dynamical affine system with n states and $n-1$ inputs is 0-flat if it is controllable.
2. Driftless systems: These are systems of the form

$$\dot{x} = \sum_{i=1}^{m} f_i(x) u_i \tag{2.28}$$

For driftless systems with two inputs, i.e.

$$\dot{x} = f_1(x) u_1 + f_2(x) u_2 \tag{2.29}$$

flatness holds, if and only if the rank of matrix $E_{k+1} := \{E_k, [E_k, E_k]\}$, $k \geq 0$ (with $E_0 := \{f_1, f_2\}$ and $[E_k, E_k] := \{[X, Y],\ X, Y \in E_k\}$ denoting a Lie bracket) is equal to $k + 2$ for $k = 0, \ldots, n-2$. It has been proven that a driftless system that is differentially flat, is also 0-flat (which means that the flat output is a function of only the elements of the state vector of the initial nonlinear system).

Moreover, for flat systems with n states and $n-2$ control inputs, i.e.

$$\dot{x} = \sum_{i=1}^{n-2} u_i f_i(x) \quad x \in R^n \tag{2.30}$$

the flatness property holds, if controllability also holds. Furthermore, the system is 0-flat if n is even.

2.2.3.3 Conditions for Applying Differential Flatness Theory

Control based on differential flatness theory can be applied to systems that admit static feedback linearization. The generic class of nonlinear systems $\dot{x} = f(x, u)$ is considered. Such a system can be transformed to the form of an affine in the input system by adding an integrator to each input [57]

$$\dot{x} = f(x) + \sum_{i=1}^{m} g_i(x) u_i \tag{2.31}$$

If the system of Eq. (2.31) can be linearized by a diffeomorphism $z = \phi(x)$ and a static state feedback $u = \alpha(x) + \beta(x)v$ into the following form

$$\begin{array}{c} \dot{z}_{i,j} = z_{i+1,j} \text{ for } 1 \leq j \leq m \text{ and } 1 \leq i \leq v_j - 1 \\ \dot{z}_{v_{i,j}} = v_j \end{array} \tag{2.32}$$

with $\sum_{j=1}^{m} v_j = n$, then $y_j = z_{1,j}$ for $1 \leq j \leq m$ are the 0-flat outputs which can be written as functions of only the elements of the state vector x. To define conditions for transforming the system of Eq. (2.31) into the canonical form described in Eq. (2.32) the following theorem holds [57]

Theorem: For the nonlinear systems described by Eq. (2.31) the following variables are defined: (i) $G_0 = \text{span}[g_1, \ldots, g_m]$, (ii) $G_1 = \text{span}[g_1, \ldots, g_m, ad_f g_1, \ldots, ad_f g_m]$, $\ldots$ (k) $G_k = \text{span}\{ad_f^j g_i \text{ for } 0 \leq j \leq k, \ 1 \leq i \leq m\}$. Then, the linearization problem for the system of Eq. (2.31) can be solved if and only if: (1). The dimension of $G_i, \ i = 1, \ldots, k$ is constant for $x \in X \subseteq R^n$ and for $1 \leq i \leq n-1$, (2). The dimension of G_{n-1} is of order n, (3). The distribution G_k is involutive for each $1 \leq k \leq n-2$.

2.2.3.4 Differential Flatness for the DFIG Model

It will be shown that the doubly-fed induction generator model is differentially flat. First the stator's magnetic flux is written in a complex number form [280, 375]

$$\psi = \psi_{s_d} + j\psi_{s_q} = \alpha e^{j\rho} \tag{2.33}$$

where $\rho = tan^{-1}(\frac{\psi_{s_b}}{\psi_{sa}})$ denotes the orientation angle of the magnetic flux vector. According to Lorentz law the torque that is developed due to the magnetic flux is given by

$$\begin{array}{c} T_e = k_1 I_m(\psi_s^* \frac{d\psi_s}{dt}) \Rightarrow T_e = k_1 I_m(\alpha e^{-j\rho}\alpha j\dot{\rho}e^{j\rho}) \Rightarrow \\ T_e = k_1 Im(j\alpha^2\dot{\rho}) \Rightarrow T_e = k_1\alpha^2\dot{\rho} \end{array} \tag{2.34}$$

The generator's equations of motion are

$$\begin{array}{c} J\dot{\omega}_r = T_m - b\dot{\omega}_r - T_e \Rightarrow \\ J\ddot{\theta} = T_m - b\dot{\theta} - k_1\alpha^2\dot{\rho} \Rightarrow \alpha = \sqrt{\frac{(T_m - b\dot{\theta} - J\ddot{\theta})}{k_1\dot{\rho}}} \end{array} \tag{2.35}$$

As flat output, the following pair of variables is considered $y = (\theta, \rho)$, where θ is the rotor's rotation angle and ρ is the orientation angle of the magnetic field. Thus, in the doubly-fed induction generator model the magnetic flux can be written as a function of the flat output y and of the associated derivatives. This is denoted as

$$\psi_s = f_g(y, \dot{y}, \ddot{y}) \tag{2.36}$$

Moreover, due to the field orientation it holds

$$\psi_{s_d} = Re\{f_g(y, \dot{y}, \ddot{y})\}, \;\; \psi_{s_q} = Im\{f_g(y, \dot{y}, \ddot{y})\} = 0 \tag{2.37}$$

It is considered that for the stator feed voltages it holds that v_{s_q} is constant (for a symmetrical grid), and v_{s_d} is associated with the grid voltage and dependant on the rotor's turn angle [148]. Moreover due to field orientation it holds $\psi_{s_q} = 0$, and from Eq. (1.3) one has

$$i_{s_d} = \frac{1}{R_s}[v_{s_d} + \frac{d\psi_{s_d}}{dt} - \omega_{dq}\psi_{s_q}] \Rightarrow i_{s_d} = \frac{1}{R_s} f_{g_1}(y, \dot{y}, \ddot{y}) \tag{2.38}$$

Moreover, from Eq. (1.4) one has

$$i_{s_q} = \frac{1}{R_s}[v_{s_q} - \frac{d\psi_{s_q}}{dt} + \omega_{d_q}\psi_{s_d}] \Rightarrow i_{s_q} = \frac{1}{R_s} f_{g_2}(y, \dot{y}, \ddot{y}) \tag{2.39}$$

From Eq. (1.7) one obtains

$$i_{r_d} = \frac{1}{M}[\psi_{s_d} - L_s i_{s_d}] \Rightarrow i_{r_d} = f_{g_3}(y, \dot{y}, \ddot{y}) \tag{2.40}$$

From Eq. (1.8) one obtains

$$i_{r_q} = \frac{1}{M}[\psi_{s_q} - L_s i_{s_q}] \Rightarrow i_{r_q} = f_{g_4}(y, \dot{y}, \ddot{y}) \tag{2.41}$$

Additionally, from Eq. (1.9) it holds

$$\psi_{r_d} = L_s i_{r_d} + M i_{s_d} \Rightarrow \psi_{r_d} = f_{g_5}(y, \dot{y}, \ddot{y}) \tag{2.42}$$

while from Eq. (1.10) one has

$$\psi_{r_q} = L_s i_{r_q} + M i_{s_q} \Rightarrow \psi_{r_q} = f_{g_6}(y, \dot{y}, \ddot{y}) \tag{2.43}$$

Since variables ψ_{r_d},ψ_{r_q}, i_{r_d} and i_{r_q} have been expressed as functions of the flat output and of the associated derivatives the same can be done for the control inputs v_{r_d} and v_{r_q}.

From Eq. (1.5) and using Eqs. (2.44) and (2.42) one obtains

$$v_{r_d} = R_r i_{r_d} + \frac{d\psi_{r_d}}{dt} - \omega_{dq}\psi_{r_q} \Rightarrow v_{r_d} = f_{g_7}(y, \dot{y}, \ddot{y}) \tag{2.44}$$

From Eq. (1.6) and using Eqs. (2.45) and (2.43) one has

$$v_{r_q} = R_r i_{r_q} + \frac{d\psi_{r_q}}{dt} - \omega_{dq}\psi_{r_d} \Rightarrow v_{r_q} = f_{g_8}(y, \dot{y}, \ddot{y}) \tag{2.45}$$

Since all elements of the state vector $x = [\theta, \omega, \psi_{s_d}, \psi_{s_q}, i_{r_d}, i_{r_q}]^T$ and the control inputs v_{r_d} and v_{r_q} of the doubly-fed induction generator can be expressed as functions of the flat output $y = [\theta, \rho]$ and of the associated derivatives, it can be concluded that the DFIG model is differentially flat.

2.2.4 Control of the Doubly-Fed Induction Generator

The differential flatness property shows that the design of a DFIG controller is possible using feed-forward control terms which are complemented by suitable error feedback terms. The design of the controller of the doubly-fed induction generator consists of two stages: (i) in the outer-loop the controller enables convergence of the stator's magnetic flux to the associated reference setpoint. The control input consists the rotor currents on the d and q reference axes (ii) in the inner-loop the controller enables convergence of the rotor's dq currents to the associated reference setpoints, where these setpoints are the control inputs used by the outer control loop (see Fig. 2.1). The control inputs for the inner loop are now the rotor voltages v_{r_d} and v_{r_q}. For the outer control loop the stator flux dynamics is given by:

$$\frac{d\psi_{s_d}}{dt} = \omega_{dq}\psi_{s_q} - \frac{1}{\tau_s}\psi_{s_d} + \frac{M}{\tau_s} i_{r_d} + v_{s_d} \tag{2.46}$$

$$\frac{d\psi_{s_q}}{dt} = -\frac{1}{\tau_s}\psi_{s_q} - \omega_{dq}\psi_{s_d} + \frac{M}{\tau_s} i_{r_q} + v_{s_q} \tag{2.47}$$

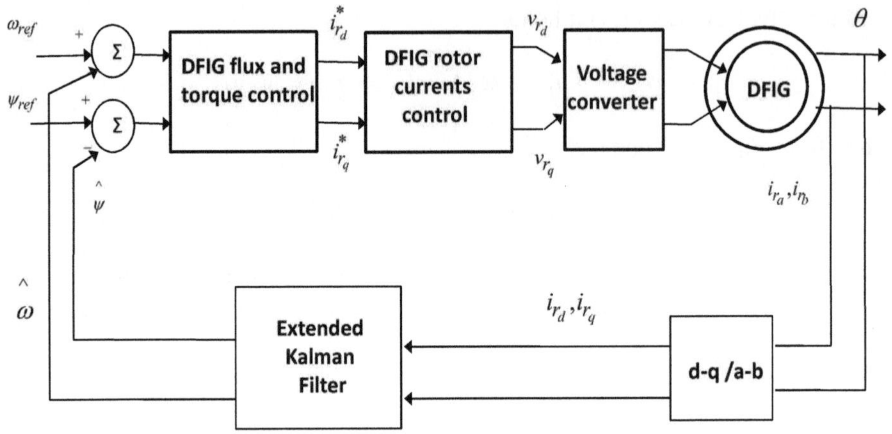

Fig. 2.1 DFIG control loop

Inner control loop of the rotor's currents:

For the inner control loop the rotor current dynamics is given by:

$$\frac{di_{r_d}}{dt} = -\beta\omega_r\psi_{s_q} + \frac{\beta}{\tau_s}\psi_{s_d} + (\omega_{dq} - \omega_r)i_{r_q} - \gamma_2 i_{r_d} - \beta v_{s_d} + \frac{1}{\sigma L_r}v_{r_d} \tag{2.48}$$

$$\frac{di_{r_q}}{dt} = \frac{\beta}{\tau_s}\psi_{s_q} - \beta\omega_r\psi_{s_d} - \gamma_2 i_{r_q} - (\omega_{dq} - \omega_r)i_{r_d} - \beta v_{s_q} + \frac{1}{\sigma L_r}v_{r_q} \tag{2.49}$$

The following control law is applied to the inner loop to make the rotor currents converge to the desirable reference setpoints

$$v_{r_d} = \sigma L_r[\beta\omega_r\psi_{s_q} - \frac{\beta}{\tau_s}\psi_{s_d} - (\omega_{dq} - \omega_r)i_{r_q} + \gamma_2 i^*_{r_d} + \beta v_{s_d} + \frac{di^*_{r_d}}{dt}] \tag{2.50}$$

$$\begin{aligned} v_{r_q} = \sigma L_r[-\frac{\beta}{\tau_s}\psi_{s_q} - \beta\omega_r\psi_{s_d} + \gamma_2 i^*_{r_q} + \\ +(\omega_{dq} - \omega_r)i_{r_d} + \beta v_{s_q} + \frac{di^*_{r_q}}{dt}] \end{aligned} \tag{2.51}$$

The inner control loop is designed so as to succeed $i_{r_q} \to i^*_{r_q}$ and $i_{r_d} \to i^*_{r_d}$, where the setpoints $i^*_{r_q}$ and $i^*_{r_d}$ are the control inputs that make the stator's flux outer control loop converge to the flux reference setpoints $\psi^*_{s_d}$ and $\psi^*_{s_q}$.

It is also noted that in DFIGs, the rotor-side converter is used to control the turbine output power and the voltage measured at the grid terminals. The power is controlled in order to follow a predefined power-speed characteristic, named tracking characteristic. The grid side converter is used to regulate the voltage of the dc bus capacitor. In addition, by using the grid side converter it is also possible to generate or absorb reactive power [62, 143].

Outer control loop of the stator's flux:

For the control of the magnetic flux, from Eqs. (2.46) and (2.47) the following matrix form is obtained:

$$\begin{pmatrix} \frac{d\psi_{s_d}}{dt} \\ \frac{d\psi_{s_q}}{dt} \end{pmatrix} = \begin{pmatrix} \omega_{dq} & -\frac{1}{\tau_s} \\ -\frac{1}{\tau_s} & -\omega_{dq} \end{pmatrix} \begin{pmatrix} \psi_{s_d} \\ \psi_{s_q} \end{pmatrix} + \begin{pmatrix} v_{s_d} \\ v_{s_q} \end{pmatrix} + \begin{pmatrix} \frac{M}{\tau_s} & 0 \\ 0 & \frac{M}{\tau_s} \end{pmatrix} \begin{pmatrix} i_{d_r} \\ i_{q_r} \end{pmatrix} \tag{2.52}$$

The magnetic flux reference setpoint is defined as $\psi_s^* = [\psi_{s_d}^*, \psi_{s_q}^*]^T$. Then the control law that assures that $\psi_s \to \psi_s^*$ is given by

$$i_{r_d} = \frac{\tau_s}{M}[-v_{s_d} + \frac{1}{\tau_s}\psi_{s_d} - \omega_{d_q}\psi_{s_q} + \dot{\psi}_{s_d}^* - r_1(\psi_{s_d} - \psi_{s_d}^*)] \tag{2.53}$$

$$i_{r_q} = \frac{\tau_s}{M}[-v_{s_q} + \frac{1}{\tau_s}\psi_{s_q} + \omega_{d_q}\psi_{s_d} + \dot{\psi}_{s_q}^* - r_2(\psi_{s_q} - \psi_{s_q}^*)] \tag{2.54}$$

To speed up convergence of the flux tracking error to zero it suffices to select appropriately coefficients r_1 and r_2 appearing in the right part of Eqs. (2.53) and (2.54).

Thus, in the outer control loop one defines the flux reference setpoint $\psi_s = [\psi_{s_d}, \psi_{s_q}]^T$ which can be attained by defining an appropriate control input i_{r_d}, i_{r_q}. The control input i_{r_d}, i_{r_q} becomes reference setpoint $i_{r_d}^*, i_{r_q}^*$ for the inner control loop. The latter control loop assures convergence of i_{r_d}, i_{r_q} to $i_{r_d}^*, i_{r_q}^*$.

Finally, considering that in the dq reference frame $\psi_{s_q} = 0$, the rotation speed of the rotor is controlled by an input associated to the magnetic flux

$$\psi_{s_d} = \frac{J}{n_p i_{s_q}}(\frac{T_m}{J} + \frac{K_m}{J}\omega_r) + \dot{\omega}_r^* - r_3(\omega_r) - \omega_r^* \tag{2.55}$$

The control signal ψ_{s_d} computed through Eq. (2.55) becomes reference setpoint for the flux control loop shown in Fig. 2.1.

2.2.5 Flux and Rotation Speed Estimator

Using Eqs. (1.26) and (1.27) one can design a suitable state observer that enables estimation of non measurable state variables through the processing of measurements of a limited number of sensors. The rotor's angular velocity ω_r as well as state variables ψ_{s_d} and ψ_{s_q} appearing in the observer's innovation can be indirectly computed using Eqs. (1.7) and (1.8) and considering that the rotor's angle θ_r and the rotor's currents i_{r_d} and i_{r_q} can be measured.

The non-measurable elements of the DFIG state vector can be estimated from the measurements of a reduced number of sensors using the Extended Kalman Filter recursion and Eqs. (1.26) and (1.27). The DFIG model is subjected to discretization and the state model is now considered [352]:

$$\begin{aligned} x(k+1) &= \phi(x(k)) + L(k)u(k) + w(k) \\ z(k) &= \gamma(x(k)) + v(k) \end{aligned} \tag{2.56}$$

where $x \in R^{m\times 1}$ is the system's state vector, and $z \in R^{p\times 1}$ is the system's output, while $w(k)$ and $v(k)$ are uncorrelated, Gaussian zero-mean noise processes with covariance matrices $Q(k)$ and $R(k)$ respectively. The operators $\phi(x)$ and $\gamma(x)$ are $\phi(x) = [\phi_1(x), \phi_2(x), \ldots, \phi_m(x)]^T$, and $\gamma(x) = [\gamma_1(x), \gamma_2(x), \ldots, \gamma_p(x)]^T$, respectively. It is assumed that ϕ and γ are sufficiently smooth in x so that each one has a valid series Taylor expansion. Following a linearization procedure, ϕ is expanded into Taylor series about $\hat{x}$:

$$\phi(x(k)) = \phi(\hat{x}(k)) + J_\phi(\hat{x}(k))[x(k) - \hat{x}(k)] + \cdots \tag{2.57}$$

where $J_\phi(x)$ is the $m\times m$ Jacobian of ϕ calculated at $\hat{x}(k)$:

$$J_\phi(x) = \frac{\partial \phi}{\partial x}|_{x=\hat{x}(k)} = \begin{pmatrix} \frac{\partial \phi_1}{\partial x_1} & \frac{\partial \phi_1}{\partial x_2} & \cdots & \frac{\partial \phi_1}{\partial x_m} \\ \frac{\partial \phi_2}{\partial x_1} & \frac{\partial \phi_2}{\partial x_2} & \cdots & \frac{\partial \phi_2}{\partial x_m} \\ \vdots & \vdots & \vdots & \vdots \\ \frac{\partial \phi_k}{\partial x_1} & \frac{\partial \phi_k}{\partial x_2} & \cdots & \frac{\partial \phi_k}{\partial x_m} \end{pmatrix} \tag{2.58}$$

Likewise, γ is expanded about $\hat{x}^-(k)$

$$\gamma(x(k)) = \gamma(\hat{x}^-(k)) + J_\gamma[x(k) - \hat{x}^-(k)] + \cdots \tag{2.59}$$

where $\hat{x}^-(k)$ is the estimation of the state vector $x(k)$ before measurement at the k-th instant to be received and $\hat{x}(k)$ is the updated estimation of the state vector after measurement at the k-th instant has been received. The $p\times m$ Jacobian $J_\gamma(x)$ is

$$J_\gamma(x) = \frac{\partial \gamma}{\partial x}|_{x=\hat{x}^-(k)} = \begin{pmatrix} \frac{\partial \gamma_1}{\partial x_1} & \frac{\partial \gamma_1}{\partial x_2} & \cdots & \frac{\partial \gamma_1}{\partial x_m} \\ \frac{\partial \gamma_2}{\partial x_1} & \frac{\partial \gamma_2}{\partial x_2} & \cdots & \frac{\partial \gamma_2}{\partial x_m} \\ \vdots & \vdots & \vdots & \vdots \\ \frac{\partial \gamma_p}{\partial x_1} & \frac{\partial \gamma_p}{\partial x_2} & \cdots & \frac{\partial \gamma_p}{\partial x_m} \end{pmatrix} \tag{2.60}$$

The resulting expressions create first order approximations of ϕ and γ. Thus the linearized version of the plant is obtained:

$$\begin{aligned} x(k+1) &= \phi(\hat{x}(k)) + J_\phi(\hat{x}(k))[x(k) - \hat{x}(k)] + w(k) \\ z(k) &= \gamma(\hat{x}^-(k)) + J_\gamma(\hat{x}^-(k))[x(k) - \hat{x}^-(k)] + v(k) \end{aligned} \tag{2.61}$$

Now, the Extended Kalman Filter (EKF) recursion is as follows: First the time update is considered: by $\hat{x}(k)$ the estimation of the state vector at instant k is denoted. Given initial conditions $\hat{x}^-(0)$ and $P^-(0)$ the recursion proceeds as:

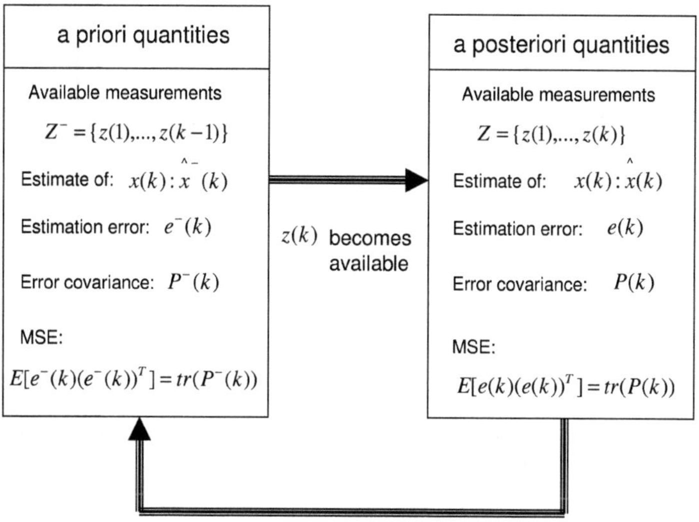

Fig. 2.2 Schematic diagram of the EKF loop

- *Measurement update*. Acquire $z(k)$ and compute:

$$\begin{aligned} K(k) &= P^{-}(k)J_{\gamma}^{T}(\hat{x}^{-}(k))\cdot[J_{\gamma}(\hat{x}^{-}(k))P^{-}(k)J_{\gamma}^{T}(\hat{x}^{-}(k)) + R(k)]^{-1} \\ \hat{x}(k) &= \hat{x}^{-}(k) + K(k)[z(k) - \gamma(\hat{x}^{-}(k))] \\ P(k) &= P^{-}(k) - K(k)J_{\gamma}(\hat{x}^{-}(k))P^{-}(k) \end{aligned} \tag{2.62}$$

- *Time update*. Compute:

$$\begin{aligned} P^{-}(k+1) &= J_{\phi}(\hat{x}(k))P(k)J_{\phi}^{T}(\hat{x}(k)) + Q(k) \\ \hat{x}^{-}(k+1) &= \phi(\hat{x}(k)) + L(k)u(k) \end{aligned} \tag{2.63}$$

The schematic diagram of the EKF loop is given in Fig. 2.2.

2.2.6 *Implementation of the EKF for Sensorless Control of the DFIG*

As analyzed, the state-space equation of the doubly-fed induction generator is given by

$$\dot{x} = f(x) + g_a v_{r_d} + g_b v_{r_q} \tag{2.64}$$

where the state vector has been defined as $x = [\theta, \omega, \psi_{s_d}, \psi_{s_q}, i_{r_d}, i_{r_q}]^T$ while functions $f(x)$, $g_a(x)$ and $g_b(x)$ have been defined as

$$f(x) = \begin{pmatrix} x_2 \\ -\frac{K_m}{J}x_2 - \frac{T_m}{J} + \frac{\eta}{J}(i_{s_q}x_3 - i_{s_d}x_4) \\ -\frac{1}{\tau_s}x_3 + \omega_{dq}x_4 + \frac{M}{\tau_s}x_5 + v_{s_d} \\ -\omega_{dq}x_3 - \frac{1}{\tau_s}x_4 + \frac{M}{\tau_s}x_6 + v_{s_q} \\ -\beta x_2 x_4 + \frac{\beta}{\tau_s}x_3 + (\omega_{dq} - x_2)x_6 - \gamma_2 x_3 \\ \frac{\beta}{\tau_s}x_4 + \beta x_2 x_3 + (\omega_{dq} - x_2)x_5 - \gamma_2 x_6 \end{pmatrix} \tag{2.65}$$

$$g_a(x) = \left(0\;0\;0\;0\;\frac{1}{\sigma L_s}\;0\right)^T \tag{2.66}$$

$$g_b(x) = \left(0\;0\;0\;0\;0\;\frac{1}{\sigma L_s}\right) \tag{2.67}$$

The measured state variables are taken to be $x_1 = \theta$, $x_5 = i_{r_d}$ and $x_6 = i_{r_q}$. Then, the associated Jacobian matrices are:

$$J_\phi = \left(J_\phi^1\; J_\phi^2\; J_\phi^3\; J_\phi^4\; J_\phi^5\; J_\phi^6\right)^T \tag{2.68}$$

where $J_\phi^1 = [0, 1, 0, 0, 0, 0]$, $J_\phi^2 = [0, \frac{K_m}{J}, \frac{\eta}{J}i_{s_q}, -\frac{\eta}{J}i_{s_d}, 0, 0]$, $J_\phi^3 = [0, 0, -\frac{1}{\tau_s}, \omega_{dq}, \frac{M}{\tau_s}, 0]$, $J_\phi^4 = [0, 0, -\omega_{dq}, -\frac{1}{\tau_s}, 0, \frac{M}{\tau_s}]$, $J_\phi^5 = [0, -\beta x_4 - x_6, -\gamma_2 + \frac{\beta}{\tau_s}, -\beta x_2, 0, \omega_{dq} - x_2]$ and $J_\phi^6 = [0, \beta x_3 - x_5, \beta x_2, \frac{\beta}{\tau_s}, \omega_{dq} - x_2, -\gamma_2]$.

$$J_\gamma = \begin{pmatrix} 1\;0\;0\;0\;0\;0 \\ 0\;0\;0\;0\;1\;0 \\ 0\;0\;0\;0\;0\;1 \end{pmatrix} \tag{2.69}$$

It is noted that: (i) the discrete-time model of Eq. (2.56) can be computed from Eqs. (2.64)–(2.67) after applying common discretization methods (e.g. first order Euler approximation of the continuous-time derivative), (ii) if a Jacobian matrix J_ϕ, associated to the drift term of the system's dynamics, is computed using the system's continuous-time description of Eq. (2.64), then in the EKF recursion of Eq. (2.62) and (2.63) it should be substituted by $I + T_s J_\phi$ where T_s is the sampling period and $I \in R^{n \times n}$ is the identity matrix.

2.2.7 *Estimation of the Wind-Generated Mechanical Torque Using EKF*

The Extended Kalman Filter can be used not only for estimation of the non-measurable state vector elements of the DFIG (such as stator's magnetic flux and rotor's angular velocity) but also for estimation of the wind generated mechanical torque. One approach to succeed this is to redefine the parameters of the state vector of the induction generator as follows: $x_1 = \theta$, $x_2 = T_m$, $x_3 = \omega_r$, $x_4 = \dot{T}_m$, $x_5 = \psi_{s_d}$,

$x_6 = \psi_{s_q}$, $x_7 = i_{r_d}$ and $x_8 = i_{r_q}$. Using the relation about the dependence of the mechanical torque on the design characteristics of the wind turbine as well as on the rotor's speed given in Eq. (1.2) and considering constant or piecewise constant wind speed v, the state equations of the DFIG can be written as:

$$f(x) = \begin{pmatrix} x_3 \\ x_4 \\ -\frac{K_m}{J}x_3 - \frac{x_2}{J} + \frac{\eta}{J}(i_{s_q}x_5 - i_{s_d}x_6) \\ \frac{1}{2}\rho\pi R^3 \frac{d^2}{dt^2}C_q(\lambda, \beta)v^2 \\ -\frac{1}{\tau_s}x_5 + \omega_{dq}x_6 + \frac{M}{\tau_s}x_7 + v_{s_d} \\ -\omega_{dq}x_5 - \frac{1}{\tau_s}x_6 + \frac{M}{\tau_s}x_8 + v_{s_q} \\ -\beta x_3 x_6 + \frac{\beta}{\tau_s}x_5 + (\omega_{dq} - x_3)x_8 - \gamma_2 x_5 \\ \frac{\beta}{\tau_s}x_6 + \beta x_3 x_5 + (\omega_{dq} - x_3)x_7 - \gamma_2 x_8 \end{pmatrix} \tag{2.70}$$

$$g_a(x) = \begin{pmatrix} 0\,0\,0\,0\,0\,0\,\frac{1}{\sigma L_s}\,0 \end{pmatrix}^T \tag{2.71}$$

$$g_b(x) = \begin{pmatrix} 0\,0\,0\,0\,0\,0\,0\,\frac{1}{\sigma L_s} \end{pmatrix} \tag{2.72}$$

Introducing to the DFIG state vector the new state variables x_2 and x_4 and defining function

$$C^*(x_3) = \frac{1}{2}\rho\pi R^3 \ddot{C}_q(\frac{\omega R}{v}, \beta)v^2 \tag{2.73}$$

the Jacobian matrices J_ϕ and J_γ associated to the mechanical torque and used in the updated EKF recursion are written as:

$$J_\phi = \left(J_\phi^1\ J_\phi^2\ J_\phi^3\ J_\phi^4\ J_\phi^5\ J_\phi^6\ J_\phi^7\ J_\phi^8\right)^T \tag{2.74}$$

where $J_\phi^1 = [0, 0, 1, 0, 0, 0, 0, 0]^T$, $J_\phi^2 = [0, 0, 0, 1, 0, 0, 0, 0]$

$J_\phi^3 = [0, -\frac{1}{J}, -\frac{K_m}{J}, 0, \frac{n_p i_{s_q}}{J}, -\frac{n_p i_{s_d}}{J}, 0, 0]$, $J_\phi^4 = [0, 0, \frac{\partial \ddot{C}^*(x_3)}{\partial x_3}, 0, 0, 0, 0, 0]$

$J_\phi^5 = [0, 0, 0, 0, -\frac{1}{\tau_s}, \omega_{dq}, \frac{M}{\tau_s}, 0]$, $J_\phi^6 = [0, 0, 0, 0, -\omega_{dq}, -\frac{1}{\tau_s}, 0, \frac{M}{\tau_s}]$

$J_\phi^7 = [0, 0, -\beta x_6 - x_8, 0, \frac{\beta}{\tau_s} - \gamma_2, -\beta x_3, 0, \omega_{dq}]$, $J_\phi^8 = [0, 0, \beta x_5 - x_7, 0, \beta x_3, \frac{\beta}{\tau_s}, \omega_{dq} - x_3, -\gamma_2]$.

$$J_\gamma = \begin{pmatrix} 1\,0\,0\,0\,0\,0\,0\,0 \\ 0\,0\,0\,0\,0\,0\,1\,0 \\ 0\,0\,0\,0\,0\,0\,0\,1 \end{pmatrix} \tag{2.75}$$

When the Jacobian matrix J_ϕ, associated to the drift term of the system's dynamics, is computed using the system's continuous-time description of Eqs. (1.26) and (1.27), then in the EKF recursion of Eqs. (2.62) and (2.63) it should be substituted by $I + T_s J_\phi$ where T_s is the sampling period and $I \in R^{n\times n}$ is the identity matrix.

The estimation of the mechanical torque T_m using the EKF recursion can be more accurate than calculating T_m with the use of Eq. (1.2) since there may be inaccuracies about the values of parameters ρ, R, λ and v as well in the modeling of the torque coefficient C_q. This type of modeling uncertainty is taken into account and compensated by the EKF, through the use of the process and measurement noise covariance matrices Q and R in the estimator's recursion described in Eqs. (2.62) and (2.63).

2.2.8 Simulation Tests

The efficiency of the proposed control scheme was tested in case of variable rotor speed setpoint assuming different wind profiles. The obtained results for the first wind torque profile are shown in Figs. 2.3, 2.4, and 2.5 while the results for the second wind torque profile are shown in Figs. 2.6, 2.7 and 2.8. The real state variable is denoted by the dashed blue line, the estimated state variable is denoted by the dashed green line, while the associated reference setpoint is denoted by the continuous red line. Although using a small number of sensors and measuring only specific elements of the DFIG state vector such as the rotor's angle and the stator's currents, the proposed control scheme succeeded accurate tracking of the reference setpoints and accurate estimation of the non-measurable parameters such as the rotation speed of the rotor, the magnetic flux of the stator and the wind-generated mechanical torque.

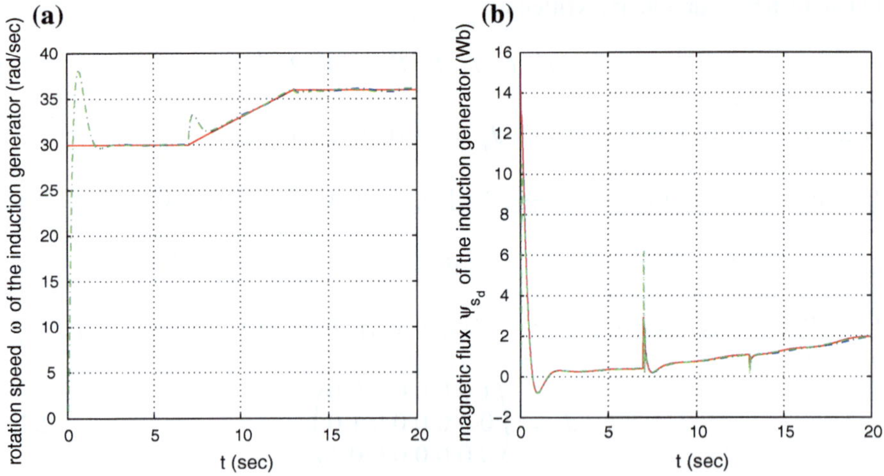

Fig. 2.3 Extended Kalman Filter-based control of the doubly-fed induction generator **a** estimation and setpoint tracking for rotor's angular velocity ω_{r_d}, **b** estimation and setpoint tracking for stator's magnetic flux ψ_{s_d}

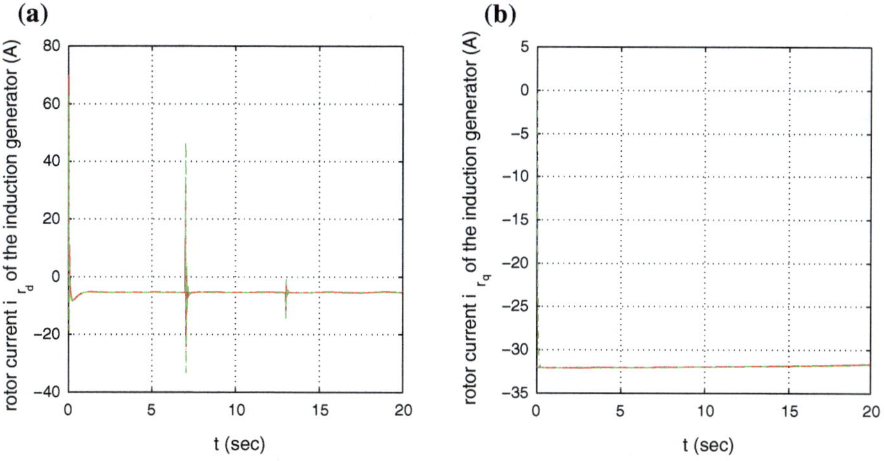

Fig. 2.4 Extended Kalman Filter-based control of the doubly-fed induction generator **a** rotor's control input current $\imath_{r_d}$, **b** rotor's control input current i_{r_q}

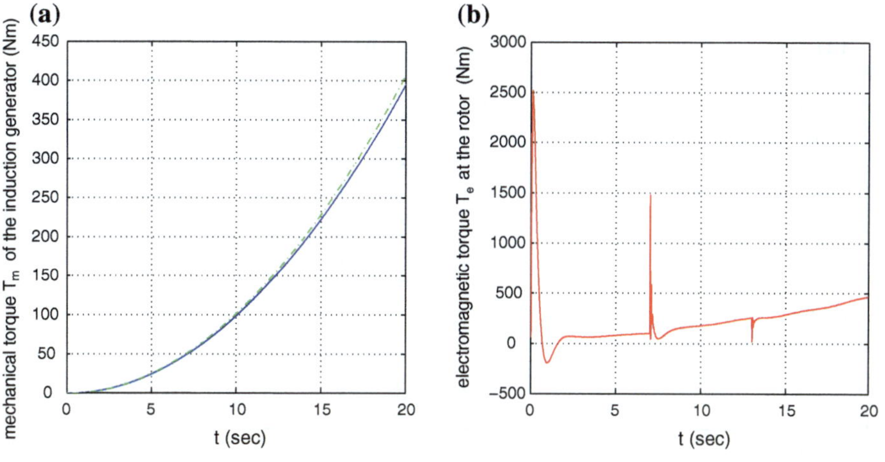

Fig. 2.5 **a** Mechanical torque T_m due to wind. **b** Electromagnetic torque T_e at the DFIG rotor

Taking into account that several variables of the induction generator's state vector (e.g. rotation speed and magnetic flux) are not directly measurable (due to sensors cost and limited reliability, sensors failures and difficulties in sensors installation) the significance of state estimation through Kalman Filtering becomes clear. It can be noticed that the Extended Kalman Filter is an efficient approach for the implementation of state estimation-based control of the sixth-order induction generator model. The Unscented Kalman Filter can be also used in place of the Extended Kalman Filter and in the latter case there will be no need to compute Jacobian matrices [202–204].

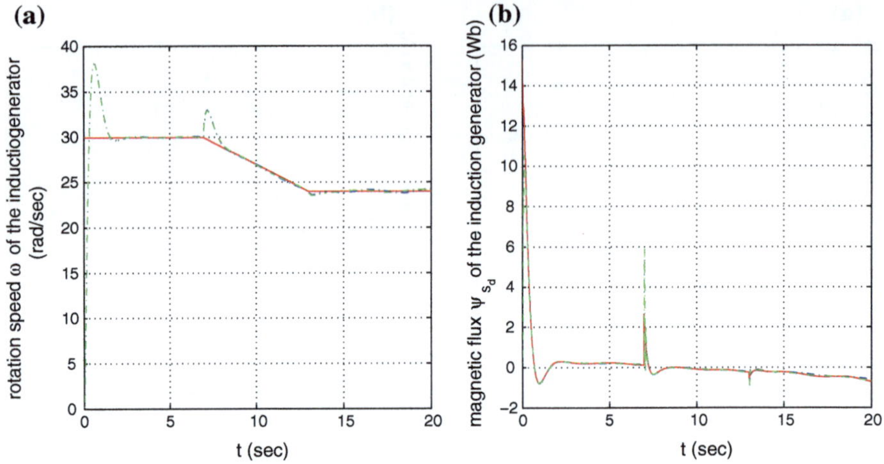

Fig. 2.6 Extended Kalman Filter-based control of the doubly-fed induction generator **a** estimation and setpoint tracking for rotor's angular velocity ω_{r_d}, **b** estimation and setpoint tracking for stator's magnetic flux ψ_{s_d}

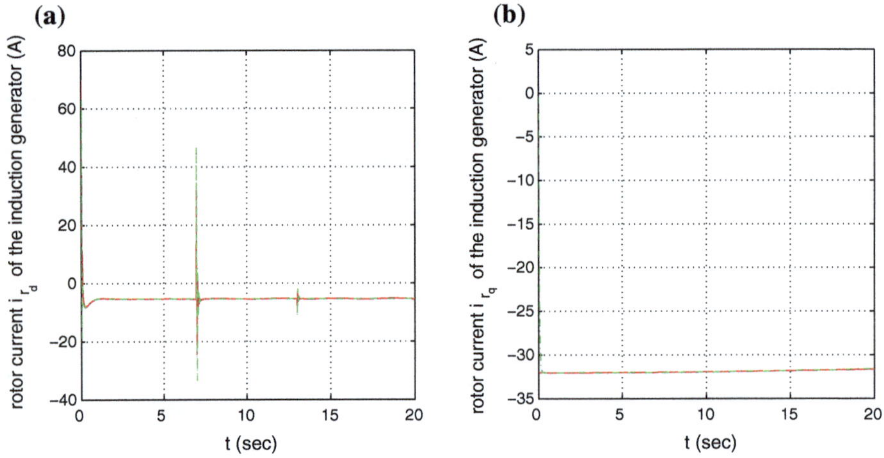

Fig. 2.7 Extended Kalman Filter-based control of the doubly-fed induction generator **a** rotor's control input current i_{r_d}, **b** rotor's control input current i_{r_q}

Finally, it is noted that to compensate for external disturbances and parameter variations in state estimation-based nonlinear control schemes the following approaches are possible (i) use of adaptive Kalman Filter, (ii) redesign of the Kalman filter in the form of a disturbance observer, (iii) redesign of the Kalman Filter in the form of a high-gain Extended Kalman Filter. Regarding (i), adaptive Kalman Filtering is implemented through the adaptation of the process and measurement noise covariance matrices in the Kalman Filter recursion so as to cope with variable noise levels

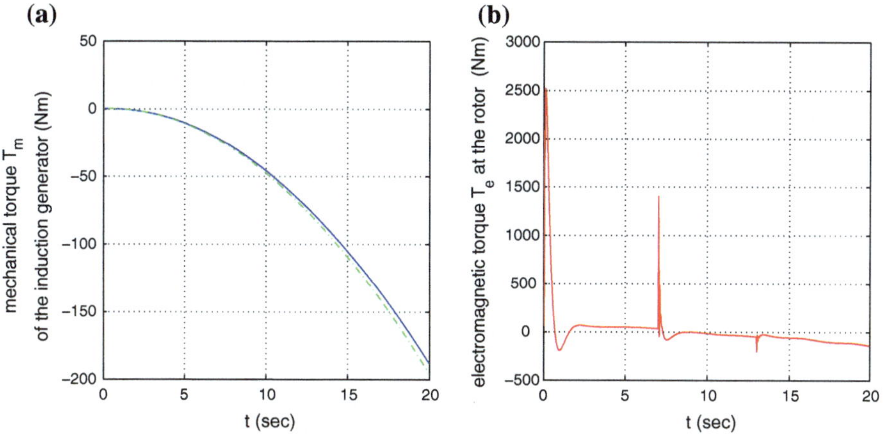

Fig. 2.8 **a** Mechanical torque T_m due to wind. **b** Electromagnetic torque T_e at the DFIG rotor

[14]. Regarding (ii), additive disturbances and parametric changes can be identified with the use of Kalman Filters that operate as disturbance observers (see work on Kalman Filtering and disturbance observers in [352]). Once the disturbance affecting the nonlinear system becomes known it can be compensated through the inclusion of an additional control term in the loop. Thus, it can be assured again that the estimation error $x - \hat{x}$ will be minimal and the performance of the control loop will be satisfactory. There are several results on disturbance observers (e.g. Extended State Observers, Unknown Input Observers or Perturbation Observers) and on their use within a Kalman Filter framework [224, 292]. Such results can be exploited so as to make Kalman Filtering for electric power generators as well as the associated state estimation-based control loop be more robust [352]. Regarding (iii) the high-gain Extended Kalman Filter can provide additional robustness to state estimation under external disturbances and parametric variations [47, 53].

2.3 Control of the DFIG Based on Global Linearization Approaches

2.3.1 Outline

In this section, flatness-based control for the doubly-fed induction generator is implemented through its transformation to the linear canonical form. As noted, differential flatness theory is currently a main direction in nonlinear dynamical systems and enables linearization and control for a wide class of systems, in a more efficient manner than Lie-algebra methods [83, 352, 388, 411, 473]. Previously, it was explained that, to find out if a dynamical system is differentially flat, the following should be examined: (i) the existence of the so-called flat output, i.e. a new variable which is expressed as a function of the system's state variables. The flat output and its derivatives should not be coupled in the form of an ordinary differential equation, (ii) the components of the system (i.e. state variables and control input) should be expressed as functions of the flat output and its derivatives [57, 137, 241, 280, 412, 456]. Differential flatness theory enables transformation to a linearized form (canonical Brunovsky form) for which the design of the controller becomes easier. By showing that a system is differentially flat, one can easily design a reference trajectory as a function of the so-called flat output and can find a control law that assures tracking of this desirable trajectory [456, 137].

This chapter is concerned with proving differential flatness of the model of the doubly-fed induction generator and its description in the Brunovksy (canonical) form [388, 411]. By defining flat outputs which are associated with the rotor's angle and with the magnetic flux of the stator, an equivalent DFIG description in the Brunovksy (linear canonical) form is obtained. It is shown that for the linearized DFIG's model it is possible to design a feedback controller. At a second stage, a novel Kalman Filtering method, the Derivative-free nonlinear Kalman Filter, is proposed for estimating the state vector elements of the linearized system which are not directly measurable. With the redesign of the proposed Kalman Filter as a disturbance observer, it becomes possible to estimate also disturbance terms affecting the DFIG model and to use these terms in the feedback controller. By avoiding linearization approximations, the proposed filtering method, improves the accuracy of estimation, and results in smooth control signal variations and in minimization of the tracking error of the associated control loop [355, 368].

2.3.2 Input-Output Linearization of the DFIG Using Lie Algebra Theory

2.3.2.1 Input-Output Linearization of the DFIG Model

The following variables are defined

$$\begin{aligned} h_1(x) &= x_1 = \theta \\ h_2(x) &= x_3^2 + x_4^2 = \psi_{s_d}^2 + \psi_{s_q}^2 \end{aligned} \tag{2.76}$$

Next, based on h_1, h_2 the following transformed state variables are defined

$$z_1 = h_1(x) = \theta \tag{2.77}$$

$$\begin{aligned} z_2 &= L_f h_1(x) \Rightarrow \\ z_2 &= f_1 \Rightarrow z_2 = x_2 \Rightarrow z_2 = \omega \end{aligned} \tag{2.78}$$

Similarly, one has

$$\begin{array}{c} z_3 = L_f^2 h_1(x) = L_f z_2 \Rightarrow \\ z_3 = f_2 \Rightarrow z_3 = -\frac{K_m}{J} x_2 - \frac{T_m}{J} + \frac{\eta}{J}(i_{r_q} x_3 - i_{r_q} x_4) \Rightarrow \\ z_3 = -\frac{K_m}{J} x_2 - \frac{T_m}{J} + \frac{\eta}{J}(x_6 x_3 - x_5 x_4) \end{array} \tag{2.79}$$

For the transformed state variable z_4 one has

$$z_4 = h_2(x) = \psi_{s_d}^2 + \psi_{s_q}^2 = x_3^2 + x_4^2 \tag{2.80}$$

and

$$\begin{array}{c} z_5 = L_f h_2(x) \Rightarrow z_5 = 2x_3 f_3 + 2x_4 f_4 \Rightarrow \\ z_5 = 2x_3[-\frac{1}{\tau_s} x_3 + \omega_{dq} x_4 + \frac{M}{\tau_s} x_5 + v_{s_d}] + \\ +2x_4[-\omega_{dq} x_3 - \frac{1}{\tau_s} x_4 + \frac{M}{\tau_s} x_6 + v_{s_q}] \end{array} \tag{2.81}$$

After the change of the state variables it holds

$$\begin{array}{c} \dot{z}_1 = z_2 \\ \dot{z}_2 = z_3 \\ \dot{z}_3 = L_f^3 h_1(x) + (L_{g_a} L_f^2 h_1(x)) u_1 + (L_{g_b} L_f^2 h_1(x)) u_2 \\ \dot{z}_4 = z_5 \\ \dot{z}_5 = L_f^2 h_2(x) + (L_{g_a} L_f h_2)(x) u_1 + (L_{g_b} L_f h_2(x)) u_2 \end{array} \tag{2.82}$$

The inputs of the above linearized and decoupled DFIG model are $u_1 = u_{r_d}$ and $u_2 = u_{r_q}$. The system of Eq. (2.82) can be written in the input-output linearized form

$$\begin{pmatrix} z_1^{(3)} \\ \ddot{z}_4 \end{pmatrix} = f_a + \tilde{M} u \tag{2.83}$$

where

$$\begin{array}{c} f_a(x) = \begin{pmatrix} L_f^3 h_1(x) \\ L_f^2 h_2(x) \end{pmatrix} \\ \tilde{M} = \begin{pmatrix} L_{g_a} L_f^2 h_1(x) & L_{g_b} L_f^2 h_1(x) \\ L_{g_a} L_f h_2(x) & L_{g_b} L_f h_2(x) \end{pmatrix} \end{array} \tag{2.84}$$

or equivalently one has the system's description in the MIMO canonical form

$$\begin{pmatrix} \dot{z}_1 \\ \dot{z}_2 \\ \dot{z}_3 \\ \dot{z}_4 \\ \dot{z}_5 \end{pmatrix} = \begin{pmatrix} 0 & 1 & 0 & 0 & 0 \\ 0 & 0 & 1 & 0 & 0 \\ 0 & 0 & 0 & 0 & 0 \\ 0 & 0 & 0 & 0 & 1 \\ 0 & 0 & 0 & 0 & 0 \end{pmatrix} \begin{pmatrix} z_1 \\ z_2 \\ z_3 \\ z_4 \\ z_5 \end{pmatrix} + \begin{pmatrix} 0 & 0 \\ 0 & 0 \\ 1 & 0 \\ 0 & 0 \\ 0 & 1 \end{pmatrix} \begin{pmatrix} v_1 \\ v_2 \end{pmatrix} \tag{2.85}$$

where

$$v_1 = L_f^3 h_1(x) + (L_{g_a}(L_f^2 h_1(x))u_1 + (L_{g_b} L_f^2 h_1(x))u_2$$
$$v_2 = L_f^2 h_2(x) + (L_{g_a} L_f h_2(x))u_1 + (L_{g_b} L_f h_2(x))u_2 \tag{2.86}$$

Returning to the compact form of Eq. (2.84) one has

$$\begin{pmatrix} z_1^{(3)} \\ \ddot{z}_4 \end{pmatrix} = \begin{pmatrix} v_1 \\ v_2 \end{pmatrix} \tag{2.87}$$

and the control signal that assures convergence of the z_1 and z_4 to the reference setpoints z_1^d and z_4^d is given by

$$v_1 = z_1^{d(3)} - k_1^{(1)}(\ddot{z}_1 - \ddot{z}_1^d) - k_2^{(1)}(\dot{z}_1 - \dot{z}_1^d) - k_3^{(1)}(z_1 - z_1^d)$$
$$v_2 = \ddot{z}_4^d - k_1^{(2)}(\dot{z}_4 - \dot{z}_4^d) - k_2^{(2)}(z_4 - z_4^d) \tag{2.88}$$

2.3.2.2 State Estimation-Based Control

For the implementation of the aforementioned control law, there is need to obtain measurements of all elements of the DFIG's state vector. The rotor's turn angle can be measured directly with the use of an encoder [33, 214]. Knowing the rotor's angle, and with the use of the decoupled induction machine's model of Eq. (2.85) it is possible to estimate the rotor's angular speed. Similarly, after obtaining measurements of the magnetic flux at the stator and with the use of the decoupled induction machine's model of Eq. (2.85) it is possible to estimate the derivatives of the magnetic flux. Due to the fact that the magnetic flux of the stator ψ_s cannot be measured directly, equations that provide indirect measurements of the flux (computed through measurements of the stator and rotor currents) will be used, that is

$$\psi_{s_d} = L_s i_{s_d} + M i_{r_d}$$
$$\psi_{s_q} = L_s i_{s_q} + M i_{r_q} \tag{2.89}$$

It is noted that the currents are measured in the ab reference frame and their computation in the dq reference frame requires the application of the associated reference frame transformation.

Using the model of Eq. (2.85) the state estimator for the DFIG is given by

$$\dot{\hat{z}} = A\hat{z} + Bv + K(z^{meas} - C\hat{z}) \tag{2.90}$$

where the estimator's gain $K \in R^{5\times 2}$

$$A = \begin{pmatrix} 0\,1\,0\,0\,0 \\ 0\,0\,1\,0\,0 \\ 0\,0\,0\,0\,0 \\ 0\,0\,0\,0\,1 \\ 0\,0\,0\,0\,0 \end{pmatrix} \quad B = \begin{pmatrix} 0\,0 \\ 0\,0 \\ 1\,0 \\ 0\,0 \\ 0\,1 \end{pmatrix}$$
$$C = \begin{pmatrix} 1\,0\,0\,0\,0 \\ 0\,0\,0\,1\,0 \end{pmatrix} \tag{2.91}$$

2.3.3 *Differential Flatness for Nonlinear Dynamical Systems*

2.3.3.1 Transformation of MIMO Nonlinear Systems into the Brunovsky Form

It is assumed now that after defining the flat outputs of the initial MIMO nonlinear model of the DFIG, and after expressing the system state variables and control inputs as functions of the flat output and of the associated derivatives, the system can be transformed in the Brunovsky canonical form:

$$\begin{array}{ll} \dot{x}_1 = x_2 & \\ \cdots & \\ \dot{x}_{r_1-1} = x_{r_1} & \\ \dot{x}_{r_1} = f_1(x) + \sum_{j=1}^{p} g_{1_j}(x)u_j + d_1 & y_1 = x_1 \\ \dot{x}_{r_1+1} = x_{r_1+2} & \cdots \\ \cdots & y_p = x_{n-r_p+1} \\ \dot{x}_{p-1} = x_p & \\ \dot{x}_p = f_p(x) + \sum_{j=1}^{p} g_{p_j}(x)u_j + d_p & \end{array} \tag{2.92}$$

where $x = [x_1, \ldots, x_n]^T$ is the state vector of the transformed system (according to the differential flatness formulation), $u = [u_1, \ldots, u_p]^T$ is the set of control inputs, $y = [y_1, \ldots, y_p]^T$ is the output vector, f_i are the drift functions and $g_{i,j}$, $i, j = 1, 2, \ldots, p$ are smooth functions corresponding to the control input gains, while d_j is a variable associated to external disturbances. In holds that $r_1 + r_2 + \cdots + r_p = n$. Having written the initial nonlinear system into the canonical (Brunovsky) form it holds

$$y_i^{(r_i)} = f_i(x) + \sum_{j=1}^{p} g_{ij}(x)u_j + d_j \tag{2.93}$$

Next the following vectors and matrices can be defined $f(x) = [f_1(x), \ldots, f_n(x)]^T$, $g(x) = [g_1(x), \ldots, g_n(x)]^T$ with $g_i(x) = [g_{1i}(x), \ldots, g_{pi}(x)]^T$, and $A =$

$diag[A_1,\dots,A_p]$, $B = diag[B_1,\dots,B_p]$, $C = diag[C_1,\dots,C_p]$, $d = [d_1,\dots,d_p]^T$, where matrix A has the MIMO canonical form, i.e. with block-diagonal elements

$$A_i = \begin{pmatrix} 0 & 1 & \cdots & 0 \\ 0 & 0 & \cdots & 0 \\ \vdots & \vdots & \cdots & \vdots \\ 0 & 0 & \cdots & 1 \\ 0 & 0 & \cdots & 0 \end{pmatrix}_{r_i \times r_i} \\ B_i^T = \begin{pmatrix} 0 & 0 & \cdots & 0 & 1 \end{pmatrix}_{1\times r_i} \\ C_i = \begin{pmatrix} 1 & 0 & \cdots & 0 & 0 \end{pmatrix}_{1\times r_i} \tag{2.94}$$

Thus, Eq. (8.112) can be written in state-space form

$$\dot{x} = Ax + Bv + B\tilde{d} \\ y = Cx \tag{2.95}$$

where the control input is written as $v = f(x) + g(x)u$.

2.3.4 Input-Output Linearization of the DFIG Using Differential Flatness Theory

2.3.4.1 Differential Flatness Properties of the DFIG

The flat outputs of the system are defined as

$$y_1 = \theta \text{ or } y = x_1 \\ y_2 = \psi_{s_d}^2 + \psi_{s_q}^2 \text{ or } y_2 = x_3^2 + x_4^2 \tag{2.96}$$

It holds that

$$\dot{y}_1 = \omega \text{ or } \dot{y}_1 = x_2 \Rightarrow \\ \ddot{y}_1 = \dot{\omega} = -\frac{K_m}{J}x_2 - \frac{T_m}{J} + \frac{\eta}{J}(x_6x_3 - x_5x_4) \Rightarrow \\ \ddot{y}_1 = \dot{\omega} = -\frac{K_m}{J}\dot{y}_1 - \frac{T_m}{J} + \frac{\eta}{J}(x_6x_3 - x_5x_4) \tag{2.97}$$

Deriving the last row of Eq. (2.97) with respect to time one obtains

$$\begin{aligned} y_1^{(3)} &= -\frac{K_m}{J}\ddot{y}_1 + \frac{\eta}{J}(\dot{x}_6x_3 + x_6\dot{x}_3 - \dot{x}_5x_4 - x_5\dot{x}_4) \Rightarrow \\ y_1^{(3)} &= -\frac{K_m}{J}\ddot{y}_1 + \frac{\eta}{J}x_3\{[\frac{\beta}{\tau_s}x_4 + \beta x_2x_3 + (\omega_{dq} - x_2)x_5 - \\ &- \gamma_2x_6 - \beta v_{s_q}] + \frac{1}{\sigma L_r}u_1\} + \frac{\eta}{J}x_6[-\frac{1}{\tau_s}x_3 + \omega_{dq}x_4 + \frac{M}{\tau_s}x_5 + v_{s_d}] \\ &- \frac{\eta}{J}x_4\{[-\beta x_2x_4 + \frac{\beta}{\tau_s}x_3 + (\omega_{dq} - x_2)x_6 - \gamma_2x_5 - \\ &- \beta v_{s_d}] + \frac{1}{\sigma L_r}u_2\} - \frac{\eta}{J}x_5[-\omega_{dq}x_3 - \frac{1}{\tau_s}x_4 + \frac{M}{\tau_s}x_6 + v_{s_q}] \end{aligned} \tag{2.98}$$

Moreover, about the second flat output it holds

$$\begin{aligned}\dot{y}_2 = 2x_3\dot{x}_3 + 2x_4\dot{x}_4 \Rightarrow \\ \dot{y}_2 = 2x_3[-\tfrac{1}{\tau_s}x_3 + \omega_{dq}x_4 + \tfrac{M}{\tau_s}x_5 + v_{s_d}] + \\ + 2x_4[-\omega_{dq}x_3 - \tfrac{1}{\tau_s}x_4 + \tfrac{M}{\tau_s}x_6 + v_{s_q}] \Rightarrow\end{aligned} \tag{2.99}$$

Consequently, it holds

$$\begin{aligned}\ddot{y}_2 = 2\dot{x}_3[-\tfrac{1}{\tau_s}x_3 + \omega_{dq}x_4 + \tfrac{M}{\tau_s}x_5 + v_{s_d}] + \\ + 2x_3[-\tfrac{1}{\tau_s}\dot{x}_3 + \omega_{dq}\dot{x}_4 + \tfrac{M}{\tau_s}\dot{x}_5] + \\ 2\dot{x}_4[-\omega_{dq}x_3 - \tfrac{1}{\tau_s}x_4 + \tfrac{M}{\tau_s}x_6 + v_{s_q}] + \\ + 2x_4[-\omega_{dq}\dot{x}_3 - \tfrac{1}{\tau_s}\dot{x}_4 + \tfrac{M}{\tau_s}\dot{x}_6]\end{aligned} \tag{2.100}$$

or equivalently

$$\begin{aligned}\ddot{y}_2 = 2[-\tfrac{1}{\tau_s}x_3 + \omega_{dq}x_4 + \tfrac{M}{\tau_s}x_5 + v_{s_d}]^2 + \\ - \tfrac{2}{\tau_s}x_3[-\tfrac{1}{\tau_s}x_3 + \omega_{dq}x_4 + \tfrac{M}{\tau_s}x_5 + v_{s_d}] \\ - 2\omega_{dq}x_3[-\omega_{dq}x_3 - \tfrac{1}{\tau_s}x_4 + \tfrac{M}{\tau_s}x_6 + v_{s_q}] \\ + \tfrac{2M}{\tau_s}x_3\{[-\beta x_2 x_4 + \tfrac{\beta}{\tau_s}x_3 + (\omega_{dq} - x_2)x_6 - \\ - \gamma_2 x_5 - \beta v_{s_d}] + \tfrac{1}{\sigma L_r}u_1\} \\ + 2[-\omega_{dq}x_3 - \tfrac{1}{\tau_s}x_4 + \tfrac{M}{\tau_s}x_6 + v_{s_q}]^2 \\ - 2\omega_{dq}x_4[-\tfrac{1}{\tau_s}x_3 + \omega_{dq}x_4 + \tfrac{M}{\tau_s}x_5 + v_{s_d}] \\ - \tfrac{2}{\tau_s}x_4[-\omega_{dq}x_3 - \tfrac{1}{\tau_s}x_4 + \tfrac{M}{\tau_s}x_6 + v_{s_q}] + \\ + 2x_4\tfrac{M}{\tau_s}\{[\tfrac{\beta}{\tau_s}x_4 + \beta x_2 x_3 + (\omega_{dq} - x_2)x_5 - \\ - \gamma_2 x_6 - \beta v_{s_q}] + \tfrac{1}{\sigma L_r}u_2\}\end{aligned} \tag{2.101}$$

It holds that $x_1 = y_1$, $x_2 = \dot{y}_1$. From the second row of Eq. (2.96) and considering that the field orientation condition requires $x_4 = \psi_{s_q} = 0$ one obtains that $x_3 = \sqrt{y_2}$. Moreover, from Eq. (2.97) it holds

$$\begin{aligned}\ddot{y}_1 = -\tfrac{K_m}{J}\dot{y}_1 - \tfrac{T_m}{J} + \tfrac{\eta}{J}\sqrt{y_2}x_6 \Rightarrow \\ x_6 = \frac{\ddot{y}_1 + \frac{K_m}{J}\dot{y}_1 + \frac{T_m}{J}}{\frac{\eta}{J}\sqrt{y_2}}, \ y_2 \neq 0\end{aligned} \tag{2.102}$$

From Eq. (2.99) one obtains

$$\begin{aligned}\dot{y}_2 = -\tfrac{2}{\tau_s}x_3^2 + \tfrac{2M}{\tau_s}x_3x_5 + 2v_{s_d}x_3 \Rightarrow \\ \dot{y}_2 + (\tfrac{2}{\tau_s}x_3 - 2v_{s_d})x_3 = \tfrac{2M}{\tau_s}x_3x_5 \Rightarrow \\ x_5 = \frac{\dot{y}_2 + (\frac{2}{\tau_s}\sqrt{y_2} - 2v_{s_d})\sqrt{y_2}}{\frac{2M}{\tau_s}\sqrt{y_2}} \ y_2 \neq 0\end{aligned} \tag{2.103}$$

Therefore, x_5 is also a function of the flat output and of its derivatives. Additionally, by solving the system of Eqs. (2.98) and (2.101) with respect to the control inputs u_1 and u_2 one obtains that the control inputs are functions of the flat output and its derivatives. Therefore, the model of the DFIG is a differentially flat one.

Next, to design the flatness-based controller for the DFIG the following transformation of the state variables is introduced: $z_1 = y_1$, $z_2 = \dot{y}_1$, $z_3 = \ddot{y}_1$, $z_4 = y_2$, $z_5 = \dot{y}_2$ for which holds

$$\begin{aligned} \dot{z}_1 &= z_2 \\ \dot{z}_2 &= z_3 \\ \dot{z}_3 &= L_f^3 h_1(x) + (L_{g_a} L_f^2 h_1(x))u_1 + (L_{g_b} L_f^2 h_1(x))u_2 \\ \dot{z}_4 &= z_5 \\ \dot{z}_5 &= L_f^2 h_2(x) + (L_{g_a} L_f h_2(x))u_1 + (L_{g_b} L_f h_2(x))u_2 \end{aligned} \tag{2.104}$$

Therefore, one obtains the decoupled and linearized representation of the system

$$\begin{pmatrix} z_1^{(3)} \\ \ddot{z}_4 \end{pmatrix} = \begin{pmatrix} L_f^3 h_1(x) \\ L_f^2 h_2(x) \end{pmatrix} + \begin{pmatrix} L_{g_a} L_f^2 h_1(x) & L_{g_b} L_f^2 h_1(x) \\ L_{g_a} L_f h_2(x) & L_{g_b} L_f h_2(x) \end{pmatrix} \begin{pmatrix} u_1 \\ u_2 \end{pmatrix} \tag{2.105}$$

or equivalently

$$\begin{pmatrix} z_1^{(3)} \\ \ddot{z}_4 \end{pmatrix} = f_a + \tilde{M}u \tag{2.106}$$

where

$$f_a = \begin{pmatrix} L_f^3 h_1(x) \\ L_f^2 h_2(x) \end{pmatrix} \quad \tilde{M} = \begin{pmatrix} L_{g_a} L_f^2 h_1(x) & L_{g_b} L_f^2 h_1(x) \\ L_{g_a} L_f h_2(x) & L_{g_b} L_f h_2(x) \end{pmatrix} \tag{2.107}$$

By defining the control inputs $v_1 = L_f^3 h_1(x) + (L_{g_a} L_f^2 h_1(x))u_1 + (L_{g_b} L_f^2 h_1(x))u_2$ and $v_2 = L_f^2 h_2(x) + (L_{g_a} L_f h_2(x))u_1 + (L_{g_b} L_f h_2(x))u_2$ one can also have the description in the MIMO canonical form

$$\begin{pmatrix} \dot{z}_1 \\ \dot{z}_2 \\ \dot{z}_3 \\ \dot{z}_4 \\ \dot{z}_5 \end{pmatrix} = \begin{pmatrix} 0 & 1 & 0 & 0 & 0 \\ 0 & 0 & 1 & 0 & 0 \\ 0 & 0 & 0 & 0 & 0 \\ 0 & 0 & 0 & 0 & 1 \\ 0 & 0 & 0 & 0 & 0 \end{pmatrix} \begin{pmatrix} z_1 \\ z_2 \\ z_3 \\ z_4 \\ z_5 \end{pmatrix} + \begin{pmatrix} 0 & 0 \\ 0 & 0 \\ 1 & 0 \\ 0 & 0 \\ 0 & 1 \end{pmatrix} \begin{pmatrix} v_1 \\ v_2 \end{pmatrix} \tag{2.108}$$

The control input for the linearized and decoupled model of the DFIG is chosen as follows

$$\begin{aligned} v_1 &= z_1^{d(3)} - k_1^{(1)}(\ddot{z}_1 - \ddot{z}_1^d) - k_2^{(1)}(\dot{z}_1 - \dot{z}_1^d) - k_3^{(1)}(z_1 - z_1^d) \\ v_2 &= \ddot{z}_4^d - k_1^{(2)}(\dot{z}_4 - \dot{z}_4^d) - k_2^{(2)}(z_4 - z_4^d) \end{aligned} \tag{2.109}$$

and finally the control input that is applied to the system is

$$u = \tilde{M}^{-1}(-f_a + v) \tag{2.110}$$

The proposed control scheme can work with the use of measurements from a small number of sensors. That is, there is need to obtain measurements of only $y_1 = \theta$ which is the turn angle of the generator's rotor, and of the magnetic flux $y_2 = \psi_s^2 = \psi_{s_d}^2 + \psi_{s_q}^2$, or due to the orientation of the magnetic field $y_2 = \psi_s^2 = \psi_{s_d}^2$. The stator flux ($\psi_s$) cannot be measured directly from a sensor (e.g. the use of Hall sensor in an electric machine with a rotating part would not be efficient), however Eq. (2.89) that relates stator flux and stator and rotor currents can be used to calculate ψ_s. Thus one has:

$$\begin{aligned} \psi_{s_d} &= L_s i_{s_d} + M i_{r_d} \\ \psi_{s_q} &= 0 \end{aligned} \tag{2.111}$$

which means that by measuring stator and rotor currents one can obtain an indirect measurement of the stator's magnetic flux ψ_{s_d}. Next, one can compute the dynamics of the magnetic flux, jointly with the dynamics of the rotor's motion through the use of the Kalman Filter. This estimation method is based on the application of the Kalman Filter recursion to the linearized equivalent of the generator's model which is given by Eq. (2.108), and is known as Derivative-free nonlinear Kalman Filter [375]. Actually, Eq. (2.108) can be written in the state-space form

$$\begin{aligned} \dot{z} &= Az + Bv \\ z^{meas} &= Cz \end{aligned} \tag{2.112}$$

where

$$A = \begin{pmatrix} 0\,1\,0\,0\,0 \\ 0\,0\,1\,0\,0 \\ 0\,0\,0\,0\,0 \\ 0\,0\,0\,0\,1 \\ 0\,0\,0\,0\,0 \end{pmatrix} \quad B = \begin{pmatrix} 0\,0 \\ 0\,0 \\ 1\,0 \\ 0\,0 \\ 0\,1 \end{pmatrix} \tag{2.113}$$

$$C = \begin{pmatrix} 1\,0\,0\,0\,0 \\ 0\,0\,0\,1\,0 \end{pmatrix}$$

The estimator's dynamics is:

$$\dot{\hat{z}} = A{\cdot}\hat{z} + B{\cdot}v + K(z^{meas} - C\hat{z}) \tag{2.114}$$

where $K \in R^{5\times2}$ is the state estimator's gain. Defining as $\tilde{A}_d$, $\tilde{B}_d$, and $\tilde{C}_d$, the discrete-time equivalents of matrices A, B and C respectively, the associated Kalman Filter-based estimator is given by [34, 209, 341, 480]

measurement update:

$$\begin{aligned} K(k) &= P^-(k)C_d^T[C_d{\cdot}P^-(k)C_d^T + R]^{-1} \\ \hat{z}(k) &= \hat{z}^-(k) + K(k)[z^{meas}(k) - C_d\hat{z}^-(k)] \\ P(k) &= P^-(k) - K(k)C_dP^-(k) \end{aligned} \tag{2.115}$$

time update:

$$\begin{aligned} P^-(k+1) &= A_d(k)P(k)A_d^T(k) + Q(k) \\ \hat{z}^-(k+1) &= A_d(k)\hat{z}(k) + B_d(k)v(k) \end{aligned} \tag{2.116}$$

2.3.5 Kalman Filter-Based Disturbance Observer for the DFIG Model

2.3.5.1 Application of a Disturbance Observer to the DFIG Model

Next, it will be considered that additive input disturbances (e.g. due to load variations) affect the DFIG model. The simultaneous estimation of the non-measurable elements of the DFIG state vector as well as the estimation of additive disturbance terms affecting the generator is possible with the use of a disturbance estimator [99, 169, 224, 292].

It is assumed that the third and fifth row of the state-space equations of the Doubly-Fed Induction Generator of Eq. (2.104) include a disturbance term

$$\begin{aligned} \dot{z}_3 &= L_f^3 h_1(x) + L_{g_a}(L_f^2 h_1(x))u_1 + L_{g_b}(L_f^2 h_2(x))u_2 + \tilde{d}_1 \\ \dot{z}_5 &= L_f^2 h_2(x) + L_{g_a}(L_f h_2(x))u_1 + L_{g_b}(L_f h_2(x))u_2 + \tilde{d}_2 \end{aligned} \tag{2.117}$$

Without loss of generality, the dynamics of the disturbance terms is described by their second order derivatives and the associated initial conditions, i.e. $\ddot{\tilde{d}}_1 = \tilde{f}_a(x)$ and $\ddot{\tilde{d}}_2 = \tilde{f}_b(x)$. Next, an extended state-space model of the system is defined that comprises as additional state variables the disturbance terms $z_6 = \tilde{d}_1$, $z_7 = \dot{\tilde{d}}_1$, while $z_8 = \tilde{d}_2$, and $z_9 = \dot{\tilde{d}}_2$. Thus, the extended state-space model is written as $\dot{z}_1 = z_2$, $\dot{z}_2 = z_3$, $\dot{z}_3 = v_1 + z_6$, $\dot{z}_4 = z_5$, and $\dot{z}_5 = v_2 + z_8$, $\dot{z}_6 = z_7$, $\dot{z}_7 = \tilde{f}_a$, $\dot{z}_8 = z_9$ and $\dot{z}_9 = \tilde{f}_b$, or in matrix form one has

$$\begin{aligned} \dot{\tilde{z}} &= \tilde{A}\tilde{z} + \tilde{B}\tilde{v} \\ \tilde{z}^{meas} &= \tilde{C}\tilde{z} \end{aligned} \tag{2.118}$$

$$\begin{pmatrix} \dot{z}_1 \\ \dot{z}_2 \\ \dot{z}_3 \\ \dot{z}_4 \\ \dot{z}_5 \\ \dot{z}_6 \\ \dot{z}_7 \\ \dot{z}_8 \\ \dot{z}_9 \end{pmatrix} = \begin{pmatrix} 0&1&0&0&0&0&0&0&0 \\ 0&0&1&0&0&0&0&0&0 \\ 0&0&0&0&0&1&0&0&0 \\ 0&0&0&0&1&0&0&0&0 \\ 0&0&0&0&0&0&0&1&0 \\ 0&0&0&0&0&0&1&0&0 \\ 0&0&0&0&0&0&0&0&0 \\ 0&0&0&0&0&0&0&0&1 \\ 0&0&0&0&0&0&0&0&0 \end{pmatrix} \begin{pmatrix} z_1 \\ z_2 \\ z_3 \\ z_4 \\ z_5 \\ z_6 \\ z_7 \\ z_8 \\ z_9 \end{pmatrix} + \begin{pmatrix} 0&0&0&0 \\ 0&0&0&0 \\ 1&0&0&0 \\ 0&0&0&0 \\ 0&1&0&0 \\ 0&0&0&0 \\ 0&0&1&0 \\ 0&0&0&0 \\ 0&0&0&1 \end{pmatrix} \begin{pmatrix} v_1 \\ v_2 \\ \tilde{f}_a \\ \tilde{f}_b \end{pmatrix} \tag{2.119}$$

$$\begin{pmatrix} z_1^{meas} \\ z_4^{meas} \end{pmatrix} = \begin{pmatrix} 1\,0\,0\,0\,0\,0\,0\,0\,0 \\ 0\,0\,0\,1\,0\,0\,0\,0\,0 \end{pmatrix} \begin{pmatrix} z_1 \\ z_2 \\ z_3 \\ z_4 \\ z_5 \\ z_6 \\ z_7 \\ z_8 \\ z_9 \end{pmatrix} \tag{2.120}$$

The associated state estimator is

$$\dot{\hat{z}} = \tilde{A}_o \hat{z} + \tilde{B}_o \tilde{v}_1 + K_o(z^{meas} - \tilde{C}\hat{z}) \tag{2.121}$$

where

$$\tilde{A}_o = \begin{pmatrix} 0\,1\,0\,0\,0\,0\,0\,0\,0 \\ 0\,0\,1\,0\,0\,0\,0\,0\,0 \\ 0\,0\,0\,0\,0\,1\,0\,0\,0 \\ 0\,0\,0\,0\,1\,0\,0\,0\,0 \\ 0\,0\,0\,0\,0\,0\,0\,1\,0 \\ 0\,0\,0\,0\,0\,0\,1\,0\,0 \\ 0\,0\,0\,0\,0\,0\,0\,0\,0 \\ 0\,0\,0\,0\,0\,0\,0\,0\,1 \\ 0\,0\,0\,0\,0\,0\,0\,0\,0 \end{pmatrix} \quad \tilde{B}_o = \begin{pmatrix} 0\,0\,0\,0 \\ 0\,0\,0\,0 \\ 1\,0\,0\,0 \\ 0\,0\,0\,0 \\ 0\,1\,0\,0 \\ 0\,0\,0\,0 \\ 0\,0\,0\,0 \\ 0\,0\,0\,0 \end{pmatrix} \tag{2.122}$$

$$\tilde{C}_o = \begin{pmatrix} 1\,0\,0\,0\,0\,0\,0\,0\,0 \\ 0\,0\,0\,1\,0\,0\,0\,0\,0 \end{pmatrix}$$

while the estimator's gain $K_o \in R^{9\times 2}$ is obtained from the standard Kalman Filter recursion [34, 209, 341, 480].

Defining as $\tilde{A}_d$, $\tilde{B}_d$, and $\tilde{C}_d$, the discrete-time equivalents of matrices $\tilde{A}_o$, $\tilde{B}_o$ and $\tilde{C}_o$ respectively, a Derivative-free nonlinear Kalman Filter can be designed for the aforementioned representation of the system dynamics [352, 355]. The associated Kalman Filter-based disturbance estimator is given by

measurement update:

$$\begin{aligned} K(k) &= P^-(k)\tilde{C}_d^T[\tilde{C}_d \cdot P^-(k)\tilde{C}_d^T + R]^{-1} \\ \hat{\tilde{z}}(k) &= \hat{\tilde{z}}^-(k) + K(k)[\tilde{C}_d\tilde{z}(k) - \tilde{C}_d\hat{\tilde{z}}^-(k)] \\ P(k) &= P^-(k) - K(k)\tilde{C}_d P^-(k) \end{aligned} \tag{2.123}$$

time update:

$$\begin{aligned} P^{-}(k+1) &= \tilde{A}_d(k)P(k)\tilde{A}_d^T(k) + Q(k) \\ \hat{\tilde{z}}^{-}(k+1) &= \tilde{A}_d(k)\hat{\tilde{z}}(k) + \tilde{B}_d(k)\tilde{v}(k) \end{aligned} \tag{2.124}$$

2.3.6 Simulation Tests

The structure of the flatness-based control scheme is depicted in Fig. 2.9. The control scheme comprises (i) the flatness-based control part which computes the control signal for the system's equivalent model that is transformed to the linear canonical form, (ii) a Kalman Filter-based disturbances estimator which provides estimates for the elements of the state vector of the DFIG, such as rotor's speed, magnetic flux at the stator as well as disturbances affecting the generator's model.

Indicative numerical values for the parameters of the considered doubly-fed induction generator model are given in Table 2.1.

Simulation tests were carried out for two different setpoints of the turn speed of the generator's rotor. The values of the generator's state vector elements are actually measured in SI units, however in the simulation results they are expressed in the per unit (p.u.) system. The results obtained for the first setpoint are depicted in Fig. 2.10

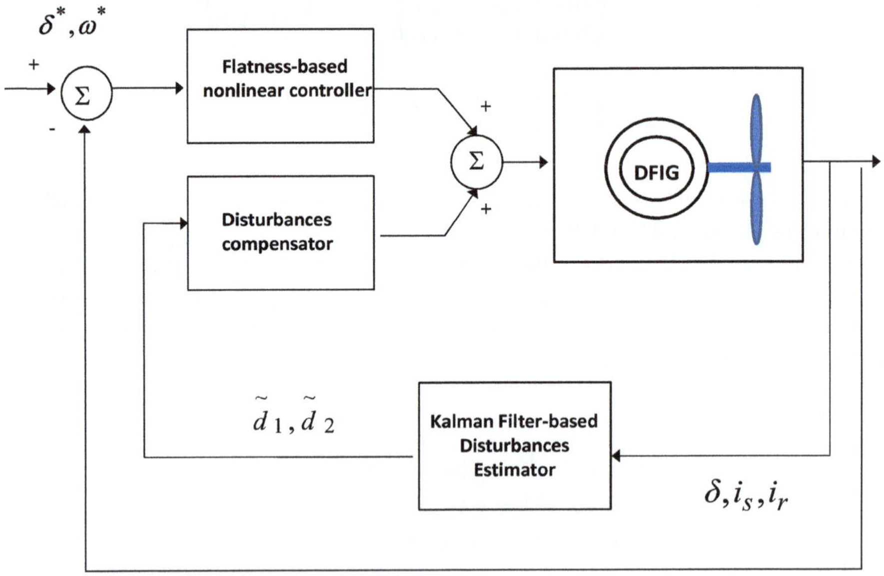

Fig. 2.9 Control loop of the DFIG comprising a flatness-based control element and an estimator for disturbances compensation

Table 2.1 Ratings of the modeled DFIG

Rated power	15.5 kW
Number of pole pairs	4
Stator resistance	0.58 Ω
Stator inductance	13 mH
Rotor resistance	1.30 Ω
Rotor inductance	3 mH
Mutual inductance	10 mH
Rotor's inertia	20.0 kg m^2

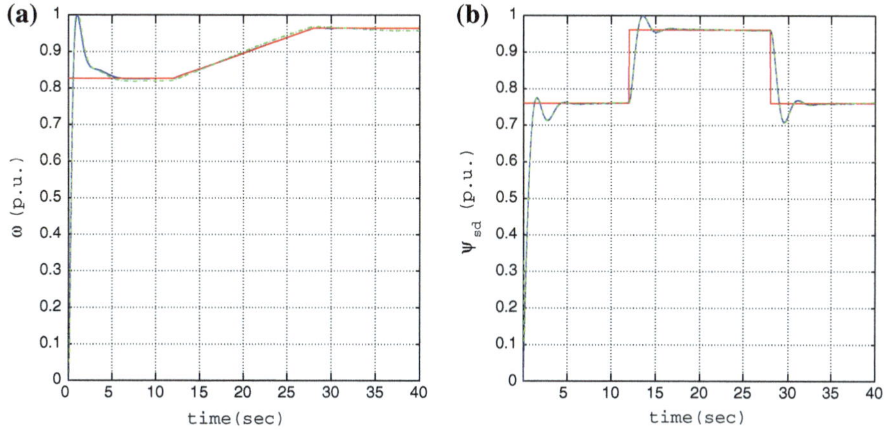

Fig. 2.10 DFIG setpoint 1: **a** Control of state variable $x_2 = \omega$. **b** Control of state variable $x_3 = \psi_{s_d}$

and Fig. 2.11. Similarly, the results obtained for the second setpoint are depicted in Figs. 2.12 and 2.13. It can be observed that the proposed control scheme assures that the rotor's turn speed follows a specific setpoint, while tracking of reference setpoints is succeeded for the components of the magnetic flux and for the rotor's currents. Several reference setpoints have been defined for the DFIG state variables, i.e. rotor's angular speed ω, rotor currents i_{r_d}, i_{r_q} and the magnetic flux ψ_{s_d} and as it can be observed from the associated diagrams, the proposed control scheme resulted in fast and accurate convergence to these setpoints. The disturbance observer that was based on the Derivative-free nonlinear Kalman Filter was capable of estimating the unknown and time-varying input disturbances affecting the DFIG model.

The selection of the magnetic flux setpoints appearing in Figs. 2.10 and 2.12 did not aim at being restricted only to the case that the DFIG is connected to a grid, which is characterized by constant voltage amplitude and frequency. The purpose of the simulation experiments was to show the capability of the proposed nonlinear control scheme to succeed convergence to time-varying magnetic flux setpoints (e.g. piecewise constant ones). Of course, the analyzed control method for the DFIG

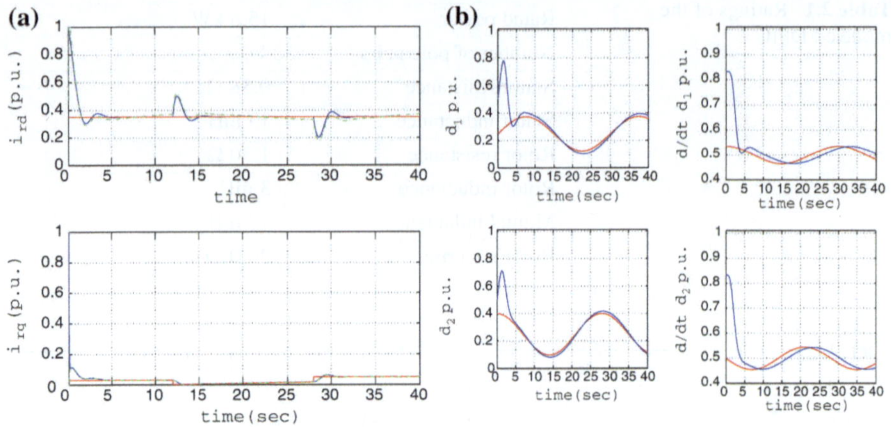

Fig. 2.11 DFIG setpoint 1: **a** Control of state variable $x_5 = i_{r_d}$ and of state variable $x_6 = i_{r_q}$. **b** Estimation of disturbance inputs d_i, $i = 1, 2$ and of their derivatives

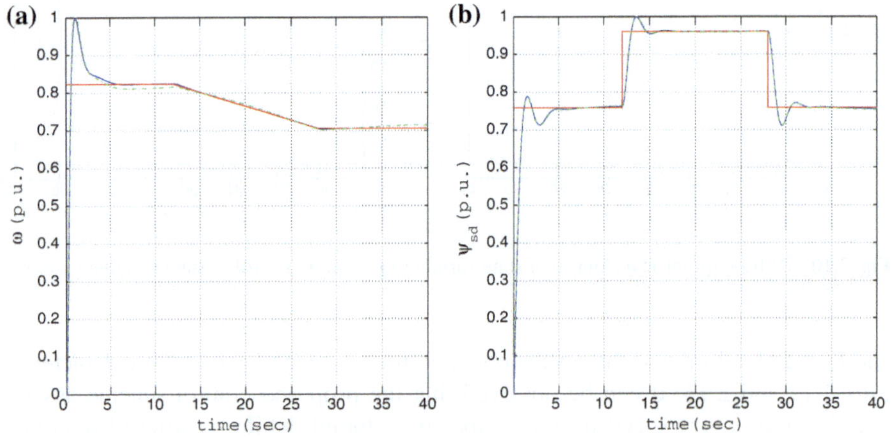

Fig. 2.12 DFIG setpoint 2: **a** Control of state variable $x_2 = \omega$. **b** Control of state variable $x_3 = \psi_{s_d}$

enables also convergence of the stator's flux to constant setpoints, but this is a subcase of what has already been presented.

The improvement in the performance of the control loop that is due to the use of a disturbance observer based on the Derivative-free nonlinear Kalman Filter is explained as follows: (i) compensation of the disturbance terms which are generated by parametric uncertainty or unknown external inputs (ii) more accurate estimation of the disturbance terms because the filtering procedure is based on an exact linearization of the system's dynamics and does not introduce numerical errors (as for example in the case of the Extended Kalman Filter). This is shown in Fig. 2.14.

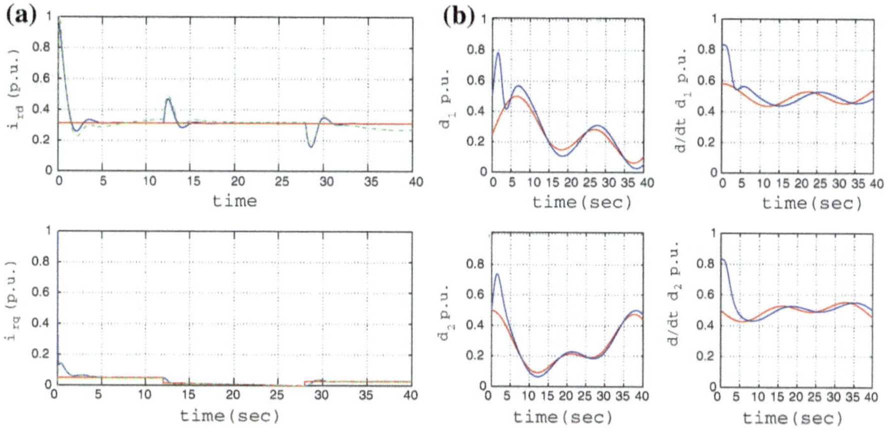

Fig. 2.13 DFIG setpoint 2: **a** Control of state variable $x_5 = i_{r_d}$ and of state variable $x_6 = i_{r_q}$. **b** Estimation of disturbance inputs d_i, $i = 1, 2$ and of their derivatives

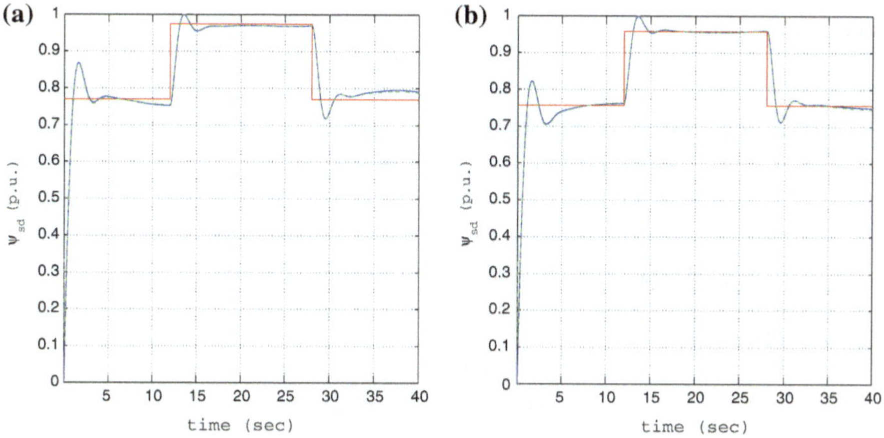

Fig. 2.14 Convergence of the stator's magnetic flux ψ_{s_d} to the reference setpoint **a** without using the disturbance observer, **b** when using the disturbance observer

2.3.7 *Input-Output Linearization of the DFIG Model with Use of Lie Algebra*

Linearization of the DFIG model can be also performed with the use of Lie algebra. The following variables have been defined

$$\begin{aligned} h_1(x) &= x_1 = \theta \\ h_2(x) &= x_3^2 + x_4^2 = \psi_{s_d}^2 + \psi_{s_q}^2 \end{aligned} \tag{2.125}$$

Next, based on h_1, h_2 the following transformed state variables are defined

$$z_1 = h_1(x) = \theta \tag{2.126}$$

$$\begin{array}{c} z_2 = L_f h_1(x) \Rightarrow \\ z_2 = f_1 \Rightarrow z_2 = x_2 \Rightarrow z_2 = \omega \end{array} \tag{2.127}$$

Similarly, one has

$$\begin{array}{c} z_3 = L_f^2 h_1(x) = L_f z_2 \Rightarrow \\ z_3 = f_2 \Rightarrow z_3 = -\frac{K_m}{J} x_2 - \frac{T_m}{J} + \frac{\eta}{J}(i_{r_q} x_3 - i_{r_q} x_4) \Rightarrow \\ z_3 = -\frac{K_m}{J} x_2 - \frac{T_m}{J} + \frac{\eta}{J}(x_6 x_3 - x_5 x_4) \end{array} \tag{2.128}$$

Moreover, one has

$$\dot{z}_3 = L_f^3 h_1(x) + (L_{g_a} L_f^2 h_1(x)) u_1 + (L_{g_b} L_f^2 h_2(x)) u_2 \tag{2.129}$$

It holds that

$$L_f^3 h_1(x) = L_f z_3 \tag{2.130}$$

$$\begin{array}{c} L_f^3 h_1(x) = -\frac{K_m}{J}[-\frac{K_m}{J} x_2 - \frac{T_m}{J} + \frac{\eta}{J}(x_6 x_3 - x_5 x_4)] \\ +\frac{\eta}{J} x_6[-\frac{1}{\tau_s} x_3 + \omega_{dq} x_4 + \frac{M}{\tau_s} x_5 + v_{s_d}] \\ -\frac{\eta}{J} x_5[-\omega_{dq} x_3 - \frac{1}{\tau_s} x_4 + \frac{M}{\tau_s} x_6 + v_{s_q}] \\ -\frac{\eta}{J} x_4[-\beta x_2 x_4 + \frac{\beta}{\tau_s} x_3 + (\omega_{dq} - x_2) x_6 - \gamma_2 x_5 - \beta v_{s_d}] \\ +\frac{\eta}{J} x_3[\frac{\beta}{\tau_s} x_4 + \beta x_2 x_3 + (\omega_{dq} - x_2) x_5 - \gamma_2 x_6 - \beta v_{s_q}] \end{array} \tag{2.131}$$

Equivalently one has

$$\begin{array}{c} L_{g_a}(L_f^2 h_1(x)) = L_{g_a} z_3 \Rightarrow \\ L_{g_a}(L_f^2 h_1(x)) = -\frac{\eta}{J}\frac{1}{\sigma L_r} x_4 \end{array} \tag{2.132}$$

and similarly

$$\begin{array}{c} L_{g_b}(L_f^2 h_1(x)) = L_{g_b} z_3 \Rightarrow \\ L_{g_b}(L_f^2 h_1(x)) = \frac{\eta}{J}\frac{1}{\sigma L_r} x_3 \end{array} \tag{2.133}$$

For the transformed state variable z_4 one has

$$z_4 = h_2(x) = \psi_{s_d}^2 + \psi_{s_q}^2 = x_3^2 + x_4^2 \tag{2.134}$$

and

$$\begin{array}{c} z_5 = L_f h_2(x) \Rightarrow z_5 = 2x_3 f_3 + 2x_4 f_4 \Rightarrow \\ z_5 = 2x_3[-\frac{1}{\tau_s} x_3 + \omega_{dq} x_4 + \frac{M}{\tau_s} x_5 + v_{s_d}] + \\ +2x_4[-\omega_{dq} x_3 - \frac{1}{\tau_s} x_4 + \frac{M}{\tau_s} x_6 + v_{s_q}] \end{array} \tag{2.135}$$

and equivalently one has

$$\dot{z}_5 = L_f^2 h_2(x) + L_{g_a}(L_f h_2(x))u_1 + L_{g_b}(L_f h_2(x))u_2 \tag{2.136}$$

It holds that

$$\begin{aligned} &L_f^2 h_2(x) = \\ &(-\tfrac{4}{\tau_s}x_3 - \tfrac{2M}{\tau_s}x_5 + 2v_{s_d})[-\tfrac{1}{\tau_s}x_3 + \omega_{dq}x_4 + \tfrac{M}{\tau_s}x_5 + v_{s_d}]+ \\ &(-\tfrac{4}{\tau_s}x_4 + \tfrac{2M}{\tau_s}x_6 + 2v_{s_q})[-\omega_{dq}x_3 - \tfrac{1}{\tau_s}x_4 + \tfrac{M}{\tau_s}x_6 + v_{s_q}]+ \\ &(\tfrac{2M}{\tau_s}x_3)[-\beta x_2 x_4 + \tfrac{\beta}{\tau_s}x_3 + (\omega_{dq} - x_2)x_6 - \gamma_2 x_5 - \beta v_{s_d}]+ \\ &(\tfrac{2M}{\tau_s}x_3)[\tfrac{\beta}{\tau_s}x_4 + \beta x_2 x_3 + (\omega_{dq} - x_2)x_5 - \gamma_2 x_6 - \beta v_{s_q}] \end{aligned} \tag{2.137}$$

Moreover, it holds that

$$L_{g_a}(L_f h_2(x)) = \tfrac{2M}{\tau_s}x_3 g_{a_5} \Rightarrow L_{g_a}(L_f h_2(x)) = \tfrac{2M}{\tau_s}\tfrac{1}{\sigma L_s}x_3 \tag{2.138}$$

and in a similar manner

$$\begin{aligned} L_{g_b}(L_f h_2(x)) &= \tfrac{2M}{\tau_s}x_4 g_{a_6} \Rightarrow \\ L_{g_a}(L_f h_2(x)) &= \tfrac{2M}{\tau_s}\tfrac{1}{\sigma L_s}x_4 \end{aligned} \tag{2.139}$$

Next, it is confirmed that after change of the state variables it holds

$$\begin{aligned} \dot{z}_1 &= z_2 \\ \dot{z}_2 &= z_3 \\ \dot{z}_3 &= L_f^3 h_1(x) + L_{g_a}(L_f^2)h_1(x)u_1 + L_{g_b}(L_f^2)h_1(x)u_2 \\ \dot{z}_4 &= z_5 \\ \dot{z}_5 &= L_f^2 h_2(x) + L_{g_a}(L_f)h_2(x)u_1 + L_{g_b}(L_f)h_2(x)u_2 \end{aligned} \tag{2.140}$$

It holds that $z_1 = \theta$, $\dot{z}_1 = \omega = z_2$, $\dot{z}_2 = \dot{\omega} = f_2(x) + g_{a_2}u_1 + g_{b_2}u_2 \Rightarrow \dot{z}_2 = f_2(x) + 0u_1 + 0u_2$ which finally gives $\dot{z}_2 = f_2(x)$. Moreover, it has been proven that $z_3 = f_2$ therefore it holds $\dot{z}_2 = z_3$. Additionally, it holds that

$$\dot{z}_3 = \tfrac{\partial z_3}{\partial x_1}\dot{x}_1 + \tfrac{\partial z_3}{\partial x_2}\dot{x}_2 + \tfrac{\partial z_3}{\partial x_3}\dot{x}_3 + \tfrac{\partial z_3}{\partial x_4}\dot{x}_4 + \tfrac{\partial z_3}{\partial x_5}\dot{x}_5 + \tfrac{\partial z_3}{\partial x_6}\dot{x}_6 \tag{2.141}$$

which in turn gives

$$\begin{aligned} \dot{z}_3 = &\tfrac{\partial z_3}{\partial x_1}f_1 + \tfrac{\partial z_3}{\partial x_2}f_2 + \tfrac{\partial z_3}{\partial x_3}f_3 + \tfrac{\partial z_3}{\partial x_4}f_4 + \\ &+ \tfrac{\partial z_3}{\partial x_5}(f_5 + \tfrac{1}{\sigma L_r}u_1) + \tfrac{\partial z_3}{\partial x_6}(f_6 + \tfrac{1}{\sigma L_r}u_2) \end{aligned} \tag{2.142}$$

that is also written as

$$\dot{z}_3 = L_f^3 h_1(x) + L_{g_a}(L_f^2 h_1(x))u_1 + L_{g_b}(L_f^2 h_2(x))u_2 \tag{2.143}$$

Similarly, one has

$$\begin{array}{c} z_4 = x_3^2 + x_4^2 \Rightarrow \\ \dot{z}_4 = 2x_3\dot{x}_3 + 2x_4\dot{x}_4 \Rightarrow \dot{z}_4 = 2x_3 f_3 + 2x_4 f_4 \\ \dot{z}_4 = 2x_3[-\frac{1}{\tau_s}x_3 + \omega_{dq}x_4 + \frac{M}{\tau_s}x_5 + v_{s_d}] + \\ +2x_4[-\omega_{dq}x_3 - \frac{1}{\tau_s}x_4 + \frac{M}{\tau_s}x_6 + v_{s_q}] \Rightarrow \dot{z}_4 = z_5 \end{array} \tag{2.144}$$

Additionally, it holds

$$\dot{z}_5 = \frac{\partial z_5}{\partial x_1}\dot{x}_1 + \frac{\partial z_5}{\partial x_2}\dot{x}_2 + \frac{\partial z_5}{\partial x_3}\dot{x}_3 + \frac{\partial z_5}{\partial x_4}\dot{x}_4 + \frac{\partial z_5}{\partial x_5}\dot{x}_5 + \frac{\partial z_6}{\partial x_6}\dot{x}_6 \tag{2.145}$$

which in turn gives

$$\begin{array}{c} \dot{z}_5 = \frac{\partial z_5}{\partial x_1}f_1 + \frac{\partial z_5}{\partial x_2}f_2 + \frac{\partial z_5}{\partial x_3}f_3 + \frac{\partial z_5}{\partial x_4}f_4 + \\ + \frac{\partial z_5}{\partial x_5}(f_5 + \frac{1}{\sigma L_r}u_1) + \frac{\partial z_5}{\partial x_6}(f_6 + \frac{1}{\sigma L_r}u_2) \end{array} \tag{2.146}$$

which subsequently gives

$$\dot{z}_5 = L_f^2 h_2(x) + L_{g_a}(L_f h_2(x))u_1 + L_{g_b}(L_f h_2(x))u_2 \tag{2.147}$$

which is the anticipated relation about $\dot{z}_5$. Consequently, Eq. (2.82) is confirmed to hold. The system of Eq. (2.82) can be written in the input-output linearized form

$$\begin{pmatrix} z_1^{(3)} \\ \ddot{z}_4 \end{pmatrix} = f_a + \tilde{M}u \tag{2.148}$$

where

$$\begin{array}{c} f_a(x) = \begin{pmatrix} L_f^3 h_1(x) \\ L_f^2 h_2(x) \end{pmatrix} \\ \tilde{M} = \begin{pmatrix} L_{g_a} L_f^2 h_1(x) & L_{g_b} L_f^2 h_2(x) \\ L_{g_a} L_f h_1(x) & L_{g_b} L_f h_2(x) \end{pmatrix} \end{array} \tag{2.149}$$

or equivalently one has the system's description in the MIMO canonical form

$$\begin{pmatrix} \dot{z}_1 \\ \dot{z}_2 \\ \dot{z}_3 \\ \dot{z}_4 \\ \dot{z}_5 \end{pmatrix} = \begin{pmatrix} 0&1&0&0&0 \\ 0&0&1&0&0 \\ 0&0&0&0&0 \\ 0&0&0&0&1 \\ 0&0&0&0&0 \end{pmatrix} \begin{pmatrix} z_1 \\ z_2 \\ z_3 \\ z_4 \\ z_5 \end{pmatrix} + \begin{pmatrix} 0&0 \\ 0&0 \\ 1&0 \\ 0&0 \\ 0&1 \end{pmatrix} \begin{pmatrix} v_1 \\ v_2 \end{pmatrix} \tag{2.150}$$

where

$$\begin{aligned} v_1 &= L_f^3 h_1(x) + L_{g_a}(L_f^2 h_1(x))u_1 + L_{g_b}(L_f^2 h_2(x))u_2 \\ v_2 &= L_f^2 h_2(x) + L_{g_a}(L_f h_1(x))u_1 + L_{g_b}(L_f h_2(x))u_2 \end{aligned} \tag{2.151}$$

2.4 Nonlinear H-Infinity Control of DFIGs

2.4.1 Outline

Control methods for DFIGs based on local linearization approaches have been developed [182, 208, 517]. In particular about nonlinear H-infinity control, the method has been primarily applied to other electromechanical systems but exhibits also the potential to be applied to DFIGs too [373, 443]. In all these methods the objective is to succeed the reliable operation of the DFIG under variable speed so as to optimize power generation and to modify the amount of the produced power in accordance to the grid's needs [161].

In this section a new control method for DFIGs is developed, based on nonlinear H-infinity control theory. The application of an approximate linearization scheme for the dynamic model of the DFIG is proposed, based on Taylor series expansion round the generator's present operating point. To perform this linearization the computation of Jacobian matrices is needed while the induced linearization error terms are treated as disturbances. For the linearized equivalent of the power generator's model an H_∞ feedback control scheme is developed. The formulation of the H_∞ control problem is based on the minimization of a quadratic cost function that comprises both the disturbance and the control input effects. The disturbance tries to maximize the cost function while the control signal tries to minimize it, within a mini-max differential game. The efficiency of the proposed nonlinear H_∞ control scheme has been tested through simulation experiments, which have shown a satisfactory performance.

Comparing to nonlinear feedback control approaches which are based on exact feedback linearization of the DFIG (as the ones based on differential flatness theory and analyzed in the previous sections and in Ref. [352, 366, 375]) the proposed H_∞ control scheme is assessed as follows: (i) it uses an approximate linearization approach of the system's dynamic model which does not follow the elaborated transformations (diffeomorphisms) of the exact linearization methods, (ii) it introduces additional disturbance error which is due to the approximate linearization of the system dynamics coming from the application of Taylor series expansion, (iii) it requires the computation of Jacobian matrices, which in the case of the sixth-order DFIG model can be also a tedious procedure, (iv) unlike exact feedback linearization, the H_∞ control term has to compensate not only for modelling uncertainties and external disturbances but needs also to annihilate the effects of the cumulative linearization error, (v) the H_∞ control approach follows an optimal control method for the computation of the control signal, however unlike exact feedback linearization control it requires the solution of Riccati equations which for the sixth-order DFIG model can be again a burdensome procedure.

2.4.2 Approximate Linearization of the Doubly-Fed Induction Generator's Dynamic Model

The nonlinear state space equation of the doubly-fed induction generator is given by

$$\dot{x} = f(x) + g_a v_{r_d} + g_b v_{r_q} \tag{2.152}$$

where the state vector has been defined as $x = [\theta, \omega, \psi_{s_d}, \psi_{s_q}, i_{r_d}, i_{r_q}]^T$ while functions $f(x)$, $g_a(x)$ and $g_b(x)$ have been defined as

$$f(x) = \begin{pmatrix} x_2 \\ -\frac{K_m}{J}x_2 - \frac{T_m}{J} + \frac{\eta}{J}(i_{r_q}x_3 - i_{r_d}x_4) \\ -\frac{1}{\tau_s}x_3 + \omega_{dq}x_4 + \frac{M}{\tau_s}x_5 + v_{s_d} \\ -\omega_{dq}x_3 - \frac{1}{\tau_s}x_4 + \frac{M}{\tau_s}x_6 + v_{s_q} \\ -\beta x_2 x_4 + \frac{\beta}{\tau_s}x_3 + (\omega_{dq} - x_2)x_6 - \gamma_2 x_3 - \beta v_{s_d} \\ \frac{\beta}{\tau_s}x_4 + \beta x_2 x_3 + (\omega_{dq} - x_2)x_5 - \gamma_2 x_6 - \beta v_{s_q} \end{pmatrix} \tag{2.153}$$

$$g_a(x) = \left(0\;0\;0\;0\;\frac{1}{\sigma L_s}\;0\right)^T \tag{2.154}$$

$$g_b(x) = \left(0\;0\;0\;0\;0\;\frac{1}{\sigma L_s}\right)^T \tag{2.155}$$

The measured state variables are taken to be $x_1 = \theta$, $x_5 = i_{r_d}$ and $x_6 = i_{r_q}$. Then, the associated Jacobian matrices are:

$$A = J_f = \begin{pmatrix} 0 & 1 & 0 & 0 & 0 & 0 \\ 0 & \frac{K_m}{J} & \frac{\eta}{J}x_6 & -\frac{\eta}{J}x_5 & -\frac{\eta}{J}x_4 & \frac{\eta}{J}x_3 \\ 0 & 0 & -\frac{1}{\tau_s} & \omega_{dq} & \frac{M}{\tau_s} & 0 \\ 0 & 0 & -\omega_{dq} & -\frac{1}{\tau_s} & 0 & \frac{M}{\tau_s} \\ 0 & -\beta x_4 - x_6 & -\gamma_2 + \frac{\beta}{\tau_s} & -\beta x_2 & 0 & \omega_{dq} - x_2 \\ 0 & \beta x_3 - x_5 & \beta x_2 & \frac{\beta}{\tau_s} & \omega_{dq} - x_2 & -\gamma_2 \end{pmatrix} \tag{2.156}$$

Moreover, linearization of the the generator's dynamics with respect to the control input variables u_1, u_2 gives

$$B = [J_{g_a} J_{g_b}] = \begin{pmatrix} 0 & 0 \\ 0 & 0 \\ 0 & 0 \\ 0 & 0 \\ \frac{1}{\sigma L_s} & 0 \\ 0 & \frac{1}{\sigma L_s} \end{pmatrix} \tag{2.157}$$

After Taylor series expansion linearization round its current operating point, the DFIG's dynamic model is written as

$$\dot{x} = Ax + Bu + d_1 \tag{2.158}$$

Parameter d_1 stands for the linearization error in the generator's dynamic model appearing in Eq. (2.158). The reference setpoints for the doubly-fed induction generator are denoted by $\mathbf{x_d} = [x_1^d, \dots, x_6^d]$. Tracking of this trajectory is succeeded after applying the control input u^*. At every time instant the control input u^* is assumed to differ from the control input u appearing in Eq. (2.158) by an amount equal to Δu, that is $u^* = u + \Delta u$

$$\dot{x}_d = Ax_d + Bu^* + d_2 \tag{2.159}$$

The dynamics of the controlled system described in Eq. (2.158) can be also written as

$$\dot{x} = Ax + Bu + Bu^* - Bu^* + d_1 \tag{2.160}$$

and by denoting $d_3 = -Bu^* + d_1$ as an aggregate disturbance term one obtains

$$\dot{x} = Ax + Bu + Bu^* + d_3 \tag{2.161}$$

By subtracting Eq. (2.159) from Eq. (2.161) one has

$$\dot{x} - \dot{x}_d = A(x - x_d) + Bu + d_3 - d_2 \tag{2.162}$$

By denoting the tracking error as $e = x - x_d$ and the aggregate disturbance term as $\tilde{d} = d_3 - d_2$, the tracking error dynamics becomes

$$\dot{e} = Ae + Bu + \tilde{d} \tag{2.163}$$

The above linearized form of the generator's model can be efficiently controlled after applying an H-infinity feedback control scheme.

2.4.3 The Nonlinear H-Infinity Control

2.4.3.1 Mini-Max Control and Disturbance Rejection

The initial nonlinear model of the induction generator is in the form

$$\dot{x} = f(x, u) \quad x \in R^n, \; u \in R^m \tag{2.164}$$

Linearization of the system (doubly-fed induction generator) is performed at each iteration of the control algorithm round its present operating point $(x^*, u^*) = (x(t), u(t - T_s))$. The linearized equivalent of the system is described by

$$\dot{x} = Ax + Bu + L \quad x \in R^n, \; u \in R^m, \; d \in R^q \tag{2.165}$$

where matrices A and B are obtained from the computation of the Jacobians

$$A = \begin{pmatrix} \frac{\partial f_1}{\partial x_1} & \frac{\partial f_1}{\partial x_2} & \cdots & \frac{\partial f_1}{\partial x_n} \\ \frac{\partial f_2}{\partial x_1} & \frac{\partial f_2}{\partial x_2} & \cdots & \frac{\partial f_2}{\partial x_n} \\ \cdots & \cdots & \cdots & \cdots \\ \frac{\partial f_n}{\partial x_1} & \frac{\partial f_n}{\partial x_2} & \cdots & \frac{\partial f_n}{\partial x_n} \end{pmatrix} |_{(x^*, u^*)} \; B = \begin{pmatrix} \frac{\partial f_1}{\partial u_1} & \frac{\partial f_1}{\partial u_2} & \cdots & \frac{\partial f_1}{\partial u_m} \\ \frac{\partial f_2}{\partial u_1} & \frac{\partial f_2}{\partial u_2} & \cdots & \frac{\partial f_2}{\partial u_m} \\ \cdots & \cdots & \cdots & \cdots \\ \frac{\partial f_n}{\partial u_1} & \frac{\partial f_n}{\partial u_2} & \cdots & \frac{\partial f_n}{\partial u_m} \end{pmatrix} |_{(x^*, u^*)} \tag{2.166}$$

and vector d denotes disturbance terms due to linearization errors. The problem of disturbance rejection for the linearized model that is described by

$$\begin{aligned} \dot{x} &= Ax + Bu + Ld \\ y &= Cx \end{aligned} \tag{2.167}$$

where $x \in R^n$, $u \in R^m$, $d \in R^q$ and $y \in R^p$, cannot be handled efficiently if the classical LQR control scheme is applied. This is because of the existence of the perturbation term d. The disturbance term d apart from modeling (parametric) uncertainty and external perturbation terms can also represent noise terms of any distribution.

In the H_∞ control approach, a feedback control scheme is designed for trajectory tracking by the system's state vector and simultaneous disturbance rejection, considering that the disturbance affects the system in the worst possible manner. The disturbances' effect are incorporated in the following quadratic cost function:

$$\begin{aligned} J(t) = \tfrac{1}{2}\int_0^T [y^T(t)y(t) + \\ + r u^T(t)u(t) - \rho^2 d^T(t)d(t)]dt, \quad r, \rho > 0 \end{aligned} \tag{2.168}$$

The significance of the negative sign in the cost function's term that is associated with the perturbation variable $d(t)$ is that the disturbance tries to maximize the cost function $J(t)$ while the control signal $u(t)$ tries to minimize it. The physical meaning of the relation given above is that the control signal and the disturbances compete to each other within a mini-max differential game. This problem of mini-max optimization can be written as

$$min_u max_d J(u,d) \tag{2.169}$$

The objective of the optimization procedure is to compute a control signal $u(t)$ which can compensate for the worst possible disturbance, the latter being externally imposed to the system. However, the solution to the mini-max optimization problem is directly related to the value of parameter ρ. This means that there is an upper bound in the disturbances magnitude that can be annihilated by the control signal.

2.4.3.2 H-Infinity Feedback Control

For the linearized system given by Eq. (2.167) the cost function of Eq. (2.168) is defined, where the coefficient r determines the penalization of the control input and the weight coefficient ρ determines the reward of the disturbances' effects. It is assumed that:

It is assumed that (i) The energy that is transferred from the disturbances signal $d(t)$ is bounded, that is $\int_0^{\infty} d^T(t)d(t)dt < \infty$, (ii) the matrices $[A/B]$ and $[A/L]$ are stabilizable, (iii) the matrix $[A/C]$ is detectable. Then, the optimal feedback control law is given by

$$u(t) = -Kx(t) \tag{2.170}$$

with

$$K = \frac{1}{r}B^T P \tag{2.171}$$

where P is a positive semi-definite symmetric matrix which is obtained from the solution of the Riccati equation

$$A^T P + PA + Q - P(\frac{1}{r}BB^T - \frac{1}{2\rho^2}LL^T)P = 0 \tag{2.172}$$

where Q is also a positive definite symmetric matrix. The worst case disturbance is given by

$$d(t) = \frac{1}{\rho^2}L^T Px(t) \tag{2.173}$$

2.4.3.3 The Role of Riccati Equation Coefficients in H_∞ Control Robustness

The parameter ρ in Eq. (2.168), is an indication of the closed-loop system robustness. If the values of $\rho > 0$ are excessively decreased with respect to r, then the solution of the Riccati equation is no longer a positive definite matrix. Consequently there is a lower bound ρ_{min} of ρ for which the H_∞ control problem has a solution. The acceptable values of ρ lie in the interval $[\rho_{min}, \infty)$. If ρ_{min} is found and used in the design of the H_∞ controller, then the closed-loop system will have increased robustness. Unlike this, if a value $\rho > \rho_{min}$ is used, then an admissible stabilizing H_∞ controller will be derived but it will be a suboptimal one. The Hamiltonian matrix

$$H = \begin{pmatrix} A & -(\frac{1}{r}BB^T - \frac{1}{\rho^2}LL^T) \\ -Q & -A^T \end{pmatrix} \tag{2.174}$$

provides a criterion for the existence of a solution of the Riccati equation Eq. (2.172). A necessary condition for the solution of the algebraic Riccati equation to be a positive semi-definite symmetric matrix is that H has no imaginary eigenvalues [352].

2.4.4 Lyapunov Stability Analysis

Through Lyapunov stability analysis it will be shown that the proposed nonlinear control scheme assures H_∞ tracking performance, and that in case of bounded disturbance terms asymptotic convergence to the reference setpoints is succeeded. The tracking error dynamics for the doubly-fed induction generator is written in the form

$$\dot{e} = Ae + Bu + L\tilde{d} \tag{2.175}$$

where in the generator's application example $L = I \in R^2$ with I being the identity matrix. The following Lyapunov equation is considered

$$V = \frac{1}{2}e^T Pe \tag{2.176}$$

where $e = x - x_d$ is the tracking error. By differentiating with respect to time one obtains

$$\begin{aligned} \dot{V} &= \tfrac{1}{2}\dot{e}^T Pe + \tfrac{1}{2}eP\dot{e} \Rightarrow \\ \dot{V} &= \tfrac{1}{2}[Ae + Bu + L\tilde{d}]^T P + \tfrac{1}{2}e^T P[Ae + Bu + L\tilde{d}] \Rightarrow \end{aligned} \tag{2.177}$$

$$\begin{aligned} \dot{V} = &\tfrac{1}{2}[e^T A^T + u^T B^T + \tilde{d}^T L^T]Pe + \\ &+ \tfrac{1}{2}e^T P[Ae + Bu + L\tilde{d}] \Rightarrow \end{aligned} \tag{2.178}$$

$$\begin{aligned} \dot{V} = &\tfrac{1}{2}e^T A^T Pe + \tfrac{1}{2}u^T B^T Pe + \tfrac{1}{2}\tilde{d}^T L^T Pe + \\ &\tfrac{1}{2}e^T PAe + \tfrac{1}{2}e^T PBu + \tfrac{1}{2}e^T PL\tilde{d} \end{aligned} \tag{2.179}$$

The previous equation is rewritten as

$$\begin{aligned} \dot{V} = &\tfrac{1}{2}e^T(A^T P + PA)e + (\tfrac{1}{2}u^T B^T Pe + \tfrac{1}{2}e^T PBu) + \\ &+ (\tfrac{1}{2}\tilde{d}^T L^T Pe + \tfrac{1}{2}e^T PL\tilde{d}) \end{aligned} \tag{2.180}$$

Assumption: For given positive definite matrix Q and coefficients r and ρ there exists a positive definite matrix P, which is the solution of the following matrix equation

$$A^T P + PA = -Q + P(\tfrac{2}{r}BB^T - \tfrac{1}{\rho^2}LL^T)P \tag{2.181}$$

Moreover, the following feedback control law is applied to the system

$$u = -\tfrac{1}{r}B^T Pe \tag{2.182}$$

By substituting Eq. (2.181) and Eq. (2.182) one obtains

$$\begin{aligned} \dot{V} &= \tfrac{1}{2}e^T[-Q + P(\tfrac{1}{r}BB^T - \tfrac{1}{2\rho^2}LL^T)P]e+ \\ &+ e^T PB(-\tfrac{1}{r}B^T Pe) + e^T PL\tilde{d} \Rightarrow \end{aligned} \tag{2.183}$$

$$\begin{aligned} \dot{V} &= -\tfrac{1}{2}e^T Qe + (\tfrac{1}{r}PBB^T Pe - \tfrac{1}{2\rho^2}e^T PLL^T)Pe \\ &- \tfrac{1}{r}e^T PBB^T Pe) + e^T PL\tilde{d} \end{aligned} \tag{2.184}$$

which after intermediate operations gives

$$\dot{V} = \tfrac{1}{2}e^T Qe - \tfrac{1}{2\rho^2}e^T PLL^T)Pe + e^T PL\tilde{d} \tag{2.185}$$

or, equivalently

$$\begin{aligned} \dot{V} &= \tfrac{1}{2}e^T Qe - \tfrac{1}{2\rho^2}e^T PLL^T)Pe+ \\ &+ \tfrac{1}{2}e^T PL\tilde{d} + \tfrac{1}{2}\tilde{d}^T L^T Pe \end{aligned} \tag{2.186}$$

Lemma: The following inequality holds

$$\tfrac{1}{2}e^T L\tilde{d} + \tfrac{1}{2}\tilde{d}L^T Pe - \tfrac{1}{2\rho^2}e^T PLL^T Pe \le \tfrac{1}{2}\rho^2\tilde{d}^T\tilde{d} \tag{2.187}$$

Proof: The binomial $(\rho\alpha - \frac{1}{\rho}b)^2$ is considered. Expanding the left part of the above inequality one gets

$$\begin{aligned} &\rho^2 a^2 + \tfrac{1}{\rho^2}b^2 - 2ab \ge 0 \Rightarrow \tfrac{1}{2}\rho^2 a^2 + \tfrac{1}{2\rho^2}b^2 - ab \ge 0 \Rightarrow \\ &ab - \tfrac{1}{2\rho^2}b^2 \le \tfrac{1}{2}\rho^2 a^2 \Rightarrow \tfrac{1}{2}ab + \tfrac{1}{2}ab - \tfrac{1}{2\rho^2}b^2 \le \tfrac{1}{2}\rho^2 a^2 \end{aligned} \tag{2.188}$$

The following substitutions are carried out: $a = \tilde{d}$ and $b = e^T PL$ and the previous relation becomes

$$\tfrac{1}{2}\tilde{d}^T L^T Pe + \tfrac{1}{2}e^T PL\tilde{d} - \tfrac{1}{2\rho^2}e^T PLL^T Pe \le \tfrac{1}{2}\rho^2\tilde{d}^T\tilde{d} \tag{2.189}$$

Equation (2.189) is substituted in Eq. (2.186) and the inequality is enforced, thus giving

$$\dot{V} \leq -\frac{1}{2}e^T Q e + \frac{1}{2}\rho^2 \tilde{d}^T \tilde{d} \tag{2.190}$$

Equation (2.190) shows that the H_∞ tracking performance criterion is satisfied. The integration of $\dot{V}$ from 0 to T gives

$$\begin{aligned} &\int_0^T \dot{V}(t)dt \leq -\frac{1}{2}\int_0^T ||e||_Q^2 dt + \frac{1}{2}\rho^2 \int_0^T ||\tilde{d}||^2 dt \Rightarrow \\ &2V(T) + \int_0^T ||e||_Q^2 dt \leq 2V(0) + \rho^2 \int_0^T ||\tilde{d}||^2 dt \end{aligned} \tag{2.191}$$

Moreover, if there exists a positive constant $M_d > 0$ such that

$$\int_0^\infty ||\tilde{d}||^2 dt \leq M_d \tag{2.192}$$

then one gets

$$\int_0^\infty ||e||_Q^2 dt \leq 2V(0) + \rho^2 M_d \tag{2.193}$$

Thus, the integral $\int_0^\infty ||e||_Q^2 dt$ is bounded. Moreover, $V(T)$ is bounded and from the definition of the Lyapunov function V in Eq. (2.176) it becomes clear that $e(t)$ will be also bounded since $e(t) \in \Omega_e = \{e | e^T P e \leq 2V(0) + \rho^2 M_d\}$. According to the above and with the use of Barbalat's Lemma one obtains $lim_{t\to\infty} e(t) = 0$.

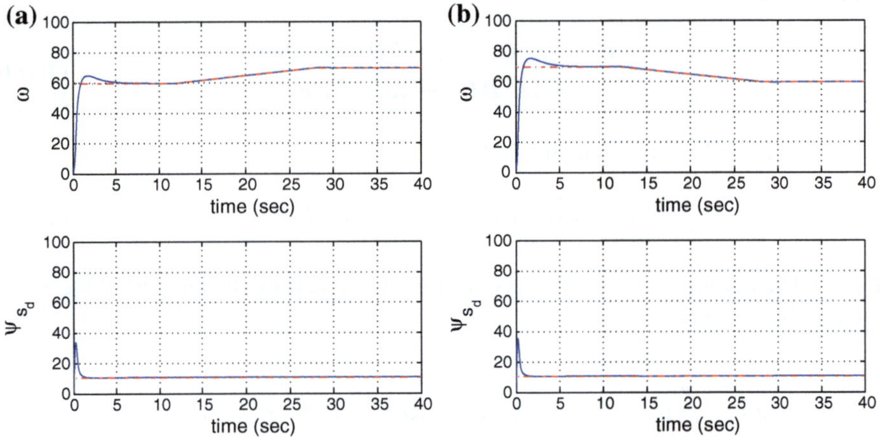

Fig. 2.15 Nonlinear H_∞ control of the doubly-fed induction generator. **a** Convergence of the rotor's angular speed and stator's magnetic flux to setpoint 1 **b** Convergence of the rotor's angular and stator's magnetic flux to setpoint 2

2.4.5 *Simulation Tests*

The performance of the proposed nonlinear H_∞ control scheme for DFIGs is tested in tracking of various setpoints. As it can be observed in Fig. 2.15 the feedback control scheme of the DFIG enabled accurate convergence to the reference setpoints. Yet simple, the considered H_∞ control law succeeded precise tracking of the reference signals. In comparison to feedback control methods for asynchronous generators which are based on exact linearization, the nonlinear H_∞ control requires the solution of an algebraic Riccati equation at each iteration of the control algorithm. The known robustness features of H_∞ control are the ones that permit to compensate for the approximation errors which were induced to the linearized model of the generator.

2.5 Flatness-Based Adaptive Fuzzy Control of DFIGs

2.5.1 *Overview*

This section presents results on model-free control of Doubly-fed Induction Generators (DFIGs) with the use of differential flatness theory and adaptive control theory. As previously explained, the problem of control of the DFIGs is a nontrivial one because the dynamic model of such electric machines is a multi-variable and nonlinear one. Moreover, under real operating conditions it exhibits parametric variations and is subjected to external perturbations. As a result empirical model-free control approaches of the PID-type lack stability and robustness, remain functional only round local operating points and in general are proven to be inefficient. DFIG control in stand-lone functioning mode has been analyzed in [56, 62, 140, 148, 279, 319, 357, 378, 379].

In particular control schemes exhibiting adaptivity features have been presented in [1, 463]. Distributed control of DFIGs and their functioning when interacting with the power grid has been studied in [473, 482]. The performance of specific DFIG control schemes under external perturbations has been presented in [8, 481, 489, 490]. Moreover, the problem of sensorless control of DFIGs has been studied in [141, 320, 486].

To cope with the DFIG control problem, in this section the differential flatness properties of the DFIG model are exploited once more. It is proven that the rotor's turn speed and the stator's magnetic flux of this electric machine are flat outputs which means that all other state variables and the control inputs of the DFIG can be expressed as differential functions of these outputs [57, 137, 241, 352, 388, 411, 412, 456]. Moreover, it is proven that the flat outputs and their derivatives are differentially independent. This comes to complete the proof about the differential flatness of the DFIG model [280, 375].

The differential flatness of the DFIG implies also that the transformation of its dynamic model to the linear canonical (Brunovsky) form is possible. After such a transformation the implementation of a stabilizing feedback controller for the generator is enabled. This solution to the control problem would be complete if no model uncertainty and parametric variations of the DFIG existed. Actually, it is assumed that only the order of the DFIG's state-space model is known while its parameters and the specific form of its state-space equations are completely unknown. In the linearized equivalent model of the generator this uncertainty takes the form of unknown functions which appear as part of the system's transformed control inputs. This unknown part of the system's dynamics is identified with the use of neurofuzzy approximators. The estimates of the DFIG's dynamics are used by a state feedback controller thus establishing an indirect adaptive control scheme. The learning procedure for these estimators is defined by the requirement to assure negative definiteness for the system's Lyapunov function [352, 366, 375]. Another issue that arises is that the state vector that is fed into the controller is partially measurable, and the missing state vector elements have to be estimated with the use of a state observer. By including such an observer in the control loop an output feedback-based control scheme is established. The stability of the control method is proven through Lyapunov stability analysis.

2.5.2 *Flatness-Based Adaptive Neurofuzzy Control*

2.5.2.1 Transformation of MIMO Nonlinear Systems into the Brunovsky Form

Following the previous procedure, that is after defining the flat outputs of the initial MIMO nonlinear model of the DFIG, that was described in Sect. 2.3 and after expressing the system's state variables and control inputs as functions of the flat output and of the associated derivatives, the DFIG is transformed into the Brunovsky canonical form of Eq. (2.108), or equivalently into the form [352, 366, 375, 378]:

$$\begin{array}{ll}
\dot{x}_1 = x_2 & \\
\dot{x}_2 = x_3 & \\
\cdots & \\
\dot{x}_{r_1-1} = x_{r_1} & \\
\dot{x}_{r_1} = f_1(x) + \sum_{j=1}^{p} g_{1_j}(x)u_j + d_1 & y_1 = x_1 \\
 & y_2 = x_{r_1-1} \\
\dot{x}_{r_1+1} = x_{r_1+2} & \cdots \\
\dot{x}_{r_1+2} = x_{r_1+3} & y_p = x_{n-r_p+1} \\
\cdots & \\
\dot{x}_{p-1} = x_p & \\
\dot{x}_p = f_p(x) + \sum_{j=1}^{p} g_{p_j}(x)u_j + d_p &
\end{array} \tag{2.194}$$

where $x = [x_1, \ldots, x_n]^T$ is the state vector of the transformed system (according to the differential flatness formulation), $u = [u_1, \ldots, u_p]^T$ is the set of control inputs, $y = [y_1, \ldots, y_p]^T$ is the output vector, f_i are the drift functions and $g_{i,j},\ i, j = 1, 2, \ldots, p$ are smooth functions corresponding to the control input gains, while d_j

is a variable associated to external disturbances. In holds that $r_1 + r_2 + \cdots + r_p = n$. Having written the initial nonlinear system into the canonical (Brunovsky) form it holds

$$y_i^{(r_i)} = f_i(x) + \sum_{j=1}^{p} g_{ij}(x)u_j + d_j \tag{2.195}$$

Equivalently, in vector form, one has the following description for the system dynamics

$$y^{(r)} = f(x) + g(x)u + d \tag{2.196}$$

where the following vectors and matrices can be defined

$$\begin{aligned}
&y^{(r)} = [y_1^{(r_1)}, \ldots, y_p^{(r_p)}] \\
&f(x) = [f_1(x), \ldots, f_p(x)]^T \\
&g(x) = [g_1(x), \ldots, g_p(x)] \\
&with\ g_i(x) = [g_{1i}(x), \ldots, g_{pi}(x)]^T \\
&A = diag[A_1, \ldots, A_p], \quad B = diag[B_1, \ldots, B_p] \\
&C^T = diag[C_1, \ldots, C_p], \quad d = [d_1, \ldots, d_p]^T
\end{aligned} \tag{2.197}$$

where matrix A has the MIMO canonical form, i.e. with elements

$$A_i = \begin{pmatrix} 0 & 1 & \cdots & 0 \\ 0 & 0 & \cdots & 0 \\ \vdots & \vdots & \cdots & \vdots \\ 0 & 0 & \cdots & 1 \\ 0 & 0 & \cdots & 0 \end{pmatrix}_{r_i \times r_i} \quad B_i = \begin{pmatrix} 0 \\ 0 \\ \cdots \\ 0 \\ 1 \end{pmatrix}_{r_i \times 1} \quad C_i^T = \begin{pmatrix} 1 \\ 0 \\ \cdots \\ 0 \\ 0 \end{pmatrix}_{r_i \times 1} \tag{2.198}$$

Thus, Eq. (2.195) can be written in state-space form

$$\begin{aligned} \dot{x} &= Ax + B[f(x) + g(x)u + \tilde{d}] \\ y &= C^T x \end{aligned} \tag{2.199}$$

which can be also written in the equivalent form:

$$\begin{aligned} \dot{x} &= Ax + Bv + B\tilde{d} \\ y &= C^T x \end{aligned} \tag{2.200}$$

where $v = f(x)+g(x)u$. The reference setpoints for the system's outputs $y_1, \ldots, y_p$ are denoted as $y_{1m}, \ldots, y_{pm}$, thus for the associated tracking errors it holds

$$\begin{aligned} e_1 &= y_1 - y_{1m} \\ e_2 &= y_2 - y_{2m} \\ &\cdots \\ e_p &= y_p - y_{pm} \end{aligned} \tag{2.201}$$

The error vector of the outputs of the transformed MIMO system is denoted as

$$\begin{aligned} E_1 &= [e_1, \ldots, e_p]^T \\ y_m &= [y_{1m}, \ldots, y_{pm}]^T \\ &\cdots \\ y_m^{(r)} &= [y_{1m}^{(r)}, \ldots, y_{pm}^{(r)}]^T \end{aligned} \tag{2.202}$$

where $y_{im}^{(r)}$ denotes the r-th order derivative of the i-th reference output of the MIMO dynamical system. Thus, one can also define the following vectors: (i) a vector containing the state variables of the system and the associated derivatives, (ii) a vector containing the reference outputs of the system and the associated derivatives

$$x = [x_1, \ldots, x_1^{r_1-1}, \ldots, x_p, \ldots, x_p^{r_p-1}]^T \tag{2.203}$$

$$Y_m = [y_{1m}, \ldots, y_{1m}^{r_1-1}, \ldots, y_{pm}, \ldots, y_{pm}^{r_p-1}]^T \tag{2.204}$$

while in a similar manner one can define a vector containing the tracking error of the system's outputs and the associated derivatives

$$e = Y_m - x = [e_1, \ldots, e_1^{r_1-1}, \ldots, e_p, \ldots, e_p^{r_p-1}]^T \tag{2.205}$$

It is assumed that matrix $g(x)$ is a nonsingular one, i.e. $g^{-1}(x)$ exists and is bounded for all $x \in U_x$, where $U_x \subset R^n$ is a compact set. In any case, the problem of singularities in matrix $g(x)$ can be handled by appropriately modifying the state feedback-based control input.

The objective of the adaptive neurofuzzy controller, denoted as $u = u(x, e|\theta)$ is: all the signals involved in the controller's design are bounded and it holds that $lim_{t \to \infty} e = 0$, (ii) the H_∞ tracking performance criterion is succeeded for a prescribed attenuation level.

In the presence of non-gaussian disturbances w_d, successful tracking of the reference signal is denoted by the H_∞ criterion [375, 379]:

$$\int_0^T e^T Q e dt \leq \rho^2 \int_0^T {w_d}^T w_d dt \tag{2.206}$$

where ρ is the attenuation level and corresponds to the maximum singular value of the transfer function $G(s)$ of the linearized model associated to Eqs. (2.199) and (2.200).

2.5.2.2 Control Law

The control signal of the MIMO nonlinear DFIG model which has been transformed into the Brunovsky form as described by Eq. (2.199) contains the unknown nonlinear functions $f(x)$ and $g(x)$. In case that the complete state vector x is measurable these unknown functions can be approximated by

$$\begin{aligned}\hat{f}(x|\theta_f) &= \Phi_f(x)\theta_f\\ \hat{g}(x|\theta_g) &= \Phi_g(x)\theta_g\end{aligned} \tag{2.207}$$

where

$$\Phi_f(x) = (\xi_f^1(x), \xi_f^2(x), \dots \xi_f^n(x))^T \tag{2.208}$$

with $\xi_f^i(x)$, $\imath = 1, \dots, n$ being the vector of kernel functions (e.g. normalized fuzzy Gaussian membership functions), where

$$\xi_f^i(x) = (\phi_f^{i,1}(x), \phi_f^{i,2}(x), \dots, \phi_f^{i,N}(x)) \tag{2.209}$$

thus giving

$$\Phi_f(x) = \begin{pmatrix} \phi_f^{1,1}(x) & \phi_f^{1,2}(x) & \cdots & \phi_f^{1,N}(x) \\ \phi_f^{2,1}(x) & \phi_f^{2,2}(x) & \cdots & \phi_f^{2,N}(x) \\ \cdots & \cdots & \cdots & \cdots \\ \phi_f^{n,1}(x) & \phi_f^{n,2}(x) & \cdots & \phi_f^{n,N}(x) \end{pmatrix} \tag{2.210}$$

while the weights vector is defined as

$$\theta_f{}^T = \left(\theta_f^1, \theta_f^2, \dots \theta_f^N\right) \tag{2.211}$$

$j = 1, \dots, N$ is the number of basis functions that is used to approximate the components of function f which are denoted as $i = 1, \dots, n$. Thus, one obtains the relation of Eq. (2.207), i.e. $\hat{f}(x|\theta_f) = \Phi_f(x)\theta_f$.

In a similar manner, for the approximation of function $g(x)$ one has

$$\Phi_g(x) = \left(\xi_g^1(x), \xi_g^2(x), \dots \xi_g^N(x)\right)^T \tag{2.212}$$

with $\xi_g^i(x)$, $\imath = 1, \dots, N$ being the vector of kernel functions (e.g. normalized fuzzy Gaussian membership functions), where

$$\xi_g^i(x) = \left(\phi_g^{i,1}(x), \phi_g^{i,2}(x), \cdots, \phi_g^{i,N}(x)\right) \tag{2.213}$$

thus giving

$$\Phi_g(x) = \begin{pmatrix} \phi_g^{1,1}(x) & \phi_g^{1,2}(x) & \cdots & \phi_g^{1,N}(x) \\ \phi_g^{2,1}(x) & \phi_g^{2,2}(x) & \cdots & \phi_g^{2,N}(x) \\ \cdots & \cdots & \cdots & \cdots \\ \phi_g^{n,1}(x) & \phi_g^{n,2}(x) & \cdots & \phi_g^{n,N}(x) \end{pmatrix} \tag{2.214}$$

while the weights vector is defined as

$$\theta_g = \left(\theta_g^1, \theta_g^2, \ldots, \theta_g^p\right) \tag{2.215}$$

where the components of matrix θ_g are defined as

$$\theta_g^j = \left(\theta_{g_1}^j, \theta_{g_2}^j, \ldots \theta_{g_N}^j\right)^T \tag{2.216}$$

$j = 1, \ldots, p$ is the number of basis functions that is used to approximate the components of function g which are denoted as $i = 1, \ldots, n$. Thus one obtains about matrix $\theta_g \in R^{N \times p}$

$$\theta_g = \begin{pmatrix} \theta_{g_1}^1 & \theta_{g_1}^2 & \cdots & \theta_{g_1}^p \\ \theta_{g_2}^1 & \theta_{g_2}^2 & \cdots & \theta_{g_2}^p \\ \cdots & \cdots & \cdots & \cdots \\ \theta_{g_N}^1 & \theta_{g_N}^2 & \cdots & \theta_{g_N}^p \end{pmatrix} \tag{2.217}$$

It holds that

$$g = \begin{pmatrix} g_1 \\ g_2 \\ \cdots \\ g_n \end{pmatrix} = \begin{pmatrix} g_1^1 & g_1^2 & \cdots & g_1^p \\ g_2^1 & g_2^2 & \cdots & g_2^p \\ \cdots & \cdots & \cdots & \cdots \\ g_n^1 & g_n^2 & \cdots & g_n^p \end{pmatrix} \tag{2.218}$$

Using the above, one finally has the relation of Eq. (2.207), i.e. $\hat{g}(x|\theta_g) = \Phi_g(x)\theta_g$. If the state variables of the system are available for measurement then a state-feedback control law can be formulated as

$$u = \hat{g}^{-1}(x|\theta_g)[-\hat{f}(x|\theta_f) + y_m^{(r)} - K^T e + u_c] \tag{2.219}$$

where $\hat{f}(x|\theta_f)$ and $\hat{g}(x|\theta_g)$ are neurofuzzy models to approximate $f(x)$ and $g(x)$, respectively. u_c is a supplementary control term, e.g. an H_∞ control term that is used to compensate for the effects of modelling inaccuracies and external disturbances. Moreover, K^T is the feedback gain matrix which assures that the characteristic polynomial of matrix $A - BK^T$ will be a Hurwitz one.

2.5.3 Estimation of the State Vector

The control of the system described by Eq. (2.196) becomes more complicated when the state vector x of the DFIG is not directly measurable and has to be reconstructed through a state observer. The following definitions are used

- error of the state vector $e = x - x_m$
- error of the estimated state vector $\hat{e} = \hat{x} - x_m$
- observation error $\tilde{e} = e - \hat{e} = (x - x_m) - (\hat{x} - x_m)$

When an observer is used to reconstruct the state vector, the control law of Eq. (2.219) is written as

$$u = \hat{g}^{-1}(\hat{x}|\theta_g)[-\hat{f}(\hat{x}|\theta_f) + y_m^{(r)} - K^T\hat{e} + u_c] \tag{2.220}$$

Applying Eq. (2.220) to the nonlinear system described by Eq. (2.196), results into $y^{(r)} = f(x)+g(x)\hat{g}^{-1}(\hat{x})[-\hat{f}(\hat{x})+y_m^{(r)} - K^T\hat{e}+u_c]+d \Rightarrow y^{(r)} = f(x)+[g(x) - \hat{g}(\hat{x}) + \hat{g}(\hat{x})]\hat{g}^{-1}(\hat{x})[-\hat{f}(\hat{x}) + y_m^{(r)} - K^T\hat{e} + u_c] + d \Rightarrow y^{(r)} = [f(x) - \hat{f}(\hat{x})] + [g(x) - \hat{g}(\hat{x})]u + y_m^{(r)} - K^T\hat{e} + u_c + d$.

It holds $e = x - x_m \Rightarrow y^{(r)} = e^{(r)} + y_m^{(r)}$. Substituting $y^{(r)}$ in the above equation gives

$$\begin{aligned} e^{(r)} + y_m^{(r)} = y_m^{(r)} - K^T\hat{e} + u_c + [f(x) - \hat{f}(\hat{x})] + \\ + [g(x) - \hat{g}(\hat{x})]u + d \end{aligned} \tag{2.221}$$

and equivalently

$$\begin{aligned} \dot{e} = Ae - BK^T\hat{e} + Bu_c + B\{[f(x) - \hat{f}(\hat{x})] + \\ + [g(x) - \hat{g}(\hat{x})]u + \tilde{d}\} \end{aligned} \tag{2.222}$$

$$e_1 = C^T e \tag{2.223}$$

where $e = [e^1, e^2, \ldots, e^p]^T$ with $e^i = [e_i, \dot{e}_i, \ddot{e}_i, \ldots, e_i^{r_i-1}]^T$, $i = 1, 2, \ldots, p$ and also $\hat{e} = [\hat{e}^1, \hat{e}^2, \ldots, \hat{e}^p]^T$ with $\hat{e}^i = [\hat{e}_i, \dot{\hat{e}}_i, \ddot{\hat{e}}_i, \ldots, \hat{e}_i^{r_i-1}]^T$, $i = 1, 2, \ldots, p$. Matrices A, B and C have been defined in Eq. (8.115).

A state observer is designed according to Eqs. (2.222) and (2.223) and is given by [375]:

$$\dot{\hat{e}} = A\hat{e} - BK^T\hat{e} + K_o[e_1 - C^T\hat{e}] \tag{2.224}$$

$$\hat{e}_1 = C^T\hat{e} \tag{2.225}$$

The feedback gain matrix is denoted as $K \in R^{n \times p}$. The observation gain matrix is denoted as $K_o \in R^{p \times n}$ and its elements are selected so as to assure the asymptotic elimination of the observation error.

2.5.4 *Application of Flatness-Based Adaptive Neurofuzzy Control to the DFIG*

2.5.4.1 Tracking Error Dynamics Under Feedback Control

If the state vector in the linearized description of the DFIG of Eq. (2.104) is constrained into $x = [x_1, x_2, x_3, x_4]^T = [\omega, \dot{\omega}, \psi_{s_d}, \dot{\psi}_{s_d}]^T$, it holds that

$$\begin{aligned} \ddot{x}_1 &= f_1(x) + g_1(x)u \\ \ddot{x}_3 &= f_2(x) + g_2(x)u \end{aligned} \tag{2.226}$$

or equivalently

$$\begin{aligned} \dot{x}_1 &= x_2 \\ \dot{x}_2 &= f_1(x) + g_1(x)u \\ \dot{x}_3 &= x_4 \\ \dot{x}_4 &= f_2(x) + g_2(x)u \end{aligned} \tag{2.227}$$

Next, taking into account also the effects of additive disturbances the dynamic model becomes

$$\begin{aligned} \ddot{x}_1 &= f_1(x,t) + g_1(x,t)u + d_1 \\ \ddot{x}_3 &= f_2(x,t) + g_2(x,t)u + d_2 \end{aligned} \tag{2.228}$$

$$\begin{pmatrix} \ddot{x}_1 \\ \ddot{x}_3 \end{pmatrix} = \begin{pmatrix} f_1(x,t) \\ f_2(x,t) \end{pmatrix} + \begin{pmatrix} g_1(x,t) \\ g_2(x,t) \end{pmatrix} u + \begin{pmatrix} d_1 \\ d_2 \end{pmatrix} \tag{2.229}$$

The following control input is defined

$$u = \begin{pmatrix} \hat{g}_1(x,t) \\ \hat{g}_2(x,t) \end{pmatrix}^{-1} \{ \begin{pmatrix} \ddot{x}_1^d \\ \ddot{x}_3^d \end{pmatrix} - \begin{pmatrix} \hat{f}_1(x,t) \\ \hat{f}_2(x,t) \end{pmatrix} - \begin{pmatrix} K_1^T \\ K_2^T \end{pmatrix} e + \begin{pmatrix} u_{c_1} \\ u_{c_2} \end{pmatrix} \} \tag{2.230}$$

where $[u_{c_1} \ u_{c_2}]^T$ is a robust control term that is used for the compensation of the model's uncertainties as well as of the external disturbances and the feedback control gain is $K_i^T = [k_1^i, k_2^i, \ldots, k_{n-1}^i, k_n^i]$. Substituting Eq. (2.230) into Eq. (2.229) the closed-loop tracking error dynamics is obtained

$$\begin{pmatrix} \ddot{x}_1 \\ \ddot{x}_3 \end{pmatrix} = \begin{pmatrix} f_1(x,t) \\ f_2(x,t) \end{pmatrix} + \begin{pmatrix} g_1(x,t) \\ g_2(x,t) \end{pmatrix} \begin{pmatrix} \hat{g}_1(x,t) \\ \hat{g}_2(x,t) \end{pmatrix}^{-1} \cdot$$
$$\cdot \{ \begin{pmatrix} \ddot{x}_1^d \\ \ddot{x}_3^d \end{pmatrix} - \begin{pmatrix} \hat{f}_1(x,t) \\ \hat{f}_2(x,t) \end{pmatrix} - \begin{pmatrix} K_1^T \\ K_2^T \end{pmatrix} e + \begin{pmatrix} u_{c_1} \\ u_{c_2} \end{pmatrix} \} + \begin{pmatrix} d_1 \\ d_2 \end{pmatrix} \tag{2.231}$$

Equation (2.231) can now be written as

$$\begin{pmatrix} \ddot{x}_1 \\ \ddot{x}_3 \end{pmatrix} = \begin{pmatrix} f_1(x,t) \\ f_2(x,t) \end{pmatrix} + \{ \begin{pmatrix} g_1(x,t) - \hat{g}_1(x,t) \\ g_2(x,t) - \hat{g}_2(x,t) \end{pmatrix} + \begin{pmatrix} \hat{g}_1(x,t) \\ \hat{g}_2(x,t) \end{pmatrix} \} \begin{pmatrix} \hat{g}_1(x,t) \\ \hat{g}_2(x,t) \end{pmatrix}^{-1} \cdot$$
$$\cdot \{ \begin{pmatrix} \ddot{x}_1^d \\ \ddot{x}_3^d \end{pmatrix} - \begin{pmatrix} \hat{f}_1(x,t) \\ \hat{f}_2(x,t) \end{pmatrix} - \begin{pmatrix} K_1^T \\ K_2^T \end{pmatrix} e + \begin{pmatrix} u_{c_1} \\ u_{c_2} \end{pmatrix} \} + \begin{pmatrix} d_1 \\ d_2 \end{pmatrix} \tag{2.232}$$

and using Eq. (2.230) this results into

$$\begin{pmatrix} \ddot{e}_1 \\ \ddot{e}_3 \end{pmatrix} = \begin{pmatrix} f_1(x,t) - \hat{f}_1(x,t) \\ f_2(x,t) - \hat{f}_2(x,t) \end{pmatrix} + \begin{pmatrix} g_1(x,t) - \hat{g}_1(x,t) \\ g_2(x,t) - \hat{g}_2(x,t) \end{pmatrix} u -$$
$$- \begin{pmatrix} K_1^T \\ K_2^T \end{pmatrix} e + \begin{pmatrix} u_{c_1} \\ u_{c_2} \end{pmatrix} + \begin{pmatrix} d_1 \\ d_2 \end{pmatrix} \tag{2.233}$$

The following description for the approximation error is defined

$$w = \begin{pmatrix} f_1(x,t) - \hat{f}_1(x,t) \\ f_2(x,t) - \hat{f}_2(x,t) \end{pmatrix} + \begin{pmatrix} g_1(x,t) - \hat{g}_1(x,t) \\ g_2(x,t) - \hat{g}_2(x,t) \end{pmatrix} u \tag{2.234}$$

Moreover, the following matrices are defined

$$A = \begin{pmatrix} 0 & 1 & 0 & 0 \\ 0 & 0 & 0 & 0 \\ 0 & 0 & 0 & 1 \\ 0 & 0 & 0 & 0 \end{pmatrix}, \quad B = \begin{pmatrix} 0 & 0 \\ 1 & 0 \\ 0 & 0 \\ 0 & 1 \end{pmatrix} \tag{2.235}$$
$$K^T = \begin{pmatrix} K_1^1 & K_2^1 & K_3^1 & K_4^1 \\ K_1^2 & K_2^2 & K_3^2 & K_4^2 \end{pmatrix}$$

Using matrices A, B, K^T, Eq. (2.233) is written in the following form

$$\dot{e} = (A - BK^T)e + Bu_c + B\{ \begin{pmatrix} f_1(x,t) - \hat{f}_1(x,t) \\ f_2(x,t) - \hat{f}_2(x,t) \end{pmatrix} +$$
$$+ \begin{pmatrix} g_1(x,t) - \hat{g}_1(x,t) \\ g_2(x,t) - \hat{g}_2(x,t) \end{pmatrix} u + \tilde{d} \} \tag{2.236}$$

When the estimated state vector $\hat{x}$ is used in the feedback control loop, equivalently to Eq. (2.222) one has

$$\dot{e} = Ae - BK^T\hat{e} + Bu_c + B\{\begin{pmatrix} f_1(x,t) - \hat{f}_1(\hat{x},t) \\ f_2(x,t) - \hat{f}_2(\hat{x},t) \end{pmatrix} + \\ + \begin{pmatrix} g_1(x,t) - \hat{g}_1(\hat{x},t) \\ g_2(x,t) - \hat{g}_2(\hat{x},t) \end{pmatrix} u + \tilde{d}\} \tag{2.237}$$

and considering that the approximation error w is now denoted as

$$w = \begin{pmatrix} f_1(x,t) - \hat{f}_1(\hat{x},t) \\ f_2(x,t) - \hat{f}_2(\hat{x},t) \end{pmatrix} + \begin{pmatrix} g_1(x,t) - \hat{g}_1(\hat{x},t) \\ g_2(x,t) - \hat{g}_2(\hat{x},t) \end{pmatrix} u \tag{2.238}$$

Equation (2.237) can be also written as

$$\dot{e} = Ae - BK^T\hat{e} + Bu_c + Bw + B\tilde{d} \tag{2.239}$$

The associated state observer will be described again by Eqs. (2.224) and (2.225).

2.5.4.2 Dynamics of the Observation Error

The observation error is defined as $\tilde{e} = e - \hat{e} = x - \hat{x}$. Subtracting Eq. (2.224) from Eq. (2.222) as well as Eq. (2.225) from Eq. (2.223) one gets

$$\dot{e} - \dot{\hat{e}} = A(e - \hat{e}) + Bu_c + B\{[f(x,t) - \hat{f}(\hat{x},t)] + \\ + [g(x,t) - \hat{g}(\hat{x},t)]u + \tilde{d}\} - K_oC^T(e - \hat{e})$$

$$e_1 - \hat{e}_1 = C^T(e - \hat{e})$$

or equivalently

$$\dot{\tilde{e}} = A\tilde{e} + Bu_c + B\{[f(x,t) - \hat{f}(\hat{x},t)] + \\ + [g(x,t) - \hat{g}(\hat{x},t)]u + \tilde{d}\} - K_oC^T\tilde{e}$$

$$\tilde{e}_1 = C^T\tilde{e}$$

which can be written as

$$\dot{\tilde{e}} = (A - K_oC^T)\tilde{e} + Bu_c + B\{[f(x,t) - \hat{f}(\hat{x},t)] + \\ + [g(x,t) - \hat{g}(\hat{x},t)]u + \tilde{d}\} \tag{2.240}$$

$$\tilde{e}_1 = C^T\tilde{e} \tag{2.241}$$

or equivalently, it can be written as

$$\dot{\tilde{e}} = (A - K_o C^T)\tilde{e} + Bu_c + Bw + \tilde{d}\} \tag{2.242}$$

$$\tilde{e}_1 = C^T \tilde{e} \tag{2.243}$$

2.5.4.3 Approximation of Functions $f(x, t)$ and $g(x, t)$

Next, the following neurofuzzy approximators of the unknown system dynamics are defined

$$\hat{f}(\hat{x}) = \begin{pmatrix} \hat{f}_1(\hat{x}|\theta_f) \ \hat{x}\in R^{4\times 1} \ \hat{f}_1(\hat{x}|\theta_f) \ \in \ R^{1\times 1} \\ \hat{f}_2(\hat{x}|\theta_f) \ \hat{x}\in R^{4\times 1} \ \hat{f}_2(\hat{x}|\theta_f) \ \in \ R^{1\times 1} \end{pmatrix} \tag{2.244}$$

with kernel functions

$$\phi_f^{i,j}(\hat{x}) = \frac{\prod_{j=1}^{n} \mu_{A_j}^{i}(\hat{x}_j)}{\sum_{i=1}^{N} \prod_{j=1}^{n} \mu_{A_j}^{i}(\hat{x}_j)} \tag{2.245}$$

where $l = 1, 2$, $\hat{x}$ is the estimate of the state vector and $\mu_{A_j^i}(\hat{x})$ is the i-th membership function associated with the l-th hidden layer node (fuzzy rule), as shown in Fig. 2.16. Similarly, the following approximators of the unknown system dynamics are defined

$$\hat{g}(\hat{x}) = \begin{pmatrix} \hat{g}_1(\hat{x}|\theta_g) \ \hat{x}\in R^{4\times 1} \ \hat{g}_1(\hat{x}|\theta_g) \ \in \ R^{1\times 2} \\ \hat{g}_2(\hat{x}|\theta_g) \ \hat{x}\in R^{4\times 1} \ \hat{g}_2(\hat{x}|\theta_g) \ \in \ R^{1\times 2} \end{pmatrix} \tag{2.246}$$

The values of the weights that result in optimal approximation are

$$\begin{aligned} \theta_f^* &= arg\ min_{\theta_f \in M_{\theta_f}} [sup_{\hat{x}\in U_{\hat{x}}} (f(x) - \hat{f}(\hat{x}|\theta_f))] \\ \theta_g^* &= arg\ min_{\theta_g \in M_{\theta_g}} [sup_{\hat{x}\in U_{\hat{x}}} (g(x) - \hat{g}(\hat{x}|\theta_g))] \end{aligned} \tag{2.247}$$

where the variation ranges for the weights are defined as

$$\begin{aligned} M_{\theta_f} &= \{\theta_f \in R^h : \ ||\theta_f|| \leq m_{\theta_f}\} \\ M_{\theta_g} &= \{\theta_g \in R^h : \ ||\theta_g|| \leq m_{\theta_g}\} \end{aligned} \tag{2.248}$$

The value of the approximation error defined in Eq. (2.234) that corresponds to the optimal values of the weights vectors θ_f^* and θ_g^* one has

$$w = \left(f(x,t) - \hat{f}(\hat{x}|\theta_f^*)\right) + \left(g(x,t) - \hat{g}(\hat{x}|\theta_g^*)\right) u \tag{2.249}$$

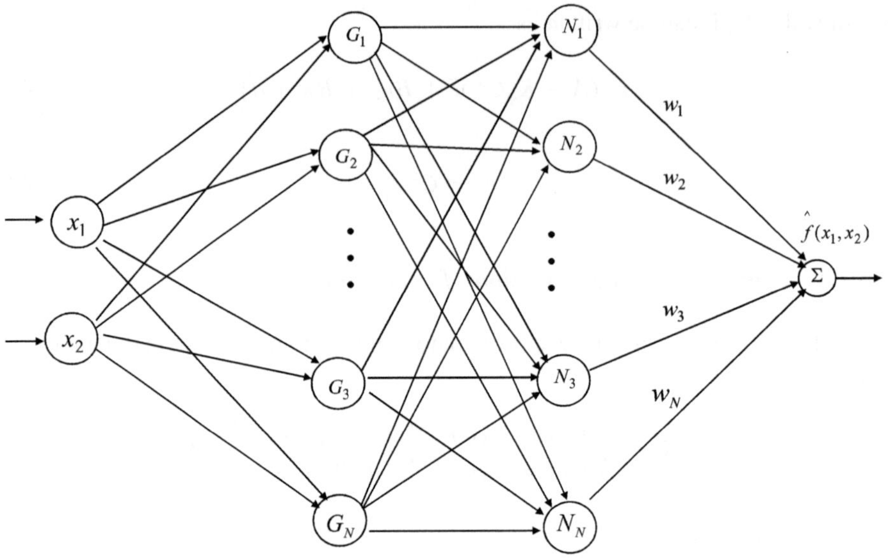

Fig. 2.16 Neurofuzzy approximator used for estimating the unknown dynamics of the DFIG: G_i Gaussian basis function, N_i: normalization unit

which is next written as

$$w = \left(f(x,t) - \hat{f}(\hat{x}|\theta_f) + \hat{f}(\hat{x}|\theta_f) - \hat{f}(\hat{x}|\theta_f^*)\right) + \\ + \left(g(x,t) - \hat{g}(\hat{x}|\theta_g) + \hat{g}(\hat{x}|\theta_g) - \hat{g}(\hat{x}|\theta_g^*)\right) u \tag{2.250}$$

which can be also written in the following form

$$w = \left(w_a + w_b\right) \tag{2.251}$$

where

$$w_a = \{[f(x,t) - \hat{f}(\hat{x}|\theta_f)] + [g(x,t) - \hat{g}(\hat{x}|\theta_g)]\}u \tag{2.252}$$

$$w_b = \{[\hat{f}(\hat{x}|\theta_f) - \hat{f}(\hat{x}|\theta_f^*)] + [\hat{g}(\hat{x},\theta_g) - \hat{g}(\hat{x}|\theta_g^*)]\}u \tag{2.253}$$

Moreover, the following weights' error vectors are defined

$$\tilde{\theta}_f = \theta_f - \theta_f^* \\ \tilde{\theta}_g = \theta_g - \theta_g^* \tag{2.254}$$

It is noted that both neural networks and neurofuzzy networks are function approximators using non-orthogonal kernel functions. This is to point out the difference to other approximators using orthogonal basis functions as for instance Fourier series expansions. The distinctive features for neurofuzzy networks are (i) they use as kernel functions in their hidden layer nodes some typical membership functions met in fuzzy logic (for instance Gaussian functions) (ii) the outputs of the hidden layer nodes are normalized so as to sum up to 1 (this is a property of most fuzzy membership functions). On the other hand in neural networks (i) it is possible to have several types of membership functions (e.g. wavelet functions, Gauss-Hermite polynomials, Bezier-Bernstein polynomials, B-splines), (ii) the outputs of the hidden layer nodes are not necessarily normalized. The learning (weights update) in both neural networks and neurofuzzy networks is performed with gradient algorithms of first or higher order (Fig. 2.16).

2.5.5 Lyapunov Stability Analysis

2.5.5.1 Design of the Lyapunov Function

The adaptation law of the neurofuzzy approximators weights θ_f and θ_g as well as the equation of the supervisory control term u_c are derived from the requirement for negative definiteness of the Lyapunov function

$$V = \frac{1}{2}\hat{e}^T P_1 \hat{e} + \frac{1}{2}\tilde{e}^T P_2 \tilde{e} + \frac{1}{2\gamma_1}\tilde{\theta}_f^T \tilde{\theta}_f + \frac{1}{2\gamma_2} tr[\tilde{\theta}_g^T \tilde{\theta}_g] \tag{2.255}$$

The selection of the Lyapunov function is based on the following principle of indirect adaptive control $\hat{e}$: $\lim_{t\to\infty} \hat{x}(t) = x_d(t)$ and $\tilde{e}$: $\lim_{t\to\infty} \hat{x}(t) = x(t)$. This yields $\lim_{t\to\infty} x(t) = x_d(t)$. Substituting Eqs. (2.224), (2.225), (2.240) and (2.241) into Eq. (2.255) and differentiating results into

$$\begin{aligned} \dot{V} = \frac{1}{2}\dot{\hat{e}}^T P_1 \hat{e} + \frac{1}{2}\hat{e}^T P_1 \dot{\hat{e}} + \frac{1}{2}\dot{\tilde{e}}^T P_2 \tilde{e} + \frac{1}{2}\tilde{e}^T P_2 \dot{\tilde{e}} + \\ + \frac{1}{\gamma_1}\dot{\tilde{\theta}}_f^T \tilde{\theta}_f + \frac{1}{\gamma_2} tr[\dot{\tilde{\theta}}_g^T \tilde{\theta}_g] \Rightarrow \end{aligned} \tag{2.256}$$

$\dot{V} = \frac{1}{2}\{(A - BK^T)\hat{e} + K_o C^T \tilde{e}\}^T P_1 \hat{e} + \frac{1}{2}\hat{e}^T P_1 \{(A - BK^T)\hat{e} + K_o C^T \tilde{e}\} + \frac{1}{2}\{(A - K_o C^T)\tilde{e} + Bu_c + B\tilde{d} + Bw\}^T P_2 \tilde{e} + \frac{1}{2}\tilde{e}^T P_2 \{(A - K_o C^T)\tilde{e} + Bu_c + B\tilde{d} + Bw\} + \frac{1}{\gamma_1}\dot{\tilde{\theta}}_f^T \tilde{\theta}_f + \frac{1}{\gamma_2} tr[\dot{\tilde{\theta}}_g^T \tilde{\theta}_g] \Rightarrow$.

or equivalently,
$\dot{V} = \frac{1}{2}\{\hat{e}^T (A - BK^T)^T + \tilde{e}^T C K_o^T\} P_1 \hat{e} + \frac{1}{2}\hat{e}^T P_1 \{(A - BK^T)\hat{e} + K_o C^T \tilde{e}\} + \frac{1}{2}\{\tilde{e}^T (A - K_o C^T)^T + u_c^T B^T + w^T B^T + \tilde{d}^T B^T\} P_2 \tilde{e} + \frac{1}{2}\tilde{e}^T P_2 \{(A - K_o C^T)\tilde{e} + Bu_c + Bw + B\tilde{d}\} + \frac{1}{\gamma_1}\dot{\tilde{\theta}}_f^T \tilde{\theta}_f + \frac{1}{\gamma_2} tr[\dot{\tilde{\theta}}_g^T \tilde{\theta}_g] \Rightarrow$

$$\begin{array}{c}\dot{V} = \frac{1}{2}\hat{e}^T(A - BK^T)^T P_1\hat{e} + \frac{1}{2}\tilde{e}^T C K_o^T P_1\hat{e} + \\ \frac{1}{2}\hat{e}^T P_1(A - BK^T)\hat{e} + \frac{1}{2}\hat{e}^T P_1 K_o C^T\tilde{e} + \frac{1}{2}\tilde{e}^T(A - K_oC^T)^T P_2\tilde{e} \\ + \frac{1}{2}(u_c^T + w^T + \tilde{d}^T)B^T P_2\tilde{e} + \frac{1}{2}\tilde{e}^T P_2(A - K_oC^T)\tilde{e} + \\ + \frac{1}{2}\tilde{e}^T P_2 B(u_c + w + \tilde{d}) + \frac{1}{\gamma_1}\dot{\tilde{\theta}}_f^T\tilde{\theta}_f + \frac{1}{\gamma_2}tr[\dot{\tilde{\theta}}_g^T\tilde{\theta}_g]\end{array} \tag{2.257}$$

Assumption 1: For given positive definite matrices Q_1 and Q_2 there exist positive definite matrices P_1 and P_2, which are the solution of the following Riccati equations [340, 352, 366, 375]

$$(A - BK^T)^T P_1 + P_1(A - BK^T) + Q_1 = 0 \tag{2.258}$$

$$\begin{array}{c}(A - K_oC^T)^T P_2 + P_2(A - K_oC^T) - \\ - P_2B(\frac{2}{r} - \frac{1}{\rho^2})B^T P_2 + Q_2 = 0\end{array} \tag{2.259}$$

The conditions given in Eq. (2.258) to (2.259) are related to the requirement that the systems described by Eqs. (2.224), (2.225), (2.240) and (2.241) have a decreasing Lyapunov function. Substituting Eq. (2.258) to (2.259) into $\dot{V}$ yields

$\dot{V} = \frac{1}{2}\hat{e}^T\{(A - BK^T)^T P_1 + P_1(A - BK^T)\}\hat{e} + \tilde{e}^T C K_o^T P_1\hat{e} + \frac{1}{2}\tilde{e}^T\{(A - K_oC^T)^T P_2 + P_2(A - K_oC^T)\}\tilde{e} + \tilde{e}^T P_2 B(u_c + w + \tilde{d}) + \frac{1}{\gamma_1}\dot{\tilde{\theta}}_f^T\tilde{\theta}_f + \frac{1}{\gamma_2}tr[\dot{\tilde{\theta}}_g^T\tilde{\theta}_g]$

that is

$\dot{V} = -\frac{1}{2}\hat{e}^T Q_1\hat{e} + \tilde{e}^T C K_o^T P_1\hat{e} - \frac{1}{2}\tilde{e}^T\{Q_2 - P_2B(\frac{2}{r} - \frac{1}{\rho^2})B^T P_2\}\tilde{e} + \tilde{e}^T P_2 B(u_c + w + \tilde{d}) + \frac{1}{\gamma_1}\dot{\tilde{\theta}}_f^T\tilde{\theta}_f + \frac{1}{\gamma_2}tr[\dot{\tilde{\theta}}_g^T\tilde{\theta}_g]$

The supervisory control u_c is decomposed in two terms, u_a and u_b.

• The control term u_a is given by

$$u_a = -\frac{1}{r}\tilde{e}^T P_2 B + \Delta u_a \tag{2.260}$$

where assuming that the measurable elements of vector $\tilde{e}$ are $\{\tilde{e}_1, \tilde{e}_3, \cdots, \tilde{e}_k\}$, the term Δu_a is such that

$$-\frac{1}{r}\tilde{e}^T P_2 B + \Delta u_a = -\frac{1}{r}\begin{pmatrix} p_{11}\tilde{e}_1 + p_{13}\tilde{e}_3 + \cdots + p_{1k}\tilde{e}_k \\ p_{13}\tilde{e}_1 + p_{33}\tilde{e}_3 + \cdots + p_{3k}\tilde{e}_k \\ \cdots\cdots\cdots \\ p_{1k}\tilde{e}_1 + p_{3k}\tilde{e}_3 + \cdots + p_{kk}\tilde{e}_k \end{pmatrix} \tag{2.261}$$

• The control term u_b is given by

$$u_b = -[(P_2B)^T(P_2B)]^{-1}(P_2B)^T C K_o^T P_1\hat{e} \tag{2.262}$$

• u_a is an H_∞ control used for the compensation of the approximation error w and the additive disturbance $\tilde{d}$. Its first component $-\frac{1}{r}\tilde{e}^T P_2 B$ has been chosen so

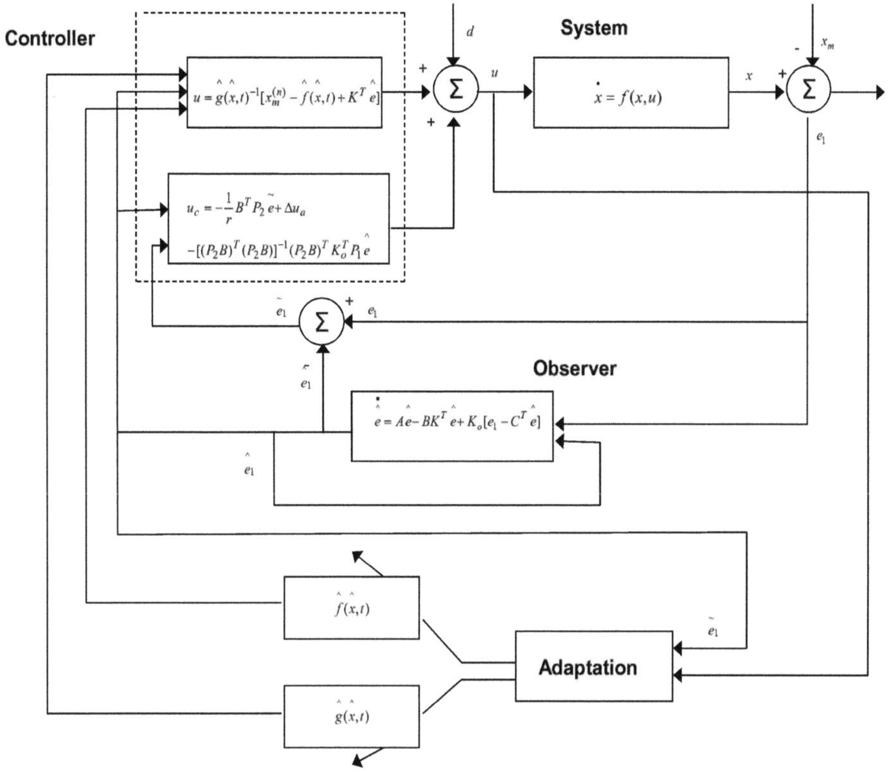

Fig. 2.17 The proposed output feedback-based adaptive fuzzy control scheme for the DFIG

as to compensate for the term $\frac{1}{r}\tilde{e}^T P_2 B B^T P_2 \tilde{e}$, which appears in the previously computed function about $\dot{V}$. By including also the second component Δu_a, one has that u_a is computed based on the feedback of only the measurable variables $\{\tilde{e}_1, \tilde{e}_3, \ldots, \tilde{e}_k\}$, out of the complete vector $\tilde{e} = [\tilde{e}_1, \tilde{e}_2, \ldots, \tilde{e}_n]$. Equation (8.182) is finally rewritten as $u_a = -\frac{1}{r}\tilde{e}^T P_2 B + \Delta u_a$.

- u_b is a control used for the compensation of the observation error (the control term u_b has been chosen so as to satisfy the condition $\tilde{e}^T P_2 B u_b = -\tilde{e}^T C K_o^T P_1 \hat{e}$.

The control scheme is depicted in Fig. 2.17
Substituting Eqs. (2.260) and (2.262) in $\dot{V}$ and assuming that Eqs. (2.258) and (2.259) hold, one gets

$$\begin{aligned} \dot{V} = &-\tfrac{1}{2}\hat{e}^T Q_1 \hat{e} + \tilde{e}^T C K_o^T P_1 \hat{e} - \tfrac{1}{2}\tilde{e}^T Q_2 \tilde{e} + \\ &+ \tfrac{1}{r}\tilde{e}^T P_2 B B^T P_2 \tilde{e} - \tfrac{1}{2\rho^2}\tilde{e}^T P_2 B B^T P_2 \tilde{e} + \\ &+ \tilde{e}^T P_2 B u_a + \tilde{e}^T P_2 B u_b + \tilde{e}^T P_2 B (w + \tilde{d}) + \\ &+ \tfrac{1}{\gamma_1}\dot{\tilde{\theta}}_f^T \tilde{\theta}_f + \tfrac{1}{\gamma_2} tr[\dot{\tilde{\theta}}_g^T \tilde{\theta}_g] \end{aligned} \tag{2.263}$$

or equivalently,

$$\begin{aligned}\dot{V} &= -\tfrac{1}{2}\hat{e}^T Q_1\hat{e} - \tfrac{1}{2}\tilde{e}^T Q_2\tilde{e} - \tfrac{1}{2\rho^2}\tilde{e}^T P_2 B B^T P_2\tilde{e}+ \\ &+\tilde{e}^T P_2 B(w+\tilde{d}+\Delta u_a) + \tfrac{1}{\gamma_1}\dot{\tilde{\theta}}_f^T\tilde{\theta}_f + \tfrac{1}{\gamma_2}tr[\dot{\tilde{\theta}}_g^T\tilde{\theta}_g]\end{aligned} \tag{2.264}$$

It holds that $\dot{\tilde{\theta}}_f = \dot{\theta}_f - \dot{\theta}_f^* = \dot{\theta}_f$ and $\dot{\tilde{\theta}}_g = \dot{\theta}_g - \dot{\theta}_g^* = \dot{\theta}_g$. The following weight adaptation laws are considered:

$$\begin{aligned}\dot{\theta}_f &= -\gamma_1\Phi(\hat{x})^T B^T P_2\tilde{e} \\ \dot{\theta}_g &= -\gamma_2\Phi(\hat{x})^T B^T P_2\tilde{e}u^T\end{aligned} \tag{2.265}$$

where assuming N fuzzy rules and associated kernel functions the matrices dimensions are $\theta_f \in R^{N\times 1}$, $\theta_g \in R^{N\times 2}$, $\Phi(x)\in R^{2\times N}$, $B\in R^{4\times 2}$, $P\in R^{4\times 4}$ and $\tilde{e}\in R^{4\times 1}$.

The update of θ_f is a gradient type algorithm [341, 344]. The update of θ_g is also a gradient type algorithm, where u_c implicitly tunes the adaptation gain γ_2. Substituting Eq. (2.265) in $\dot{V}$ gives

$$\begin{aligned}\dot{V} &= -\tfrac{1}{2}\hat{e}^T Q_1\hat{e} - \tfrac{1}{2}\tilde{e}^T Q_2\tilde{e} - \tfrac{1}{2\rho^2}\tilde{e}^T P_2 B B^T P_2\tilde{e} + B^T P_2\tilde{e}(w+d+\Delta u_a) \\ &+\tfrac{1}{\gamma_1}(-\gamma_1)\tilde{e}^T P_2 B\Phi(\hat{x})(\theta_f-\theta_f^*) + \tfrac{1}{\gamma_2}(-\gamma_2)tr[u\tilde{e}^T P_2 B\Phi(\hat{x})(\theta_g-\theta_g^*)]\end{aligned}$$

or equivalently

$$\begin{aligned}\dot{V} &= -\tfrac{1}{2}\hat{e}^T Q_1\hat{e} - \tfrac{1}{2}\tilde{e}^T Q_2\tilde{e} - \tfrac{1}{2\rho^2}\tilde{e}^T P_2 B B^T P_2\tilde{e} + B^T P_2\tilde{e}(w+\tilde{d}+\Delta u_a) \\ &+\tfrac{1}{\gamma_1}(\gamma_1)\tilde{e}^T P_2 B\Phi(\hat{x})(\theta_f-\theta_f^*) + \tfrac{1}{\gamma_2}(\gamma_2)tr[u\tilde{e}^T P_2 B(\hat{g}(\hat{x}|\theta_g) - \hat{g}(\hat{x}|\theta_g^*)]\end{aligned}$$

Taking into account that $u\in R^{2\times 1}$ and $\tilde{e}^T PB(\hat{g}(x|\theta_g) - \hat{g}(x|\theta_g^*)) \in R^{1\times 2}$ it holds

$$\begin{aligned}\dot{V} &= -\tfrac{1}{2}\hat{e}^T Q_1\hat{e} - \tfrac{1}{2}\tilde{e}^T Q_2\tilde{e} - \tfrac{1}{2\rho^2}\tilde{e}^T P_2 B B^T P_2\tilde{e} + B^T P_2\tilde{e}(w+\tilde{d}+\Delta u_a)+ \\ &+\tfrac{1}{\gamma_1}(-\gamma_1)\tilde{e}^T P_2 B\Phi(\hat{x})(\theta_f-\theta_f^*) + \tfrac{1}{\gamma_2}(-\gamma_2)tr[\tilde{e}^T P_2 B(\hat{g}(\hat{x}|\theta_g) - \hat{g}(\hat{x}|\theta_g^*))u]\end{aligned}$$

Since $\tilde{e}^T P_2 B(\hat{g}(\hat{x}|\theta_g) - \hat{g}(\hat{x}|\theta_g^*))u\in R^{1\times 1}$ it holds

$$\begin{aligned}&tr(\tilde{e}^T P_2 B(\hat{g}(x|\theta_g) - \hat{g}(x|\theta_g^*)u) = \\ &= \tilde{e}^T P_2 B(\hat{g}(x|\theta_g) - \hat{g}(x|\theta_g^*))u\end{aligned} \tag{2.266}$$

Therefore, one finally obtains

$$\begin{aligned}\dot{V} &= -\tfrac{1}{2}\hat{e}^T Q_1\hat{e} - \tfrac{1}{2}\tilde{e}^T Q_2\tilde{e} - \tfrac{1}{2\rho^2}\tilde{e}^T P_2 B B^T P_2\tilde{e} + B^T P_2\tilde{e}(w+\tilde{d}+\Delta u_a)+ \\ &+\tfrac{1}{\gamma_1}(-\gamma_1)\tilde{e}^T P_2 B\Phi(\hat{x})(\theta_f-\theta_f^*) + \tfrac{1}{\gamma_2}(-\gamma_2)\tilde{e}^T P_2 B(\hat{g}(\hat{x}|\theta_g) - \hat{g}(\hat{x}|\theta_g^*))u\end{aligned}$$

Next, the following approximation error is defined

$$w_\alpha = [\hat{f}(\hat{x}|\theta_f^*) - \hat{f}(\hat{x}|\theta_f)] + [\hat{g}(\hat{x}|\theta_g^*) - \hat{g}(\hat{x}|\theta_g)]u \tag{2.267}$$

Thus, one obtains

$$\begin{aligned}\dot{V} = & -\tfrac{1}{2}\hat{e}^T Q_1 \hat{e} - \tfrac{1}{2}\tilde{e}^T Q_2 \tilde{e} - \tfrac{1}{2\rho^2}\tilde{e}^T P_2 B B^T P_2 \tilde{e} + \\ & + B^T P_2 \tilde{e}(w + \tilde{d}) + \tilde{e}^T P_2 B w_\alpha\end{aligned} \tag{2.268}$$

Denoting the aggregate approximation error and disturbances vector as

$$w_1 = w + \tilde{d} + w_\alpha + \Delta u_a \tag{2.269}$$

the derivative of the Lyapunov function becomes

$$\dot{V} = -\tfrac{1}{2}\hat{e}^T Q_1 \hat{e} - \tfrac{1}{2}\tilde{e}^T Q_2 \tilde{e} - \tfrac{1}{2\rho^2}\tilde{e}^T P_2 B B^T P_2 \tilde{e} + \tilde{e}^T P_2 B w_1 \tag{2.270}$$

which in turn is written as

$$\begin{aligned}\dot{V} = & -\tfrac{1}{2}\hat{e}^T Q_1 \hat{e} - \tfrac{1}{2}\tilde{e}^T Q_2 \tilde{e} - \tfrac{1}{2\rho^2}\tilde{e}^T P_2 B B^T P_2 \tilde{e} + \\ & + \tfrac{1}{2}\tilde{e}^T P B w_1 + \tfrac{1}{2} w_1^T B^T P_2 \tilde{e}\end{aligned} \tag{2.271}$$

Lemma: The following inequality holds

$$\tfrac{1}{2}\tilde{e}^T P_2 B w_1 + \tfrac{1}{2} w_1^T B^T P_2 \tilde{e} - \tfrac{1}{2\rho^2}\tilde{e}^T P_2 B B^T P_2 \tilde{e} \leq \tfrac{1}{2}\rho^2 w_1^T w_1 \tag{2.272}$$

Proof: The binomial $(\rho a - \frac{1}{\rho} b)^2 \geq 0$ is considered. Expanding the left part of the above inequality one gets

$$\begin{aligned}& \rho^2 a^2 + \tfrac{1}{\rho^2} b^2 - 2ab \geq 0 \Rightarrow \tfrac{1}{2}\rho^2 a^2 + \tfrac{1}{2\rho^2} b^2 - ab \geq 0 \Rightarrow \\ & ab - \tfrac{1}{2\rho^2} b^2 \leq \tfrac{1}{2}\rho^2 a^2 \Rightarrow \tfrac{1}{2}ab + \tfrac{1}{2}ab - \tfrac{1}{2\rho^2} b^2 \leq \tfrac{1}{2}\rho^2 a^2\end{aligned} \tag{2.273}$$

The following substitutions are carried out: $a = w_1$ and $b = \tilde{e}^T P_2 B$ and the previous relation becomes

$$\tfrac{1}{2} w_1^T B^T P_2 \tilde{e} + \tfrac{1}{2}\tilde{e}^T P_2 B w_1 - \tfrac{1}{2\rho^2}\tilde{e}^T P_2 B B^T P_2 \tilde{e} \leq \tfrac{1}{2}\rho^2 w_1^T w_1 \tag{2.274}$$

The above relation is used in $\dot{V}$, and the right part of the associated inequality is enforced

$$\dot{V} \leq -\frac{1}{2}\hat{e}^T Q_1 \hat{e} - \frac{1}{2}\tilde{e}^T Q_2 \tilde{e} + \frac{1}{2}\rho^2 w_1^T w_1 \tag{2.275}$$

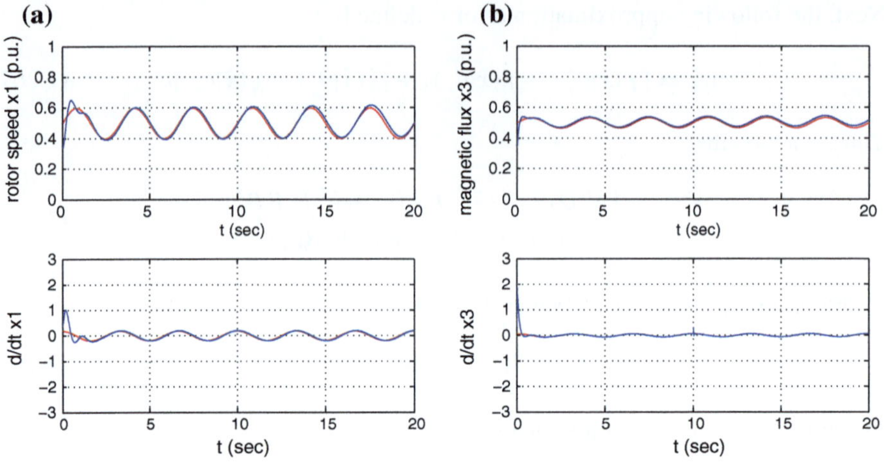

Fig. 2.18 Tracking of setpoint 1 (*red lines*) from state variables of the DFIG (*blue lines*): **a** rotor's speed x_1 and its derivative x_2, **b** magnetic flux of the stator x_3 and its derivative x_4

Thus, Eq. (2.275) can be written as

$$\dot{V} \leq -\frac{1}{2}E^T Q E + \frac{1}{2}\rho^2 w_1^T w_1 \tag{2.276}$$

where

$$E = \begin{pmatrix} \hat{e} \\ \tilde{e} \end{pmatrix}, \quad Q = \begin{pmatrix} Q_1 & 0 \\ 0 & Q_2 \end{pmatrix} = diag[Q_1, Q_2] \tag{2.277}$$

Hence, the H_∞ performance criterion is derived. For ρ sufficiently small Eq. (2.275) will be true and the H_∞ tracking criterion will be satisfied. In that case, the integration of $\dot{V}$ from 0 to T gives

$$\begin{array}{c} \int_0^T \dot{V}(t)dt \leq -\frac{1}{2}\int_0^T ||E||^2 dt + \frac{1}{2}\rho^2 \int_0^T ||w_1||^2 dt \Rightarrow \\ 2V(T) - 2V(0) \leq -\int_0^T ||E||_Q^2 dt + \rho^2 \int_0^T ||w_1||^2 dt \Rightarrow \\ 2V(T) + \int_0^T ||E||_Q^2 dt \leq 2V(0) + \rho^2 \int_0^T ||w_1||^2 dt \end{array} \tag{2.278}$$

It is assumed that there exists a positive constant $M_w > 0$ such that $\int_0^\infty ||w_1||^2 dt \leq M_w$. Therefore for the integral $\int_0^T ||E||_Q^2 dt$ one gets

$$\int_0^\infty ||E||_Q^2 dt \leq 2V(0) + \rho^2 M_w \tag{2.279}$$

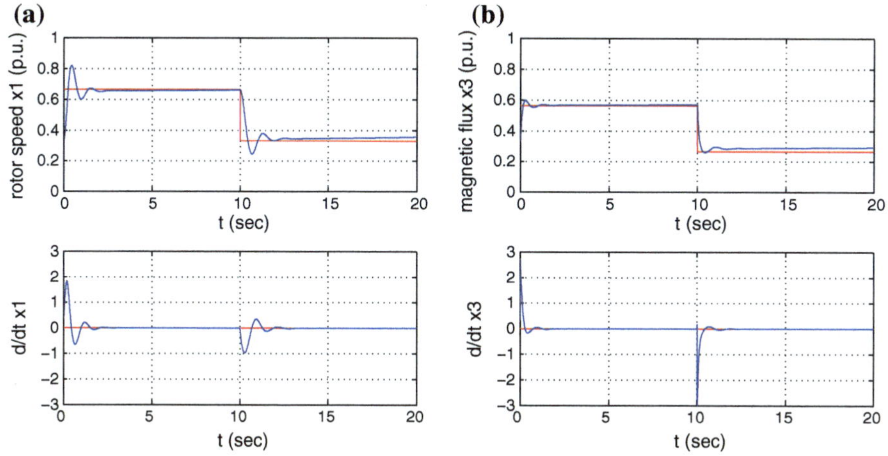

Fig. 2.19 Tracking of setpoint 2 (*red lines*) from state variables of the DFIG (*blue lines*): **a** rotor's speed x_1 and its derivative x_2, **b** magnetic flux of the stator x_3 and its derivative x_4

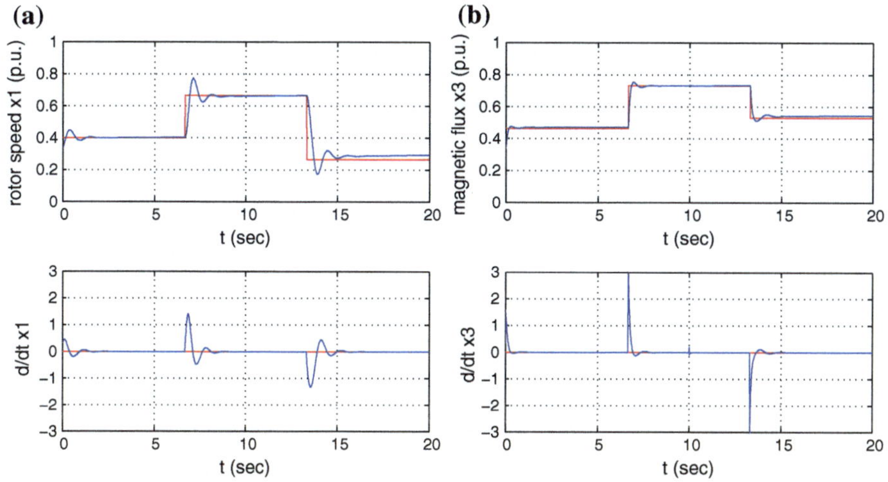

Fig. 2.20 Tracking of setpoint 3 (*red lines*) from state variables of the DFIG (*blue lines*): **a** rotor's speed x_1 and its derivative x_2, **b** magnetic flux of the stator x_3 and its derivative x_4

Thus, the integral $\int_0^{\infty}||E||_Q^2 dt$ is bounded and according to Barbalat's Lemma

$$\begin{array}{c} \lim_{t\to\infty} E(t) = 0 \Rightarrow \\ \lim_{t\to\infty} \hat{e}(t) = 0 \\ \lim_{t\to\infty} \tilde{e}(t) = 0 \end{array} \tag{2.280}$$

Therefore $\lim_{t\to\infty} e(t) = 0$.

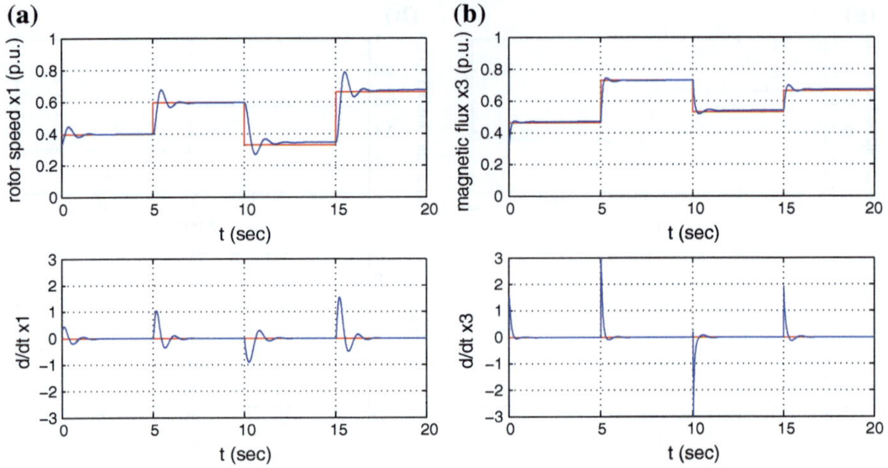

Fig. 2.21 Tracking of setpoint 4 (*red lines*) from state variables of the DFIG (*blue lines*): **a** rotor's speed x_1 and its derivative x_2, **b** magnetic flux of the stator x_3 and its derivative x_4

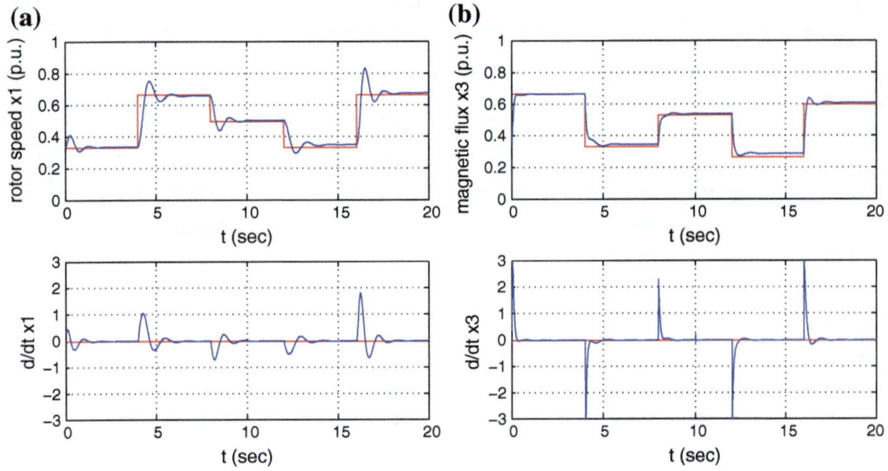

Fig. 2.22 Tracking of setpoint 5 (*red lines*) from state variables of the DFIG (*blue lines*): **a** rotor's speed x_1 and its derivative x_2, **b** magnetic flux of the stator x_3 and its derivative x_4

2.5.6 Simulation Tests

The performance of the proposed adaptive neurofuzzy control scheme for the Doubly-fed induction generator, using only output feedback, has been evaluated through simulation experiments. Indicative numerical values for the parameters of the considered doubly-fed induction generator model are given in Table 2.1, appearing in Sect. 2.3.6. The turn speed of the rotor ω was considered to be

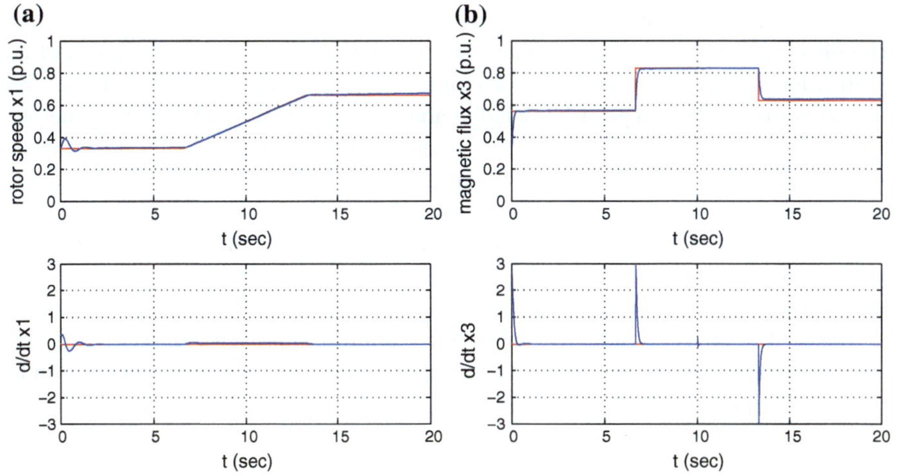

Fig. 2.23 Tracking of setpoint 6 (*red lines*) from state variables of the DFIG (*blue lines*): **a** rotor's speed x_1 and its derivative x_2, **b** magnetic flux of the stator x_3 and its derivative x_4

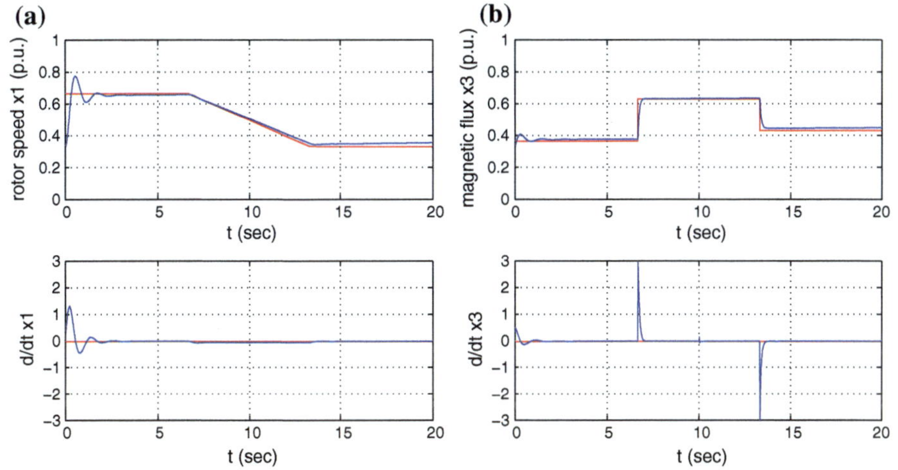

Fig. 2.24 Tracking of setpoint 7 (*red lines*) from state variables of the DFIG (*blue lines*): **a** rotor's speed x_1 and its derivative x_2, **b** magnetic flux of the stator x_3 and its derivative x_4

measurable. Moreover, the magnetic flux of the stator ψ_{s_d} was considered to be computed from measurements of stator and rotor currents, following the previously explained procedure. The obtained experimental results are presented in Figs. 2.18, 2.19, 2.20, 2.21, 2.22, 2.23 and 2.24. It can be noticed that the proposed control scheme succeeded in all cases fast and accurate tracking of the reference setpoints.

The treated control problem, that is (i) simultaneous estimation of the unknown dynamics of the generator and estimation of the machine's state vector, and

(ii) use of this information by an adaptive control scheme, is of elevated difficulty. The provided Lyapunov analysis has come to confirm the stability properties of the control loop. The presented adaptive control method of the DFIG is a model-free one. This means that no prior knowledge about the dynamic model of the generator is needed. Moreover, no knowledge is needed about the mechanical torque exerted on the generator's rotor (e.g. due to wind, steam or water fall). Through the identification procedure that was analyzed in Sect. 2.5.4 it becomes possible to estimate in real-time the unknown dynamical model of the generator and subsequently to use this information in the control loop. Besides, this adaptive control method does not need to measure the complete state vector of the DFIG and is based only on output feedback. The analyzed adaptive neurofuzzy control approach exhibits robustness to external perturbation thus assuring the reliable functioning of the generator under harsh conditions.

Chapter 3
Control of the Functioning of Synchronous Generators

3.1 Outline

The chapter analyzes different approaches for nonlinear control of synchronous generators: (i) flatness-based control through transformation to the canonical form, (ii) flatness-based control in cascading loops, (iii) differential geometry-based control, (iv) stabilizing control of synchronous power generators in the s-frequency domain.

First, the chapter is concerned with proving differential flatness of the permanent magnet synchronous generator model and its resulting description in the Brunovksy (canonical) form. At a second stage and with the use of the linearized model, Kalman Filtering is proposed for estimating the non directly measurable elements of the state vector of the power generator. To enable efficient operation of the power generator under variable mechanical input power (torque) the Kalman Filter is extended towards a disturbances estimator. This enables to estimate simultaneously both the state vector of the generator and the unknown mechanical input power (torque). Finally, state estimation-based control is applied to assure that the rotation speed of the generator tracks a predefined reference setpoint. The aggregate control input is generated by including in the state-feedback control law a term that is based on the estimation of the unknown input power and of its derivatives and which compensates for the effects of variation of this input power.

Next, the chapter proposes a new method for the control of the nonlinear dynamics of synchronous generators which is based on differential flatness theory and which is implemented in successive loops. The method assumes that the generator's state-space model can be transformed to the so-called triangular form. The controller design proceeds by showing that each row of the state-space model of the synchronous generator stands for a differentially flat system, where the flat output is chosen to be the associated state variable. Next, for each subsystem which is linked with a row of the state-space model a virtual control input is computed, that can invert the

G. Rigatos, *Intelligent Renewable Energy Systems*,
Green Energy and Technology, DOI 10.1007/978-3-319-39156-4_3

subsystem's dynamics and can eliminate the subsystem's tracking error. From the last row of the state-space description, the control input that is actually applied to the synchronous generator is found. This control input contains recursively all virtual control inputs which were computed for the individual subsystems associated with the previous rows of the state-space equation. Thus, by tracing the rows of the state-space model backwards, at each iteration of the control algorithm, one can finally obtain the control input that should be applied to the synchronous generator so as to assure that all its state vector elements will converge to the desirable setpoints. Indicative evaluation results of the control method are presented in this section in the form of simulation experiments.

Finally, the chapter analyzes a robust power system stabilizer (PSS) which can damp-out oscillations in synchronous generators connected to the electricity grid. Oscillations of small magnitude and low frequency, linked with the electromechanical models of power systems, often persist for long periods of time and in some cases present limitations on the power transfer capability. The proposed PSS is designed according to Kharitonov's extremal gain margin theory. It has the following advantages: (i) it is based on simultaneous stabilization of limited number of extreme plants, (ii) the control design can be based on frequency response analysis techniques (root locus diagrams or Nyquist plots), (iii) the resulting controller is a low-order phase lead compensator, which is robust to the change of operating points. The proposed stabilizer for power generators is tested through simulation experiments.

3.2 Flatness-Based Control of Synchronous Generators

3.2.1 Outline

In the design of nonlinear controllers for power generators it is important to measure all state variables needed to generate a feedback control law. In case of state variables for which direct measurement is technically difficult or costly, estimation can be performed with nonlinear filtering methods [20, 112, 197, 217, 249, 266, 304]. The chapter proposes derivative-free nonlinear Kalman Filtering for sensorless control of Permanent Magnet Synchronous Generators.

For nonlinear electric power generators, and under Gaussian noise, the Extended Kalman Filter (EKF) is frequently applied for estimating the non-measurable state variables through the processing of input and output sequences [34, 174, 209, 310]. The Extended Kalman Filter is based on linearization of the dynamical system using Taylor series expansion [334, 341, 342, 343, 480]. Although EKF is efficient in several estimation problems, it is characterized by cumulative errors due to the local linearization assumption and this may affect the accuracy of the state estimation or even risk the stability of the observer-based control loop.

It is also known that one can attempt transformation of the permanent magnet synchronous generator model into the canonical (Brunovsky) form through the application of the Lie-algebra theory. By using such differential geometric methods it is possible to arrive at a description of the system in the linear canonical form if the relative degree of the system is equal to the order of the system. After transformation to the canonical form, state estimation with the use of a linear system is possible. However, this linearization procedure requires the computation of Lie derivatives (partial derivatives on the vector fields describing the system dynamics) which can be a complicated computation procedure.

Aiming at finding more efficient state estimation methods for the model of the permanent magnet synchronous generator (with reference to the Lie algebra-based state estimator design), in this chapter a derivative-free approach to Kalman filtering is introduced. In the proposed derivative-free Kalman Filtering method the generator is subjected to a linearization transformation that is based on the differential flatness theory and next state estimation is performed by applying the standard Kalman Filter recursion to the linearized model. Unlike the Lie algebra-based estimator design method, the proposed approach provides estimates of the state vector of the permanent magnet synchronous generator without the need for derivatives and Jacobians calculation. By avoiding linearization approximations, the proposed filtering method improves the accuracy of estimation of the system state variables, and results in smooth control signal variations and in minimization of the tracking error of the associated control loop [348, 353, 354].

As previously noted, differential flatness theory has been widely used for developing nonlinear control systems [352, 388, 411]. To find out if a dynamical system is differentially flat, the following should be examined: (i) the existence of the so-called flat output, i.e. a new variable which is expressed as a function of the system's state variables. The flat output and its derivatives should not be coupled in the form of an ordinary differential equation, (ii) the components of the system (i.e. state variables and control input) should be expressed as functions of the flat output and its derivatives [137, 241, 229]. As explained, differential flatness theory enables transformation to a linearized form (canonical Brunovsky form) for which the design of the controller becomes easier. Moreover by showing that a system is differentially flat one can easily design a reference trajectory as a function of the so-called flat output and can find a control law that assures tracking of this desirable trajectory [137, 456].

This chapter is concerned with proving differential flatness of the permanent magnet synchronous generator model and its resulting description in the Brunovksy (canonical) form [281]. At a second stage and with the use of the linearized model, Kalman Filtering is proposed for estimating the non directly measurable elements of the state vector of the power generator. To enable efficient operation of the power generator under variable mechanical input power (torque) the Kalman Filter is extended towards a disturbances estimator. This enables to estimate simultaneously both the state vector of the generator and the unknown mechanical input power (torque). Finally, state estimation-based control is applied to assure that the rotation speed of the generator tracks a predefined reference setpoint. The aggregate control input is generated by including in it the state-feedback control law a term that is based on the estimation of the unknown input power and of its derivatives and which compensates for the effects of variation of this input power.

3.2.2 Lie Algebra-Based Design of Nonlinear State Estimators

3.2.2.1 Relative Degree for Nonlinear Systems

The nonlinear model of the PMSG given in Eq. (1.31) is in an affine in-the-input form, i.e.

$$\begin{aligned} \dot{x}(t) &= f(x) + g(x)u \\ y(t) &= h(x) \end{aligned} \tag{3.1}$$

where $x \in R^n$ is the state vector, $u \in R$ is the control vector, $y \in R$ is the output vector, $f(x)$ and $g(x)$ vector fields that belong both in R^n and $h(x)$ is the scalar function of x. It will be shown that a state estimator for the nonlinear model of the PMSG can be derived using Lie algebra, which introduces a change of coordinates (diffeomorphism) that enables to write a nonlinear system into an equivalent linear form. Next, the following definitions from Lie algebra theory are used [215]:

Lie Derivative: For a given differentiable scalar function $h(x)$ of $x = [x_1, x_2, x_3, \ldots x_n]^T$ and a vector field $f(x) = [f_1, f_2, f_3, \ldots, f_n]^T$, the Lie derivative of function $h(x)$ along the vector field $f(x)$, is a new scalar function defined by $L_f h(x)$ is obtained as follows

$$L_f h(x) = \frac{\partial h(x)}{\partial x} f(x) = \sum_{i=1}^{n} \frac{\partial h(x)}{\partial x_i} f_i(x) \tag{3.2}$$

Relative degree: If the Lie derivative of the function $L_f^{r-1} h(x)$ along vector field $g(x)$ is not equal to zero in a neighborhood Ω, i.e.

$$L_g L_f^{r-1} h(x) \neq 0 \tag{3.3}$$

then it is said that the system has relative degree r in Ω.

The relative degree of the system is a parameter to take into account in the design of controllers or observers for nonlinear dynamical systems. If the relative degree of a system is equal to its order n, then the system is exactly linearizable. If $r < n$ then the system is partially linearizable.

3.2.3 Nonlinear Observer Design for Exactly Linearizable Systems

Under the condition $r = n$ one has [215]:

$$\begin{array}{c} L_g L_f^{1-1} h(x) = L_g L_f^{2-1} h(x) = \cdots = L_g L_f^{n-2} h(x) = 0 \\ L_g L_f^{n-1} h(x) \neq 0 \end{array} \tag{3.4}$$

Next a change of coordinates is performed as follows:

$$z_1 = y = h(x) = L_f^{1-1} h(x) \tag{3.5}$$

Moreover, it holds

$$\dot{z}_1 = \frac{\partial h(x)}{\partial x} \dot{x} \tag{3.6}$$

Substituting the system's dynamics into Eq. (3.6) one gets

$$\begin{aligned} \dot{z}_1 &= \frac{\partial h(x)}{\partial x} f(x) + \frac{\partial h(x)}{\partial x} g(x) \cdot u \\ &= L_f^{2-1} h(x) + L_g L_f^{1-1} h(x) u \end{aligned} \tag{3.7}$$

According to the relative degree condition it holds $L_g L_f^{1-1} h(x) = 0$ and from the Eq. (3.7) one gets

$$\dot{z}_1 = L_f h(x) = z_2 \tag{3.8}$$

In a similar way

$$\begin{array}{c} \dot{z}_2 = L_f^2 h(x) = z_3 \\ \cdots \\ \cdots \\ \dot{z}_{n-1} = L_f^{n-1} h(x) = z_n \end{array} \tag{3.9}$$

Again, using the property of the system's relative degree one has $L_g L_f^{n-1} h(x) \neq 0$ one obtains

$$\dot{z}_n = L_f^n h(x) + L_g L_f^{n-1} h(x) u = \alpha(x) + b(x) u = v \tag{3.10}$$

where $\alpha(x) = L_f^n h(x)$, $b(x) = L_g L_f^{n-1} h(x)$ and $v = \alpha(x) + b(x)u$. After this change of coordinates the system can be written as

$$\dot{z} = Az + Bu \tag{3.11}$$

where

$$z = \phi(x) = \begin{pmatrix} h(x) \\ L_f h(x) \\ \cdots \\ \cdots \\ L_f^{n-1} h(x) \end{pmatrix} \tag{3.12}$$

while matrices A, B appearing in the previous state-space equation are given by

$$A = \begin{pmatrix} 0 & 1 & 0 & \cdots & 0 \\ 0 & 0 & 1 & \cdots & 0 \\ \vdots & \vdots & \vdots & \cdots & \vdots \\ 0 & 0 & 0 & \cdots & 1 \\ 0 & 0 & 0 & \cdots & 0 \end{pmatrix} \\ B = \begin{pmatrix} 0 & 0 & \cdots & 0 & 1 \end{pmatrix}^T \tag{3.13}$$

The state observer for the transformed system of Eq. (3.11) is

$$\dot{\hat{z}} = A\hat{z} + Bv + K(y - h(\hat{x})) \text{ or} \\ \dot{\hat{z}} = A\hat{z} + B[\alpha(\hat{x}) + b(\hat{x})u] + K(y - h(\hat{x})) \text{ or} \\ \dot{\hat{z}} = A\hat{z} + B[\alpha(\phi^{-1}(\hat{z})) + b(\phi^{-1}(\hat{z}))u] + K(y - h(\phi^{-1}(\hat{z}))) \tag{3.14}$$

It is also possible to express the state observer using a nonlinear model, i.e.

$$\dot{\hat{x}} = f(\hat{x}) + g(\hat{x})u + L(y - h(\hat{x})) \tag{3.15}$$

It will be shown that the nonlinear observer's gain L is now given by

$$L = (J_\phi(\hat{x}))^{-1} K \tag{3.16}$$

where matrix $J_\phi(\hat{x})$ is the Jacobian of the new coordinates $z_1, \ldots, z_n$ which is obtained after the nonlinear change of coordinates and K is the observer's gain computed for the linearized equivalent of the system of Eq. (3.14). The observer gain K for the linearized system can be obtained through the Kalman Filter recursion. One has that

$$d\hat{z} = \begin{pmatrix} dh(\hat{x}) \\ dL_f h(\hat{x}) \\ \cdots \\ dL_f^{n-1} h(\hat{x}) \end{pmatrix} = J_\phi(\hat{x}) d\hat{x} \tag{3.17}$$

or equivalently

$$\dot{\hat{z}} = J_\phi(\hat{x}) \dot{\hat{x}} \tag{3.18}$$

It holds that the Jacobian matrix of $\phi(\hat{x})$ with respect to $\hat{x}$ can be written as

$$J_\phi(\hat{x}) = \frac{\partial \phi(\hat{x})}{\partial \hat{x}} \tag{3.19}$$

Using the state observer dynamics described in Eq. (3.15) one has

$$\frac{\partial \phi(\hat{x})}{\partial \hat{x}}\dot{\hat{x}} = \frac{\partial \phi(\hat{x})}{\partial \hat{x}} f(\hat{x}) + \frac{\partial \phi(\hat{x})}{\partial \hat{x}} g(\hat{x})u + \frac{\partial \phi(\hat{x})}{\partial \hat{x}} L(\hat{x})(y - h(\hat{x})) \tag{3.20}$$

Considering that for the first row of the Jacobian matrix it holds $\phi(\hat{x}) = h(\hat{x})$ one has

$$\frac{\partial h(\hat{x})}{\partial \hat{x}}\dot{\hat{x}} = \frac{\partial h(\hat{x})}{\partial \hat{x}} f(\hat{x}) + \frac{\partial h(\hat{x})}{\partial \hat{x}} g(\hat{x})u + K(y - h(\hat{x})) \tag{3.21}$$

or equivalently

$$\dot{\hat{z}}_1 = L_f h(\hat{x}) + L_g h(\hat{x})u + K_1(y - h(\hat{x})) \tag{3.22}$$

Moreover, it holds $L_g h(\hat{x}) = 0$ and $L_f h(\hat{x}) = \hat{z}_2$ and thus one obtains

$$\dot{\hat{z}}_1 = \hat{z}_2 + K_1(y - h(\hat{x})) \tag{3.23}$$

In a similar manner one has

$$\begin{aligned} \dot{\hat{z}}_2 &= \hat{z}_3 + K_2(y - h(\hat{x})) \\ &\cdots \\ \dot{\hat{z}}_{n-1} &= \hat{z}_n + K_{n-1}(y - h(\hat{x})) \end{aligned} \tag{3.24}$$

Additionally, using that $L_g L_f^{n-1} \neq 0$ one can finally write

$$\dot{\hat{z}}_n = \alpha(\phi^{-1}(\hat{z})) + b(\phi^{-1}(\hat{z}))u + K_n(y - h(\hat{x})) \tag{3.25}$$

where $\alpha(\phi^{-1}(\hat{z})) = L_f^{n-1} h(\hat{x}) = L_f^{n-1} h(\phi^{-1}(\hat{z}))$, and $b(\phi^{-1}(\hat{z})) = L_g L_f^{n-1} h(\hat{x}) = L_g L_f^{n-1} h(\phi^{-1}(\hat{z}))$. Using the previous notation one obtains the formulation of the nonlinear estimator's gain L as a function of the observation gain K for the linearized equivalent of the system described in Eq. (3.14). Thus one finally arrives at Eq. (3.16)

$$(J_\phi(\hat{x}))L = K \Rightarrow L = (J_\phi(\hat{x}))^{-1} K \tag{3.26}$$

3.2.3.1 Linearization of PMSG Dynamics Using Lie-Algebra

The nonlinear dynamical model of the PMSG was described in Eqs. (1.32)–(1.35). The linearization procedure with the use of the Lie algebra gives:

$$z_1 = L_f^0 h(x) = h(x) = \Delta\delta \tag{3.27}$$

$$z_2 = L_f^1 h(x) = \left(\frac{\partial h}{\partial x_1} \ \frac{\partial h}{\partial x_2} \ \frac{\partial h}{\partial x_3}\right) \cdot \begin{pmatrix} f_1 \\ f_2 \\ f_3 \end{pmatrix} \Rightarrow$$
$$z_2 = x_2 = \Delta\omega \tag{3.28}$$

$$z_3 = L_f^2 h(x) = \left(\frac{\partial z_2}{\partial x_1} \ \frac{\partial z_2}{\partial x_2} \ \frac{\partial z_2}{\partial x_3}\right) \cdot \begin{pmatrix} f_1 \\ f_2 \\ f_3 \end{pmatrix} \Rightarrow$$
$$z_3 = f_2 = -\frac{D}{2J}(\omega - \omega_0) + \omega_0 \frac{P_m}{2J} - \frac{\omega_0}{2J}\frac{V_s E_q'}{x_{d\Sigma}'} sin(\Delta\delta) \tag{3.29}$$

Moreover, according to the previous analysis it holds

$$\dot{z}_3 = L_f^3 h(x) + L_g L_f^2 h(x) u \tag{3.30}$$

where

$$L_f^3 h(x) = \left(\frac{\partial z_3}{\partial x_1} \ \frac{\partial z_3}{\partial x_2} \ \frac{\partial z_3}{\partial x_3}\right) \cdot \begin{pmatrix} f_1 \\ f_2 \\ f_3 \end{pmatrix} \Rightarrow$$
$$L_f^3 h(x) = \left(-\frac{\omega_0}{2J}\frac{V_s}{x_{d\Sigma}'} x_3 cos(x_1)\dot{x}_1\right) -$$
$$-\frac{D}{2J}\left(\frac{-D}{2J}x_2 + \omega_0 \frac{P_m}{2J} - \frac{\omega_0}{2J}\frac{V_s}{x_{d\Sigma}'} x_3 sin(x_1)\right)\left(\frac{1}{T_d'}x_3 + \frac{1}{T_{do}}\frac{x_d - x_d'}{x_{d\Sigma}'} cos(x_1)\right) \Rightarrow \tag{3.31}$$
$$L_f^3 h(x) = \left(\frac{D}{2J}\right)^2 \dot{x}_1 - \omega_0 \frac{D}{2J}\frac{P_m}{2J} + \frac{D}{(2J)^2}\omega_0 \frac{V_s}{x_{d\Sigma}'} x_3 sin(x_1) + \frac{\omega_0}{2J}\frac{V_s}{x_{d\Sigma}'}\frac{1}{T_d'} x_3 sin(x_1) -$$
$$-\frac{\omega_0}{2J}\frac{V_s}{x_{d\Sigma}'}\frac{1}{T_{do}}\frac{x_d - x_d'}{x_{d\Sigma}'} V_s cos(x_1) sin(x_1) - \frac{\omega_0}{2H}\frac{V_s}{x_{d\Sigma}'} x_3 cos(x_1) x_2$$

Finally, in an equivalent manner one obtains

$$L_g L_f^2 h(x) = \left(\frac{\partial z_3}{\partial x_1} \ \frac{\partial z_3}{\partial x_2} \ \frac{\partial z_3}{\partial x_3}\right) \cdot \begin{pmatrix} g_1 \\ g_2 \\ g_3 \end{pmatrix} \Rightarrow L_g L_f^2 h(x) = -\frac{\omega_0}{2J}\frac{1}{T_{do}}\frac{V_s}{x_{d\Sigma}'} sin(x_1) \tag{3.32}$$

Defining the control input for the linearized system $v = L_f^3 h(x) + L_g L_f^2 h(x) u$, the state-space model can be written in the following linear canonical (Brunovsky) form

$$\begin{pmatrix} \dot{z}_1 \\ \dot{z}_2 \\ \dot{z}_3 \end{pmatrix} = \begin{pmatrix} 0 & 1 & 0 \\ 0 & 0 & 1 \\ 0 & 0 & 0 \end{pmatrix} \begin{pmatrix} z_1 \\ z_2 \\ z_3 \end{pmatrix} + \begin{pmatrix} 0 \\ 0 \\ 1 \end{pmatrix} v \tag{3.33}$$

3.2.4 Differential Flatness for Nonlinear Dynamical Systems

3.2.4.1 Overview of Differential Flatness Properties

As explained before, differential flatness is a structural property of a class of nonlinear systems, denoting that all system variables (such as state vector elements and control inputs) can be written in terms of a set of specific variables (the so-called flat outputs) and their derivatives. The following nonlinear system is considered:

$$\dot{x}(t) = f(x(t), u(t)) \tag{3.34}$$

The time is $t \in R$, the state vector is $x(t) \in R^n$ with initial conditions $x(0) = x_0$, and the input is $u(t) \in R^m$. Next, the properties of differentially flat systems are given [137, 229, 241, 352, 388, 411, 456]:

The finite dimensional system of Eq. (3.34) can be written in the general form of an ordinary differential equation (ODE), i.e. $S_i(w, \dot{w}, \ddot{w}, \dots, w^{(i)}),\ \ i = 1, 2, \dots, q$. The term w is a generic notation for the system variables (these variables are for instance the elements of the system's state vector $x(t)$ and the elements of the control input $u(t)$) while $w^{(i)}$, $i = 1, 2, \dots, q$ are the associated derivatives. Using the definitions of Chap. 2, such a system is differentially flat if there are m functions $y = (y_1, \dots, y_m)$ of the system variables and of their time-derivatives, i.e. $y_i = \phi(w, \dot{w}, \ddot{w}, \dots, w^{(\alpha_i)}),\ \ i = 1, \dots, m$ satisfying the following two conditions [137, 281, 342]:

1. There does not exist any differential relation of the form $R(y, \dot{y}, \dots, y^{(\beta)}) = 0$ which implies that the derivatives of the flat output are not coupled in the sense of an ODE, or equivalently it can be said that the flat output is differentially independent.
2. All system variables (i.e. the elements of the system's state vector w and the control input) can be expressed using only the flat output y and its time derivatives $w_i = \psi_i(y, \dot{y}, \dots, y^{(\gamma_i)}),\ \ i = 1, \dots, s$.

Next, an example is given to explain the design of a differentially flat controller for finite dimensional systems of known parameters.

Example: Flatness-based control for a nonlinear system of known parameters. Consider the following model [229]:

$$\begin{aligned} \dot{x}_1 &= x_3 - x_2 u \\ \dot{x}_2 &= -x_2 + u \\ \dot{x}_3 &= x_2 - x_1 + 2x_2(u - x_2) \end{aligned} \tag{3.35}$$

The candidate flat output is $y_1 = x_1 + \frac{x_2^2}{2}$. Thus one gets:

$$\begin{aligned} y_1 &= x_1 + \frac{x_2^2}{2} \\ y_2 &= \dot{y}_1 = (x_3 - x_2 u) + x_2(u - x_2) = x_3 - x_2^2 \\ y_3 &= \dot{y}_2 = \ddot{y}_1 = x_2 - x_1 + 2x_2(u - x_2) - 2x_2(u - x_2) = -x_1 + x_2 \\ v &= \dot{y}_3 = y_1^{(3)} = -x_3 + x_2 u - x_2 + u = -x_2 - x_3 + u(1 + x_2) \end{aligned} \tag{3.36}$$

It can be verified that property (1) holds, i.e. there does not exist any differential relation of the form $R(y, \dot{y}, \ldots, y^{(\beta)}) = 0$, and this implies that the derivatives of the flat output are not coupled. Moreover, it can be shown that property (2) also holds i.e. the components w of the system (elements of the system's state vector and control input) can be expressed using only the flat output y and its time derivatives $w_i = \psi_i(y, \dot{y}, \ldots, y^{(\gamma_i)}), i = i, \ldots, s$.

For instance to calculate x_1 with respect to $y_1, \dot{y}_1, \ddot{y}_1$ and $y_1^{(3)}$ the relation of $\ddot{y}_1$ is used, i.e.:

$$x_1^2 + 2x_1(1 + \ddot{y}_1) + \ddot{y}_1^2 - 2y_1 = 0 \tag{3.37}$$

from which two possible solutions are derived, i.e.: $x_1 = -(1 + \ddot{y}_1 - \sqrt{1 + 2(y_1 + \ddot{y}_1)})$ and $x_1 = -(1 + \ddot{y}_1 + \sqrt{1 + 2(y_1 + \ddot{y}_1)})$. Keeping the biggest of these two solutions one obtains:

$$\begin{aligned} x_1 &= -(1 + \ddot{y}_1) + \sqrt{1 + 2(y_1 + \ddot{y}_1)} \\ x_2 &= \ddot{y}_1 + x_1 \\ x_3 &= \dot{y}_1 + \ddot{y}_1^2 + 2x_1\ddot{y}_1 + x_1^2 \\ u &= \frac{y_1^3 + \ddot{y}_1^2 + \ddot{y}_1 + \dot{y}_1 + x_1 + 2x_1\ddot{y}_1 + x_1^2}{1 + x_1 + \ddot{y}_1} \end{aligned} \tag{3.38}$$

The computation of the equivalent model of the system in the linear canonical form is summarized as follows: by finding the derivatives of the flat output one gets a set of equations which can be solved with respect to the state variables and the control input of the initial state-space description of the system. First, the binomial of variable x_1 given in Eq. (3.37) is solved providing x_1 as a function of the flat output and its derivatives. Next, using the expression for x_1 and Eq. (3.38), state variable x_2 is also written as a function of the flat output and its derivatives. Finally, using the expressions for both x_1 and x_2 and Eq. (3.38), state variable x_3 is written as a function of the flat output and its derivatives. Thus one can finally express the state vector elements and the control input as function of the flat output and its derivatives, which completes the proof about differential flatness of the system.

From Eq. (3.38) it can be concluded that the initial system of Eq. (3.35) is indeed differentially flat. Using the flat output and its derivatives, the system of Eq. (3.35) can be written in Brunovsky (canonical) form:

$$\begin{pmatrix} \dot{y}_1 \\ \dot{y}_2 \\ \dot{y}_3 \end{pmatrix} = \begin{pmatrix} 0 & 1 & 0 \\ 0 & 0 & 1 \\ 0 & 0 & 0 \end{pmatrix} \begin{pmatrix} y_1 \\ y_2 \\ y_3 \end{pmatrix} + \begin{pmatrix} 0 \\ 0 \\ 1 \end{pmatrix} v \tag{3.39}$$

Therefore, a transformation of the system into a linear equivalent is obtained and then it is straightforward to design a a controller based on linear control theory. Thus, given the reference trajectory $[x_1^*, x_2^*, x_3^*]^T$ one can find the transformed reference trajectory $[y_1^*, \dot{y}_1^*, \ddot{y}_1^*]^T$ and select the appropriate control input v that succeeds tracking. Knowing v, the control u of the initial system can be found.

3.2.5 *Differential Flatness and Transformation into the Canonical Form*

The classes of systems for which the application of differential flatness theory results into a canonical form are overviewed. Transformation into the Brunovksy form can be succeeded for systems that admit static feedback linearization (i.e. a change of coordinates for both the system state variables and the system's control input). Single input differentially flat systems admit static feedback linearization and therefore can be finally written in the Brunovsky form. For flat multi-input systems necessary and sufficient conditions that allow an endogenous transformation into Brunovsky coordinates can be also stated [281, 360].

In particular, for the case of a single-input differentially flat dynamical system one has:

$$\dot{x} = f_s(x, t) + g_s(x, t)(u + \tilde{d}), \ \ x \in R^n, \ \ u \in R, \ \ \tilde{d} \in R \tag{3.40}$$

where $f_s(x, t)$, $g_s(x, t)$ are nonlinear vector fields defining the system's dynamics, u denotes the control input and $\tilde{d}$ denotes additive input disturbances. Knowing that the system of Eq. (3.40) is differentially flat, the next step is to try to write it into a Brunovsky form. The selected flat output is again denoted by y. Then, as analyzed in Sect. 3.2.4, for the state variables x_i of the system of Eq. (3.40) it holds

$$x_i = \phi_i(y, \dot{y}, \cdots, y^{(r-1)}), i = 1, \ldots, n \tag{3.41}$$

while for the control input it holds

$$u = \psi(y, \dot{y}, \ldots, y^{(r-1)}, y^{(r)}) \tag{3.42}$$

Introducing the new state variables $y_1 = y$ and $y_i = y^{(i-1)}, \ i = 2, \ldots, n$, the initial system of Eq. (3.40) can be written in the Brunovsky form:

$$\begin{pmatrix} \dot{y}_1 \\ \dot{y}_2 \\ \cdots \\ \cdots \\ \dot{y}_{n-1} \\ \dot{y}_n \end{pmatrix} = \begin{pmatrix} 0 & 1 & 0 & \cdots & 0 \\ 0 & 0 & 1 & \cdots & 0 \\ \cdots & \cdots & \cdots & \cdots & \cdots \\ \cdots & \cdots & \cdots & \cdots & \cdots \\ 0 & 0 & 0 & \cdots & 1 \\ 0 & 0 & 0 & \cdots & 0 \end{pmatrix} \begin{pmatrix} y_1 \\ y_2 \\ \cdots \\ \cdots \\ y_{n-1} \\ y_n \end{pmatrix} + \begin{pmatrix} 0 \\ 0 \\ \cdots \\ \cdots \\ 0 \\ 1 \end{pmatrix} v \tag{3.43}$$

where $v = f(x,t) + g(x,t)(u + \tilde{d})$ is the control input for the linearized model, and $\tilde{d}$ denotes additive input disturbances.

3.2.6 Differential Flatness of the Synchronous Generator

It will be shown that the dynamic model of the PMSG is a differentially flat one, i.e. it holds that all state variables and its control inputs can be written as functions of the flat output and its derivatives. Moreover, it will be shown that by expressing the elements of the state vector as functions of the flat output and its derivatives one obtains a transformation of the PMSG model into the linear canonical (Brunovsky) form.

Defining the state vector of the PMSG as $x = [\Delta\delta, \Delta\omega, E'_q]^T$ one has that

$$\begin{aligned} \dot{x}_1 &= x_2 \\ \dot{x}_2 &= -\tfrac{D}{2J}x_2 + \omega_0\tfrac{P_m}{2J} - \tfrac{\omega_0}{2J}\tfrac{V_s}{x'_{d\Sigma}}x_3 sin(x_1) \\ \dot{x}_3 &= -\tfrac{1}{T'_d}x_3 + \tfrac{1}{T_{do}}\tfrac{x_d - x'_d}{x'_{d\Sigma}}V_s cos(x_1) + \tfrac{1}{T_{do}}u \end{aligned} \tag{3.44}$$

The flat output is chosen to be $y = x_1$. Therefore it holds

$$x_1 = y \tag{3.45}$$

$$x_2 = \dot{y} \tag{3.46}$$

while from the second row of Eq. (3.44) one has

$$\ddot{y} = -\frac{D}{2J}\dot{y} + \omega_0\frac{P_m}{2J} - \frac{\omega_0}{2J}\frac{V_s x_3}{x'_{d\Sigma}}sin(y) \tag{3.47}$$

Thus, for $x_1 \neq \pm n\pi$ (where $n = 0, 1, 2, \ldots$) one obtains

$$\begin{aligned} x_3 &= \frac{\omega_0\frac{P_m}{2J} - \ddot{y} - \frac{D}{2J}\dot{y}}{\frac{\omega_0}{2J}\frac{V_s x_3}{x'_{d\Sigma}}sin(y)}, \text{ or} \\ x_3 &= f_a(y, \dot{y}, \ddot{y}) \end{aligned} \tag{3.48}$$

From the third row of Eq. (3.44) one has

$$u = T_{do}[\dot{x}_3 + \frac{1}{T'_d}x_3 \frac{1}{T_{do}} \frac{x_d - x'_d}{x'_{d\Sigma}} V_s cos(x_1)], \text{ or}$$
$$u = f_b(y, \dot{y}, \ddot{y}) \quad (3.49)$$

Therefore, all state variables and the control input of the PMSG can be written as functions of the flat output and its derivatives, and the PMSG model is a differentially flat one.

Next, the following change of variables is performed: $y_1 = y$, $y_2 = \dot{y}$, $y_3 = \ddot{y}$. It also holds

$$\dot{y}_3 = y^{(3)} = \ddot{x}_2 \Rightarrow$$
$$y^{(3)} = -\frac{D}{2J}\dot{x}_2 - \frac{\omega_0}{2H}\frac{V_s}{x_{d\Sigma}}'\dot{x}_3 sin(x_1) - \frac{\omega_0}{2J}\frac{V_s}{x_{d\Sigma}}' x_3 cos(x_1)\dot{x}_1 \quad (3.50)$$

By substituting $\dot{x}_2$ and $\dot{x}_3$ from the second and third rows of Eq. (3.44), respectively, and after intermediate operations one obtains

$$y^{(3)} = f_c(y, \dot{y}, \ddot{y}) + g_c(y, \dot{y}, \ddot{y})u \quad (3.51)$$

where

$$f_c(y, \dot{y}, \ddot{y}) = (\frac{D}{2J^2})\dot{y} - \omega_0 \frac{D}{2J}\frac{P_m}{2J} + \omega_0 \frac{D}{(2J)^2}\frac{V_s}{x'_{d\Sigma}} x_3 sin(\dot{y}) + \frac{\omega_0}{2J}\frac{V_s}{x'_{d\Sigma}}\frac{1}{T'_d} x_3 sin(y) -$$
$$- \frac{\omega_0}{2J}\frac{V_s}{x'_{d\Sigma}}\frac{1}{T_{do}}\frac{x_d - x'_d}{x'_{d\Sigma}} V_s cos(y) sin(y) - \frac{\omega_0}{2J}\frac{V_s}{x'_{d\Sigma}} x_3 cos(y)\dot{y} \quad (3.52)$$

$$g_c(y, \dot{y}, \ddot{y}) = -\frac{\omega_0}{2J}\frac{1}{T_{do}}\frac{V_s}{x'_{d\Sigma}} sin(y) \quad (3.53)$$

Thus, the system can be written in the following linear canonical (Brunovsky) form

$$\begin{pmatrix} \dot{y}_1 \\ \dot{y}_2 \\ \dot{y}_3 \end{pmatrix} = \begin{pmatrix} 0 & 1 & 0 \\ 0 & 0 & 1 \\ 0 & 0 & 0 \end{pmatrix} \begin{pmatrix} y_1 \\ y_2 \\ y_3 \end{pmatrix} + \begin{pmatrix} 0 \\ 0 \\ 1 \end{pmatrix} v \quad (3.54)$$

while the new control input for the linearized system is $v = f_c(y, \dot{y}, \ddot{y}) + g_c(y, \dot{y}, \ddot{y})u$. It can be noticed that the linearized equivalent of the system that is obtained after applying differential flatness theory, is the same with one obtained after applying the Lie algebra-based method.

The controller design for the linearized system described in Eq. (3.54) is carried out using common pole placement methods. Since the overall system dynamics is described by the differential equation

$$y^{(3)} = v \tag{3.55}$$

a suitable feedback control law that makes the flat output track a desirable setpoint $y_d(t)$ is given by

$$v = y_d^{(3)}(t) - k_1(\ddot{y}(t) - \ddot{y}_d(t)) - k_2(\dot{y}(t) - \dot{y}_d(t)) - k_3(y(t) - y_d(t))) \tag{3.56}$$

The control input that is finally applied to the PMSG is given by

$$u(t) = g_c^{-1}(t)[v(t) - f_c(t)] \tag{3.57}$$

The control law of Eq. (3.56) results in the closed-loop dynamics

$$e^{(3)}(t) + k_1\ddot{e}(t) + k_2\dot{e}(t) + k_3e(t) = 0 \tag{3.58}$$

By selecting the feedback control gains k_1,k_2 and k_3 such that the associated characteristic polynomial $p(s) = s^3 + k_1s^2 + k_2s + k_3$ is a Hurwitz one, i.e. it has poles in the left complex semi-plane one has that the tracking error converges asympotically to zero

$$lim_{t\to\infty}e(t) = 0 \Rightarrow lim_{t\to\infty}y(t) = y_d(t) \tag{3.59}$$

Since the flat output converges to the desirable setpoint and since all state variables x_1,x_2 and x_3 are expressed as functions of the flat output and its derivatives one has that the state variables also converge to the desirable setpoints and therefore efficient control of the PMSG is achieved.

Moreover, for the linearized equivalent of the system one can perform state estimation with the use of the standard Kalman Filter recursion and can compute also the state vector elements which are not directly measurable (i.e. ω and E_q').

3.2.7 *Robust State Estimation-Based Control of the PMSG*

3.2.7.1 Unknown Input Observers

The previous method for PMSG controller design was based on the assumption that the mechanical input power (torque) P_m that is applied to the PMSG can be directly measured and that this input torque is a constant or piecewise constant variable. However, this is not always the case since the measurement of P_m may be technically difficult while P_m maybe a time-varying signal (e.g. mechanical excitation due to wind, steam or water flow). In the latter case one can consider variable P_m as an external disturbance to the generator's model. The simultaneous estimation of the non-measurable elements of the PMSG state vector (i.e. ω and E_q') as well as the estimation of the disturbance term (input torque P_m) is possible if using a disturbance estimator [85, 98, 99, 169, 292].

A first type of disturbance estimator is the Unknown Input Observer. This is applied to dynamical systems of the form

$$\begin{aligned} \dot{x} &= Ax + B(u + w_e) \\ z &= Cx \end{aligned} \tag{3.60}$$

while the disturbance dynamics is given by

$$\begin{aligned} \dot{d} &= A_f d \\ w_e &= C_f d \end{aligned} \tag{3.61}$$

Then, the unknown input observer provides a state estimate of the extended state vector

$$\begin{pmatrix} \dot{\hat{x}} \\ \dot{\hat{d}} \end{pmatrix} = \begin{pmatrix} A & BC_f \\ 0 & A_f \end{pmatrix} \begin{pmatrix} \hat{x} \\ \hat{d} \end{pmatrix} + \begin{pmatrix} B \\ 0 \end{pmatrix} u + K(z - C\hat{x}) \tag{3.62}$$

In the generic case one can assume that the disturbances vector w_e varies dynamically in time. However, in several cases it suffices to assume a constant or piecewise constant disturbance $\dot{w}_e(z) = 0$ where $A_f = 0$ and $C_f = 1$. The observer's gain can be obtained through the standard Kalman Filter recursion.

3.2.7.2 Perturbation Observer

The perturbation observer is an extension of the unknown inputs observer which takes into account not only external disturbances but also parametric uncertainties. In the discrete-time form, the system dynamics is given by

$$\begin{aligned} x(k+1) &= Ax(k) + Bu(k) + w_f(k) \\ z(k) &= Cx(k) \end{aligned} \tag{3.63}$$

while the disturbance dynamics is given by

$$\begin{aligned} d(k) &= A_f d(k-1) + B_f(B^{+}(\hat{x}(k) - A\hat{x}(k-1)) - u(k-1)) \\ \hat{w}_f(k) &= C_f d(k) \\ \hat{x}(k+1) &= A\hat{x}(k) + B(u(k) + \hat{w}_f(k) + L(z(k) - C\hat{x}(k)) \end{aligned} \tag{3.64}$$

where B^{+} is the Moore-Penrose pseudo-inverse of matrix B. The unknown input can represent traditional external disturbances and model uncertainties, i.e. $w_f = w_e + \Delta A x_k + \Delta B u_k$.

3.2.7.3 Extended State Observer

The Extended State Observer uses a canonical form so the unmodelled dynamics appear at the disturbance estimation part. The system's description in the canonical form is given by

$$\begin{aligned} x_1^{(n)} &= f(x, t, u, w_f) + b_m u \\ z &= x_1 \\ x &= \left(x_1 \; \dot{x}_1 \; \ldots \; x_1^{(n-1)}\right)^T \end{aligned} \tag{3.65}$$

$$\begin{pmatrix} \dot{\hat{x}}_1 \\ \ldots \\ \dot{\hat{x}}_{n-1} \\ \dot{\hat{x}}_n \\ \dot{\hat{f}} \end{pmatrix} = \begin{pmatrix} \hat{x}_2 \\ \ldots \\ \hat{x}_n \\ \hat{f} + b_m u \\ 0 \end{pmatrix} + K(x_1 - \hat{x}_1) \tag{3.66}$$

The Extended State Observer can be also modified to take into account derivatives of the disturbance

$$\begin{aligned} x_1^{(n)} &= f(x, t, u, w_f) + b_m u \\ z &= x_1 \\ x &= \left(x_1 \; \dot{x}_1 \; \ldots \; x_1^{(n-1)}\right)^T \\ F &= \left(f \; \dot{f} \; \ldots f^{(h-1)}\right)^T \end{aligned} \tag{3.67}$$

and now the state and disturbance observer takes the form

$$\begin{pmatrix} \dot{\hat{x}}_1 \\ \ldots \\ \dot{\hat{x}}_{n-1} \\ \dot{\hat{x}}_n \\ \dot{\hat{F}}_1 \\ \ldots \\ \dot{\hat{F}}_{h-1} \\ \dot{\hat{F}}_h \end{pmatrix} = \begin{pmatrix} \hat{x}_2 \\ \ldots \\ \hat{x}_n \\ \hat{f} + b_m u \\ \hat{F}_2 \\ \ldots \\ \hat{F}_h \\ 0 \end{pmatrix}^T + K(x_1 - \hat{x}_1) \tag{3.68}$$

The latter form of the Extended State Observer described in Eq. (3.68) enables to track various types of disturbances. For example, $h = 1$ allows estimation of disturbance dynamics defined by its first order derivative, and $h = 2$ allows estimation of disturbance dynamics defined by its second order derivative.

3.2.8 Estimation of PMSG Disturbance Input with Kalman Filtering

3.2.8.1 State Estimation with the Derivative-Free Nonlinear Kalman Filter

It was shown that using differential flatness theory the initial nonlinear model of the PMSG can be written in the canonical form

$$\begin{pmatrix} \dot{y}_1 \\ \dot{y}_2 \\ \dot{y}_3 \end{pmatrix} = \begin{pmatrix} 0 & 1 & 0 \\ 0 & 0 & 1 \\ 0 & 0 & 0 \end{pmatrix} \begin{pmatrix} y_1 \\ y_2 \\ \dot{y}_3 \end{pmatrix} + \begin{pmatrix} 0 \\ 0 \\ 1 \end{pmatrix} v \tag{3.69}$$

Thus one has a linear model of the form

$$\begin{aligned} \dot{y}_f &= A_f y_f + B_f v \\ z_f &= C_f y_f \end{aligned} \tag{3.70}$$

where $y_f = [y_1, y_2, y_3]^T$ and matrices A_f,B_f,C_f are in the canonical form

$$A_f = \begin{pmatrix} 0 & 1 & 0 \\ 0 & 0 & 1 \\ 0 & 0 & 0 \end{pmatrix} B_f = \begin{pmatrix} 0 \\ 0 \\ 1 \end{pmatrix} C_f^T = \begin{pmatrix} 1 \\ 0 \\ 0 \end{pmatrix} \tag{3.71}$$

where the measurable variable $y_1 = \delta$ is associated with the turn angle of the rotor. For the previous model, and after carrying out discretization of matrices A_f, B_f and C_f with common discretization methods one can perform linear Kalman filtering using Eqs. (3.82) and (3.83). This is *Derivative-free nonlinear Kalman filtering* for the model of the generator which, unlike EKF, is performed without the need to compute Jacobian matrices and does not introduce numerical errors due to approximative linearization with Taylor series expansion [353, 354, 361].

3.2.8.2 Kalman Filter-Based Estimation of Disturbances

Up to now the mechanical input torque of the generator, which has been denoted as P_m was considered to be constant or piecewise constant while it has been also considered that it is possible to measure it. Now it is assumed that the mechanical input torque varies in time, and in that case the aggregate disturbance input exerted on the generator's model is

$$T_m = -\omega_0 \frac{D}{(2J)^2} P_m + \frac{\omega_0}{2J} \dot{P}_m \tag{3.72}$$

It is also assumed that the dynamics of the disturbance term T_m is defined by its n-th order derivative $T_m^{(n)}$. Considering now that after expressing the system's state variables as functions of the flat outputs and their derivatives the PMSG's dynamics is given by

$$\begin{aligned} y^{(3)} &= f_c(y,\dot{y},\ddot{y}) + g_c(y,\dot{y},\ddot{y})u + \frac{1}{2H}\dot{P}_m \text{ or} \\ y^{(3)} &= v - \omega_0\frac{D}{(2J)}^2 P_m + \frac{\omega_0}{2J}\dot{P}_m \text{ or} \\ y^{(3)} &= v - T_m \end{aligned} \tag{3.73}$$

where

$$\begin{aligned} f_c(y,\dot{y},\ddot{y}) = (\frac{D}{2J^2})\dot{y} - \omega_0\frac{D}{2J}\frac{P_m}{2J} + \frac{D}{(2J)^2}\omega_0\frac{V_s}{x'_{d\Sigma}}x_3 sin(\dot{y}) + \frac{\omega_0}{2J}\frac{V_s}{x'_{d\Sigma}}\frac{1}{T'_d}x_3 sin(y) - \\ -\frac{\omega_0}{2J}\frac{V_s}{x'_{d\Sigma}}\frac{1}{T_{do}}\frac{x_d - x'_d}{x'_{d\Sigma}}V_s cos(y)sin(y) - \frac{\omega_0}{2H}\frac{V_s}{x'_{d\Sigma}}x_3 cos(y)\dot{y} \end{aligned} \tag{3.74}$$

$$g_c(y,\dot{y},\ddot{y}) = -\frac{D}{2M}\frac{\omega_0}{T_{do}}\frac{V_s}{x'_{d\Sigma}}sin(y) \tag{3.75}$$

$$\begin{pmatrix}\dot{y}_1\\ \dot{y}_2\\ \dot{y}_3\end{pmatrix} = \begin{pmatrix}0&1&0\\0&0&1\\0&0&0\end{pmatrix}\begin{pmatrix}y_1\\ y_2\\ y_3\end{pmatrix} + \begin{pmatrix}0\\0\\1\end{pmatrix}\left(v - \omega_0\frac{D}{(2J)^2}P_m + \frac{\omega_0}{2J}\dot{P}_m\right) \tag{3.76}$$

Next, the state vector of the model of Eq. (8.41) is extended to include as additional state variables the disturbance input T_m. Then, in the new state-space description one has $z_1 = y_1$, $z_2 = y_2$, $z_3 = y_3$, $z_4 = T_m = -\omega_0\frac{D}{(2J)^2}P_m + \frac{\omega_0}{2J}\dot{P}_m$, $z_5 = \dot{T}_m$, and $z_6 = \ddot{T}_m$. Without loss of generality, the disturbance input dynamics is assumed to be described by its third order derivative $\dot{z}_6 = T_m^{(3)}$. Using the previous definition of state variables one has the matrix equations

$$\dot{z} = \tilde{A}{\cdot}z + \tilde{B}{\cdot}\tilde{v} \tag{3.77}$$

where the control input is

$$\tilde{v} = \left(v \; T_m^{(3)}\right)^T \tag{3.78}$$

with

$$\tilde{A}=\begin{pmatrix}0&1&0&0&0&0\\0&0&1&0&0&0\\0&0&0&1&0&0\\0&0&0&0&1&0\\0&0&0&0&0&1\\0&0&0&0&0&0\end{pmatrix}\quad \tilde{B}=\begin{pmatrix}0&0\\0&0\\1&0\\0&0\\0&0\\0&1\end{pmatrix}\quad \tilde{C}^T=\begin{pmatrix}1\\0\\0\\0\\0\\0\end{pmatrix} \tag{3.79}$$

where the measurable state variable is z_1. Since the dynamics of the disturbance input are taken to be unknown in the design of the associated disturbances' estimator one has the following dynamics:

$$\dot{z}_o=\tilde{A}_o\cdot z+\tilde{B}_o\cdot\tilde{v}+K(C_oz-C_o\hat{z}) \tag{3.80}$$

where $K\in R^{6\times 1}$ is the state estimator's gain and

$$\tilde{A}_o=\begin{pmatrix}0&1&0&0&0&0\\0&0&1&0&0&0\\0&0&0&1&0&0\\0&0&0&0&1&0\\0&0&0&0&0&1\\0&0&0&0&0&0\end{pmatrix}\quad \tilde{B}_o=\begin{pmatrix}0&0\\0&0\\1&0\\0&0\\0&0\\0&0\end{pmatrix}\quad \tilde{C}_o^T=\begin{pmatrix}1\\0\\0\\0\\0\\0\end{pmatrix} \tag{3.81}$$

It can be confirmed that the disturbance observer model of the PMSG defined in Eq. (3.81) is observable. Defining as $\tilde{A}_d$, $\tilde{B}_d$, and $\tilde{C}_d$, the discrete-time equivalents of matrices $\tilde{A}_o$, $\tilde{B}_o$ and $\tilde{C}_o$ respectively, a Derivative-free nonlinear Kalman Filter can be designed for the aforementioned representation of the system dynamics. The associated Kalman Filter-based disturbance estimator is depicted in Fig. 3.1 and is given by [363].

measurement update:

$$\begin{aligned}K(k)&=P^-(k)\tilde{C}_d^T[\tilde{C}_d\cdot P^-(k)\tilde{C}_d^T+R]^{-1}\\ \hat{x}(k)&=\hat{x}^-(k)+K(k)[z(k)-\tilde{C}_d\hat{x}^-(k)]\\ P(k)&=P^-(k)-K(k)\tilde{C}_dP^-(k)\end{aligned} \tag{3.82}$$

time update:

$$\begin{aligned}P^-(k+1)&=\tilde{A}_d(k)P(k)\tilde{A}_d^T(k)+Q(k)\\ \hat{x}^-(k+1)&=\tilde{A}_d(k)\hat{x}(k)+\tilde{B}_d(k)\tilde{v}(k)\end{aligned} \tag{3.83}$$

To compensate for the disturbance terms it suffices to use in the control loop the modified control input, which actually removes the effects of the external disturbance variable T_m.

$$v^*=v-\hat{T}_m \text{ or } v^*=v-\hat{z}_4 \tag{3.84}$$

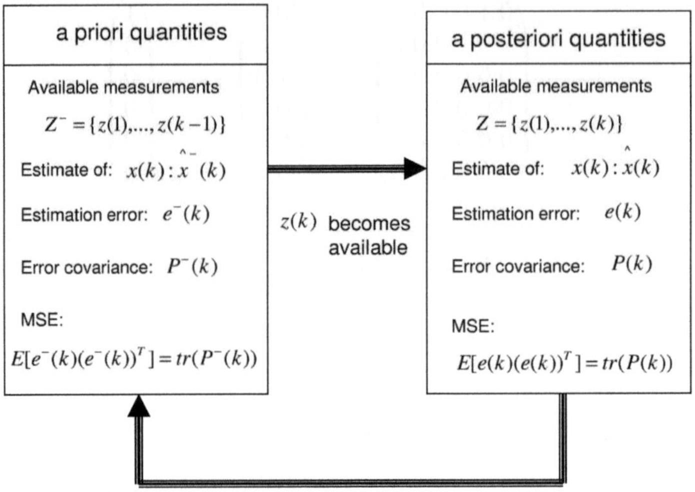

Fig. 3.1 Schematic diagram of the Kalman Filter loop

3.2.9 Simulation Experiments

To evaluate the performance of the proposed nonlinear control scheme, that uses Kalman Filtering to estimate the nonmeasurable state vector elements of the PMSG and the external disturbances, simulation experiments have been carried out. Different rotation speed setpoints had been assumed. Moreover, different input torques (mechanical input power profiles) have been considered to affect the PMSG dynamic model (Single Machine Infinite Bus model). The control loop used in the PMSG control is shown in Fig. 3.2.

First, the case of measurable piecewise constant mechanical input power (torque) was examined. It was assumed that the input torque could be measured and could be directly used in the controller's design. The derivative-free nonlinear Kalman Filter enabled the estimation of specific elements of the state vector, such as ω and E'_q. The associated results about the tracking performance of the control loop are shown in Figs. 3.3 and 3.4. The desirable setpoint is denoted by a red line, the real value of the state variable is denoted by a blue line, while the estimated value of the state variable in printed in green. The units of the PMSG state variables have been expressed in p.u. (per unit system). On can observe that the proposed sensorless control scheme for the PMSG succeeds fast and accurate convergence to the desirable setpoints.

Next, the case of PMSG operation under unknown input power (torque) was examined. The input power was considered to be a disturbance input to the PMSG model and it was assumed that its change in time was defined by the third derivative of the associated variable, i.e. $T_m^{(3)}$ where $T_m = -\omega_0 \frac{D}{(2J)^2} P_m + \frac{\omega_0}{2J} \dot{P}_m$. The disturbance dynamics was completely unknown to the controller and its identification was performed in real time by the disturbance estimator. It is shown that the derivative-free nonlinear Kalman Filter redesigned as a disturbance observer is capable of

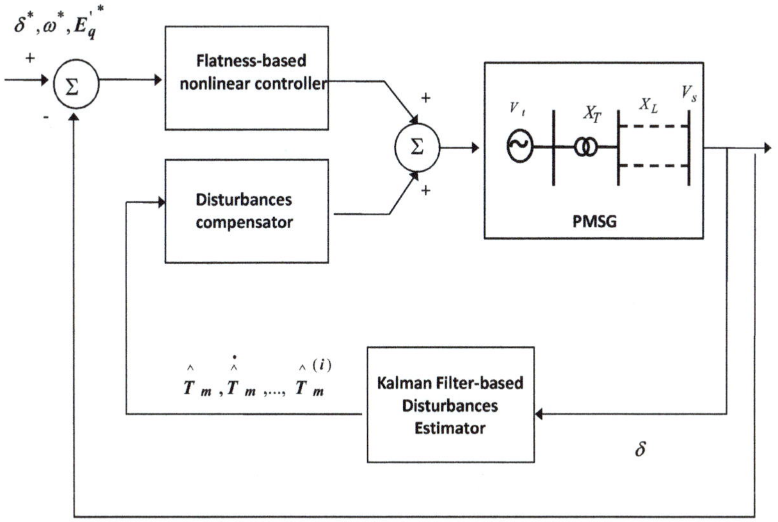

Fig. 3.2 Control loop for the PSMG comprising a flatness-based nonlinear controller and a Kalman Filter-based disturbances estimator

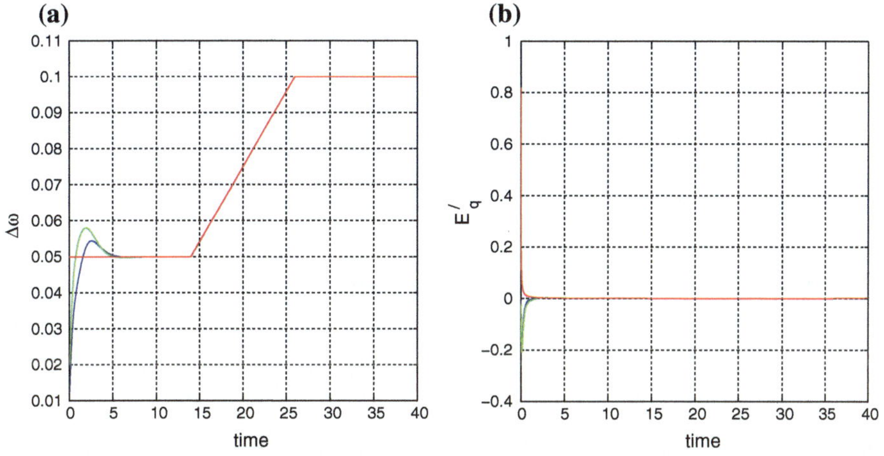

Fig. 3.3 Sensorless control of the PMSG under measurable mechanical input torque in case of speed reference setpoint 1: **a** convergence of the real and estimated values of the angular speed difference $\Delta\omega$, **b** convergence of the real and estimated values of the quadrature axis voltage E_q'

estimating simultaneously the nonmeasurable state variables of the generator (ω and E_q'), as well as the unknown disturbance input T_m. A first set of results is concerned with the tracking performance of the control loop in case of piecewise constant disturbance input, as shown in Figs. 3.5 and 3.6. The estimation of the piecewise constant disturbance input is shown in Fig. 3.7. Additionally, results about the

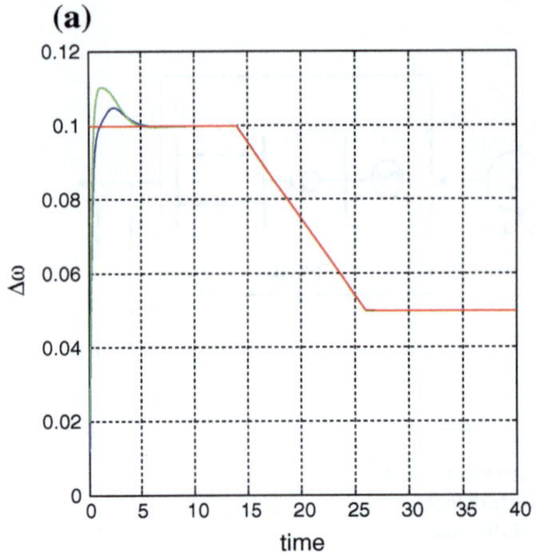

Fig. 3.4 Sensorless control of the PMSG under measurable mechanical input torque in case of speed reference setpoint 2: **a** convergence of the real and estimated values of the angular speed difference $\Delta\omega$, **b** convergence of the real and estimated values of the quadrature axis voltage E_q'

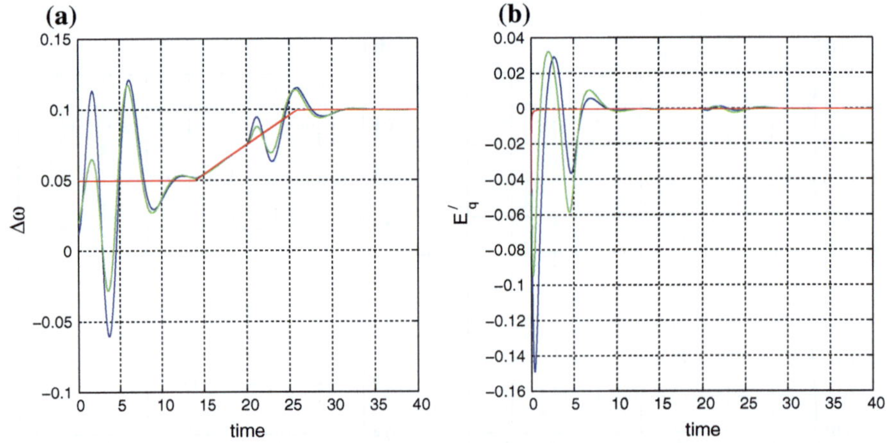

Fig. 3.5 Sensorless control of the PMSG under non-measurable (piecewise constant) mechanical input torque in case of speed reference setpoint 1: **a** convergence of the real and estimated values of the angular speed difference $\Delta\omega$, **b** convergence of the real and estimated values of the quadrature axis voltage E_q'

tracking performance of the control loop in case of time-varying disturbance input are shown in Figs. 3.8 and 3.9, while the estimation of the time-varying disturbance input is shown in Fig. 3.10. The units of the PMSG state variables have been expressed again in p.u. (per unit system).

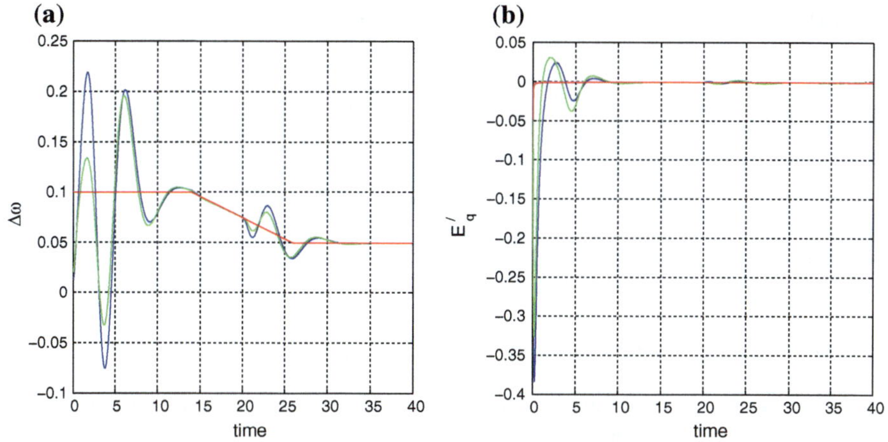

Fig. 3.6 Sensorless control of the PMSG under non-measurable (piecewise constant) mechanical input torque in case of speed reference setpoint 2: **a** convergence of the real and estimated values of the angular speed difference $\Delta\omega$, **b** convergence of the real and estimated values of the quadrature axis voltage E_q'

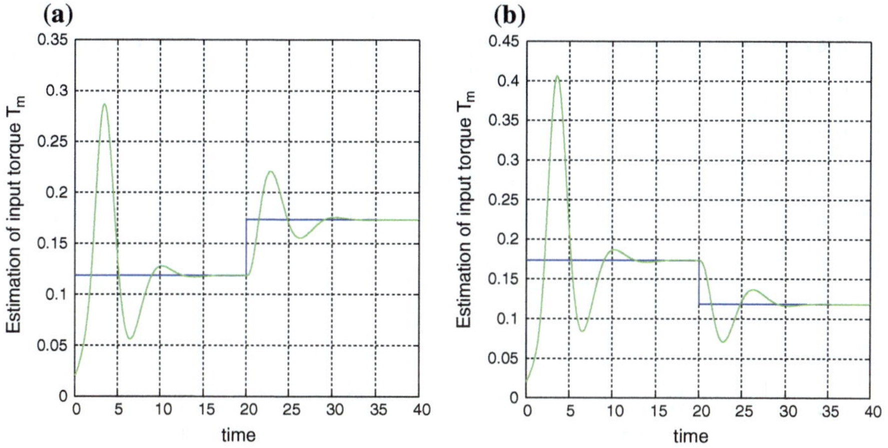

Fig. 3.7 Estimation of the non-measurable (piecewise constant) input torque in sensorless control of the PMSG through the processing of rotor angle measurements: **a** in case of speed reference setpoint 1 **b** in case of speed reference setpoint 2

The simulation experiments have confirmed that the proposed state estimation-based control scheme not only enables implementation of PMSG control through the measurement of a small number of variables (e.g. of only the rotor's turn angle) but also improves the robustness of the PMSG control loop in case of varying speed setpoints and varying mechanical input torque.

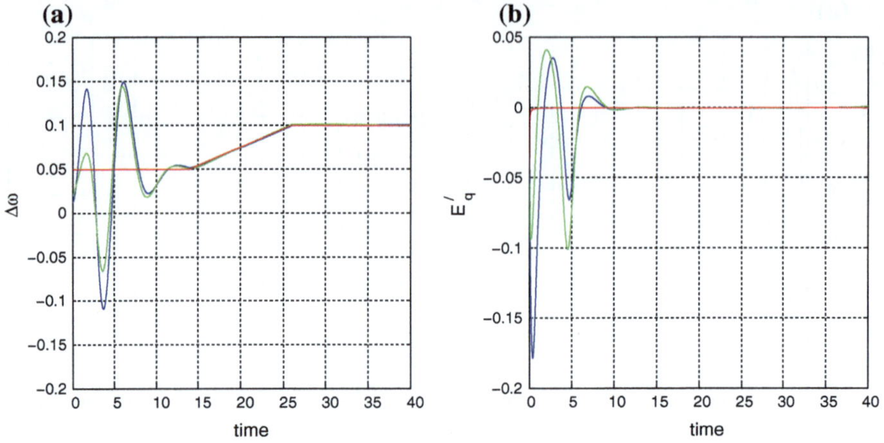

Fig. 3.8 Sensorless control of the PMSG under non-measurable (time varying) mechanical input torque in case of speed reference setpoint 1: **a** convergence of the real and estimated values of the angular speed difference $\Delta\omega$, **b** convergence of the real and estimated values of the quadrature axis voltage $E_q^{\prime}$

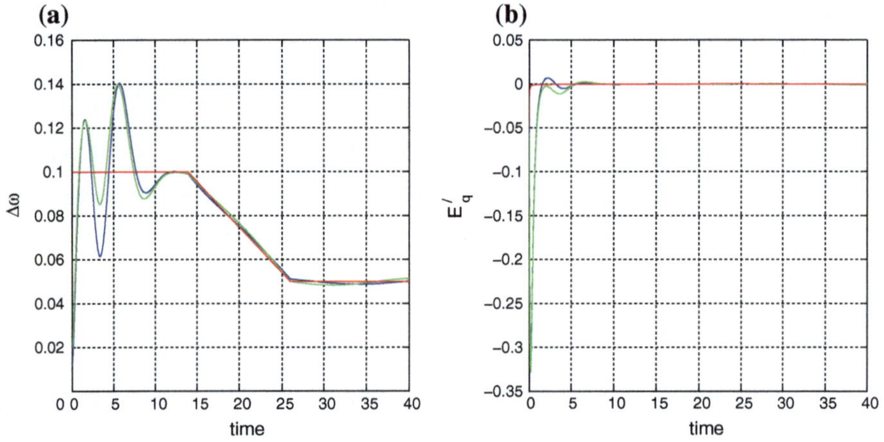

Fig. 3.9 Sensorless control of the PMSG under non-measurable (time varying) mechanical input torque in case of speed reference setpoint 2: **a** convergence of the real and estimated values of the angular speed difference $\Delta\omega$, **b** convergence of the real and estimated values of the quadrature axis voltage $E_q^{\prime}$

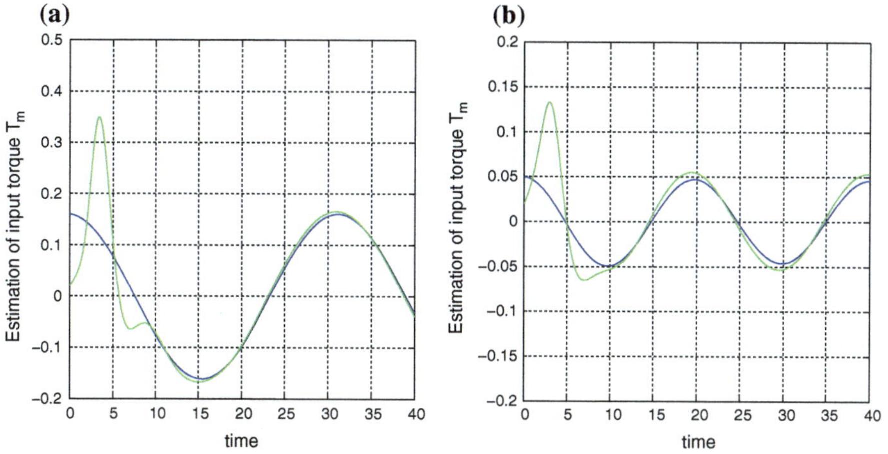

Fig. 3.10 Estimation of the non-measurable (time varying) input torque in sensorless control of the PMSG: **a** in case of speed reference setpoint 1, **b** in case of speed reference setpoint 2

3.3 Flatness-Based Control of Synchronous Generators in Successive Loops

3.3.1 Outline

Next, a different implementation of flatness-based control is proposed for the model of the synchronous generator. This approach comes to complement flatness-based control implemented through transformation of the system's dynamics into the canonical form [352, 366, 375]. As noted, a system is considered to be differentially flat if all its state variables and its control inputs can be expressed as functions of one single algebraic variable which is the so-called flat output, and also as functions of the flat-output's derivatives [137, 240, 241, 280, 456]. The differential flatness property enables the transformation of the nonlinear system's dynamics in the linear canonical form and the design of a state feedback controller through the application of pole placement techniques in the linearized equivalent model of the system [57, 412, 516]. In this section, a different approach is developed for controller design in nonlinear dynamical systems which exhibit the differential flatness property. The method makes use of the initial nonlinear model of the generator and of its decomposition into a set of nonlinear subsystems for which the differential flatness property holds.

The proposed control method is directly applicable to nonlinear systems of the so-called triangular form, or to system's which can be transformed into such a form. Each row of the state-space model of the nonlinear system is considered to be a subsystem, which satisfies the differential flatness properties. For each subsystem which is associated with a row of the state-space model a virtual control input is computed, capable of inverting the subsystem's dynamics and of eliminating the subsystem's tracking error. The control input that is actually applied to the nonlinear generator is found from the last row of the state-space description. This control input incorporates in a recursive manner all virtual control inputs which were computed for the individual subsystems associated with the previous rows of the state-space equation. The control input that should be applied to the synchronous generator model so as to assure that all its state vector elements will converge to the desirable setpoints, is obtained at each iteration of the control algorithm, by tracing the rows of the state-space model backwards. The performance of the proposed control method is evaluated through simulation experiments, in the case of nonlinear robotic models and in the case of nonlinear power generators.

3.3.2 Flatness-Based Control Through Transformation into the Canonical Form

Example: To remind of flatness-based control through linearization transformations, the following state-space equation of a nonlinear electric machine is considered [354]:

$$\begin{aligned} \dot{x}_1 &= x_2 \\ \dot{x}_2 &= k_1 x_2 + k_2 x_3 + k_3 x_3^2 + k_4 T_1 \\ \dot{x}_3 &= k_5 x_2 + k_6 x_2 x_3 + k_7 x_3 + k_8 u \end{aligned} \tag{3.85}$$

where, $x_1 = \theta$ is the rotor's turn angle, $x_2 = \dot{\theta}$ is the rotor's angular speed and $x_3 = I$ is the armature's current. Moreover, T_1 the load torque, u is the field voltage and k_i are coefficients. From the second row of Eq. (3.85) one obtains,

$$\begin{aligned} \ddot{x}_2 &= k_1 \dot{x}_2 + k_2 \dot{x}_3 + 2k_3 x_3 \dot{x}_3 \Rightarrow \\ \ddot{x}_2 &= k_1 \dot{x}_2 + k_2 \dot{x}_3 + 2k_3 k_5 x_2 x_3 + 2k_3 k_6 x_2 x_3^2 + \\ &\quad + 2k_3 k_7 x_3^2 + 2k_3 k_8 x_3 u \end{aligned} \tag{3.86}$$

Thus the input-output relation can be written as

$$\ddot{x}_2 = \bar{f}(x) + \bar{g}(x) u \tag{3.87}$$

where, $\bar{f}(x) = k_1 \dot{x}_2 + k_2 \dot{x}_3 + 2k_3 k_5 x_2 x_3 + 2k_3 k_6 x_2 x_3^2 + 2k_3 k_7 x_3^2$, and $\bar{g}(x) = 2k_3 k_8 x_3$. Taking $k_1 = 0$ and considering the torque T_1 as external disturbance, the nonlinear DC motor model of Eq. (3.85) becomes

$$\begin{aligned} \dot{x}_1 &= x_2 \\ \dot{x}_2 &= k_2 x_3 + k_3 x_3^2 \\ \dot{x}_3 &= k_5 x_2 + k_6 x_2 x_3 + k_7 x_3 + k_8 u \end{aligned} \tag{3.88}$$

Then, setting the flat output to be $y = x_1$ one can see that all state variables x_i, $i = 1, 2, 3$ and the control input u can be expressed as functions of this output and its derivatives. Indeed it holds

$$\begin{aligned} x_1 &= y \\ x_2 &= \dot{y} \\ x_3 &= \frac{-k_2 + \sqrt{|k_2^2 + 4k_3\ddot{y}|}}{2k_3} \\ u &= \frac{1}{\bar{g}(x)}[y^{(3)} - \bar{f}(x)] \end{aligned} \tag{3.89}$$

The aforementioned system of Eq. (3.88) can be written in the Brunovsky (canonical) form:

$$\begin{pmatrix} \dot{y}_1 \\ \dot{y}_2 \\ \dot{y}_3 \end{pmatrix} = \begin{pmatrix} 0 & 1 & 0 \\ 0 & 0 & 1 \\ 0 & 0 & 0 \end{pmatrix} \begin{pmatrix} y_1 \\ y_2 \\ y_3 \end{pmatrix} + \begin{pmatrix} 0 \\ 0 \\ 1 \end{pmatrix} v \tag{3.90}$$

where $v = \bar{f}(x,t) + \bar{g}(x,t)u$. Using a control input of the form $u = \bar{g}(x,t)^{-1}$ $[v - \bar{f}(x,t)]$ it is possible to make the electric machine's angle track any desirable setpoint.

Therefore, a transformation of the system into a linear equivalent is obtained and then it is straightforward to design a a controller based on linear control theory. Thus, given the reference trajectory $[x_1^*, x_2^*, x_3^*]^T$ one can find the transformed reference trajectory $[y_1^*, \dot{y}_1^*, \ddot{y}_1^*]^T$ and select the appropriate control input v that succeeds tracking. Knowing v, the control u of the initial system can be found.

For the linearized description of the system given above, the state feedback controller is $v = y_d^{(3)} - k_{f_1}(\ddot{y} - \ddot{y}_d) - k_{f_2}(\dot{y} - \dot{y}_d) - k_{f_3}(y - y_d)$. This results in tracking error dynamics of the form $e^{(3)}(t) + k_{f_1}\ddot{e}(t) + k_{f_2}\dot{e}(t) + k_{f_3}e(t) = 0$. By selecting the feedback gains k_{f_i}, $i = 1, 2, 3$ such that the previous characteristic polynomial is a Hurwitz one, it assured that $lim_{t\to\infty}e(t) = 0$.

3.3.3 A New Approach to Flatness-Based Control for Nonlinear Power Systems

The following nonlinear power system, of the synchronous generator, is now examined:

$$\begin{aligned} \dot{x} &= f(x) + g(x)u \quad x \in R^n \quad u \in R \\ y &= h(x) \end{aligned} \tag{3.91}$$

Moreover, it is considered that the power system can be written in the so-called triangular form (this is also succeeded through transformation to the canonical form):

$$
\begin{aligned}
\dot{x}_1 &= f_1(x_1) + g_1(x_1)x_2 \\
\dot{x}_2 &= f_2(x_1, x_2) + g_2(x_1, x_2)x_3 \\
\dot{x}_3 &= f_3(x_1, x_2, x_3) + g_3(x_1, x_2, x_3)x_4 \\
&\cdots \\
\dot{x}_i &= f_i(x_1, x_2, \ldots, x_i) + g_i(x_1, x_2, \ldots, x_i)x_{i+1} \\
&\cdots \\
\dot{x}_{n-1} &= f_{n-1}(x_1, x_2, \ldots, x_{n-1}) + g_{n-1}(x_1, x_2, \ldots, x_{n-1})x_n \\
\dot{x}_n &= f_n(x_1, x_2, \ldots, x_n) + g_n(x_1, x_2, \ldots, x_n)u
\end{aligned}
\tag{3.92}
$$

The following virtual control inputs $\alpha_i = x_{i+1}$ are defined per row of the state-space model of Eq. (3.92)

$$
\begin{aligned}
\dot{x}_1 &= f_1(x_1) + g_1(x_1)\alpha_1 \\
\dot{x}_2 &= f_2(x_1, x_2) + g_2(x_1, x_2)\alpha_2 \\
\dot{x}_3 &= f_3(x_1, x_2, x_3) + g_3(x_1, x_2, x_3)\alpha_3 \\
&\cdots \\
\dot{x}_i &= f_i(x_1, x_2, \ldots, x_i) + g_i(x_1, x_2, \ldots, x_i)\alpha_i \\
&\cdots \\
\dot{x}_{n-1} &= f_{n-1}(x_1, x_2, \ldots, x_{n-1}) + g_{n-1}(x_1, x_2, \ldots, x_{n-1})\alpha_{n-1} \\
\dot{x}_n &= f_n(x_1, x_2, \ldots, x_n) + g_n(x_1, x_2, \ldots, x_n)u
\end{aligned}
\tag{3.93}
$$

The system of Eq. (3.93) is a differentially flat one. It is considered that $y = x_1$ is the flat output of system. It can be easily shown that each virtual control input $\alpha_i = x_{i+1}$, $i = 1, 2, \ldots$ can be expressed as a function of the flat output and its derivatives, since it holds

$$
\alpha_i = \frac{1}{g_i(x_1, x_2, \ldots, x_i)}(\dot{x}_i - f(x_1, x_2, \ldots, x_i))
\tag{3.94}
$$

For $i = 1$ one has

$$
\alpha_1 = \frac{1}{g_1(x_1)}(\dot{x}_1 - f(x_1))
\tag{3.95}
$$

which means that α_1 is a function of the flat output and its derivative. For $i = 2$ one has

$$
\alpha_2 = \frac{1}{g_2(x_1, x_2)}(\dot{x}_2 - f_2(x_1, x_2))
\tag{3.96}
$$

which means that $\alpha_2 = x_3$ is a function of the flat output $y = x_1$ and its derivatives. Continuing in a similar manner one has that $\alpha_{n-1} = x_n$ and consequently $\alpha_n = u$ is also a function of the flat output $y = x_1$ and its derivatives. According to the above, one has a nonlinear dynamical system in which, all its state variables and the control input can be written as functions of the flat output and its derivatives. Therefore, such a system is differentially flat.

Additionally, by considering each row of the model of Eq. (3.93), one has a set of n subsystems of the form

$$\dot{x}_i = f_i(x_1, x_2, \ldots, x_i) + g_i(x_1, x_2, \ldots, x_i)\alpha_i \tag{3.97}$$

where each subsystem describes the dynamics of the single state variable x_i. For each one of these subsystems one can consider the state variable x_i as the flat output. Obviously, the virtual control input α_i is a function of this flat output and its derivatives. Therefore, each local subsystem is also differentially flat.

Next, one can compute the virtual inputs which are applied to each subsystem. For the first subsystem, which is associated with the first row of Eq. (3.92), and by defining $z_i = x_i - x_i^* = x_1 - \alpha_{i-1}$, the virtual control input is given by

$$\begin{aligned} \alpha_1 = x_2^* &= \frac{1}{g_1(x_1)}(\dot{x}_1^* - f(x_1) - K_1^1(x_1 - x_1^*)) \Rightarrow \\ \alpha_1 = x_2^* &= \frac{1}{g_1(x_1)}(\dot{x}_1^* - f(x_1) - K_1^1 z_1) \end{aligned} \tag{3.98}$$

From the second row of Eq. (3.92), and using that $z_2 = x_2 - x_2^* = x_2 - \alpha_1$ one has

$$\begin{aligned} \alpha_2 = x_3^* &= \frac{1}{g_2(x_1, x_2)}(\dot{x}_2^* - f_2(x_1, x_2) - K_1^2(x_2 - x_2^*)) \Rightarrow \\ \alpha_2 = x_3^* &= \frac{1}{g_2(x_1, x_2)}(\dot{\alpha}_1 - f_2(x_1, x_2) - K_1^2 z_2) \end{aligned} \tag{3.99}$$

From the third row of Eq. (3.92), and using that $z_3 = x_3 - x_3^* = x_3 - \alpha_2$ one has

$$\begin{aligned} \alpha_3 = x_4^* &= \frac{1}{g_3(x_1, x_2, x_3)}(\dot{x}_3^* - f_3(x_1, x_2, x_3) - K_1^3(x_3 - x_3^*)) \Rightarrow \\ \alpha_3 = x_4^* &= \frac{1}{g_3(x_1, x_2, x_3)}(\dot{\alpha}_2 - f_3(x_1, x_2, x_3) - K_1^3 z_3) \end{aligned} \tag{3.100}$$

Continuing in a similar manner and from the i-th row of the state-space description of the system given in Eq. (3.92), and while also using that $z_i = x_i - x_i^* = x_i - \alpha_{i-1}$ one obtains

$$\begin{aligned} \alpha_i = x_{i+1}^* &= \frac{1}{g_i(x_1, x_2, \ldots, x_i)}(\dot{x}_i^* - f_i(x_1, x_2, \ldots, x_i) - K_1^i(x_i - x_i^*)) \Rightarrow \\ \alpha_i = x_{i+1}^* &= \frac{1}{g_i(x_1, x_2, \ldots, x_i)}(\dot{\alpha}_{i-1} - f_i(x_1, x_2, \ldots, x_3) - K_1^i z_i) \end{aligned} \tag{3.101}$$

Equivalently, from the $n-1$-th row of the state-space model of Eq. (3.92) and using that $z_{n-1} = x_{n-1} - x_{n-1}^* = x_{n-1} - \alpha_{n-2}$ one has

$$\alpha_{n-1}=x_n^*=\frac{1}{g_{n-1}(x_1,x_2,\dots,x_{n-1})}(\dot{x}_{n-1}^*-f_{n-1}(x_1,x_2,\dots,x_{n-1})-K_1^{n-1}(x_{n-1}-x_{n-1}^*))\Rightarrow$$
$$\alpha_{n-1}=x_n^*=\frac{1}{g_{n-1}(x_1,x_2,\dots,x_{n-1})}(\dot{\alpha}_{n-2}-f_{n-1}(x_1,x_2,\dots,x_{n-1})-K_1^{n-1}z_{n-1}) \tag{3.102}$$

Finally, from the n-th row of the state-space model of Eq. (3.92) and using that $z_n=x_n-x_n^*=x_n-\alpha_{n-1}$ one has

$$\alpha_n=u=\frac{1}{g_n(x_1,x_2,\dots,x_n)}(\dot{x}_n^*-f_n(x_1,x_2,\dots,x_n)-K_1^n(x_n-x_n^*))\Rightarrow$$
$$\alpha_n=u=\frac{1}{g_n(x_1,x_2,\dots,x_n)}(\dot{\alpha}_{n-1}-f_n(x_1,x_2,\dots,x_n)-K_1^n z_n) \tag{3.103}$$

The computation of the control input u that should be actually applied to the nonlinear system is performed in a recursive manner by processing backwards the virtual control inputs described in Eqs. (3.98)–(3.103).

Thus, from the last row of the state-space description the control input that is actually applied to the nonlinear system is found. This control input contains recursively all virtual control inputs which were computed for the individual subsystems associated with the previous rows of the state-space equation. Thus, by tracing the rows of the state-space model backwards, at each iteration of the control algorithm, one can finally obtain the control input that should be applied to the nonlinear system so as to assure that all its state vector elements will converge to the desirable setpoints.

3.3.4 Closed-Loop Dynamics

By substituting Eq. (3.103) into the last row of the state space model of Eq. (3.92), and using the definition $x_n-a_{n-1}=z_n$, one obtains:

$$\begin{gathered}\dot{x}_n=\dot{a}_{n-1}-K_1^n(x_n-\alpha_{n-1})\Rightarrow\\(\dot{x}_n-\dot{a}_{n-1})+K_1^n(x_n-\alpha_{n-1})=0\Rightarrow\\\dot{z}_n+K_1^n z_n=0\end{gathered} \tag{3.104}$$

By substituting Eq. (3.102) into the last row of the state space model of Eq. (3.92), and using the definition $x_{n-1}-a_{n-2}=z_{n-1}$, one obtains:

$$\begin{gathered}\dot{x}_{n-1}=\dot{a}_{n-2}-K_1^{n-1}(x_{n-1}-\alpha_{n-2})\Rightarrow\\(\dot{x}_{n-1}-\dot{a}_{n-2})+K_1^{n-1}(x_{n-1}-\alpha_{n-2})=0\Rightarrow\\\dot{z}_{n-1}+K_1^{n-1}z_{n-1}=0\end{gathered} \tag{3.105}$$

By substituting Eq. (3.101) into the last row of the state space model of Eq. (3.92), and using the definition $x_i-a_{i-1}=z_i$, one obtains:

$$\begin{gathered}\dot{x}_i = \dot{a}_{i-1} - K_1^{n-1}(x_i - \alpha_{i-1}) \Rightarrow \\ (\dot{x}_i - \dot{a}_{i-1}) + K_1^i(x_i - \alpha_{i-1}) = 0 \Rightarrow \\ \dot{z}_i + K_1^i z_i = 0\end{gathered} \tag{3.106}$$

while continuing backwards and by substituting Eq. (3.99) into the second row of the state space model of Eq. (3.92), and using the definition $x_2 - a_1 = z_2$, one gets:

$$\begin{gathered}\dot{x}_2 = \dot{a}_1 - K_1^2(x_2 - \alpha_1) \Rightarrow \\ (\dot{x}_2 - \dot{a}_1) + K_1^2(x_2 - \alpha_1) = 0 \Rightarrow \\ \dot{z}_2 + K_1^2 z_2 = 0\end{gathered} \tag{3.107}$$

Finally, by substituting Eq. (3.98) into the first row of the state space model of Eq. (3.92), one has:

$$\begin{gathered}\dot{x}_1 = \dot{x}_1 - K_1^1(x_1 - x_1^d) \Rightarrow \\ (\dot{x}_1 - \dot{x}_1^d) + K_1^1(x_1 - x_1^d) = 0 \Rightarrow \\ \dot{z}_1 + K_1^1 z_1 = 0\end{gathered} \tag{3.108}$$

Therefore, after the application of the feedback control law, the closed-loop dynamics becomes $\dot{z}_1 + K_1^1 z_1 = 0$, $\dot{z}_2 + K_1^2 z_2 = 0$, $\dot{z}_3 + K_1^3 z_3 = 0$, ..., $\dot{z}_i + K_1^i z_i = 0$, ..., $\dot{z}_{n-1} + K_1^{n-1} z_{n-1} = 0$, $\dot{z}_n + K_1^n z_n = 0$.

In matrix form, the closed-loop dynamics is written as

$$\begin{pmatrix} \dot{z}_1 \\ \dot{z}_2 \\ \dot{z}_3 \\ \cdots \\ \dot{z}_i \\ \cdots \\ \dot{z}_{n-1} \\ \dot{z}_n \end{pmatrix} = \begin{pmatrix} -K_1^1 \, 0 \, 0 \cdots 0 \cdots 0 \, 0 \\ 0 \, -K_1^2 \, 0 \cdots 0 \cdots 0 \, 0 \\ 0 \, 0 \, -K_1^3 \cdots 0 \cdots 0 \, 0 \\ \cdots \cdots \cdots \cdots \cdots \\ 0 \, 0 \, 0 \cdots -K_1^i \cdots 0 \, 0 \\ \cdots \cdots \cdots \cdots \cdots \\ 0 \, 0 \, 0 \cdots 0 \cdots -K_1^{n-1} \, 0 \\ 0 \, 0 \, 0 \cdots 0 \cdots 0 \, -K_1^n \end{pmatrix} \begin{pmatrix} z_1 \\ z_2 \\ z_3 \\ \cdots \\ z_i \\ \cdots \\ z_{n-1} \\ z_n \end{pmatrix} \tag{3.109}$$

or equivalently

$$\dot{Z} = KZ \tag{3.110}$$

By selecting the eigenvalues of matrix K to be in the left complex semiplane, one has that

$$lim_{t\to\infty} Z = 0_{n\times 1} \tag{3.111}$$

which also implies that $lim_{t\to\infty}x_1 = x_1^d$, $lim_{t\to\infty}x_2 = \alpha_1 = x_2^d$, $lim_{t\to\infty}x_3 = \alpha_2 = x_3^d$, ..., $lim_{t\to\infty}x_i = \alpha_{i-1} = x_i^d$, ..., $lim_{t\to\infty}x_{n-1} = \alpha_{n-2} = x_{n-1}^d$, and $lim_{t\to\infty}x_n = \alpha_{n-1} = x_n^d$.

To prove asymptotic stability for the proposed control scheme the following Lyapunov function can be defined

$$V = \sum_{i=1}^{N} \frac{1}{2} z_i^2 \tag{3.112}$$

The time derivative of the aforementioned Lyapunov function is

$$\dot{V} = \sum_{i=1}^{N} z_i \dot{z}_i \Rightarrow \dot{V} = -\sum_{i=1}^{N} K_1^i z_i^2 \Rightarrow \dot{V} < 0 \tag{3.113}$$

By selecting the feedback control gains K_1^i, $i = 1, \ldots, n$ to be $K_1^i > 0$, the asymptotic stability of the control loop is assured.

3.3.5 Comparison to Backstepping Control

Backstepping control can be applied to nonlinear dynamical systems in the triangular form. A backstepping control law can be derived for systems of the form [215, 352, 353]:

$$\begin{aligned} \dot{x} &= f(x) + g(x)u \\ y &= h(x) \end{aligned} \tag{3.114}$$

for which holds

$$\dot{y} = \frac{\partial h(x)}{\partial x}\dot{x} = \frac{\partial h(x)}{\partial x}[f(x) + g(x)u] = L_f h(x) + L_g h(x)u \tag{3.115}$$

where the Lie derivatives are defined as

$$L_f h(x) = \frac{\partial}{\partial x} h(x) f(x), \ L_g h(x) = \frac{\partial}{\partial x} h(x) g(x) \tag{3.116}$$

The system of Eq. (3.114) can be written in cascading form

$$\begin{aligned} \dot{x}_1 &= f_1(x_1) + g_1(x_1)x_2 \\ \dot{x}_2 &= f_2(x_1, x_2) + g_2(x_1, x_2)x_3 \\ \dot{x}_3 &= f_3(x_1, x_2, x_3) + g_3(x_1, x_2, x_3)x_4 \\ &\cdots \cdots \\ &\cdots \cdots \\ \dot{x}_{n_1} &= f_{n-1}(x_1, x_2, \ldots, x_{n-1}) + g_{n-1}(x_1, x_2, \ldots, x_{n-1})x_n \\ \dot{x}_n &= f_n(x_1, x_2, \ldots, x_n) + g_{n-1}(x_1, x_2, \ldots, x_n)u \\ y &= h(x_1) \end{aligned} \tag{3.117}$$

Then, the n-th order backstepping SISO controller is given by the recursive relation

$$\begin{aligned}
\alpha_1 &= \frac{1}{L_{g_1}h(x_1)}[\dot{y}_d - L_{f_1}h(x_1) - k_1z_1 - n_1(z_1)z_1] \\
\alpha_2 &= \frac{1}{L_{g_2}h(x_1,x_2)}[\dot{\alpha}_1 - f_2(x_1,x_2) - L_{g_1}h(x_1)z_1 - k_2z_2 - n_2(z_2)z_2] \\
&\cdots \\
\alpha_i &= \frac{1}{L_{g_i}h(x_1,x_2,\dots,x_i)}[\dot{\alpha}_{i-1} - f_i(x_1,\dots,x_i) - g_{i-1}(x_1,x_2,\dots,x_{i-1})z_{i-1} - \\
&\quad -k_iz_1 - n_i(z_i)z_i] \\
&\cdots \\
\alpha_n &= \frac{1}{L_{g_n}h(x_1,x_2,\dots,x_n)}[\dot{\alpha}_{n-1} - f_n(x_1,\dots,x_n) - g_{n-1}(x_1,x_2,\dots,x_{n-2})z_{n-1} - \\
&\quad -k_nz_n - n_n(z_n)z_n] \\
u &= \alpha_n
\end{aligned} \tag{3.118}$$

with $z_1 = h(x_1) - y_d$ and $z_i = x_i - \alpha_{i-1}$. Such a backstepping controller results in closed-loop dynamics given by $\dot{z} = -K(z)z + S(x)z$, with

$$K(z) = \begin{pmatrix} k_1 + n_1(z_1) & 0 & \cdots & 0 \\ 0 & k_2 + n_2(z_2) & \cdots & 0 \\ \cdots & \cdots & \cdots & \cdots \\ 0 & 0 & \cdots & k_n + n_n(z_n) \end{pmatrix} \tag{3.119}$$

$$S(x) = \begin{pmatrix} 0 & L_{g_1}(x_1) & 0 & \cdots & 0 & 0 & 0 \\ -L_{g_1}(x_1) & 0 & g_2(x_1,x_2) & 0 & 0 & 0 & \\ 0 & -g_2(x_1,x_2) & 0 & \cdots & 0 & 0 & 0 \\ \cdots & \cdots & \cdots & \cdots & \cdots & \cdots & \cdots \\ 0 & 0 & 0 & \cdots & 0 & g_{n-1}(x_1,\cdots,x_{n-1}) & 0 \\ 0 & 0 & 0 & \cdots & -g_{n-1}(x_1,\cdots,x_{n-1}) & 0 & 0 \\ 0 & 0 & 0 & \cdots & 0 & -g_n(x_1,\cdots,x_n & 0) \end{pmatrix} \tag{3.120}$$

From the above it can be noticed that the so-called backstepping control, which is based on the recursive computation of the control signal of the system after applying virtual control inputs to the individual rows of the state-space model, can be completely substituted by the proposed flatness-based control method. A backstepping control law can be derived for systems of the triangular form [215, 337, 353]. However, as it was previously analyzed, by showing that each row of the state-space model stands for a subsystem that satisfies differential flatness properties one can apply effectively to each subsystem the controller design stages found in input-output linearizing flatness-based control methods.

3.3.6 Simulation Tests

The above noted flatness-based control method is applied to the model of a Permanent Magnet Synchronous Generator (PMSG). The single-machine infinite-bus model of the PMSG (Fig. 3.11) is described by a nonlinear state space model of the form

$$\dot{x} = f(x) + g(x)u \tag{3.121}$$

where the state vector x is defined as $x = \left(\Delta\delta \ \Delta\omega \ E_q'\right)^T$. The vector fields $f(x)$ and $g(x)$ are defined as

$$f(x) = \begin{pmatrix} \omega - \omega_0 \\ -\frac{D}{2J}(\omega - \omega_0) + \omega_0\frac{P_m}{2J} - \omega_0\frac{1}{2J}\frac{V_s E_q'}{x_{d\sigma}'}sin(\Delta\delta) \\ -\frac{1}{T_d'}E_q' + \frac{1}{T_{do}}\frac{x_d - x_d'}{x_{d\sigma}'}V_s cos(\Delta\delta) \end{pmatrix} \tag{3.122}$$

$$g(x) = \left(0 \ 0 \ \frac{1}{T_{do}}\right)^T \tag{3.123}$$

with control input $u = E_f$ being the field voltage (equivalent voltage in the excitation coil) and measurable output the rotation speed of the rotor

$$y = h(x) = \delta - \delta_0 \tag{3.124}$$

The obtained results are depicted in Figs. 3.12 and 3.13. It can be noticed that the proposed control method succeeds fast and accurate tracking of the reference setpoints for the state variables of the generator. The associated variations of the control input remain also smooth. This comes to confirm the previously given theoretical analysis about the stability and convergence of the proposed control scheme and to show its potential application to electric power generation systems.

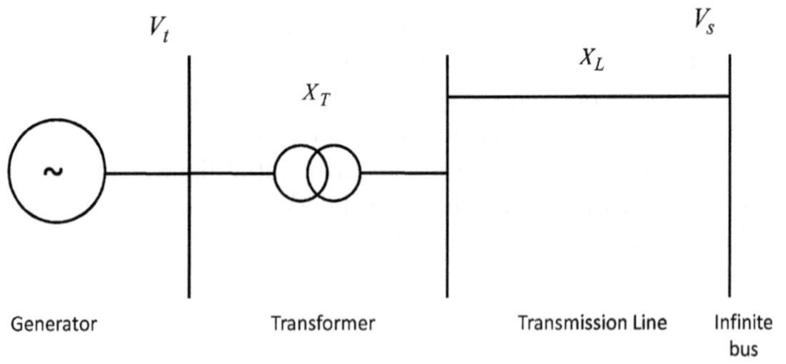

Fig. 3.11 Model of the permanent magnet synchronous generator connected to an infinite bus

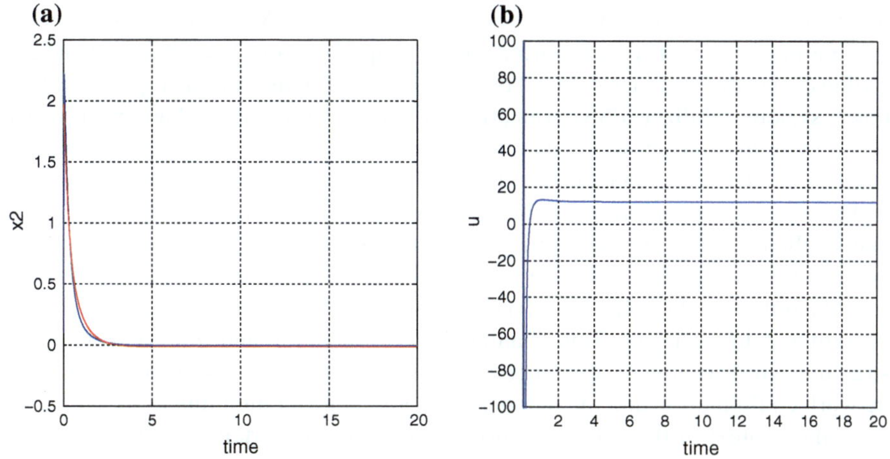

Fig. 3.12 **a** The deviation $\Delta\omega$ of the PMSG rotor angular velocity from the synchronous speed converges to 0, **b** associated control input E_f

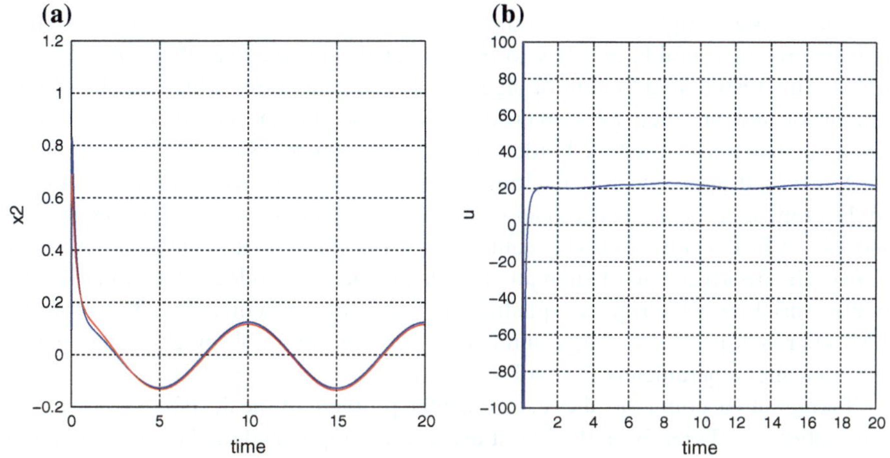

Fig. 3.13 **a** The deviation $\Delta\omega$ of the PMSG rotor angular velocity from the synchronous speed is made to track a reference setpoint, **b** associated control input E_f

3.4 Stabilizing Control of Synchronous Generators Using Interval Polynomials Theory

3.4.1 Outline

The objective of this section is to develop robust power system stabilizers that will be able to suppress the oscillatory behavior of the power systems over a wide range of operating conditions, thus assuring their secure operation. Modern large-scale power systems have commonly experienced adverse impacts on the their operation and cascading events due to under-damped oscillations [225]. This section proposes a robust power system stabilizer (PSS) as an effective way to damp-out oscillations in electric power systems. Since power generation systems are actually nonlinear, conventional fixed parameter, PSS cannot cope with great changes in operating conditions.

Due to frequent variations in the power system's operating conditions it is difficult to obtain a precise model of its dynamics and consequently to design a power system stabilizer [383]. The most common causes of uncertainty about power system dynamics are the following: (i) changes in power consumption, (ii) changes in operating states resulting from changes in the generation and transmission device structure, (iii) variation of operating conditions of the generators, (iv) changes in the network configuration and the number of operating generation units.

In the context of the above mentioned parametric uncertainties, the problem of designing a power system stabilizer using a nonlinear dynamical model is a non-trivial one. A first approach would be to linearize the system's model using state feedback and for the linearized model to solve a Riccati equation that will finally provide the feedback gain of an H_∞ controller [10, 125, 176, 243]. The problem becomes more complicated when the complete state vector of the power system is not directly measurable and has to be estimated with the use of a state observer [341]. This results in a system of algebraic Riccati equations which can be of high dimension and thus their solution needs elaborated computations. To reach a solution often one has to use a reduced order model of the power system.

A different approach for obtaining a robust power system stabilizer would be to decompose the nonlinear model of the power system dynamics into local linear models and to design local H_∞ controllers using linear matrix inequalities (LMI) [273]. The disadvantage of such an approach is that it introduces a number of Lyapunov variables which grows quadratically with the system's size. Therefore the LMI approach involves a large number of extra variables, although the number of the controller's parameters is small.

Taking into account the above, to solve the problem of the design of a power system stabilizer efficiently, and without using any complicated mathematical manipulations, Kharitonov's theorem can be proposed. Kharitonov's theorem is particularly useful for synthesizing fixed and low order controllers that simultaneously stabilize a given interval plant family in the parametric robust control area. Using Kharitonov's theorem, there have been many developments in the field of parametric robust control related to the stability and performance of uncertain systems presented as interval plants [185, 434, 442].

The model of a single-machine infinite-bus power system is considered in this section and a stabilizer is designed for this power system using Kharitonov's extremal gain margin theorem. According to Kharitonov's theory the nonlinear power system can be modeled by a set of transfer functions, where each transfer function is described by the ratio of two interval polynomials. Such polynomials have the form $p(s) = a_n s^n + a_{n-1} s^{n-1} + \cdots + a_0$, where each coefficient a_i is independent of the others and varies within an interval of known lower and upper bounds, i.e. $a_i \in [a_i^-, a_i^+]$. Thus, the problem of stabilization of a nonlinear power system is transformed into stabilization of a set of linear systems and can be achieved with the use of a simple phase-lead compensator, and frequency response techniques, such as the root-locus diagram [48, 422, 446].

3.4.2 *Stabilization for the Single-Machine Infinite-Bus Model*

3.4.2.1 Problem Statement

In a renewable power generation system, the electromechanical coupling between the rotor and the rest of the system components results in a behavior similar to that of a damped oscillator. Disturbances such as sudden changes in loads, changes in transmission line parameters, or fluctuations in the output of turbine and faults, cause low frequency oscillations around the equilibrium state. Furthermore, the use of fast acting high gain Automated Voltage Regulators (AVR) and the development of interconnected power systems with transfer of power across transmission lines have introduced more uncertainty in the power system model and have resulted in further degradation of the power system's oscillatory behavior. These oscillations limit the power transmission capability of the network and, sometimes, even cause a loss of generators' synchronism and a subsequent breakdown of the entire system [345]. The application of Power System Stabilizer (PSS) can help in damping out these oscillations and improve the system stability.

Usually, there are two approaches to achieve efficient operation of power systems in variable conditions: (i) adaptive control, and (ii) robust control. Adaptive control is based on the idea of continuously adapting the controller's parameters according to recent measurements. However, the performance of adaptive controllers maybe unsatisfactory during the learning phase, particularly when they are improperly initialized. Successful operation of adaptive controllers requires the measurements to satisfy persistence of excitation conditions, otherwise the adjustment of the controller's parameters fails. Moreover, the design of nonlinear adaptive controllers can be complicated, and the selection of the adaptation gains which assure closed stability may require extensive calculations [341]. Robust control can be an effective approach to deal with uncertainties introduced by variations of the power system operating conditions. Among many techniques available in robust control theory H_∞ control has received significant attention in applications of power system stabilization [86]. The H_∞ controller is selected so that the system's closed-loop response lies between pre-specified bounds. The design of such H_∞ controllers requires the solution of Riccati equations.

Taking into account the flaws and limitations of the aforementioned power stabilization approaches, this chapter proposes a simple, yet robust controller for power system stabilization, using Kharitonov's stability theory. Diagrams of the single-machine infinite-bus systems are given in Figs. 3.14 and 3.15. The PSS receives as input the machine's speed deviation signal $\Delta\omega$. To select the parameters of the PSS,

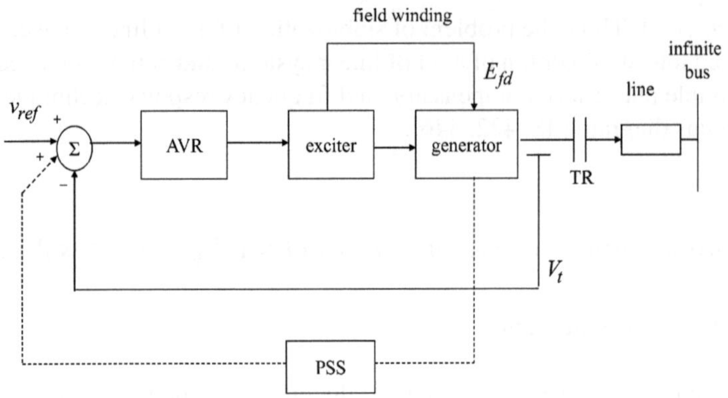

Fig. 3.14 Basic components of the single-machine infinite bus power system

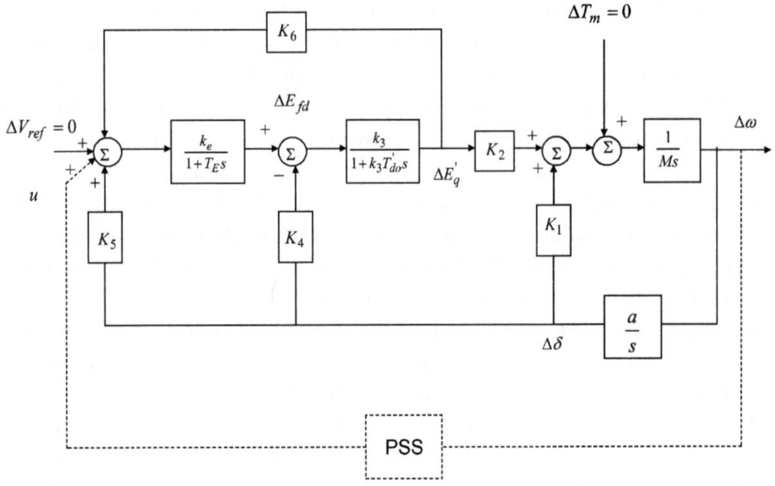

Fig. 3.15 Block diagram of the linearized single-machine infinite bus power system

usually the power system dynamics is linearized at an operating point, a transfer function is obtained and linear control techniques are applied. However, in practice the parameters of such a linear model are load dependent and thus the PSS has to be adjusted at different loads. This may include calculations over a large number of operating points. A different approach is to consider robust control methods in which the PSS parameters will be chosen so as to assure stability of the power system model and suppression of the oscillatory behavior for a wide range of operating conditions. In the sequel Kharitonov's theory will be applied to the design of a robust power system stabilizer.

3.4.2.2 Transfer Function of the Single-Machine Infinite-Bus Model

The fourth-order transfer function for the single-machine infinite-bus system is given by [422]

$$\frac{\Delta\omega}{U} = \frac{b_s}{a_4 s^4 + a_3 s^3 + a_2 s^2 + a_1 s + a_0} \tag{3.125}$$

The transfer function coefficients are

$$\begin{aligned}
\alpha_4 &= M \cdot T \cdot T_E \\
\alpha_3 &= M(T + T_E) \\
\alpha_2 &= M + 314 k_1 T \cdot T_E + k_E k_3 k_6 M \\
\alpha_1 &= 314(k_1 - k_2 k_3 k_4 - k_E k_2 k_3 k_5 + k_E k_1 k_3 k_6) \\
\alpha_0 &= 314(k_1 = k_2 k_3 k_4 - k_E k_2 k_3 k_5 + k_E k_1 k_3 k_6) \\
b &= k_E k_2 k_3
\end{aligned} \tag{3.126}$$

where $T = k_3 T'_{do}$. The parameters of the single machine infinite-bus system are as follows: $k_1 \ldots k_6$ are the parameters of the power system block diagram, T is the mechanical torque, T'_{do} is the open-circuit d-axis transient time constant, M is the inertia coefficient, k_E and T_E are the exciter gain and time constant respectively, $\Delta\omega$ is the machine's speed deviation and U is the stabilizing signal (PSS output). The paraments of the power system model are summarized in Table 3.1.

Table 3.1 Parameters of single-machine infinite-bus power system

$\Delta\omega$	Machine's speed deviation
δ	Angle between the q-axis and the infinite busbar
U	PSS output
V_t	Terminal voltage
E_{fd}	Generator field voltage
$k_1 \ldots k_6$	Constant coefficients
T'_{d0}	Open circuit d-axis time constant
M	Inertia coefficient
k_E	Exciter gain
T_E	Exciter time constant
T	Input mechanical torque
ΔT_m	Disturbance mechanical torque

3.4.3 *Kharitonov's Stability Theory*

3.4.3.1 Basics of Kharitonov's Theory

The stability theory developed by Kharitonov deals with the robust Hurwitz stability of interval polynomials and can be used in the design of robust power system stabilizers. The main points of Kharitonov's theory are as follows [446]:

Let $I(s)$ be the set of the n-th order real polynomials of the form $\delta x = \delta_0 + \delta_1 s + \delta_2 s^2 + \cdots + \delta_n s^n$ where the coefficients $\delta_i,\ i = 1, \ldots, n$ take values in the intervals $\delta_0 \in [x_0, y_0]$, $\delta_1 \in [x_1, y_1]$, ..., $\delta_n \in [x_n, y_n]$ (Fig. 3.16). The coefficient vector δ is defined as $\delta = [\delta_0, \delta_1, \ldots, \delta_n]$. Therefore, δ lies in the hyper-rectangle box of coefficients, as shown in Fig. 3.17 (see [48])

$$\Pi = \delta : \delta \in R^{n+1},\ x_i \le p_i \le y_i \tag{3.127}$$

It is assumed that 0 is not in $[x_n, y_n]$. This kind of polynomials is called 'interval polynomials' and $I(s)$ is the family of 'interval polynomials'. Kharitonov's theorem provides a simple necessary and sufficient condition for the family of interval polynomials $I(s)$.

Theorem: Each polynomial of the family $I(s)$ is Hurwitz if and only if the following four extreme polynomials are Hurwitz (i.e. they have all their roots in the left complex semi-plane)

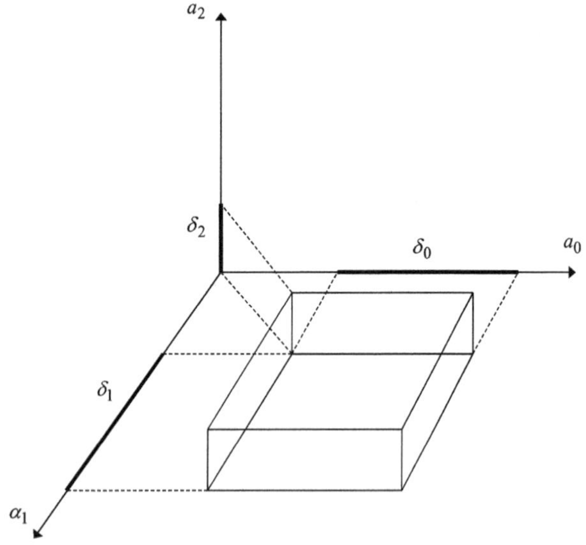

Fig. 3.16 The hyperrectangle defined by the parameters of the parameters of the interval characteristic polynomial

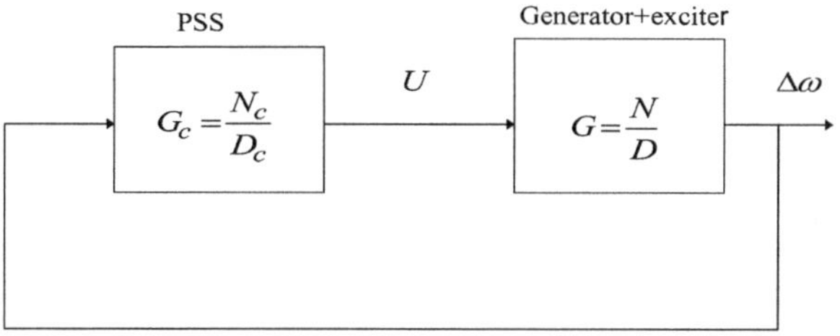

Fig. 3.17 Closed-loop for power system stabilization

$$
\begin{aligned}
K^1(s) &= x_0 + x_1 s + y_2 s^2 + y_3 s^3 + x_4 s^4 + x_5 s^5 + x_6 s^6 + \cdots \\
K^2(s) &= x_0 + y_1 s + y_2 s^2 + x_3 s^3 + x_4 s^4 + y_5 s^5 + y_6 s^6 + \cdots \\
K^3(s) &= y_0 + x_1 s + x_2 s^2 + y_3 s^3 + y_4 s^4 + x_5 s^5 + x_6 s^6 + \cdots \\
K^4(s) &= y_0 + y_1 s + x_2 s^2 + x_3 s^3 + y_4 s^4 + y_5 s^5 + x_6 s^6 + \cdots
\end{aligned}
\tag{3.128}
$$

A weakness of Kharitonov's theorem is the assumption that the coefficients perturb independently. Obviously, in many problems this assumption does not hold, since the characteristic polynomial coefficients perturb inter-dependently through other primary parameters. However, even in these cases Kharitonov's theorem can give computationally simple answers by overbounding the actual perturbations by an axis parallel box Π in the coefficient space.

3.4.3.2 Extremal Properties of Kharitonov's Polynomials

The case of a closed-loop control system with unity feedback is considered, where the forward transfer function $G(s) = \frac{n(s)}{d(s)}$ contains parameter uncertainty. The polynomials $n(s)$ and $d(s)$ vary in independent polynomial families $N(s)$ and $D(s)$ respectively. Moreover, the family of transfer functions of Eq. (3.129) in considered:

$$G(s) := \{G(s) = \frac{n(s)}{d(s)} : n(s) \in N(s),\ d(s) \in D(s)\} \tag{3.129}$$

where each interval polynomial of degree n can be written as

$$\delta(s) = \delta_0 + \delta_1 s + \delta_2 s^2 + \cdots + \delta_n s^n \tag{3.130}$$

with the coefficients lying in the ranges $\delta_0 \in [x_0, y_0]$, $\delta_1 \in [x_1, y_1]$, $\cdots$, $\delta_n \in [x_n, y_n]$.

It is assumed that the closed-loop system with unity feedback containing $G(s)$ is robustly stable. Thus, the characteristic polynomial of the closed-loop system given by $P(s) = d(s) + n(s)$ is Hurwitz for all $(n(s), d(s)) \in (N(s) \times D(s))$. Let $K_N^i(s),\ i = 1, 2, 3, 4$, and $K_D^j(s),\ j = 1, 2, 3, 4$ denote the Kharitonov polynomials associated with $N(s)$ and $D(s)$ respectively. The positive set of Kharitonov systems $G_K^+(s)$ which are associated with the interval family $G(s)$ is now defined as:

$$G_K^+(s) = \{\frac{K_N^i(s)}{K_D^j(s)}, \; i,j = 1,2,3,4\} \tag{3.131}$$

The following two theorems hold:

Theorem 1: The closed-loop system with unity feedback containing $G(s)$ is robustly stable if and only if each of the positive Kharitonov systems in $G_K^+(s)$ is robustly stable.

Theorem 2: (Extremal Gain Margin Theorem) The worst case gain margin of a system at the point P over the interval family $G(s)$ is the minimum gain margin corresponding to the positive Kharitonov systems $G_K^+(s)$.

Finally, the Kharitonov plants (linear system descriptions) are defined as follows: the interval plant family P is considered consisting of all plants of the form:

$$P(s,q,r) = \frac{N(s,q)}{D(s,r)} = \frac{q_m s^m + q_{m-1}s^{m-1} + \cdots + q_1 s + q_0}{s^n + r_{n-1}s^{n-1} + \cdots + r_1 s + r_0} \tag{3.132}$$

where $N(s,q)$ and $D(s,r)$ are interval polynomials, with coefficients which vary in the rectangles Q and R as defined by the cartesian products of $[q_i^-, q_i^+]$, $i = 0, 1, \ldots, m$ and $[r_i^-, r_i^+]$, $i = 0, 1, \ldots, n-1$. It is assumed that $N(s,q)$ and $D(s,r)$ do not have common roots and that $D(s,r)$ does not have roots on the imaginary axis for all $r \in R$. Given $c \in C$, the polynomials family

$$N(s,q) - cD(s,r) \tag{3.133}$$

is said to be *robustly stable* if all its roots are in the left complex semi-plane, $\forall\, q \in Q$ and $r \in R$. The Kharitonov polynomials for the numerator are defined as

$$\begin{aligned}
N_1(s) &= q_0^+ + q_1^+ s + q_2^- s^2 + q_3^- s^3 + q_4^- s^4 + q_5^+ s^5 + \cdots \\
N_2(s) &= q_0^+ + q_1^- s + q_2^+ s^2 + q_3^+ s^3 + q_4^- s^4 + q_5^- s^5 + \cdots \\
N_3(s) &= q_0^- + q_1^+ s + q_2^+ s^2 + q_3^- s^3 + q_4^- s^4 + q_5^+ s^5 + \cdots \\
N_4(s) &= q_0^+ + q_1^- s + q_2^- s^2 + q_3^+ s^3 + q_4^+ s^4 + q_5^- s^5 + \cdots
\end{aligned} \tag{3.134}$$

The interval polynomial $N(s,q)$ is Hurwitz $\forall\, q \in Q$ if and only if the 4 Kharitonov polynomials $N_i(s)$, $i = 1, 2, 3, 4$ are Hurwitz. The Kharitonov polynomials for the denominator are defined as

$$\begin{aligned}
D_1(s) &= r_0^+ + r_1^+ s + r_2^- s^2 + r_3^- s^3 + r_4^- s^4 + r_5^+ s^5 + \cdots \\
D_2(s) &= r_0^+ + r_1^- s + r_2^+ s^2 + r_3^+ s^3 + r_4^- s^4 + r_5^- s^5 + \cdots \\
D_3(s) &= r_0^- + r_1^+ s + r_2^+ s^2 + r_3^- s^3 + r_4^- s^4 + r_5^+ s^5 + \cdots \\
D_4(s) &= r_0^+ + r_1^- s + r_2^- s^2 + r_3^+ s^3 + r_4^+ s^4 + r_5^- s^5 + \cdots
\end{aligned} \tag{3.135}$$

The interval polynomial $D(s,r)$ is Hurwitz $\forall\, r \in R$ if and only if the 4 Kharitonov polynomials $D_k(s)$, $k = 1, 2, 3, 4$ are Hurwitz.

3.4.4 Design of the Power System Stabilizer

The basic control loop for power system stabilization is shown in the following diagram

Depending on the closed-loop specifications various types of compensators can be chosen (e.g. lead, lag, lead-lad, PD, PI, PID, or controllers based on state-space representation of the power system) [434]. In the system under study a first order lead compensator is chosen

$$G_c = \frac{N_c}{D_c} = K_c \frac{s + \mu_c}{s + \pi_c} \tag{3.136}$$

The characteristic polynomial of the closed-loop thus becomes

$$A(s + \mu_c)N + (s + \pi_c)D = 0 \tag{3.137}$$

The objective is to find the controller's parameters K_c, μ_c, π_c that can stabilize the characteristic polynomial of Eq. (3.137). To succeed this one has the following options: (i) to examine the four Kharitonov polynomials which are associated with the characteristic polynomial of the closed loop which are given in Eq. (3.128) (ii) to study the stability of the 16 associated Kharitonov plants which are given by

$$G_i = \frac{N_{i1}(s)}{D_{i2}(s)}, \quad i_1 = 1, \ldots, 4, \ i_2 = 1, \ldots, 4 \tag{3.138}$$

The closed-loop system is robustly stable if the controller can stabilize all the 16 closed-loop polynomials

$$K_c(s + \mu_c)N_{i1}(s) + (s + \pi_c)D_{i2}(s) = 0 \tag{3.139}$$

The stability analysis of the Kharitonov plants, or of the positive Kharitonov systems G_k^+ can be carried out using either algebraic methods (Routh-Hurwitz stability criterion) or graphical methods, such as the Nyquist plot and the root locus. The steps for the PSS design are summarized as follows:

1. The linearized model of the power system is obtained, as for example in Eq. (3.125). The compensator's zero μ_c and pole π_c are obtained so as to achieve the desirable performance for the nominal plant.
2. The range of variation of the model parameters α_i and b_j is obtained.
3. Using Eqs. (3.138) and (3.139) the Kharitonov extremal plants and the associated characteristic polynomials are obtained.
4. Using the root-locus technique for the 16 Kharitonov plants and assuming variable gain K_c, the compensator that stabilizes all Kharitonov plants is obtained. The range of K_c that stabilizes each Kharitonov plant is first calculated. The gain K_c that stabilizes the system over the whole range of operating points is chosen in the intersection of the ranges of K_c found for each Kharitonov plant.

3.4.5 Simulation Tests

The transfer function that describes the single-machine infinite-bus model is:

$$G(s) = \frac{b \cdot s}{a_4 s^4 + a_3 s^3 + a_2 s^2 + a_1 s + a_0} \tag{3.140}$$

Using [422] indicative ranges of the parameters of the model are $b \in [21, 96.1], a_4 = 8.34, a_3 = 167.3, a_2 \in [427.1, 744.4], a_1 \in [2941, 7500], a_0 \in [1093, 1258]$. The nominal values of the model are found in the middle of the aforementioned variation ranges. The control-loop of the nominal system is first considered to contain a lead-lag compensator of the form

$$G_c(s) = K_c \frac{s + \mu_1}{s + \pi_1} \frac{s + \mu_2}{s + \pi_2} \tag{3.141}$$

where the μ_1 and π_1 are the zeros/poles of the phase-lead part while μ_2, π_2 are the poles/zeros of the phase-lag part. Using the stability conditions of the closed loop

$$||G_c(s)G(s) = 1|| \quad \angle G_c(s)G(s) = -180^\circ \tag{3.142}$$

and the requirement that the steady-state error of the control loop should be minimized one obtains the compensator

$$G_c(s) = 63.58 \frac{s + 2}{s + 2.33} \frac{s + 1.75}{s + 0.05} \tag{3.143}$$

Assuming step input to the control loop, the machine's angle difference $\Delta\delta$ is depicted in Fig. 3.18. It can be observed that by introducing the compensator $G_c(s)$ in the control-loop the desirable transient and steady-state and transient-phase performance are achieved. It can be seen that the steady-state becomes practically 0, oscillations are suppressed and settling time is reduced.

Next, the design of a robust phase-lead compensator was performed, assuming that the model's parameters where defined within the uncertainty ranges described after Eq. 3.140. The phase lead compensator was $G_c(s) = K_c \frac{s+\mu_1}{s+\pi_1}$. According to Kharitonov's theorem, to find the stabilizing gain K_c one has to calculate the gain K_c which stabilizes each one of the 16 Kharitonov plant's G_K^+ defined in Eq. (3.138). However, since coefficient b has two extreme values b_{min} and b_{max} the 16 Kharitonov plants of Eq. (3.138) are reduced to only 8 models. Thus one obtains the following extreme Kharitonov plants

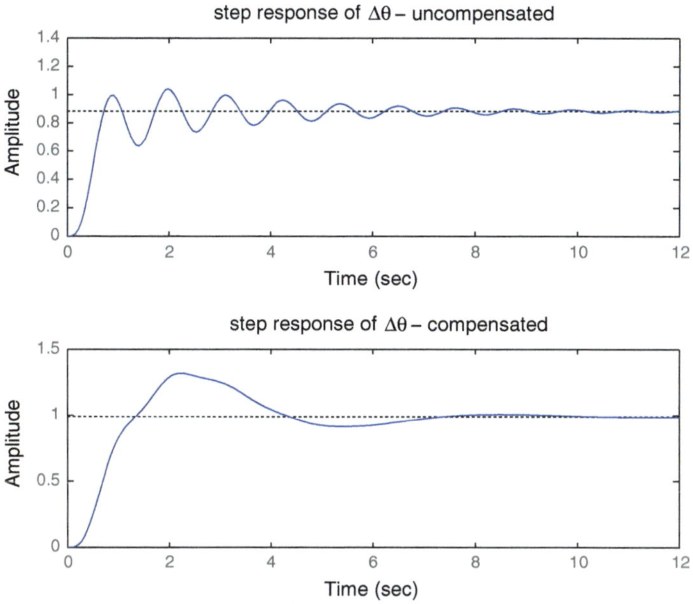

Fig. 3.18 The inclusion of the compensator $G_c(s)$ in the control-loop results in the desirable transient and steady-state performance of the nominal power system model

$$\begin{array}{ll} G_k^1(s) = \frac{21s}{8.34s^4+167.3s^3+744.4s^2+2941s+1093} & G_k^2(s) = \frac{96.1s}{8.34s^4+167.3s^3+744.4s^2+2941s+1093} \\ G_k^3(s) = \frac{21s}{8.34s^4+167.3s^3+427.1s^2+7500s+1258} & G_k^4(s) = \frac{96.1s}{8.34s^4+167.3s^3+744.4s^2+2941s+1258} \\ G_k^5(s) = \frac{21s}{8.34s^4+167.3s^3+427.1s^2+2941s+1258} & G_k^6(s) = \frac{96.1s}{8.34s^4+167.3s^3+427.1s^2+2941s+1258} \\ G_k^7(s) = \frac{21s}{8.34s^4+167.3s^3+744.4s^2+7500s+1093} & G_k^8(s) = \frac{96.1s}{8.34s^4+167.3s^3+744.4s^2+7500s+1093} \end{array} \tag{3.144}$$

The root-locus diagrams which correspond to the models $G_c(s)G_k^i$, $i = 1, \ldots, 8$ are depicted in Fig. 3.19. From these diagrams one obtains the variation range $\Delta K_c^i = [K_{c_{min}}^i, K_{c_{max}}^i]$, $i = 1, \ldots, 8$ of the gain K_c which assures stability to each one of the associated loops, while the overall stabilizing gain belongs to the intersection of ΔK_c^i. A suitable stabilizing gain is thus taken to be $K_c = 2$.

The impulse response of the 8 unit-feedback closed loops which contain the stabilizing gain K_c, the phase lead compensator $G_c(s) = \frac{s+\mu_1}{s+\pi_1}$ and each one of the extreme Kharitonov polynomials G_k^i is shown in Figs. 3.20a and 3.21a.

Comparing to the response of the unit-feedback closed-loop without phase lead compensator and with gain K_c which has been selected according to the stability ranges of the nominal plant, as depicted in Figs. 3.20b and 3.21b, one can observe

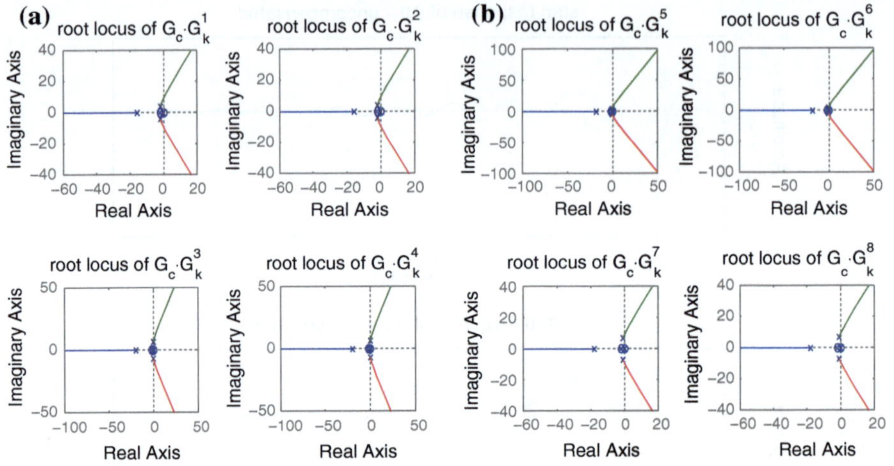

Fig. 3.19 **a** Root-locus plots of the Kharitonov extremal plants $G_c(s)G_k^1$ to $G_c(s)G_k^4$, **b** root-locus plots of the Kharitonov extremal plants $G_c(s)G_k^5$ to $G_c(s)G_k^8$

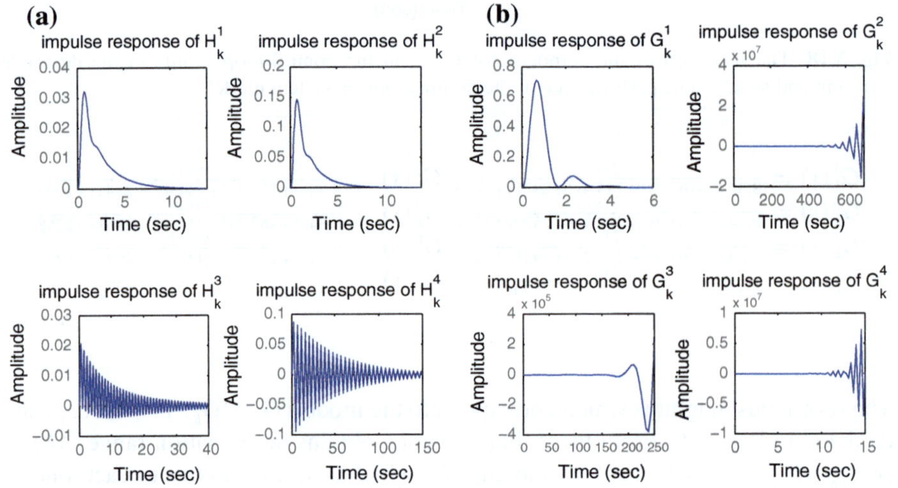

Fig. 3.20 **a** Impulse response of the Kharitonov extremal plants G_k^1 to G_k^4, in a unit-feedback closed-loop which includes a phase-lead compensator, **b** impulse response of the Kharitonov extremal plants G_k^1 to G_k^4, in a unit-feedback closed-loop with proportional gain K_c chosen according to the nominal plant's stability ranges

the improvements in the closed-loop performance which is due to the use of the power system stabilizer. It can be seen that the proposed robust power system stabilizer results in fast elimination of oscillations and suppression of the oscillations'

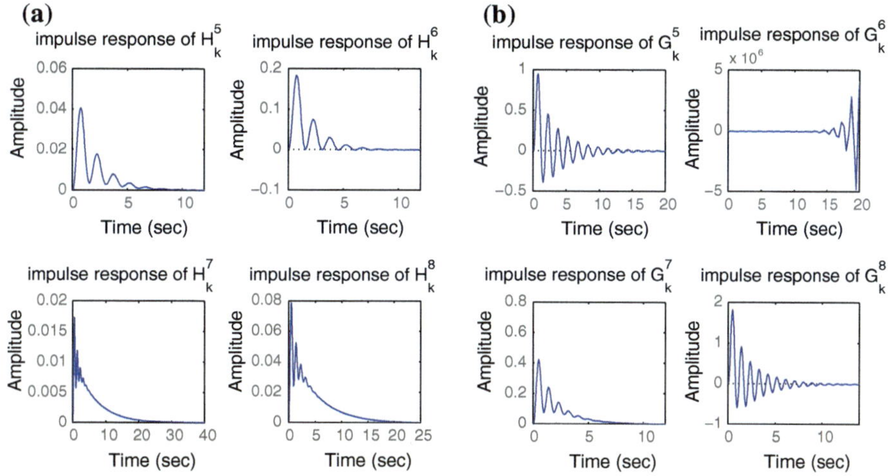

Fig. 3.21 **a** Impulse response of the Kharitonov extremal plants G_k^5 to G_k^8, in a unit-feedback closed-loop which includes a phase-lead compensator, **b** impulse response of the Kharitonov extremal plants G_k^5 to G_k^8, in a unit-feedback closed-loop with proportional gain K_c chosen according to the nominal plant's stability ranges

amplitude. On the other hand the design of a control-loop which does not take into account the parametric uncertainty in the single-machine infinite-bus model results in undesirable oscillations and may even lead the power system to instability as shown in Figs. 3.20b and 3.21b.

Chapter 4
Control of the Functioning of Multiphase Electric Machines

4.1 Outline

The chapter analyzes different approaches for nonlinear control of multi-phase electric machines: (i) H-infinity control of multi-phase synchronous electric machines, (ii) H-infinity control of doubly-fed reluctance machines, (iii) flatness-based adaptive control of doubly-fed reluctance machines.

First, in this chapter power generation with the use of multi-phase synchronous machines is considered. The dynamic model of the 6-phase synchronous electric machine undergoes an approximate linearisation, through Taylor series expansion. The linearization is performed round local operating points which are defined at each time instant by the present value of the system's state vector and the last value of the control input that was exerted on it. The linearisation procedure requires the computation of Jacobian matrices at the aforementioned operating points. The modelling error, which is due to the truncation of higher order terms in the Taylor series expansion is perceived as a perturbation that should be compensated by the robustness of the control loop. Next, for the linearized equivalent model of the 6-phase synchronous electric machine, an H-infinity feedback control loop is designed. This approach, is based on the concept of a differential game that takes place between the control input (which tries to minimize the deviation of the state vector from the reference setpoints) and the disturbance input (that tries to maximize it). In such a case, the computation of the optimal control input requires the solution of an algebraic Riccati equation at each iteration of the control algorithm. The known robustness properties of H-infinity control enable compensation of model uncertainty and rejection of the perturbation terms that affect the 6-phase synchronous machine. The stability of the control loop is proven through Lyapunov analysis. Actually, it is shown that H-infinity tracking performance is succeeded, while conditionally the asymptotic stability of the control loop is also assured. The efficiency of the proposed control scheme for the 6-phase synchronous machine is further confirmed through simulation experiments.

G. Rigatos, *Intelligent Renewable Energy Systems*,
Green Energy and Technology, DOI 10.1007/978-3-319-39156-4_4

Next, the chapter develops an H-infinity approach for the problem of nonlinear optimal control of doubly-fed reluctance machines. The dynamic model of the machines is subjected to linearization round local operating points, through Taylor series expansion and the computation of Jacobian matrices. For the linearized model an H-infinity feedback controller is designed, capable of compensating for the modelling error of the approximate linearization as well as for external perturbations affecting the machine. The computation of the feedback control gain is based on the solution of an algebraic Riccati equation that is performed at each iteration of the control algorithm. Lyapunov stability analysis for the control loop of the reluctance machine arrives at an H-infinity tracking performance criterion, and finally the asymptotic stability of the control loop is demonstrated. The excellent tracking performance of the H-infinity control method is confirmed through simulation experiments.

Finally, the chapter proposes an adaptive control approach that is capable of compensating for model uncertainty and parametric changes of the doubly-fed reluctance machines (DFRMs), as well as for the lack of measurements about the DFRM's state vector elements. First it is proven that the DFRM's model is a differentially flat one. This means that all its state variables and its control inputs can be written as differential functions of key state variables which are the so-called flat outputs. Moreover, this implies that the flat outputs and their derivatives are not connected through an ordinary differential equation. By exploiting differential flatness properties it is shown that the DFRM model can be transformed into the linear canonical form. For the latter description, the new control inputs comprise unknown nonlinear functions which can be identified with the use of neurofuzzy approximators. The estimated dynamics of the generator is used by a feedback controller thus establishing an indirect adaptive control scheme. Moreover, to robustify the control loop a supplementary control term is computed using H-infinity control theory. Another problem that has to be dealt with comes from the inability to measure the complete state vector of the generator. Thus, a state observer is implemented in the control loop. The stability of the considered observer-based adaptive control approach is proven using Lyapunov analysis. Moreover, the performance of the control scheme is evaluated through simulation experiments.

4.2 Nonlinear H-infinity Control of Multi-phase Electric Machines

4.2.1 Overview

The use of multi-phase electric machines in power generation applications, exhibits specific advantages, such as improved torque and power characteristics and robustness to failures taking place in the electrical part of such machines [23, 54, 139, 170, 199, 239, 250]. Aiming at achieving the reliable functioning of multi-phase machines under variable operating conditions, several control methods have been developed in the recent years [2, 199, 285, 286, 389, 400]. This section presents a new nonlinear H-infinity control method for multi-phase electric machines, using as application example the 6-phase Permanent Magnet Synchronous Machine (PMSM). The method is based on an approximate linearization of the multi-phase machine's dynamics, following Taylor series expansion round local equilibria. At each time instant the linearizarion point is defined by the present value of the state vector and the most recent value of the control inputs vector. This linearization requires the computation of the system's Jacobian matrices [34, 341, 344]. The modelling error which is due to the truncation of higher order terms in the Taylor series expansion is perceived as perturbation which is compensated by the robustness of the H-infinity control loop. The stability of the control method is proven through Lyapunov analysis, following the procedure that has been developed for the case of the induction generators.

4.2.2 Dynamic Model of the 6-Phase Synchronous Machine

4.2.2.1 Dynamics of the 6-Phase PMSM

The dynamic model of a six-phase Permanent Magnet Synchronous Machine (PMSM) will be analyzed next. There are two three-phase systems, the first one is denoted as *abc* and the second one is denoted as *xyz* and results from the first system by a 30° rotation (Fig. 4.1) [250]. By applying Kirchhoff's law in these phase systems one obtains:

$$v_{abc} = \bar{R}_s i_{abc} + \frac{d}{dt}\lambda_{abc} \tag{4.1}$$

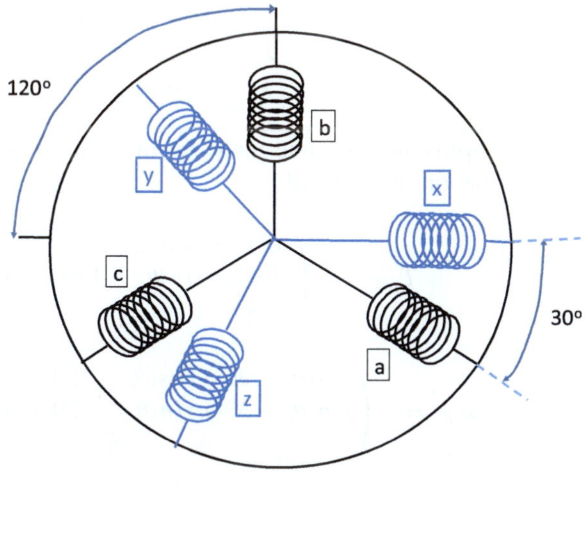

Fig. 4.1 A six-phase electric machine

$$\lambda_{abc} = \bar{L}_{11} i_{abc} + \bar{L}_{12} i_{xyz} + \lambda'_{mabc} \tag{4.2}$$

$$v_{xyz} = \bar{R}_s i_{xyz} + \frac{d}{dt}\lambda_{xyz} \tag{4.3}$$

$$\lambda_{xyz} = \bar{L}_{22} i_{xyz} + \bar{L}_{21} i_{abc} + \lambda_{mxyz} \tag{4.4}$$

where $\bar{R}_s = diag[R_s, R_s, R_s]^T$ is the stator resistance matrix, $v_{abc} = [v_a, v_b, v_c]^T$ is the phase voltage vector of the abc winding, $i_{abc} = [i_a, i_b, i_c]^T$ is the current vector of the abc winding, $v_{xyz} = [v_x, v_y, v_z]^T$ is the phase voltage vector of the xyz winding, $i_{xyz} = [i_x, i_y, i_z]^T$ is the current vector of the xyz winding, $\lambda_{abc} = [\lambda_a, \lambda_b, \lambda_c]^T$ is the stator flux linkage vector of the abc winding, $\lambda_{xyz} = [\lambda_x, \lambda_y, \lambda_z]^T$ is the stator flux linkage vector of the xyz winding, $\bar{L}_{11} = diag[L_{11}, L_{11}, L_{11}]^T$ is the stator inductance vector of the abc winding, $\bar{L}_{12} = diag[L_{12}, L_{12}, L_{12}]^T$ is a mutual inductance vector, $\bar{L}_{22} = diag[L_{22}, L_{22}, L_{22}]^T$ is the stator inductance vector of the xyz winding, and $\bar{L}_{21} = diag[L_{21}, L_{21}, L_{21}]^T$ is also a mutual inductance vector. Moreover λ'_{mabc} is the permanent-magnet flux linkage vector of the abc winding and λ'_{mxyz} is the permanent-magnet flux linkage vector of the xyz winding.

In aggregate one has the following equations

$$\begin{aligned} v_a &= R_s i_a + \frac{d}{dt}\lambda_a \\ v_b &= R_s i_b + \frac{d}{dt}\lambda_b \\ v_c &= R_s i_c + \frac{d}{dt}\lambda_c \end{aligned} \tag{4.5}$$

$$\begin{aligned} \lambda_a &= L_{11} i_a + L_{12} i_x + \lambda'_{ma} \\ \lambda_b &= L_{11} i_b + L_{12} i_y + \lambda'_{mb} \\ \lambda_c &= L_{11} i_c + L_{12} i_z + \lambda'_{mc} \end{aligned} \tag{4.6}$$

$$\begin{aligned} v_x &= R_s i_x + \frac{d}{dt}\lambda_x \\ v_y &= R_s i_y + \frac{d}{dt}\lambda_y \\ v_z &= R_s i_z + \frac{d}{dt}\lambda_z \end{aligned} \tag{4.7}$$

$$\begin{aligned} \lambda_x &= L_{22} i_x + L_{21} i_a + \lambda'_{mx} \\ \lambda_y &= L_{22} i_y + L_{21} i_b + \lambda'_{my} \\ \lambda_z &= L_{22} i_z + L_{21} i_c + \lambda'_{mz} \end{aligned} \tag{4.8}$$

To simplify the dynamic model of the multi-phase PMSM the following Park transformations are applied:

$$T_{qd1} = \frac{2}{3} \begin{pmatrix} cos(\theta_e) & cos(\theta_e - 120°) & cos(\theta_e + 120°) \\ sin(\theta_e) & sin(\theta_e - 120°) & sin(\theta_e + 120°) \\ \frac{1}{2} & \frac{1}{2} & \frac{1}{2} \end{pmatrix} \tag{4.9}$$

$$T_{qd2} = \frac{2}{3} \begin{pmatrix} cos(\theta_e - 30°) & cos(\theta_e - 150°) & cos(\theta_e + 90°) \\ sin(\theta_e - 30°) & sin(\theta_e - 150°) & sin(\theta_e + 90°) \\ \frac{1}{2} & \frac{1}{2} & \frac{1}{2} \end{pmatrix} \tag{4.10}$$

where T_{dq1} is the transformation leading from the abc reference frame to the $dq1$ reference frame, T_{dq2} is the transformation leading from the xyz reference frame to the $dq2$ reference frame, and θ_e is the rotor flux angle. In the new reference frame, the six-phase PMSM model becomes

$$\begin{aligned} v_{q_1} &= R_s i_{q_1} + L_{q11}\tfrac{d}{dt}i_{q_1} + \omega_e(L_{d11}i_{d1}) + \lambda_m^{'} \\ v_{d_1} &= R_s i_{d_1} + L_{d11}\tfrac{d}{dt}i_{d_1} - \omega_r L_{q11}i_{q_1} \\ v_{q_2} &= R_s i_{q_2} + L_{q22}\tfrac{d}{dt}i_{q_2} + \omega_e(L_{d22}i_{d2}) + \lambda_m^{'} \\ v_{d_2} &= R_s i_{d_2} + L_{d22}\tfrac{d}{dt}i_{d_2} - \omega_r L_{q22}i_{q_2} \\ \omega_e &= \tfrac{p}{2}\omega_r \end{aligned} \tag{4.11}$$

where v_{d_1} and v_{q_1} are the stator abc windings' voltages expressed in the $dq1$ reference frame, v_{d_2} and v_{q_2} are the stator xyz windings' voltages expressed in the $dq2$ reference frame, i_{d_1} and i_{q_1} are the stator abc windings' currents expressed in the $dq1$ reference frame, i_{d_2} and i_{q_2} are the stator xyz windings' currents expressed in the $dq2$ reference frame, L_{d11} and L_{q11} are the stator abc windings' inductances expressed in the $dq1$ reference frame, L_{d22} and L_{q22} are the stator xyz windings' inductances expressed in the $dq2$ reference frame, ω_r is the rotor's angular velocity, ω_e is the rotation speed of the dq reference frame, $\lambda_m^{'}$ is the permanent magnet flux linkage, and p is the number of poles of the electric machine.

The electric torque of the six-phase PMSM is given by

$$\begin{aligned} T_e = \tfrac{3}{2}\tfrac{p}{2}([(L_{d11}i_{d1} + \lambda_m^{'})i_{q1} + (L_{d11} - L_{q11})i_{d1}i_{q1}] + \\ + [(L_{d22} + \lambda_m^{'})i_{q2} + (L_{d22} - L_{q22})i_{d2}i_{q2}]) \end{aligned} \tag{4.12}$$

Moreover, about the rotational motion of the rotor of this electric motor one has

$$J\tfrac{d}{dt}\omega_r + B\omega_r = T_L - T_e \tag{4.13}$$

where ω_r is the rotor's turn speed, J is the moment of inertia, B is a friction coefficient, T_L is an external load's torque and T_e is the electric torque exerted on the rotor due to magnetic field developed at the stator.

4.2.3 State-Space Description of the 6-Phase PMSM

4.2.3.1 Dynamic Model of the 6-Phase Machine

The following state variables are defined for the 6-phase PMSM machine. These are: $x_1 = \omega_r, x_2 = i_{d1}, x_3 = i_{q1}, x_4 = i_{d2}, x_5 = i_{q2}$. Moreover, the following control inputs are defined: $u_1 = v_{d1}$, $u_2 = v_{q1}$, $u_3 = v_{d2}$, $u_4 = v_{q2}$. Using this notation, the state-space model of the PMSM is written as:

$$\dot{x}_1 = -\frac{B}{J}x_1 - \frac{3}{2}\frac{p}{2J}([(L_{d11}x_2 + \lambda_m^{'})x_3 + (L_{d11} - L_{q11})x_2x_3] + \\ + [(L_{d22} + \lambda_m^{'})x_5 + (L_{d22} - L_{q22})x_4x_5]) + \frac{T_L}{J} \tag{4.14}$$

$$\dot{x}_2 = -\frac{R_s}{L_{d11}}x_2 - \frac{p}{2}x_1\frac{L_{q11}}{L_{d11}}x_3 + \frac{1}{L_{d11}}u_1 \tag{4.15}$$

$$\dot{x}_3 = -\frac{R_s}{L_{q11}}x_3 - \frac{p}{2}x_1(\frac{L_{d11}}{L_{q11}}x_2 + \frac{1}{L_{q11}}\lambda_m^{'}) + \frac{1}{L_{q11}}u_2 \tag{4.16}$$

$$\dot{x}_4 = -\frac{R_s}{L_{d22}}x_4 - \frac{p}{2}x_1\frac{L_{q22}}{L_{d22}}x_5 + \frac{1}{L_{d22}}u_3 \tag{4.17}$$

$$\dot{x}_5 = -\frac{R_s}{L_{q22}}x_5 - \frac{p}{2}x_1(\frac{L_{d22}}{L_{q22}}x_4 + \frac{1}{L_{q22}}\lambda_m^{'}) + \frac{1}{L_{q22}}u_2 \tag{4.18}$$

It can be noticed that the equations that constitute the dynamics of the electric part of the 6-phase machine for the two different frames are decoupled, while coupling appears in the part describing the mechanical part of the machine and the associated rotational motion.

Using the previous notation one can also arrive at a state-space description of the 6-phase machine in matrix form

$$\dot{x} = f(x) + g(x)u \tag{4.19}$$

where $x \in R^{5\times1}$, $u \in R$ and $f(x) \in R^{5\times1}$, $g(x) \in R^{5\times4}$ and

$$f(x) = \begin{pmatrix} -\frac{B}{J}x_1 + \frac{3}{2}\frac{p}{2J}([(L_{d11}x_2 + \lambda_m^{'})x_3 + (L_{d11} - L_{q11})x_2x_3] + \\ + [(L_{d22}x_4 + \lambda_m^{'})x_5 + (L_{d22} - L_{q22})x_4x_5]) + \frac{T_L}{J} \\ -\frac{R_s}{L_{d11}}x_2 - \frac{p}{2}x_1\frac{L_{q11}}{L_{d11}}x_3 \\ -\frac{R_s}{L_{q11}}x_3 - \frac{p}{2}x_1(\frac{L_{d11}}{L_{q11}}x_2 + \frac{1}{L_{q11}}\lambda_m^{'}) \\ -\frac{R_s}{L_{d22}}x_4 - \frac{p}{2}x_1\frac{L_{q22}}{L_{d22}}x_5 \\ -\frac{R_s}{L_{q22}}x_5 - \frac{p}{2}x_1(\frac{L_{d22}}{L_{q22}}x_4 + \frac{1}{L_{q22}}\lambda_m^{'}) \end{pmatrix} \tag{4.20}$$

$$g(x) = \begin{pmatrix} 0 & 0 & 0 & 0 \\ \frac{1}{L_{d11}} & 0 & 0 & 0 \\ 0 & \frac{1}{L_{q11}} & 0 & 0 \\ 0 & 0 & \frac{1}{L_{d22}} & 0 \\ 0 & 0 & 0 & \frac{1}{L_{q22}} \end{pmatrix} \tag{4.21}$$

Next, linearization of the 6-phase electric machine is carried out by computing the associated Jacobian matrices. It holds that

$$\begin{aligned} \nabla_x [f(x) + g(x)u] \mid_{(x^*,u^*)} = \nabla_x f(x) \\ \nabla_u [f(x) + g(x)u] \mid_{(x^*,u^*)} = g(x^*) \end{aligned} \tag{4.22}$$

where

$$\nabla_x f(x) = \begin{pmatrix} \frac{\partial f_1}{\partial x_1} & \frac{\partial f_1}{\partial x_2} & \frac{\partial f_1}{\partial x_3} & \frac{\partial f_1}{\partial x_4} & \frac{\partial f_1}{\partial x_5} \\ \frac{\partial f_2}{\partial x_1} & \frac{\partial f_2}{\partial x_2} & \frac{\partial f_2}{\partial x_3} & \frac{\partial f_2}{\partial x_4} & \frac{\partial f_2}{\partial x_5} \\ \frac{\partial f_3}{\partial x_1} & \frac{\partial f_3}{\partial x_2} & \frac{\partial f_3}{\partial x_3} & \frac{\partial f_3}{\partial x_4} & \frac{\partial f_3}{\partial x_5} \\ \frac{\partial f_4}{\partial x_1} & \frac{\partial f_4}{\partial x_2} & \frac{\partial f_4}{\partial x_3} & \frac{\partial f_4}{\partial x_4} & \frac{\partial f_4}{\partial x_5} \\ \frac{\partial f_5}{\partial x_1} & \frac{\partial f_5}{\partial x_2} & \frac{\partial f_5}{\partial x_3} & \frac{\partial f_5}{\partial x_4} & \frac{\partial f_5}{\partial x_5} \end{pmatrix} \tag{4.23}$$

where, for the first row of the Jacobian matrix one has $\frac{\partial f_1}{\partial x_1} = -\frac{B}{J}$, $\frac{\partial f_1}{\partial x_2} = -\frac{3}{2}\frac{P}{2J}[L_{d11}x_3 + (L_{d11} - L_{q11})x_3]$, $\frac{\partial f_1}{\partial x_3} = -\frac{3}{2}\frac{P}{2J}[(L_{d11}x_2 + \lambda_m)' + (L_{d11} - L_{q11})x_2]$, $\frac{\partial f_1}{\partial x_4} = -\frac{3}{2}\frac{P}{2J}[L_{d22}x_5 + (L_{d22} - L_{q22})x_5]$, and $\frac{\partial f_1}{\partial x_5} = -\frac{3}{2}\frac{P}{2J}[(L_{d22}x_4 + \lambda_m^{'}) + (L_{d22} - L_{q22})x_4]$.

For the second row of the Jacobian matrix one gets $\frac{\partial f_2}{\partial x_1} = -\frac{p}{2}\frac{L_{q11}}{L_{d11}}x_3$, $\frac{\partial f_2}{\partial x_2} = -\frac{R_s}{L_{d11}}$, $\frac{\partial f_2}{\partial x_3} = -\frac{p}{2}\frac{L_{q11}}{L_{d11}}x_1$ $\frac{\partial f_2}{\partial x_4} = 0$, and $\frac{\partial f_2}{\partial x_5} = 0$.

For the third row of Jacobian matrix one has $\frac{\partial f_3}{\partial x_1} = -\frac{p}{2}(\frac{L_{d11}}{L_{q11}}x_2 + \frac{1}{L_{q11}}\lambda_m^{'})$, $\frac{\partial f_3}{\partial x_2} = -\frac{p}{2}x_1(\frac{L_{d11}}{L_{q11}})$, $\frac{\partial f_3}{\partial x_3} = -\frac{R_s}{L_{q11}}$, $\frac{\partial f_3}{\partial x_4} = 0$, and $\frac{\partial f_3}{\partial x_5} = 0$.

For the fourth row of the Jacobian matrix one gets $\frac{\partial f_4}{\partial x_1} = -\frac{p}{2}\frac{L_{q22}}{L_{d22}}x_5$, $\frac{\partial f_4}{\partial x_2} = 0$, $\frac{\partial f_4}{\partial x_3} = 0$, $\frac{\partial f_4}{\partial x_4} = -\frac{R_s}{L_{d22}}$, and $\frac{\partial f_4}{\partial x_5} = -\frac{p}{2}x_1\frac{L_{g22}}{L_{d22}}$.

For the fifth row of the Jacobian matrix one gets $\frac{\partial f_5}{\partial x_1} = -\frac{p}{2}(\frac{L_{d22}}{L_{q22}}x_4 + \frac{1}{L_{q22}}\lambda_m^{'})$, $\frac{\partial f_5}{\partial x_2} = 0$, $\frac{\partial f_5}{\partial x_3} = 0$, $\frac{\partial f_5}{\partial x_4} = -\frac{p}{2}\frac{L_{d22}}{L_{q22}}x_1$, and $\frac{\partial f_5}{\partial x_5} = -\frac{R_s}{L_{q22}}$.

4.2.3.2 Tracking Error Dynamics

The stability analysis for the multi-phase electric machines follows the stages presented in the case of the induction generator. After linearization round its current operating point, the six-phase machine's dynamic model is written as

$$\dot{x} = Ax + Bu + d_1 \tag{4.24}$$

Parameter d_1 stands for the linearization error in the six-phase machine's dynamic model appearing in Eq. (4.24). The reference setpoints for the multi-phase motor are denoted by $\mathbf{x_d} = [x_1^d, \ldots, x_6^d]$. Tracking of this trajectory is succeeded after applying the control input u^*. At every time instant the control input u^* is assumed to differ from the control input u appearing in Eq. (4.24) by an amount equal to Δu, that is $u^* = u + \Delta u$

$$\dot{x}_d = Ax_d + Bu^* + d_2 \tag{4.25}$$

The dynamics of the controlled system described in Eq. (4.24) can be also written as

$$\dot{x} = Ax + Bu + Bu^* - Bu^* + d_1 \tag{4.26}$$

and by denoting $d_3 = -Bu^* + d_1$ as an aggregate disturbance term one obtains

$$\dot{x} = Ax + Bu + Bu^* + d_3 \tag{4.27}$$

By subtracting Eq. (4.25) from Eq. (4.27) one has

$$\dot{x} - \dot{x}_d = A(x - x_d) + Bu + d_3 - d_2 \tag{4.28}$$

By denoting the tracking error as $e = x - x_d$ and the aggregate disturbance term as $\tilde{d} = d_3 - d_2$, the tracking error dynamics becomes

$$\dot{e} = Ae + Bu + \tilde{d} \tag{4.29}$$

The above linearized form of the six-phase machine's model can be efficiently controlled after applying an H-infinity feedback control scheme.

4.2.4 The Nonlinear H-infinity Control

4.2.4.1 Mini-Max Control and Disturbance Rejection

The initial nonlinear model of the six-phase machine is in the form

$$\dot{x} = f(x, u) \;\; x \in R^n, \;\; u \in R^m \tag{4.30}$$

Linearization of the system (multi-phase motor) is performed at each iteration of the control algorithm, through Taylor series expansion, round its present operating point $(x^*, u^*) = (x(t), u(t - T_s))$. The linearized equivalent of the system is described by

$$\dot{x} = Ax + Bu + L\tilde{d} \quad x \in R^n, \ u \in R^m, \ \tilde{d} \in R^q \tag{4.31}$$

where matrices A and B are obtained from the computation of the Jacobians

$$A = \begin{pmatrix} \frac{\partial f_1}{\partial x_1} & \frac{\partial f_1}{\partial x_2} & \cdots & \frac{\partial f_1}{\partial x_n} \\ \frac{\partial f_2}{\partial x_1} & \frac{\partial f_2}{\partial x_2} & \cdots & \frac{\partial f_2}{\partial x_n} \\ \cdots & \cdots & \cdots & \cdots \\ \frac{\partial f_n}{\partial x_1} & \frac{\partial f_n}{\partial x_2} & \cdots & \frac{\partial f_n}{\partial x_n} \end{pmatrix} |_{(x^*,u^*)} \ B = \begin{pmatrix} \frac{\partial f_1}{\partial u_1} & \frac{\partial f_1}{\partial u_2} & \cdots & \frac{\partial f_1}{\partial u_m} \\ \frac{\partial f_2}{\partial u_1} & \frac{\partial f_2}{\partial u_2} & \cdots & \frac{\partial f_2}{\partial u_m} \\ \cdots & \cdots & \cdots & \cdots \\ \frac{\partial f_n}{\partial u_1} & \frac{\partial f_n}{\partial u_2} & \cdots & \frac{\partial f_n}{\partial u_m} \end{pmatrix} |_{(x^*,u^*)} \tag{4.32}$$

and vector $\tilde{d}$ denotes disturbance terms due to linearization errors. As already explained in the case of the induction generator, the problem of disturbance rejection for the linearized model that is described by

$$\begin{aligned} \dot{x} &= Ax + Bu + L\tilde{d} \\ y &= Cx \end{aligned} \tag{4.33}$$

where $x \in R^n$, $u \in R^m$, $\tilde{d} \in R^q$ and $y \in R^p$, cannot be handled efficiently if the classical LQR control scheme is applied. This is because of the existence of the perturbation term $\tilde{d}$. Again, the disturbance term $\tilde{d}$ apart from modeling (parametric) uncertainty and external perturbation terms can also represent noise terms of any distribution.

In the H_∞ control approach, a feedback control scheme is designed for trajectory tracking by the system's state vector and simultaneous disturbance rejection, considering that the disturbance affects the system in the worst possible manner. The disturbances' effect are incorporated in the following quadratic cost function:

$$\begin{aligned} J(t) = \tfrac{1}{2}\int_0^T [y^T(t)y(t) + \\ + ru^T(t)u(t) - \rho^2\tilde{d}^T(t)\tilde{d}(t)]dt, \quad r, \rho > 0 \end{aligned} \tag{4.34}$$

The significance of the negative sign in the cost function's term that is associated with the perturbation variable $\tilde{d}(t)$ is that the disturbance tries to maximize the cost function $J(t)$ while the control signal $u(t)$ tries to mininize it. The physical meaning of the relation given above is that the control signal and the disturbances compete to each other within a mini-max differential game. This problem of mini-max optimization can be written as

$$min_u max_{\tilde{d}} J(u, \tilde{d}) \tag{4.35}$$

Also in the case of multi-phase electric machines, the objective of the optimization procedure is to compute a control signal $u(t)$ which can compensate for the worst possible disturbance, that is externally imposed to the system. However, the solution to the mini-max optimization problem is directly related to the value of the parameter ρ. This means that there is an upper bound in the disturbances magnitude that can be annihilated by the control signal.

4.2.4.2 H-infinity Feedback Control

For the linearized system of the multi-phase machine given by Eq. (4.33) the cost function of Eq. (4.34) is defined, where the coefficient r determines the penalization of the control input and the weight coefficient ρ determines the reward of the disturbances' effects. It is assumed that:

It is assumed that (i) The energy that is transferred from the disturbances signal $\tilde{d}(t)$ is bounded, that is $\int_0^\infty \tilde{d}^T(t)\tilde{d}(t)dt < \infty$, (ii) the matrices $[A, B]$ and $[A, L]$ are stabilizable, (iii) the matrix $[A, C]$ is detectable. Then, the optimal feedback control law is given by

$$u(t) = -Kx(t) \tag{4.36}$$

with

$$K = \frac{1}{r}B^T P \tag{4.37}$$

where P is a positive semi-definite symmetric matrix which is obtained from the solution of the Riccati equation

$$A^T P + PA + Q - P(\tfrac{1}{r}BB^T - \tfrac{1}{2\rho^2}LL^T)P = 0 \tag{4.38}$$

where Q is also a positive definite symmetric matrix. The worst case disturbance is given by

$$\tilde{d}(t) = \frac{1}{\rho^2}L^T Px(t) \tag{4.39}$$

The diagram of the considered control loop is depicted in Fig. 4.2.

4.2.5 *Lyapunov Stability Analysis*

Through Lyapunov stability analysis it will be shown that the proposed nonlinear control scheme assures H_∞ tracking performance for the six-phase machine, and that in case of bounded disturbance terms asymptotic convergence to the reference setpoints is succeeded.

The tracking error dynamics for the multi-phase motor is written in the form

$$\dot{e} = Ae + Bu + L\tilde{d} \tag{4.40}$$

where in the six-phase machine's case $L = I \in R^2$ with I being the identity matrix. Variable $\tilde{d}$ denotes model uncertainties and external disturbances of the motor's model. The following Lyapunov equation is considered

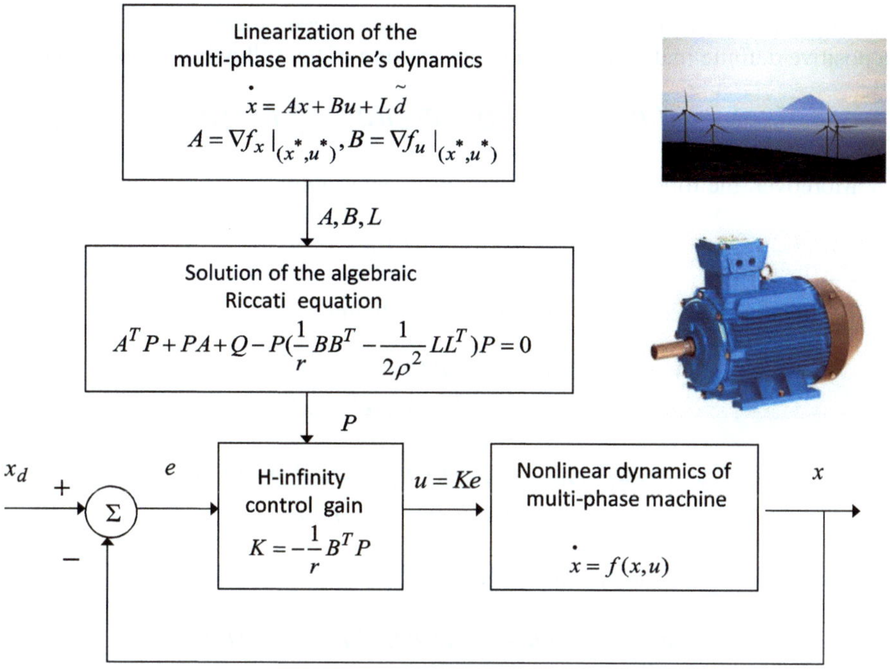

Fig. 4.2 Diagram of the control scheme for the six-phase synchronous machine

$$V = \tfrac{1}{2}e^T P e \tag{4.41}$$

where $e = x - x_d$ is the tracking error. By differentiating with respect to time one obtains

$$\begin{aligned}\dot{V} &= \tfrac{1}{2}\dot{e}^T P e + \tfrac{1}{2} e P \dot{e} \Rightarrow \\ \dot{V} &= \tfrac{1}{2}[Ae + Bu + L\tilde{d}]^T P + \tfrac{1}{2}e^T P[Ae + Bu + L\tilde{d}] \Rightarrow\end{aligned} \tag{4.42}$$

$$\begin{aligned}\dot{V} = \tfrac{1}{2}[e^T A^T + u^T B^T + \tilde{d}^T L^T]Pe + \\ + \tfrac{1}{2}e^T P[Ae + Bu + L\tilde{d}] \Rightarrow\end{aligned} \tag{4.43}$$

$$\begin{aligned}\dot{V} = \tfrac{1}{2}e^T A^T P e + \tfrac{1}{2}u^T B^T P e + \tfrac{1}{2}\tilde{d}^T L^T P e + \\ \tfrac{1}{2}e^T P A e + \tfrac{1}{2}e^T P B u + \tfrac{1}{2}e^T P L \tilde{d}\end{aligned} \tag{4.44}$$

The previous equation is rewritten as

$$\begin{aligned}\dot{V} = \tfrac{1}{2}e^T(A^T P + PA)e + (\tfrac{1}{2}u^T B^T P e + \tfrac{1}{2}e^T P B u) + \\ + (\tfrac{1}{2}\tilde{d}^T L^T P e + \tfrac{1}{2}e^T P L \tilde{d})\end{aligned} \tag{4.45}$$

Assumption: For given positive definite matrix Q and coefficients r and ρ there exists a positive definite matrix P, which is the solution of the following matrix equation

$$A^T P + PA = -Q + P(\tfrac{2}{r}BB^T - \tfrac{1}{\rho^2}LL^T)P \tag{4.46}$$

Moreover, the following feedback control law is applied to the system

$$u = -\tfrac{1}{r}B^T Pe \tag{4.47}$$

By substituting Eqs. (4.46) and (4.47) one obtains

$$\begin{aligned}\dot{V} = \tfrac{1}{2}e^T[-Q + P(\tfrac{2}{r}BB^T - \tfrac{1}{2\rho^2}LL^T)P]e+ \\ +e^T PB(-\tfrac{1}{r}B^T Pe) + e^T PL\tilde{d} \Rightarrow\end{aligned} \tag{4.48}$$

$$\begin{aligned}\dot{V} = -\tfrac{1}{2}e^T Qe + (\tfrac{2}{r}PBB^T Pe - \tfrac{1}{2\rho^2}e^T PLL^T)Pe \\ -\tfrac{1}{r}e^T PBB^T Pe) + e^T PL\tilde{d}\end{aligned} \tag{4.49}$$

which after intermediate operations gives

$$\dot{V} = -\tfrac{1}{2}e^T Qe - \tfrac{1}{2\rho^2}e^T PLL^T Pe + e^T PL\tilde{d} \tag{4.50}$$

or, equivalently

$$\begin{aligned}\dot{V} = -\tfrac{1}{2}e^T Qe - \tfrac{1}{2\rho^2}e^T PLL^T Pe+ \\ +\tfrac{1}{2}e^T PL\tilde{d} + \tfrac{1}{2}\tilde{d}^T L^T Pe\end{aligned} \tag{4.51}$$

Lemma: The following inequality holds

$$\tfrac{1}{2}e^T L\tilde{d} + \tfrac{1}{2}\tilde{d}L^T Pe - \tfrac{1}{2\rho^2}e^T PLL^T Pe \le \tfrac{1}{2}\rho^2\tilde{d}^T\tilde{d} \tag{4.52}$$

Proof: The binomial $(\rho\alpha - \frac{1}{\rho}b)^2$ is considered. Expanding the left part of the above inequality one gets

$$\begin{aligned}\rho^2 a^2 + \tfrac{1}{\rho^2}b^2 - 2ab \ge 0 \Rightarrow \tfrac{1}{2}\rho^2 a^2 + \tfrac{1}{2\rho^2}b^2 - ab \ge 0 \Rightarrow \\ ab - \tfrac{1}{2\rho^2}b^2 \le \tfrac{1}{2}\rho^2 a^2 \Rightarrow \tfrac{1}{2}ab + \tfrac{1}{2}ab - \tfrac{1}{2\rho^2}b^2 \le \tfrac{1}{2}\rho^2 a^2\end{aligned} \tag{4.53}$$

The following substitutions are carried out: $a = \tilde{d}$ and $b = e^T PL$ and the previous relation becomes

$$\tfrac{1}{2}\tilde{d}^T L^T Pe + \tfrac{1}{2}e^T PL\tilde{d} - \tfrac{1}{2\rho^2}e^T PLL^T Pe \le \tfrac{1}{2}\rho^2\tilde{d}^T\tilde{d} \tag{4.54}$$

Equation (4.54) is substituted in Eq. (4.51) and the inequality is enforced, thus giving

$$\dot{V} \leq -\tfrac{1}{2}e^T Q e + \tfrac{1}{2}\rho^2 \tilde{d}^T \tilde{d} \tag{4.55}$$

Equation (4.55) shows that the H_∞ tracking performance criterion is satisfied. The integration of $\dot{V}$ from 0 to T gives

$$\begin{array}{c} \int_0^T \dot{V}(t)dt \leq -\tfrac{1}{2}\int_0^T ||e||_Q^2 dt + \tfrac{1}{2}\rho^2 \int_0^T ||\tilde{d}||^2 dt \Rightarrow \\ 2V(T) + \int_0^T ||e||_Q^2 dt \leq 2V(0) + \rho^2 \int_0^T ||\tilde{d}||^2 dt \end{array} \tag{4.56}$$

Moreover, if there exists a positive constant $M_d > 0$ such that

$$\int_0^\infty ||\tilde{d}||^2 dt \leq M_d \tag{4.57}$$

then one gets

$$\int_0^\infty ||e||_Q^2 dt \leq 2V(0) + \rho^2 M_d \tag{4.58}$$

Thus, the integral $\int_0^\infty ||e||_Q^2 dt$ is bounded. Moreover, $V(T)$ is bounded and from the definition of the Lyapunov function V in Eq. (4.41) it becomes clear that $e(t)$ will be also bounded since $e(t) \in \Omega_e = \{e | e^T P e \leq 2V(0) + \rho^2 M_d\}$. According to the above and with the use of Barbalat's Lemma one obtains $lim_{t\rightarrow\infty} e(t) = 0$.

4.2.6 Robust State Estimation with the Use of the H_∞ Kalman Filter

Another problem that has to be dealt with in the design of a state feedback controller for the 6-phase PMSM is that in several operating conditions the complete state vector might not be measurable. To reconstruct the missing information about the state vector of the multi-phase machine it is proposed to apply a filtering scheme and based on it to apply state estimation-based control using the H-infinity Kalman Filter [60, 112, 197, 376]. The recursion of the H_∞ Kalman Filter, for the model of the six-phase machine given in Eq. (4.95), can be formulated in terms of a *measurement update* and a *time update* part.

Measurement update:

$$\begin{array}{l} D(k) = [I - \theta W(k) P^-(k) + C^T(k) R(k)^{-1} C(k) P^-(k)]^{-1} \\ K(k) = P^-(k) D(k) C^T(k) R(k)^{-1} \\ \hat{x}(k) = \hat{x}^-(k) + K(k)[y(k) - C\hat{x}^-(k)] \end{array} \tag{4.59}$$

Time update:

$$\begin{array}{l} \hat{x}^-(k+1) = A(k)x(k) + B(k)u(k) \\ P^-(k+1) = A(k)P^-(k)D(k)A^T(k) + Q(k) \end{array} \tag{4.60}$$

where it is assumed that parameter θ is sufficiently small to assure that the prior to measurement update of the covariance matrix $P^{-}(k) - \theta W(k) + C^T(k)R(k)^{-1}C(k)$ will be positive definite. When $\theta = 0$ the H_∞ Kalman Filter becomes equivalent to the standard Kalman Filter.

4.2.7 Simulation Tests

The efficiency of the proposed control method was evaluated through simulation experiments in several test cases. The associated results have been presented in Figs. 4.3, 4.4, 4.5, 4.6 and 4.7. The developed control scheme was a state estimation-based one, since it was assumed that only specific state vector elements of the 6-phase PMSM could be directly measured. Actually, only measurements about state variables $x_2 = i_{d1}$ and $x_4 = i_{d2}$ were directly used, while the rest of the state vector elements of the multi-phase machine (that is the rotor's turn speed $x_1 = \omega_r$ and the currents $x_3 = i_{q1}$ and $x_5 = i_{q2}$) were estimated through filtering. It is also noted that the permanent magnet flux linkage $\lambda_m^{'}$ that appears in Eq. (4.11), can be dynamically estimated by the H-infinity Kalman Filter if it is considered as additional state variable in the motor's model. In the latter case the H-infinity Kalman Filter functions as an extended state observer.

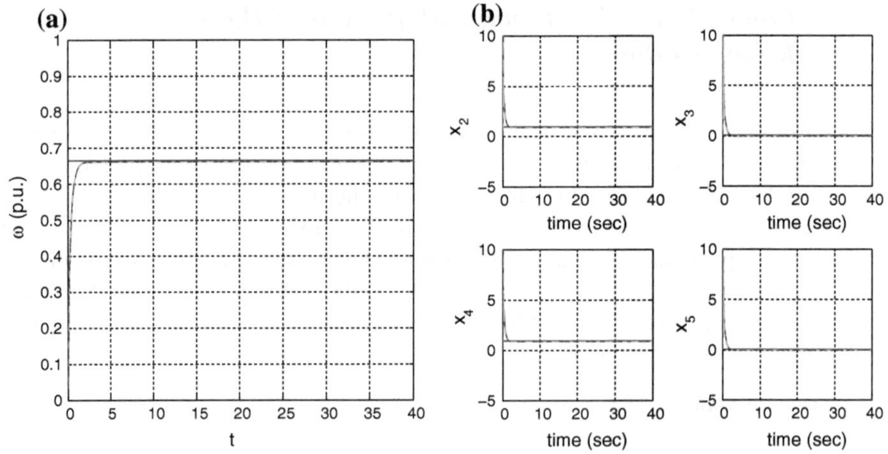

Fig. 4.3 **a** Convergence of the rotation speed $x_1 = \omega$ of the 6-phase PMSM motor to setpoint 1, **b** convergence of state variables x_2 to x_5 of the multi-phase machine to the reference signals of setpoints 1

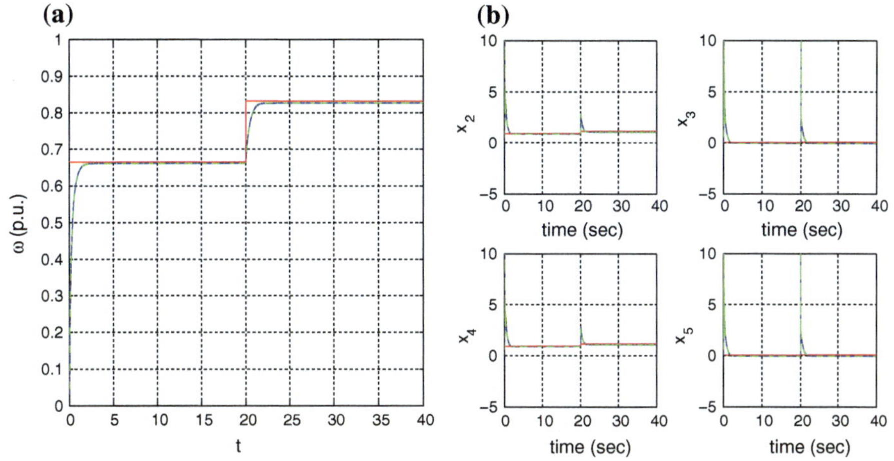

Fig. 4.4 **a** Convergence of the rotation speed $x_1 = \omega$ of the 6-phase PMSM motor to setpoint 2, **b** convergence of state variables x_2 to x_5 of the multi-phase machine to the reference signals of setpoints 2

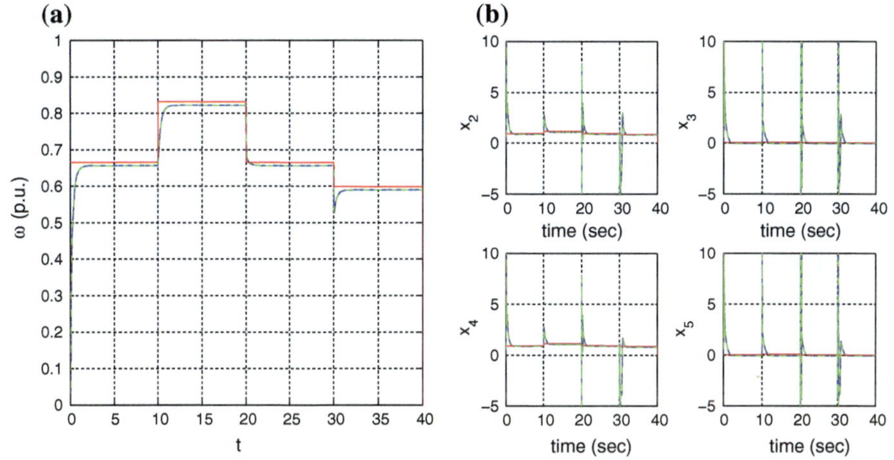

Fig. 4.5 **a** Convergence of the rotation speed $x_1 = \omega$ of the 6-phase PMSM motor to setpoint 3, **b** convergence of state variables x_2 to x_5 of the multi-phase machine to the reference signals of setpoints 3

At each iteration of the control algorithm an estimate of the state vector was provided by the H-infinity Kalman Filter and this was used next by the H-infinity controller. The feedback gain of the H-infinity controller was based on the solution of the previously described algebraic Riccati equation. It can be noticed, that in all cases the rotor's turn speed $x_1 = \omega$ converged fast to its setpoint while for all other

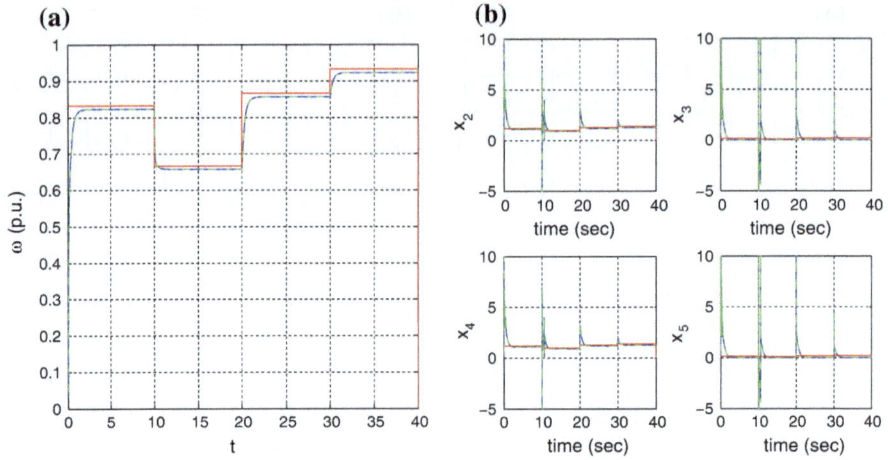

Fig. 4.6 **a** Convergence of the rotation speed $x_1 = \omega$ of the 6-phase PMSM motor to setpoint 4, **b** convergence of state variables x_2 to x_5 of the multi-phase machine to the reference signals of setpoints 4

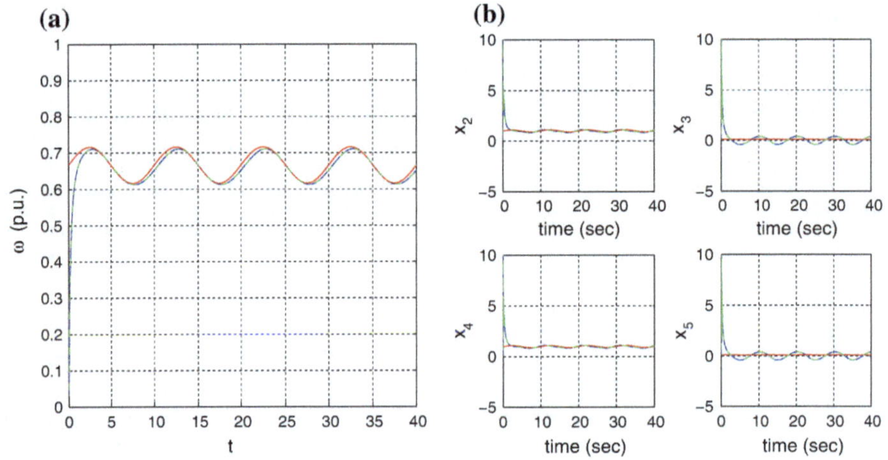

Fig. 4.7 **a** Convergence of the rotation speed $x_1 = \omega$ of the 6-phase PMSM motor to setpoint 5, **b** convergence of state variables x_2 to x_5 of the multi-phase machine to the reference signals of setpoints 5

stsate variables of the machine x_2 to x_5 fast and accurate tracking of the reference setpoints was also achieved. Yet, simple in concept and implementation the proposed nonlinear H-infinity control method has given very satisfactory results.

4.3 An H-infinity Approach to Optimal Control of Doubly-Fed Reluctance Machines

4.3.1 Overview

As pointed out, the deployment of renewable energy systems and the growing need for electric power has motivated research on new and more efficient electric power generators. Comparing to synchronous and induction machines, brushless doubly-fed reluctance machines achieve high power and torque while keeping low their power losses [318, 409, 425, 502]. In such machines there is no winding or permanent magnets on the rotor and all the windings are located on the stator. The machine has no brushes and slip rings and therefore it can be considered as a machine that is less prone to failures. To solve the control problem of reluctance machines several approaches have been proposed, among which vector control methods [5, 6, 46, 77, 179, 201] as well as some robust and adaptive control methods [232, 449, 460]. This section proposes a new nonlinear optimal control method for reluctance machines which is based on H-infinity control theory [151, 346, 352, 375, 376]. First, the dynamic model of the reluctance machines is analyzed and it is explained that a primary three-phase winding at the stator is directly connected to the grid while a secondary winding is connected to the grid through an AC/DC/AC converter and is used to control the stator's magnetic field and consequently the rotor's motion. Next, a nonlinear control algorithm is developed which can be applied to the machine through a PWM scheme.

The nonlinear dynamical model of the doubly-fed reluctance machine is linearized round a local operating point which is defined by the present value of the machine's state vector and the last value of the control inputs vector that was exerted on it. This linearization procedure requires a Taylor series expansion and the computation of the associated Jacobian matrices [34, 341, 344]. The linearization error which is due to the truncation of higher order terms in the Taylor series is viewed as a perturbation that is compensated by the robustness of the control algorithm. Next, a stabilizing state feedback controller is designed for the linearized model of the reluctance machine using H-infinity control theory. A cost function is defined comprising a quadratic term based on the system's state vector error. The control problem is viewed as a mini-max differential game between the control input (which tries to minimize the cost function) and the disturbances that try to maximize the cost function.

The feedback gain of the H-infinity controller is computed from the solution of an algebraic Riccati equation. This solution is recalculated at each iteration of the control algorithm. By applying Lyapunov stability analysis it is shown that an H-infinity tracking performance criterion is satisfied for the control loop. This signifies that the control system remains robust to modelling uncertainties and external disturbances. Moreover, under moderate conditions it is proven that global asymptotic stability holds and that the tracking error of the state vector is eliminated. To cope with the case of a partially measurable state vector in the reluctance machine, state estimation-based control is applied [60, 112, 181, 197]. Moreover, to assure robust state estimation the use of the H-infinity Kalman Filter is proposed [160, 408]. The performance of the H-infinity nonlinear control scheme is confirmed through simulation experiments.

4.3.2 Dynamic Model of the Doubly-Fed Reluctance Machine

4.3.2.1 Dynamics of the Electrical and Mechanical Part of the Machine

The brushless doubly-fed reluctance machine has two separate stator windings where the first one is noted as stator winding while the second one is noted as control winding (Fig. 4.8). The power winding is directly connected to the 3-phase grid and using the dq reference notation one has two voltage components v_d and v_q. The frequency of current at the primary winding is the one of the grid and is denoted as ω_1. The secondary winding is connected to the grid through an AC/DC/AC converter thus it can have current at a frequency different from the one of the grid, which is denoted as ω_2. The power winding has p_1 poles while the secondary (control) winding has $p_{@}$ poles. The rotor of the machine has no windings and the number of poles in it is usually chosen to be $p_r = (p_1 + p_2)/2$. About the electrical angular speed of the reluctance rotor one has $\omega_r = \omega_1 - \omega_2$ and that $p_r \omega_r = \omega_m$

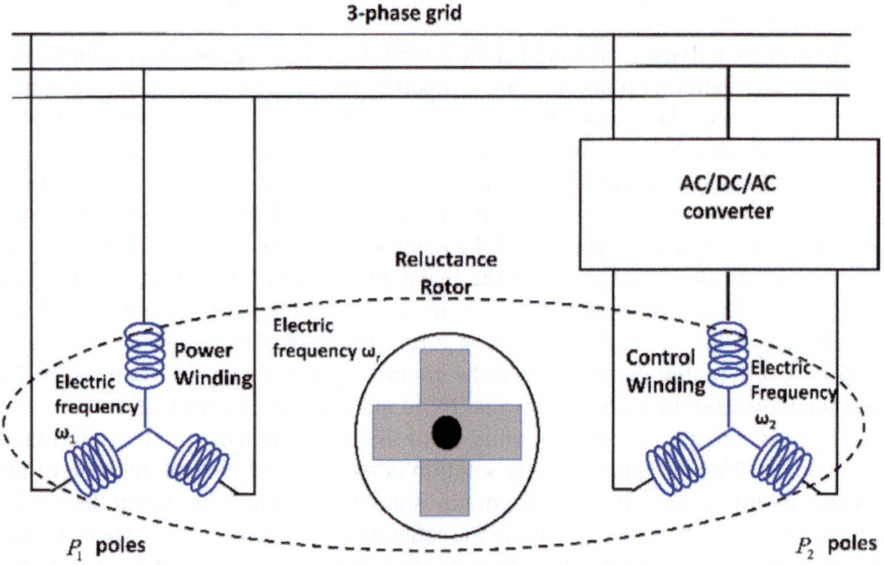

Fig. 4.8 Doubly-fed reluctance machine and its windings

Next, by applying Kirchhoff's laws at the primary and the secondary winding of the reluctance machine one obtains the dynamic model of its electrical part. Thus, at the primary winding one has [449]

$$\dot{\lambda}_{1d} = -R_1 i_{1d} + \omega_1 \lambda_{1q} + v_{1d} \tag{4.61}$$

$$\dot{\lambda}_{1q} = -R_1 i_{1q} - \omega_1 \lambda_{1d} + v_{1q} \tag{4.62}$$

while at the secondary winding it holds

$$\dot{\lambda}_{2d} = -R_2 i_{2d} + \omega_2 \lambda_{2q} + v_{2d} \tag{4.63}$$

$$\dot{\lambda}_{2q} = -R_2 i_{2q} - \omega_2 \lambda_{2d} + v_{2q} \tag{4.64}$$

The magnetic flux at the primary winding is the result of the inductance of this winding and of the mutual inductance (coupling) with the secondary winding

$$\lambda_{1d} = L_1 i_{1d} + L_{12} i_{2d} \tag{4.65}$$

$$\lambda_{1q} = L_1 i_{1q} - L_{12} i_{2q} \tag{4.66}$$

Equivalently, the magnetic flux at the secondary winding is the result of the inductance of this winding and of the mutual inductance (coupling) with the primary winding

$$\lambda_{2d} = L_2 i_{2d} + L_{12} i_{1d} \tag{4.67}$$

$$\lambda_{2q} = L_2 i_{2q} - L_{12} i_{1q} \tag{4.68}$$

In the above equations variables v_{1d}, v_{1q} are the grid voltage components in the dq reference frame which are exerted as inputs to the circuit of the primary winding, while v_{2d}, v_{2q} are the voltage components at the output of the AC/DC/AC converter, expressed in the dq reference frame, which are exerted as inputs to the circuit of the secondary winding.

Equivalently i_{1d}, i_{1q} are the currents of the primary winding expressed in the dq reference frame, while i_{2d}, i_{2q} are the currents of the secondary winding represented again in the dq reference frame.

Moreover, λ_{1d}, λ_{1q} are the components of the magnetic flux of the primary winding written in the dq reference frame, while λ_{2d}, λ_{2q} are components of the magnetic flux of the secondary winding expressed again in the dq reference frame.

Additionally, R_1 and L_1 is the resistance and the inductance of the primary winding, while R_2 and L_2 is the resistance and the inductance of the secondary winding. Moreover, L_{12} is the mutual inductance (coupling) between the primary and the secondary winding.

The electromagnetic torque of the machine is a function of the currents and the magnetic flux expressed in the dq reference frame. It also depends on the number of rotor's poles. Thus, one has

$$T_e = \frac{3}{2}\frac{L_{12}}{L_1}p_r(\lambda_{1d}i_{2q} + \lambda_{1q}i_{2d}) \tag{4.69}$$

Next, the dynamics of the mechanical part of the reluctance machine (generator) can be formulated. Considering that the moment of inertia of the rotor is J_r, and denoting the rotation speed of the rotor as ω one has that

$$\begin{array}{c}\dot{\omega} = \frac{1}{J_r}(-T_e + T_m) \Rightarrow \\ \dot{\omega} = \frac{1}{J_r}[-\frac{3}{2}\frac{L_{12}}{L_1}p_r(\lambda_{1d}i_{2q} + \lambda_{1q}i_{2d}) + T_m]\end{array} \tag{4.70}$$

Moreover, the active power of the reluctance machine is computed as follows [449]

$$P^* = \frac{3}{2}(v_{1d}i_{1d} + v_{1q}i_{1d}) \tag{4.71}$$

or an equivalent relation can be obtained from $P = T_e\omega$ that is

$$P^* = \frac{3}{2}\frac{L_{12}}{L_1}p_r(\lambda_{1d}i_{2q} + \lambda_{1q}i_{2d})\omega \tag{4.72}$$

whereas the reactive power of the machine is given by

$$Q^* = 3/2(v_{1d}i_{1q} - v_{1q}i_{1d}) \tag{4.73}$$

4.3.2.2 State-Space Model of the Reluctance Machine

By substituting the relations about the magnetic flux in the primary and the secondary winding which are given in Eqs. (4.66)–(4.68) into the Eqs. (4.61)–(4.64) and after solving with respect to $\dot{i}_{1d}$, $\dot{i}_{1q}$, $\dot{i}_{2d}$, and $\dot{i}_{2q}$, one arrives at the following description:

$$\begin{pmatrix} L_1 & 0 & L_{12} & 0 \\ 0 & L_1 & 0 & -L_{12} \\ L_{12} & 0 & L_2 & 0 \\ 0 & -L_{12} & 0 & L_2 \end{pmatrix} \begin{pmatrix} d/dt i_{1d} \\ d/dt i_{1q} \\ d/dt i_{2d} \\ d/dt i_{2q} \end{pmatrix} = \begin{pmatrix} -R_1 i_{1d} + \omega_1(L_1 i_{1q} - L_{12}i_{2q}) + v_{1d} \\ -R_1 i_{1q} - \omega_1(L_1 i_{1d} + L_{12}i_{2d}) + v_{1q} \\ -R_2 i_{2d} + \omega_2(L_2 i_{2q} - L_{12}i_{1q}) + v_{2d} \\ -R_2 i_{2q} - \omega_2(L_2 i_{2d} + L_{12}i_{1d}) + v_{1q} \end{pmatrix} \tag{4.74}$$

By denoting $\sigma = 1/(L_1L_2 - L_{12}^2)$, and by using that about the grid voltage holds that v_{1d} is constant and v_{1q} is zero, the previous relation provides finally the following equations about the electrical dynamics of the machine:

$$\begin{array}{c}\frac{d}{dt}i_{1d} = \sigma\{-R_1L_2i_{1d} + \omega_1L_1L_2i_{1q} - \omega_1L_2L_{12}i_{2q} + L_2V_s + \\ + R_2L_{12}i_{2d} - \omega_2L_2L_{12}i_{2q} + \omega_2L_{12}^2i_{1q} - L_{12}v_{2d}\}\end{array} \tag{4.75}$$

$$\begin{aligned}\frac{d}{dt}i_{1q} &= \sigma\{-R_1L_2i_{1q} + \omega_1L_1L_2i_{1d} + \omega_1L_2L_{12}i_{2d} - \\ &- R_2L_{12}i_{2q} - \omega_2L_2L_{12}i_{2d} + \omega_2L_{12}^2i_{1d} - L_{12}v_{2q}\}\end{aligned} \tag{4.76}$$

$$\begin{aligned}\frac{d}{dt}i_{2d} &= \sigma\{-R_1L_{12}i_{2d} - \omega_1L_1L_{12}i_{1q} + \omega_1L_{12}^2i_{2q} - L_{12}V_s - \\ &- R_2L_1i_{2d} + \omega_2L_1L_2i_{2q} - \omega_2L_1L_{12}i_{1q} + L_1v_{2d}\}\end{aligned} \tag{4.77}$$

$$\begin{aligned}\frac{d}{dt}i_{2q} &= \sigma\{-R_1L_{12}i_{1q} - \omega_1L_1L_{12}i_{1d} - \omega_1L_{12}^2i_{2d} - \\ &- R_2L_1i_{2q} - \omega_2L_1L_2i_{2d} - \omega_2L_1L_{12}i_{1d} + L_1v_{2q}\}\end{aligned} \tag{4.78}$$

Moreover, from Eq. (4.70) about the rotor's turn speed one has

$$\dot{\omega} = \frac{1}{J_r}\left(\frac{3}{2}\frac{L_{12}}{L_1}p_r(L_1i_{1d}i_{2q} + L_{12}i_{2d}i_{2q} + L_1i_{1q}i_{2d} + L_{12}i_{2q}i_{2d})\right) \tag{4.79}$$

Additionally, by considering damping (friction) effects in the rotor's motion one has

$$\dot{\omega} = -\frac{b}{J}\omega + \frac{1}{J_r}\left(\frac{3}{2}\frac{L_{12}}{L_1}p_r(L_1i_{1d}i_{2q} + L_{12}i_{2d}i_{2q} + L_1i_{1q}i_{2d} + L_{12}i_{2q}i_{2d})\right) \tag{4.80}$$

Next, by defining the state variables of the model as $x_1 = \omega$, $x_2 = i_{1d}$, $x_3 = i_{1q}$, $x_4 = i_{2d}$, $x_5 = i_{2q}$ and the control inputs $u_1 = v_{2d}$ and $u_2 = v_{2q}$ the following state-space description for the reluctance machine is obtained:

$$\dot{x}_1 = -\frac{b}{J}x_1 - \frac{3}{2J}\frac{L_{12}}{L_1}p_r(L_1x_2x_5 + L_{12}x_3x_5 + L_1x_4x_5 + L_{12}x_3x_5) + \frac{T_m}{J} \tag{4.81}$$

$$\begin{aligned}\dot{x}_2 &= -\sigma R_1L_2x_2 - \sigma\omega_1L_1L_2x_3 - \sigma\omega_1L_2L_{12}x_5 + \sigma L_2V_s - \\ &- \sigma R_2L_{12}x_4 - \sigma\omega_2L_2L_{12}x_5 + \sigma\omega_2L_{12}^2x_3 - \sigma L_{12}u_1\end{aligned} \tag{4.82}$$

$$\begin{aligned}\dot{x}_3 &= -\sigma R_1L_2x_3 - \sigma\omega_1L_1L_2x_2 + \sigma\omega_1L_2L_{12}x_4 - \\ &- \sigma R_2L_{12}x_5 - \sigma\omega_2L_2L_{12}x_4 + \sigma\omega_2L_{12}^2x_2 + \sigma L_{12}u_2\end{aligned} \tag{4.83}$$

$$\begin{aligned}\dot{x}_4 &= -\sigma R_1L_{12}x_2 - \sigma\omega_1L_1L_{12}x_3 + \sigma\omega_1L_{12}^2x_5 - \sigma L_{12}V_s - \\ &- \sigma R_2L_1x_4 + \sigma\omega_2L_1L_2x_5 - \sigma\omega_2L_1L_{12}x_3 + \sigma L_1u_1\end{aligned} \tag{4.84}$$

$$\begin{aligned}\dot{x}_5 &= -\sigma R_1L_{12}x_3 - \sigma\omega_1L_1L_{12}x_2 - \sigma\omega_1L_{12}^2x_4 - \\ &- \sigma R_2L_1x_5 - \sigma\omega_2L_1L_2x_4 - \sigma\omega_2L_1L_{12}x_2 + \sigma L_1u_2\end{aligned} \tag{4.85}$$

Consequently, the state-space description of the system becomes:

$$\dot{x} = f(x) + g(x)u \tag{4.86}$$

where $x \in R^{5\times1}$, $u \in R^{2\times1}$, $f(x) \in R^{5\times1}$ and $g(x) \in R^{5\times2}$, with the drift matrix to be given by

$$f(x)=\begin{pmatrix} -\frac{b}{J}x_1+\frac{3}{2J}\frac{L_{12}}{L_1}p_r(L_1x_2x_5+L_{12}x_3x_5+L_1x_4x_5+L_{12}x_3x_5)+\frac{T_m}{J} \\ -\sigma R_1L_2x_2-\sigma\omega_1L_1L_2x_3-\sigma\omega_1L_2L_{12}x_5+\sigma L_2V_s- \\ -\sigma R_2L_{12}x_4-\sigma\omega_2L_2L_{12}x_5+\sigma\omega_2L_{12}^2x_3 \\ -\sigma R_1L_2x_3-\sigma\omega_1L_1L_2x_2+\sigma\omega_1L_2L_{12}x_4- \\ -\sigma R_2L_{12}x_5-\sigma\omega_2L_2L_{12}x_4+\sigma\omega_2L_{12}^2x_2 \\ -\sigma R_1L_{12}x_2-\sigma\omega_1L_1L_{12}x_3+\sigma\omega_1L_{12}^2x_5-\sigma L_{12}V_s- \\ -\sigma R_2L_1x_4+\sigma\omega_2L_1L_2x_5-\sigma\omega_2L_1L_{12}x_3 \\ -\sigma R_1L_{12}x_3-\sigma\omega_1L_1L_{12}x_2-\sigma\omega_1L_{12}^2x_4- \\ -\sigma R_2L_1x_5-\sigma\omega_2L_1L_2x_4-\sigma\omega_2L_1L_{12}x_2 \end{pmatrix} \tag{4.87}$$

and the inputs gain matrix to be given by

$$g(x)=\begin{pmatrix} 0 & 0 \\ -\sigma L_{12} & 0 \\ 0 & \sigma L_{12} \\ \sigma L_1 & 0 \\ 0 & \sigma L_2 \end{pmatrix} \tag{4.88}$$

4.3.3 Linearization of the Reluctance Machine's State-Space Models

Local linearization is performed for the reluctance machine's state-space model, round the operating point (x^*, u^*) where x^* is the present value of the system's state vector and u^* is the last value of the control input that was exerted on the machine. Thus, one obtains the linearized description

$$\dot{x}=Ax+Bu+\tilde{d} \tag{4.89}$$

For the previous description of the reluctance machine by the state-space equation of Eq. (4.123) it holds that $A=\nabla_x f(x)$ and $B=g(x)$ with

$$A=\nabla_x[f(x)+g(x)u]=\begin{pmatrix} \frac{\partial f_1}{\partial x_1} & \frac{\partial f_1}{\partial x_2} & \frac{\partial f_1}{\partial x_3} & \frac{\partial f_1}{\partial x_4} & \frac{\partial f_1}{\partial x_5} \\ \frac{\partial f_2}{\partial x_1} & \frac{\partial f_2}{\partial x_2} & \frac{\partial f_2}{\partial x_3} & \frac{\partial f_2}{\partial x_4} & \frac{\partial f_2}{\partial x_5} \\ \frac{\partial f_3}{\partial x_1} & \frac{\partial f_3}{\partial x_2} & \frac{\partial f_3}{\partial x_3} & \frac{\partial f_3}{\partial x_4} & \frac{\partial f_3}{\partial x_5} \\ \frac{\partial f_4}{\partial x_1} & \frac{\partial f_4}{\partial x_2} & \frac{\partial f_4}{\partial x_3} & \frac{\partial f_4}{\partial x_4} & \frac{\partial f_4}{\partial x_5} \\ \frac{\partial f_5}{\partial x_1} & \frac{\partial f_5}{\partial x_2} & \frac{\partial f_5}{\partial x_3} & \frac{\partial f_5}{\partial x_4} & \frac{\partial f_5}{\partial x_5} \end{pmatrix} \tag{4.90}$$

About the first row of the Jacobian matrix one has $\frac{\partial f_1}{\partial x_1}=-b$, $\frac{\partial f_1}{\partial x_2}=-\frac{3}{2J}\frac{L_{12}}{L_1}p_r$ (L_1x_5), $\frac{\partial f_1}{\partial x_3}=-\frac{3}{2J}\frac{L_{12}}{L_1}p_r(2L_{12}x_5)$, $\frac{\partial f_1}{\partial x_4}=-\frac{3}{2J}\frac{L_{12}}{L_1}p_r(L_1x_5)$, $\frac{\partial f_1}{\partial x_5}=-\frac{3}{2J}\frac{L_{12}}{L_1}p_r$ $(L_1x_2+2L_{12}x_3+L_1x_4)$.

About the second row of the Jacobian matrix one has $\frac{\partial f_2}{\partial x_1}=0$, $\frac{\partial f_2}{\partial x_2}=-\sigma R_1L_2$, $\frac{\partial f_2}{\partial x_3}=-\sigma\omega_1L_1L_2+\sigma\omega_2L_{12}^2$, $\frac{\partial f_2}{\partial x_4}=-\sigma R_2L_{12}$, $\frac{\partial f_2}{\partial x_5}=-\sigma\omega_1L_1L_{12}-\sigma\omega_2L_2L_{12}$.

About the third row of the Jacobian matrix one has $\frac{\partial f_3}{\partial x_1}=0$, $\frac{\partial f_3}{\partial x_2}=-\sigma\omega_1L_1L_2+\sigma\omega_2L_{12}^2$, $\frac{\partial f_3}{\partial x_3}=-\sigma R_1L_2$, $\frac{\partial f_3}{\partial x_4}=\sigma\omega_1L_2L_{12}-\sigma\omega_2L_2L_{12}$, $\frac{\partial f_3}{\partial x_5}=-\sigma R_1L_2-\sigma R_2L_{12}$.

About the fourth row of the Jacobian matrix one has $\frac{\partial f_4}{\partial x_1}=0$, $\frac{\partial f_4}{\partial x_2}=-\sigma R_1L_{12}$, $\frac{\partial f_4}{\partial x_3}=-\sigma\omega_1L_1L_{12}-\sigma L_1L_{12}$, $\frac{\partial f_4}{\partial x_4}=-\sigma R_2L_1$, $\frac{\partial f_4}{\partial x_5}=\sigma\omega_1L_{12}^2+\sigma\omega_2L_1L_2$.

About the fifth row of the Jacobian matrix one has $\frac{\partial f_5}{\partial x_1}=0$, $\frac{\partial f_5}{\partial x_2}=-\sigma\omega_1L_1L_{12}-\sigma\omega_2L_1L_{12}$,4 $\frac{\partial f_5}{\partial x_3}=-\sigma R_1L_{12}$, $\frac{\partial f_5}{\partial x_4}=-\sigma\omega_1L_{12}^2-\sigma\omega_2L_1L_2$, $\frac{\partial f_5}{\partial x_5}=-\sigma R_2L_1$, while one also has

$$B=\nabla_u[f(x)+g(x)u]=\begin{pmatrix}0&0\\ \sigma L_{12}&0\\ 0&\sigma L_{12}\\ \sigma L_1&0\\ 0&\sigma L_2\end{pmatrix} \tag{4.91}$$

4.3.4 The Nonlinear H-infinity Control

4.3.4.1 Mini-Max Control and Disturbance Rejection

The concept of nonlinear H-infinity control was explained in the case of the 6-phase synchronous machine, and will be also applied to the model of the doubly-fed reluctance machine. The initial nonlinear model of the doubly-fed reluctance machine is in the form

$$\dot{x}=f(x,u)\ \ x\in R^n,\ \ u\in R^m \tag{4.92}$$

Linearization of the system (doubly-fed reluctance machine) is performed at each iteration of the control algorithm, through Taylor series expansion, round its present operating point $(x^*,u^*)=(x(t),u(t-T_s))$. The linearized equivalent of the system is described by

$$\dot{x}=Ax+Bu+L\tilde{d}\ \ x\in R^n,\ \ u\in R^m,\ \tilde{d}\in R^q \tag{4.93}$$

where matrices A and B are obtained from the computation of the Jacobians

$$A = \begin{pmatrix} \frac{\partial f_1}{\partial x_1} & \frac{\partial f_1}{\partial x_2} & \cdots & \frac{\partial f_1}{\partial x_n} \\ \frac{\partial f_2}{\partial x_1} & \frac{\partial f_2}{\partial x_2} & \cdots & \frac{\partial f_2}{\partial x_n} \\ \cdots & \cdots & \cdots & \cdots \\ \frac{\partial f_n}{\partial x_1} & \frac{\partial f_n}{\partial x_2} & \cdots & \frac{\partial f_n}{\partial x_n} \end{pmatrix} |_{(x^*,u^*)} \quad B = \begin{pmatrix} \frac{\partial f_1}{\partial u_1} & \frac{\partial f_1}{\partial u_2} & \cdots & \frac{\partial f_1}{\partial u_m} \\ \frac{\partial f_2}{\partial u_1} & \frac{\partial f_2}{\partial u_2} & \cdots & \frac{\partial f_2}{\partial u_m} \\ \cdots & \cdots & \cdots & \cdots \\ \frac{\partial f_n}{\partial u_1} & \frac{\partial f_n}{\partial u_2} & \cdots & \frac{\partial f_n}{\partial u_m} \end{pmatrix} |_{(x^*,u^*)} \tag{4.94}$$

and vector $\tilde{d}$ denotes disturbance terms due to linearization errors. The problem of disturbance rejection for the linearized model that is described by

$$\begin{aligned} \dot{x} &= Ax + Bu + L\tilde{d} \\ y &= Cx \end{aligned} \tag{4.95}$$

where $x \in R^n$, $u \in R^m$, $\tilde{d} \in R^q$ and $y \in R^p$, cannot be handled efficiently if the classical LQR control scheme is applied. This is because of the existence of the perturbation term $\tilde{d}$. The disturbance term $\tilde{d}$ apart from modeling (parametric) uncertainty and external perturbation terms can also represent noise terms of any distribution.

In the H_∞ control approach, a feedback control scheme is designed for trajectory tracking by the system's state vector and simultaneous disturbance rejection, considering that the disturbance affects the system in the worst possible manner. The disturbances' effect are incorporated in the following quadratic cost function:

$$J(t) = \frac{1}{2}\int_0^T [y^T(t)y(t) + ru^T(t)u(t) - \rho^2\tilde{d}^T(t)\tilde{d}(t)]dt, \quad r, \rho > 0 \tag{4.96}$$

The significance of the negative sign in the cost function's term that is associated with the perturbation variable $\tilde{d}(t)$ is that the disturbance tries to maximize the cost function $J(t)$ while the control signal $u(t)$ tries to mininize it. The physical meaning of the relation given above is that the control signal and the disturbances compete to each other within a mini-max differential game. This problem of mini-max optimization can be written as

$$min_u max_{\tilde{d}} J(u, \tilde{d}) \tag{4.97}$$

The objective of the optimization procedure is again to compute a control signal $u(t)$ which can compensate for the worst possible disturbance, that is externally imposed to the system. However, the solution to the mini-max optimization problem is directly related to the value of the parameter ρ. This means that there is an upper bound in the disturbances magnitude that can be annihilated by the control signal.

4.3.4.2 H-infinity Feedback Control

For the linearized system given by Eq. (4.95) the cost function of Eq. (4.96) is defined, where the coefficient r determines the penalization of the control input and the weight coefficient ρ determines the reward of the disturbances' effects.

It is assumed that (i) The energy that is transferred from the disturbances signal $\tilde{d}(t)$ is bounded, that is $\int_0^{\infty}\tilde{d}^T(t)\tilde{d}(t)dt < \infty$, (ii) the matrices $[A, B]$ and $[A, L]$ are stabilizable, (iii) the matrix $[A, C]$ is detectable. Then, the optimal feedback control law is given by

$$u(t) = -Kx(t) \tag{4.98}$$

with

$$K = \frac{1}{r}B^T P \tag{4.99}$$

where P is a positive semi-definite symmetric matrix which is obtained from the solution of the Riccati equation

$$A^T P + PA + Q - P(\frac{1}{r}BB^T - \frac{1}{2\rho^2}LL^T)P = 0 \tag{4.100}$$

where Q is also a positive definite symmetric matrix. The worst case disturbance is given by

$$\tilde{d}(t) = \frac{1}{\rho^2}L^T Px(t) \tag{4.101}$$

The diagram of the considered control loop is depicted in Fig. 4.9.

4.3.5 Lyapunov Stability Analysis

Through Lyapunov stability analysis it will be shown that the proposed nonlinear control scheme assures H_∞ tracking performance for the doubly-fed reluctance machine, and that in case of bounded disturbance terms asymptotic convergence to the reference setpoints is succeeded.

The tracking error dynamics for the doubly-fed reluctance machine is written in the form

$$\dot{e} = Ae + Bu + L\tilde{d} \tag{4.102}$$

where in the doubly-fed reluctance machine's case $L = I \in R^{5\times 5}$ with I being the identity matrix. Variable $\tilde{d}$ denotes model uncertainties and external disturbances of the motor's model. The following Lyapunov equation is considered

$$V = \frac{1}{2}e^T Pe \tag{4.103}$$

where $e = x - x_d$ is the tracking error. By differentiating with respect to time one obtains

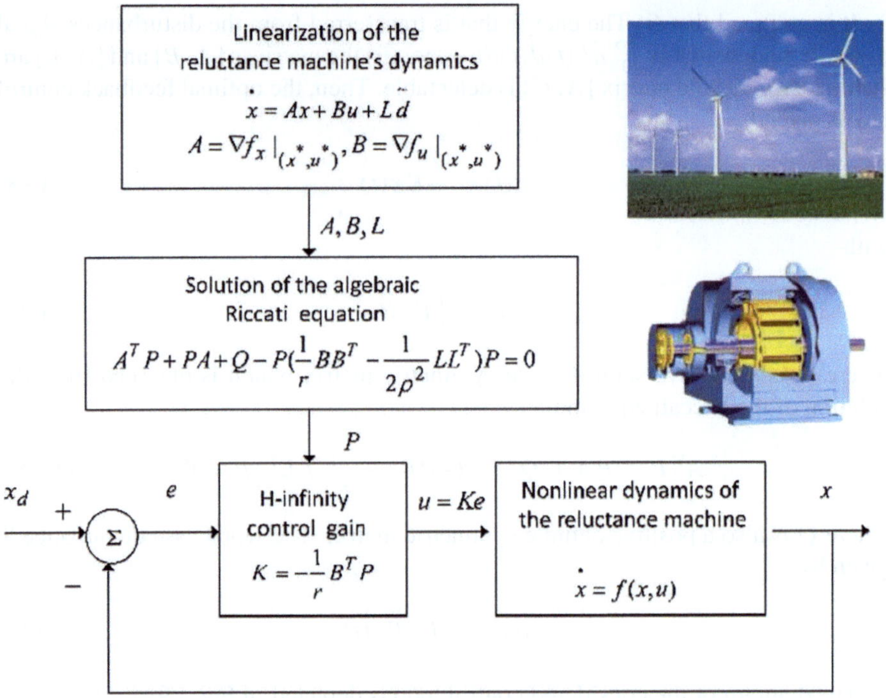

Fig. 4.9 Diagram of the control scheme for the doubly-fed reluctance machine

$$
\begin{aligned}
\dot{V} &= \tfrac{1}{2}\dot{e}^T Pe + \tfrac{1}{2}eP\dot{e} \Rightarrow \\
\dot{V} &= \tfrac{1}{2}[Ae + Bu + L\tilde{d}]^T P + \tfrac{1}{2}e^T P[Ae + Bu + L\tilde{d}] \Rightarrow
\end{aligned}
\tag{4.104}
$$

$$
\begin{aligned}
\dot{V} = &\tfrac{1}{2}[e^T A^T + u^T B^T + \tilde{d}^T L^T]Pe + \\
&+ \tfrac{1}{2}e^T P[Ae + Bu + L\tilde{d}] \Rightarrow
\end{aligned}
\tag{4.105}
$$

$$
\begin{aligned}
\dot{V} = &\tfrac{1}{2}e^T A^T Pe + \tfrac{1}{2}u^T B^T Pe + \tfrac{1}{2}\tilde{d}^T L^T Pe + \\
&\tfrac{1}{2}e^T PAe + \tfrac{1}{2}e^T PBu + \tfrac{1}{2}e^T PL\tilde{d}
\end{aligned}
\tag{4.106}
$$

The previous equation is rewritten as

$$
\begin{aligned}
\dot{V} = &\tfrac{1}{2}e^T(A^T P + PA)e + (\tfrac{1}{2}u^T B^T Pe + \tfrac{1}{2}e^T PBu) + \\
&+ (\tfrac{1}{2}\tilde{d}^T L^T Pe + \tfrac{1}{2}e^T PL\tilde{d})
\end{aligned}
\tag{4.107}
$$

Assumption: For given positive definite matrix Q and coefficients r and ρ there exists a positive definite matrix P, which is the solution of the following matrix equation

$$A^T P + PA = -Q + P(\tfrac{2}{r}BB^T - \tfrac{1}{\rho^2}LL^T)P \tag{4.108}$$

Moreover, the following feedback control law is applied to the system

$$u = -\tfrac{1}{r}B^T Pe \tag{4.109}$$

By substituting Eqs. (4.108) and (4.109) one obtains

$$\begin{aligned}\dot{V} = &\tfrac{1}{2}e^T[-Q + P(\tfrac{2}{r}BB^T - \tfrac{1}{2\rho^2}LL^T)P]e + \\ &+ e^T PB(-\tfrac{1}{r}B^T Pe) + e^T PL\tilde{d} \Rightarrow\end{aligned} \tag{4.110}$$

$$\begin{aligned}\dot{V} = &-\tfrac{1}{2}e^T Qe + (\tfrac{2}{r}PBB^T Pe - \tfrac{1}{2\rho^2}e^T PLL^T)Pe \\ &-\tfrac{1}{r}e^T PBB^T Pe) + e^T PL\tilde{d}\end{aligned} \tag{4.111}$$

which after intermediate operations gives

$$\dot{V} = -\tfrac{1}{2}e^T Qe - \tfrac{1}{2\rho^2}e^T PLL^T Pe + e^T PL\tilde{d} \tag{4.112}$$

or, equivalently

$$\begin{aligned}\dot{V} = &-\tfrac{1}{2}e^T Qe - \tfrac{1}{2\rho^2}e^T PLL^T Pe + \\ &+ \tfrac{1}{2}e^T PL\tilde{d} + \tfrac{1}{2}\tilde{d}^T L^T Pe\end{aligned} \tag{4.113}$$

Lemma: The following inequality holds

$$\tfrac{1}{2}e^T L\tilde{d} + \tfrac{1}{2}\tilde{d}L^T Pe - \tfrac{1}{2\rho^2}e^T PLL^T Pe \leq \tfrac{1}{2}\rho^2\tilde{d}^T\tilde{d} \tag{4.114}$$

Proof: The binomial $(\rho\alpha - \frac{1}{\rho}b)^2$ is considered. Expanding the left part of the above inequality one gets

$$\begin{aligned}&\rho^2 a^2 + \tfrac{1}{\rho^2}b^2 - 2ab \geq 0 \Rightarrow \tfrac{1}{2}\rho^2 a^2 + \tfrac{1}{2\rho^2}b^2 - ab \geq 0 \Rightarrow \\ &ab - \tfrac{1}{2\rho^2}b^2 \leq \tfrac{1}{2}\rho^2 a^2 \Rightarrow \tfrac{1}{2}ab + \tfrac{1}{2}ab - \tfrac{1}{2\rho^2}b^2 \leq \tfrac{1}{2}\rho^2 a^2\end{aligned} \tag{4.115}$$

The following substitutions are carried out: $a = \tilde{d}$ and $b = e^T PL$ and the previous relation becomes

$$\tfrac{1}{2}\tilde{d}^T L^T Pe + \tfrac{1}{2}e^T PL\tilde{d} - \tfrac{1}{2\rho^2}e^T PLL^T Pe \leq \tfrac{1}{2}\rho^2\tilde{d}^T\tilde{d} \tag{4.116}$$

Equation (4.116) is substituted in Eq. (4.113) and the inequality is enforced, thus giving

$$\dot{V} \leq -\tfrac{1}{2}e^T Qe + \tfrac{1}{2}\rho^2\tilde{d}^T\tilde{d} \tag{4.117}$$

Equation (4.117) shows that the H_∞ tracking performance criterion is satisfied. The integration of $\dot{V}$ from 0 to T gives

$$\begin{array}{c}\int_0^T \dot{V}(t)dt \leq -\frac{1}{2}\int_0^T ||e||_Q^2 dt + \frac{1}{2}\rho^2\int_0^T ||\tilde{d}||^2 dt \Rightarrow \\ 2V(T) + \int_0^T ||e||_Q^2 dt \leq 2V(0) + \rho^2\int_0^T ||\tilde{d}||^2 dt\end{array} \quad (4.118)$$

Moreover, if there exists a positive constant $M_d > 0$ such that

$$\int_0^\infty ||\tilde{d}||^2 dt \leq M_d \quad (4.119)$$

then one gets

$$\int_0^\infty ||e||_Q^2 dt \leq 2V(0) + \rho^2 M_d \quad (4.120)$$

Thus, the integral $\int_0^\infty ||e||_Q^2 dt$ is bounded. Moreover, $V(T)$ is bounded and from the definition of the Lyapunov function V in Eq. (4.103) it becomes clear that $e(t)$ will be also bounded since $e(t) \in \Omega_e = \{e|e^T Pe \leq 2V(0) + \rho^2 M_d\}$. According to the above and with the use of Barbalat's Lemma one obtains $lim_{t\to\infty} e(t) = 0$.

4.3.6 Robust State Estimation with the Use of the H_∞ Kalman Filter

The control loop for the doubly-fed reluctance machine has to be implemented with the use of information provided by a small number of sensors and by processing only a small number of state variables. To reconstruct the missing information about the state vector of the doubly-fed reluctance machine it is proposed to use a filtering scheme and based on it to apply state estimation-based control [60, 112, 197]. The recursion of the H_∞ Kalman Filter, for the model of the DFRM, can be formulated in terms of a *measurement update* and a *time update* part.

Measurement update:

$$\begin{array}{l}D(k) = [I - \theta W(k)P^-(k) + C^T(k)R(k)^{-1}C(k)P^-(k)]^{-1} \\ K(k) = P^-(k)D(k)C^T(k)R(k)^{-1} \\ \hat{x}(k) = \hat{x}^-(k) + K(k)[y(k) - C\hat{x}^-(k)]\end{array} \quad (4.121)$$

Time update:

$$\begin{array}{l}\hat{x}^-(k+1) = A(k)x(k) + B(k)u(k) \\ P^-(k+1) = A(k)P^-(k)D(k)A^T(k) + Q(k)\end{array} \quad (4.122)$$

where it is assumed that parameter θ is sufficiently small to assure that the covariance matrix $P^-(k) - \theta W(k) + C^T(k)R(k)^{-1}C(k)$ will be positive definite.When $\theta = 0$

the H_∞ Kalman Filter becomes equivalent to the standard Kalman Filter. One can measure only a part of the state vector of the DFRM, such as state variables $x_2 = i_{1d}$ and $x_4 = i_{2d}$ and can estimate through filtering the rest of the state vector elements (that is the rotor's turn speed $x_1 = \omega_r$ and the currents $x_3 = i_{1q}$ and $x_5 = i_{2q}$). Moreover, the proposed Kalman filtering method can be used for sensor fusion purposes.

4.3.7 Simulation Tests

The efficiency of the proposed control method was evaluated through simulation experiments in several test cases. The associated results have been presented in Figs. 4.10, 4.11, 4.12, 4.13, 4.14 and 4.15. The developed control scheme was a state estimation-based one, since it was assumed that only specific state vector elements of the doubly-fed reluctance machine could be directly measured. Actually, only measurements about state variables $x_2 = i_{1d}$ and $x_4 = i_{2d}$ were directly used, while the rest of the state vector elements of the multi-phase machine (that is the rotor's turn speed $x_1 = \omega_r$ and the currents $x_3 = i_{1q}$ and $x_5 = i_{2q}$) were estimated through filtering.

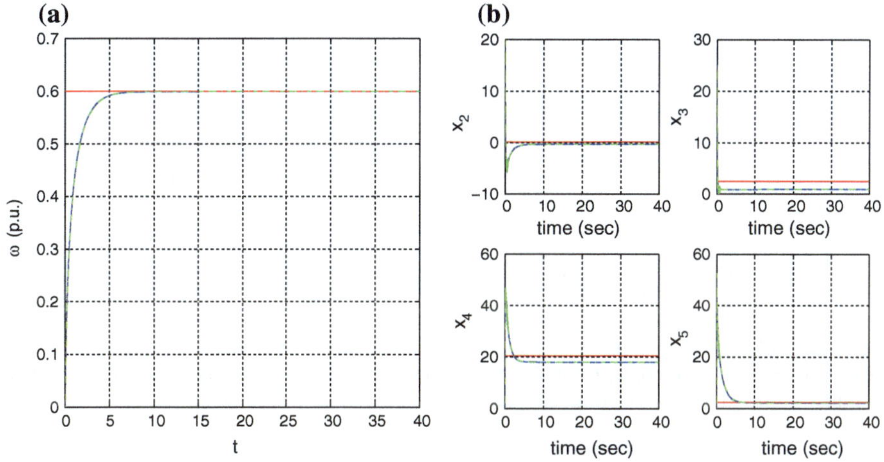

Fig. 4.10 Tracking of setpoint 1 for the doubly-fed reluctance machine. **a** Convergence of state variable $x_1 = \omega$ to the reference setppoint, **b** tracking of reference setpoints by state variables (stator currents) x_2 to x_5 (*red line* setpoint, *blue line* real value, *green line* estimated value)

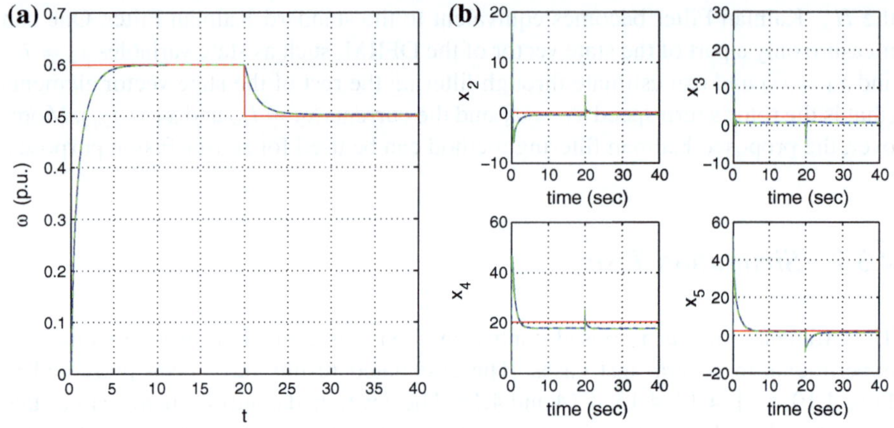

Fig. 4.11 Tracking of setpoint 2 for the doubly-fed reluctance machine. **a** Convergence of state variable $x_1 = \omega$ to the reference setppoint, **b** tracking of reference setpoints by state variables (stator currents) x_2 to x_5 (*red line* setpoint, *blue line* real value, *green line* estimated value)

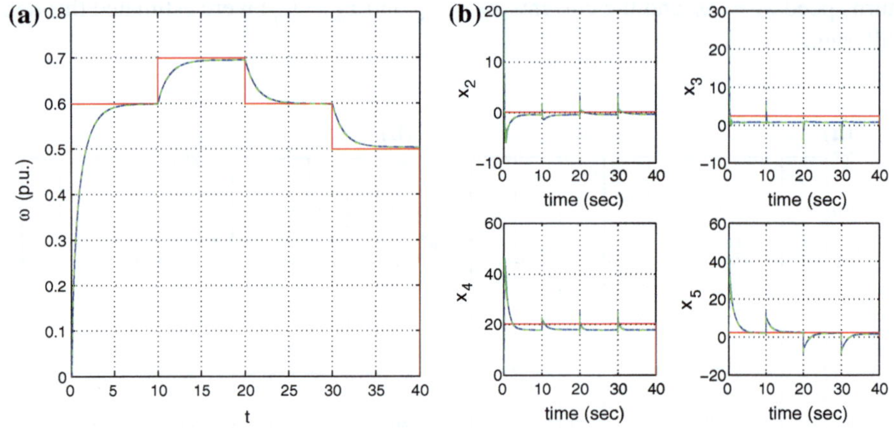

Fig. 4.12 Tracking of setpoint 3 for the doubly-fed reluctance machine. **a** Convergence of state variable $x_1 = \omega$ to the reference setppoint, **b** tracking of reference setpoints by state variables (stator currents) x_2 to x_5 (*red line* setpoint, *blue line* real value, *green line* estimated value)

At each iteration of the control algorithm an estimate of the state vector was provided by the H-infinity Kalman Filter and this was used next by the H-infinity controller. The feedback gain of the H-infinity controller was based on the solution of the previously described algebraic Riccati equation. It can be noticed, that in all cases the rotor's turn speed $x_1 = \omega$ converged fast to its setpoint while for all other

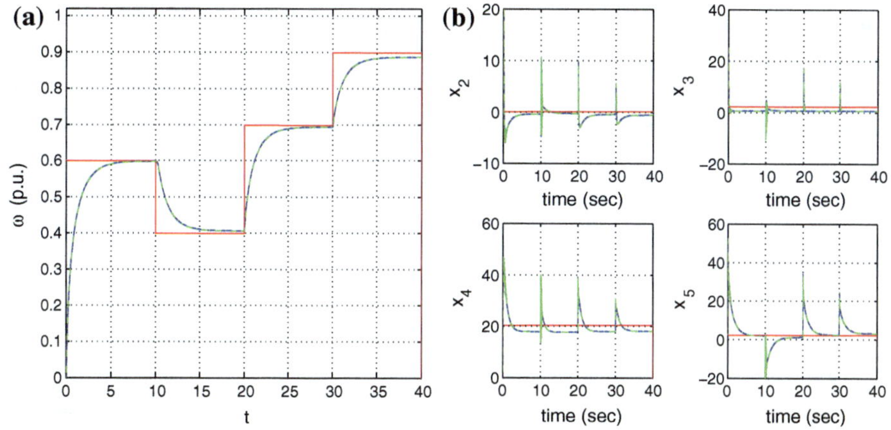

Fig. 4.13 Tracking of setpoint 4 for the doubly-fed reluctance machine. **a** Convergence of state variable $x_1 = \omega$ to the reference setppoint, **b** tracking of reference setpoints by state variables (stator currents) x_2 to x_5 (*red line* setpoint, *blue line* real value, *green line* estimated value)

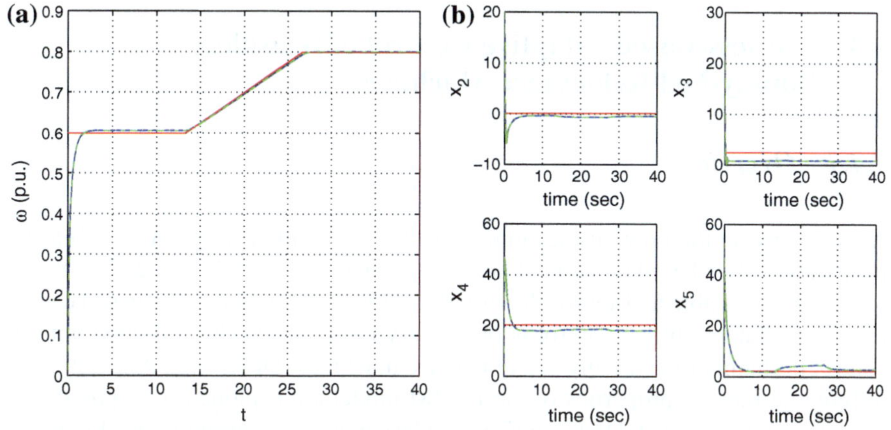

Fig. 4.14 Tracking of setpoint 5 for the doubly-fed reluctance machine. **a** Convergence of state variable $x_1 = \omega$ to the reference setppoint, **b** tracking of reference setpoints by state variables (stator currents) x_2 to x_5 (*red line* setpoint, *blue line* real value, *green line* estimated value)

state variables of the machine x_2 to x_5 fast and accurate tracking of the reference setpoints was also achieved. Yet, simple in concept and implementation the proposed nonlinear H-infinity control method has given very satisfactory results.

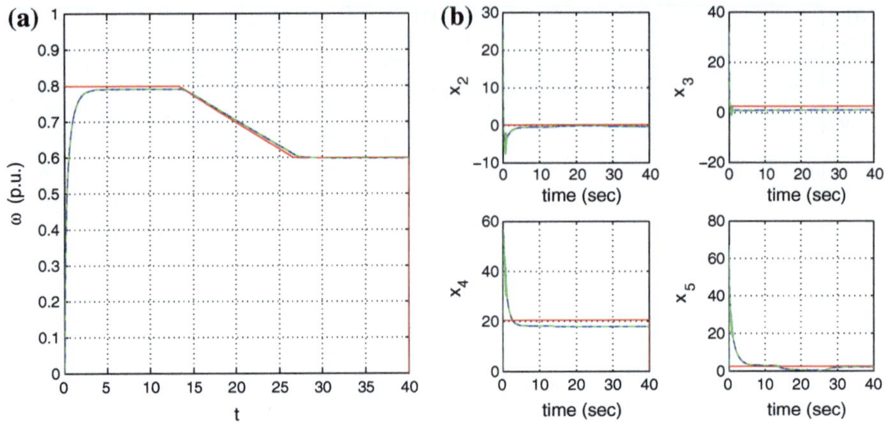

Fig. 4.15 Tracking of setpoint 6 for the doubly-fed reluctance machine. **a** Convergence of state variable $x_1 = \omega$ to the reference setppoint, **b** tracking of reference setpoints by state variables (stator currents) x_2 to x_5 (*red line* setpoint, *blue line* real value, *green line* estimated value)

4.4 Flatness-Based Adaptive Control of Brushless Doubly-Fed Reluctance Machines

4.4.1 Overview

As explained in the previous section, control of the nonlinear dynamics of electric machines is often based on global linearization methods [52, 92, 238, 277, 278, 381, 451]. Among established approaches for global linearization-based control of electric machines, flatness-based control is of primary importance. Thus, to cope with the control problem of the doubly-fed reluctance machines (DFRM), in this section the differential flatness properties of the DFRM model are exploited. It is proven that the rotor's turn speed and the stator's secondary winding's current in the doubly-fed reluctance machine are flat outputs which means that all other state variables and the control inputs of the DFRM can be expressed as differential functions of these outputs [57, 137, 241, 280, 388, 411, 412, 456]. Moreover, it is proven that the flat outputs and their derivatives are not connected through an ordinary differential equation. This comes to complete the proof about the differential flatness of the DFRM model.

The differential flatness of the DFRM implies also that the transformation of its dynamic model to the linear canonical (Brunovsky) form is possible. After such a transformation the implementation of a stabilizing feedback controller for the reluctance generator is enabled. This solution to the control problem would be complete if no model uncertainty and parametric variations of the DFRM existed. Actually, it is assumed that only the order of the DFRM's state-space model is known while its parameters and the specific form of its state-space equations are completely unknown. In the linearized equivalent model of the generator this uncertainty takes the form of unknown functions which appear as part of the system's transformed control inputs.

This unknown part of the system's dynamics is identified with the use of neurofuzzy approximators. The estimates of the DFRM's dynamics are used by a state feedback controller thus establishing an indirect adaptive control scheme. The adaptation of the weights in the neurofuzzy approximators is based on gradient algorithms [34, 341, 344]. The learning procedure for these estimators is defined by the requirement to assure that the first derivative of the system's Lyapunov function is negative [352, 375]. Another issue that arises is that the state vector that is fed into the controller is partially measurable, and the missing state vector elements have to be estimated with the use of a state observer. By including such an observer in the control loop an output feedback-based control scheme is established. The stability of the control method is proven through Lyapunov analysis.

4.4.2 Outline of the Dynamic Model of the DFRM

As it was previously explained, by defining the state variables of the model as $x_1 = \omega$, $x_2 = i_{1d}$, $x_3 = i_{1q}$, $x_4 = i_{2d}$, $x_5 = i_{2q}$ and the control inputs $u_1 = v_{2d}$ and $u_2 = v_{2q}$ the following state-space description for the reluctance machine is obtained:

$$\dot{x} = f(x) + g(x)u \tag{4.123}$$

where $x \in R^{5\times 1}$, $u \in R^{2\times 1}$, $f(x) \in R^{5\times 1}$ and $g(x) \in R^{5\times 2}$, with the drift matrix to be given by

$$f(x) = \begin{pmatrix} -\frac{b}{J}x_1 + \frac{3}{2J}\frac{L_{12}}{L_1}p_r(L_1x_2x_5 + L_{12}x_3x_5 + \\ +L_1x_4x_5 + L_{12}x_3x_5) + \frac{T_m}{J} \\ \\ -\sigma R_1L_2x_2 - \sigma\omega_1L_1L_2x_3 - \sigma\omega_1L_2L_{12}x_5 + \sigma L_2V_s - \\ -\sigma R_2L_{12}x_4 - \sigma\omega_2L_2L_{12}x_5 + \sigma\omega_2L_{12}^2x_3 \\ \\ -\sigma R_1L_2x_3 - \sigma\omega_1L_1L_2x_2 + \sigma\omega_1L_2L_{12}x_4 - \\ -\sigma R_2L_{12}x_5 - \sigma\omega_2L_2L_{12}x_4 + \sigma\omega_2L_{12}^2x_2 \\ \\ -\sigma R_1L_{12}x_2 - \sigma\omega_1L_1L_{12}x_3 + \sigma\omega_1L_{12}^2x_5 - \sigma L_{12}V_s - \\ -\sigma R_2L_1x_4 + \sigma\omega_2L_1L_2x_5 - \sigma\omega_2L_1L_{12}x_3 \\ \\ -\sigma R_1L_{12}x_3 - \sigma\omega_1L_1L_{12}x_2 - \sigma\omega_1L_{12}^2x_4 - \\ -\sigma R_2L_1x_5 - \sigma\omega_2L_1L_2x_4 - \sigma\omega_2L_1L_{12}x_2 \end{pmatrix} \tag{4.124}$$

and the inputs gain matrix to be given by

$$g(x)^T = \begin{pmatrix} 0 & -\sigma L_{12} & 0 & \sigma L_1 & 0 \\ 0 & 0 & \sigma L_{12} & 0 & \sigma L_2 \end{pmatrix} \tag{4.125}$$

4.4.3 Differential Flatness Properties of the Reluctance Machine

4.4.3.1 Differential Flatness of the Reluctance Machine

The flat output of the model is chosen to be $Y = [y_1, y_2]$ where $y_1 = x_1 = \omega$ and $y_2 = x_4 = i_{2d}$. It will be shown that all state variables and the control inputs of the model are differential functions of its flat output and consequently the model of the reluctance machine is a differentially flat one.

From Eq. (4.81) one solves with respect to x_2, thus obtaining

$$\begin{aligned} x_2 &= \frac{1}{\frac{3}{2J}\frac{L_{12}}{L_1}p_r(L_1x_5)}\cdot[\dot{x}_1 + bx_1 - \tfrac{3}{2J}\tfrac{L_{12}}{L_1}p_r(L_{12}x_3x_5 + L_1x_4x_5 + L_{12}x_3x_5 - \tfrac{T_m}{J})] \Rightarrow \\ x_2 &= \frac{1}{\frac{3}{2J}\frac{L_{12}}{L_1}p_r(L_1x_5)}\cdot[\dot{y}_1 + by_1 - \tfrac{3}{2J}\tfrac{L_{12}}{L_1}p_r(L_{12}x_3x_5 + L_1y_2x_5 + L_{12}x_3x_5 - \tfrac{T_m}{J})] \Rightarrow \\ x_2 &= \tilde{f}_2(y_1, \dot{y}_1, x_3, x_5) \end{aligned} \tag{4.126}$$

Moreover, from Eq. (4.126) and by differentiating its last row with respect to time one gets

$$\dot{x}_2 = \tilde{f}_2'(\dot{Y}, \ddot{Y}, \dot{x}_5, \dot{x}_3, x_5, \dot{x}_5) \tag{4.127}$$

From Eqs. (4.82) and (4.84) one obtains

$$\begin{aligned} L_1\dot{x}_2 + L_{12}\dot{x}_4 = L_1[\sigma R_1L_2x_2 - \sigma\omega_1L_1L_2x_3 - \\ -\sigma\omega_1L_2L_{12}x_5 + \sigma L_2V_s - \sigma R_2L_{12}x_4 - \sigma\omega_2L_2L_{12}x_5 + \sigma\omega_2L_{12}^2x_3] + \\ +L_{12}[\sigma R_1L_{12}x_2 - \sigma\omega_1L_1L_{12}x_3 + \sigma\omega_1L_{12}^2x_5 - \sigma L_{12}V_s - \sigma R_2L_1x_4 + \\ +\sigma\omega_2L_1L_2x_5 - \sigma\omega_2L_1L_{12}x_3] \end{aligned} \tag{4.128}$$

From Eqs. (4.83) and (4.85) one obtains

$$\begin{aligned} L_2\dot{x}_3 - L_{12}\dot{x}_5 = L_2[-\sigma R_1L_2x_3 - \sigma\omega_1L_1L_2x_2 + \sigma\omega_1L_2L_{12}x_4 - \\ -\sigma R_2L_{12}x_5 - \sigma\omega_2L_2L_{12}x_4 + \sigma\omega_2L_{12}^2x_2] - \\ L_{12}[-\sigma R_1L_{12}x_3 - \sigma\omega_1L_1L_{12}x_2 - \sigma\omega_1L_{12}^2x_4 - \\ -\sigma R_2L_1x_5 - \sigma\omega_2L_1L_2x_4 - \sigma\omega_2L_1L_{12}x_2] \end{aligned} \tag{4.129}$$

By substituting Eqs. (4.126) and (4.127) into Eq. (4.128) one obtains

$$a_1x_3 + a_2\dot{x}_3 + a_3x_5 + a_4\dot{x}_5 = h_1(Y, \dot{Y}, \ddot{Y}) \tag{4.130}$$

By substituting Eqs. (4.126) and (4.127) into Eq. (4.129) one obtains

$$b_1x_3 + b_2\dot{x}_3 + b_3x_5 + b_4\dot{x}_5 = h_2(Y, \dot{Y}, \ddot{Y}) \tag{4.131}$$

By solving Eqs. (4.130) and (4.131) with respect to x_3 and $\dot{x}_3$ one gets

$$x_3 = m_1(Y, \dot{Y}, \ddot{Y}, x_5, \dot{x}_5) \tag{4.132}$$

By substituting Eq. (4.132) into Eq. (4.126) one has

$$c_1x_5 + c_2\dot{x}_5 = n_1(Y, \dot{Y}, \ddot{Y}) \tag{4.133}$$

By substituting Eq. (4.132) into Eq. (4.129) one has

$$d_1x_5 + d_2\dot{x}_5 = n_2(Y, \dot{Y}, \ddot{Y}) \tag{4.134}$$

From Eqs. (4.133) and (4.134) one finally gets

$$x_5 = p_1(Y, \dot{Y}, \ddot{Y}) \tag{4.135}$$

Substituting Eq. (4.135) into Eq. (4.132) one gets that

$$x_3 = m_1(Y, \dot{Y}, \ddot{Y}, \dots) \tag{4.136}$$

Moreover, by substituting Eqs. (4.135), and (4.136) into Eq. (4.126) one gets

$$x_2 = m_1(Y, \dot{Y}, \ddot{Y}, \dots) \tag{4.137}$$

Consequently, one has that the state variables x_i, $i = 1, \dots, 5$ are differential functions of the flat output. Moreover, by solving Eqs. (4.84) and (4.85) with respect to u_1 and u_2 one has also that the control inputs of the model are differential functions of the flat output, that is

$$u_1 = \tilde{f}_{u_1}(Y, \dot{Y}, \ddot{Y}, \dots) \tag{4.138}$$

$$u_2 = \tilde{f}_{u_2}(Y, \dot{Y}, \ddot{Y}, \dots) \tag{4.139}$$

This completes the proof about the differential flatness of the model of the doubly-fed reluctance machine.

4.4.3.2 Transformation of the Dynamic Model into the Canonical Form

Next, the model is transformed into the canonical form. By differentiating Eq. (4.81) with respect to time one obtains

$$\ddot{x}_1=-\frac{b}{J}x_1-\frac{3}{2J_r}\frac{L_{12}}{L_1}p_r[L_1\dot{x}_2x_5+L_1x_2\dot{x}_5+2L_{12}\dot{x}_3x_5+2L_{12}x_3\dot{x}_5+L_1\dot{x}_4x_5+L_1x_4\dot{x}_5] \tag{4.140}$$

and by substituting in the above equation $\dot{x}_4$ from Eq. (4.81) and $\dot{x}_5$ from Eq. (4.85) one arrives at

$$\begin{aligned}\ddot{x}_1=-\frac{b}{J}x_1&+\frac{3}{2J_r}\frac{L_{12}}{L_1}p_r(L_1x_5)[-\sigma R_1L_2x_2-\sigma\omega_1L_1L_2x_3-\sigma\omega_1L_2L_{12}x_5+\sigma L_2V_s-\\&-\sigma R_2L_{12}x_4-\sigma\omega_2L_2L_{12}x_5+\sigma\omega_2L_{12}^2x_3-\sigma L_{12}u_1]+\\&+\frac{3}{2J_r}\frac{L_{12}}{L_1}p_r(2L_{12}x_5)[-\sigma R_1L_2x_3-\sigma\omega_1L_1L_2x_2+\sigma\omega_1L_2L_{12}x_4-\\&-\sigma R_2L_{12}x_5-\sigma\omega_2L_2L_{12}x_4+\sigma\omega_2L_{12}^2x_2+\sigma L_{12}u_2]\\&+\frac{3}{2J_r}\frac{L_{12}}{L_1}p_r(L_1x_5)[\sigma R_1L_{12}x_2-\sigma\omega_1L_1L_{12}x_3+\sigma\omega_1L_{12}^2x_5-\sigma L_{12}V_s-\\&-\sigma R_2L_1x_4+\sigma\omega_2L_1L_2x_5-\sigma\omega_2L_1L_{12}x_3+\sigma L_1u_1]+\\&+\frac{3}{2J_r}\frac{L_{12}}{L_1}p_r(2L_{12}x_3+L_1x_4)[-\sigma R_1L_{12}x_3-\sigma\omega_1L_1L_{12}x_2-\sigma\omega_1L_{12}^2x_4-\\&-\sigma R_2L_1x_5-\sigma\omega_2L_1L_2x_4-\sigma\omega_2L_1L_{12}x_2+\sigma L_1u_2]\end{aligned} \tag{4.141}$$

The previous equation is also written in the form

$$\ddot{x}_1=f_1(x)+g_{11}(x)u_1+g_{12}(x)u_2 \tag{4.142}$$

with

$$\begin{aligned}f_1(x)=-\frac{b}{J}x_1&-\frac{3}{2J_r}\frac{L_{12}}{L_1}p_r(L_1x_5)[-\sigma R_1L_2x_2-\sigma\omega_1L_1L_2x_3-\sigma\omega_1L_2L_{12}x_5+\sigma L_2V_s-\\&-\sigma R_2L_{12}x_4-\sigma\omega_2L_2L_{12}x_5+\sigma\omega_2L_{12}^2x_3]+\\&+\frac{3}{2J_r}\frac{L_{12}}{L_1}p_r(2L_{12}x_5)[-\sigma R_1L_2x_3-\sigma\omega_1L_1L_2x_2+\sigma\omega_1L_2L_{12}x_4-\\&-\sigma R_2L_{12}x_5-\sigma\omega_2L_2L_{12}x_4+\sigma\omega_2L_{12}^2x_2]\\&+\frac{3}{2J_r}\frac{L_{12}}{L_1}p_r(L_1x_5)[\sigma R_1L_{12}x_2-\sigma\omega_1L_1L_{12}x_3+\sigma\omega_1L_{12}^2x_5-\sigma L_{12}V_s-\\&-\sigma R_2L_1x_4+\sigma\omega_2L_1L_2x_5-\sigma\omega_2L_1L_{12}x_3]+\\&+\frac{3}{2J_r}\frac{L_{12}}{L_1}p_r(2L_{12}x_3+L_1x_4)[-\sigma R_1L_{12}x_3-\sigma\omega_1L_1L_{12}x_2-\sigma\omega_1L_{12}^2x_4-\\&-\sigma R_2L_1x_5-\sigma\omega_2L_1L_2x_4-\sigma\omega_2L_1L_{12}x_2]\end{aligned} \tag{4.143}$$

while also

$$g_{11}(x)=\frac{3}{2J_r}\frac{L_{12}}{L_1}p_r(L_1x_5)(\sigma L_1-\sigma L_{12}) \tag{4.144}$$

$$g_{12}(x)=\frac{3}{2J_r}\frac{L_{12}}{L_1}p_r[(2L_{12}x_5)(\sigma L_{12})+(2L_{12}x_3+L_1x_4)(\sigma L_1)] \tag{4.145}$$

Moreover, from Eq. (4.85) one gets

$$\dot{x}_4=f_2(x)+g_{21}(x)u_1+g_{22}(x)u_2 \tag{4.146}$$

with

$$f_2(x) = \sigma R_1 L_{12} x_2 - \sigma\omega_1 L_1 L_{12} x_3 + \sigma\omega_1 L_{12}^2 x_5 - \sigma L_{12} V_s - \\ - \sigma R_2 L_1 x_4 + \sigma\omega_2 L_1 L_2 x_5 - \sigma\omega_2 L_1 L_{12} x_3 \tag{4.147}$$

and also

$$g_{21}(x) = \sigma L_1 \tag{4.148}$$

$$g_{22}(x) = 0 \tag{4.149}$$

Thus, by defining the state variables $z_1 = x_1$ and $z_2 = x_4$ one arrives at the input-output linearized form of the system

$$\begin{pmatrix} \dot{z}_1 \\ \dot{z}_2 \\ \dot{z}_3 \end{pmatrix} = \begin{pmatrix} 0 & 1 & 0 \\ 0 & 0 & 0 \\ 0 & 0 & 0 \end{pmatrix} \begin{pmatrix} z_1 \\ z_2 \\ z_3 \end{pmatrix} + \begin{pmatrix} 0 & 0 \\ 1 & 0 \\ 0 & 1 \end{pmatrix} \begin{pmatrix} v_1 \\ v_2 \end{pmatrix} \tag{4.150}$$

where the new inputs to the linearized system are

$$v_1 = f_1(x) + g_{11}(x)u_1 + g_{12}(x)u_2 \\ v_2 = f_2(x) + g_{21}(x)u_1 + g_{22}(x)u_2 \tag{4.151}$$

The stabilizing feedback control for the system is

$$v_1 = \ddot{z}_1^d - K_1^1(\dot{z}_1 - \dot{z}_1^d) - K_2^1(z_1 - z_1^d) \\ v_2 = \dot{z}_2^d - K_2^2(z_2 - z_2^d) \tag{4.152}$$

where the feedback gains are chosen such that the characteristic polynomials associated with the tracking error to be Hurwitz stable. Using that

$$\begin{pmatrix} v_1 \\ v_2 \end{pmatrix} = \begin{pmatrix} f_1(x) \\ f_2(x) \end{pmatrix} + \begin{pmatrix} g_{11}(x) & g_{12}(x) \\ g_{21}(x) & g_{22}(x) \end{pmatrix} \begin{pmatrix} u_1 \\ u_2 \end{pmatrix} \tag{4.153}$$

the control input that is really exerted on the system is

$$\begin{pmatrix} u_1 \\ u_2 \end{pmatrix} = \begin{pmatrix} g_{11}(x) & g_{12}(x) \\ g_{21}(x) & g_{22}(x) \end{pmatrix}^{-1} \{ \begin{pmatrix} v_1 \\ v_2 \end{pmatrix} - \begin{pmatrix} f_1(x) \\ f_2(x) \end{pmatrix} \} \tag{4.154}$$

4.4.4 Flatness-Based Adaptive Neurofuzzy Control

4.4.4.1 Transformation of MIMO Nonlinear Systems into the Brunovsky Form

Following the previous procedure, that is after defining the flat outputs of the initial MIMO nonlinear model of the Doubly-Fed Reluctance Machine (DFRM), and after expressing the system's state variables and control inputs as functions of the flat output and of the associated derivatives, the DFRM is transformed into the Brunovsky canonical form of Eq. (4.150), or equivalently into the form [352, 366, 375]:

$$\begin{array}{ll}\dot{x}_1 = x_2 & \\ \dot{x}_2 = x_3 & \\ \cdots & \\ \dot{x}_{r_1-1} = x_{r_1} & y_1 = x_1 \\ \dot{x}_{r_1} = f_1(x) + \sum_{j=1}^{p} g_{1_j}(x)u_j + d_1 & y_2 = x_{r_1-1} \\ \dot{x}_{r_1+1} = x_{r_1+2} & \cdots \\ \dot{x}_{r_1+2} = x_{r_1+3} & y_p = x_{n-r_p+1} \\ \cdots & \\ \dot{x}_{p-1} = x_p & \\ \dot{x}_p = f_p(x) + \sum_{j=1}^{p} g_{p_j}(x)u_j + d_p & \end{array} \tag{4.155}$$

where $x = [x_1, \ldots, x_n]^T$ is the state vector of the transformed system (according to the differential flatness formulation), $u = [u_1, \ldots, u_p]^T$ is the set of control inputs, $y = [y_1, \ldots, y_p]^T$ is the output vector, f_i are the drift functions and $g_{i,j}$, $i, j = 1, 2, \ldots, p$ are smooth functions corresponding to the control input gains, while d_j is a variable associated to external disturbances. In holds that $r_1 + r_2 + \cdots + r_p = n$. Having written the initial nonlinear system into the canonical (Brunovsky) form it holds

$$y_i^{(r_i)} = f_i(x) + \sum_{j=1}^{p} g_{ij}(x)u_j + d_j \tag{4.156}$$

Equivalently, in vector form, one has the following description for the system dynamics

$$y^{(r)} = f(x) + g(x)u + d \tag{4.157}$$

and the following vectors and matrices can be defined

$$\begin{array}{l} y^{(r)} = [y_1^{(r_1)}, \ldots, y_p^{(r_p)}] \\ f(x) = [f_1(x), \ldots, f_p(x)]^T \\ g(x) = [g_1(x), \ldots, g_p(x)] \\ \text{with } g_i(x) = [g_{1i}(x), \ldots, g_{pi}(x)]^T \\ A = diag[A_1, \ldots, A_p], \;\; B = diag[B_1, \ldots, B_p] \\ C^T = diag[C_1, \ldots, C_p], \;\; d = [d_1, \ldots, d_p]^T \end{array} \tag{4.158}$$

where matrix A has the MIMO canonical form, i.e. with elements

$$A_i = \begin{pmatrix} 0 & 1 & \cdots & 0 \\ 0 & 0 & \cdots & 0 \\ \vdots & \vdots & \cdots & \vdots \\ 0 & 0 & \cdots & 1 \\ 0 & 0 & \cdots & 0 \end{pmatrix}_{r_i \times r_i} \quad B_i = \begin{pmatrix} 0 \\ 0 \\ \cdots \\ 0 \\ 1 \end{pmatrix}_{r_i \times 1} \quad C_i^T = \begin{pmatrix} 1 \\ 0 \\ \cdots \\ 0 \\ 0 \end{pmatrix}_{r_i \times 1} \tag{4.159}$$

Thus, Eq. (4.156) can be described in state-space form

$$\begin{aligned} \dot{x} &= Ax + B[f(x) + g(x)u + \tilde{d}] \\ y &= C^T x \end{aligned} \tag{4.160}$$

which can be also written in the equivalent form:

$$\begin{aligned} \dot{x} &= Ax + Bv + B\tilde{d} \\ y &= C^T x \end{aligned} \tag{4.161}$$

where $v = f(x) + g(x)u$. The reference setpoints for the system's outputs $y_1, \ldots, y_p$ are denoted as $y_{1m}, \ldots, y_{pm}$, thus for the associated tracking errors it holds

$$\begin{aligned} e_1 &= y_1 - y_{1m} \\ e_2 &= y_2 - y_{2m} \\ &\cdots \\ e_p &= y_p - y_{pm} \end{aligned} \tag{4.162}$$

The error vector of the outputs of the transformed MIMO system is denoted as

$$\begin{aligned} E_1 &= [e_1, \ldots, e_p]^T \\ y_m &= [y_{1m}, \ldots, y_{pm}]^T \\ &\cdots \\ y_m^{(r)} &= [y_{1m}^{(r)}, \ldots, y_{pm}^{(r)}]^T \end{aligned} \tag{4.163}$$

where $y_{im}^{(r)}$ denotes the r-th order derivative of the i-th reference output of the MIMO dynamical system. Thus, one can also define the following vectors: (i) a vector containing the state variables of the system and the associated derivatives, (ii) a vector containing the reference outputs of the system and the associated derivatives

$$x = [x_1, \ldots, x_1^{r_1-1}, \ldots, x_p, \cdots, x_p^{r_p-1}]^T \tag{4.164}$$

$$Y_m = [y_{1m}, \ldots, y_{1m}^{r_1-1}, \ldots, y_{pm}, \ldots, y_{pm}^{r_p-1}]^T \tag{4.165}$$

while in a similar manner one can define a vector containing the tracking error of the system's outputs and the associated derivatives

$$e = Y_m - x = [e_1, \ldots, e_1^{r_1-1}, \ldots, e_p, \ldots, e_p^{r_p-1}]^T \tag{4.166}$$

It is assumed that matrix $g(x)$ is a nonsingular one, i.e. $g^{-1}(x)$ exists and is bounded for all $x \in U_x$, where $U_x \subset R^n$ is a compact set. In any case, the problem of singularities in matrix $g(x)$ can be handled by appropriately modifying the state feedback-based control input.

The objective of the adaptive neurofuzzy controller, denoted as $u = u(x, e|\theta)$ is: all the signals involved in the controller's design remain bounded and it holds that $lim_{t\to\infty} e = 0$, (ii) the H_∞ tracking performance criterion is succeeded for a prescribed attenuation level.

In the presence of non-Gaussian disturbances w_d, successful tracking of the reference signal is denoted by the H_∞ criterion [352, 366, 375]:

$$\int_0^T e^T Qedt \leq \rho^2 \int_0^T {w_d}^T w_d dt \tag{4.167}$$

where ρ is the attenuation level and corresponds to the maximum singular value of the transfer function $G(s)$ of the linearized model associated to Eqs. (4.160) and (4.161).

4.4.4.2 Control Law

The control signal of the MIMO nonlinear DFRM model which has been transformed into the Brunovsky form as described by Eq. (4.161) contains the unknown nonlinear functions $f(x)$ and $g(x)$. In case that the complete state vector x is measurable these unknown functions can be approximated by

$$\begin{aligned}\hat{f}(x|\theta_f) &= \Phi_f(x)\theta_f \\ \hat{g}(x|\theta_g) &= \Phi_g(x)\theta_g\end{aligned} \tag{4.168}$$

where

$$\Phi_f(x) = (\xi_f^1(x), \xi_f^2(x), \dots \xi_f^n(x))^T \tag{4.169}$$

with $\xi_f^i(x), \imath = 1, \dots, n$ being the vector of kernel functions (e.g. normalized fuzzy Gaussian membership functions), where

$$\xi_f^i(x) = (\phi_f^{i,1}(x), \phi_f^{i,2}(x), \dots, \phi_f^{i,N}(x)) \tag{4.170}$$

thus giving

$$\Phi_f(x) = \begin{pmatrix} \phi_f^{1,1}(x) & \phi_f^{1,2}(x) & \dots & \phi_f^{1,N}(x) \\ \phi_f^{2,1}(x) & \phi_f^{2,2}(x) & \dots & \phi_f^{2,N}(x) \\ \dots & \dots & \dots & \dots \\ \phi_f^{n,1}(x) & \phi_f^{n,2}(x) & \dots & \phi_f^{n,N}(x) \end{pmatrix} \tag{4.171}$$

while the weights vector is defined as

$$\theta_f{}^T = \left(\theta_f^1, \theta_f^2, \dots \theta_f^N\right) \tag{4.172}$$

$j = 1, \ldots, N$ is the number of basis functions that is used to approximate the components of function f which are denoted as $i = 1, \ldots, n$. Thus, one obtains the relation of Eq. (4.168), i.e. $\hat{f}(x|\theta_f) = \Phi_f(x)\theta_f$.

In a similar manner, for the approximation of function g one has

$$\Phi_g(x) = \left(\xi_g^1(x), \xi_g^2(x), \ldots \xi_g^N(x)\right)^T \tag{4.173}$$

with $\xi_g^i(x)$, $\imath = 1, \ldots, N$ being the vector of kernel functions (e.g. normalized fuzzy Gaussian membership functions), where

$$\xi_g^i(x) = \left(\phi_g^{i,1}(x), \phi_g^{i,2}(x), \ldots, \phi_g^{i,N}(x)\right) \tag{4.174}$$

thus giving

$$\Phi_g(x) = \begin{pmatrix} \phi_g^{1,1}(x) & \phi_g^{1,2}(x) & \ldots & \phi_g^{1,N}(x) \\ \phi_g^{2,1}(x) & \phi_g^{2,2}(x) & \ldots & \phi_g^{2,N}(x) \\ \ldots & \ldots & \ldots & \ldots \\ \phi_g^{n,1}(x) & \phi_g^{n,2}(x) & \ldots & \phi_g^{n,N}(x) \end{pmatrix} \tag{4.175}$$

while the weights vector is defined as

$$\theta_g = \left(\theta_g^1, \theta_g^2, \ldots, \theta_g^p\right) \tag{4.176}$$

where the components of matrix θ_g are defined as

$$\theta_g^j = \left(\theta_{g1}^j, \theta_{g2}^j, \ldots \theta_{gN}^j\right)^T \tag{4.177}$$

$j = 1, \ldots, p$ is the number of basis functions that is used to approximate the components of function g which are denoted as $i = 1, \ldots, n$. Thus one obtains about matrix $\theta_g \in R^{N \times p}$

$$\theta_g = \begin{pmatrix} \theta_{g1}^1 & \theta_{g1}^2 & \ldots & \theta_{g1}^p \\ \theta_{g2}^1 & \theta_{g2}^2 & \ldots & \theta_{g2}^p \\ \ldots & \ldots & \ldots & \ldots \\ \theta_{gN}^1 & \theta_{gN}^2 & \ldots & \theta_{gN}^p \end{pmatrix} \tag{4.178}$$

It holds that

$$g = \begin{pmatrix} g_1 \\ g_2 \\ \ldots \\ g_n \end{pmatrix} = \begin{pmatrix} g_1^1 & g_1^2 & \ldots & g_1^p \\ g_2^1 & g_2^2 & \ldots & g_2^p \\ \ldots & \ldots & \ldots & \ldots \\ g_n^1 & g_n^2 & \ldots & g_n^p \end{pmatrix} \tag{4.179}$$

Using the above, one finally has the relation of Eq. (4.168), i.e. $\hat{g}(x|\theta_g) = \Phi_g(x)\theta_g$. If the state variables of the system are available for measurement then a state-feedback control law can be formulated as

$$u = \hat{g}^{-1}(x|\theta_g)[-\hat{f}(x|\theta_f) + y_m^{(r)} - K^T e + u_c] \tag{4.180}$$

where $\hat{f}(x|\theta_f)$ and $\hat{g}(x|\theta_g)$ are neurofuzzy models to approximate $f(x)$ and $g(x)$, respectively. u_c is a supplementary control term, e.g. an H_∞ control term that is used to compensate for the effects of modelling inaccuracies and external disturbances. Moreover, K^T is the feedback gain matrix that assures that the characteristic polynomial of matrix $A - BK^T$ will be a Hurwitz one.

4.4.4.3 Estimation of the State Vector

As in the case of adaptive control for induction machines presented in Chap. 2, in the DFRM case control of the system described by Eq. (4.157) becomes more complicated when the state vector x of the DFRM is not directly measurable and has to be reconstructed through a state observer. The following definitions are used

- error of the state vector $e = x - x_m$
- error of the estimated state vector $\hat{e} = \hat{x} - x_m$
- observation error $\tilde{e} = e - \hat{e} = (x - x_m) - (\hat{x} - x_m)$

When an observer is used to reconstruct the state vector, the control law of Eq. (4.180) is written as

$$u = \hat{g}^{-1}(\hat{x}|\theta_g)[-\hat{f}(\hat{x}|\theta_f) + y_m^{(r)} - K^T \hat{e} + u_c] \tag{4.181}$$

Applying Eq. (4.181) to the nonlinear system described by Eq. (4.157), results into
$y^{(r)} = f(x) + g(x)\hat{g}^{-1}(\hat{x})[-\hat{f}(\hat{x}) + y_m^{(r)} - K^T\hat{e} + u_c] + d \Rightarrow y^{(r)} = f(x) + [g(x) - \hat{g}(\hat{x}) + \hat{g}(\hat{x})]\hat{g}^{-1}(\hat{x})[-\hat{f}(\hat{x}) + y_m^{(r)} - K^T\hat{e} + u_c] + d \Rightarrow y^{(r)} = [f(x) - \hat{f}(\hat{x})] + [g(x) - \hat{g}(\hat{x})]u + y_m^{(r)} - K^T\hat{e} + u_c + d.$

It holds $e = x - x_m \Rightarrow y^{(r)} = e^{(r)} + y_m^{(r)}$. Substituting $y^{(r)}$ in the above equation gives

$$\begin{aligned} e^{(r)} + y_m^{(r)} = y_m^{(r)} - K^T\hat{e} + u_c + [f(x) - \hat{f}(\hat{x})] + \\ +[g(x) - \hat{g}(\hat{x})]u + d \end{aligned} \tag{4.182}$$

and equivalently

$$\begin{aligned} \dot{e} = Ae - BK^T\hat{e} + Bu_c + B\{[f(x) - \hat{f}(\hat{x})] + \\ +[g(x) - \hat{g}(\hat{x})]u + \tilde{d}\} \end{aligned} \tag{4.183}$$

$$e_1 = C^T e \tag{4.184}$$

where $e = [e^1, e^2, \dots, e^p]^T$ with $e^i = [e_i, \dot{e}_i, \ddot{e}_i, \dots, e_i^{r_i-1}]^T$, $i = 1, 2, \dots, p$ and also $\hat{e} = [\hat{e}^1, \hat{e}^2, \dots, \hat{e}^p]^T$ with $\hat{e}^i = [\hat{e}_i, \dot{\hat{e}}_i, \ddot{\hat{e}}_i, \dots, \hat{e}_i^{r_i-1}]^T$, $i = 1, 2, \dots, p$. Matrices A,B and C have been defined in Eq. (4.159).

A state observer is designed according to Eqs. (4.183) and (4.184) and is given by [375]:

$$\dot{\hat{e}} = A\hat{e} - BK^T\hat{e} + K_o[e_1 - C^T\hat{e}] \tag{4.185}$$

$$\hat{e}_1 = C^T\hat{e} \tag{4.186}$$

The feedback gain matrix is denoted as $K \in R^{n \times p}$. The observation gain matrix is denoted as $K_o \in R^{p \times n}$ and its elements are selected so as to assure the asymptotic elimination of the observation error.

4.4.5 *Application of Flatness-Based Adaptive Neurofuzzy Control to the DFRM*

4.4.5.1 Tracking Error Dynamics Under Feedback Control

In the state vector in the linearized description of the DFRM of Eqs. (4.142) and (4.146), where the state vector is constrained into $x = [x_1, x_2, x_3]^T = [\omega, \dot{\omega}, i_{2_d}]^T$, and $g_1(x) = [g_{11}(x)\ g_{12}(x)]$, $g_2(x) = [g_{21}(x)\ g_{22}(x)]$, $u = [u_1\ u_2]^T$ it holds that

$$\begin{aligned} \ddot{x}_1 &= f_1(x) + g_1(x)u \\ \dot{x}_3 &= f_2(x) + g_2(x)u \end{aligned} \tag{4.187}$$

or equivalently

$$\begin{aligned} \dot{x}_1 &= x_2 \\ \dot{x}_2 &= f_1(x) + g_1(x)u \\ \dot{x}_3 &= f_2(x) + g_2(x)u \end{aligned} \tag{4.188}$$

Next, taking into account also the effects of additive disturbances the dynamic model becomes

$$\begin{aligned} \ddot{x}_1 &= f_1(x,t) + g_1(x,t)u + d_1 \\ \dot{x}_3 &= f_2(x,t) + g_2(x,t)u + d_2 \end{aligned} \tag{4.189}$$

$$\begin{pmatrix} \ddot{x}_1 \\ \dot{x}_3 \end{pmatrix} = \begin{pmatrix} f_1(x,t) \\ f_2(x,t) \end{pmatrix} + \begin{pmatrix} g_1(x,t) \\ g_2(x,t) \end{pmatrix} u + \begin{pmatrix} d_1 \\ d_2 \end{pmatrix} \tag{4.190}$$

The following control input is defined

$$u = \begin{pmatrix} \hat{g}_1(x,t) \\ \hat{g}_2(x,t) \end{pmatrix}^{-1} \{ \begin{pmatrix} \ddot{x}_1^d \\ \dot{x}_3^d \end{pmatrix} - \begin{pmatrix} \hat{f}_1(x,t) \\ \hat{f}_2(x,t) \end{pmatrix} - \begin{pmatrix} K_1^T \\ K_2^T \end{pmatrix} e + \begin{pmatrix} u_{c_1} \\ u_{c_2} \end{pmatrix} \} \tag{4.191}$$

where $[u_{c_1}\ u_{c_2}]^T$ is a robust control term that is used for the compensation of the model's uncertainties as well as of the external disturbances and the feedback gains vector is given by $K_i^T = [k_1^i, k_2^i, \ldots, k_{n-1}^i, k_n^i]$. Substituting Eq. (4.191) into Eq. (4.190) the closed-loop tracking error dynamics is obtained

$$\begin{pmatrix} \ddot{x}_1 \\ \dot{x}_3 \end{pmatrix} = \begin{pmatrix} f_1(x,t) \\ f_2(x,t) \end{pmatrix} + \begin{pmatrix} g_1(x,t) \\ g_2(x,t) \end{pmatrix} \begin{pmatrix} \hat{g}_1(x,t) \\ \hat{g}_2(x,t) \end{pmatrix}^{-1} \cdot \\ \cdot \{ \begin{pmatrix} \ddot{x}_1^d \\ \dot{x}_3^d \end{pmatrix} - \begin{pmatrix} \hat{f}_1(x,t) \\ \hat{f}_2(x,t) \end{pmatrix} - \begin{pmatrix} K_1^T \\ K_2^T \end{pmatrix} e + \begin{pmatrix} u_{c_1} \\ u_{c_2} \end{pmatrix} \} + \begin{pmatrix} d_1 \\ d_2 \end{pmatrix} \tag{4.192}$$

Equation (4.192) can now be written as

$$\begin{pmatrix} \ddot{x}_1 \\ \dot{x}_3 \end{pmatrix} = \begin{pmatrix} f_1(x,t) \\ f_2(x,t) \end{pmatrix} + \{ \begin{pmatrix} g_1(x,t) - \hat{g}_1(x,t) \\ g_2(x,t) - \hat{g}_2(x,t) \end{pmatrix} + \begin{pmatrix} \hat{g}_1(x,t) \\ \hat{g}_2(x,t) \end{pmatrix} \} \begin{pmatrix} \hat{g}_1(x,t) \\ \hat{g}_2(x,t) \end{pmatrix}^{-1} \cdot \\ \cdot \{ \begin{pmatrix} \ddot{x}_1^d \\ \dot{x}_3^d \end{pmatrix} - \begin{pmatrix} \hat{f}_1(x,t) \\ \hat{f}_2(x,t) \end{pmatrix} - \begin{pmatrix} K_1^T \\ K_2^T \end{pmatrix} e + \begin{pmatrix} u_{c_1} \\ u_{c_2} \end{pmatrix} \} + \begin{pmatrix} d_1 \\ d_2 \end{pmatrix} \tag{4.193}$$

and using Eq. (4.191) this results into

$$\begin{pmatrix} \ddot{e}_1 \\ \dot{e}_3 \end{pmatrix} = \begin{pmatrix} f_1(x,t) - \hat{f}_1(x,t) \\ f_2(x,t) - \hat{f}_2(x,t) \end{pmatrix} + \begin{pmatrix} g_1(x,t) - \hat{g}_1(x,t) \\ g_2(x,t) - \hat{g}_2(x,t) \end{pmatrix} u - \\ - \begin{pmatrix} K_1^T \\ K_2^T \end{pmatrix} e + \begin{pmatrix} u_{c_1} \\ u_{c_2} \end{pmatrix} + \begin{pmatrix} d_1 \\ d_2 \end{pmatrix} \tag{4.194}$$

The following description for the approximation error is defined

$$w = \begin{pmatrix} f_1(x,t) - \hat{f}_1(x,t) \\ f_2(x,t) - \hat{f}_2(x,t) \end{pmatrix} + \begin{pmatrix} g_1(x,t) - \hat{g}_1(x,t) \\ g_2(x,t) - \hat{g}_2(x,t) \end{pmatrix} u \tag{4.195}$$

Moreover, the following matrices are defined

$$A = \begin{pmatrix} 0 & 1 & 0 \\ 0 & 0 & 0 \\ 0 & 0 & 0 \end{pmatrix}, \quad B = \begin{pmatrix} 0 & 0 \\ 1 & 0 \\ 0 & 1 \end{pmatrix} \\ K^T = \begin{pmatrix} K_1^1 & K_2^1 & K_3^1 \\ K_1^2 & K_2^2 & K_3^2 \end{pmatrix} \tag{4.196}$$

Using matrices A, B, K^T, Eq. (4.194) is written in the following form

$$\dot{e} = (A - BK^T)e + Bu_c + B\{\begin{pmatrix} f_1(x,t) - \hat{f}_1(x,t) \\ f_2(x,t) - \hat{f}_2(x,t) \end{pmatrix} + \\ + \begin{pmatrix} g_1(x,t) - \hat{g}_1(x,t) \\ g_2(x,t) - \hat{g}_2(x,t) \end{pmatrix} u + \tilde{d}\} \tag{4.197}$$

When the estimated state vector $\hat{x}$ is used in the feedback control loop, equivalently to Eq. (4.183) one has

$$\dot{e} = Ae - BK^T\hat{e} + Bu_c + B\{\begin{pmatrix} f_1(x,t) - \hat{f}_1(\hat{x},t) \\ f_2(x,t) - \hat{f}_2(\hat{x},t) \end{pmatrix} + \\ + \begin{pmatrix} g_1(x,t) - \hat{g}_1(\hat{x},t) \\ g_2(x,t) - \hat{g}_2(\hat{x},t) \end{pmatrix} u + \tilde{d}\} \tag{4.198}$$

and considering that the approximation error w is now denoted as

$$w = \begin{pmatrix} f_1(x,t) - \hat{f}_1(\hat{x},t) \\ f_2(x,t) - \hat{f}_2(\hat{x},t) \end{pmatrix} + \begin{pmatrix} g_1(x,t) - \hat{g}_1(\hat{x},t) \\ g_2(x,t) - \hat{g}_2(\hat{x},t) \end{pmatrix} u \tag{4.199}$$

Equation (4.198) can be also written as

$$\dot{e} = Ae - BK^T\hat{e} + Bu_c + Bw + B\tilde{d} \tag{4.200}$$

The associated state observer will be described again by Eqs. (4.185) and (4.186).

4.4.5.2 Dynamics of the Observation Error

The observation error is defined as $\tilde{e} = e - \hat{e} = x - \hat{x}$. Subtracting Eq. (4.185) from Eq. (4.183) as well as Eq. (4.186) from Eq. (4.184) one gets

$$\begin{gathered} \dot{e} - \dot{\hat{e}} = A(e - \hat{e}) + Bu_c + B\{[f(x,t) - \hat{f}(\hat{x},t)] + \\ + [g(x,t) - \hat{g}(\hat{x},t)]u + \tilde{d}\} - K_oC^T(e - \hat{e}) \\ e_1 - \hat{e}_1 = C^T(e - \hat{e}) \end{gathered}$$

or equivalently

$$\begin{gathered} \dot{\tilde{e}} = A\tilde{e} + Bu_c + B\{[f(x,t) - \hat{f}(\hat{x},t)] + \\ + [g(x,t) - \hat{g}(\hat{x},t)]u + \tilde{d}\} - K_oC^T\tilde{e} \\ \tilde{e}_1 = C^T\tilde{e} \end{gathered}$$

which can be written as

$$\dot{\tilde{e}} = (A - K_o C^T)\tilde{e} + Bu_c + B\{[f(x,t) - \hat{f}(\hat{x},t)] + \\ +[g(x,t) - \hat{g}(\hat{x},t)]u + \tilde{d}\} \tag{4.201}$$

$$\tilde{e}_1 = C^T \tilde{e} \tag{4.202}$$

or equivalently, it can be written as

$$\dot{\tilde{e}} = (A - K_o C^T)\tilde{e} + Bu_c + Bw + \tilde{d}\} \tag{4.203}$$

$$\tilde{e}_1 = C^T \tilde{e} \tag{4.204}$$

4.4.5.3 Approximation of Functions $f(x,t)$ and $g(x,t)$

Next, the following neurofuzzy approximators of the unknown system dynamics are defined

$$\hat{f}(\hat{x}) = \begin{pmatrix} \hat{f}_1(\hat{x}|\theta_f) \; \hat{x} \in R^{3\times 1} \; \hat{f}_1(\hat{x}|\theta_f) \; \in \; R^{1\times 1} \\ \hat{f}_2(\hat{x}|\theta_f) \; \hat{x} \in R^{3\times 1} \; \hat{f}_2(\hat{x}|\theta_f) \; \in \; R^{1\times 1} \end{pmatrix} \tag{4.205}$$

with kernel functions

$$\phi_f^{i,j}(\hat{x}) = \frac{\prod_{j=1}^{n} \mu_{A_j}^i(\hat{x}_j)}{\sum_{i=1}^{N} \prod_{j=1}^{n} \mu_{A_j}^i(\hat{x}_j)} \tag{4.206}$$

where $l = 1, 2$, $\hat{x}$ is the estimate of the state vector and $\mu_{A_j^i}(\hat{x})$ is the i-th membership function associated with the l-th hidden layer node (fuzzy rule), as shown in Fig. 4.16. Similarly, the following approximators of the unknown system dynamics are defined

$$\hat{g}(\hat{x}) = \begin{pmatrix} \hat{g}_1(\hat{x}|\theta_g) \; \hat{x} \in R^{3\times 1} \; \hat{g}_1(\hat{x}|\theta_g) \; \in \; R^{1\times 2} \\ \hat{g}_2(\hat{x}|\theta_g) \; \hat{x} \in R^{3\times 1} \; \hat{g}_2(\hat{x}|\theta_g) \; \in \; R^{1\times 2} \end{pmatrix} \tag{4.207}$$

The values of the weights that result in optimal approximation are

$$\theta_f^* = arg \; min_{\theta_f \in M_{\theta_f}} [sup_{\hat{x} \in U_{\hat{x}}} (f(x) - \hat{f}(\hat{x}|\theta_f))] \\ \theta_g^* = arg \; min_{\theta_g \in M_{\theta_g}} [sup_{\hat{x} \in U_{\hat{x}}} (g(x) - \hat{g}(\hat{x}|\theta_g))] \tag{4.208}$$

where the variation ranges for the weights are defined as

$$M_{\theta_f} = \{\theta_f \in R^h : \; ||\theta_f|| \leq m_{\theta_f}\} \\ M_{\theta_g} = \{\theta_g \in R^h : \; ||\theta_g|| \leq m_{\theta_g}\} \tag{4.209}$$

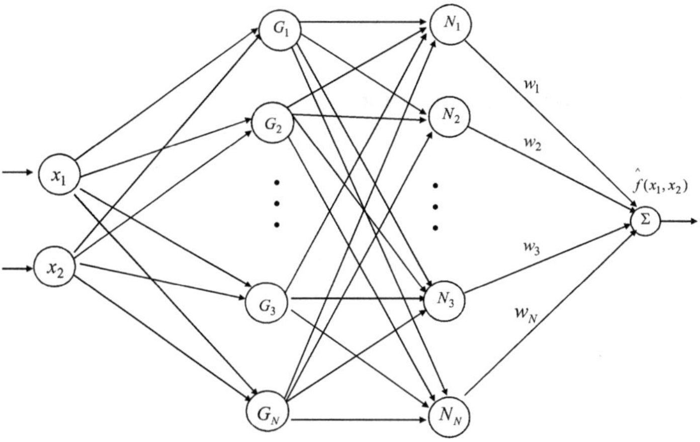

Fig. 4.16 Neurofuzzy approximator used for estimating the unknown dynamics of the DFRM

The value of the approximation error defined in Eq. (4.195) that corresponds to the optimal values of the weights vectors θ_f^* and θ_g^* one has

$$w = \left(f(x,t) - \hat{f}(\hat{x}|\theta_f^*)\right) + \left(g(x,t) - \hat{g}(\hat{x}|\theta_g^*)\right) u \tag{4.210}$$

which is next written as

$$\begin{aligned} w = &\left(f(x,t) - \hat{f}(\hat{x}|\theta_f) + \hat{f}(\hat{x}|\theta_f) - \hat{f}(\hat{x}|\theta_f^*)\right) + \\ &+ \left(g(x,t) - \hat{g}(\hat{x}|\theta_g) + \hat{g}(\hat{x}|\theta_g) - \hat{g}(\hat{x}|\theta_g^*)\right) u \end{aligned} \tag{4.211}$$

which can be also written in the following form

$$w = \left(w_a + w_b\right) \tag{4.212}$$

where

$$w_a = \{[f(x,t) - \hat{f}(\hat{x}|\theta_f)] + [g(x,t) - \hat{g}(\hat{x}|\theta_g)]\}u \tag{4.213}$$

$$w_b = \{[\hat{f}(\hat{x}|\theta_f) - \hat{f}(\hat{x}|\theta_f^*)] + [\hat{g}(\hat{x},\theta_g) - \hat{g}(\hat{x}|\theta_g^*)]\}u \tag{4.214}$$

Moreover, the following weights' error vectors are defined

$$\begin{aligned} \tilde{\theta}_f &= \theta_f - \theta_f^* \\ \tilde{\theta}_g &= \theta_g - \theta_g^* \end{aligned} \tag{4.215}$$

4.4.6 Lyapunov Stability Analysis

4.4.6.1 Design of the Lyapunov Function

The stability analysis for the DFRM adaptive control follows the cases presented in the case of DFIG adaptive control. The adaptation law of the neurofuzzy approximators weights θ_f and θ_g as well as the equation of the supervisory control term u_c are derived from the requirement for a negative first order derivative of the Lyapunov function

$$V = \frac{1}{2}\hat{e}^T P_1 \hat{e} + \frac{1}{2}\tilde{e}^T P_2 \tilde{e} + \frac{1}{2\gamma_1}\tilde{\theta}_f^T \tilde{\theta}_f + \frac{1}{2\gamma_2} tr[\tilde{\theta}_g^T \tilde{\theta}_g] \tag{4.216}$$

The selection of the Lyapunov function is based on the following principle of indirect adaptive control $\hat{e} : \lim_{t\to\infty} \hat{x}(t) = x_d(t)$ and $\tilde{e} : \lim_{t\to\infty} \hat{x}(t) = x(t)$. This yields $\lim_{t\to\infty} x(t) = x_d(t)$. Substituting Eqs. (4.185), (4.186), (4.201), (4.202) into Eq. (4.216) and differentiating results into

$$\begin{aligned} \dot{V} = \frac{1}{2}\dot{\hat{e}}^T P_1 \hat{e} + \frac{1}{2}\hat{e}^T P_1 \dot{\hat{e}} + \frac{1}{2}\dot{\tilde{e}}^T P_2 \tilde{e} + \frac{1}{2}\tilde{e}^T P_2 \dot{\tilde{e}} + \\ + \frac{1}{\gamma_1}\dot{\tilde{\theta}}_f^T \tilde{\theta}_f + \frac{1}{\gamma_2} tr[\dot{\tilde{\theta}}_g^T \tilde{\theta}_g] \Rightarrow \end{aligned} \tag{4.217}$$

$$\begin{aligned} \dot{V} = \frac{1}{2}\{(A - BK^T)\hat{e} + K_o C^T \tilde{e}\}^T P_1 \hat{e} + \\ + \frac{1}{2}\hat{e}^T P_1 \{(A - BK^T)\hat{e} + K_o C^T \tilde{e}\} + \frac{1}{2}\{(A - K_o C^T)\tilde{e} + Bu_c + B\tilde{d} + Bw\}^T P_2 \tilde{e} + \\ + \frac{1}{2}\tilde{e}^T P_2 \{(A - K_o C^T)\tilde{e} + Bu_c + B\tilde{d} + Bw\} + \\ \frac{1}{\gamma_1}\dot{\tilde{\theta}}_f^T \tilde{\theta}_f + \frac{1}{\gamma_2} tr[\dot{\tilde{\theta}}_g^T \tilde{\theta}_g] \Rightarrow . \end{aligned} \tag{4.218}$$

$$\dot{V} = \frac{1}{2}\{\hat{e}^T (A - BK^T)^T + \tilde{e}^T C K_o^T\} P_1 \hat{e} + \frac{1}{2}\hat{e}^T P_1 \{(A - BK^T)\hat{e} + K_o C^T \tilde{e}\} + \frac{1}{2}\{\tilde{e}^T (A - K_o C^T)^T + u_c^T B^T + w^T B^T + \tilde{d}^T B^T\} P_2 \tilde{e} + \frac{1}{2}\tilde{e}^T P_2 \{(A - K_o C^T)\tilde{e} + Bu_c + Bw + B\tilde{d}\} + \frac{1}{\gamma_1}\dot{\tilde{\theta}}_f^T \tilde{\theta}_f + \frac{1}{\gamma_2} tr[\dot{\tilde{\theta}}_g^T \tilde{\theta}_g] \Rightarrow$$

$$\dot{V} = \frac{1}{2}\hat{e}^T (A - BK^T)^T P_1 \hat{e} + \frac{1}{2}\tilde{e}^T C K_o^T P_1 \hat{e} + \frac{1}{2}\hat{e}^T P_1 (A - BK^T)\hat{e} + \frac{1}{2}\hat{e}^T P_1 K_o C^T \tilde{e} + \frac{1}{2}\tilde{e}^T (A - K_o C^T)^T P_2 \tilde{e} + \frac{1}{2}(u_c^T + w^T + \tilde{d}^T) B^T P_2 \tilde{e} + \frac{1}{2}\tilde{e}^T P_2 (A - K_o C^T)\tilde{e} + \frac{1}{2}\tilde{e}^T P_2 B(u_c + w + \tilde{d}) + \frac{1}{\gamma_1}\dot{\tilde{\theta}}_f^T \tilde{\theta}_f + \frac{1}{\gamma_2} tr[\dot{\tilde{\theta}}_g^T \tilde{\theta}_g]$$

Assumption 1: For given positive definite matrices Q_1 and Q_2 there exist positive definite matrices P_1 and P_2, which are the solution of the following Riccati equations [352, 375]

$$(A - BK^T)^T P_1 + P_1 (A - BK^T) + Q_1 = 0 \tag{4.219}$$

$$\begin{aligned} (A - K_o C^T)^T P_2 + P_2 (A - K_o C^T) - \\ - P_2 B(\tfrac{2}{r} - \tfrac{1}{\rho^2}) B^T P_2 + Q_2 = 0 \end{aligned} \tag{4.220}$$

The conditions given in Eqs. (4.219) to (4.220) are related to the requirement that the systems described by Eqs. (4.185), (4.186) and Eqs. (4.201), (4.202) have a decreasing Lyapunov functions. Substituting Eqs. (4.219) to (4.220) into $\dot{V}$ yields

$\dot{V} = \frac{1}{2}\hat{e}^T\{(A-BK^T)^T P_1 + P_1(A-BK^T)\}\hat{e} + \tilde{e}^T C K_o^T P_1\hat{e} + \frac{1}{2}\tilde{e}^T\{(A-K_oC^T)^T$
$P_2 + P_2(A-K_oC^T)\}\tilde{e} + \tilde{e}^T P_2 B(u_c + w + \tilde{d}) + \frac{1}{\gamma_1}\dot{\tilde{\theta}}_f^T\tilde{\theta}_f + \frac{1}{\gamma_2}tr[\dot{\tilde{\theta}}_g^T\tilde{\theta}_g]$

that is
$\dot{V} = -\frac{1}{2}\hat{e}^T Q_1\hat{e} + \tilde{e}^T C K_o^T P_1\hat{e} - \frac{1}{2}\tilde{e}^T\{Q_2 - P_2B(\frac{2}{r} - \frac{1}{\rho^2})B^T P_2\}\tilde{e} + \tilde{e}^T P_2 B(u_c + w +$
$\tilde{d}) + \frac{1}{\gamma_1}\dot{\tilde{\theta}}_f^T\tilde{\theta}_f + \frac{1}{\gamma_2}tr[\dot{\tilde{\theta}}_g^T\tilde{\theta}_g]$

The supervisory control u_c is decomposed in two terms, u_a and u_b.

- The control term u_a is given by

$$u_a = -\frac{1}{r}\tilde{e}^T P_2 B + \Delta u_a \tag{4.221}$$

where assuming that the measurable elements of vector $\tilde{e}$ are $\{\tilde{e}_1, \tilde{e}_3, \ldots, \tilde{e}_k\}$, the term Δu_a is such that

$$-\frac{1}{r}\tilde{e}^T P_2 B + \Delta u_a = -\frac{1}{r}\begin{pmatrix} p_{11}\tilde{e}_1 + p_{13}\tilde{e}_3 + \cdots + p_{1k}\tilde{e}_k \\ p_{13}\tilde{e}_1 + p_{33}\tilde{e}_3 + \cdots + p_{3k}\tilde{e}_k \\ \cdots\cdots\cdots \\ p_{1k}\tilde{e}_1 + p_{3k}\tilde{e}_3 + \cdots + p_{kk}\tilde{e}_k \end{pmatrix} \tag{4.222}$$

- The control term u_b is given by

$$u_b = -[(P_2B)^T(P_2B)]^{-1}(P_2B)^T C K_o^T P_1\hat{e} \tag{4.223}$$

- u_a is an H_∞ control used for the compensation of the approximation error w and the additive disturbance $\tilde{d}$. Its first component $-\frac{1}{r}\tilde{e}^T P_2 B$ has been chosen so as to compensate for the term $\frac{1}{r}\tilde{e}^T P_2 B B^T P_2\tilde{e}$, which appears in the previously computed function about $\dot{V}$. By including also the second component Δu_a, one has that u_a is computed based on the feedback only the measurable variables $\{\tilde{e}_1, \tilde{e}_3, \ldots, \tilde{e}_k\}$, out of the complete vector $\tilde{e} = [\tilde{e}_1, \tilde{e}_2, \ldots, \tilde{e}_n]$. Equation (4.221) is finally rewritten as $u_a = -\frac{1}{r}\tilde{e}^T P_2 B + \Delta u_a$.
- u_b is a control used for the compensation of the observation error (the control term u_b has been chosen so as to satisfy the condition $\tilde{e}^T P_2 B u_b = -\tilde{e}^T C K_o^T P_1\hat{e}$.

The control scheme is depicted in Fig. 4.17.

Substituting Eqs. (4.221) and (4.223) in $\dot{V}$ and assuming that Eqs. (4.219) and (4.220) hold, one gets

$$\begin{aligned}\dot{V} = &-\tfrac{1}{2}\hat{e}^T Q_1\hat{e} + \tilde{e}^T C K_o^T P_1\hat{e} - \tfrac{1}{2}\tilde{e}^T Q_2\tilde{e} + \\ &+\tfrac{1}{r}\tilde{e}^T P_2 B B^T P_2\tilde{e} - \tfrac{1}{2\rho^2}\tilde{e}^T P_2 B B^T P_2\tilde{e} + \\ &+\tilde{e}^T P_2 B u_a + \tilde{e}^T P_2 B u_b + \tilde{e}^T P_2 B(w + \tilde{d}) + \\ &+\tfrac{1}{\gamma_1}\dot{\tilde{\theta}}_f^T\tilde{\theta}_f + \tfrac{1}{\gamma_2}tr[\dot{\tilde{\theta}}_g^T\tilde{\theta}_g]\end{aligned} \tag{4.224}$$

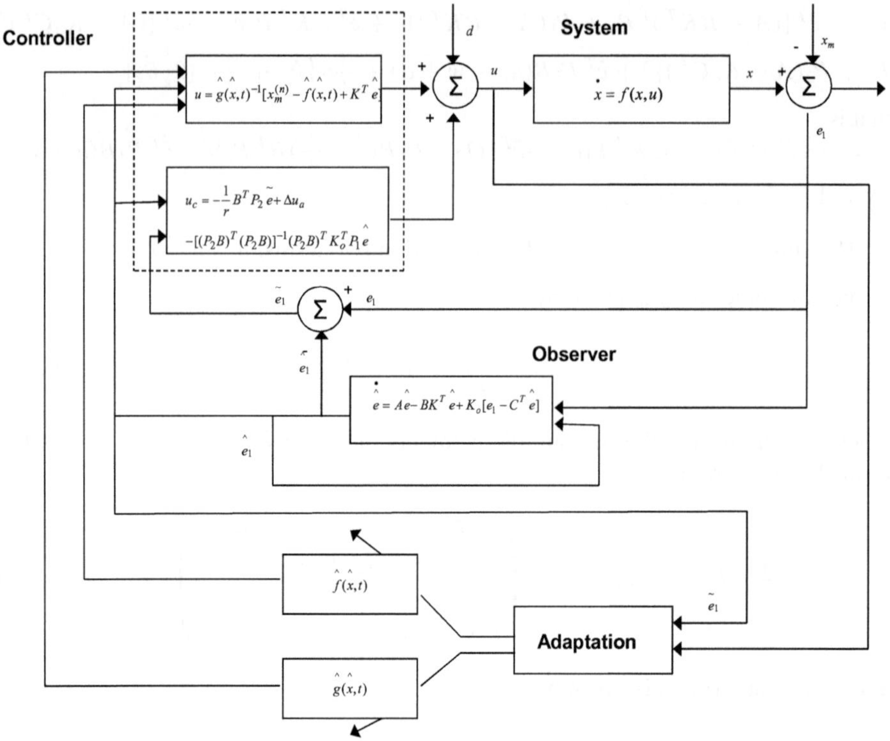

Fig. 4.17 The proposed adaptive fuzzy control scheme based on output feedback for the DFRM

or equivalently,

$$\begin{aligned}\dot V = & -\tfrac{1}{2}\hat e^T Q_1 \hat e - \tfrac{1}{2}\tilde e^T Q_2 \tilde e - \tfrac{1}{2\rho^2}\tilde e^T P_2 B B^T P_2 \tilde e + \\ & + \tilde e^T P_2 B(w + \tilde d + \Delta u_a) + \tfrac{1}{\gamma_1}\dot{\tilde\theta}_f^T \tilde\theta_f + \tfrac{1}{\gamma_2} tr[\dot{\tilde\theta}_g^T \tilde\theta_g]\end{aligned} \tag{4.225}$$

It holds that $\dot{\tilde\theta}_f = \dot\theta_f - \dot\theta_f^* = \dot\theta_f$ and $\dot{\tilde\theta}_g = \dot\theta_g - \dot\theta_g^* = \dot\theta_g$. The following weight adaptation laws are considered:

$$\begin{aligned}\dot\theta_f &= -\gamma_1 \Phi(\hat x)^T B^T P_2 \tilde e \\ \dot\theta_g &= -\gamma_2 \Phi(\hat x)^T B^T P_2 \tilde e u^T\end{aligned} \tag{4.226}$$

where assuming N fuzzy rules and associated kernel functions the matrices dimensions are $\theta_f \in R^{N\times 1}$, $\theta_g \in R^{N\times 2}$, $\Phi(x) \in R^{2\times N}$, $B \in R^{3\times 2}$, $P \in R^{3\times 3}$ and $\tilde e \in R^{3\times 1}$.

The update of θ_f is a gradient type algorithm [341, 344]. The update of θ_g is also a gradient type algorithm, where u_c implicitly tunes the adaptation gain γ_2. Substituting Eq. (4.226) in $\dot V$ gives

$$\dot{V} = -\frac{1}{2}\hat{e}^T Q_1\hat{e} - \frac{1}{2}\tilde{e}^T Q_2\tilde{e} - \frac{1}{2\rho^2}\tilde{e}^T P_2 B B^T P_2\tilde{e} + B^T P_2\tilde{e}(w + d + \Delta u_a) + \\ + \frac{1}{\gamma_1}(-\gamma_1)\tilde{e}^T P_2 B\Phi(\hat{x})(\theta_f - \theta_f^*) + \frac{1}{\gamma_2}(-\gamma_2)tr[u\tilde{e}^T P_2 B\Phi(\hat{x})(\theta_g - \theta_g^*)] \tag{4.227}$$

or equivalently

$$\dot{V} = -\frac{1}{2}\hat{e}^T Q_1\hat{e} - \frac{1}{2}\tilde{e}^T Q_2\tilde{e} - \frac{1}{2\rho^2}\tilde{e}^T P_2 B B^T P_2\tilde{e} + B^T P_2\tilde{e}(w + \tilde{d} + \Delta u_a) + \\ + \frac{1}{\gamma_1}(-\gamma_1)\tilde{e}^T P_2 B\Phi(\hat{x})(\theta_f - \theta_f^*) + \frac{1}{\gamma_2}(-\gamma_2)tr[u\tilde{e}^T P_2 B(\hat{g}(\hat{x}|\theta_g) - \hat{g}(\hat{x}|\theta_g^*)] \tag{4.228}$$

Taking into account that $u \in R^{2\times 1}$ and $\tilde{e}^T PB(\hat{g}(x|\theta_g) - \hat{g}(x|\theta_g^*)) \in R^{1\times 2}$ it holds

$$\dot{V} = -\frac{1}{2}\hat{e}^T Q_1\hat{e} - \frac{1}{2}\tilde{e}^T Q_2\tilde{e} - \frac{1}{2\rho^2}\tilde{e}^T P_2 B B^T P_2\tilde{e} + B^T P_2\tilde{e}(w + \tilde{d} + \Delta u_a) + \\ + \frac{1}{\gamma_1}(-\gamma_1)\tilde{e}^T P_2 B\Phi(\hat{x})(\theta_f - \theta_f^*) + \frac{1}{\gamma_2}(-\gamma_2)tr[\tilde{e}^T P_2 B(\hat{g}(\hat{x}|\theta_g) - \hat{g}(\hat{x}|\theta_g^*))u] \tag{4.229}$$

Since $\tilde{e}^T P_2 B(\hat{g}(\hat{x}|\theta_g) - \hat{g}(\hat{x}|\theta_g^*))u \in R^{1\times 1}$ it holds

$$tr(\tilde{e}^T P_2 B(\hat{g}(x|\theta_g) - \hat{g}(x|\theta_g^*)u) = \\ = \tilde{e}^T P_2 B(\hat{g}(x|\theta_g) - \hat{g}(x|\theta_g^*))u \tag{4.230}$$

Therefore, one finally obtains

$$\dot{V} = -\frac{1}{2}\hat{e}^T Q_1\hat{e} - \frac{1}{2}\tilde{e}^T Q_2\tilde{e} - \frac{1}{2\rho^2}\tilde{e}^T P_2 B B^T P_2\tilde{e} + B^T P_2\tilde{e}(w + \tilde{d} + \Delta u_a) + \\ + \frac{1}{\gamma_1}(-\gamma_1)\tilde{e}^T P_2 B\Phi(\hat{x})(\theta_f - \theta_f^*) + \frac{1}{\gamma_2}(-\gamma_2)\tilde{e}^T P_2 B(\hat{g}(\hat{x}|\theta_g) - \hat{g}(\hat{x}|\theta_g^*))u \tag{4.231}$$

Next, the following approximation error is defined

$$w_\alpha = [\hat{f}(\hat{x}|\theta_f^*) - \hat{f}(\hat{x}|\theta_f)] + [\hat{g}(\hat{x}|\theta_g^*) - \hat{g}(\hat{x}|\theta_g)]u \tag{4.232}$$

Thus, one obtains

$$\dot{V} = -\frac{1}{2}\hat{e}^T Q_1\hat{e} - \frac{1}{2}\tilde{e}^T Q_2\tilde{e} - \frac{1}{2\rho^2}\tilde{e}^T P_2 B B^T P_2\tilde{e} + \\ + B^T P_2\tilde{e}(w + \tilde{d}) + \tilde{e}^T P_2 B w_\alpha \tag{4.233}$$

Denoting the aggregate approximation error and disturbances vector as

$$w_1 = w + \tilde{d} + w_\alpha + \Delta u_a \tag{4.234}$$

the derivative of the Lyapunov function becomes

$$\dot{V} = -\frac{1}{2}\hat{e}^T Q_1\hat{e} - \frac{1}{2}\tilde{e}^T Q_2\tilde{e} - \frac{1}{2\rho^2}\tilde{e}^T P_2 B B^T P_2\tilde{e} + \tilde{e}^T P_2 B w_1 \tag{4.235}$$

which in turn is written as

$$\dot{V} = -\frac{1}{2}\hat{e}^T Q_1\hat{e} - \frac{1}{2}\tilde{e}^T Q_2\tilde{e} - \frac{1}{2\rho^2}\tilde{e}^T P_2 B B^T P_2\tilde{e} + \\ + \frac{1}{2}\tilde{e}^T P B w_1 + \frac{1}{2}w_1^T B^T P_2\tilde{e} \tag{4.236}$$

Lemma: The following inequality holds

$$\tfrac{1}{2}\tilde{e}^T P_2 B w_1 + \tfrac{1}{2}w_1^T B^T P_2 \tilde{e} - \tfrac{1}{2\rho^2}\tilde{e}^T P_2 B B^T P_2 \tilde{e} \le \tfrac{1}{2}\rho^2 w_1^T w_1 \tag{4.237}$$

Proof: The binomial $(\rho a - \frac{1}{\rho}b)^2 \ge 0$ is considered. Expanding the left part of the above inequality one gets

$$\begin{aligned} \rho^2 a^2 + \tfrac{1}{\rho^2}b^2 - 2ab \ge 0 \Rightarrow \tfrac{1}{2}\rho^2 a^2 + \tfrac{1}{2\rho^2}b^2 - ab \ge 0 \Rightarrow \\ ab - \tfrac{1}{2\rho^2}b^2 \le \tfrac{1}{2}\rho^2 a^2 \Rightarrow \tfrac{1}{2}ab + \tfrac{1}{2}ab - \tfrac{1}{2\rho^2}b^2 \le \tfrac{1}{2}\rho^2 a^2 \end{aligned} \tag{4.238}$$

The following substitutions are carried out: $a = w_1$ and $b = \tilde{e}^T P_2 B$ and the previous relation becomes

$$\tfrac{1}{2}w_1^T B^T P_2 \tilde{e} + \tfrac{1}{2}\tilde{e}^T P_2 B w_1 - \tfrac{1}{2\rho^2}\tilde{e}^T P_2 B B^T P_2 \tilde{e} \le \tfrac{1}{2}\rho^2 w_1^T w_1 \tag{4.239}$$

The above relation is used in $\dot{V}$, and the right part of the associated inequality is enforced

$$\dot{V} \le -\frac{1}{2}\hat{e}^T Q_1 \hat{e} - \frac{1}{2}\tilde{e}^T Q_2 \tilde{e} + \frac{1}{2}\rho^2 w_1^T w_1 \tag{4.240}$$

Thus, Eq. (4.240) can be written as

$$\dot{V} \le -\frac{1}{2}E^T Q E + \frac{1}{2}\rho^2 w_1^T w_1 \tag{4.241}$$

where

$$E = \begin{pmatrix} \hat{e} \\ \tilde{e} \end{pmatrix}, \quad Q = \begin{pmatrix} Q_1 & 0 \\ 0 & Q_2 \end{pmatrix} = diag[Q_1, Q_2] \tag{4.242}$$

Hence, the H_∞ performance criterion is derived. For ρ sufficiently small Eq. (4.240) will be true and the H_∞ tracking criterion will be satisfied. In that case, the integration of $\dot{V}$ from 0 to T gives

$$\begin{aligned} \int_0^T \dot{V}(t)dt \le -\tfrac{1}{2}\int_0^T ||E||^2 dt + \tfrac{1}{2}\rho^2 \int_0^T ||w_1||^2 dt \Rightarrow \\ 2V(T) - 2V(0) \le -\int_0^T ||E||_Q^2 dt + \rho^2 \int_0^T ||w_1||^2 dt \Rightarrow \\ 2V(T) + \int_0^T ||E||_Q^2 dt \le 2V(0) + \rho^2 \int_0^T ||w_1||^2 dt \end{aligned} \tag{4.243}$$

It is assumed that there exists a positive constant $M_w > 0$ such that $\int_0^\infty ||w_1||^2 dt \le M_w$.

Therefore for the integral $\int_0^T ||E||_Q^2 dt$ one gets

$$\int_0^\infty ||E||_Q^2 dt \le 2V(0) + \rho^2 M_w \tag{4.244}$$

Thus, the integral $\int_0^{\infty}||E||_Q^2 dt$ is bounded and according to Barbalat's Lemma

$$\begin{array}{c} \lim_{t\to\infty} E(t) = 0 \Rightarrow \\ \lim_{t\to\infty} \hat{e}(t) = 0 \ \lim_{t\to\infty} \tilde{e}(t) = 0 \end{array} \tag{4.245}$$

Therefore $\lim_{t\to\infty} e(t) = 0$.

4.4.7 Simulation Tests

The efficiency of the proposed flatness-based control method for doubly-fed reluctance machines has been confirmed with the use of simulation experiments. In the simulation tests, the dynamic model of the DFRM was taken to be completely unknown. The system's dynamics were identified with the used of the previously analyzed neurofuzzy approximators. The estimates about the unknown dynamics of the electric machine were used for the computation of the feedback control input, and in this manner an indirect adaptive control scheme was established. To estimate for the non-measurable state vector elements of the machine a state observer was used in the control loop. The measured state variables were the rotation speed of the DFRM and the currents of the secondary winding of the stator. To improve the robustness of the control scheme an H-infinity control term was included in the control input. This was computed through the solution of the previously analyzed algebraic Riccati equation, at each iteration of the control algorithm. The obtained results are depicted in Fig. 4.18.

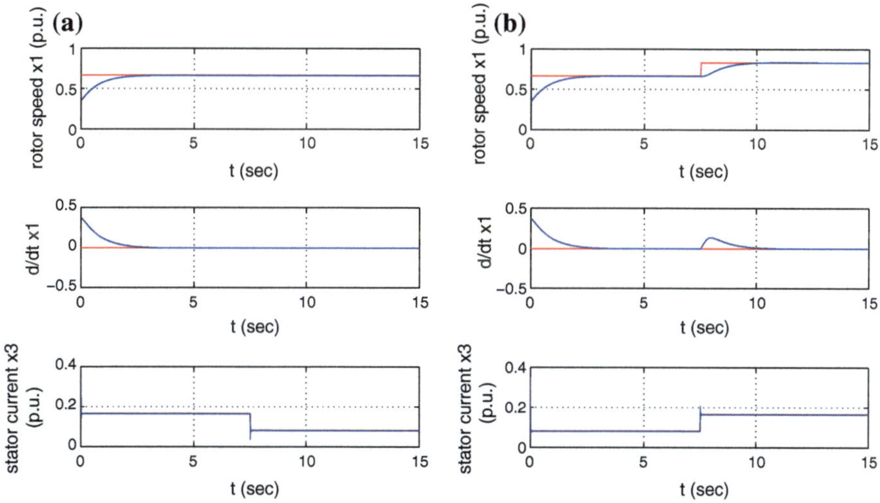

Fig. 4.18 **a** Tracking of setpoint 1 for the doubly-fed reluctance machine, **b** tracking of setpoint 2 for the doubly-fed reluctance machine

Through the previous simulation experiments, the satisfactory performance of the flatness-based adaptive fuzzy control for the DFRM has been confirmed. It can be noticed that fast and accurate tracking of the reference setpoints was achieved. The transients of the state variables did not exhibit abrupt changes while the variations of the control input remained also smooth. The proposed adaptive control method functions well although no prior information is used about the dynamics of the machine and although only output feedback can be used from the DFRM. Both these features are important, because one cannot rely on the assumption that the dynamics of the controlled system is precisely known and remains unaltered after a long time of functioning of the machine under hard conditions. Moreover, one cannot rely on the assumption that the entire state vector of the DFRM can be measured, because this requires the installation and faultless use of several sensors, and this is not always technically feasible.

Chapter 5
Control of the Functioning of DC to DC and AC to DC Converters

5.1 Outline

The chapter analyzes different approaches for nonlinear control of AC to DC converters (rectifiers): (i) flatness-based control through transformation to the canonical form, (ii) differential geometry-based control, (iii) nonlinear H-infinity control. Moreover, the chapter presents flatness-based control for the HVDC-VSC transmission system.

First, the chapter proposes a method for nonlinear control of the dynamical system that is formed by a DC-DC converter and a DC motor, making use of differential flatness theory. First it is proven that the aforementioned system is differentially flat one which means that all its state vector elements and its control inputs can be expressed as differential functions of primary state variables which are defined to be the system's flat outputs. By exploiting the differential flatness properties of the model its transformation to a linearized canonical (Brunovsky) form becomes possible. For the latter description of the system one can design a stabilizing feedback controller. Moreover, estimation of the nonmeasurable state vector elements of the system is achieved by applying the Derivative-free nonlinear Kalman Filter. This filter consists of the Kalman Filter recursion applied to the linearized equivalent model of the system and of an inverse transformation that is based on differential flatness theory and which enables to obtain estimates of the initial nonlinear state-space model. Moreover, to compensate for parametric uncertainties and external perturbations the filter is redesigned as a disturbance observer. By estimating the perturbation inputs that affect the joint model consisting of the DC-DC converter and of the DC motor their compensation becomes possible.

Next, the chapter is concerned with proving differential flatness of the three-phase voltage source converter (VSC) model and its resulting description in the Brunovksy (canonical) form. For the linearized canonical model of the converter a feedback controller is designed. At a second stage, the previously noted Kalman Filtering method (derivative-free nonlinear Kalman Filtering) is introduced. The proposed Kalman Filter is redesigned as disturbance observer for estimating additive input

G. Rigatos, *Intelligent Renewable Energy Systems*,
Green Energy and Technology, DOI 10.1007/978-3-319-39156-4_5

disturbances to the VSC model. These estimated disturbance terms are finally used by a feedback controller that enables the DC output voltage track desirable setpoints.

Next, the chapter introduces a new method for feedback control of the nonlinear dynamics of the three-phase voltage source converters. The control method consists of a repetitive solution of an H-infinity control problem for the voltage source converter, based on a locally linearized model of the converter and takes place at each iteration of the control algorithm. The converter's model is locally linearized round its current operating point through the computation of the associated Jacobian matrices. Using the linearized model of the converter an H-infinity feedback control law is computed. The known robustness features of H-infinity control enable to compensate for the errors of the approximate linearization, as well as to eliminate the effects of external perturbations. The efficiency of the proposed control scheme is shown analytically and is confirmed through simulation experiments. The method can be also applied to other power electronics models.

Finally, the chapter develops a control method for the VSC-HVDC system, that is for an AC to DC voltage source converter connected to the electricity grid through a high voltage DC transmission line terminating at an inverter. By showing that the VSC-HVDC system is a differentially flat one, its transformation to the linear canonical form becomes possible. This is a global input-output linearization procedure that results into an equivalent dynamic model of the VSC-HVDC system for which the design of a state feedback controller becomes possible. Moreover, to estimate and compensate for modeling uncertainty terms and perturbation inputs exerted on the VSC-HVDC model it is proposed to include in the control loop a disturbance observer that is based on the Derivative-free nonlinear Kalman Filter. This filtering method makes use of the linearized equivalent model of the VSC-HVDC system and of an inverse transformation which is based on differential flatness theory and which finally provides estimates of the state variables of the initial nonlinear model.

5.2 Control of DC to DC Converters

5.2.1 Overview

DC to DC converters are needed for administering the power generated by DC power units, such as photovoltaics or fuel cells. A nonlinear control method, based on differential flatness theory, is developed for the model that is formed after connecting a DC motor to a DC-DC converter. This scheme can be used for the exploitation of the power produced by photovoltaic units, since the DC-DC converter controls the level of the produced output voltage. In turn the DC output voltage can be fed into DC motors, as for instance in the case of actuators, mechatronic devices and pumps [166, 260, 313]. Nonlinear control approaches for DC-DC converters connected to DC motors have been presented in [16, 18, 129, 284, 519]. In particular global linearization approaches and flatness-based control of such systems have been presented in [21, 22, 252, 253, 406, 410, 438]. In this section, a state-space description is obtained first for the system consisting of the DC-DC converter and the DC motor. Next, the differential flatness properties of this model are proven using as flat output the rotation angle of the motor. Differential flatness means that all state variables and the control input of the system can be expressed as functions of the flat output and its derivatives [241, 352, 366, 375, 388, 411].

By proving that the system is a differentially flat one, it is confirmed that it can be transformed to the linear canonical (Brunovsky) form [57, 137, 229, 240, 288, 387, 516]. After such a transformation, the computation of a stabilizing feedback control input becomes possible. To estimate the non-measurable state variables of the system the Derivative-free nonlinear Kalman Filter is proposed [352, 366, 375]. This consists of the Kalman Filter recursion applied on the equivalent linearized model of the system [34, 341, 344]. Moreover, an inverse transformation is used, being based on differential flatness theory and which allows to obtain estimates of the state variables of the initial nonlinear model. Next, the Derivative-free nonlinear Kalman Filter is designed as a disturbance observer. This enables simultaneous estimation of the state variables of the system and of disturbance terms affecting it [355, 374]. By identifying such disturbance inputs, their compensation becomes possible with the inclusion of an additional control term in the aggregate feedback control signal.

5.2.2 *Differential Flatness of the Model of a DC-DC Converter Connected to a DC Motor*

DC-DC converters controlling DC electric machines can be found in several applications, as for instance in photovoltaic powered pumps or in desalination units. Control is implemented through a pulse-width-modulation (PWM) approach (Fig. 5.1). The dynamic model of the DC-DC converter connected to a DC motor has been analyzed in Chap. 1, and is overviewed as follows:

$$\begin{aligned}
\dot{x}_1 &= x_2 \\
\dot{x}_2 &= -\tfrac{B}{J}x_2 + \tfrac{K_a}{J}x_5 + \tfrac{1}{J}\tau_d \\
\dot{x}_3 &= -\tfrac{1}{L}x_4 + \tfrac{1}{L}u \\
\dot{x}_4 &= \tfrac{1}{C}x_3 - \tfrac{1}{RC}x_4 + \tfrac{1}{C}x_5 \\
\dot{x}_5 &= \tfrac{1}{L_m}x_4 - \tfrac{R_m}{L_m}x_5 + \tfrac{K_e}{L_m}x_2
\end{aligned} \tag{5.1}$$

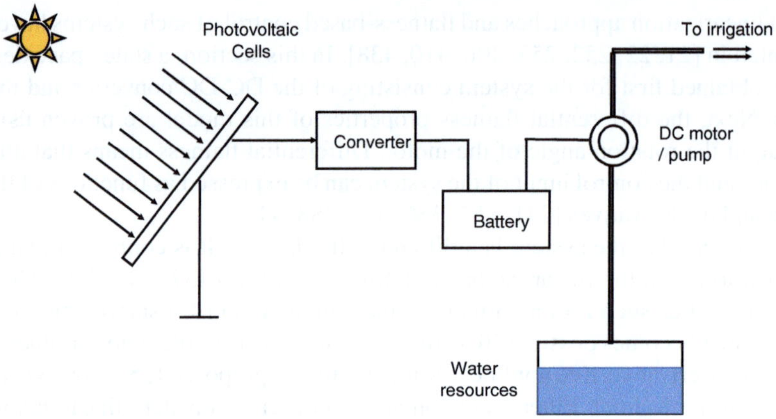

Fig. 5.1 Photovoltaic powered pump

Next, the differential flatness properties of the model are proven. The flat output of the system is taken to be $y = x_1$, that is the turn angle of the rotor. From the first row of the model one gets

$$x_2 = \dot{x}_1 = \dot{y} \tag{5.2}$$

From the second row of the model one gets

$$x_5 = \frac{\dot{x}_2 + \frac{B}{J}x_2 - \frac{1}{J}mglsin(x_1)}{\frac{K_a}{J}} \Rightarrow x_5 = f_a(y, \dot{y}, \ddot{y}) \tag{5.3}$$

From the fifth row of the model one gets

$$x_4 = \frac{\dot{x}_5 + \frac{K_e}{L_m}x_2 + \frac{R_m}{L_m}x_5}{\frac{1}{L_m}} \Rightarrow x_4 = f_b(y, \dot{y}, \ddot{y}, y^{(3)}) \tag{5.4}$$

From the fourth row of the model one gets

$$x_3 = \frac{\dot{x}_4 + \frac{1}{RC}x_4 + \frac{1}{C}x_5}{\frac{1}{C}} \Rightarrow x_3 = f_c(y, \dot{y}, \ddot{y}, y^{(3)}, y^{(4)}) \tag{5.5}$$

From the third row of the model one gets

$$u = L(\dot{x}_3 + \frac{1}{L}x_4) \Rightarrow u = f_d(y, \dot{y}, \ddot{y}, y^{(3)}, y^{(4)}, y^{(5)}) \tag{5.6}$$

Consequently, all state variables and control inputs of the model are differential functions of the flat output and the model is a differentially flat one.

5.2.3 Transformation of the Dynamic Model into the Canonical Form

Knowing that the dynamic model of the system that is formed by the DC-DC converter and the DC motor is a differentially flat, its transformation to the linear canonical (Brunovsky) form is assured. Indeed by differentiating the first row of the state-space model with respect to time one obtains:

$$\dot{x}_1 = x_2 \Rightarrow \ddot{x}_1 = \dot{x}_2 \Rightarrow \ddot{x}_1 = -\frac{B}{J}x_2 + \frac{K_m}{J}x_5 + mglsin(x_1) \tag{5.7}$$

By differentiating once more with respect to time one gets:

$$\begin{aligned} x_1^{(3)} &= -\frac{B}{J}\dot{x}_2 + \frac{K_m}{J}\dot{x}_5 + mglcos(x_1)\dot{x}_1 \Rightarrow \\ x_1^{(3)} &= -\frac{B}{J}[-\frac{B}{J}x_2 + \frac{K_a}{J}x_5 + mglsin(x_1)] + \\ &\frac{K_m}{J}(\frac{K_e}{L_m}x_2 + \frac{1}{L_m}x_4 - \frac{R_m}{L_m}x_5) + \\ &mglcos(x_1)\dot{x}_1 \end{aligned} \tag{5.8}$$

After intermediate operations one obtains:

$$\begin{aligned} x_1^{(3)} &= [(\frac{B}{J})^2 + (\frac{K_aK_e}{JK_m})]x_2 + [-\frac{Bmgl}{J}sin(x_1) + mglcos(x_1)x_2] \\ &+ \\ &\frac{K_a}{JL_m}x_4 + [-\frac{BK_a}{J^2} - \frac{K_aR_m}{JL_m}]x_5 \end{aligned} \tag{5.9}$$

By differentiating once more with respect to time one arrives at:

$$\begin{aligned} x_1^{(4)} &= [(\frac{B}{J})^2 + (\frac{K_aK_e}{JL_m})]\dot{x}_2 \\ &+[-\frac{Bmgl}{J}cos(x_1)\dot{x}_1 - mglsin(x_1)\dot{x}_1x_2 + mglcos(x_1)\dot{x}_2] + \\ &\frac{K_a}{JL_m}\dot{x}_4 + [-\frac{BK_a}{J^2} - \frac{K_aR_m}{JL_m}]\dot{x}_5 \end{aligned} \tag{5.10}$$

which after intermediate operations gives

$$\begin{aligned} x_1^{(4)} &= [(\frac{B}{J})^2(\frac{K_aK_e)}{JL_m} + mglcos(x_1)] \cdot [-\frac{B}{J}x_2 + \frac{K_a}{J}x_5 + mglcos(x_1)] + \\ &+[-\frac{Bmgl}{J}cos(x_1)x_2 - mglsin(x_1)x_2^2] + \frac{K_a}{JL_m}[\frac{1}{c}x_3 - \frac{1}{RC}x_4 + \frac{1}{C}x_5] + \\ &+[-\frac{BK_a}{J^2} - \frac{K_aR_m}{JL_m}] \cdot [\frac{1}{c}x_3 - \frac{1}{RC}x_4 + \frac{1}{C}x_5] \end{aligned} \tag{5.11}$$

By differentiating once more with respect to time one gets:

$$\begin{aligned} x_1^{(5)} &= -[mglsin(x_1)\dot{x}_1] \cdot [-\frac{B}{J}\dot{x}_2 + \frac{K_a}{J}\dot{x}_5 + mglcos(x_1)] + \\ &[(\frac{B}{J})^2 + \frac{K_aK_e]}{JL_m} + mglcos(x_1) \cdot [-\frac{B}{J}\dot{x}_2 + \frac{K_a}{J}\dot{x}_5 + mglcos(x_1)\dot{x}_1] + \\ &[\frac{Bmgl}{J}sin(x_1)\dot{x}_1x_2 - \frac{Bmgl}{J}cos(x_1)\dot{x}_2 - mglcos(x_1)\dot{x}_1x_2^2 - mglsin(x_1)2x_2\dot{x}_2] + \\ &\frac{K_a}{JL_mC}(-\frac{1}{L})x_4 + \frac{K_a}{JL_mC}(\frac{1}{L})u \end{aligned} \tag{5.12}$$

The input-output linearized system is also written in the concise form:

$$y^{(5)} = f(y, \ldots, y^{(5)}) + g(y, \ldots, y^{(5)})u \quad (5.13)$$

and by defining $v = f(y, \ldots, y^{(5)}) + g(y, \ldots, y^{(5)})u$ one gets the equivalent description

$$y^{(5)} = v \quad (5.14)$$

For the previous description, the design of a stabilizing feedback controller gives:

$$u = y_d^{(5)} - k_1(y^{(4)} - y_d^{(4)}) - k_2(y^{(3)} - y_d^{(3)}) - k_3(\ddot{y} - \ddot{y}_d) - k_4(\dot{y} - \dot{y}_d) - k_5(y - y_d) \quad (5.15)$$

while the control input that is actually exerted on the system is

$$u = \frac{1}{g(y, \ldots, y^{(5)})}[v - f(y, \ldots, y^{(5)})] \quad (5.16)$$

with the previous control law the closed-loop dynamics becomes

$$e^{(5)} + k_1 e^{(4)} + k_2 e^{(3)} + k_3 \ddot{e} + k_4 \dot{e} + k_5 e = 0 \quad (5.17)$$

Thus, through suitable selection of the feedback control gains k_i $i = 1, \ldots, 5$ so as the associated characteristic polynomial to be a Hurwitz stable one, it is assured that

$$lim_{t \Rightarrow \infty} e(t) = 0 \quad (5.18)$$

5.2.4 *Disturbances Compensation with the Derivative-Free Nonlinear Kalman Filter*

Next, the problem of disturbances compensation is treated. Without loss of generality it is assumed that additive input disturbances $\tilde{d}$ affect the linearized model of the converter

$$y^{(5)} = v + \tilde{d} \quad (5.19)$$

and that these disturbances are described by their time derivatives up to order 3, that is $\tilde{d}^{(3)} = f_d$. Actually, every signal $\tilde{d}$ can be equivalently described by its time derivatives up to order n and the associated initial conditions. However, since estimation of the signal $\tilde{d}$ and of its derivatives is going to be performed with the use of Kalman Filtering, there is no prior constraint about knowing these initial conditions.

The state vector of the DC-DC converter's model is extended by introducing as additional state variables the disturbance inputs and their derivatives. Thus, $z_1 = y$, $z_2 = \dot{y}$, $z_3 = \ddot{y}$, $z_4 - y^{(3)}$, $z_5 - y^{(4)}$, $z_6 = \tilde{d}$, $z_7 = \dot{\tilde{d}}$ and $z_8 = \ddot{\tilde{d}}$. In this manner, and by defining the extended state vector $Z = [z_1, z_2, \dots, z_8]^T$ one arrives at a state-space description into the linear canonical (Brunovsky) form:

$$\begin{pmatrix} \dot{z}_1 \\ \dot{z}_2 \\ \dot{z}_3 \\ \dot{z}_4 \\ \dot{z}_5 \\ \dot{z}_6 \\ \dot{z}_7 \\ \dot{z}_8 \end{pmatrix} = \begin{pmatrix} 0 & 1 & 0 & 0 & 0 & 0 & 0 & 0 \\ 0 & 0 & 1 & 0 & 0 & 0 & 0 & 0 \\ 0 & 0 & 0 & 1 & 0 & 0 & 0 & 0 \\ 0 & 0 & 0 & 0 & 1 & 0 & 0 & 0 \\ 0 & 0 & 0 & 0 & 0 & 1 & 0 & 0 \\ 0 & 0 & 0 & 0 & 0 & 0 & 1 & 0 \\ 0 & 0 & 0 & 0 & 0 & 0 & 0 & 1 \\ 0 & 0 & 0 & 0 & 0 & 0 & 0 & 0 \end{pmatrix} \begin{pmatrix} z_1 \\ z_2 \\ z_3 \\ z_4 \\ z_5 \\ z_6 \\ z_7 \\ z_8 \end{pmatrix} + \begin{pmatrix} 0 & 0 \\ 0 & 0 \\ 0 & 0 \\ 0 & 0 \\ 1 & 0 \\ 0 & 0 \\ 0 & 0 \\ 0 & 1 \end{pmatrix} \begin{pmatrix} v \\ f_d \end{pmatrix} \tag{5.20}$$

$$z^m = \begin{pmatrix} 1 & 0 & 0 & 0 & 0 & 0 & 0 & 0 \end{pmatrix} Z \tag{5.21}$$

The previous description of the system is written in the linear state-space form:

$$\begin{aligned} \dot{z} &= Az + Bv \\ z^m &= Cz \end{aligned} \tag{5.22}$$

To perform simultaneous estimation of the non-measurable state vector elements and of the disturbance inputs the following disturbance observer is defined

$$\begin{aligned} \dot{\hat{z}} &= A_o \hat{z} + B_o v + K_f (z_m - \hat{z}_m) \\ \hat{z}_m &= C_o \hat{z} \end{aligned} \tag{5.23}$$

where $A_o = A$, $C_o = C$ and

$$B_o = \begin{pmatrix} 0 & 0 & 0 & 0 & 1 & 0 & 0 & 0 \end{pmatrix}^T \tag{5.24}$$

The observer's gain is found through the Kalman Filter's recursion [34, 341, 344]. To apply Kalman Filtering on the linearized equivalent model of the system, which is also known as Derivative-free nonlinear Kalman Filter, matrices A_o, B_o, C_o are subjected to discretization with the use of common discretization methods. The filter also comprises an inverse transformation based on differential flatness theory and on Eqs. (5.2)–(5.5), which enables to obtain state estimates of the initial nonlinear model. The discrete-time equivalents of the aforementioned matrices are denoted as A_d, B_d and C_d. The Kalman Filter's algorithm consists of a measurement update and a time-update stage. By denoting as Q, R the process and measurement noise covariance matrices, as P^- the state vector error covariance matrix prior to measurement taking and by P the state vector error after measurement taking one has.

measurement update:

$$
\begin{aligned}
K_f(k) &= P^-(k)C_d^T[C_d P^-(k)C_d^T + R]^{-1} \\
\hat{x}(k) &= \hat{x}^-(k) + K_f(k)[z_m - \hat{z}_m] \\
P(k) &= P^-(k) - K_f(k)C_d P^-(k)
\end{aligned}
\tag{5.25}
$$

time update:

$$
\begin{aligned}
P^-(k+1) &= A_d P(k) A_d^T + Q \\
\hat{x}^-(k+1) &= A_d \hat{x}(k) + B_d v(k)
\end{aligned}
\tag{5.26}
$$

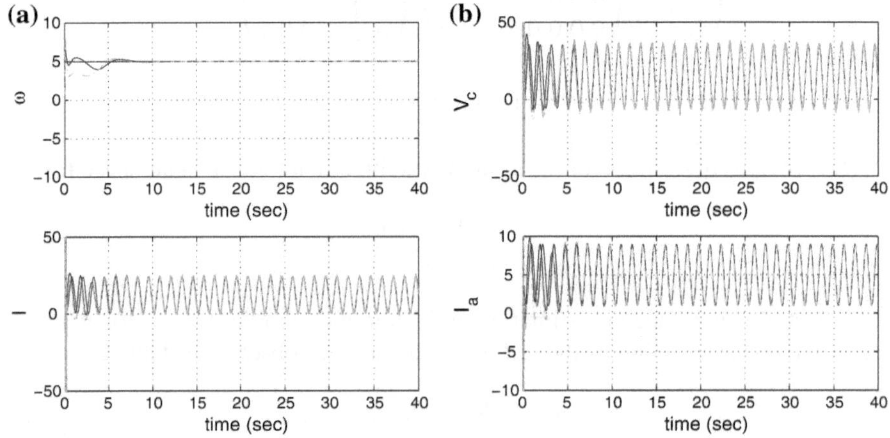

Fig. 5.2 Tracking of setpoint 1 for the system consisting of the DC-DC converter connected to the DC motor. **a** State variables $x_2 = \omega$ and $x_3 = I$, **b** state variables $x_4 = V_c$ and $x_5 = I_a$

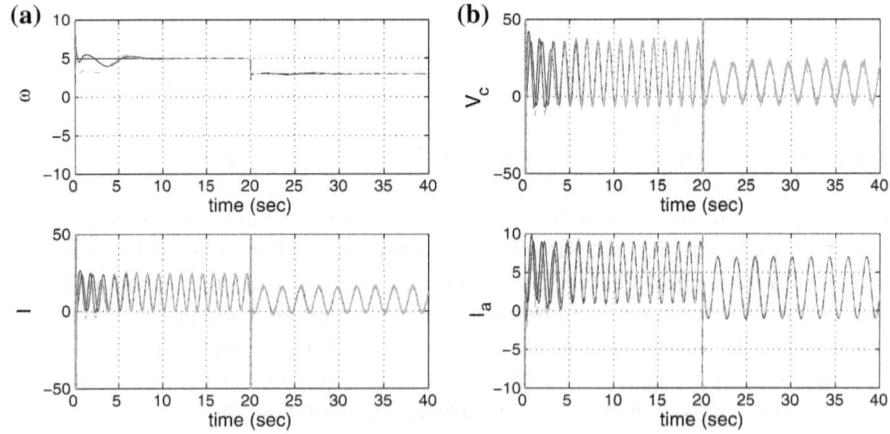

Fig. 5.3 Tracking of setpoint 2 for the system consisting of the DC-DC converter connected to the DC motor. **a** State variables $x_2 = \omega$ and $x_3 = I$, **b** state variables $x_4 = V_c$ and $x_5 = I_a$

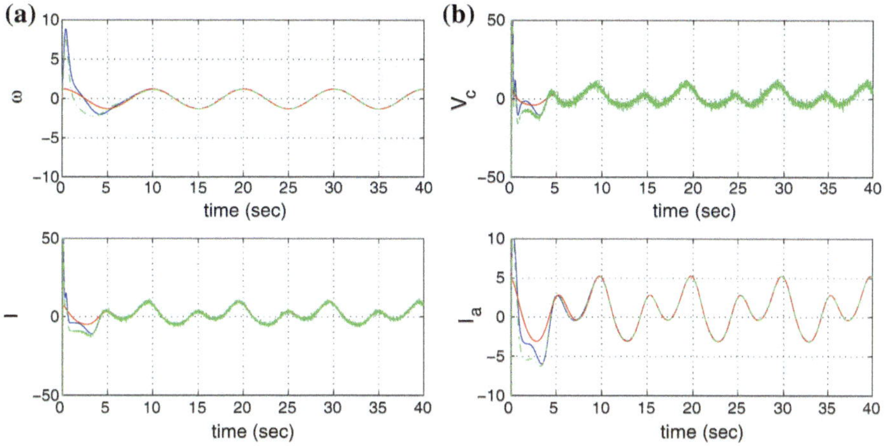

Fig. 5.4 Tracking of setpoint 3 for the system consisting of the DC-DC converter connected to the DC motor. **a** State variables $x_2 = \omega$ and $x_3 = I$, **b** state variables $x_4 = V_c$ and $x_5 = I_a$

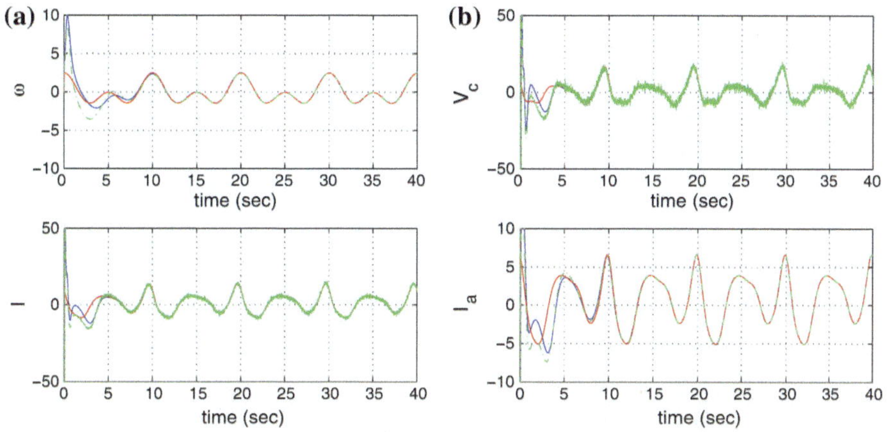

Fig. 5.5 Tracking of setpoint 2 for the system consisting of the DC-DC converter connected to the DC motor. **a** State variables $x_2 = \omega$ and $x_3 = I$, **b** state variables $x_4 = V_c$ and $x_5 = I_a$

5.2.5 *Simulation Tests*

The efficiency of the proposed control scheme was tested through simulation experiments. Flatness-based control was applied to the system consisting of the DC-DC converter and of the DC motor. The only measurable state variable of the system was taken to be the rotor's turn angle $x_1 = \theta$. The Derivative-free nonlinear Kalman Filter was used as a disturbance observer, for estimating simultaneously the non-measurable state variables of the model and the additive disturbance inputs that were affecting it. The obtained results are depicted in Figs. 5.2, 5.3, 5.4, 5.5, 5.6 and 5.7.

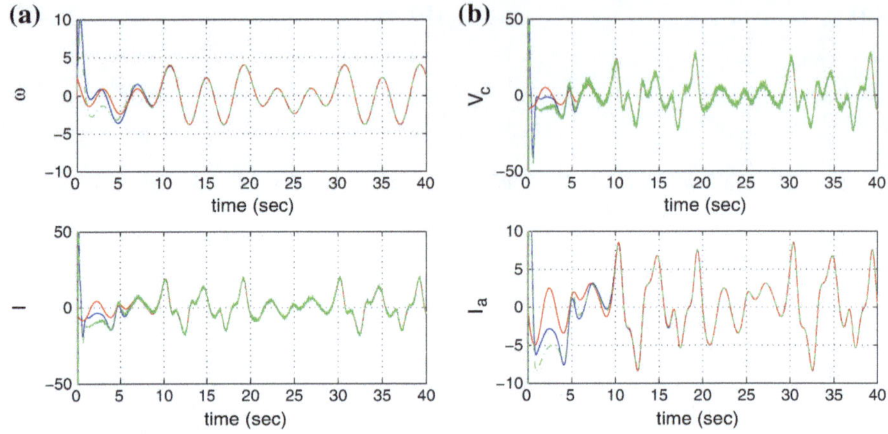

Fig. 5.6 Tracking of setpoint 5 for the system consisting of the DC-DC converter connected to the DC motor. **a** State variables $x_2 = \omega$ and $x_3 = I$, **b** state variables $x_4 = V_c$ and $x_5 = I_a$

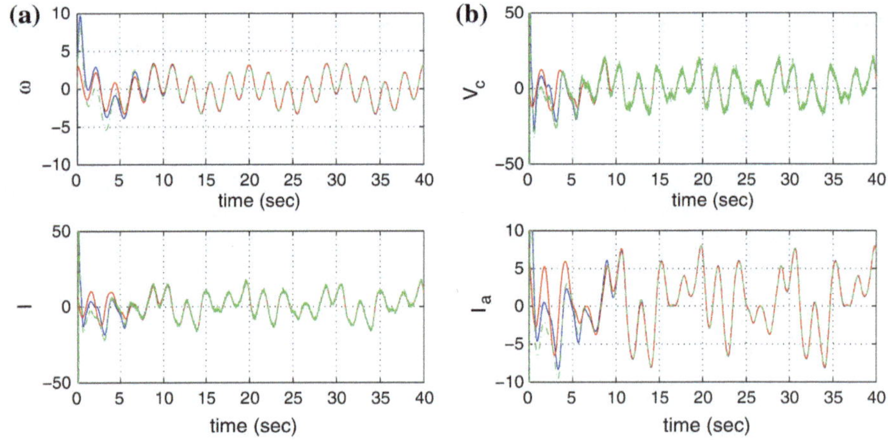

Fig. 5.7 Tracking of setpoint 6 for the system consisting of the DC-DC converter connected to the DC motor. **a** State variables $x_2 = \omega$ and $x_3 = I$, **b** state variables $x_4 = V_c$ and $x_5 = I_a$

The real value of the state variable is depicted with the blue line, the estimated value of it is depicted with the green line while the reference setpoint is depicted with the red line. It can be noticed that, despite implementation of a state estimation-based control scheme and despite the presence of external disturbances, the flatness-based control succeeded fast and accurate tracking to the reference setpoints for all state variables of the system.

The control inputs exerted on the system which consists of the DC-DC converter, are shown in Figs. 5.8a, 5.9a and 5.10a. These diagrams depict the ratio between the control input that is applied to the DC-DC converter and the constant voltage E

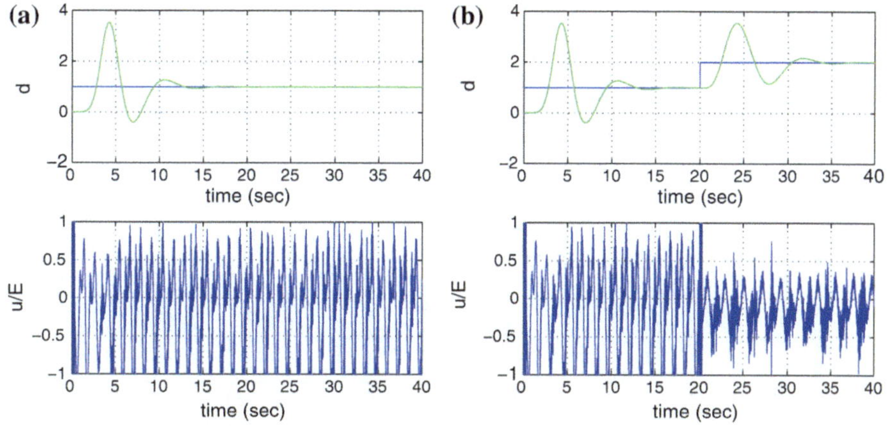

Fig. 5.8 *Top row* Kalman Filter-based estimation of disturbance inputs affecting the model of the DC-DC converter and the DC motor, *bottom row* Ratio u/E between the time-varying control input u applied to the DC-DC converter and the constant DC input E, in case of **a** tracking setpoint 1, **b** tracking setpoint 2

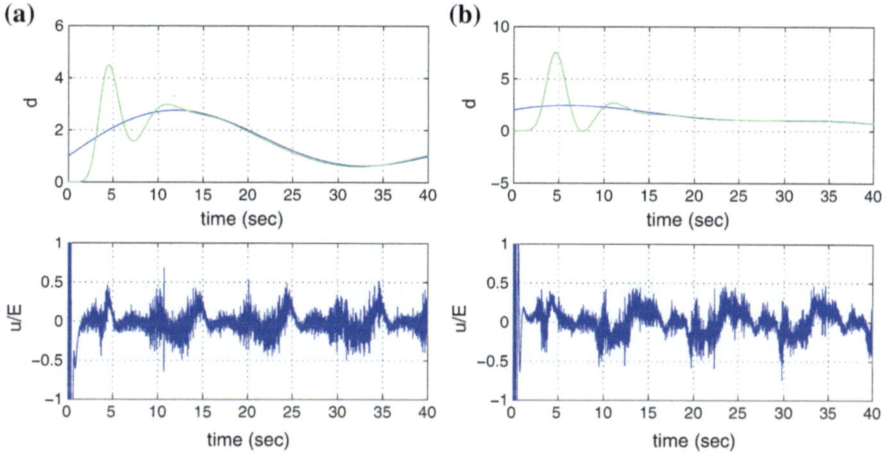

Fig. 5.9 *Top row* Kalman Filter-based estimation of disturbance inputs affecting the model of the DC-DC motor and the DC motor, *bottom row* Ratio u/E between the time-varying control input u applied to the DC-DC converter and the constant DC input E, in case of **a** tracking setpoint 3, **b** tracking setpoint 4

(Fig. 5.1). This ratio provides information about the implementation of PWM control of the converter and the associated duty cycle. Moreover, in Figs. 5.8b, 5.9b and 5.10b one can see that the Kalman Filter-based disturbance observer provides accurate estimates of the additive disturbance inputs that affected the converter's and motor's model. By identifying in real-time these perturbation terms their compensation became possible, after including an additional input to the feedback control signal.

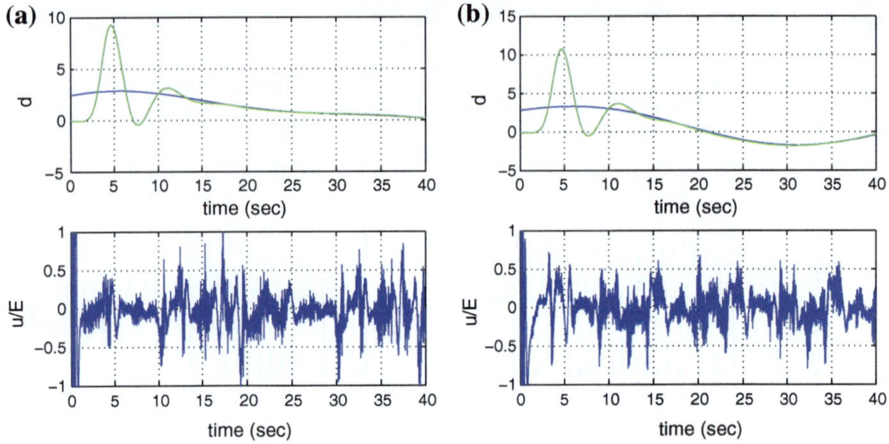

Fig. 5.10 *Top row* Kalman Filter-based estimation of disturbance inputs affecting the model of the DC-DC motor and the DC motor, *bottom row* Ratio u/E between the time-varying control input u applied to the DC-DC converter and the constant DC input E, in case of **a** tracking setpoint 5, **b** tracking setpoint 6

5.3 Control of Three-Phase AC to DC Converters

5.3.1 Overview

AC to DC converters are necessary for connecting DC electric machines to the AC grid, as well as for connecting AC sectors of the grid through HVDC lines. This section proposes a nonlinear control scheme for three-phase voltage source converters where estimation of disturbances and variations of the load current is performed with the use of a derivative-free implementation of nonlinear Kalman Filtering. Voltage source converters, are three-phase filtered rectifiers, that are widely used in the electric power grid (mainly for power flow control). Voltage source converters are the main building blocks of power flow controllers in transmission lines. For example, VSCs are contained in Unified Power Flow Controllers (UPFCs), or distribution-static synchronous compensators (D-STATCOMs). VSCs enable control of the amplitude and phase angle of the AC terminal voltages. Moreover, their bidirectional power flow capabilities allow VSCs to perform real and/or reactive power flow control in AC transmission lines [488, 506].

The dynamic model of voltage source converters is a nonlinear one and several nonlinear control methods have been proposed for it. Linearization round certain operating points with the computation of Jacobian matrices and control with the use of linear feedback control methods has been one of the proposed solutions [51, 444, 506]. Operation range can be restricted and a relatively big capacitor may be needed for keeping a constant DC-voltage in presence of a varying load. Initial non-linear control approaches consider the representation of the voltage source converter dynamics in the dq reference frame and use PI compensators. Passivity-based control methods have been proposed in [128]. Neural/fuzzy control methods for VSCs have been analyzed in [13, 73–76]. Back-stepping control approaches have been proposed in [13]. State estimation-based control for voltage source converters has been studied in [187, 234, 236]. The Power Matching Modulation approach has been proposed in [66, 67], while the Virtual Flux control method has been analyzed in [271, 272].

Other control approaches consider input-output linearization, as well as input-state linearization [287]. The latter methods transform the nonlinear system into a decoupled linear one. It is also known that one can attempt transformation of the nonlinear VSC model into the linear canonical (Brunovsky) form through the application of Lie-algebra theory. With the application of Lie-algebra methods it is possible to arrive at a description of the system in the linear canonical form if the relative degree of the system is equal to the order of the system. However, this linearization procedure requires the computation of Lie derivatives (partial derivatives on the vector fields describing the system dynamics) and this can be a tedious procedure.

Moreover, differential flatness theory has been proposed for VSC control [149, 150, 424]. As previously explained, differential flatness theory enables linearization for a wider class of systems than the one achieved with Lie-algebra methods [352, 388, 411]. The criteria for defining if a system is differentially flat have been previously defined: (i) the existence of the so-called flat output, i.e. a new variable which is expressed as a function of the system's state variables. The flat output and its derivatives should not be coupled in the form of an ordinary differential equation, (ii) the components of the system (i.e. state variables and control input) should be expressed as functions of the flat output and its derivatives [137, 229, 241, 456]. Differential flatness theory enables transformation to a linearized form (canonical Brunovsky form) for which the design of the controller becomes easier. Moreover, by showing that a system is differentially flat one can easily design a reference trajectory as a function of the so-called flat output and can find a control law that assures tracking of this desirable trajectory [137, 456].

This section is concerned with proving differential flatness of the three-phase voltage source converter model and its resulting description in the Brunovksy (canonical) form [281]. It is shown that for the linearized converter's model it is possible to design a feedback controller. At a second stage, a novel Kalman Filtering method, the Derivative-free nonlinear Kalman Filter, is proposed for estimating the non directly measurable elements of the state vector of the linearized system. With the redesign of the proposed Kalman Filter as a disturbance observer, it becomes possible to estimate disturbance terms which are due to variations of the load current and next to use these terms in the feedback controller. By avoiding linearization approximations, the proposed filtering method, improves the accuracy of estimation, and results in smooth control signal variations and in minimization of the tracking error of the associated control loop [57, 352, 355, 368].

5.3.2 *Linearization of the Converter's Model Using Lie Algebra*

5.3.2.1 Outcome of the Lie Algebra-Based Linearization

The electric circuit of the three-phase voltage source converter is depicted again in Fig. 5.11. Linearization of the converter's model will be performed using Lie-algebra and with the computation of the associated Lie derivatives [215, 266]. The following state variables are defined: $z_1 = h_1(x)$, $z_2 = L_f h_1(x)$, and $z_3 = h_2(x)$. After intermediate computations (which will be explained in the following section) one arrives at the VSC description:

$$\begin{aligned} \dot{z}_1 &= z_2 \\ \dot{z}_2 &= v_1 \\ \dot{z}_3 &= v_2 \end{aligned} \tag{5.27}$$

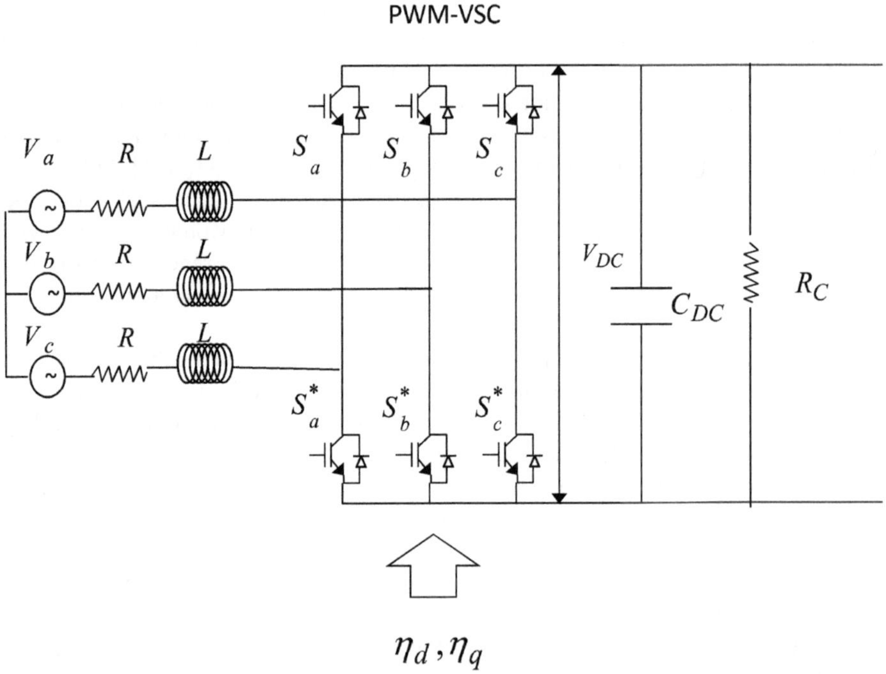

Fig. 5.11 Electrical circuit of the voltage source converter

or at the state-space description

$$\begin{pmatrix}\dot{z}_1\\ \dot{z}_2\\ \dot{z}_3\end{pmatrix} = \begin{pmatrix}0&1&0\\0&0&0\\0&0&0\end{pmatrix}\begin{pmatrix}z_1\\ z_2\\ z_3\end{pmatrix} + \begin{pmatrix}0&0\\1&0\\0&1\end{pmatrix}\begin{pmatrix}v_1\\ v_2\end{pmatrix}$$
$$\begin{pmatrix}y_1\\ y_2\end{pmatrix} = \begin{pmatrix}1&0&0\\0&0&1\end{pmatrix}\begin{pmatrix}z_1\\ z_2\\ z_3\end{pmatrix} \tag{5.28}$$

where the new control inputs are defined as

$$\begin{aligned} v_1 &= L_f^2 h_1(x) + L_{g_1}L_f h_1(x)u_1 + L_{g_2}L_f h_2(x)u_2 \\ v_2 &= f_{21} + g_{22}u_2 \end{aligned} \tag{5.29}$$

Therefore, the initial nonlinear system of the voltage source converter is transformed into the linear canonical form. The linearized system is controllable and observable.

5.3.2.2 Computation of Lie Derivatives for the VSC Model

Linearization of the converter's model can be performed using Lie-algebra and with the computation of the associated Lie derivatives [215, 266]. The following state variables are defined: $z_1 = h_1(x)$, $z_2 = L_f h_1(x)$, and $z_3 = h_2(x)$. Thus one gets

$$\begin{aligned} L_f h_1(x) &= \frac{\partial h_1}{\partial x_1}f_1 + \frac{\partial h_1}{\partial x_2}f_2 + \frac{\partial h_1}{\partial x_3}f_3 \Rightarrow L_f h_1(x) = \frac{3}{2}Lx_1(-\frac{R}{L}x_1 + \omega x_2 + \frac{v_d}{L}) \\ &+ \frac{3}{2}Lx_2(-\omega x_1 - \frac{R}{L}x_2) + C_{dc}x_3(-\frac{x_3}{C_{dc}R_c}) \\ &\Rightarrow z_2^1 = L_f h_1(x) = -\frac{3}{2}R(x_1^2 + x_2^2) + \frac{3}{2}v_d x_1 - \frac{1}{R_c}x_3^2. \end{aligned}$$

Similarly, one has

$$\dot{z}_2 = L_f^2 h_1(x) + L_{g_1}L_f h_1(x)u_1 + L_{g_2}L_f h_2(x)u_2 \tag{5.30}$$

where

$$\begin{aligned} L_f^2 h_1(x) &= \frac{\partial z_2^1}{\partial x_1}f_1 + \frac{\partial z_2^1}{\partial x_2}f_2 + \frac{\partial z_2^1}{\partial x_3}f_3 \Rightarrow L_f^2 h_1(x) = (-3Rx_1 + \frac{3}{2}v_d)(-\frac{R}{L}x_1 + \omega x_2 + \frac{v_d}{L}) \\ &+ (-3Rx_2)(-\omega x_1 - \frac{R}{L}x_2) + (-\frac{2}{R_c}x_3)(-\frac{x_3}{C_{dc}R_c}) \Rightarrow L_f^2 h_1(x) \\ &= \frac{3R^2}{L}(x_1^2 + x_2^2) - \frac{3R}{L}v_d x_1 + \frac{2x_3^2}{C_{dc}R_c^2} \end{aligned}$$

while it also holds

$$L_{g_1}L_f h_1(x) = \frac{\partial z_2^1}{\partial x_1}g_{11} + \frac{\partial z_2^1}{\partial x_2}g_{21} + \frac{\partial z_2^1}{\partial x_3}g_{31} \Rightarrow L_{g_1}L_f h_1(x)$$
$$= (-3Rx_1 + \frac{3}{2}v_d)(-\frac{x_3}{2L}) + (-3Rx_2)0 + (-\frac{2}{R_c}x_3)(\frac{3x_1}{4C_{dc}}) \Rightarrow L_{g_1}L_f h_1(x)$$
$$= \frac{3R}{2L}x_1x_3 - \frac{3}{4L}v_d x_3 - \frac{3x_1x_3}{2R_cC_{dc}}$$

and also

$$L_{g_2}L_f h_1(x) = \frac{\partial z_2^1}{\partial x_1}g_{12} + \frac{\partial z_2^1}{\partial x_2}g_{22} + \frac{\partial z_2^1}{\partial x_3}g_{32} \Rightarrow L_{g_2}L_f h_1(x)$$
$$= (-3Rx_1 + \frac{3}{2}v_d)0 + (-3Rx_2)(-\frac{x_3}{2L}) + (-\frac{2}{R_c}x_3)(\frac{3x_2}{4C_{dc}}) \Rightarrow L_{g_2}L_f h_1(x)$$
$$= \frac{3Rx_2x_3}{2L} - \frac{3x_2x_3}{2R_cC_{dc}}.$$

Therefore, one has

$$\begin{aligned} \dot{z}_1 &= z_2 \\ \dot{z}_2 &= L_f^2 h_1(x) + L_{g_1}L_f h_1(x)u_1 + L_{g_2}L_f h_1(x)u_2 \end{aligned} \tag{5.31}$$

It can be confirmed that it holds $\dot{z}_1 = z_2$. Indeed one has $z_1 = h_1(x)$ therefore

$$\dot{z}_1 = \frac{\partial z_1^1}{\partial x_1}\dot{x}_1 + \frac{\partial z_1^1}{\partial x_2}\dot{x}_2 + \frac{\partial z_1^1}{\partial x_3}\dot{x}_3 \Rightarrow \dot{z}_1 = \frac{3}{2}Lx_1(-\frac{R}{L}x_1 + \omega x_2 + \frac{v_d}{L})$$
$$+\frac{3}{2}Lx_2(-\omega x_1 - \frac{R}{L}x_2) + C_{dc}x_3(-\frac{x_3}{C_{dc}R_c}) \Rightarrow \dot{z}_1$$
$$= -\frac{3}{2}Rx_1^2 + \frac{3}{2}L\omega x_1x_2 + \frac{3}{2}v_dx_1 - \frac{3}{2}Rx_2^2 - \frac{3}{2}L\omega x_1x_2 - \frac{1}{R_C}x_3^2.$$

Thus, one finally obtains

$$\dot{z}_1 = -\tfrac{3}{2}R(x_1^2 + x_2^2) + \tfrac{3}{2}v_dx_1 - \tfrac{1}{R_c}x_3^2 \tag{5.32}$$

Equation (5.30) can be also confirmed. Indeed, using that $z_2 = L_f h_1(x) = -\frac{3}{2}R(x_1^2 + x_2^2) + \frac{3}{2}v_dx_1 - \frac{1}{R_c}x_3^2$ one obtains

$$\dot{z}_2 = \frac{\partial z_2^1}{\partial x_1}\dot{x}_1 + \frac{\partial z_2^1}{\partial x_2}\dot{x}_2 + \frac{\partial z_2^1}{\partial x_3}\dot{x}_3 \Rightarrow \dot{z}_2 = (-3Rx_1 + \frac{3}{2}v_d)(-\frac{R}{L}x_1 + \omega x_2 + \frac{v_d}{L} - \frac{x_3}{2L}u_1)$$
$$+(-3Rx_2)(-\omega x_1 - \frac{R}{L}x_2 - \frac{x_3}{2L}u_2) + (-\frac{2}{R_c}x_3)(-, \frac{x_3}{C_{dc}R_c} + \frac{3x_1}{4C_{dc}}u_1 + \frac{3x_2}{4C_{dc}}u_2) \Rightarrow$$

$$\begin{aligned}\dot{z}_2 = \tfrac{3R^2}{L}(x_1^2+x_2^2) - \tfrac{3R}{L}v_d x_1 + \tfrac{2x_3^2}{C_{dc}R_c^2} \\ +(\tfrac{3R}{2L}x_1x_3 - \tfrac{3}{4L}v_d x_3 - \tfrac{3x_1x_3}{2R_cC_{dc}})u_1 + (\tfrac{3Rx_2x_3}{2L} - \tfrac{3x_2x_3}{2R_cC_{dc}})u_2\end{aligned} \tag{5.33}$$

which through the previous relations about $L_f^2h_1(x)$, $L_{g_1}L_fh_1(x)$ and $L_{g_2}L_fh_1(x)$ confirms finally Eq. (5.30).

The third state variable is also examined, that is $z_3 = h_2(x) = i_q$. It holds that

$$\begin{aligned}L_fh_2(x) = \tfrac{\partial h_2}{\partial x_1}f_1 + \tfrac{\partial h_2}{\partial x_2}f_2 + \tfrac{\partial h_2}{\partial x_3}f_3 \Rightarrow \\ L_fh_2(x) = 0f_1 + 1f_2 + 0f_3 \Rightarrow \\ L_fh_2(x) = -\omega x_1 - \tfrac{R}{L}x_2\end{aligned} \tag{5.34}$$

Equivalently, one gets

$$\begin{aligned}L_{g_1}h_2(x) = \tfrac{\partial h_2}{\partial x_1}g_{11} + \tfrac{\partial h_2}{\partial x_2}g_{21} + \tfrac{\partial h_2}{\partial x_3}g_{31} \Rightarrow \\ L_{g_1}h_2(x) = 0g_{11} + 1g_{21} + 0g_{31} \Rightarrow \\ L_{g_1}h_2(x) = 0\end{aligned} \tag{5.35}$$

and similarly one has

$$\begin{aligned}L_{g_2}h_2(x) = \tfrac{\partial h_2}{\partial x_1}g_{12} + \tfrac{\partial h_2}{\partial x_2}g_{22} + \tfrac{\partial h_2}{\partial x_3}g_{32} \Rightarrow \\ L_{g_2}h_2(x) = 0g_{12} + 1g_{22} + 0g_{32} \Rightarrow \\ L_{g_2}h_2(x) = -\tfrac{x_3}{2L}\end{aligned} \tag{5.36}$$

Therefore, it holds

$$\begin{aligned}\dot{z}_3 = L_fh_2(x) + L_{g_1}h_2(x)u_1 + L_{g_2}h_2(x)u_2 \Rightarrow \\ \dot{z}_3 = (-\omega x_1 - \tfrac{R}{L}x_2) + 0u_1 - \tfrac{x_3}{2L}u_2 \\ \dot{z}_3 = (-\omega x_1 - \tfrac{R}{L}x_2) - \tfrac{x_3}{2L}u_2\end{aligned} \tag{5.37}$$

Equation (5.37) corresponds to the second line of the state-space equations of V_{dc} and therefore confirms the relation

$$\dot{z}_3 = L_fh_2(x) + L_{g_1}h_2(x)u_1 + L_{g_1}h_2(x)u_2 \tag{5.38}$$

Therefore one arrives at the following equations

$$\begin{aligned}\dot{z}_1 = z_2 \\ \dot{z}_2 = L_f^2h_1(x) + L_{g_1}L_fh_1(x)u_1 + L_{g_2}L_fh_2(x)u_2 \\ \dot{z}_3 = f_{21} + g_{22}u_2\end{aligned} \tag{5.39}$$

where the new control inputs are defined as

$$\begin{aligned}v_1 = L_f^2h_1(x) + L_{g_1}L_fh_1(x)u_1 + L_{g_2}L_fh_2(x)u_2 \\ v_2 = f_{21} + g_{22}u_2\end{aligned} \tag{5.40}$$

Consequently, one gets

$$\begin{aligned}\dot{z}_1 &= z_2\\ \dot{z}_2 &= v_1\\ \dot{z}_3 &= v_2\end{aligned} \tag{5.41}$$

or in state-space description

$$\begin{pmatrix}\dot{z}_1\\ \dot{z}_2\\ \dot{z}_3\end{pmatrix} = \begin{pmatrix}0 & 1 & 0\\ 0 & 0 & 0\\ 0 & 0 & 0\end{pmatrix}\begin{pmatrix}z_1\\ z_2\\ z_3\end{pmatrix} + \begin{pmatrix}0 & 0\\ 1 & 0\\ 0 & 1\end{pmatrix}\begin{pmatrix}v_1\\ v_2\end{pmatrix}$$

$$\begin{pmatrix}y_1\\ y_2\end{pmatrix} = \begin{pmatrix}1 & 0 & 0\\ 0 & 0 & 1\end{pmatrix}\begin{pmatrix}z_1\\ z_2\\ z_3\end{pmatrix} \tag{5.42}$$

Therefore, the initial nonlinear system of the voltage source converter is transformed into the linear canonical form.

5.3.3 *Differential Flatness of the Voltage Source Converter*

5.3.3.1 Differential Flatness of the Voltage Source Converter's Model

It will be shown that the dynamic model of the voltage source converter is a differentially flat one, i.e. it holds that all state variables and its control inputs can be written as functions of the flat outputs and their derivatives. Moreover, it will be shown that by expressing the elements of the state vector as functions of the flat outputs and their derivatives one obtains a transformation of the VSC model into the linear canonical (Brunovsky) form.

The following flat outputs are defined'

$$y_{f_1} = h_1 = \tfrac{3L}{4}(x_1^2 + x_2^2) + \tfrac{C_{dc_2}}{2}x_3^2 \tag{5.43}$$

$$y_{f_2} = h_2 = i_q \tag{5.44}$$

It holds that

$$\dot{y}_{f_1} = -\tfrac{3}{2}R(x_1^2 + x_2^2) + \tfrac{3}{2}v_d x_1 - \tfrac{1}{R_c}x_3^2 \tag{5.45}$$

while deriving once more with respect to time one gets

$$\begin{aligned}\ddot{y}_{f_1} &= \tfrac{3R^2}{2}(x_1^2 + x_2^2) - \tfrac{3R}{L}v_d x_1 + \tfrac{2x_3^2}{C_{dc}R_c^2} + \\ &(-\tfrac{3R}{2L}x_1x_3 - \tfrac{3}{4L}v_d x_3 - \tfrac{3x_1x_3}{2R_cC_{dc}})u_1 + (\tfrac{3Rx_2x_3}{2L} - \tfrac{3x_2x_3}{2R_cC_{dc}})u_2\end{aligned} \tag{5.46}$$

Moreover,

$$\dot{y}_{f_2} = (-\omega x_1 - \tfrac{R}{L}x_2) - \tfrac{x_3}{2L}u_2 \tag{5.47}$$

It can be confirmed that all state variables and the control inputs of the VSC model can be written as functions of the flat outputs y_{f_1}, y_{f_2} and of their derivatives. It holds that $y_{f_2} = x_2 = i_q$. Moreover, using the definition of the flat outputs it holds

$$\dot{y}_{f_1} = -\tfrac{3}{2}R(x_1^2 + x_2^2) + \tfrac{3}{2}v_d x_1 - \tfrac{1}{R_c}x_3^2 \tag{5.48}$$

$$y_{f_2} = x_2 \tag{5.49}$$

Solving Eq. (5.48) with respect to x_3^2 one obtains

$$x_3^2 = R_c[-\tfrac{3}{2}Rx_1^2 - \tfrac{3}{2}Ry_{f_2}^2 + \tfrac{3}{2}v_d x_1 - \dot{y}_{f_1}] \tag{5.50}$$

By substituting Eq. (5.50) into Eq. (5.43) one gets

$$y_{f_1} = \frac{3}{4}Lx_1^2 + \frac{3}{4}Ly_{f_2}^2 + \frac{C_{dc}}{2}R_c[-\frac{3}{2}Rx_1^2 - \frac{3}{2}Ry_2^2 + \frac{3}{2}v_d x_1 - \dot{y}_{f_1}] \Rightarrow (\frac{3L - 3C_{dc}R_cR}{4})x_1^2 + \frac{3C_{dc}R_cv_d}{4}x_1$$
$$+(\frac{3}{4}Ly_{f_2}^2 - \frac{3C_{dc}R_cR}{4}y_{f_2}^2 - \frac{C_{dc}R_c}{2}\dot{y}_{f_1} - y_{f_1}) = 0$$

By computing the roots of the binomial given in the above equation it becomes possible to express state variable y_1 as a function of the flat outputs and their derivatives. Thus, by keeping the maximum of the binomial's roots one obtains

$$x_1 = f_a(y_{f_1}, \dot{y}_{f_1}, y_{f_2}) \tag{5.51}$$

Using the relation for x_1 described in Eq. (5.51) into Eq. (5.50) one has

$$\begin{gathered} x_3^2 = R_c[-\tfrac{3}{2}Rf_a(y_{f_1}, \dot{y}_{f_1}, y_{f_2}) - \tfrac{3}{2}Ry_{f_2}^2 + \tfrac{3}{2}v_d f_a(y_{f_1}, \dot{y}_{f_1}, y_{f_2}) - \dot{y}_1] \Rightarrow \\ x_3 = R_c[-\tfrac{3}{2}Rf_a(y_{f_1}, \dot{y}_{f_1}, y_{f_2}) - \tfrac{3}{2}Ry_{f_2}^2 + \\ \tfrac{3}{2}v_d f_a(y_{f_1}, \dot{y}_{f_1}, y_{f_2}) - \dot{y}_1]^{\frac{1}{2}} \Rightarrow x_3 = f_b(y_{f_1}, \dot{y}_{f_1}, y_{f_2}) \end{gathered} \tag{5.52}$$

From the first line of the state-space description of the system given in Eqs. (5.77)–(5.79) it holds that

$$\begin{gathered} \dot{x}_1 = (-\tfrac{R}{L}x_1 + \omega x_2 + \tfrac{v_d}{L}) - \tfrac{x_3}{2L}u_1 \Rightarrow \\ u_1 = \frac{(-\frac{R}{L}x_1 + \omega x_2 + \frac{v_d}{L}) - \dot{x}_1}{\frac{x_3}{2L}} \Rightarrow \\ u_1 = f_c(y_{f_1}, \dot{y}_{f_1}, y_{f_2}) \end{gathered} \tag{5.53}$$

Similarly, from the second line of the state-space description of the system given in Eqs. (5.77)–(5.79) it holds that

$$\dot{x}_2 = (-\omega x_1 - \tfrac{R}{L}x_2) - \tfrac{x_3}{2L}u_2 \Rightarrow$$
$$u_2 = \frac{-\omega x_1 - \frac{R}{L}x_2 - \dot{x}_2}{\frac{x_3}{2L}} \Rightarrow \qquad (5.54)$$
$$u_2 = f_d(y_{f_1}, \dot{y}_{f_1}, y_{f_2})$$

Consequently, all state variables and the control inputs in the model of the voltage source converter can be written as functions of the flat outputs and their derivatives. Thus, the VSC model is a differentially flat one.

5.3.3.2 Linearization of the Converter's Model Using Differential Flatness Theory

Using the definitions of the flat outputs y_{f_1} and y_{f_2} for the VSC model, and considering the new state variables $z_1 = y_{f_1}$, $z_2 = \dot{y}_{f_1}$ and $z_3 = y_{f_2}$ one obtains

$$\dot{z}_1 = z_2$$
$$\dot{z}_2 = \tfrac{3R^2}{2}(x_1^2 + x_2^2) - \tfrac{3R}{L}v_d x_1 + \tfrac{2x_3^2}{C_{dc}R_c^2} + (\tfrac{3R}{2L}x_1x_3 - \tfrac{3}{4L}v_d x_3 - \tfrac{3x_1x_3}{2R_cC_{dc}})u_1 + (\tfrac{3Rx_2x_3}{2L} - \tfrac{3x_2x_3}{2R_cC_{dc}})u_2 \qquad (5.55)$$
$$\dot{z}_3 = (-\omega x_1 - \tfrac{R}{L}x_2) - \tfrac{x_3}{2L}u_2$$

or equivalently

$$\dot{z}_1 = z_2$$
$$\dot{z}_2 = v_1 \qquad (5.56)$$
$$\dot{z}_3 = v_2$$

which can be also written in state-space form

$$\begin{pmatrix}\dot{z}_1\\ \dot{z}_2\\ \dot{z}_3\end{pmatrix} = \begin{pmatrix}0&1&0\\0&0&0\\0&0&0\end{pmatrix}\begin{pmatrix}z_1\\ z_2\\ z_3\end{pmatrix} + \begin{pmatrix}0&0\\1&0\\0&1\end{pmatrix}\begin{pmatrix}v_1\\ v_2\end{pmatrix}$$
$$\begin{pmatrix}\dot{y}_1\\ \dot{y}_2\end{pmatrix} = \begin{pmatrix}1&0&0\\0&0&1\end{pmatrix}\begin{pmatrix}z_1\\ z_2\\ z_3\end{pmatrix} \qquad (5.57)$$

As already noted, the linearized system is controllable and observable. The new control inputs $v_i,\ i=1,2$ are the same as the ones defined in the Lie algebra-based approach, as shown in Eq. (5.29). The linearized model of the VSC can be described by the following two equations

$$\begin{aligned}\ddot{z}_1 &= v_1\\ \dot{z}_3 &= v_2\end{aligned} \tag{5.58}$$

The control inputs which enable convergence to the desirable setpoints are

$$\begin{aligned}v_1 &= \ddot{z}_1^d - k_1^1(\dot{z}_1^1 - \dot{z}_1^d) - k_2^1(z_1 - z_1^d)\\ v_2 &= \dot{z}_3^d - k_1^2(z_3 - z_3^d)\end{aligned} \tag{5.59}$$

The control law of Eq. (5.59) succeeds the following tracking error dynamics

$$\begin{aligned}\ddot{e}_1 + k_1^1\dot{e}_1 + k_2^1 e_1 &= 0\\ \dot{e}_3 + k_1^2 e_3 &= 0\end{aligned} \tag{5.60}$$

which results into $lim_{t\Rightarrow\infty}e_1(t)=0$ and $lim_{t\Rightarrow\infty}e_3(t)=0$. It can be noticed that the linearized model obtained after the application of differential flatness theory is the same with the one obtained with the use of the Lie derivative approach. It holds that

$$\begin{aligned}v_1 &= L_f^2h_1(x) + L_{g_1}L_fh_1(x)u_1 + L_{g_2}L_fh_1(x)u_2\\ v_2 &= L_fh_2(x) + L_{g_1}h_2(x)u_1 + L_{g_2}h_2(x)u_2\end{aligned} \tag{5.61}$$

or, in matrix form

$$\begin{pmatrix}v_1\\ v_2\end{pmatrix} = \begin{pmatrix}L_f^2h_1(x)\\ L_fh_2(x)\end{pmatrix} + \begin{pmatrix}L_{g_1}L_fh_1(x) & L_{g_2}L_fh_1(x)\\ L_{g_1}h_2(x) & L_{g_2}h_2(x)\end{pmatrix}\begin{pmatrix}u_1\\ u_2\end{pmatrix} \tag{5.62}$$

It holds that

$$\begin{aligned}\begin{pmatrix}v_1\\ v_2\end{pmatrix} &= \begin{pmatrix}\frac{3R^2}{2}(x_1^2+x_2^2) - \frac{3R}{L}v_dx_1 + \frac{2x_3^2}{C_{dc}R_c^2}\\ -\omega x_1 - \frac{R}{L}x_2\end{pmatrix} +\\ &+ \begin{pmatrix}\frac{3R}{2L}x_1x_3 - \frac{3}{4L}v_dx_3 - \frac{3x_1x_3}{2R_cC_{dc}} & \frac{3Rx_2x_3}{2L} - \frac{3x_2x_3}{2R_cC_{dc}}\\ 0 & -\frac{x_3}{2L}\end{pmatrix}\begin{pmatrix}u_1\\ u_2\end{pmatrix}\end{aligned} \tag{5.63}$$

According to the above, the VSC dynamic model can be also written in a more compact form as

$$\begin{aligned}\tilde{v} &= \tilde{f} + M\tilde{u} \Rightarrow\\ \tilde{u} &= M^{-1}(\tilde{v} - \tilde{f})\end{aligned} \tag{5.64}$$

5.3.4 Kalman Filter-Based Disturbance Observer for the VSC Model

The simultaneous estimation of the non-measurable elements of the VSC state vector (e.g. $\dot{y}_{f_1}$) as well as the estimation of additive disturbance terms (e.g. associated with variations of the load current i_L) is possible with the use of a disturbance estimator [85, 98, 99, 169, 292].

Next, it will be considered that additive input disturbances (e.g. due to load variations) affect the VSC model. Thus, it is assumed that the third row of the state-space equations of the voltage source converter of Eq. (1.86) includes a disturbance term

$$\dot{x}_3 = -\frac{x_3}{C_{dc}R_c} + \frac{3x_1}{4C_{dC}}u_1 + \frac{3x_2}{4C_{dC}}u_2 + \tilde{d} \tag{5.65}$$

By describing the system's dynamics using differential flatness theory and the definition of the input v_1 given in Eq. (5.63), the disturbances' effects appear as follows

$$\ddot{y}_{f_1} = v_1 + (-\frac{2}{R_c}x_3\tilde{d}) \tag{5.66}$$

which means that the additive disturbance term is now described by

$$\tilde{T}_d = -\frac{2}{R_c}x_3\tilde{d} \tag{5.67}$$

The disturbance $\tilde{T}_d$ may also comprise additional uncertainty terms associated with the numerical values of the parameters of the VSC model. It is assumed that the aggregate dynamics of term $\tilde{T}_d$ is described by its third order derivative

$$\tilde{T}_d^{(3)} = f_L(y_{f_1}, \dot{y}_{f_1}, y_{f_2}) \tag{5.68}$$

Thus, it holds that $z_1 = y_{f_1}$, $z_2 = \dot{y}_{f_1}$, $z_3 = y_{f_2}$, $z_4 = \tilde{T}_d$, $z_5 = \dot{\tilde{T}}_d$ and $z_6 = \ddot{\tilde{T}}_d$. The dynamics of the extended state-space model is written as

$$\begin{array}{lll} \dot{z}_1 = z_2 \; \dot{z}_2 = v_1 + \tilde{T}_d & & \dot{z}_3 = v_2 \\ \dot{z}_4 = z_5 & \dot{z}_5 = z_6 & \dot{z}_6 = f_L(y_{f_1}, \dot{y}_{f_1}, y_{f_2}) \end{array} \tag{5.69}$$

or in matrix form one has

$$\begin{array}{c} \dot{\tilde{z}} = \tilde{A}\tilde{z} + \tilde{B}\tilde{v} \\ \tilde{y} = \tilde{C}\tilde{z} \end{array} \tag{5.70}$$

$$\begin{pmatrix} \dot{z}_1 \\ \dot{z}_2 \\ \dot{z}_3 \\ \dot{z}_4 \\ \dot{z}_5 \\ \dot{z}_6 \end{pmatrix} = \begin{pmatrix} 0 & 1 & 0 & 0 & 0 & 0 \\ 0 & 0 & 0 & 1 & 0 & 0 \\ 0 & 0 & 0 & 0 & 0 & 0 \\ 0 & 0 & 0 & 0 & 1 & 0 \\ 0 & 0 & 0 & 0 & 0 & 1 \\ 0 & 0 & 0 & 0 & 0 & 0 \end{pmatrix} \begin{pmatrix} z_1 \\ z_2 \\ z_3 \\ z_4 \\ z_5 \\ z_6 \end{pmatrix} + \begin{pmatrix} 0 & 0 & 0 \\ 1 & 0 & 0 \\ 0 & 1 & 0 \\ 0 & 0 & 0 \\ 0 & 0 & 0 \\ 0 & 0 & 1 \end{pmatrix} \begin{pmatrix} v_1 \\ v_2 \\ f_L(y_{f_1}, \dot{y}_{f_1}, y_{f_2}) \end{pmatrix}$$

$$\begin{pmatrix} y_1 \\ y_2 \end{pmatrix} = \begin{pmatrix} 1 & 0 & 0 & 0 & 0 & 0 \\ 0 & 0 & 1 & 0 & 0 & 0 \end{pmatrix} \begin{pmatrix} z_1 \\ z_2 \\ z_3 \\ z_4 \\ z_5 \\ z_6 \end{pmatrix} \tag{5.71}$$

The associated state estimator is

$$\dot{\hat{z}} = \tilde{A}_o \hat{z} + \tilde{B}_o \tilde{v}_1 + K_o(\tilde{y} - \tilde{C}\hat{z}) \tag{5.72}$$

where

$$\tilde{A}_o = \begin{pmatrix} 0 & 1 & 0 & 0 & 0 & 0 \\ 0 & 0 & 0 & 1 & 0 & 0 \\ 0 & 0 & 0 & 0 & 0 & 0 \\ 0 & 0 & 0 & 0 & 1 & 0 \\ 0 & 0 & 0 & 0 & 0 & 1 \\ 0 & 0 & 0 & 0 & 0 & 0 \end{pmatrix} \quad \tilde{B}_o = \begin{pmatrix} 0 & 0 & 0 \\ 1 & 0 & 0 \\ 0 & 1 & 0 \\ 0 & 0 & 0 \\ 0 & 0 & 0 \\ 0 & 0 & 0 \end{pmatrix} \tag{5.73}$$

$$\tilde{C}_o = \begin{pmatrix} 1 & 0 & 0 & 0 & 0 & 0 \\ 0 & 0 & 1 & 0 & 0 & 0 \end{pmatrix}$$

while the estimator's gain $K_o \in R^{6\times 2}$ is obtained from the standard Kalman Filter recursion [174, 209].

Defining as $\tilde{A}_d$, $\tilde{B}_d$, and $\tilde{C}_d$, the discrete-time equivalents of matrices $\tilde{A}_o$, $\tilde{B}_o$ and $\tilde{C}_o$ respectively, a Derivative-free nonlinear Kalman Filter can be designed for the aforementioned representation of the system dynamics [352, 355]. The associated Kalman Filter-based disturbance estimator is given by

measurement update:

$$\begin{aligned} K(k) &= P^-(k)\tilde{C}_d^T[\tilde{C}_d \cdot P^-(k)\tilde{C}_d^T + R]^{-1} \\ \hat{\tilde{z}}(k) &= \hat{z}^-(k) + K(k)[\tilde{C}_d\tilde{z}(k) - \tilde{C}_d\hat{\tilde{z}}^-(k)] \\ P(k) &= P^-(k) - K(k)\tilde{C}_d P^-(k) \end{aligned} \tag{5.74}$$

time update:

$$\begin{aligned} P^-(k+1) &= \tilde{A}_d(k)P(k)\tilde{A}_d^T(k) + Q(k) \\ \hat{\tilde{z}}^-(k+1) &= \tilde{A}_d(k)\hat{\tilde{z}}(k) + \tilde{B}_d(k)\tilde{v}(k) \end{aligned} \tag{5.75}$$

5.3.5 Simulation Tests

To evaluate the performance of the proposed nonlinear control scheme, that uses Kalman Filtering to estimate the nonmeasurable disturbances of the VSC model, simulation experiments have been carried out. Different DC voltage setpoints V_{dc} have been assumed. Moreover, different external disturbance terms $\tilde{d}$ (e.g. due to load perturbations) have been considered to affect the VSC dynamic model. The control loop used in the VSC control is shown in Fig. 5.12.

Several cases of VSC functioning under different perturbation terms have been presented. The disturbance dynamics was completely unknown to the controller and its identification was performed in real time by the disturbance estimator. It is shown that the derivative-free nonlinear Kalman Filter redesigned as a disturbance observer is capable of estimating with accuracy the unknown disturbance input $\tilde{d}$. The associated results are presented in Figs. 5.13, 5.14, 5.15 and 5.16. Several reference setpoints have been defined for the VSC state variables, i.e. currents i_d, i_q and the output voltage V_{dc} and as it can be observed from the associated diagrams, the proposed control scheme resulted in fast and accurate convergence to these setpoints. The disturbance observer that was based on the Derivative-free nonlinear Kalman Filter was capable of estimating the unknown and time-varying input disturbances affecting the VSC model.

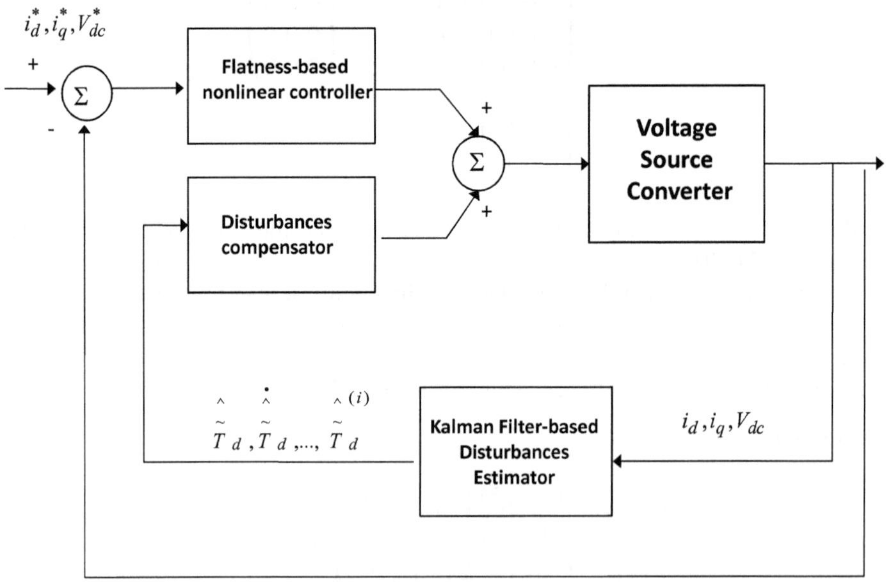

Fig. 5.12 Control loop for the VSC comprising a flatness-based nonlinear controller and a Kalman Filter-based disturbances estimator

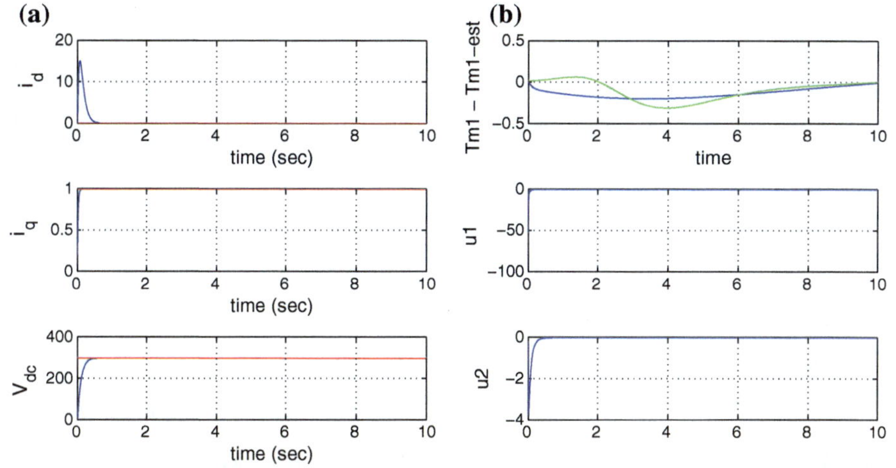

Fig. 5.13 **a** Control of the state variables of the VSC in case of reference setpoint 1, **b** estimation of disturbance input and variation of the control inputs

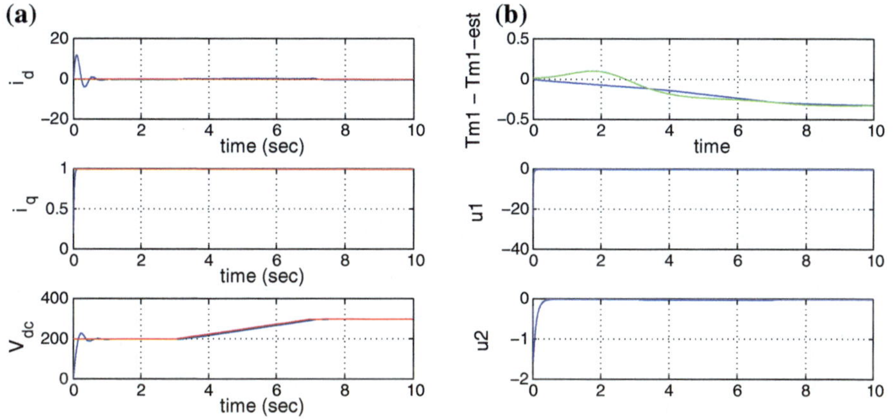

Fig. 5.14 **a** Control of the state variables of the VSC in case of reference setpoint 2, **b** estimation of disturbance input and variation of the control inputs

The simulation experiments have confirmed that the proposed state estimation-based control scheme not only enables implementation of VSC control through the measurement of a small number of variables (e.g. the ones appearing in the flat outputs y_{f_1} and y_{f_2}) but also improves the robustness of the VSC control loop in case of disturbances.

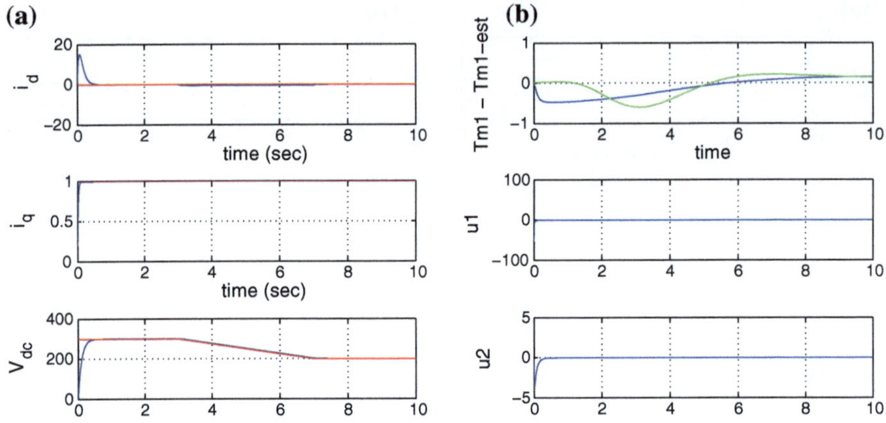

Fig. 5.15 **a** Control of the state variables of the VSC in case of reference setpoint 3, **b** estimation of disturbance input and variation of the control inputs

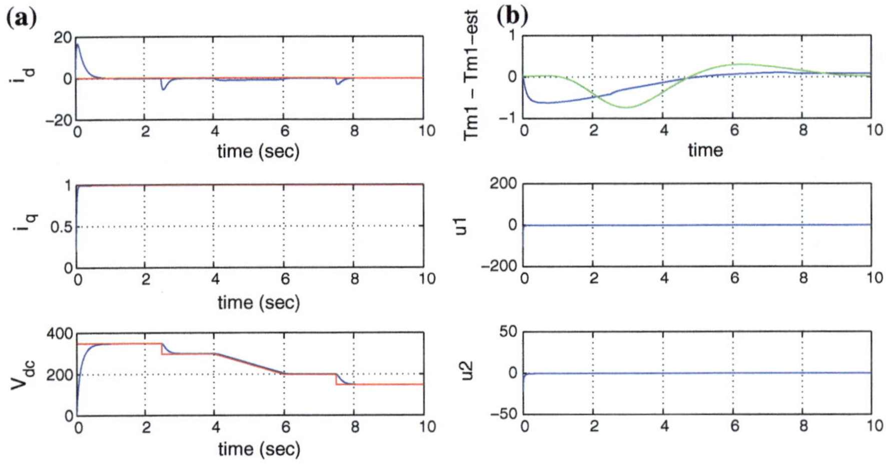

Fig. 5.16 **a** Control of the state variables of the VSC in case of reference setpoint 4, **b** estimation of disturbance input and variation of the control inputs

The improvement in the performance of the control loop that is due to the use of a disturbance observer based on the Derivative-free nonlinear Kalman Filter is explained as follows: (i) compensation of the disturbance terms which are generated by parametric uncertainty or unknown external inputs (ii) more accurate estimation of the disturbance terms because the filtering procedure is based on an exact linearization of the system's dynamics and does not introduce numerical errors (as for example in the case of the Extended Kalman Filter). This is shown in Fig. 5.17.

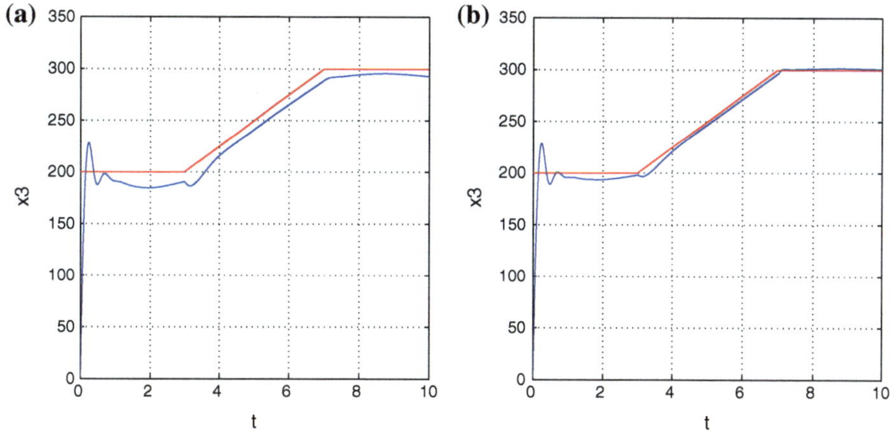

Fig. 5.17 Convergence of the output voltage V_{dc} to the reference setpoint **a** without using the disturbance observer, **b** when using the disturbance observer

5.4 Nonlinear H-infinity Control of VSC

5.4.1 Outline

The H-infinity control method has been already applied to power generators and multi-phase machines. This chapter proposes the application of an approximate linearization scheme for the dynamic model of the voltage source converter that is based on Taylor series expansion round the converter's temporal operating point. [34, 341]. To perform this linearization, the computation of Jacobian matrices is needed while the induced linearization error terms are treated as disturbances. For the linearized equivalent of the converter's model an H_∞ feedback control scheme is developed. The formulation of the H_∞ control problem is based on the minimization of a quadratic cost function that comprises both the disturbance and the control input effects. The disturbance tries to maximize the cost function while the control signal tries to minimize, within a mini-max differential game.

On the one side the proposed control method is characterized by its simplicity and its efficient performance in several operating conditions. On the other side, comparing to nonlinear feedback control approaches which are based on exact feedback linearization (as the ones based on differential flatness theory and analyzed in Refs. [352, 366, 375]), the proposed H_∞ control scheme for the VSC model is assessed as follows: (i) it uses an approximate linearization approach of the system's dynamic model which does not follow the elaborated transformations (diffeomorphisms) of the exact linearization methods, (ii) it introduces additional disturbance error which is due to the approximate linearization of the converter's dynamics dynamics coming from the application of Taylor series expansion, (iii) it requires the computation of Jacobian matrices, which in the case of electric power systems of high dimensionality can be also a burdensome procedure, (iv) unlike exact feedback linearization, the H_∞ control term has to compensate not only for modelling uncertainties and external disturbances but needs also to annihilate the effects of the cumulative linearization error, (v) the H_∞ control approach follows optimal control methods for the computation of the control signal, however unlike exact feedback linearization control it requires the solution of Riccati equations which for systems of high dimensionality can be again a demanding procedure.

5.4.2 Linearization of the Voltage Source Converter's Dynamic Model

The previously defined model of the AC to DC voltage source converter is examined. A three-phase voltage-source converter is considered (Fig. 5.11). The voltage source converter model in the rotating dq reference frame is given by [236, 424]:

$$\begin{aligned} L\dot{i}_d &= -Ri_d + L\omega i_q + v_d - \frac{V_{dc}}{2}u_1 \\ L\dot{i}_q &= -L\omega i_d - Ri_q + v_q - \frac{V_{dc}}{2}u_2 \\ C_{dc}\dot{V}_{dc} &= -\frac{1}{R_c}V_{dc} + \frac{3}{4}i_d u_1 + \frac{3}{4}i_q u_2 \end{aligned} \tag{5.76}$$

where i_d, i_q are the line currents (i_a, i_b, i_c) after transformation in the dq reference frame, and equivalently v_d, v_q are the phase voltages v_a, v_b, v_c after transformation in the dq reference frame. Variable V_{dc} denotes the DC voltage output of the converter, $u_1 = \eta_d$ and $u_2 = \eta_q$ stand for control inputs. The line losses and the transformer conduction losses are modelled by R and the inverter switching losses are modeled by R_c. Moreover, v_q is taken to be 0.

In Fig. 5.11 the electric circuit of the voltage source converter is depicted. The state vector is denoted as $x = [i_d, i_q, V_{dc}]^T$, while it is considered that all state vector elements are measurable thus constituting also the system's output $y = [i_d, i_q, V_{dc}]^T$, The control input is denoted by the vector $u = [\eta_d, \eta_q, v_d]^T$. Thus the MIMO nonlinear model of the VSC is written in the state-space form

$$\begin{aligned} \dot{x} &= f(x) + G(x)u \\ y &= h(x) \end{aligned} \tag{5.77}$$

where

$$f = \begin{pmatrix} -\frac{R}{L}x_1 + \omega x_2 \\ -\omega x_1 - \frac{R}{L}x_2 \\ -\frac{x_3}{C_{dc}R_c} \end{pmatrix} \quad G = [g_1 \; g_2] = \begin{pmatrix} -\frac{x_3}{2L} & 0 & \frac{1}{L} \\ 0 & -\frac{x_3}{2L} & 0 \\ \frac{3x_1}{4C_{dc}} & \frac{3x_1}{4C_{dc}} & 0 \end{pmatrix} \tag{5.78}$$

$$h = \begin{pmatrix} h_1 \\ h_2 \\ h_3 \end{pmatrix} = \begin{pmatrix} i_d \\ i_q \\ V_{dc} \end{pmatrix} \tag{5.79}$$

Considering linearization of the model, through Taylor series expansion, round the present value of the state vector $x(t)$ of the voltage source converter and round the present value of the input vector $u(t-T_s)$, where T_S is the sampling period, the Jacobian matrices of the converter's model are:

$$A=\begin{pmatrix} -\frac{R}{L} & \omega & -\frac{1}{2L}u_1(t-T_s) \\ -\omega & -\frac{R}{L} & -\frac{1}{2L}u_2(t-T_s) \\ \frac{3u_1(t-T_s)}{4C_{dc}} & \frac{3u_2(t-T_s)}{4C_{dc}} & -\frac{1}{C_{dc}R_c} \end{pmatrix} \tag{5.80}$$

$$B=\begin{pmatrix} -\frac{x_3}{2L} & 0 & \frac{1}{L} \\ 0 & -\frac{x_3}{2L} & 0 \\ \frac{3x_1}{4C_{dc}} & \frac{3x_2}{4C_{dc}} & 0 \end{pmatrix} \tag{5.81}$$

After linearization round its current operating point, the converter's dynamic model is written as

$$\dot{x}=Ax+Bu+d_1 \tag{5.82}$$

Parameter d_1 stands for the linearization error in the converter's dynamic model appearing in Eq. (5.82). The reference setpoints for the voltage source converter are denoted by $\mathbf{x_d}=[x_1^d, x_2^d, x_3^d]^T$. Tracking of this trajectory is succeeded after applying the control input u^*. At every time instant the control input u^* is assumed to differ from the control input u appearing in Eq. (5.82) by an amount equal to Δu, that is $u^*=u+\Delta u$

$$\dot{x}_d=Ax_d+Bu^*+d_2 \tag{5.83}$$

The dynamics of the controlled VSC system described in Eq. (5.82) can be also written as

$$\dot{x}=Ax+Bu+Bu^*-Bu^*+d_1 \tag{5.84}$$

and by denoting $d_3=-Bu^*+d_1$ as an aggregate disturbance term one obtains

$$\dot{x}=Ax+Bu+Bu^*+d_3 \tag{5.85}$$

By subtracting Eq. (5.83) from Eq. (5.85) one has

$$\dot{x}-\dot{x}_d=A(x-x_d)+Bu+d_3-d_2 \tag{5.86}$$

By denoting the tracking error as $e=x-x_d$ and the aggregate disturbance term as $\tilde{d}=d_3-d_2$, the tracking error dynamics becomes

$$\dot{e}=Ae+Bu+\tilde{d} \tag{5.87}$$

The above linearized form of the converter's model can be efficiently controlled after applying an H-infinity feedback control scheme. Since the third element of the control input v_d is taken to be known and constant (grid's voltage) the H-infinity control problem will be solved with respect to the first two elements of the control vector that is (u_1, u_2) using a new B matrix that consist of the first two columns of the initial B matrix.

5.4.3 *Nonlinear H-infinity Control for the Three-Phase VSC*

5.4.3.1 Mini-Max Control and Disturbance Rejection

The initial nonlinear system of the VSC is in the form

$$\dot{x} = f(x,u) \;\; x \in R^n, \;\; u \in R^m \tag{5.88}$$

Linearization of the system is performed at each iteration of the control algorithm round its present operating point $(x^*, u^*) = (x(t), u(t - T_s))$. The linearized equivalent of the system is described by

$$\dot{x} = Ax + Bu + Ld \;\; x \in R^n, \;\; u \in R^m, \;\; d \in R^q \tag{5.89}$$

where matrices A and B are obtained from the computation of the Jacobians

$$A = \begin{pmatrix} \frac{\partial f_1}{\partial x_1} & \frac{\partial f_1}{\partial x_2} & \cdots & \frac{\partial f_1}{\partial x_n} \\ \frac{\partial f_2}{\partial x_1} & \frac{\partial f_2}{\partial x_2} & \cdots & \frac{\partial f_2}{\partial x_n} \\ \cdots & \cdots & \cdots & \cdots \\ \frac{\partial f_n}{\partial x_1} & \frac{\partial f_n}{\partial x_2} & \cdots & \frac{\partial f_n}{\partial x_n} \end{pmatrix} |_{(x^*,u^*)} \;\; B = \begin{pmatrix} \frac{\partial f_1}{\partial u_1} & \frac{\partial f_1}{\partial u_2} & \cdots & \frac{\partial f_1}{\partial u_m} \\ \frac{\partial f_2}{\partial u_1} & \frac{\partial f_2}{\partial u_2} & \cdots & \frac{\partial f_2}{\partial u_m} \\ \cdots & \cdots & \cdots & \cdots \\ \frac{\partial f_n}{\partial u_1} & \frac{\partial f_n}{\partial u_2} & \cdots & \frac{\partial f_n}{\partial u_m} \end{pmatrix} |_{(x^*,u^*)} \tag{5.90}$$

and vector d denotes disturbance terms due to linearization errors. The problem of disturbance rejection for the linearized model that is described by

$$\begin{aligned} \dot{x} &= Ax + Bu + Ld \\ y &= Cx \end{aligned} \tag{5.91}$$

where $x \in R^n$, $u \in R^m$, $d \in R^q$ and $y \in R^p$, cannot be handled efficiently if the classical LQR control scheme is applied. This is because of the existence of the perturbation term d. The disturbance term d apart from modeling (parametric) uncertainty and external perturbation terms can also represent noise terms of any distribution.

In the H_∞ control approach, a feedback control scheme is designed for trajectory tracking by the converter's state vector and simultaneous disturbance rejection, considering that the disturbance affects the VSC system in the worst possible manner. The disturbances' effect are incorporated in the following quadratic cost function:

$$J(t) = \frac{1}{2}\int_0^T [y^T(t)y(t) + ru^T(t)u(t) - \rho^2 d^T(t)d(t)]dt, \quad r, \rho > 0 \tag{5.92}$$

As in the case of H-infinity control previously applied to power generators, the significance of the negative sign in the cost function's term that is associated with the perturbation variable $d(t)$ is that the disturbance tries to maximize the cost function $J(t)$ while the control signal $u(t)$ tries to minimize it. The physical meaning of the relation given above is that the control signal and the disturbances compete to each other within a mini-max differential game. This problem of mini-max optimization can be written as

$$min_u max_d J(u, d) \tag{5.93}$$

The objective of the optimization procedure is to compute a control signal $u(t)$ which can compensate for the worst possible disturbance, that is externally imposed to the VSC system. Again, the solution to the mini-max optimization problem is directly related to the value of the parameter ρ. This means that there is an upper bound in the disturbances magnitude that can be annihilated by the control signal.

5.4.3.2 H-Infinity Feedback Control

For the linearized VSC system given by Eq. (5.91) the cost function of Eq. (5.92) is defined, where the coefficient r determines the penalization of the control input and the weight coefficient ρ determines the reward of the disturbances' effects. It is assumed that:

The energy that is transferred from the disturbances signal $d(t)$ is considered to be bounded, that is $\int_0^\infty d^T(t)d(t)dt < \infty$, (ii) the matrices $[A \; B]$ and $[A \; L]$ are stabilizable, (iii) the matrix $[A \; C]$ is detectable. Then, the optimal feedback control law is given by

$$u(t) = -Kx(t) \tag{5.94}$$

with

$$K = \frac{1}{r}B^T P \tag{5.95}$$

where P is a positive semi-definite symmetric matrix which is obtained from the solution of the Riccati equation

$$A^T P + PA + Q - P(\frac{1}{r}BB^T - \frac{1}{2\rho^2}LL^T)P = 0 \tag{5.96}$$

where Q is also a positive definite symmetric matrix. The worst case disturbance is given by $d(t) = \frac{1}{\rho^2}L^T Px(t)$.

5.4.4 Lyapunov Stability Analysis

Lyapunov stability analysis for the VSC model follows the approach which has been already analyzed in the case of H-infinity control of induction generators and multi-phase machines. Through Lyapunov stability analysis it will be shown that the proposed nonlinear control scheme assures H_∞ tracking performance, and that in case of bounded disturbance terms asymptotic convergence to the reference setpoints is succeeded. The tracking error dynamics for the three-phase voltage source converter is written in the form

$$\dot{e} = Ae + Bu + L\tilde{d} \tag{5.97}$$

where in the converter's application example $L = I \in R^2$ with I being the identity matrix. The following Lyapunov equation is considered

$$V = \frac{1}{2}e^T Pe \tag{5.98}$$

where $e = x - x_d$ is the tracking error. By differentiating with respect to time one obtains

$$\begin{gathered} \dot{V} = \frac{1}{2}\dot{e}^T Pe + \frac{1}{2}eP\dot{e} \Rightarrow \\ \dot{V} = \frac{1}{2}[Ae + Bu + L\tilde{d}]^T P + \frac{1}{2}e^T P[Ae + Bu + L\tilde{d}] \Rightarrow \end{gathered} \tag{5.99}$$

$$\begin{gathered} \dot{V} = \frac{1}{2}[e^T A^T + u^T B^T + \tilde{d}^T L^T]Pe + \\ + \frac{1}{2}e^T P[Ae + Bu + L\tilde{d}] \Rightarrow \end{gathered} \tag{5.100}$$

$$\begin{gathered} \dot{V} = \frac{1}{2}e^T A^T Pe + \frac{1}{2}u^T B^T Pe + \frac{1}{2}\tilde{d}^T L^T Pe + \\ + \frac{1}{2}e^T PAe + \frac{1}{2}e^T PBu + \frac{1}{2}e^T PL\tilde{d} \end{gathered} \tag{5.101}$$

The previous equation is rewritten as

$$\begin{gathered} \dot{V} = \frac{1}{2}e^T(A^T P + PA)e + (\frac{1}{2}u^T B^T Pe + \frac{1}{2}e^T PBu) + \\ + (\frac{1}{2}\tilde{d}^T L^T Pe + \frac{1}{2}e^T PL\tilde{d}) \end{gathered} \tag{5.102}$$

Assumption: For given positive definite matrix Q and coefficients r and ρ there exists a positive definite matrix P, which is the solution of the following matrix equation

$$A^T P + PA = -Q + P(\frac{1}{r}BB^T - \frac{1}{\rho^2}LL^T)P \tag{5.103}$$

Moreover, the following feedback control law is applied to the system

$$u = -\frac{1}{r}B^T Pe \tag{5.104}$$

By substituting Eqs. (5.103) and (5.104) one obtains

$$\begin{aligned}\dot{V} = \tfrac{1}{2}e^T[-Q + P(\tfrac{1}{r}BB^T - \tfrac{1}{2\rho^2}LL^T)P]e + \\ + e^T PB(-\tfrac{1}{r}B^T Pe) + e^T PL\tilde{d} \Rightarrow\end{aligned} \tag{5.105}$$

$$\begin{aligned}\dot{V} = -\tfrac{1}{2}e^T Qe + (\tfrac{1}{r}PBB^T Pe - \tfrac{1}{2\rho^2}e^T PLL^T)Pe \\ -\tfrac{1}{r}e^T PBB^T Pe + e^T PL\tilde{d}\end{aligned} \tag{5.106}$$

which after intermediate operations gives

$$\dot{V} = -\tfrac{1}{2}e^T Qe - \tfrac{1}{2\rho^2}e^T PLL^T Pe + e^T PL\tilde{d} \tag{5.107}$$

or, equivalently

$$\begin{aligned}\dot{V} = -\tfrac{1}{2}e^T Qe - \tfrac{1}{2\rho^2}e^T PLL^T Pe + \\ + \tfrac{1}{2}e^T PL\tilde{d} + \tfrac{1}{2}\tilde{d}^T L^T Pe\end{aligned} \tag{5.108}$$

Lemma: The following inequality holds

$$\tfrac{1}{2}e^T L\tilde{d} + \tfrac{1}{2}\tilde{d}L^T Pe - \tfrac{1}{2\rho^2}e^T PLL^T Pe \leq \tfrac{1}{2}\rho^2\tilde{d}^T\tilde{d} \tag{5.109}$$

Proof: The binomial $(\rho\alpha - \frac{1}{\rho}b)^2$ is considered. Expanding the left part of the above inequality one gets

$$\begin{aligned}\rho^2a^2 + \tfrac{1}{\rho^2}b^2 - 2ab \geq 0 \Rightarrow \tfrac{1}{2}\rho^2a^2 + \tfrac{1}{2\rho^2}b^2 - ab \geq 0 \Rightarrow \\ ab - \tfrac{1}{2\rho^2}b^2 \leq \tfrac{1}{2}\rho^2a^2 \Rightarrow \tfrac{1}{2}ab + \tfrac{1}{2}ab - \tfrac{1}{2\rho^2}b^2 \leq \tfrac{1}{2}\rho^2a^2\end{aligned} \tag{5.110}$$

The following substitutions are carried out: $a = \tilde{d}$ and $b = e^T PL$ and the previous relation becomes

$$\tfrac{1}{2}\tilde{d}^T L^T Pe + \tfrac{1}{2}e^T PL\tilde{d} - \tfrac{1}{2\rho^2}e^T PLL^T Pe \leq \tfrac{1}{2}\rho^2\tilde{d}^T\tilde{d} \tag{5.111}$$

Equation (5.111) is substituted in Eq. (5.108) and the inequality is enforced, thus giving

$$\dot{V} \leq -\tfrac{1}{2}e^T Qe + \tfrac{1}{2}\rho^2\tilde{d}^T\tilde{d} \tag{5.112}$$

For sufficiently small ρ the right part of Eq. (5.112) will remain upper bounded by zero. Equation (5.112) shows that the H_∞ tracking performance criterion is satisfied. The integration of $\dot{V}$ from 0 to T gives

$$\begin{aligned}\int_0^T \dot{V}(t)dt \leq -\tfrac{1}{2}\int_0^T ||e||_Q^2 dt + \tfrac{1}{2}\rho^2\int_0^T ||\tilde{d}||^2 dt \Rightarrow \\ 2V(T) + \int_0^T ||e||_Q^2 dt \leq 2V(0) + \rho^2\int_0^T ||\tilde{d}||^2 dt\end{aligned} \tag{5.113}$$

Moreover, if there exists a positive constant $M_d > 0$ such that

$$\int_0^{\infty} ||\tilde{d}||^2 dt \leq M_d \tag{5.114}$$

then one gets

$$\int_0^{\infty} ||e||_Q^2 dt \leq 2V(0) + \rho^2 M_d \tag{5.115}$$

Thus, the integral $\int_0^{\infty} ||e||_Q^2 dt$ is bounded. Moreover, $V(T)$ is bounded and from the definition of the Lyapunov function V in Eq. (5.98) it becomes clear that $e(t)$ will be also bounded since $e(t) \in \Omega_e = \{e | e^T Pe \leq 2V(0) + \rho^2 M_d\}$. According to the above and with the use of Barbalat's Lemma one obtains $lim_{t\to\infty} e(t) = 0$. The latter stands for an asymptotic stability condition for the control loop.

5.4.5 Simulation Tests

The performance of the proposed nonlinear H_{∞} control scheme is tested in tracking of various setpoints. Indicative values of the converter's parameters are as follows: $R = 4\,\Omega$, $L = 165\,\text{mH}$, $C_{dc} = 2.1\,\text{mF}$, $w = 120\pi$ rad/sec, $v_d = 380\,\text{V}$, and $R_c = 3\,\text{k}\Omega$. It is noted that precise knowledge of the converter's parameters is not a restrictive condition. The nonlinear H-infinity controller is robust to uncertainty of the model's parameters therefore, the functioning of the control loop remains reliable even if the nominal parameters' values used in the design of the controller differ from the real parameter values.

In all simulation cases the performance of the proposed controller was satisfactory, with minimum tracking error and fast convergence to the reference setpoints (Fig. 5.18). The obtained results are depicted in Figs. 5.19 and 5.20. In can be observed that for the particular choice of the controller parameters, the time needed for the output of the voltage source converter to reach the desirable DC voltage levels was a few hundred milliseconds (less than 0.2 s).

Moreover, additional simulation experiments have been performed to test the nonlinear H-infinity control loop under parametric uncertainty and external perturbations for the converter's model. As it can be observed in Figs. 5.21 and 5.22 the nonlinear controller succeeded convergence of the converter's state vector elements to the desirable setpoints. The perturbations signals affecting the converter's control inputs are depicted in Fig. 5.23a, b, respectively.

Finally, diagrams are provided about the three-phase voltage variables v_a, v_b, v_c of the electricity grid to which the AC to DC voltage converter is connected. These are depicted in Fig. 5.24a. These diagrams were obtained by considering that the voltage variables v_d and v_q expressed in the dq reference frame are known and by applying to them the Park transform (this finally provided the three-phase voltage variables v_a, v_b and v_c). Additionally, diagrams are provided about the variation of the active power p_f and the reactive power q_f produced by the three-phase voltage source

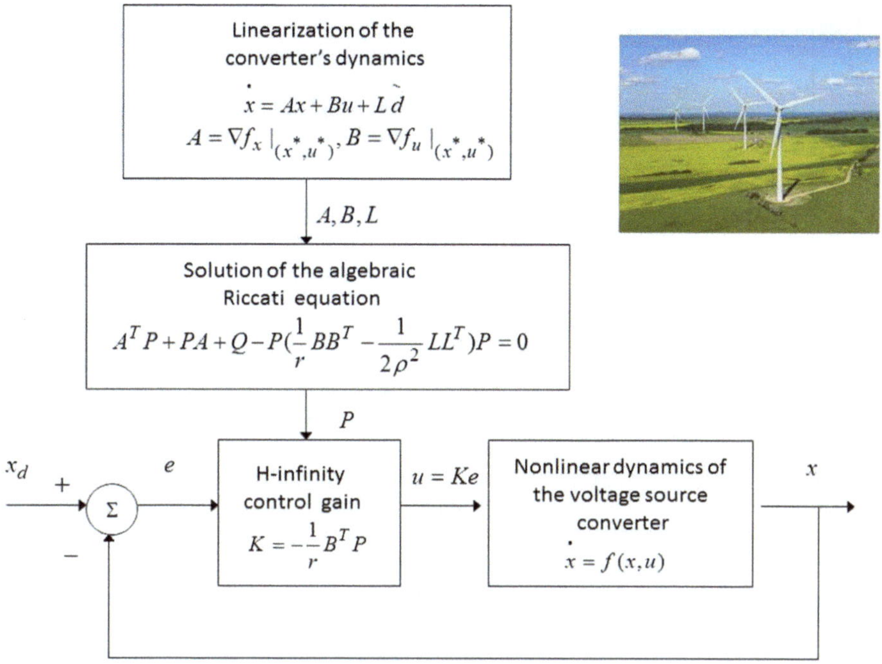

Fig. 5.18 Diagram of the H-infinity nonlinear control for the voltage source converter

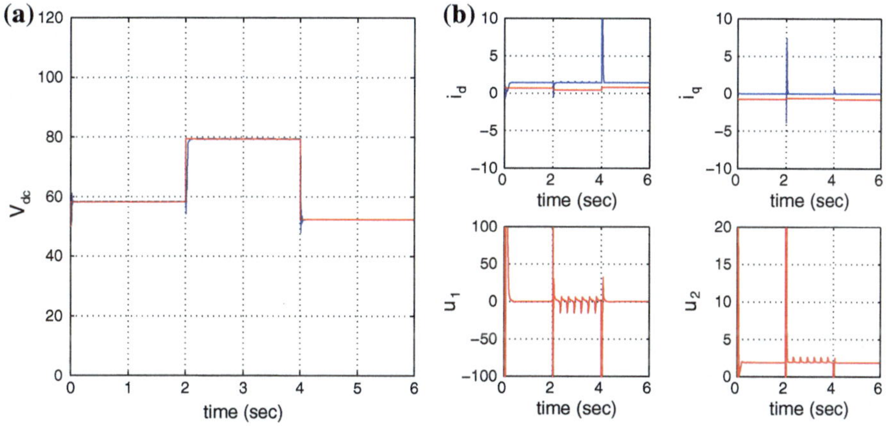

Fig. 5.19 Functioning of the converter without external (grid) perturbations: **a** Tracking of reference setpoint 1 by the DC voltage output of the three-phase voltage source converter, **b** variation of the converter's state vector elements and control inputs

converter. The active power of the converter was computed by $p_f = v_d i_d + v_q i_q$, while the reactive power was computed by $q_f = v_d i_q - v_q i_d$. The associated results are depicted in Fig. 5.24b. Using the values of p_f and q_f it can be found that the converter's power factor is about 95 %.

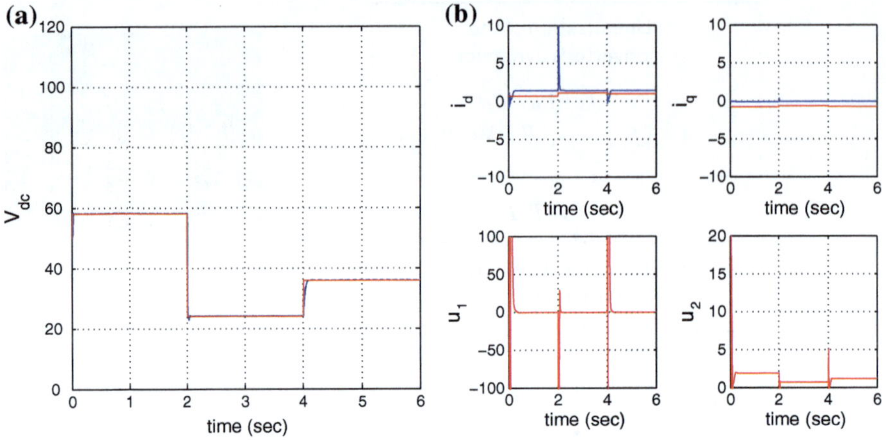

Fig. 5.20 Functioning of the converter without external (grid) perturbations: **a** Tracking of reference setpoint 2 by the DC voltage output of the three-phase voltage source converter, **b** variation of the converter's state vector elements and control inputs

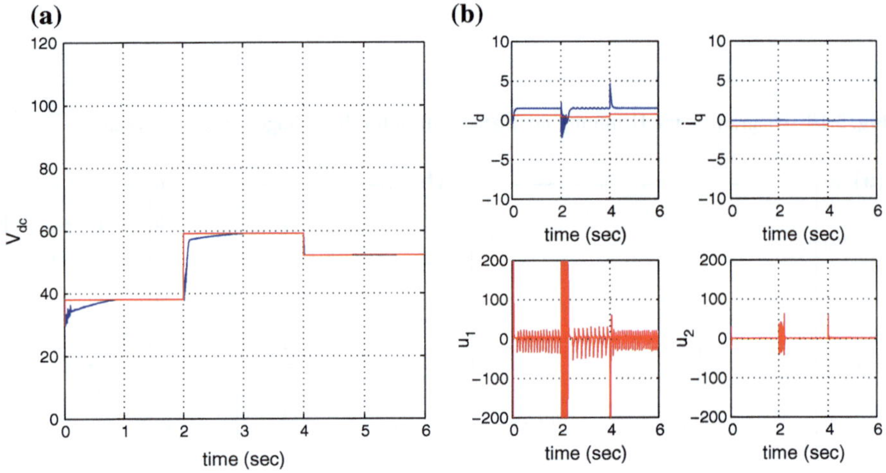

Fig. 5.21 Functioning of the converter under external (grid) perturbations: **a** Tracking of reference setpoint 1 by the DC voltage output, **b** variation of the converter's state vector elements and control inputs

The accuracy of setpoint tracking that was succeeded with the proposed control scheme is given in the Table 5.1, for the case of the two reference setpoints, examined above. It can be noticed that despite the rapid variations of the reference setpoint the tracking error for the state variables $x_i, \ i = 1, 2, 3$ of the VSC remained small.

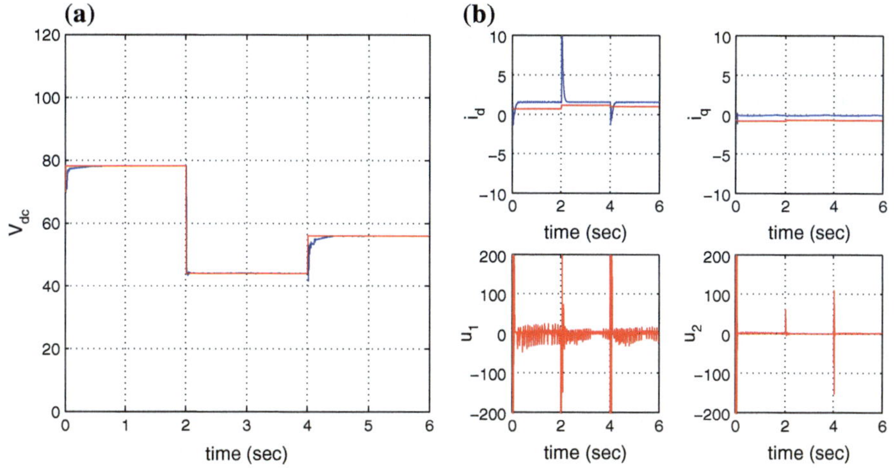

Fig. 5.22 Functioning of the converter under external (grid) perturbations: **a** Tracking of reference setpoint 2 by the DC voltage output, **b** variation of the converter's state vector elements and control inputs

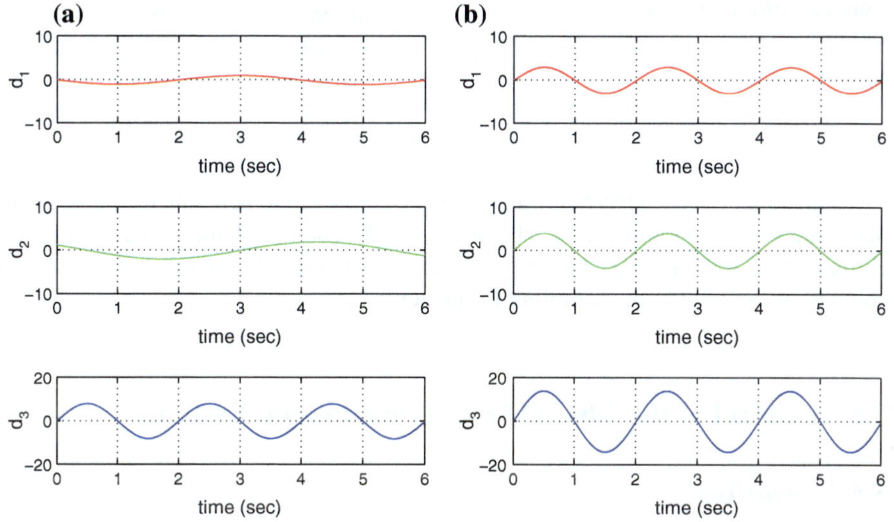

Fig. 5.23 Functioning of the converter under external (grid) perturbations: **a** Disturbance inputs affecting the converter when tracking setpoint 1, **b** Disturbance inputs affecting the converter when tracking setpoint 2

Regarding, the practical implementation of the proposed control scheme for three-phase voltage source converters, a pulse width modulation (PWM) approach can be used. This means that instead of giving to the converter control inputs of varying amplitude, the control input's amplitude is kept constant while the parameter that determines the control input's effect is the duration of the control pulse (pulse width).

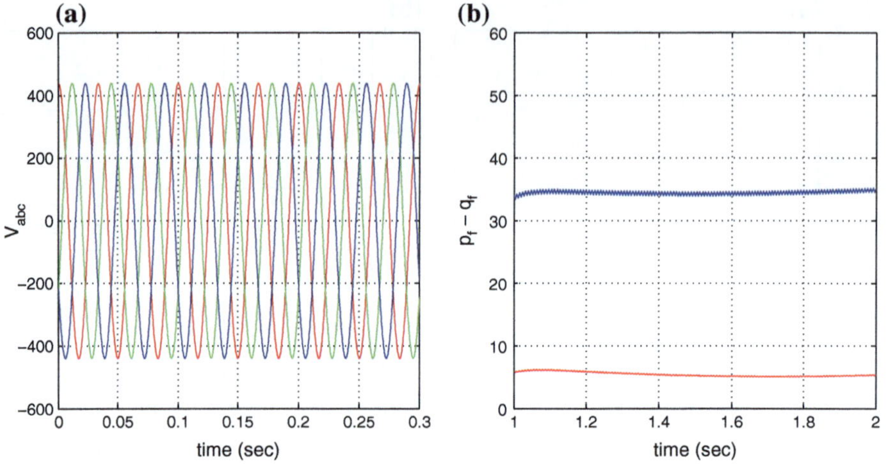

Fig. 5.24 **a** Three-phase voltage variables of the electric power grid to which the converter is connected, **b** Variation in time of the active power (*blue line*) and of the reactive power (*red line*) produced by the converter

Table 5.1 RMSE for setpoint tracking

	$RMSE_1$	$RMSE_2$
x_1	0.0269	0.0245
x_2	0.0268	0.0268
x_3	0.0050	0.0042

The sinusoidal PWM (SPWM) strategy can be employed to generate the pulse width modulated control input [235]. The switching frequency of the carrier is usually chosen to be several times larger than the frequency of the modulated signal and equal to or larger than the sampling frequency.

5.5 Control of the VSC-HVDC Transmission System

5.5.1 *Outline*

Another application for VSC converters is found in the case of VSC-HVDC transmission lines. The integration of renewable energy sources in the electricity grid has imposed in several cases the development of methods for power transmission through AC to DC and DC to AC conversion, and with the intervention of high voltage DC (HVDC) lines. Towards the design of more efficient control schemes for VSC-HVDC grid connections the following results can be noted: in [25, 26, 303] H_∞ control has been proposed for high voltage direct current links. In [89, 90, 104] input-output linearization of the voltage converter (VSC) and HVDC link is performed and accordingly a state feedback controller is developed. In [126, 127] feedback linearization is performed to model of multiple HVDC links and subsequently optimal control is applied. In [496] a decentralized PI passivity-based controller is developed for high-voltage direct current transmission systems. Moreover, in [264] model predictive control and decentralized PID control is applied to the system of multiple HVDC links. In [331] a sliding-mode control method has been developed for the VSC-HVDC model. Finally, in [395] flatness-based control is proposed for the HVDC dynamics considering that the latter is described by a PDE model, while in [459] a feedforward-feedback control scheme is implemented for active power control in VSC-HVDC.

The present section proposes a global linearization approach, through differential flatness theory, for the problem of nonlinear feedback control and state estimation of a Voltage Source Converter—HVDC line system. First, the dynamic model of this electric power transmission system is considered. This comprises the state-space equations of an AC to DC voltage source converter, that is connected to the rest of the grid through an HVDC line terminating at an inverter. This dynamic model describes the stages of transmission of electric power from a power generator to the rest of the grid, through the intervention of an HVDC line. Next, it is shown that this model is a differentially flat one, which means that all its state variables and its control inputs can be expressed as differential functions of one specific state variable of the model, which the so-called flat output. By proving that differential flatness properties hold for the electric power transmission system it is also shown that the initial nonlinear VSC-HVDC state-space model can be written into the linear canonical (Brunovsky) form. This is a global (exact) linearization procedure, that is valid throughout the entire state-space of the VSC-HVDC system and which is not based on approximate truncation of nonlinearities. For the latter description of the system's dynamics the design of a feedback controller becomes possible, thus permitting to make the voltage output at the inverter's side track any reference setpoint [352, 366, 375].

Another problem that has to be dealt with in the design of the feedback control scheme is that the control loop is subject to modelling uncertainty and external disturbances. To estimate and compensate for these perturbation terms it is proposed to include in the control loop a disturbance observer that is based on the Derivative-free nonlinear Kalman Filter. Actually, the Derivative-free nonlinear Kalman Filter consists of the Kalman Filter recursion applied to the linearized equivalent model of the VSC-HVDC and of an inverse transformation based again on differential flatness theory which enables to obtain estimates for the state variables of the initial nonlinear model. The inclusion of an additional term in the control input, based on the identification of these exogenous or endogenous perturbations, makes possible their compensation and improves the robustness of the control loop [352, 366, 375]. The performance of the considered control scheme for the VSC-HVDC system is evaluated through simulation experiments.

5.5.2 Lie Algebra-Based Linearization of the VSC-HVDC Dynamics

First, it will be shown that input-output linearization of the VSC-HVDC system can be succeeded with the use of Lie algebra methods. Using Lie derivatives the following state variables are defined $z_1 = h_1(x), z_2 = L_f h_1(x), z_3 = h_2(x)$ and $z_4 = L_f h_2(x)$. It holds that:

$$z_2 = L_f h_1(x) \Rightarrow z_2 = \frac{\partial h_1}{\partial x_1} f_1 + \frac{\partial h_1}{\partial x_2} f_2 + \frac{\partial h_1}{\partial x_3} f_3 + \frac{\partial h_1}{\partial x_4} f_4 \tag{5.116}$$

Thus, one gets

$$\begin{aligned} L_f h_1(x) = \tfrac{3L}{2} x_1[-\tfrac{R}{L}x_1 + \omega x_2 + \tfrac{v_d}{L}] + \\ + \tfrac{3L}{2} x_2[-\omega x_1 - \tfrac{R}{L}x_2 + \tfrac{v_q}{L}] \\ + 0[\tfrac{1}{Z_L C_{DC_2}} x_4 - \tfrac{1}{Z_L C_{DC_2}} x_3] \\ + C_{DC_1} x_4[-\tfrac{1}{Z_L C_{DC_1}} x_4 + \tfrac{1}{Z_L C_{DC_1}} x_3] \end{aligned} \tag{5.117}$$

which after intermediate computations gives

$$L_f h_1(x) = -\tfrac{3R}{2}(x_1^2 + x_2^2) + \tfrac{3v_d}{2} x_1 - \tfrac{x_4^2}{Z_L} + \tfrac{x_3 x_4}{Z_L} \tag{5.118}$$

Moreover, it holds that

$$\begin{aligned} L_{g_1} h_1(x) = \tfrac{\partial h_1}{\partial x_1} g_{11} + \tfrac{\partial h_1}{\partial x_2} g_{12} + \tfrac{\partial h_1}{\partial x_3} g_{13} + \tfrac{\partial h_1}{\partial x_4} g_{14} \Rightarrow \\ L_{g_1} h_1(x) = \tfrac{3L}{2} x_1 \cdot (-\tfrac{x_4}{2L}) + \tfrac{3L}{2} x_2 \cdot 0 + 0 \cdot 0 + (C_{DC_1} x_4) \cdot (\tfrac{3x_1}{4C_{DC_1}}) \Rightarrow \\ L_{g_1} h_1(x) = -\tfrac{3x_1 x_4}{4} + \tfrac{3x_1 x_4}{4} \Rightarrow L_{g_1} h_1(x) = 0 \end{aligned} \tag{5.119}$$

Similarly, one obtains

$$\begin{aligned} L_{g_2} h_1(x) = \tfrac{\partial h_1}{\partial x_1} g_{21} + \tfrac{\partial h_1}{\partial x_2} g_{22} + \tfrac{\partial h_1}{\partial x_3} g_{23} + \tfrac{\partial h_1}{\partial x_4} g_{24} \Rightarrow \\ L_{g_2} h_1(x) = \tfrac{3L}{2} x_1 \cdot 0 + \tfrac{3L}{2} x_2 \cdot (-\tfrac{x_4}{2L}) + 0 \cdot 0 + C_{DC_1} x_4 (\tfrac{3x_2}{4C_{DC_1}}) \Rightarrow \\ L_{g_2} h_1(x) = -\tfrac{3x_2 x_4}{4} + \tfrac{3x_2 x_4}{4} \Rightarrow L_{g_2} h_1(x) = 0 \end{aligned} \tag{5.120}$$

Next, the Lie derivatives of the second flat output are computed:

$$z_4 = L_f h_2(x) \Rightarrow z_4 = \frac{\partial h_2}{\partial x_1} f_1 + \frac{\partial h_2}{\partial x_2} f_2 + \frac{\partial h_2}{\partial x_3} f_3 + \frac{\partial h_2}{\partial x_4} f_4 \tag{5.121}$$

which gives

$$\begin{aligned} L_f h_2(x) = 0 \cdot f_1 + 0 \cdot f_2 + 1 \cdot f_3 + 0 \cdot f_4 \Rightarrow \\ L_f h_2(x) = \tfrac{1}{Z_L C_{DC_2}} x_4 - \tfrac{1}{Z_L C_{DC_2}} x_3 \end{aligned} \tag{5.122}$$

and in a similar manner one gets

$$\begin{array}{c} L_{g_1}h_2(x) = \frac{\partial h_2}{\partial x_1}g_{11} + \frac{\partial h_2}{\partial x_2}g_{12} + \frac{\partial h_2}{\partial x_3}g_{13} + \frac{\partial h_2}{\partial x_4}g_{14} \Rightarrow \\ L_{g_1}h_2(x) = 0 \cdot g_{11} + 0 \cdot g_{12} + 1 \cdot g_{13} + 0 \cdot g_{14} \Rightarrow \\ L_{g_1}h_2(x) = 0 \end{array} \tag{5.123}$$

while it also holds

$$\begin{array}{c} L_{g_2}h_2(x) = \frac{\partial h_2}{\partial x_1}g_{21} + \frac{\partial h_2}{\partial x_2}g_{22} + \frac{\partial h_2}{\partial x_3}g_{23} + \frac{\partial h_2}{\partial x_4}g_{24} \Rightarrow \\ L_{g_2}h_2(x) = 0 \cdot g_{21} + 0 \cdot g_{22} + 1 \cdot g_{23} + 0 \cdot g_{24} \Rightarrow \\ L_{g_2}h_2(x) = 0 \end{array} \tag{5.124}$$

In the following, higher order Lie derivatives of the system are computed:

$$\begin{array}{c} L_f^2 h_1(x) = (-3Rx_1 + \frac{3v_d}{2})[-\frac{R}{L}x_1 + \omega x_2 + \frac{v_d}{L}] - \\ -3Rx_2[-\omega x_1 - \frac{R}{L}x_2 + \frac{v_q}{L}] + \\ + \frac{x_4}{Z_L}[\frac{1}{Z_L C_{DC_2}}x_4 - \frac{1}{Z_L C_{DC_2}}x_3] + \\ (-\frac{2x_4}{Z_L} + \frac{x_3}{Z_L})[-\frac{1}{Z_L C_{DC_1}}x_4 + \frac{1}{Z_L C_{DC_1}}x_3] \end{array} \tag{5.125}$$

After performing intermediate computations, the previous relation gives

$$\begin{array}{c} L_f^2 h_1(x) = \frac{3R^2}{L}(x_1^2 + x_2^2) - \frac{6Rx_1v_d}{L} + \frac{3\omega v_d x_2}{2} + \frac{3v_d^2}{4} + \\ + \frac{2x_4^2}{Z_L^2 C_{DC_1}} - \frac{2x_1x_4}{Z_L^2 C_{DC_1}} + \frac{x_4^2}{Z_L^2 C_{DC_2}} - \frac{x_3x_4}{Z_L^2 C_{DC_2}} - \\ - \frac{x_3x_4}{Z_L^2 C_{DC_1}} + -\frac{x_3^2}{Z_L^2 C_{DC_1}} \end{array} \tag{5.126}$$

In a similar manner one computes:

$$\begin{array}{c} L_{g_1}L_fh_1(x) = L_{g_1}z_2 \Rightarrow L_{g_1}L_fh_1(x) = \frac{\partial z_2}{\partial x_1}g_{11} + \frac{\partial z_2}{\partial x_2}g_{12} + \frac{\partial z_2}{\partial x_3}g_{13} + \frac{\partial z_2}{\partial x_4}g_{14} \Rightarrow \\ L_{g_1}L_fh_1(x) = (-3Rx_1 + \frac{3v_d}{2}) \cdot (-\frac{x_4}{2L}) + (-3Rx_2 \cdot 0) + \frac{x_4}{Z_L} \cdot 0 + (-\frac{2x_4}{Z_L} + \frac{x_3}{Z_L})\frac{3x_1}{4C_{DC_1}} \Rightarrow \\ L_{g_1}L_fh_1(x) = \frac{3Rx_1x_4}{2L} - \frac{3v_dx_4}{4L} - \frac{6x_1x_4}{4Z_LC_{DC_1}} + \frac{3x_1x_3}{4Z_LC_{DC_1}} \end{array} \tag{5.127}$$

while it also holds

$$\begin{array}{c} L_{g_2}L_fh_1(x) = L_{g_2}z_2 \Rightarrow L_{g_2}L_fh_2(x) = \frac{\partial z_2}{\partial x_1}g_{21} + \frac{\partial z_2}{\partial x_2}g_{22} + \frac{\partial z_2}{\partial x_3}g_{23} + \frac{\partial z_2}{\partial x_4}g_{24} \Rightarrow \\ L_{g_2}L_fh_1(x) = (-3Rx_1 + \frac{3v_d}{2}) \cdot 0 - (3Rx_2) \cdot (-\frac{x_4}{2L}) + \frac{x_4}{Z_L} \cdot 0 + (-\frac{2x_4}{Z_L} + \frac{x_3}{Z_L}) \cdot (\frac{3x_2}{4C_{DC_1}}) \end{array} \tag{5.128}$$

which finally gives

$$L_{g_2}L_fh_1(x) = \frac{3Rx_2x_4}{2L} - \frac{6x_2x_4}{4Z_LC_{DC_1}} + \frac{3x_2x_3}{4Z_LC_{DC_1}} \tag{5.129}$$

Following the same approach one computes

$$\begin{array}{c}L_f^2h_2(x)=L_fz_4\Rightarrow L_f^2h_2(x)=\frac{\partial z_4}{\partial x_1}f_1+\frac{\partial z_4}{\partial x_2}f_2+\frac{\partial z_4}{\partial x_3}f_3+\frac{\partial z_4}{\partial x_4}f_4\Rightarrow\\ L_f^2h_2(x)=0\cdot f_1+0\cdot f_2-\frac{1}{Z_LC_{DC_2}}[\frac{1}{Z_LC_{DC_2}}x_4-\frac{1}{Z_LC_{DC_2}}x_3]+\\ +\frac{1}{Z_LC_{DC_2}}[-\frac{1}{Z_LC_{DC_1}}x_4+\frac{1}{Z_LC_{DC_1}}x_3]\end{array} \tag{5.130}$$

which after intermediate operations gives

$$L_f^2h_2(x)=-\frac{x_4}{Z_L^2C_{DC_2}^2}+\frac{x_3}{Z_L^2C_{DC_2}^2}-\frac{x_4}{Z_L^2C_{DC_1}C_{DC_2}}+\frac{x_3}{Z_L^2C_{DC_1}C_{DC_2}} \tag{5.131}$$

In a similar manner one computes

$$\begin{array}{c}L_{g_1}L_fh_2(x)=L_{g_1}z_4\Rightarrow L_{g_1}L_fh_2(x)=\frac{\partial z_4}{\partial x_1}g_{11}+\frac{\partial z_4}{\partial x_2}g_{12}+\frac{\partial z_4}{\partial x_3}g_{13}+\frac{\partial z_4}{\partial x_4}g_{14}\Rightarrow\\ L_{g_1}L_fh_2(x)=0\cdot g_{11}+0\cdot g_{12}-\frac{1}{Z_LC_{DC_2}}\cdot 0+\frac{1}{Z_LC_{DC_2}}\cdot(\frac{3x_1}{4C_{DC_1}})\Rightarrow\\ L_{g_1}L_fh_2(x)=\frac{3x_1}{4Z_LC_{DC_1}C_{DC_2}}\end{array} \tag{5.132}$$

Finally, by applying the same method one computes

$$\begin{array}{c}L_{g_2}L_fh_2(x)=L_{g_2}z_4\Rightarrow L_{g_2}L_fh_2(x)=\frac{\partial z_4}{\partial x_1}g_{21}+\frac{\partial z_4}{\partial x_2}g_{22}+\frac{\partial z_4}{\partial x_3}g_{23}+\frac{\partial z_4}{\partial x_4}g_{24}\Rightarrow\\ L_{g_2}L_fh_2(x)=0\cdot g_{21}+0\cdot g_{22}-\frac{1}{Z_LC_{DC_2}}\cdot 0+\frac{1}{Z_LC_{DC_2}}\cdot(\frac{3x_2}{4C_{DC_1}})\Rightarrow\\ L_{g_2}L_fh_2(x)=\frac{3x_2}{4Z_LC_{DC_1}C_{DC_2}}\end{array} \tag{5.133}$$

It can be confirmed that it holds

$$\begin{array}{c}\dot{z}_1=z_2\\ \dot{z}_2=L_f^2h_1(x)+L_{g_1}L_fh_1(x)u_1+L_{g_2}L_fh_1(x)u_2\\ \dot{z}_3=z_4\\ \dot{z}_4=L_f^2h_2(x)+L_{g_1}L_fh_2(x)u_1+L_{g_2}L_fh_2(x)u_2\end{array} \tag{5.134}$$

The input-output linearized model takes the form

$$\begin{array}{c}\dot{z}_1=L_f^2h_1(x)+L_{g_1}L_fh_1(x)u_1+L_{g_2}L_fh_1(x)u_2\\ \dot{z}_2=L_f^2h_2(x)+L_{g_1}L_fh_2(x)u_1+L_{g_2}L_fh_2(x)u_2\end{array} \tag{5.135}$$

By defining the new control inputs

$$\begin{array}{c}v_1=L_f^2h_1(x)+L_{g_1}L_fh_1(x)u_1+L_{g_2}L_fh_1(x)u_2\\ v_2=L_f^2h_1(x)+L_{g_1}L_fh_1(x)u_1+L_{g_2}L_fh_1(x)u_2\end{array} \tag{5.136}$$

the VSC-HVDC system is written in the chain of integrators form

$$\begin{aligned} \ddot{z}_1 &= v_1 \\ \ddot{z}_1 &= v_2 \end{aligned} \quad (5.137)$$

Moreover, by defining the feedback control inputs

$$\begin{aligned} v_1 = \ddot{z}_1^d - k_d^1(\dot{z}_1 - \dot{z}_1^d) - k_p^1(z_1 - z_1^d) \\ v_2 = \ddot{z}_2^d - k_d^2(\dot{z}_2 - \dot{z}_2^d) - k_p^2(z_2 - z_2^d) \end{aligned} \quad (5.138)$$

The tracking error dynamics for the closed loop becomes $\ddot{e}_1 + k_d^1\dot{e}_1 + k_p^1 e_1 = 0$ and $\ddot{e}_2 + k_d^2\dot{e}_2 + k_p^2 e_2 = 0$. By selecting the feedback gains k_p^1, k_d^1 and k_p^2, k_d^2 such as the associated characteristic polynomials to be Hurwitz ones, one has the tracking error dynamics $lim_{t \Rightarrow \infty} e_1(t) = 0$ and $lim_{t \Rightarrow \infty} e_2(t) = 0$.

To compute the control input that is actually applied to the VSC-HVDC model one proceeds as follows: Eq. (5.135) is rewritten in the matrix form

$$\ddot{\tilde{y}}_f = \tilde{f} + \tilde{M}\tilde{u} \quad (5.139)$$

where

$$\tilde{y}_f = \begin{pmatrix} y_{f_1} \\ y_{f_2} \end{pmatrix} \; \tilde{f} = \begin{pmatrix} L_f^2 h_1(x) \\ L_f^2 h_2(x) \end{pmatrix} \; \tilde{M} = \begin{pmatrix} L_{g_1} L_f h_1(x) & L_{g_2} L_f h_1(x) \\ L_{g_1} L_f h_2(x) & L_{g_2} L_f h_2(x) \end{pmatrix} \; \tilde{v} = \begin{pmatrix} v_1 \\ v_2 \end{pmatrix} \quad (5.140)$$

Using the above, the control input that should be exerted on the VSC-HVDC model is for eliminating the tracking error is

$$\tilde{u} = \tilde{M}^{-1}[\tilde{v} - \tilde{f}] \quad (5.141)$$

5.5.3 *Differential Flatness of the VSC-HVDC System*

Next, it will be shown that the dynamic model of the voltage source converter is a differentially flat one, i.e. it holds that all state variables and its control inputs can be written as functions of the flat outputs and their derivatives [57, 137, 229, 241, 387, 388, 411]. Moreover, it will be shown that by expressing the elements of the state vector as functions of the flat outputs and their derivatives one obtains a transformation of the VSC-HVDC model, depicted in Fig. 5.25, into the linear canonical (Brunovsky) form.

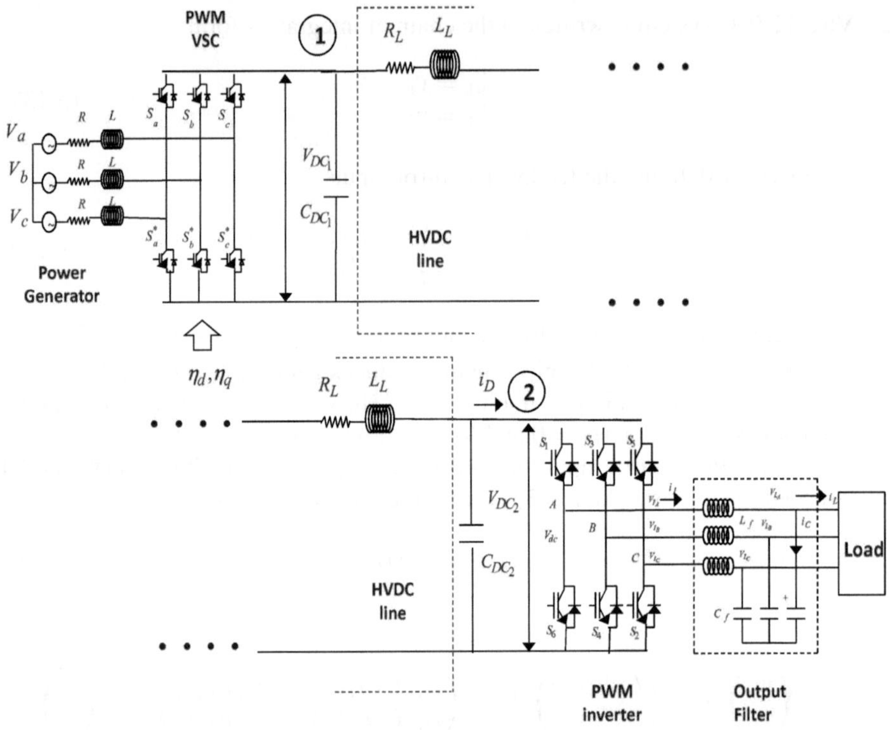

Fig. 5.25 Transmission system comprising a voltage source converter (VSC), a high voltage DC line (HVDC) and an inverter

The dynamic model of the joint VSC-HVDC dynamics has been defined as

$$\dot{x}_1 = -\frac{R}{L}x_1 + \omega x_2 + \frac{v_d}{L} - \frac{x_4}{2L}u_1 \tag{5.142}$$

$$\dot{x}_2 = -\omega x_1 - \frac{R}{L}x_2 + \frac{v_q}{L} - \frac{x_4}{2L}u_2 \tag{5.143}$$

$$\dot{x}_3 = \frac{1}{Z_L C_{DC_2}}x_4 - \frac{1}{Z_L C_{DC_2}}x_3 \tag{5.144}$$

$$\dot{x}_4 = -\frac{1}{Z_L C_{DC_1}}x_4 + \frac{1}{Z_L C_{DC_1}}x_3 + \frac{3x_1}{4C_{DC_1}}u_1 + \frac{3x_2}{4C_{DC_1}}u_2 \tag{5.145}$$

The flat output of the system is taken to be $y_f = [y_{f_1}, y_{f_2}]^T$, where

$$y_{f_1} = \frac{3L}{4}(x_1^2 + x_2^2) + \frac{C_{DC_1}}{2}x_4^2 \tag{5.146}$$

$$y_{f_2} = x_3 \tag{5.147}$$

By deriving Eq. (5.146) with respect to time one obtains

$$\begin{array}{c}\dot{y}_{f_1}=\frac{3L}{2}x_1\dot{x}_1+\frac{3L}{2}x_2\dot{x}_2+C_{DC_1}x_4\dot{x}_4\Rightarrow\\ \dot{y}_{f_1}=\frac{3L}{2}x_1[-\frac{R}{L}x_1+\omega x_2+\frac{v_d}{L}-\frac{x_4}{2L}u_1]+\\ +\frac{3L}{2}x_2[-\frac{R}{L}x_2-\omega x_1+\frac{v_q}{L}-\frac{x_4}{2L}u_2]+\\ +C_{DC_1}x_4[-\frac{1}{Z_LC_{DC_1}}x_4+\frac{1}{Z_LC_{DC_1}}x_3+\frac{3x_1}{4C_{DC_1}}u_1+\frac{3x_2}{4C_{DC_1}}u_2]\end{array} \tag{5.148}$$

Using that v_d is constant (d-axis component of the grid voltage) and $v_q=0$, and after intermediate computations one obtains

$$\dot{y}_{f_1}=-\frac{3R}{2}(x_1^2+x_2^2)+\frac{3x_1}{2}v_d-\frac{x_4^2}{Z_L}+\frac{x_3x_4}{Z_L} \tag{5.149}$$

From Eq. (5.144) one has

$$\begin{array}{c}x_4=(Z_LC_{DC_2})[\dot{x}_3+\frac{1}{Z_LC_{DC_2}}x_3]\Rightarrow\\ x_4=(Z_LC_{DC_2})[\dot{y}_{f_2}+\frac{1}{Z_LC_{DC_2}}y_{f_2}]\Rightarrow\\ x_4=f_a(y_f,\dot{y}_f)\end{array} \tag{5.150}$$

which means that state variable x_4 is a function of the flat output and of its derivatives. Moreover, from Eq. (5.146) one gets

$$\begin{array}{c}(x_1^2+x_2^2)=\frac{4}{3L}[y_{f_1}-\frac{C_{DC_1}}{2}x_4^2]\Rightarrow\\ (x_1^2+x_2^2)=f_b(y_f,\dot{y}_f)\end{array} \tag{5.151}$$

Next, using Eq. (5.149) one has

$$\begin{array}{c}x_1=\frac{2}{3v_d}\{\dot{y}_{f_1}+\frac{3R}{2}(x_1^2+x_2^2)+\frac{x_4^2}{Z_L}-\frac{x_3x_4}{Z_L}\}\Rightarrow\\ x_1=\frac{2}{3v_d}\{\dot{y}_{f_1}+\frac{3R}{2}f_b(y_f,\dot{y}_f)+\frac{x_4^2}{Z_L}-\frac{y_{f_2}x_4}{Z_L}\}\end{array} \tag{5.152}$$

Using Eq. (5.150) in Eq. (5.152) one finally gets that state variable x_1 is a function of the flat output and its derivatives

$$x_1=f_c(y_f,\dot{y}_f) \tag{5.153}$$

Moreover, from Eqs. (5.151) and (5.153) it holds $(x_1^2+x_2^2)=f_b(y_f,\dot{y}_f)$ and $x_1=f_c(y_f,\dot{y}_f)$. Thus by solving with respect to x_2 one gets

$$\begin{array}{c}x_2^2=f_b(y_f,\dot{y}_f)-f_c^2(y_f,\dot{y}_f)\Rightarrow\\ x_2=\sqrt{|f_b(y_f,\dot{y}_f)-f_c^2(y_f.\dot{y}_f)|}\Rightarrow\\ x_2=f_d(y_f,\dot{y}_f)\end{array} \tag{5.154}$$

which means that state variable x_2 is also a function of the flat output and of its derivatives. Next by solving Eq. (5.142) with respect to the control input u_1 and Eq. (5.143) with respect to the control input u_2 one obtains

$$u_1 = \frac{2L}{x_4}\{\dot{x}_1 + \frac{R}{L}x_1 - \omega x_2 - \frac{v_d}{L}\} \tag{5.155}$$

$$u_2 = \frac{2L}{x_4}\{\dot{x}_2 + \omega x_1 + \frac{R}{L}x_2 - \frac{v_q}{L}\} \tag{5.156}$$

Using Eqs. (5.150), (5.153) and (5.154) in Eq. (5.54) it can be concluded that

$$u_1 = f_e(y_f, \dot{y}_f) \tag{5.157}$$

Similarly, using Eqs. (5.150), (5.153) and (5.154) in Eq. (5.156) it can be concluded that

$$u_2 = f_g(y_f, \dot{y}_f) \tag{5.158}$$

Consequently, all state variables and the control inputs of the VSC-HVDC model can be written as differential functions of the flat output $y_f = [y_{f_1}, y_{f_2}]^T$. Thus, the VSC-HVDC system is a differentially flat one.

5.5.4 *Flatness-Based Control of the VSC-HVDC System*

Next, a flatness-based controller will be designed for the VSC-HVDC system. Using Eq. (5.148) and deriving $\dot{y}_{f_1}$ once more with respect to time one gets

$$\begin{array}{c} \ddot{y}_{f_1} = -\frac{3R}{2}(2x_1\dot{x}_1 + 2x_2\dot{x}_2) + 3\frac{\dot{x}_1}{2}v_d - \frac{2x_4\dot{x}_4}{Z_L} + \frac{\dot{x}_3x_4}{Z_L} + \frac{x_3\dot{x}_4}{Z_L} \Rightarrow \\ \ddot{y}_{f_1} = (-3Rx_1 + \frac{3v_d}{2})[-\frac{R}{L}x_1 + \omega x_2 + \frac{v_d}{L} - \frac{x_4}{2L}u_1] - \\ -3Rx_2[-\omega x_1 + \frac{R}{L}x_2 + \frac{v_q}{L} - \frac{x_4}{2L}u_2] + \\ +\frac{x_4}{Z_L}(\frac{x_4}{Z_LC_{DC_2}} - \frac{x_3}{Z_LC_{DC_2}}) + \\ +(-\frac{2x+4}{Z_L} + \frac{x_3}{Z_L})[-\frac{x_4}{Z_LC_{DC_1}} + \frac{x_3}{Z_LC_{DC_1}} + \frac{3x_1}{4C_{DC_1}}u_1 + \frac{3x_2}{4C_{DC_1}}u_2] \end{array} \tag{5.159}$$

which after intermediate computations gives

$$\begin{array}{c} \ddot{y}_{f_1} = [\frac{3R^2}{L}(x_1^2 + x_2^2) - \frac{6Rx_1v_d}{L} + \frac{3\omega v_dx_2}{2} + \frac{3v_d^2}{4} \\ + \frac{2x_4^2}{Z_L^2C_{DC_1}} - \frac{2x_1x_4}{Z_L^2C_{DC_1}} + \frac{x_4^2}{Z_L^2C_{DC_2}} - \frac{x_3x_4}{Z_L^2C_{DC_2}} - \\ - \frac{x_3x_4}{Z_L^2C_{DC_1}} + -\frac{x_3^2}{Z_L^2C_{DC_1}}] + \\ +[\frac{3Rx_1x_4}{2L} - \frac{3v_dx_4}{4L} - \frac{6x_1x_4}{4Z_LC_{DC_1}} + \frac{3x_1x_3}{4Z_LC_{DC_1}}]u_1 + \\ +[\frac{3Rx_2x_4}{2L} - \frac{6x_2x_4}{4Z_LC_{DC_1}} + \frac{3x_2x_3}{4Z_LC_{DC_1}}]u_2 \end{array} \tag{5.160}$$

Equation (5.160) can be also written in the concise form

$$\ddot{y}_{f_1} = L_f^2h_1(x) + L_{g_1}L_fh_1(x)u_1 + L_{g_2}L_fh_1(x)u_2 \tag{5.161}$$

Functions $L_f^2h_1(x)$, $L_{g_1}L_fh_1(x)$ and $L_{g_2}L_fh_1(x)$ are defined in detail in Eqs. (5.126), (5.127) and (5.129) respectively, which have been provided in the analysis of the Lie algebra-based linearization of the VSC-HVDC system. Additionally, from the second component of the flat outputs vector given in Eq. (5.147) one gets

$$\begin{aligned} y_{f_2} &= x_2 \Rightarrow \dot{y}_{f_3} = \dot{x}_3 \Rightarrow \\ \dot{y}_{f_2} &= \frac{1}{Z_L C_{DC_2}} x_4 - \frac{1}{Z_L C_{DC_2}} x_3 \end{aligned} \tag{5.162}$$

By deriving once more the above relation with respect to time one gets

$$\ddot{y}_{f_2} = \frac{1}{Z_L C_{DC_2}} \dot{x}_4 - \frac{1}{Z_L C_{DC_2}} \dot{x}_3 \tag{5.163}$$

Next, by substituting Eqs. (5.144) and (5.145) into Eq. (5.163) one arrives at

$$\begin{aligned} \ddot{y}_{f_2} = \frac{1}{Z_L C_{DC_2}} [-\frac{1}{Z_L C_{DC_1}} x_4 + \frac{1}{Z_L C_{DC_1}} x_3 + \frac{3x_1}{4C_{DC_1}} u_1 + \frac{3x_2}{4C_{DC_1}} u_2] - \\ - \frac{1}{Z_L C_{DC_2}} [\frac{1}{Z_L C_{DC_2}} x_4 - \frac{1}{Z_L C_{DC_2}} x_3] \end{aligned} \tag{5.164}$$

which after intermediate operations gives

$$\begin{aligned} \ddot{y}_{f_2} = [-\frac{x_4}{Z_L^2 C_{DC_1} C_{DC_2}} + \frac{x_3}{Z_L^2 C_{DC_1} C_{DC_2}} - \frac{x_4}{Z_L^2 C_{DC_2}{}^2} + \frac{x_3}{Z_L^2 C_{DC_2}{}^2}] + \\ + \frac{3x_1}{4Z_L C_{DC_1} C_{DC_2}} u_1 + \frac{3x_2}{4Z_L C_{DC_1} C_{DC_2}} u_2 \end{aligned} \tag{5.165}$$

which can be also written in the more concise form

$$\ddot{y}_{f_2} = L_f^2 h_2(x) + L_{g_1} L_f h_2(x) u_1 + L_{g_2} L_f h_2(x) u_2 \tag{5.166}$$

Functions $L_f^2h_2(x)$, $L_{g_1}L_fh_2(x)$ and $L_{g_2}L_fh_2(x)$ are defined in detail in Eqs. (5.131)–(5.133) respectively, which have been provided in the analysis of the Lie algebra-based linearization of the VSC-HVDC system. Next, Eqs. (5.160) and (5.161) are written in matrix form

$$\begin{pmatrix} \ddot{y}_{f_1} \\ \ddot{y}_{f_1} \end{pmatrix} = \begin{pmatrix} L_f^2 h_1(x) \\ L_f^2 h_2(x) \end{pmatrix} + \begin{pmatrix} L_{g_1} L_f h_1(x) & L_{g_2} L_f h_1(x) \\ L_{g_1} L_f h_2(x) & L_{g_2} L_f h_2(x) \end{pmatrix} \begin{pmatrix} u_1 \\ u_2 \end{pmatrix} \tag{5.167}$$

By defining the new control inputs $v_1 = L_f^2h_1(x) + L_{g_1}L_fh_1(x)u_1 + L_{g_2}L_fh_1(x)u_2$ and $v_2 = L_f^2h_2(x) + L_{g_1}L_fh_2(x)u_1 + L_{g_2}L_fh_2(x)u_2$ one arrives at the linearized and decoupled description of the dynamics of the VSC-HVDC model

$$\begin{aligned} \ddot{y}_{f_1} &= v_1 \\ \ddot{y}_{f_2} &= v_2 \end{aligned} \tag{5.168}$$

Moreover, by defining the following vectors and matrices

$$\tilde{y}_f = \begin{pmatrix} y_{f_1} \\ y_{f_2} \end{pmatrix} \; \tilde{f} = \begin{pmatrix} L_f^2 h_1(x) \\ L_f^2 h_2(x) \end{pmatrix} \; \tilde{M} = \begin{pmatrix} L_{g_1} L_f h_1(x) & L_{g_2} L_f h_1(x) \\ L_{g_1} L_f h_2(x) & L_{g_2} L_f h_2(x) \end{pmatrix} \; \tilde{v} = \begin{pmatrix} v_1 \\ v_2 \end{pmatrix} \tag{5.169}$$

one obtains the following matrix-form description of the system's dynamics

$$\ddot{\tilde{y}}_f = \tilde{f} + \tilde{M}\tilde{u} \Rightarrow \tilde{u} = \tilde{M}^{-1}[\tilde{v} - f] \tag{5.170}$$

where $\tilde{u}$ is the control input that is actually exerted on the VSC-HVDC system, while the transformed control input $\tilde{v}$ is given by

$$\tilde{v} = \ddot{\tilde{y}}_f^d - K_d(\dot{\tilde{y}}_f - \dot{\tilde{y}}_f^d) - K_p(\tilde{y}_f - \tilde{y}_f^d) \tag{5.171}$$

For the linearized model of the VSC-HVDC dynamics one can also obtain a description in the linear canonical (Brunovsky) form. The following state variables are defined: $z_1 = y_{f_1}$, $z_2 = \dot{y}_{f_1}$, $z_3 = y_{f_2}$ and $z_4 = \dot{y}_{f_2}$. This results in the following state-space description

$$\begin{gathered} \dot{z}_1 = z_2 \\ \dot{z}_2 = L_f^2 h_1(x) + L_{g_1} L_f h_1(x) u_1 + L_{g_2} L_f h_1(x) u_2 \\ \dot{z}_3 = z_4 \\ \dot{z}_4 = L_f^2 h_2(x) + L_{g_1} L_f h_2(x) u_1 + L_{g_2} L_f h_2(x) u_2 \end{gathered} \tag{5.172}$$

The previous state-space description results also in the linear canonical (Brunovsky) form:

$$\begin{pmatrix} \dot{z}_1 \\ \dot{z}_2 \\ \dot{z}_3 \\ \dot{z}_4 \end{pmatrix} = \begin{pmatrix} 0 & 1 & 0 & 0 \\ 0 & 0 & 0 & 0 \\ 0 & 0 & 0 & 1 \\ 0 & 0 & 0 & 0 \end{pmatrix} \begin{pmatrix} z_1 \\ z_2 \\ z_3 \\ z_4 \end{pmatrix} + \begin{pmatrix} 0 & 0 \\ 1 & 0 \\ 0 & 0 \\ 0 & 1 \end{pmatrix} \begin{pmatrix} v_1 \\ v_2 \end{pmatrix} \tag{5.173}$$

and since the measurable variables for this system are taken to be $z_1 = y_{f_1}$ and $z_3 = y_{f_2}$ the associated measurement equation becomes

$$\begin{pmatrix} z_1^m \\ z_3^m \end{pmatrix} = \begin{pmatrix} 1 & 0 & 0 & 0 \\ 0 & 0 & 1 & 0 \end{pmatrix} \begin{pmatrix} z_1 \\ z_2 \\ z_3 \\ z_4 \end{pmatrix} \tag{5.174}$$

By defining the new control inputs

$$\begin{aligned} v_1 &= L_f^2 h_1(x) + L_{g_1} L_f h_1(x) u_1 + L_{g_2} L_f h_1(x) u_2 \\ v_2 &= L_f^2 h_1(x) + L_{g_1} L_f h_1(x) u_1 + L_{g_2} L_f h_1(x) u_2 \end{aligned} \tag{5.175}$$

the VSC-HVDC system is written in the chain of integrators form

$$\begin{aligned} \ddot{z}_1 &= v_1 \\ \ddot{z}_1 &= v_2 \end{aligned} \tag{5.176}$$

Moreover, by defining the feedback control inputs

$$\begin{aligned} v_1 &= \ddot{z}_1^d - k_d^1(\dot{z}_1 - \dot{z}_1^d) - k_p^1(z_1 - z_1^d) \\ v_2 &= \ddot{z}_2^d - k_d^2(\dot{z}_2 - \dot{z}_2^d) - k_p^2(z_2 - z_2^d) \end{aligned} \tag{5.177}$$

The tracking error dynamics for the closed loop becomes $\ddot{e}_1 + k_d^1\dot{e}_1 + k_p^1 e_1 = 0$ and $\ddot{e}_2 + k_d^2\dot{e}_2 + k_p^2 e_2 = 0$. By selecting the feedback gains k_p^1, k_d^1 and k_p^2, k_d^2 such as the associated characteristic polynomials to be Hurwitz ones, one has the tracking error dynamics $lim_{t \Rightarrow \infty} e_1(t) = 0$ and $lim_{t \Rightarrow \infty} e_2(t) = 0$. It is possible to experiment with different values of the gains k_p^1, k_d^1 and k_p^2, k_d^2 within the ranges that assure the stability of the control loop, so as to achieve certain performance indexes (such as overshoot or settling time).

To compute the control input that is actually applied to the VSC-HVDC model one proceeds as follows: Eqs. (5.175) and (5.176) are rewritten in the matrix form

$$\ddot{\tilde{y}}_f = \tilde{f} + \tilde{M}\tilde{u} \tag{5.178}$$

where

$$\tilde{y}_f = \begin{pmatrix} y_{f_1} \\ y_{f_2} \end{pmatrix} \; \tilde{f} = \begin{pmatrix} L_f^2 h_1(x) \\ L_f^2 h_2(x) \end{pmatrix} \; \tilde{M} = \begin{pmatrix} L_{g_1} L_f h_1(x) & L_{g_2} L_f h_1(x) \\ L_{g_1} L_f h_2(x) & L_{g_2} L_f h_2(x) \end{pmatrix} \; \tilde{v} = \begin{pmatrix} v_1 \\ v_2 \end{pmatrix} \tag{5.179}$$

Using the above, the control input that should be exerted on the VSC-HVDC model for eliminating the tracking error is

$$\tilde{u} = \tilde{M}^{-1}[\tilde{v} - \tilde{f}] \tag{5.180}$$

5.5.5 Compensation of Disturbances Using the Derivative-Free Nonlinear Kalman Filter

Next, it is assumed that the VSC-HVDC model is subjected to additive input disturbances, that is

$$\begin{aligned}\ddot{y}_{f_1} &= L_f^2 h_1(x) + L_{g_1} L_f h_1(x) u_1 + L_{g_2} L_f h_1(x) u_2 + \tilde{d}_1 \\ \ddot{y}_{f_2} &= L_f^2 h_2(x) + L_{g_1} L_f h_2(x) u_1 + L_{g_2} L_f h_2(x) u_2 + \tilde{d}_2\end{aligned} \tag{5.181}$$

or equivalently

$$\begin{aligned}\ddot{y}_{f_1} &= v_1 + \tilde{d}_1 \\ \ddot{y}_{f_2} &= v_2 + \tilde{d}_2\end{aligned} \tag{5.182}$$

Moreover, it is considered that the disturbance inputs can be described by their n-th order derivative. This is because each function can be described either by its mathematical equations or equivalently by its derivatives and the associated initial conditions. However, in the studied case there will be finally no use of initial conditions because the disturbance function will be estimated by the Kalman Filter which is not dependent on the processing of initial conditions. Without loss of generality it is assumed that the additive disturbance terms $\tilde{d}_i,\ i = 1, 2$ are described by the signal's second order derivative, that is

$$\ddot{\tilde{d}}_1 = f_{d_1} \quad \ddot{\tilde{d}}_2 = f_{d_2} \tag{5.183}$$

Then by defining the new state variables $z_1 = y_{f_1}$, $z_2 = \dot{y}_{f_1}$, $z_3 = y_{f_2}$, $z_4 = \dot{y}_{f_2}$, $z_5 = \tilde{d}_1$, $z_6 = \dot{\tilde{d}}_1$, $z_7 = \tilde{d}_2$, $z_8 = z_6 = \dot{\tilde{d}}_2$ one obtains the following extended state-space description of the system

$$\begin{pmatrix}\dot{z}_1\\ \dot{z}_2\\ \dot{z}_3\\ \dot{z}_4\\ \dot{z}_5\\ \dot{z}_6\\ \dot{z}_7\\ \dot{z}_8\end{pmatrix} = \begin{pmatrix}0&1&0&0&0&0&0&0\\ 0&0&0&0&1&0&0&0\\ 0&0&0&1&0&0&0&0\\ 0&0&0&0&0&0&1&0\\ 0&0&0&0&0&1&0&0\\ 0&0&0&0&0&0&0&0\\ 0&0&0&0&0&0&0&1\\ 0&0&0&0&0&0&0&0\end{pmatrix}\begin{pmatrix}z_1\\ z_2\\ z_3\\ z_4\\ z_5\\ z_6\\ z_7\\ z_8\end{pmatrix} + \begin{pmatrix}0&0\\ 1&0\\ 0&0\\ 0&1\\ 0&0\\ 0&0\\ 0&0\\ 0&0\end{pmatrix}\begin{pmatrix}v_1\\ v_2\end{pmatrix} \tag{5.184}$$

while by denoting the extended state vector $z_e = [z_1, z_2, z_3, z_4, z_5, z_6, z_7, z_8]^T$, the measurement equation for the extended system becomes

$$\begin{pmatrix}z_1^m\\ z_3^m\end{pmatrix} = \begin{pmatrix}1&0&0&0&0&0&0&0\\ 0&0&1&0&0&0&0&0\end{pmatrix} z_e \tag{5.185}$$

According to the above the extended state-space model of the VSC-HVDC system is written in the linear matrix from

$$\begin{aligned}\dot{z}_e &= A z_e + B v \\ z_e^m &= C z_e\end{aligned} \tag{5.186}$$

For the extended state-space model of the VSC-HVDC system it is possible to perform simultaneously states and disturbances estimation using the Derivative-free nonlinear Kalman Filter. The latter consists of the Kalman Filter recursion on the linearized equivalent model of the VSC-HVDC system and of an inverse transformation, based on the results of Sect. 5.5.3 which allows to obtain estimates of the state variables of the initial nonlinear model described in Eq. (5.142) to Eq. (5.145). The filter's algorithm uses the discrete time equivalents of the previously defined matrices A, B and C, which are denoted as A_d, B_d and C_d respectively. The proposed filtering method consists of a *measurement update* and a *time update* stage [34, 341, 344]:

measurement update:

$$\begin{aligned} K(k) &= P^-(k)C_d^T[C_dP(k)C_d^T + R(k)]^{-1} \\ \hat{z}_e(k) &= \hat{z}_e^-(k) + K(k)[z_e^m(k) - C_d\hat{z}_e^-(k)] \\ P(k) &= P^-(k) - K(k)C_dP^-(k) \end{aligned} \tag{5.187}$$

time update:

$$\begin{aligned} P^-(k+1) &= A_dP(k)A_d^T + Q(k) \\ \hat{z}^-(k+1) &= A_d\hat{z}(k) + B_dv(k) \end{aligned} \tag{5.188}$$

where $Q(k)$ and $R(k)$ stand for process and measurement noise covariance matrices. After estimating the disturbance inputs, that is $\hat{z}_5 = \hat{\tilde{d}}_1$ and $\hat{z}_7 = \hat{\tilde{d}}_2$ the control input that is applied to VSC-HVDC system is modified as follows:

$$\begin{aligned} v_1^* &= v_1 - \hat{z}_5 \\ v_2^* &= v_2 - \hat{z}_7 \end{aligned} \tag{5.189}$$

Thus, with respect to Eq. (5.182), the closed-loop dynamics becomes

$$\begin{aligned} \ddot{y}_{f_1} &= v_1 - \hat{z}_5 + \tilde{d}_1 \\ \ddot{y}_{f_2} &= v_2 - \hat{z}_7 + \tilde{d}_2 \end{aligned} \tag{5.190}$$

5.5.6 Simulation Tests

The performance of the control loop for the VSC-HVDC system was tested through simulation experiments. The associated results are presented in Figs. 5.26, 5.27, 5.28, 5.29 and 5.30. Actually, in Fig. 5.26a, 5.27a, 5.28a, 5.29a and 5.30a the convergence of the real state variables of the system $x_i,\ i = 1, \ldots, 4$ to their reference setpoints is presented. Additionally, in Fig. 5.26b, 5.27b, 5.28b, 5.29b and 5.30b the estimated disturbance inputs of the VSC-HVDC model are plotted. Moreover, diagrams of the variation of the system's control inputs are provided. In the aforementioned diagrams, the per unit (p.u.) measurement system has been used. It can be observed that through the proposed control scheme, the state variables of the VSC-HVDC

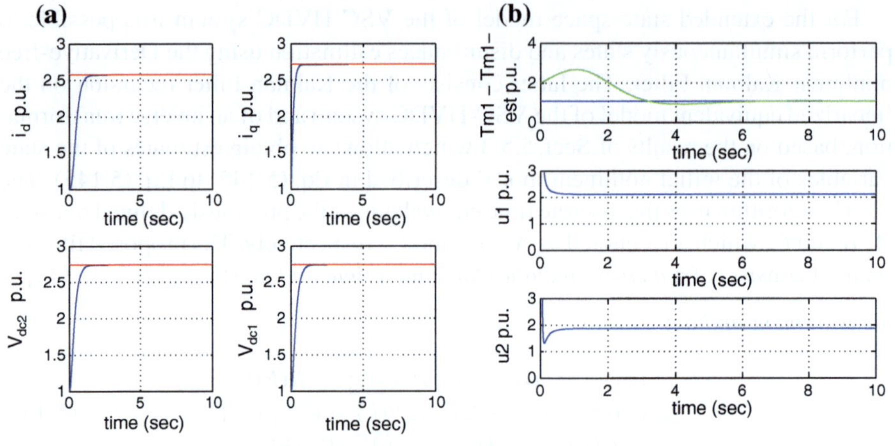

Fig. 5.26 Convergence of the state vector elements $x_i,\ i=1,\ldots,4$ to the associated reference setpoints No 1 for the initial nonlinear model of the VSC-HVDC system, **b** estimation of disturbance input $\tilde{d}$ on the VSC-HVDC model and variation of control inputs $u_i,\ i=1,2$

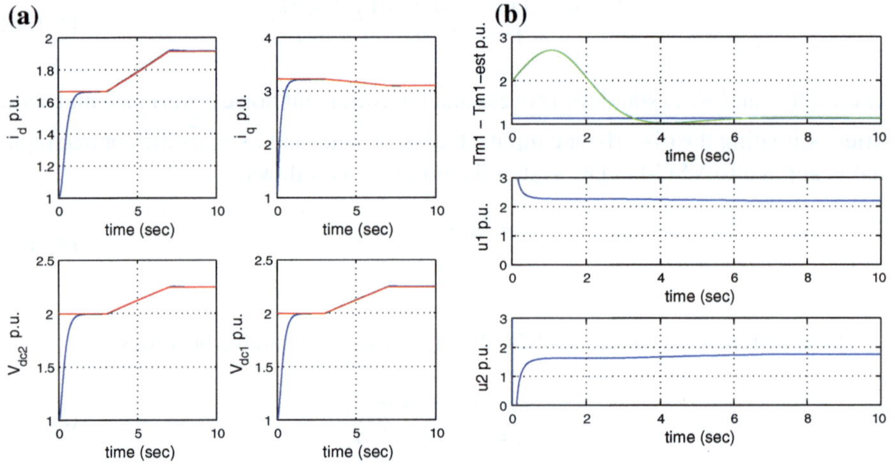

Fig. 5.27 **a** Convergence of the state vector elements $x_i,\ i=1,\ldots,4$ to the associated reference setpoints No 2 for the initial nonlinear model of the VSC-HVDC system, **b** estimation of disturbance input $\tilde{d}$ on the VSC-HVDC model and variation of control inputs $u_i,\ i=1,2$

system converge fast to the reference setpoints. Furthermore, the performance of the control loop is improved against external disturbances by including in it the Derivative-free nonlinear Kalman Filter. By functioning as a disturbance observer the filter enables to estimate in real-time additive disturbance inputs that affect the VSC-HVDC model and subsequently to compensate for them.

In is noted that the vector control approach which is implemented through PID controllers is unsuitable, because it is not of proven stability while its robustness

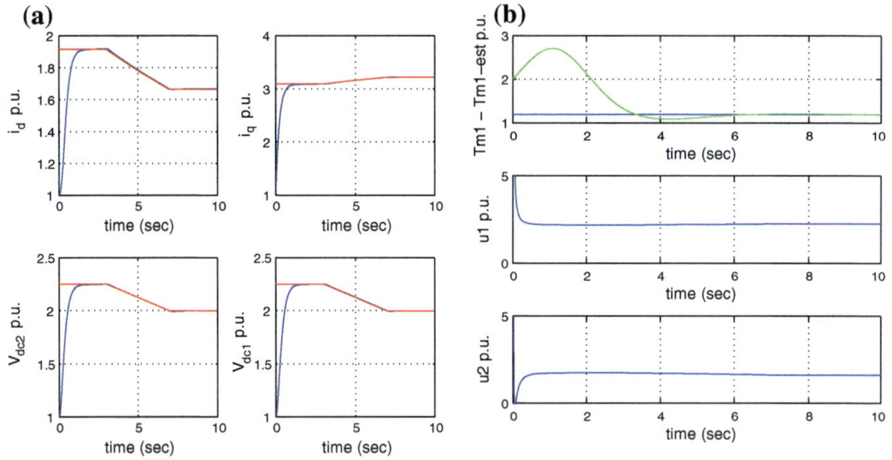

Fig. 5.28 Convergence of the state vector elements $x_i,\ i = 1, \ldots, 4$ to the associated reference setpoints No 3 for the initial nonlinear model of the VSC-HVDC system, **b** estimation of disturbance input $\tilde{d}$ on the VSC-HVDC model and variation of control inputs $u_i,\ i = 1, 2$

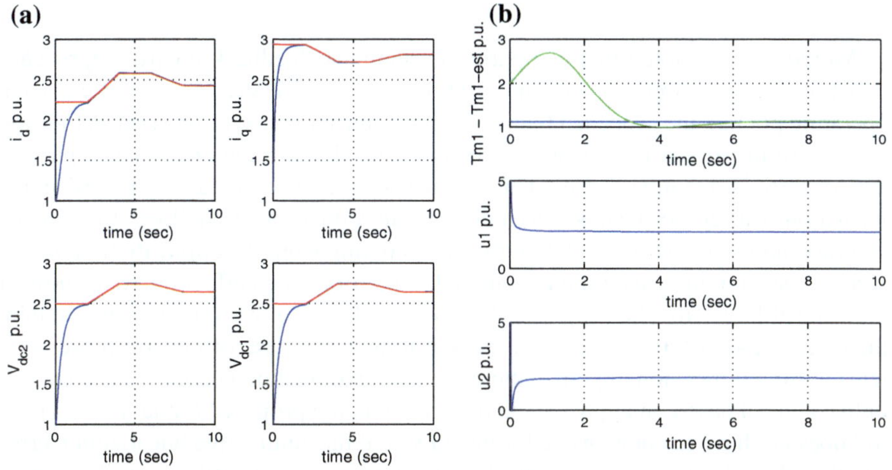

Fig. 5.29 Convergence of the state vector elements $x_i,\ i = 1, \ldots, 4$ to the associated reference setpoints No 4 for the initial nonlinear model of the VSC-HVDC system, **b** estimation of disturbance input $\tilde{d}$ on the VSC-HVDC model and variation of control inputs $u_i,\ i = 1, 2$

cannot be assured. Taking into account that the dynamic model of the VSC-HVDC system is characterized by parametric uncertainties and external perturbations and that it functions under varying conditions, the performance of vector control becomes questionable and consequently such a controller is not recommended for the VSC-HVDC model.

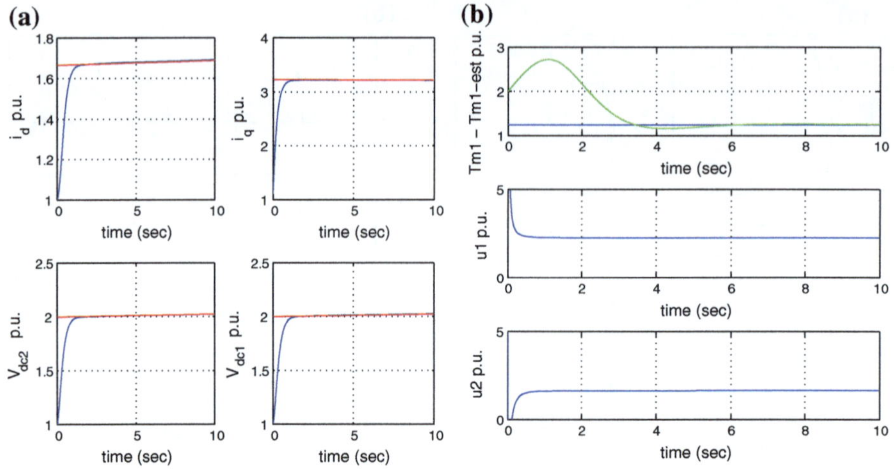

Fig. 5.30 Convergence of the state vector elements $x_i,\ i = 1, \ldots, 4$ to the associated reference setpoints No 5 for the initial nonlinear model of the VSC-HVDC system, **b** estimation of disturbance input $\tilde{d}$ on the VSC-HVDC model and variation of control inputs $u_i,\ i = 1, 2$

Moreover, it is noted that comparing to nonlinear feedback control approaches which are based on approximate linearization of the VSC-HVDC system (round local operating points), the proposed flatness-based control method is assessed as follows: (i) it performs global linearization which is valid through the entire state-space of the converter's dynamic model. This is done with the use of state variables' transformations (diffeomorphisms), (ii) it does not introduce any modelling error. On the contrary, approximate linearization methods coming from the application of Taylor series expansion induce cumulative modelling errors which affect the robustness of the control loop, (iii) the proposed flatness-based control method does not require the computation of Jacobian matrices, and thus in the case of electric power systems of high dimensionality an elaborated computational procedure is avoided, (iv) unlike control based on approximate linearization, the proposed flatness-based control does not have to compensate for the effects of the cumulative linearization error and thus the robustness of the control loop is improved, (v) unlike control based on approximate linearization, the proposed flatness-based control does not require the solution of Riccati equations, which means that for systems of high dimensionality a complicated computational procedure is again avoided.

Chapter 6
Control of the Functioning of DC to AC Converters

6.1 Outline

The chapter analyzes different approaches for nonlinear control of DC to AC converters (inverters): (i) flatness-based control through transformation to the canonical form, (ii) differential geometry-based control. Moreover, the chapter presents flatness-based control and nonlinear H-infinity control for DC to DC converters. Finally the chapter analyzes adaptive control of active power filters.

First, the chapter exploits the use of three-phase voltage inverters (DC to AC converters) in electric power system, such as photovoltaics connected with the rest of the grid. The chapter proposes a nonlinear feedback control method for three-phase inverters, which is based on differential flatness theory and a derivative-free implementation of nonlinear Kalman Filtering. This means that all its state variables and the control inputs can be written as functions of a single algebraic variable which is the flat output. By exploiting differential flatness properties it is shown that the inverter's model can be transformed to the linear canonical (Brunovsky's) form. For the latter description the design of a state feedback controller becomes possible, e.g. using pole placement methods. Moreover, to estimate the non-measurable state variables of the linearized equivalent of the inverter, the Derivative-free nonlinear Kalman Filter is used. This consists of the Kalman Filter recursion applied on the linearized inverter's model and of an inverse transformation that is based on differential flatness theory, which enables to compute estimates of the state variables of the initial nonlinear system. Furthermore, by redesigning the aforementioned filter as a disturbance observer it becomes also possible to estimate disturbance terms that affect the inverter and subsequently to compensate for them. The performance and disturbance rejection capability of the proposed nonlinear feedback control scheme is evaluated through simulation experiments.

Next, the chapter proposes a differential flatness theory-based approach for adaptive fuzzy control of active power filters. Aiming at maintaining the quality of the produced electric power within desirable levels and at compensating for harmonics distortions, the use of active power filters has become necessary in several parts of the

G. Rigatos, *Intelligent Renewable Energy Systems*,
Green Energy and Technology, DOI 10.1007/978-3-319-39156-4_6

electricity grid. An active power filter consists of an inverter which injects current to the power grid that annihilates the effects of specific harmonics. By eliminating such harmonics, it is achieved to maintain specified quality features for the grid's current and voltage. To this end, a new method of adaptive control for active power filters is developed in this section. By proving that the active power filter is a differentially flat system, its transformation to the linear canonical (Brunovsky) form becomes possible. In this new description the control input of the active power filter comprises unknown nonlinear terms which are identified with the use of neurofuzzy networks and through an adaptation/learning procedure. These estimated parts of the system's dynamics are used in an indirect adaptive control scheme, which finally makes the outputs of the active power filter converge to the desirable setpoints. The learning rate in the aforementioned adaptation procedure is given a value which assures that a suitably chosen Lyapunov function will always have a negative first-order derivative. Under the proposed control method, the closed loop of the active power filter is shown to satisfy the H-infinity tracking criterion, which implies a maximum capability for rejection of external perturbations as well as of modelling errors. The performance of the control scheme is tested through simulation experiments.

6.2 Flatness-Based Control of Inverters

6.2.1 Outline

Inverters, that is DC to AC converters are used for connecting DC power units such as photovoltaics, to the AC grid. A nonlinear control scheme is developed for three-phase inverters (DC to AC converters) using differential flatness theory and the Derivative-free nonlinear Kalman Filter [352, 366, 375]. Efficient control of inverters is important for the reliable operation of the electric power grid and the connection to it of distributed DC power generation units (such as photovoltaic arrays providing power to the main grid or local photovoltaic cells connected to a microgrid infrastructure) [32, 70, 74, 80, 82, 115, 216, 248, 261, 271, 464, 475]. Moreover, inverters find extensive use in HVDCs (High Voltage DC) lines and in their connection with the rest of the AC electric power network [183, 257, 437]. Indicative results on nonlinear control of inverters with the use of feedback linearization methods can be found in [49, 266, 458]. In the same direction, the present chapter proposes a method for inverters control, based on differential flatness theory and on a new nonlinear filtering method under the name Derivative-free nonlinear Kalman Filter.

The dynamic model of the inverter is a nonlinear multi-variable one [115, 183]. The control inputs are the voltages generated by an H-bridge circuit while the state vector comprises as elements the voltages at the load's side and the currents at the H-bridge circuit output. It is shown that this model is a differentially flat one, which means that all state variables and the control inputs can be written as functions of a single algebraic variable which is the so-called flat output (actually for the inverter's model the flat output is the vector of the voltages at the load side expressed in the *dq* reference frame) [183, 352]. By applying differential flatness theory it is proven that the inverter's model can be transformed into the linear canonical (Brunowsky) form [57, 137, 229, 241, 281, 387, 388, 411]. This transformation enables the design of a state feedback controller, e.g. using pole placement methods.

Other problems that have to be dealt with in the design of the feedback control scheme, are the estimation of nonmeasurable variables of the inverter's state vector (such as currents and rate of change of voltages), as well as the compensation of disturbance terms that affect the inverter's model. To solve the first problem, the Derivative-free nonlinear Kalman Filter is used. This consists of the standard Kalman Filter recursion applied on the linearized equivalent model of the system and of an inverse transformation that is based on differential flatness theory and which enables to obtain the estimates of the state variables of the initial nonlinear model [355, 359, 367, 370]. To solve the second problem, the Derivative-free nonlinear Kalman Filter is redesigned as a disturbance observer. This enables to obtain simultaneously estimates of the inverter's state vector and of the additive perturbation terms that affect its inputs. Since the Derivative-free nonlinear Kalman Filter keeps the optimality properties of the linear Kalman Filter algorithm, it succeeds fast and accurate estimation of the perturbation and modelling uncertainty terms [34, 341, 344, 356, 375].

6.2.2 Lie Algebra-Based Control of the Inverter's Model

The inverter's model was defined in Chap. 1 and is outlined next. The state-space description of the inverter has been defined as:

$$\frac{d}{dt}\begin{pmatrix} V_{L_d} \\ V_{L_q} \\ i_{I_d} \\ i_{I_q} \end{pmatrix} = \begin{pmatrix} \omega V_{L_q} + \frac{1}{C_f} i_{I_d} - \frac{1}{C_f}\frac{p_f V_{L_d} + q_f V_{L_q}}{V_{L_d}^2 + V_{L_q}^2} + \omega C_f V_{L_q} - \frac{\omega L_f V_{L_q}(i_{I_d}^2 + i_{I_q}^2)}{(V_{L_d}^2 + V_{L_q}^2)} \\ -\omega V_{L_d} + \frac{1}{C_f} i_{I_q} - \frac{1}{C_f}\frac{p_f V_{L_q} - q_f V_{L_d}}{V_{L_d}^2 + V_{L_q}^2} - \omega C_f V_{L_d} + \frac{\omega L_f V_{L_d}(i_{I_d}^2 + i_{I_q}^2)}{(V_{L_d}^2 + V_{L_q}^2)} \\ \omega i_{I_q} - \frac{1}{L_f} V_{L_d} \\ -\omega i_{I_d} - \frac{1}{L_f} V_{L_q} \end{pmatrix} + \begin{pmatrix} 0 & 0 \\ 0 & 0 \\ \frac{1}{L_f} & 0 \\ 0 & \frac{1}{L_f} \end{pmatrix} \begin{pmatrix} V_{I_d} \\ V_{I_q} \end{pmatrix} \tag{6.1}$$

while the measurement equation of the inverter's model is

$$\begin{pmatrix} y_1 \\ y_2 \end{pmatrix} = \begin{pmatrix} V_{L_d} \\ V_{L_q} \end{pmatrix} = \begin{pmatrix} 1 & 0 & 0 & 0 \\ 0 & 1 & 0 & 0 \end{pmatrix} \begin{pmatrix} V_{L_d} \\ V_{L_q} \\ i_{L_d} \\ i_{L_q} \end{pmatrix} \tag{6.2}$$

and by using the state variables notation $x_1 = V_{L_d}$, $x_2 = V_{L_q}$, $x_3 = i_{L_d}$ and $x_4 = i_{L_q}$ one has

$$\frac{d}{dt}\begin{pmatrix} x_1 \\ x_2 \\ x_3 \\ x_4 \end{pmatrix} = \begin{pmatrix} \omega x_2 + \frac{1}{C_f}x_3 - \frac{1}{C_f}\frac{p_f x_1 + q_f x_2}{x_1^2+x_2^2} + \omega C_f x_2 - \frac{\omega L_f x_2(x_3^2+x_4^2)}{(x_1^2+x_2^2)} \\ -\omega x_1 + \frac{1}{C_f}x_4 - \frac{1}{C_f}\frac{p_f x_2 - q_f x_1}{x_1^2+x_2^2} - \omega C_f x_1 + \frac{\omega L_f x_1(x_3^2+x_4^2)}{(x_1^2+x_2^2)} \\ \omega x_4 - \frac{1}{L_f}x_1 \\ -\omega x_3 - \frac{1}{L_f}x_2 \end{pmatrix} + \tag{6.3}$$

$$+ \begin{pmatrix} 0 & 0 \\ 0 & 0 \\ \frac{1}{L_f} & 0 \\ 0 & \frac{1}{L_f} \end{pmatrix}\begin{pmatrix} u_1 \\ u_2 \end{pmatrix}$$

while the measurement equation of the inverter's model is

$$\begin{pmatrix} y_1 \\ y_2 \end{pmatrix} = \begin{pmatrix} V_{L_d} \\ V_{L_q} \end{pmatrix} = \begin{pmatrix} 1\,0\,0\,0 \\ 0\,1\,0\,0 \end{pmatrix}\begin{pmatrix} x_1 \\ x_2 \\ x_3 \\ x_4 \end{pmatrix} \tag{6.4}$$

thus, the inverter's model is written in the nonlinear state-space form

$$\begin{matrix} \dot{x} = f(x) + G(x)u \\ y = h(x) \end{matrix} \tag{6.5}$$

where $f(x)\in R^{4\times 1}$, $G(x)\in R^{4\times 1}$ and $h(x)\in R^{2\times 4}$.

The following functions are defined next

$$\begin{matrix} y_1 = h_1(x) = x_1 \\ y_2 = h_2(x) = x_2 \end{matrix} \tag{6.6}$$

It holds that $z_1 = h_1(x) = x_1$, $z_2 = L_f h_1(x)$ and $\dot{z}_2 = L_f^2 h_1(x) + L_{g_a} L_f h_1(x) u_1 + L_{g_b} L_f h_1(x) u_2$ and also $z_3 = h_2(x) = x_2$, $z_4 = L_f h_2(x)$ and $\dot{z}_4 = L_f^2 h_2(x) + L_{g_a} L_f h_2(x) u_1 + L_{g_b} L_f h_2(x) u_2$

Using the above, one has

$$\begin{matrix} z_2 = L_f h_1(x) \Rightarrow z_2 = \frac{\partial h_1}{\partial x_1}f_1 + \frac{\partial h_1}{\partial x_2}f_2 + \frac{\partial h_1}{\partial x_3}f_3 + \frac{\partial h_1}{\partial x_4}f_4 \Rightarrow \\ z_2 = 1 f_1 \Rightarrow z_2 = \omega x_2 + \frac{1}{C_f}x_3 - \frac{1}{C_f}\{\frac{p_f x_1 + g_f x_2}{x_1^2+x_2^2} + \omega C_f x_2 - \frac{\omega L_f x_2(x_3^2+x_4^2)}{(x_1^2+x_2^2)}\} \end{matrix} \tag{6.7}$$

In a similar manner one computes

$$L_f^2h_1(x) = L_f z_2 \Rightarrow L_f^2h_1(x) = \frac{\partial z_1}{\partial x_1}f_1 + \frac{\partial z_1}{\partial x_2}f_2 + \frac{\partial z_1}{\partial x_3}f_3 + \frac{\partial z_1}{\partial x_4}f_4 \Rightarrow \quad (6.8)$$

Thus, it holds that

$$L_f^2h_1(x) = -\frac{1}{C_f}\{\frac{p_f(x_1^2+x_2^2)-(p_f x_1+g_f x_2)(2x_1)}{(x_1^2+x_2^2)^2} + \frac{\omega L_f x_2(x_3^2+x_4^2)(2x_1)}{(x_1^2+x_2^2)^2}\}\dot{x}_1 +$$
$$+[\omega - \frac{1}{C_f}\{\frac{g_f(x_1^2+x_2^2)-(p_f x_1+g_f x_2)2x_2}{(x_1^2+x_2^2)^2} + \omega C_f - \frac{\omega L_f(x_3^2+x_4^2)(x_1^2+x_2^2)+\omega L_f x_2(x_3^2+x_4^2)2x_2}{(x_1^2+x_2^2)^2}\}]\dot{x}_2$$
$$[\frac{1}{C_f} - \frac{1}{C_f}\{-\frac{\omega L_f x_2 2x_3}{(x_1^2+x_2^2)^2}\}](\omega x_4 - \frac{1}{L_f}x_1) + [-\frac{1}{C_f}\{\frac{-\omega L_f x_2 2x_4}{(x_1^2+x_2^2)}\}](-\omega x_3 - \frac{1}{L_f}x_2) \quad (6.9)$$

and after intermediate operations one arrives at the same formulation as in the case of differential flatness theory. Thus, one has

$$L_f^2h_1(x) = \omega\dot{x}_2 + \frac{1}{C_f}(\omega x_4 - \frac{1}{L_f}x_1) - \frac{1}{C_f}\{\frac{(p_f\dot{x}_1+q_f\dot{x}_2)(x_1^2+x_2^2)-(p_f x_1+q_f x_2)(2x_1\dot{x}_1+2x_2\dot{x}_2)}{(x_1^2+x_2^2)^2}\}$$
$$-\omega\dot{x}_2 + \frac{1}{C_f}\{\frac{\omega L_f\dot{x}_2(x_3^2+x_4^2)(x_1^2+x_2^2)-\omega L_f x_2(x_3^2+x_4^2)(2x_1\dot{x}_1+2x_2\dot{x}_2)}{(x_1^2+x_2^2)^2}\}$$
$$+\frac{\omega L_f x_2}{C_f}\cdot\frac{2x_3(\omega x_4-\frac{1}{L_f}x_1)}{(x_1^2+x_2^2)} + \frac{\omega L_f x_2}{C_f}\cdot\frac{2x_1(-\omega x_3-\frac{1}{L_f}x_2)}{(x_1^2+x_2^2)} \quad (6.10)$$

In a similar manner one computes

$$L_{g_a}L_f h_1(x) = L_{g_a}z_1 \Rightarrow L_{g_a}L_f h_1(x) = \frac{\partial z_2}{\partial x_1}g_{a_1} + \frac{\partial z_2}{\partial x_2}g_{a_2} + \frac{\partial z_2}{\partial x_3}g_{a_3} + \frac{\partial z_2}{\partial x_4}g_{a_4} \Rightarrow$$
$$L_{g_a}L_f h_1(x) = \frac{1}{L_f}\frac{\partial z_2}{\partial x_3} \Rightarrow \quad (6.11)$$
$$L_{g_a}L_f h_1(x) = \frac{1}{C_f}\{\frac{\omega L_f x_2(2x_3)}{(x_1^2+x_2^2)L_f} + \frac{1}{L_f}\}$$

and also

$$L_{g_b}L_f h_1(x) = L_{g_a}z_2 \Rightarrow L_{g_b}L_f h_2(x) = \frac{\partial z_2}{\partial x_1}g_{b_1} + \frac{\partial z_2}{\partial x_2}g_{b_2} + \frac{\partial z_2}{\partial x_3}g_{b_3} + \frac{\partial z_2}{\partial x_4}g_{b_4} \Rightarrow$$
$$L_{g_b}L_f h_1(x) = \frac{1}{L_f}\frac{\partial z_2}{\partial x_4} \Rightarrow \quad (6.12)$$
$$L_{g_b}L_f h_1(x) = \frac{\omega L_f x_2}{C_f}\frac{2x_4}{(x_1^2+x_2^2)}\frac{1}{L_f}$$

Following a similar procedure one obtains

$$z_4 = L_f h_2(x) \Rightarrow z_4 = \frac{\partial h_2}{\partial x_1}f_1 + \frac{\partial h_2}{\partial x_2}f_2 + \frac{\partial h_2}{\partial x_3}f_3 + \frac{\partial h_2}{\partial x_4}f_4 \Rightarrow$$
$$z_4 = f_2 \Rightarrow z_4 = -\omega x_1 + \frac{1}{C_f}x_4 - \frac{1}{C_f}\{\frac{p_f x_2 - q_f x_1}{(x_1^2+x_2^2)} - \omega C_f + \frac{\omega L_f x_1(x_3^2+x_4^2)}{(x_1^2+x_2^2)}\} \quad (6.13)$$

Moreover, one computes

$$L_f^2h_2(x) = L_fz_4 \Rightarrow L_f^2h_2(x) = \frac{\partial z_4}{\partial x_1}f_1 + \frac{\partial z_4}{\partial x_2}f_2 + \frac{\partial z_4}{\partial x_3}f_3 + \frac{\partial z_4}{\partial x_4}f_4 \Rightarrow$$
$$L_f^2h_2(x) = \frac{\partial z_4}{\partial x_1}\dot{x}_1 + \frac{\partial z_4}{\partial x_2}\dot{x}_2 + \frac{\partial z_4}{\partial x_3}\omega x_4 + \frac{\partial z_4}{\partial x_4} - \omega x_3 \qquad (6.14)$$

which is also written as

$$L_f^2h_2(x) = [-\omega - \frac{1}{C_f}\{\frac{-g_f(x_1^2+x_2^2)-(p_fx_2-q_fx_1)(2x_1)}{(x_1^2+x_2^2)^2} - \omega C_f\} +$$
$$+\omega L_f\frac{(x_3^2+x_4^2)(x_1^2+x_2^2)-\omega L_f x_1(x_3^2+x_4^2)2x_1}{(x_1^2+x_2^2)^2}]\dot{x}_1$$
$$[-\frac{1}{C_f}\{\frac{p_f(x_1^2+x_2^2)-(p_fx_2-q_fx_4)2x_2}{x_1^2+x_2^{2\,2}} - \frac{\omega L_f x_1(x_3^2+x_4^2)2x_2}{(x_1^2+x_2^2)^2}\}]\dot{x}_2 \qquad (6.15)$$
$$[-\frac{1}{C_f}\{\frac{\omega L_f x_1 2x_3}{(x_1^2+x_2^2)}\}](\omega x_4 - \frac{1}{L_f}x_1)$$
$$[\frac{1}{C_f} - \frac{1}{C_f}\{\frac{\omega L_f x_1 2x_4}{(x_1^2+x_2^2)}\}](-\omega x_3 - \frac{1}{L_f}x_2)$$

which after intermediate operations gives

$$L_f^2h_2(x) = -\omega\dot{x}_1 + \frac{1}{C_f}(-\omega x_3) - \frac{1}{C_f}\{\frac{(p_f\dot{x}_2-g_f\dot{x}_1)(x_1^2+x_2^2)-(p_fx_2-g_fx_1)(2x_1\dot{x}_1+2x_2\dot{x}_2)}{(x_1^2+x_2^2)^2}\}+$$
$$\omega\dot{x}_1 - \frac{1}{C_f}\{\frac{\omega L_f\dot{x}_1(x_3^2+x_4^2)(x_1^2+x_2^2)-\omega L_f x_1(x_3^2+x_4^2)(2x_1\dot{x}_1+2x_2\dot{x}_2)}{(x_1^2+x_2^2)^2}\}$$
$$-\frac{1}{C_f}\frac{\omega L_f x_1 2x_3(\omega x_4-\frac{1}{L_f}x_1)}{(x_1^2+x_2^2)} - \frac{1}{C_f}\frac{\omega L_f x_1 2x_4(-\omega x_3-\frac{1}{L_f}x_2)}{(x_1^2+x_2^2)} \qquad (6.16)$$

Following a similar procedure one finds

$$L_{g_a}L_fh_2(x) = L_{g_a}z_4 \Rightarrow L_{g_a}L_fh_2(x) = \frac{\partial z_4}{\partial x_1}g_{a_1} + \frac{\partial z_4}{\partial x_2}g_{a_2} + \frac{\partial z_4}{\partial x_3}g_{a_3} + \frac{\partial z_4}{\partial x_4}g_{a_4} \Rightarrow$$
$$L_{g_b}L_fh_2(x) = \frac{1}{L_f}\frac{\partial z_4}{\partial x_3} \Rightarrow \qquad (6.17)$$
$$L_{g_a}L_fh_2(x) = -\frac{1}{C_f}\frac{\omega L_f x_1 2x_3}{(x_1^2+x_2^2)}\frac{1}{L_f}$$

and equivalently

$$L_{g_b}L_fh_2(x) = L_{g_b}z_4 \Rightarrow L_{g_b}L_fh_2(x) = \frac{\partial z_4}{\partial x_1}g_{b_1} + \frac{\partial z_4}{\partial x_2}g_{b_2} + \frac{\partial z_4}{\partial x_3}g_{b_3} + \frac{\partial z_4}{\partial x_4}g_{b_4} \Rightarrow$$
$$L_{g_b}L_fh_2(x) = \frac{-1}{L_f}\frac{\partial z_4}{\partial x_4} \Rightarrow \qquad (6.18)$$
$$L_{g_b}L_fh_2(x) = -\frac{1}{C_f}\{\frac{\omega L_f x_1 2x_4}{(x_1^2+x_2^2)}\frac{1}{L_f} + \frac{1}{L_f}\}$$

After the Lie algebra-based linearization procedure, the dynamics of the inverter is written in the form

$$\begin{aligned}\ddot{x}_1 &= L_f^2 h_1(x) + L_{g_a} L_f h_1(x) u_1 + L_{g_b} L_f h_1(x) u_2 \\ \ddot{x}_2 &= L_f^2 h_2(x) + L_{g_a} L_f h_2(x) u_1 + L_{g_b} L_f h_2(x) u_2\end{aligned} \tag{6.19}$$

The following new control inputs are defined

$$\begin{aligned}v_1 &= L_f^2 h_1(x) + L_{g_a} L_f h_1(x) u_1 + L_{g_b} L_f h_1(x) u_2 \\ v_2 &= L_f^2 h_2(x) + L_{g_a} L_f h_2(x) u_1 + L_{g_b} L_f h_2(x) u_2\end{aligned} \tag{6.20}$$

This results in the linearized system's dynamics

$$\begin{aligned}\ddot{x}_1 &= v_1 \\ \ddot{x}_2 &= v_2\end{aligned} \tag{6.21}$$

which is also written in a state-space form, after defining the state variables $z_1 = x_1$, $z_2 = \dot{x}_1$, $z_3 = x_2$ and $z_4 = \dot{x}_2$

$$\begin{pmatrix}\dot{z}_1 \\ \dot{z}_2 \\ \dot{z}_3 \\ \dot{z}_4\end{pmatrix} = \begin{pmatrix}0 & 1 & 0 & 0 \\ 0 & 0 & 0 & 0 \\ 0 & 0 & 0 & 1 \\ 0 & 0 & 0 & 0\end{pmatrix} \begin{pmatrix}z_1 \\ z_2 \\ z_3 \\ z_4\end{pmatrix} + \begin{pmatrix}0 & 0 \\ 1 & 0 \\ 0 & 0 \\ 0 & 1\end{pmatrix} \begin{pmatrix}v_1 \\ v_2\end{pmatrix} \tag{6.22}$$

$$\begin{pmatrix}z_1^m \\ z_2^m\end{pmatrix} = \begin{pmatrix}1 & 0 & 0 & 0 \\ 0 & 0 & 1 & 0\end{pmatrix} \begin{pmatrix}z_1 \\ z_2 \\ z_3 \\ z_4\end{pmatrix} \tag{6.23}$$

6.2.3 *Differential Flatness of the Inverter's Model*

The flat output of the inverter is taken to be the vector

$$y = [y_1, y_2] = [V_{L_d}, V_{L_q}] \tag{6.24}$$

It will be shown that the system is a differentially flat one, which means that all its state variables and the control inputs can be written as functions of the flat output and their derivatives. It holds that

$$\begin{aligned}x_1 &= V_{L_d} \Rightarrow x_1 = y_1 \\ x_2 &= V_{L_q} \Rightarrow x_2 = y_2\end{aligned} \tag{6.25}$$

The first row of the state-space equations is

$$\dot{x}_1 = \omega x_2 + \frac{1}{C_f}x_3 - \frac{1}{C_f}\frac{p_f x_1 + q_f x_2}{x_1^2 + x_2^2} - \omega x_2 + \frac{1}{C_f}\frac{\omega L_f x_2 (x_3^2 + x_4^2)}{x_1^2 + x_2^2} \tag{6.26}$$

The second row of the state-space equations is

$$\dot{x}_2 = \omega x_1 + \frac{1}{C_f}x_4 - \frac{1}{C_f}\frac{p_f x_2 + q_f x_1}{x_1^2 + x_2^2} + \omega x_1 + \frac{1}{C_f}\frac{\omega L_f x_1 (x_3^2 + x_4^2)}{x_1^2 + x_2^2} \tag{6.27}$$

Both equations are solved with respect to their first term

$$\frac{1}{C_f}\frac{\omega L_f x_2 (x_3^2 + x_4^2)}{x_1^2 + x_2^2} = \dot{x}_1 - \omega x_2 - \frac{1}{C_f}x_3 + \frac{1}{C_f}\frac{p_f x_1 + q_f x_2}{(x_1^2 + x_2^2)} + \omega x_2 \tag{6.28}$$

$$\frac{1}{C_f}\frac{\omega L_f x_1 (x_3^2 + x_4^2)}{x_1^2 + x_2^2} = \dot{x}_2 + \omega x_1 - \frac{1}{C_f}x_4 + \frac{1}{C_f}\frac{p_f x_2 - q_f x_1}{(x_1^2 + x_2^2)} - \omega x_1 \tag{6.29}$$

By dividing the above two equations one obtains

$$-\frac{x_2}{x_1} = \frac{\dot{x}_1 - \omega x_2 - \frac{1}{C_f}x_3 + \frac{1}{C_f}\frac{p_f x_1 + q_f x_2}{(x_1^2 + x_2^2)} + \omega x_2}{\dot{x}_2 + \omega x_1 - \frac{1}{C_f}x_4 + \frac{1}{C_f}\frac{p_f x_2 - q_f x_1}{(x_1^2 + x_2^2)} - \omega x_1} \tag{6.30}$$

or using in the notation the elements of the flat output vector

$$-\frac{y_2}{y_1} = \frac{\dot{y}_1 - \omega y_2 - \frac{1}{C_f}x_3 + \frac{1}{C_f}\frac{p_f y_1 + q_f y_2}{(y_1^2 + y_2^2)} + \omega y_2}{\dot{y}_2 + \omega y_1 - \frac{1}{C_f}x_4 + \frac{1}{C_f}\frac{p_f y_2 - q_f y_1}{(y_1^2 + y_2^2)} - \omega y_1} \tag{6.31}$$

which equivalently gives

$$\begin{aligned} -\frac{y_2}{y_1}\dot{y}_2 - \omega y_2 + \frac{1}{C_f}\left(\frac{y_2}{y_1}\right)\frac{p_f y_2 - q_f y_1}{(y_1^2 + y_2^2)} + \omega y_2 = \\ = \dot{y}_1 - \omega y_2 - \frac{1}{C_f}x_3 + \frac{1}{C_f}\frac{p_f y_1 + q_f y_2}{(y_1^2 + y_2^2)} + \omega y_2 \end{aligned} \tag{6.32}$$

Equation (6.32) is solved with respect to x_3. It holds that

$$\begin{aligned} x_3 = C_f\{(\tfrac{y_2}{y_1})\dot{y}_2 + \omega y_2 - \frac{1}{C_f}(\tfrac{y_2}{y_1})x_4 + \frac{1}{C_f}(\tfrac{y_2}{y_1})\frac{p_f y_2 - q_f y_1}{y_1^2 + y_2^2} - \omega y_2 \\ + \dot{y}_1 - \omega \dot{y}_2 + \frac{1}{C_f}\frac{p_f y_1 + q_f y_2}{y_1^2 + y_2^2} + \omega y_2\} \end{aligned} \tag{6.33}$$

which equivalently gives

$$\begin{aligned} x_3 = -\frac{y_2}{y_1}x_4 + C_f\{\frac{y_2}{y_1}\dot{y}_2 + \omega y_2 + \frac{1}{C_f}(\tfrac{y_2}{y_1})\frac{p_f y_2 - q_f y_1}{(y_1^2 + y_2^2)} - \omega y_2 \\ + \dot{y}_1 - \omega y_2 + \frac{1}{C_f}\frac{p_f y_1 + q_f y_2}{(y_1^2 + y_2^2)} + \omega y_2\} \end{aligned} \tag{6.34}$$

which is also written as

$$x_3 = -(\tfrac{y_2}{y_1})x_4 + f_a(y_1, \dot{y}_1, y_2, \dot{y}_2) \tag{6.35}$$

Next, Eq. (6.35) is substituted into Eq. (6.26). Thus, one obtains

$$\begin{aligned}\dot{x}_2 = -\omega x_1 + \tfrac{1}{C_f}x_4 - \tfrac{1}{C_f}\tfrac{p_f x_2 - q_f x_1}{(x_1^2+x_2^2)} + \omega x_1 - \\ -\tfrac{1}{C_f}\tfrac{\omega L_f x_1\{[-(\frac{y_2}{y_1})x_4 + f_a(y_1,\dot{y}_1,y_2,\dot{y}_2)]^2 + x_4^2\}}{(x_1^2+x_2^2)}\end{aligned} \tag{6.36}$$

or equivalently

$$\begin{aligned}\dot{y}_2 = -\omega y_1 + \tfrac{1}{C_f}x_4 - \tfrac{1}{C_f}\tfrac{p_f y_2 - q_f y_1}{(y_1^2+y_2^2)} + \omega y_1 + \\ -\tfrac{1}{C_f}\tfrac{\omega L_f y_1\{[-(\frac{y_2}{y_1})x_4 + f_a(y_1,\dot{y}_1,y_2,\dot{y}_2)]^2 + x_4^2\}}{(y_1^2+y_2^2)}\end{aligned} \tag{6.37}$$

Equation (6.37) is a binomial which can be solved with respect to x_4. Thus, one obtains that

$$x_4 = f_b(y_1, \dot{y}_1, y_2, \dot{y}_2) \tag{6.38}$$

Substituting Eq. (6.39) into Eq. (6.35) one gets

$$x_3 = -(\tfrac{y_2}{y_1})f_b(y_1, \dot{y}_1, y_2, \dot{y}_2) + f_4(y_1, \dot{y}_1, y_2, \dot{y}_2) \tag{6.39}$$

From Eqs. (6.38) and (6.39) on has that state variables x_3 and x_4 can be written as functions of the flat outputs and their derivatives. Moreover, from the last two rows of the state-space equations one has that

$$\dot{x}_3 = \omega x_4 - \tfrac{1}{L_f}x_1 + \tfrac{1}{L_f}u_1 \tag{6.40}$$

$$\dot{x}_4 = -\omega x_3 - \tfrac{1}{L_f}x_2 + \tfrac{1}{L_f}u_2 \tag{6.41}$$

From Eq. (6.40) one gets

$$u_1 = L_f\{\dot{x}_3 - \omega x_4 + \tfrac{1}{L_f}x_1\} \Rightarrow u_1 = f_c(y_1, \dot{y}_1, y_2, \dot{y}_2)\} \tag{6.42}$$

From Eq. (6.41) one gets

$$u_2 = L_f\{\dot{x}_4 - \omega x_3 + \tfrac{1}{L_f}x_2\} \Rightarrow u_2 = f_d(y_1, \dot{y}_1, y_2, \dot{y}_2)\} \tag{6.43}$$

Consequently, the control inputs are also functions of the flat outputs and their derivatives. Thus, the inverter's model is a differentially flat one.

6.2.4 Flatness-Based Control of the Inverter

Next, a flatness-based controller will be designed. To this end Eqs. (6.26) and (6.27) are used again. By considering p_f, q_f as piecewise constant and by deriving Eq. (6.26) with respect to time, one has

$$\begin{aligned}\ddot{x}_1 = \omega\dot{x}_2 + \tfrac{1}{C_f}\dot{x}_3 - \tfrac{1}{C_f}\tfrac{\{(p_f\dot{x}_1+q_f\dot{x}_2)(x_1^2+x_2^2)-(p_f x_1+q_f x_2)(2x_1\dot{x}_1+2x_2\dot{x}_2)}{(x_1^2+x_2^2)^2\}} - \\ -\omega\dot{x}_2 + \tfrac{1}{C_f}\{\tfrac{\omega L_f\dot{x}_2(x_3^2+x_4^2)(x_1^2+x_2^2)+\omega L_f x_2(2x_3\dot{x}_3+2x_4\dot{x}_4)(x_1^2+x_2^2)}{(x_1^2+x_2^2)^2}\} \\ -\tfrac{\omega L_f x_2(x_3^2+x_4^2)(2x_1\dot{x}_1+2x_2\dot{x}_2)}{(x_1^2+x_2^2)^2}\end{aligned} \tag{6.44}$$

The time derivatives are substituted from the associated rows of the state-space equations. Next, substitutions of the terms $\dot{x}_3$ and $\dot{x}_4$ are performed. It holds that

$$\begin{aligned}\ddot{x}_1 = \omega\dot{x}_2 + \tfrac{1}{C_f}(\omega x_4 - \tfrac{1}{L_f}x_1 + \tfrac{1}{L_f}u_1) - \tfrac{1}{C_f}\tfrac{\{(p_f\dot{x}_1+q_f\dot{x}_2)(x_1^2+x_2^2)-(p_f x_1+q_f x_2)(2x_1\dot{x}_1+2x_2\dot{x}_2)}{(x_1^2+x_2^2)^2\}} - \\ -\omega\dot{x}_2 + \tfrac{1}{C_f}\{\tfrac{\omega L_f\dot{x}_2(x_3^2+x_4^2)(x_1^2+x_2^2)+\omega L_f x_2(x_3^2+x_4^2)(2x_1\dot{x}_1+2x_2\dot{x}_2)}{(x_1^2+x_2^2)^2}\} \\ +\tfrac{\omega L_f x_2}{C_f}\tfrac{2x_3(x_1^2+x_2^2)}{(x_1^2+x_2^2)^2}(\omega x_4 - \tfrac{1}{L_f}x_1 + \tfrac{1}{L_f}u_1) + \tfrac{\omega L_f x_2}{C_f}\tfrac{2x_4(x_1^2+x_2^2)}{(x_1^2+x_2^2)^2}(-\omega x_3 - \tfrac{1}{L_f}x_2 + \tfrac{1}{L_f}u_2)\end{aligned} \tag{6.45}$$

The previous relation is rewritten by regrouping terms which multiply the control inputs u_1 and u_2. Thus one obtains

$$\begin{aligned}\ddot{x}_1 = \omega\dot{x}_2 + \tfrac{1}{C_f}(\omega x_4 - \tfrac{1}{L_f}x_1) - \tfrac{1}{C_f}\{\tfrac{(p_f\dot{x}_1+q_f\dot{x}_2)(x_1^2+x_2^2)-(p_f x_1+q_f x_2)(2x_1\dot{x}_1+2x_2\dot{x}_2)}{(x_1^2+x_2^2)^2}\} \\ -\omega\dot{x}_2 + \tfrac{1}{C_f}\{\tfrac{\omega L_f\dot{x}_2(x_3^2+x_4^2)(x_1^2+x_2^2)-\omega L_f x_2(x_3^2+x_4^2)(2x_1\dot{x}_1+2x_2\dot{x}_2)}{(x_1^2+x_2^2)^2}\} \\ +\tfrac{\omega L_f x_2}{C_f}\cdot\tfrac{2x_3(\omega x_4-\frac{1}{L_f}x_1)}{(x_1^2+x_2^2)} + \tfrac{\omega L_f x_2}{C_f}\cdot\tfrac{2x_4(-\omega x_3-\frac{1}{L_f}x_2)}{(x_1^2+x_2^2)} + \\ +\tfrac{1}{C_f}\{\tfrac{\omega L_f x_2(2x_3)}{(x_1^2+x_2^2)L_f} + \tfrac{1}{L_f}\}u_1 + \tfrac{\omega L_f x_2}{C_f}\tfrac{2x_4}{(x_1^2+x_2^2)}\tfrac{1}{L_f}u_2\end{aligned} \tag{6.46}$$

The previous relation can be also written in the form

$$\ddot{x}_1 = L_f^2 h_1(x) + L_{g_a}L_f h_1(x)u_1 + L_{g_b}L_f h_1(x)u_2 \tag{6.47}$$

In a similar manner, by differentiating the second row of the state-space equations with respect to time one has

$$\ddot{x}_2 = -\omega\dot{x}_1 + \frac{1}{C_f}\dot{x}_4 - \frac{1}{C_f}\frac{(p_f\dot{x}_2 - g_f\dot{x}_1)(x_1^2+x_2^2) - (p_f x_2 - g_f x_1(2x_1\dot{x}_1 + 2x_2\dot{x}_2))}{(x_1^2+x_2^2)^2} +$$
$$+\omega\dot{x}_1 - \frac{1}{C_f}\{\frac{\omega L_f\dot{x}_1(x_3^2+x_4^2)(x_1^2+x_2^2) + \omega L_f x_1(2x_3\dot{x}_3 + 2x_4\dot{x}_4)(x_1^2+x_2^2)}{(x_1^2+x_2^2)^2} - \tag{6.48}$$
$$\frac{-\omega L_f x_1(x_3^2+x_4^2)(2x_1\dot{x}_1+2x_2\dot{x}_2)}{(x_1^2+x_2^2)^2}\}$$

By substituting the derivative terms for x_3 and x_4, according to the third and fourth rows of the state-space equations one has

$$\ddot{x}_2 = -\omega\dot{x}_1 + \frac{1}{C_f}(-\omega x_3 - \frac{1}{L_f}x_2 + \frac{1}{L_f}u_2) -$$
$$-\frac{1}{C_f}\{\frac{(p_f\dot{x}_2 - g_f\dot{x}_1)(x_1^2+x_2^2) - (p_f x_2 - g_f x_1)(2x_1\dot{x}_1 + 2x_2\dot{x}_2)}{(x_1^2+x_2^2)^2}\} +$$
$$+\omega\dot{x}_1 - \frac{1}{C_f}\{\frac{\omega L_f\dot{x}_1(x_3^2+x_4^2)(x_1^2+x_2^2) - \omega L_f x_1(x_3^2+x_4^2)(2x_1\dot{x}_1+2x_2\dot{x}_2)}{(x_1^2+x_2^2)^2}\} \tag{6.49}$$
$$-\frac{1}{C_f}\frac{\omega L_f x_1 2x_3(x_1^2+x_2^2)}{(x_1^2+x_2^2)^2}(\omega x_4 - \frac{1}{L_f}x_1 + \frac{1}{L_f}u_1) -$$
$$-\frac{1}{C_f}\frac{\omega L_f x_1 2x_4(x_1^2+x_2^2)}{(x_1^2+x_2^2)^2}(-\omega x_3 - \frac{1}{L_f}x_2 + \frac{1}{L_f}u_2)$$

By grouping the terms that multiply the control inputs u_1 and u_2 one obtains

$$\ddot{x}_2 = -\omega\dot{x}_1 + \frac{1}{C_f}(-\omega x_3) - \frac{1}{C_f}\{\frac{(p_f\dot{x}_2 - g_f\dot{x}_1)(x_1^2+x_2^2) - (p_f x_2 - g_f x_1)(2x_1\dot{x}_1 + 2x_2\dot{x}_2)}{(x_1^2+x_2^2)^2}\} +$$
$$\omega\dot{x}_1 - \frac{1}{C_f}\{\frac{\omega L_f\dot{x}_1(x_3^2+x_4^2)(x_1^2+x_2^2) - \omega L_f x_1(x_3^2+x_4^2)(2x_1\dot{x}_1+2x_2\dot{x}_2)}{(x_1^2+x_2^2)^2}\}$$
$$-\frac{1}{C_f}\frac{\omega L_f x_1 2x_3(\omega x_4 - \frac{1}{L_f}x_1)}{(x_1^2+x_2^2)} - \frac{1}{C_f}\frac{\omega L_f x_1 2x_4(-\omega x_3 - \frac{1}{L_f}x_2)}{(x_1^2+x_2^2)} - \tag{6.50}$$
$$-\frac{1}{C_f}\frac{\omega L_f x_1 2x_3}{(x_1^2+x_2^2)}\frac{1}{L_f}u_1 - \frac{1}{C_f}\{\frac{\omega L_f x_1 2x_4}{(x_1^2+x_2^2)}\frac{1}{L_f} + \frac{1}{L_f}\}u_2$$

The previous equation can be also written in the form

$$\ddot{x}_2 = L_f^2 h_2(x) + L_{g_a}L_f h_2(x)u_1 + L_{g_b}L_f h_2(x)u_2 \tag{6.51}$$

Thus, one obtains an input-output linearized description of the inverter in the form

$$\begin{aligned}\ddot{x}_1 &= L_f^2 h_1(x) + L_{g_a}L_f h_1(x)u_1 + L_{g_b}L_f h_1(x)u_2\\ \ddot{x}_2 &= L_f^2 h_2(x) + L_{g_a}L_f h_2(x)u_1 + L_{g_b}L_f h_2(x)u_2\end{aligned} \tag{6.52}$$

The following control inputs are defined

$$
\begin{aligned}
v_1 &= L_f^2 h_1(x) + L_{g_a} L_f h_1(x) u_1 + L_{g_b} L_f h_1(x) u_2 \\
v_2 &= L_f^2 h_2(x) + L_{g_a} L_f h_2(x) u_1 + L_{g_b} L_f h_2(x) u_2
\end{aligned}
\tag{6.53}
$$

Thus, the inverter's dynamics becomes

$$
\begin{aligned}
\ddot{x}_1 &= v_1 \\
\ddot{x}_2 &= v_2
\end{aligned}
\tag{6.54}
$$

For this form of the system's dynamics the design of a state feedback controller is easy. This takes the form:

$$
\begin{aligned}
v_1 &= \ddot{x}_1 - k_d^1(\dot{x}_1 - \dot{x}_1^d) - k_p^1(x_1 - x_1^d) \\
v_2 &= \ddot{x}_2 - k_d^2(\dot{x}_2 - \dot{x}_2^d) - k_p^2(x_2 - x_2^d)
\end{aligned}
\tag{6.55}
$$

The control input that is actually applied to the system is computed as:

$$
\tilde{v} = \tilde{f} + \tilde{M} u \tag{6.56}
$$

or equivalently

$$
\begin{pmatrix} v_1 \\ v_2 \end{pmatrix} = \begin{pmatrix} L_f^2 h_1(x) \\ L_f^2 h_2(x) \end{pmatrix} + \begin{pmatrix} L_{g_a} L_f h_1(x) & L_{g_b} L_f h_1(x) \\ L_{g_a} L_f h_2(x) & L_{g_b} L_f h_2(x) \end{pmatrix} \begin{pmatrix} u_1 \\ u_2 \end{pmatrix} \tag{6.57}
$$

which means that the control input that is finally applied to the system is

$$
\tilde{u} = \tilde{M}^{-1}(\tilde{v} - \tilde{f}) \tag{6.58}
$$

To estimate the non-measurable state variables of the input-output linearized system the Derivative-free nonlinear Kalman Filter can be used. By defining the new state variables $z_1 = x_1, z_2 = \dot{x}_1, z_3 = x_2$ and $z_4 = \dot{x}_2$, the following state-space description of the system can be used

$$
\begin{pmatrix} \dot{z}_1 \\ \dot{z}_2 \\ \dot{z}_3 \\ \dot{z}_4 \end{pmatrix} = \begin{pmatrix} 0 & 1 & 0 & 0 \\ 0 & 0 & 0 & 0 \\ 0 & 0 & 0 & 1 \\ 0 & 0 & 0 & 0 \end{pmatrix} \begin{pmatrix} z_1 \\ z_2 \\ z_3 \\ z_4 \end{pmatrix} + \begin{pmatrix} 0 & 0 \\ 1 & 0 \\ 0 & 0 \\ 0 & 1 \end{pmatrix} \begin{pmatrix} v_1 \\ v_2 \end{pmatrix} \tag{6.59}
$$

$$
\begin{pmatrix} z_1^m \\ z_2^m \end{pmatrix} = \begin{pmatrix} 1 & 0 & 0 & 0 \\ 0 & 0 & 1 & 0 \end{pmatrix} \begin{pmatrix} z_1 \\ z_2 \\ z_3 \\ z_4 \end{pmatrix} \tag{6.60}
$$

The associated state estimator has the form

$$\dot{\hat{z}} = A\hat{z} + Bv + K_f(z^m - C\hat{z}) \tag{6.61}$$

For the linearized equivalent of the system it is possible to perform state estimation using the Derivative-free nonlinear Kalman Filter. The previously defined matrices A, B and C are substituted by their discrete-time equivalents A_d, B_d and C_d. This is done through common discretization methods. The recursion of the Kalman Filter is [344, 375]:

measurement update:

$$\begin{aligned} K_f(k) &= P^-(k)C_d^T[C_dP^-(k)C_d^T + R(k)]^{-1} \\ \hat{x}(k) &= \hat{x}^-(k) + K_f(k)[C_dz(k) - C_d\hat{z}(k)] \\ P(k) &= P^-(k) - K(k)C_dP^-(k) \end{aligned} \tag{6.62}$$

time update:

$$\begin{aligned} P^-(k+1) &= A_dP(k)A_d^T + Q(k) \\ \hat{x}^-(k+1) &= A_d\hat{x}(k) + B_dv(k) \end{aligned} \tag{6.63}$$

The linearized equivalent model of the inverter can thus provide the state estimates $\hat{x}_1, \dot{\hat{x}}_1, \hat{x}_2$ and $\dot{\hat{x}}_2$. Finally using Eqs. (6.38) and (6.39) one can obtain estimates for all state variables of the initial nonlinear model.

6.2.5 State and Disturbances Estimation with Nonlinear Kalman Filtering

The estimator can be also redesigned in the form of a disturbance observer. It is considered that the linearized model of the inverter is affected by additive input disturbances. The disturbance terms can describe both modelling uncertainties and external perturbations.

$$\begin{aligned} \ddot{x}_1 &= v_1 + \tilde{d}_1 \\ \ddot{x}_2 &= v_2 + \tilde{d}_2 \end{aligned} \tag{6.64}$$

These take the form $\tilde{d}_1^{(n)} = f_{d_1}$ and $\tilde{d}_2^{(n)} = f_{d_2}$. The disturbances' dynamics is represented by the n-th order derivative of the disturbances variables together with the associated initial conditions (however, since the disturbances are going to be estimated by the Kalman Filter, the knowledge of the initial conditions finally becomes obsolete).

Without loss of generality it is assumed that $n = 2$. The state vector is extended by including as additional state variables the disturbances and their derivatives. Thus, one has $z_1 = x_1$, $z_2 = \dot{x}_1$, $z_3 = x_2$, $z_4 = \dot{x}_2$, $z_5 = \tilde{d}_1$, $z_6 = \dot{\tilde{d}}_1$, $z_7 = \tilde{d}_2$ and $z_8 = \dot{\tilde{d}}_2$. Therefore, one has the extended state estimator of the form

$$\dot{\hat{z}}_e = A_e \hat{z}_e + B_e v + K_{f_e}(z_e^m - C_e \hat{z}_e) \tag{6.65}$$

where

$$A_e = \begin{pmatrix} 0\,1\,0\,0\,0\,0\,0\,0 \\ 0\,0\,0\,0\,1\,0\,0\,0 \\ 0\,0\,0\,1\,0\,0\,0\,0 \\ 0\,0\,0\,0\,0\,0\,1\,0 \\ 0\,0\,0\,0\,0\,1\,0\,0 \\ 0\,0\,0\,0\,0\,0\,0\,0 \\ 0\,0\,0\,0\,0\,0\,0\,1 \\ 0\,0\,0\,0\,0\,0\,0\,0 \end{pmatrix} \quad B_e = \begin{pmatrix} 0\,0 \\ 1\,0 \\ 0\,0 \\ 0\,1 \\ 0\,0 \\ 0\,0 \\ 0\,0 \\ 0\,0 \end{pmatrix} \quad C_e^T = \begin{pmatrix} 1\,0 \\ 0\,0 \\ 0\,1 \\ 0\,0 \\ 0\,0 \\ 0\,0 \\ 0\,0 \\ 0\,0 \end{pmatrix} \tag{6.66}$$

For the extended state-space description of the system it is possible again to perform state estimation using the Derivative-free nonlinear Kalman Filter. The previously defined matrices A_e, B_e and C_e are substituted by their discrete-time equivalents A_{e_d}, B_{e_d} and C_{e_d}. This is done again through common discretization methods. The recursion of the Kalman Filter in this case is:

measurement update:

$$\begin{aligned} K_f(k) &= P^-(k)C_d^T[C_{e_d}P^-(k)C_{e_d}^T + R(k)]^{-1} \\ \hat{x}(k) &= \hat{x}^-(k) + K_f(k)[C_{e_d}z(k) - C_{e_d}\hat{z}(k)] \\ P(k) &= P^-(k) - K(k)C_{e_d}P^-(k) \end{aligned} \tag{6.67}$$

time update:

$$\begin{aligned} P^-(k+1) &= A_{e_d}P(k)A_{e_d}^T + Q(k) \\ \hat{x}^-(k+1) &= A_{e_d}\hat{x}(k) + B_{e_d}v(k) \end{aligned} \tag{6.68}$$

After identifying the disturbance terms, the control input of the inverter is modified as follows:

$$\begin{aligned} v_1 &= \ddot{x}_1 - k_d^1(\dot{x}_1 - \dot{x}_1^d) - k_p^1(x_1 - x_1^d) - \hat{z}_5 \\ v_2 &= \ddot{x}_2 - k_d^2(\dot{x}_2 - \dot{x}_2^d) - k_p^2(x_2 - x_2^d) - \hat{z}_7 \end{aligned} \tag{6.69}$$

6.2.6 Simulation Tests

The efficiency of the inverter's nonlinear control scheme was tested in the tracking of several time-varying setpoints. The associated results are presented in Figs. 6.1, 6.2 and 6.3. It can be noticed that through the proposed control approach, fast and

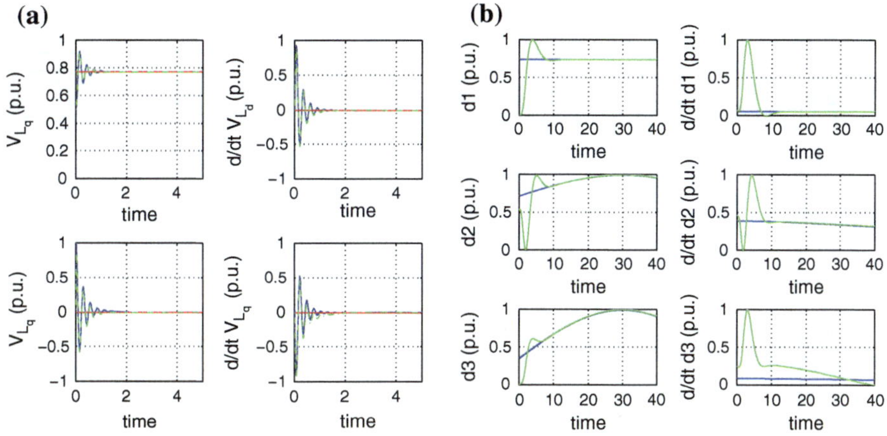

Fig. 6.1 **a** Convergence of the state vector elements $V_{L,d}$ and $V_{L,q}$ of the three-phase inverter to the associated reference setpoints 1. **b** Estimation of disturbance inputs affecting the inverter's model, with the use of the Derivative-free nonlinear Kalman Filter

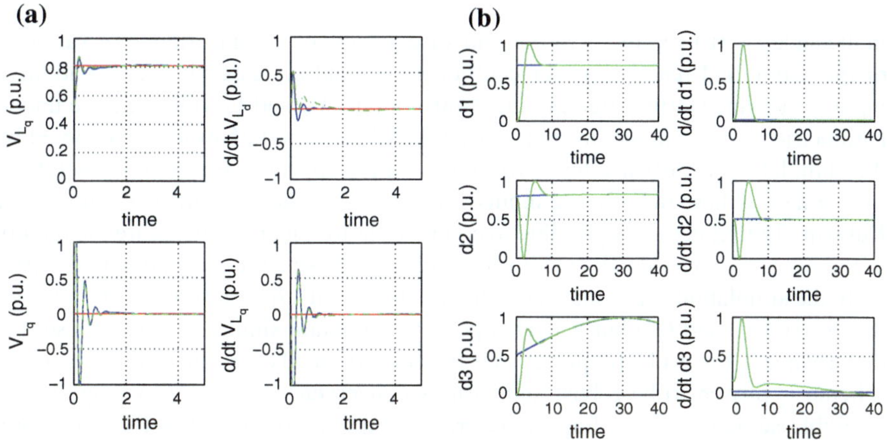

Fig. 6.2 **a** Convergence of the state vector elements $V_{L,d}$ and $V_{L,q}$ of the three-phase inverter to the associated reference setpoints 2. **b** Estimation of disturbance inputs affecting the inverter's model, with the use of the Derivative-free nonlinear Kalman Filter

accurate tracking of the AC voltage reference setpoints can be succeeded. This is shown in Figs. 6.1a, 6.2a and 6.3a. Moreover, it can be noticed that the Derivative-free nonlinear Kalman Filter enables precise estimation and compensation of the modelling uncertainties and the external perturbation terms that affect the inverter's model. This is shown in Fig. 6.1b, 6.2b and 6.3b.

Using the state-space model of Eq. (6.3) as well as the linearized equivalent of Eq. (6.64), the disturbance terms affecting the inverter's model are as follows: (i) parametric uncertainty associated with the values of the capacitance C_f and of the

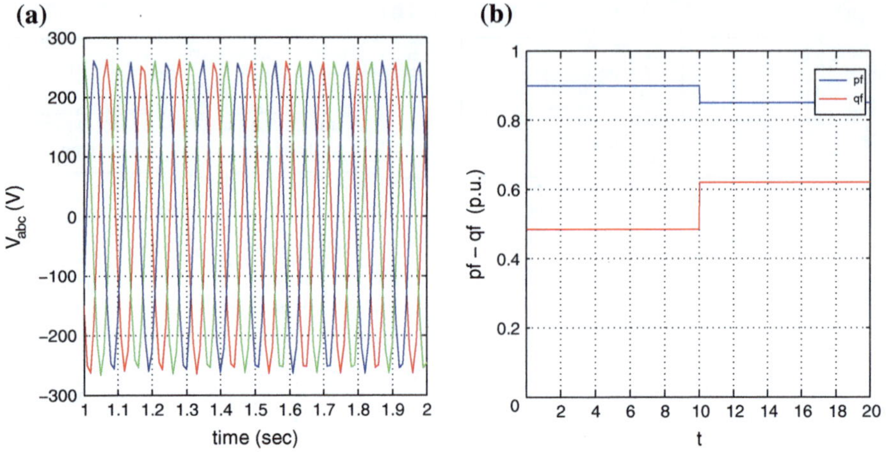

Fig. 6.3 **a** Three-phase voltage variables at the output of the inverter. **b** Active and reactive power of the inverter

inductance L_f, (ii) variations of the grid's frequency ω, (iii) uncertainty or variation of the active and reactive power coefficients p_f and q_f respectively, (iv) additive input disturbance terms affecting the control inputs V_{i_d} and V_{i_q} (these can be due to parametric changes in the inverter's circuit). All these perturbations are denoted by the cumulative disturbance terms $\tilde{d}_1$ and $\tilde{d}_2$ which appear in Eq. (6.64). Therefore, by estimating these disturbance inputs with the use of the Derivative-free nonlinear Kalman Filter, it becomes possible to compensate for them and to assure the quality of the AC power provided by the inverter either to the stand-alone load or to the grid.

In the simulation experiments the disturbance inputs $\tilde{d}_1$ and $\tilde{d}_2$ of Eq. (6.64) were due to (i), (iii) and (iv) but it is also possible to obtain similar results in case that disturbance in the grid's frequency ω is introduced. In all cases the proposed control scheme is proven to be robust and assures convergence of the AC voltage to the desirable setpoints. Regarding the reference setpoints presented in the simulation tests these were sinusoidal voltage signals of variable amplitude and frequency. The disturbance inputs contained also sinusoidal terms of variable amplitude and frequency. It was confirmed that in all cases the control scheme succeeded elimination of the setpoints' tracking error.

From the stability and convergence analysis of the proposed state estimation-based control scheme, it can be assured that both the steady-state and the transient response of the inverter's control loop are satisfactory. The selection of the feedback control gains given in Eq. (6.69) are the ones that determine the closed-loops transient characteristics. Moreover, by choosing the Kalman Filter for estimating the state vector elements which are used by the state-feedback controller, one can assure that undesirable transients will be avoided (unlike for instance the transients generated by sliding-mode observer) [123, 420]. Actually, the Kalman Filter assures fast and smooth elimination of the estimation error (thus also avoiding chattering that usually appears when sliding-mode observers are used in the control loop).

The proposed differential flatness theory-based control and filtering method for three-phase inverters can be generalized towards synchronizing control of parallel inverters connecting distributed DC power units to the grid. This will be analyzed in Chap. 8 The latter problem is particularly significant for integrating into the electricity network renewable energy sources of the DC type such as photovoltaics, while also assuring that frequency and amplitude characteristics of the grid's voltage will not be distorted and that the grid's stability will be also assured [49, 378, 397, 503, 512].

6.3 Flatness-Based Adaptive Control of Active Power Filters

6.3.1 *Overview*

Next, the use of inverters is examined in the problem of improvement of the quality features of the voltage transmitted and distributed through the electricity grid. Due to the connection of nonlinear loads to the power grid, disturbances such as distortion of harmonics are induced to voltage and current waveforms. For the compensation of such perturbations the use of active power filters in the electricity grid has been proposed. Thus, control of active power filters, permitting the current and voltage signals of the power grid to converge to their desirable waveforms, has become a field of intensive research. Identification of voltage harmonics with neural or other signal processing methods has been presented in [95, 142, 173] while identification of harmonics with the use of Kalman Filtering has been analyzed in [210]. The estimated harmonics are used for generating a control input that is re-injected to the grid and which finally eliminates the undesirable harmonics. There have been also several results on adaptive control of shunt active power filters. For instance model reference adaptive control has been presented in [93, 133, 405], while neuro-fuzzy adaptive control has been analyzed in [132–134, 136]. There have been also examples of robust control methods such as sliding mode control [256] and H-infinity control based on output feedback [78]. Additionally, Lyapunov theory-based methods for controller design of active power filters have been shown in [93, 329]. Other types of controllers for this problem are implemented in cascading loops [330], or in multiple SISO loops [314]. Finally one can note control methods in s-frequency domain making use of transfer functions of the linearized dynamics of the active power filter [19].

Aiming at rejecting harmonics' distortions and at preserving the desirable waveforms for the transmitted voltage and currents signals in the electricity grid, the present chapter proposes adaptive fuzzy control of active power filters. The developed control method is based on differential flatness theory [352, 366, 375]. With the use of differential flatness theory the dynamics of the active power filter is transformed into the linear canonical form. After this transformation, the resulting control inputs are shown to contain nonlinear elements which depend on the active power filter's parameters. The nonlinear terms which appear in the control inputs are approximated by neuro-fuzzy networks. It is shown that a suitable learning law can be defined for the aforementioned neuro-fuzzy approximators so as to preserve the closed-loop system stability. With the use of Lyapunov stability analysis it is proven that the proposed adaptive fuzzy control scheme for active power filters results in H_∞ tracking performance.

6.3.2 Dynamic Model of the Active Power Filter

6.3.2.1 Differential Flatness of the Active Power Filter

As previously explained, the flat output of a system is a function of the its state variables and control input variables, thus being written in the form $y_i = h(x, u, \dot{u}, \dots, u^{(r)})$. The MIMO system of the active power filter is said to be differentially flat if there exists a set of flat outputs $y = (y_1, \dots, y_m)$, satisfying the following two conditions [137, 241, 388, 411, 456]:

1. There does not exist any differential relation of the form $R(y, \dot{y}, \dots, y^{(\beta)}) = 0$ which implies that the derivatives of the flat output are not coupled in the sense of an ODE, or equivalently it can be said that the flat output is differentially independent.
2. All system variables (i.e. the elements of the system's state vector w and the control input) can be expressed using only the flat output y and its time derivatives $w_i = \psi_i(y, \dot{y}, \dots, y^{(\gamma_i)}),\ i = 1, \dots, s$.

In the case of the active power filter, the flat output of the system is taken to be the vector

$$y = [y_1, y_2, y_3] = [x_1, x_2, x_3] \tag{6.70}$$

By omitting the effects of disturbances to the model of the active power filter, its state-space description is rewritten as

$$\begin{aligned}
\dot{y}_1 &= -\tfrac{R_c}{L_c}y_1 + \tfrac{V_1}{V_c} - \tfrac{1}{L_c}u_1\\
\dot{y}_2 &= -\tfrac{R_c}{L_c}y_2 + \tfrac{V_2}{V_c} - \tfrac{1}{L_c}u_2\\
\dot{y}_3 &= -\tfrac{R_c}{L_c}y_3 + \tfrac{V_3}{V_c} - \tfrac{1}{L_c}u_3
\end{aligned} \tag{6.71}$$

Obviously, it holds that $u_1 = f_a(y, \dot{y})$, $u_2 = f_b(y, \dot{y})$ and $u_3 = f_c(y, \dot{y})$, therefore all state variables and the control inputs can be written as functions of the flat output and its derivatives. Therefore, the system is a differentially flat one.

Next, by defining the following control inputs

$$\begin{aligned} v_1 &= -\tfrac{R_c}{L_c}y_1 + \tfrac{V_1}{V_c} - \tfrac{1}{L_c}u_1 \\ v_2 &= -\tfrac{R_c}{L_c}y_2 + \tfrac{V_2}{V_c} - \tfrac{1}{L_c}u_2 \\ v_3 &= -\tfrac{R_c}{L_c}y_3 + \tfrac{V_3}{V_c} - \tfrac{1}{L_c}u_3 \end{aligned} \tag{6.72}$$

one has the equivalent dynamics of the system

$$\dot{y}_1 = v_1 \quad \dot{y}_2 = v_2 \quad \dot{y}_3 = v_3 \tag{6.73}$$

Equation (6.73) stands for the MIMO canonical form description of the system's dynamics. Considering that there is no modelling uncertainty or external perturbations, the feedback control input that assures convergence to the reference setpoints $y^d = [y_1^d, y_2^d, y_3^d]$ is

$$\begin{aligned} v_1 &= \dot{y}_1^d - k_1(y_1 - y_1^d) \\ v_2 &= \dot{y}_2^d - k_2(y_2 - y_2^d) \\ v_3 &= \dot{y}_3^d - k_3(y_3 - y_3^d) \end{aligned} \tag{6.74}$$

Defining, $e_i = y_i - y_i^d,\ i = 1, 2, 3$, and for suitable choice of the feedback gains $k_i,\ i = 1, 2, 3$ the previous control input results in the tracking error dynamics

$$\begin{aligned} \dot{e}_1 + k_1 e_1 &= 0 \Rightarrow lim_{t\to\infty} e_1 = 0 \\ \dot{e}_2 + k_1 e_2 &= 0 \Rightarrow lim_{t\to\infty} e_2 = 0 \\ \dot{e}_3 + k_3 e_3 &= 0 \Rightarrow lim_{t\to\infty} e_3 = 0 \end{aligned} \tag{6.75}$$

Knowing that $v = f(x, t) + g(x, t)u$, where $f(x, t) = [f_1, f_2, f_3]^T$ and using that the control inputs gain matrix is $g(x, t) = diag[g_1, g_2, g_3]$, the control input that is actually applied to the active power filter is $u = g^{-1}(x, t)[v - f(x, t)]$.

The problem that arises in the previous analysis is that neither the dynamic model of the active power filter is accurately known (the values of the circuit parameters are not precise), nor the disturbances affecting this model (harmonics distortion) can be directly measured. Therefore, the recommended solution for making the outputs of the active power filter converge to reference setpoints is to apply adaptive control.

6.3.3 *Application if Flatness-Based Adaptive Fuzzy Control to Inverters*

The control signal of the MIMO nonlinear system which has been transformed into the Brunovsky form as described by Eq. (6.73). This also takes the form of Eq. (4.157)

$$y^{(r)} = f(x) + g(x)u + d \tag{6.76}$$

An equivalent states-space description of the system has been shown to be

$$\dot{x} = Ax + B[f(x) + g(x)u + \tilde{d}] \\ y = C^T x \tag{6.77}$$

The control signal $v = f(x) + g(x)u$ contains the unknown nonlinear functions $f(x)$ and $g(x)$ which can be approximated by

$$\hat{f}(x|\theta_f) = \Phi_f(x)\theta_f \\ \hat{g}(x|\theta_g) = \Phi_g(x)\theta_g \tag{6.78}$$

where

$$\Phi_f(x) = (\xi_f^1(x), \xi_f^2(x), \dots \xi_f^n(x))^T \tag{6.79}$$

with $\xi_f^i(x), \imath = 1, \dots, n$ being the vector of kernel functions (e.g. normalized fuzzy Gaussian membership functions), where

$$\xi_f^i(x) = (\phi_f^{i,1}(x), \phi_f^{i,2}(x), \dots, \phi_f^{i,N}(x)) \tag{6.80}$$

thus giving

$$\Phi_f(x) = \begin{pmatrix} \phi_f^{1,1}(x) & \phi_f^{1,2}(x) & \cdots & \phi_f^{1,N}(x) \\ \phi_f^{2,1}(x) & \phi_f^{2,2}(x) & \cdots & \phi_f^{2,N}(x) \\ \cdots & \cdots & \cdots & \cdots \\ \phi_f^{n,1}(x) & \phi_f^{n,2}(x) & \cdots & \phi_f^{n,N}(x) \end{pmatrix} \tag{6.81}$$

while the weights vector is defined as

$$\theta_f{}^T = \left(\theta_f^1, \theta_f^2, \dots \theta_f^N\right) \tag{6.82}$$

$j = 1, \dots, N$ is the number of basis functions that is used to approximate the components of function f which are denoted as $i = 1, \dots, n$. Thus, one obtains the relation of Eq. (6.78), i.e. $\hat{f}(x|\theta_f) = \Phi_f(x)\theta_f$
In a similar manner, for the approximation of function g one has

$$\Phi_g(x) = \left(\xi_g^1(x), \xi_g^2(x), \dots \xi_g^N(x)\right)^T \tag{6.83}$$

with $\xi_g^i(x), \imath = 1, \dots, N$ being the vector of kernel functions (e.g. normalized fuzzy Gaussian membership functions), where

$$\xi_g^i(x) = \left(\phi_g^{i,1}(x), \phi_g^{i,2}(x), \dots, \phi_g^{i,N}(x)\right) \tag{6.84}$$

thus giving

$$\Phi_g(x) = \begin{pmatrix} \phi_g^{1,1}(x) & \phi_g^{1,2}(x) & \cdots & \phi_g^{1,N}(x) \\ \phi_g^{2,1}(x) & \phi_g^{2,2}(x) & \cdots & \phi_g^{2,N}(x) \\ \cdots & \cdots & \cdots & \cdots \\ \phi_g^{n,1}(x) & \phi_g^{n,2}(x) & \cdots & \phi_g^{n,N}(x) \end{pmatrix} \tag{6.85}$$

while the weights vector is defined as

$$\theta_g = \left(\theta_g^1, \theta_g^2, \ldots, \theta_g^p\right)^T \tag{6.86}$$

where the components of matrix θ_g are defined as

$$\theta_g^j = \left(\theta_{g_1}^j, \theta_{g_2}^j, \ldots \theta_{g_N}^j\right) \tag{6.87}$$

$j = 1, \ldots, N$ is the number of basis functions that is used to approximate the components of function g which are denoted as $i = 1, \ldots, n$. Thus one obtains about matrix $\theta_g \in R^{N \times p}$

$$\theta_g = \begin{pmatrix} \theta_{g_1}^1 & \theta_{g_1}^2 & \cdots & \theta_{g_1}^p \\ \theta_{g_2}^1 & \theta_{g_2}^2 & \cdots & \theta_{g_2}^p \\ \cdots & \cdots & \cdots & \cdots \\ \theta_{g_N}^1 & \theta_{g_N}^2 & \cdots & \theta_{g_N}^p \end{pmatrix} \tag{6.88}$$

It holds that

$$g = \begin{pmatrix} g_1 \\ g_2 \\ \cdots \\ g_n \end{pmatrix} = \begin{pmatrix} g_1^1 & g_1^2 & \cdots & g_1^p \\ g_2^1 & g_2^2 & \cdots & g_2^p \\ \cdots & \cdots & \cdots & \cdots \\ g_n^1 & g_n^2 & \cdots & g_n^p \end{pmatrix} \tag{6.89}$$

Using the above, one finally has the relation of Eq. (6.78), i.e. $\hat{g}(x|\theta_g) = \Phi_g(x)\theta_g$. If the state variables of the system are available for measurement then a state-feedback control law can be formulated as

$$u = \hat{g}^{-1}(x|\theta_g)[-\hat{f}(x|\theta_f) + y_m^{(r)} - K^T e + u_c] \tag{6.90}$$

where $\hat{f}(x|\theta_f)$ and $\hat{g}(x|\theta_g)$ are fuzzy models to approximate $f(x)$ and $g(x)$, respectively. u_c is a supervisory control term, e.g. H_∞ control term that is used to compensate for the effects of modelling inaccuracies and external disturbances. Using the system's state-space description of Eq. (6.77) the control term u_c is defined as

$$u_c = -\frac{1}{r} B^T P e \tag{6.91}$$

Moreover, K^T is the feedback gain matrix that assures that the characteristic polynomial of the resulting closed-loop dynamics will be a Hurwitz one.

6.3.4 *Flatness-Based Adaptive Control for Active Power Filters*

Adaptive neurofuzzy control for MIMO nonlinear dynamical systems has been studied in [372, 441]. About the dynamics of the active power filter described in Sect. 6.3.2 and using that its control input is $u \in R^{3\times 1}$ it holds that

$$\begin{aligned}\dot{x}_1 &= f_1(x) + [g_1(x)\ 0\ 0]u\\ \dot{x}_2 &= f_2(x) + [0\ g_2(x)\ 0]u\\ \dot{x}_3 &= f_3(x) + [0\ 0\ g_3(x)]u\end{aligned} \tag{6.92}$$

Next, taking into account also the effects of additive disturbances to the control input of the active filter the dynamic model becomes

$$\begin{aligned}\dot{x}_1 &= f_1(x,t) + [g_1(x,t)\ 0\ 0]u + \tilde{d}_1\\ \dot{x}_2 &= f_2(x,t) + [0\ g_2(x,t)\ 0]u + \tilde{d}_2\\ \dot{x}_3 &= f_3(x,t) + [0\ 0\ g_3(x,t)]u + \tilde{d}_3\end{aligned} \tag{6.93}$$

Next, the following notation is introduced

$$\begin{aligned}g_A &= [g_1(x,t)\ 0\ 0] & \hat{g}_A &= [\hat{g}_1(x,t)\ 0\ 0]\\ g_B &= [0\ g_2(x,t)\ 0] & \hat{g}_B &= [0\ \hat{g}_2(x,t)\ 0]\\ g_C &= [0\ 0\ g_3(x,t)] & \hat{g}_C &= [0\ 0\ \hat{g}_3(x,t)]\end{aligned} \tag{6.94}$$

and thus the state-space description of the active power filter becomes

$$\begin{pmatrix}\dot{x}_1\\ \dot{x}_2\\ \dot{x}_3\end{pmatrix} = \begin{pmatrix}f_1(x,t)\\ f_2(x,t)\\ f_3(x,t)\end{pmatrix} + \begin{pmatrix}g_A\\ g_B\\ g_C\end{pmatrix} u + \begin{pmatrix}\tilde{d}_1\\ \tilde{d}_2\\ \tilde{d}_3\end{pmatrix} \tag{6.95}$$

Considering the vector of the tracking error $e = [x_1 - x_1^d,\ x_2 - x_2^d,\ x_3 - x_3^d]^T$, the following control input is defined

$$u = \begin{pmatrix}\hat{g}_A\\ \hat{g}_B\\ \hat{g}_C\end{pmatrix}^{-1} \cdot \left\{ \begin{pmatrix}\dot{x}_1^d\\ \dot{x}_2^d\\ \dot{x}_3^d\end{pmatrix} - \begin{pmatrix}\hat{f}_1(x,t)\\ \hat{f}_2(x,t)\\ \hat{f}_3(x,t)\end{pmatrix} - \begin{pmatrix}K_1^T\\ K_2^T\\ K_3^T\end{pmatrix} e + \begin{pmatrix}u_{c_1}\\ u_{c_2}\\ u_{c_3}\end{pmatrix} \right\} \tag{6.96}$$

where $[u_{c_1}\ u_{c_2}\ u_{c_3}]^T$ is the robust control term of Eq. (6.91) that is used for the compensation of the model's uncertainties as well as of the external disturbances and the vector of the feedback gain is $K_i^T = [K_1^i, K_2^i, K_3^i]$. Substituting Eq. (6.96) into Eq. (6.95) the closed-loop tracking error dynamics is obtained

$$\begin{pmatrix}\dot{x}_1\\ \dot{x}_2\\ \dot{x}_3\end{pmatrix}=\begin{pmatrix}f_1(x,t)\\ f_2(x,t)\\ f_3(x,t)\end{pmatrix}+\begin{pmatrix}g_A\\ g_B\\ g_C\end{pmatrix}\begin{pmatrix}\hat{g}_A\\ \hat{g}_B\\ \hat{g}_C\end{pmatrix}^{-1}\cdot$$
$$\{\begin{pmatrix}\dot{x}_1^d\\ \dot{x}_2^d\\ \dot{x}_3^d\end{pmatrix}-\begin{pmatrix}\hat{f}_1(x,t)\\ \hat{f}_2(x,t)\\ \hat{f}_3(x,t)\end{pmatrix}-\begin{pmatrix}K_1^T\\ K_2^T\\ K_3^T\end{pmatrix}e+\begin{pmatrix}u_{c_1}\\ u_{c_2}\\ u_{c_3}\end{pmatrix}\}+\begin{pmatrix}\tilde{d}_1\\ \tilde{d}_2\\ \tilde{d}_3\end{pmatrix} \quad (6.97)$$

Equation (6.97) can now be written as

$$\begin{pmatrix}\dot{x}_1\\ \dot{x}_2\\ \dot{x}_3\end{pmatrix}=\begin{pmatrix}f_1(x,t)\\ f_2(x,t)\\ f_3(x,t)\end{pmatrix}+\{\begin{pmatrix}g_A-\hat{g}_A\\ g_B-\hat{g}_B\\ g_C-\hat{g}_C\end{pmatrix}+\begin{pmatrix}\hat{g}_A\\ \hat{g}_B\\ \hat{g}_C\end{pmatrix}\}\cdot$$
$$\cdot\begin{pmatrix}\hat{g}_A\\ \hat{g}_B\\ \hat{g}_C\end{pmatrix}^{-1}\cdot\{\begin{pmatrix}\dot{x}_1^d\\ \dot{x}_2^d\\ \dot{x}_3^d\end{pmatrix}-\begin{pmatrix}\hat{f}_1(x,t)\\ \hat{f}_2(x,t)\\ \hat{f}_3(x,t)\end{pmatrix}-\begin{pmatrix}K_1^T\\ K_2^T\\ K_3^T\end{pmatrix}e+\begin{pmatrix}u_{c_1}\\ u_{c_2}\\ u_{c_3}\end{pmatrix}\}+\begin{pmatrix}\tilde{d}_1\\ \tilde{d}_2\\ \tilde{d}_3\end{pmatrix} \quad (6.98)$$

and using Eq. (6.96) this results into

$$\begin{pmatrix}\dot{e}_1\\ \dot{e}_2\\ \dot{e}_3\end{pmatrix}=\begin{pmatrix}f_1(x,t)-\hat{f}_1(x,t)\\ f_2(x,t)-\hat{f}_2(x,t)\\ f_3(x,t)-\hat{f}_3(x,t)\end{pmatrix}+\begin{pmatrix}g_A-\hat{g}_A\\ g_B-\hat{g}_B\\ g_C-\hat{g}_C\end{pmatrix}\begin{pmatrix}\hat{g}_A\\ \hat{g}_B\\ \hat{g}_C\end{pmatrix}^{-1}u-$$
$$-\begin{pmatrix}K_1^T\\ K_2^T\\ K_3^T\end{pmatrix}e+\begin{pmatrix}u_{c_1}\\ u_{c_2}\\ u_{c_3}\end{pmatrix}+\begin{pmatrix}\tilde{d}_1\\ \tilde{d}_2\\ \tilde{d}_3\end{pmatrix} \quad (6.99)$$

The following description for the approximation error is defined

$$w=\begin{pmatrix}f_1(x,t)-\hat{f}_1(x,t)\\ f_2(x,t)-\hat{f}_2(x,t)\\ f_3(x,t)-\hat{f}_3(x,t)\end{pmatrix}+\begin{pmatrix}g_A-\hat{g}_A\\ g_B-\hat{g}_B\\ g_C-\hat{g}_C\end{pmatrix}\begin{pmatrix}\hat{g}_A\\ \hat{g}_B\\ \hat{g}_C\end{pmatrix}^{-1}u \quad (6.100)$$

Moreover, the following matrices are defined

$$A=\begin{pmatrix}0&0&0\\ 0&0&0\\ 0&0&0\end{pmatrix},\quad B=\begin{pmatrix}1&0&0\\ 0&1&0\\ 0&0&1\end{pmatrix}\quad K^T=\begin{pmatrix}K_1^1&0&0\\ 0&K_2^2&0\\ 0&0&K_3^3\end{pmatrix} \quad (6.101)$$

Using matrices A, B, K^T, Eq. (6.99) is written in the following form

$$\dot{e} = (A - BK^T)e + Bu_c + B\{\begin{pmatrix} f_1(x,t) - \hat{f}_1(x,t) \\ f_2(x,t) - \hat{f}_2(x,t) \\ f_3(x,t) - \hat{f}_3(x,t) \end{pmatrix} + \begin{pmatrix} g_A - \hat{g}_A \\ g_B - \hat{g}_B \\ g_C - \hat{g}_C \end{pmatrix} \begin{pmatrix} \hat{g}_A \\ \hat{g}_B \\ \hat{g}_C \end{pmatrix}^{-1} u + \tilde{d}\} \tag{6.102}$$

Next, the following approximators of the unknown system dynamics are defined

$$\hat{f}(x) = \begin{pmatrix} \hat{f}_1(x|\theta_f) \ x \in R^{3\times 1} \ \hat{f}_1(x|\theta_f) \ \in \ R^{1\times 1} \\ \hat{f}_2(x|\theta_f) \ x \in R^{3\times 1} \ \hat{f}_2(x|\theta_f) \ \in \ R^{1\times 1} \\ \hat{f}_3(x|\theta_f) \ x \in R^{3\times 1} \ \hat{f}_3(x|\theta_f) \ \in \ R^{1\times 1} \end{pmatrix} \tag{6.103}$$

with kernel functions

$$\phi_f^{i,j}(x) = \frac{\prod_{j=1}^{n} \mu_{A_j}^{i}(x_j)}{\sum_{i=1}^{N} \prod_{j=1}^{n} \mu_{A_j}^{i}(x_j)} \tag{6.104}$$

where $l = 1, 2$ and $\mu_{A_j^i}(x)$ is the i-th membership function of the antecedent (IF) part of the l-th fuzzy rule. Similarly, the following approximators of the unknown system dynamics are defined

$$\hat{g}(x) = \begin{pmatrix} \hat{g}_A(x|\theta_g) \ x \in R^{3\times 1} \ \hat{g}_A(x|\theta_g) \ \in \ R^{1\times 3} \\ \hat{g}_B(x|\theta_g) \ x \in R^{3\times 1} \ \hat{g}_B(x|\theta_g) \ \in \ R^{1\times 3} \\ \hat{g}_C(x|\theta_g) \ x \in R^{3\times 1} \ \hat{g}_C(x|\theta_g) \ \in \ R^{1\times 3} \end{pmatrix} \tag{6.105}$$

or equivalently

$$\hat{g}(x) = \begin{pmatrix} \hat{g}_1(x|\theta_g) & 0 & 0 \\ 0 & \hat{g}_2(x|\theta_g) & 0 \\ 0 & 0 & \hat{g}_3(x|\theta_g) \end{pmatrix} \tag{6.106}$$

with $x \in R^{3\times 1}$. The values of the weights that result in optimal approximation are

$$\begin{aligned} \theta_f^* &= arg\ min_{\theta_f \in M_{\theta_f}} [sup_{x \in U_x} (f(x) - \hat{f}(x|\theta_f))] \\ \theta_g^* &= arg\ min_{\theta_g \in M_{\theta_g}} [sup_{x \in U_x} (g(x) - \hat{g}(x|\theta_g))] \end{aligned} \tag{6.107}$$

where the variation ranges for the weights are defined as

$$\begin{aligned} M_{\theta_f} &= \{\theta_f \in R^h : \ ||\theta_f|| \le m_{\theta_f}\} \\ M_{\theta_g} &= \{\theta_g \in R^h : \ ||\theta_g|| \le m_{\theta_g}\} \end{aligned} \tag{6.108}$$

For the value of the approximation error defined in Eq. (6.100) that corresponds to the optimal values of the weights vectors θ_f^* and θ_g^* one has

$$w = \left(f(x,t) - \hat{f}(x|\theta_f^*)\right) + \left(g(x,t) - \hat{g}(x|\theta_g^*)\right) \left(\hat{g}(x,t)\right)^{-1} u \tag{6.109}$$

which is next written as

$$w = \left(f(x,t) - \hat{f}(x|\theta_f) + \hat{f}(x|\theta_f) - \hat{f}(x|\theta_f^*)\right) + \\ + \left(g(x,t) - \hat{g}(x|\theta_g) + \hat{g}(x|\theta_g) - \hat{g}(x|\theta_g^*)\right)\left(\hat{g}(x,t)\right)^{-1} u \tag{6.110}$$

which can be also written in the following form

$$w = \left(w_a + w_b\right) \tag{6.111}$$

where

$$w_a = \{[\hat{f}(x,\theta_f) - \hat{f}(x|\theta_f^*)] + [\hat{g}(x,\theta_g) - \hat{g}(x|\theta_g^*)]\}\cdot\left(\hat{g}(x,t)\right)^{-1} u \tag{6.112}$$

$$w_b = \{[f(x,t) - \hat{f}(x|\theta_f)] + [g(x,t) - \hat{g}(x|\theta_g)]\} \\ \left(\hat{g}(x,t)\right)^{-1} u \tag{6.113}$$

Moreover, the following weights error vectors are defined

$$\tilde{\theta}_f = \theta_f - \theta_f^* \\ \tilde{\theta}_g = \theta_g - \theta_g^* \tag{6.114}$$

6.3.5 Lyapunov Stability Analysis for the Active Power Filter

The following quadratic Lyapunov function for active power systems' control is defined

$$V = \frac{1}{2}e^T P e + \frac{1}{2\gamma_1}\tilde{\theta}_f^T\tilde{\theta}_f + \frac{1}{2\gamma_2}tr[\tilde{\theta}_g^T\tilde{\theta}_g] \tag{6.115}$$

Parameter γ_1 is the learning rate used in the adaptation of the weights of the neurofuzzy approximator for $f(x)$, while parameter γ_2 is the learning rate used in the adaptation of the weights of the neurofuzzy approximation for $g(x)$. It holds that

$$\dot{V} = \tfrac{1}{2}\dot{e}^T P e + \tfrac{1}{2}e^T P\dot{e} + \tfrac{1}{\gamma_1}\dot{\tilde{\theta}}_f^T\tilde{\theta}_f + \tfrac{1}{\gamma_2}tr[\dot{\tilde{\theta}}_g^T\tilde{\theta}_g] \tag{6.116}$$

The tracking error dynamics is described by

$$\dot{e} = (A - BK^T)e + Bu_c + B\{\begin{pmatrix} f_1(x,t) - \hat{f}_1(x,t) \\ f_2(x,t) - \hat{f}_2(x,t) \\ f_3(x,t) - \hat{f}_3(x,t)\end{pmatrix}\} + \begin{pmatrix} g_A - \hat{g}_A \\ g_B - \hat{g}_B \\ g_C - \hat{g}_C\end{pmatrix}\begin{pmatrix}\hat{g}_A \\ \hat{g}_B \\ \hat{g}_C\end{pmatrix}^{-1} u + \tilde{d}\} \tag{6.117}$$

and defining the approximation error

$$w = \begin{pmatrix} f_1(x,t) - \hat{f}_1(x,t) \\ f_2(x,t) - \hat{f}_2(x,t) \\ f_3(x,t) - \hat{f}_3(x,t) \end{pmatrix} + \begin{pmatrix} g_A - \hat{g}_A \\ g_B - \hat{g}_B \\ g_C - \hat{g}_C \end{pmatrix} \begin{pmatrix} \hat{g}_A \\ \hat{g}_B \\ \hat{g}_C \end{pmatrix}^{-1} u \quad (6.118)$$

the previous relation can be also written as

$$\dot{e} = (A - BK^T)e + Bu_c + B(w + \tilde{d}) \quad (6.119)$$

From Eq. (6.116) one obtains

$$\begin{aligned} \dot{V} &= \tfrac{1}{2}\{e^T(A - BK^T)^T + u_c^T B^T + \\ &+ (w + \tilde{d})^T B^T\}Pe + \tfrac{1}{2}e^T P\{(A - BK^T)e + \\ &+ Bu_c + B(w + \tilde{d})\} + \tfrac{1}{\gamma_1}\tilde{\dot{\theta}}_f^T\tilde{\theta}_f + \tfrac{1}{\gamma_2}tr[\dot{\tilde{\theta}}_g^T\tilde{\theta}_g] \end{aligned} \quad (6.120)$$

which in turn gives

$$\begin{aligned} \dot{V} &= \tfrac{1}{2}e^T\{(A - BK^T)^T P + P(A - BK^T)\}e + \\ &\tfrac{1}{2}2e^T PBu_c + \tfrac{1}{2}2B^T Pe(w + \tilde{d}) + \\ &+ \tfrac{1}{\gamma_1}\dot{\tilde{\theta}}_f^T\tilde{\theta}_f + \tfrac{1}{\gamma_2}tr[\dot{\tilde{\theta}}_g^T\tilde{\theta}_g] \end{aligned} \quad (6.121)$$

Assumption 1: For given positive definite matrix Q there exists a positive definite matrix P, which is the solution of the following matrix equation

$$(A - BK^T)^T P + P(A - BK^T) - PB(\tfrac{2}{r} - \tfrac{1}{\rho^2})B^T P + Q = 0 \quad (6.122)$$

Substituting Eqs. (6.122) and (6.91) into $\dot{V}$ yields after some operations

$$\begin{aligned} \dot{V} &= \tfrac{1}{2}e^T\{-Q + PB(\tfrac{2}{r} - \tfrac{1}{\rho^2})B^T P\}e + \\ e^T PB\{-\tfrac{1}{r}B^T Pe\} &+ B^T P(w + \tilde{d}) + \tfrac{1}{\gamma_1}\dot{\tilde{\theta}}_f^T\tilde{\theta}_f + \tfrac{1}{\gamma_2}tr[\dot{\tilde{\theta}}_g^T\tilde{\theta}_g] \end{aligned} \quad (6.123)$$

Therefore it holds

$$\begin{aligned} \dot{V} &= -\tfrac{1}{2}e^T Qe - \tfrac{1}{2\rho^2}e^T PBB^T Pe + e^T PB(w + \tilde{d}) + \\ &\tfrac{1}{\gamma_1}\dot{\tilde{\theta}}_f^T\tilde{\theta}_f + \tfrac{1}{\gamma_2}tr[\dot{\tilde{\theta}}_g^T\tilde{\theta}_g] \end{aligned} \quad (6.124)$$

It also holds that

$$\begin{aligned} \dot{\tilde{\theta}}_f &= \dot{\theta}_f - \dot{\theta}_f^* = \dot{\theta}_f \\ \dot{\tilde{\theta}}_g &= \dot{\theta}_g - \dot{\theta}_g^* = \dot{\theta}_g \end{aligned} \quad (6.125)$$

The following weights adaptation law is used

$$\begin{aligned}\dot{\theta}_f &= -\gamma_1 \Phi(x)^T B^T P e \\ \dot{\theta}_g &= -\gamma_2 \Phi(x)^T B^T P e u^T\end{aligned} \tag{6.126}$$

which stands for a gradient-type law for updating the neurofuzzy approximator's parameters [34, 344]. Assuming N fuzzy rules and associated kernel functions the matrices dimensions are $\theta_f \in R^{N\times 1}$, $\theta_g \in R^{N\times 3}$, $\Phi(x) \in R^{3\times N}$, $B \in R^{3\times 3}$, $P \in R^{3\times 3}$ and $e \in R^{3\times 1}$. Therefore it holds

$$\begin{aligned}\dot{V} = &-\tfrac{1}{2}e^T Q e - \tfrac{1}{2\rho^2}e^T PBB^T P e + e^T PB(w + \tilde{d}) + \\ &+ \tfrac{1}{\gamma_1}(-\gamma_1)e^T PB\Phi(x)(\theta_f - \theta_f^*) + \\ &+ \tfrac{1}{\gamma_2}(-\gamma_2)tr[ue^T PB\Phi(x)(\theta_g - \theta_g^*)]\end{aligned} \tag{6.127}$$

or

$$\begin{aligned}\dot{V} = &-\tfrac{1}{2}e^T Q e - \tfrac{1}{2\rho^2}e^T PBB^T P e + e^T PB(w + \tilde{d}) + \\ &+ \tfrac{1}{\gamma_1}(-\gamma_1)e^T PB\Phi(x)(\theta_f - \theta_f^*) + \\ &+ \tfrac{1}{\gamma_2}(-\gamma_2)tr[ue^T PB(\hat{g}(x|\theta_g) - \hat{g}(x|\theta_g^*)]\end{aligned} \tag{6.128}$$

Taking into account that $u \in R^{2\times 1}$ and $e^T PB(\hat{g}(x|\theta_g) - \hat{g}(x|\theta_g^*)) \in R^{1\times 2}$ it holds

$$\begin{aligned}\dot{V} = &-\tfrac{1}{2}e^T Q e - \tfrac{1}{2\rho^2}e^T PBB^T P e + e^T PB(w + \tilde{d}) + \\ &+ \tfrac{1}{\gamma_1}(-\gamma_1)e^T PB\Phi(x)(\theta_f - \theta_f^*) + \\ &+ \tfrac{1}{\gamma_2}(-\gamma_2)tr[e^T PB(\hat{g}(x|\theta_g) - \hat{g}(x|\theta_g^*))u]\end{aligned} \tag{6.129}$$

Since $e^T PB(\hat{g}(x|\theta_g) - \hat{g}(x|\theta_g^*))u \in R^{1\times 1}$ it holds

$$\begin{aligned}&tr(e^T PB(\hat{g}(x|\theta_g) - \hat{g}(x|\theta_g^*)u) = \\ &= e^T PB(\hat{g}(x|\theta_g) - \hat{g}(x|\theta_g^*))u\end{aligned} \tag{6.130}$$

Therefore, one finally obtains

$$\begin{aligned}\dot{V} = &-\tfrac{1}{2}e^T Q e - \tfrac{1}{2\rho^2}e^T PBB^T P e + e^T PB(w + \tilde{d}) + \\ &+ \tfrac{1}{\gamma_1}(-\gamma_1)e^T PB\Phi(x)(\theta_f - \theta_f^*) + \\ &+ \tfrac{1}{\gamma_2}(-\gamma_2)e^T PB(\hat{g}(x|\theta_g) - \hat{g}(x|\theta_g^*))u\end{aligned} \tag{6.131}$$

Next the following approximation error is defined

$$w_\alpha = [\hat{f}(x|\theta_f) - \hat{f}(x|\theta_f^*)] + [\hat{g}(x|\theta_g) - \hat{g}(x|\theta_g^*)]u \tag{6.132}$$

Thus, one obtains

$$\begin{aligned}\dot{V} = &-\tfrac{1}{2}e^T Q e - \tfrac{1}{2\rho^2}e^T PBB^T P e + \\ &+ e^T PB(w + \tilde{d}) + e^T PB w_\alpha\end{aligned} \tag{6.133}$$

Denoting the aggregate approximation error and disturbances vector as

$$w_1 = w + \tilde{d} + w_\alpha \tag{6.134}$$

the derivative of the Lyapunov function becomes

$$\dot{V} = -\frac{1}{2}e^T Qe - \frac{1}{2\rho^2}e^T PBB^T Pe + e^T PBw_1 \tag{6.135}$$

which in turn is written as

$$\begin{aligned} \dot{V} = -\frac{1}{2}e^T Qe - \frac{1}{2\rho^2}e^T PBB^T Pe + \\ + \frac{1}{2}e^T PBw_1 + \frac{1}{2}w_1^T B^T Pe \end{aligned} \tag{6.136}$$

Next, a Lemma is introduced:

Lemma: The following inequality holds:

$$\frac{1}{2}e^T PBw_1 + \frac{1}{2}w_1^T B^T Pe - \frac{1}{2\rho^2}e^T PBB^T Pe \leq \frac{1}{2}\rho^2 w_1^T w_1 \tag{6.137}$$

Proof: The binomial $(\rho a - \frac{1}{\rho}b)^2 \geq 0$ is considered. Expanding the left part of the above inequality one gets

$$\begin{aligned} \rho^2 a^2 + \frac{1}{\rho^2}b^2 - 2ab \geq 0 \Rightarrow \frac{1}{2}\rho^2 a^2 + \frac{1}{2\rho^2}b^2 - ab \geq 0 \Rightarrow \\ ab - \frac{1}{2\rho^2}b^2 \leq \frac{1}{2}\rho^2 a^2 \Rightarrow \frac{1}{2}ab + \frac{1}{2}ab - \frac{1}{2\rho^2}b^2 \leq \frac{1}{2}\rho^2 a^2 \end{aligned} \tag{6.138}$$

The following substitutions are carried out: $a = w_1$ and $b = e^T PB$ and the previous relation becomes

$$\frac{1}{2}w_1^T B^T Pe + \frac{1}{2}e^T PBw_1 - \frac{1}{2\rho^2}e^T PBB^T Pe \leq \frac{1}{2}\rho^2 w_1^T w_1 \tag{6.139}$$

The previous inequality is used in $\dot{V}$, and the right part of the associated inequality is enforced

$$\dot{V} \leq -\frac{1}{2}e^T Qe + \frac{1}{2}\rho^2 w_1^T w_1 \tag{6.140}$$

The attenuation coefficient ρ can be chosen such that the right part of Eq. (6.140) is always upper bounded by 0. For instance, it suffices at every iteration of the control algorithm to have

$$\begin{aligned} -\frac{1}{2}e^T Qe + \frac{1}{2}\rho^2 ||w_1||^2 \leq 0 \Rightarrow -\frac{1}{2}||e||_Q^2 + \frac{1}{2}\rho^2 ||w_1||^2 \leq 0 \Rightarrow \\ \frac{1}{2}\rho^2 ||w_1||^2 \leq \frac{1}{2}||e||_Q^2 + \Rightarrow \rho^2 \leq \frac{||e||_Q^2}{||w_1||^2} \end{aligned} \tag{6.141}$$

Again without knowledge of the uncertainties and disturbance term $||w_1||$ a sufficiently small value of ρ can assure that the above inequality holds and thus that the loop's stability is assured. Actually, ρ should be given the least value which permits to obtain a solution of the Riccati equation, given in Eq. (6.122).

Equation (6.140) can be used to show that the H_∞ performance criterion is satisfied. The integration of $\dot{V}$ from 0 to T gives

$$\begin{aligned}&\int_0^T \dot{V}(t)dt \leq -\frac{1}{2}\int_0^T ||e||^2dt + \frac{1}{2}\rho^2\int_0^T ||w_1||^2dt \Rightarrow \\ &2V(T) + \int_0^T ||e||_Q^2dt \leq 2V(0) + \rho^2\int_0^T ||w_1||^2dt\end{aligned} \tag{6.142}$$

Moreover, if there exists a positive constant $M_w > 0$ such that

$$\int_0^\infty ||w_1||^2dt \leq M_w \tag{6.143}$$

then one gets

$$\int_0^\infty ||e||_Q^2dt \leq 2V(0) + \rho^2M_w \tag{6.144}$$

Thus, the integral $\int_0^\infty ||e||_Q^2dt$ is bounded and according to Barbalat's Lemma one obtains $lim_{t\rightarrow\infty}e(t) = 0$.

In the presented adaptive fuzzy control approach there is no prior assumption about bounded weights of the neuro-fuzzy approximators. It is the Lyapunov stability analysis that assures that the tracking error of the control loop will follow the H_∞ tracking performance criterion and that the weights values will remain bounded. The only assumption needed for the design of the controller and for succeeding H_∞ tracking performance for the control loop is that there exists a solution for a Riccati equation associated to the linearized error dynamics of the differentially flat model. This assumption is quite reasonable for several nonlinear systems, thus providing a systematic approach to the design of robust flatness-based controllers for dynamical systems characterized by model uncertainty and subjected to external disturbances [171, 172].

6.3.6 Simulation Tests

The efficiency of the proposed adaptive fuzzy control scheme for active power filters has been evaluated through simulation experiments. In the performed tests it has been considered that the dynamic model of the active power filter, as well as the external disturbances that affected the control loop were completely unknown. The obtained results are depicted in Figs. 6.4 and 6.5. The variation of the grid currents is depicted with the blue line, while the associated reference setpoint is depicted with the red line. The simulation experiments have confirmed the findings of the Lyapunov stability analysis and have shown that by rejecting harmonics distortions the lines' currents converged fast and accurately to the reference waveforms.

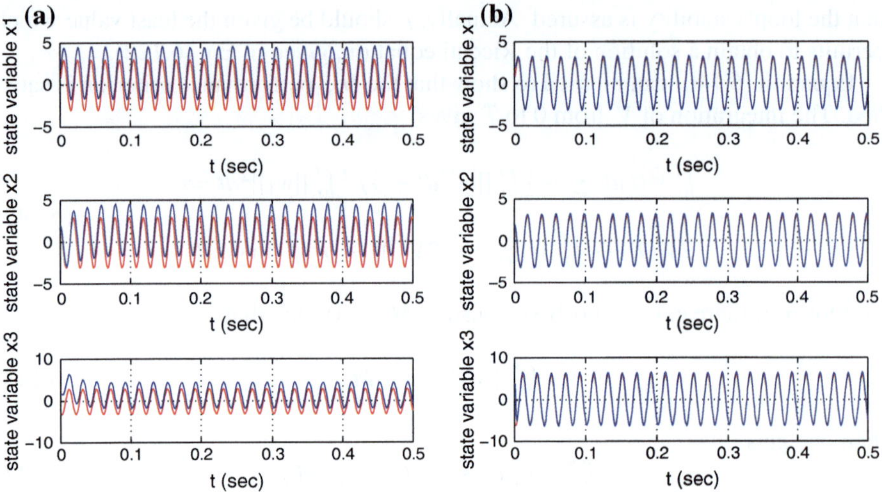

Fig. 6.4 **a** Grid currents I_1, I_2 and I_3 at nominal frequency of 50 Hz and subjected to harmonics distortion. **b** Grid currents I_1, I_2 and I_3 at nominal frequency of 50 Hz after activation of the active power filter and rejection of harmonics distortion

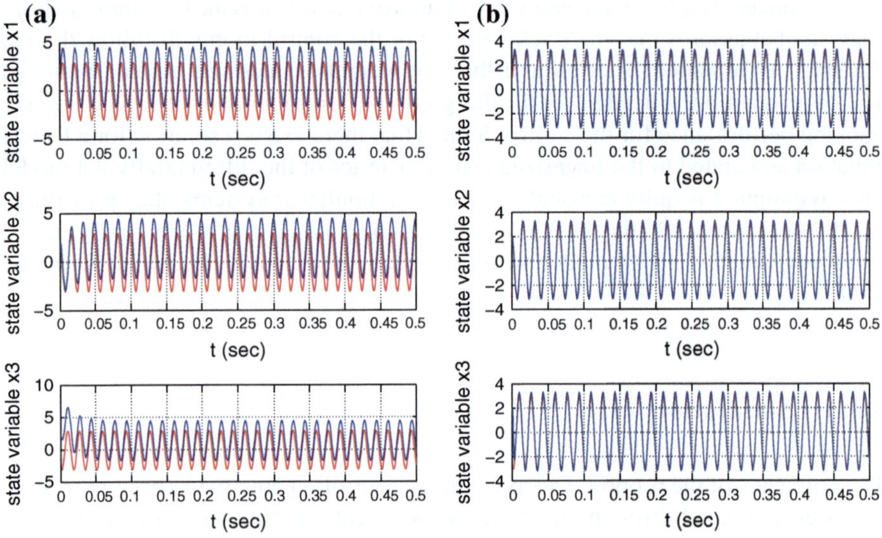

Fig. 6.5 **a** Grid currents I_1, I_2 and I_3 at nominal frequency of 60 Hz and subjected to harmonics distortion. **b** Grid currents I_1, I_2 and I_3 at nominal frequency of 60 Hz after activation of the active power filter and rejection of harmonics distortion

Chapter 7
Control of Fuel Cells and Batteries

7.1 Outline

The chapter is concerned with electrochemical systems used for power generation as well as for power storage. First, the chapter analyzes flatness-based control for proton exchange membrane (PEM) fuel cells. Additionally it gives results on nonlinear H-infinity control of PEM fuel cells. Moreover, it proposes flatenss-based control for the PDE model of the Li-ions batteries.

First, the chapter presents an approach to nonlinear control of fuel cells using differential flatness theory and Kalman Filtering. It is proven that the dynamic model of fuel cells is a differentially flat one which means that all its state variables and control inputs can be expressed as differential functions of specific state vector elements which are the so-called flat outputs of the system. By exploiting the differential flatness properties of the model its transformation to an equivalent linear form (canonical Brunovsky form) becomes possible. For the latter description of the system's dynamics the design of a state-feedback controller is achieved. This control scheme should be also robust to model uncertainties and external perturbations. To cope with this problem the state-space description of the PEM fuel cells is extended by considering as additional state variables the derivatives of the aggregate disturbance input. Next, a Kalman Filter-based disturbance observer is applied to the linearized extended model of the fuel cells. This estimation method enables to identify the disturbance and model uncertainty terms that affect the system and to introduce a complementary control element that compensates for the perturbations' effects.

Moreover, the chapter applies a new nonlinear H-infinity control approach to PEM fuel cells. First, the dynamic model of the PEM fuel cells undergoes approximate linearisation, through Taylor series expansion, round local operating points which are defined at each time instant by the present value of the system's state vector and the last value of the control input that was exerted on it. The linearisation procedure requires the computation of Jacobian matrices at the aforementioned operating points. The modelling error, which is due to the truncation of higher order terms in the Taylor series expansion is perceived as a perturbation that should be compensated by the

G. Rigatos, *Intelligent Renewable Energy Systems*,
Green Energy and Technology, DOI 10.1007/978-3-319-39156-4_7

robustness of the control loop. Next, for the linearized equivalent model of the PEM fuel cells, an H-infinity feedback control loop is designed. This approach, is based on a differential game that takes place between the control input (which tries to minimize the deviation of the state vector from the reference setpoints) and the disturbance input (that tries to maximize it). In such a case, the computation of the optimal control input requires the solution of an algebraic Riccati equation at each iteration of the control algorithm. The known robustness properties of H-infinity control enable compensation of model uncertainty and rejection of the perturbation terms that affect the PEM fuel cells system. The stability of the control loop is proven through Lyapunov analysis. Actually, it is shown that H-infinity tracking performance is achieved, while under moderate conditions the asymptotic stability of the control loop is also demonstrated.

Finally, the chapter analyzes estimation and control for the PDE dynamics describing particles diffusion in Li-ion batteries. It is proven that this PDE dynamics satisfies differential flatness properties and this enables to solve the associated state estimation problem and to design a stabilizing feedback controller. First, by applying semi-discretization and the finite differences method the particles' diffusion PDE model is decomposed into an equivalent set of nonlinear coupled ODEs and a state-space description is obtained. Next, by defining specific state variables as virtual control inputs it is shown that each row of the state-space model is a differentially flat subsystem for which a feedback control law can be found that eliminates the output's tracking error. From the last row of the state-space model, the control input that is finally exerted on the PDE system is computed and this contains recursively all virtual control inputs of the previous rows of the state-space description. Moreover, by exploiting differential flatness properties the state estimation problem for the diffusion PDE model can be solved which consequently means that feedback control can be implemented with the use of a small number of measurements. It is shown that, being differentially flat, the state-space model of the PDE can be transformed to the canonical form and that state-estimation can be performed with the use of the Derivative-free nonlinear Kalman Filter. The filter consists of the Kalman Filter recursion, applied to the canonical form state-space description of the PDE and of a transformation based on differential flatness theory which provides estimates for the initial state variables of the diffusion model. The stability of the PDE control method is proven through Lyapunov analysis.

7.2 Flatness-Based Control of PEM Fuel Cells

7.2.1 Outline

Fuel cells are a form of rapidly expanding renewable energy sources. A fuel cell is an electrochemical energy device that converts the chemical energy of the reaction between hydrogen and oxygen into electricity and heat giving also water as byproduct of the reaction [131, 244, 298, 299, 399, 332, 432, 385]. Fuel cells can significantly contribute to renewable power generation and their use gets widely deployed in the smart grid [100, 450, 227, 283, 257]. In this section a nonlinear feedback control method that is based on differential flatness theory is developed for Proton Exchange Membrane (PEM) Fuel Cells [352, 366, 375]. First it is proven that the dynamic model of the fuel cells is a differentially flat one [137, 240, 241, 388, 411]. This means that all its state variables and its control inputs can be expressed as differential functions of a primary variable which is the so-called flat output. Differential (linear) independence is another property that holds between the flat output and its derivatives. By exploiting differential flatness properties the fuel cells' model can be transformed into an equivalent linearized description which is the canonical Brunovsky form [57, 229, 288, 387, 516]. The latter representation of the system the design of a stabilizing feedback controller becomes possible.

Another problem that has to be dealt with in the deign of the fuel cells' nonlinear controller is that the system is subjected to model uncertainties and external perturbations. To compensate for these disturbances it is proposed to use a Kalman Filter-based disturbance observer in the control loop [355, 374]. The state-space description of the PEM fuel cells is extended by considering as additional state variables the derivatives of the aggregate disturbance input. Through the Kalman Filter recursion and by processing exclusively measurements of the system's output it becomes also possible to identify the perturbation input. The applied Kalman Filter method, also known as Derivative-free nonlinear Kalman Filter consists of the Kalman Filter algorithm applied on the input-output linearized model of the PEM fuel cells. Next, the feedback control law for the system is modified with the inclusion of an additional element which annihilates the aggregate disturbances effects.

7.2.2 Linearization of the Fuel Cells Dynamics

7.2.2.1 Differential Flatness of the PEM Fuel-Cells Model

As already noted in Chap. 1, the dynamics of the fuel cells system is given through the following two sets of differential equations [299]:

Anode mole conservation:

$$\frac{dP_{H_2}}{dt} = \frac{RT}{V_a}[H_{2in} - H_{2used} - H_{2out}] \tag{7.1}$$

$$\frac{dP_{H_2}O_A}{dt} = \frac{RT}{V_a}[H_2O_{in} - H_2O_{used} - H_2O_{out}] \tag{7.2}$$

Cathode mole conservation:

$$\frac{dP_{O_2}}{dt}=\frac{RT}{V_c}[O_{2in}-O_{2used}-O_{2out}] \tag{7.3}$$

$$\frac{dP_{N_2}}{dt}=\frac{RT}{V_c}[N_{2in}-N_{2out}] \tag{7.4}$$

$$\frac{dP_{H_2O_c}}{dt}=\frac{RT}{V_c}[H_2O_{cin}-H_2O_{cproduced}-H_2O_{cout}+H_2O_{cmbr}] \tag{7.5}$$

In the above state equations H_{2in}, O_{2in}, H_2O_{Ain}, N_{2in}, and H_2O_{cin} are the inlet flow rates of hydrogen, oxygen, nitrogen, anode-side water and cathode-side water. Moreover, H_{2out}, O_{2out}, H_2O_{Aout}, N_{2out} and H_2O_{cout} are the outlet flow rates of each reactant. Additionally, H_{2used}, O_{2used} and $H_2O_{cproduced}$ are the usage and produced concentrations of the reactants. Furthermore, H_2O_{mbr} is the water concentration transferred through the membrane and is a function of the stack current and of the humidity (which is assumed to remain constant). It is also noted that V_a is the anode's volume and V_c is the cathode's volume (multiplied by the reactant's mass concentration in mole). The state-space model of PEM fuel-celss has been written in the form

$$\dot{x}_1=c_1(x_4-x_1-x_2-c_2)-\frac{c_3x_1W_{co,out}}{c_4x_1+c_5x_2+c_6}-c_7\zeta \tag{7.6}$$

$$\dot{x}_2=c_8(x_4-x_1-x_2-c_2)-\frac{c_3x_2W_{co,out}}{c_4x_1+c_5x_2+c_6} \tag{7.7}$$

$$\dot{x}_3=-c_9x_3-c_{10}[(\frac{x_4}{c_{11}})^{c_{12}}-1]+c_{13}u \tag{7.8}$$

$$\dot{x}_4=c_{14}\{1+c_{15}[(\frac{x_4}{c_{11}})^{c_{12}}-1]\}\cdot[W_{cp}-c_{16}(x_4-x_1-x_2-c_2)] \tag{7.9}$$

or equivalently in matric form

$$\dot{x}=f(x)+g(x)u\Rightarrow$$

$$\begin{pmatrix}\dot{x}_1\\ \dot{x}_2\\ \dot{x}_3\\ \dot{x}_4\end{pmatrix}=\begin{pmatrix}c_1(x_4-x_1-x_2-c_2)-\frac{c_3x_1W_{co,out}}{c_4x_1+c_5x_2+c_6}-c_7\zeta\\ \dot{x}_2=c_8(x_4-x_1-x_2-c_2)-\frac{c_3x_2W_{co,out}}{c_4x_1+c_5x_2+c_6}\\ \dot{x}_3=-c_9x_3-c_{10}[(\frac{x_4}{c_{11}})^{c_{12}}-1]\\ c_{14}\{1+c_{15}[(\frac{x_4}{c_{11}})^{c_{12}}-1]\}\cdot[W_{cp}-c_{16}(x_4-x_1-x_2-c_2)]\end{pmatrix}+\begin{pmatrix}0\\ 0\\ c_{13}\\ 0\end{pmatrix}u \tag{7.10}$$

It is proven that the dynamic model of the PEM fuel cells given in Eqs. (7.6)–(7.9) is a differentially flat one, which means that all its state variables and its control inputs can be written as differential functions of the flat output and its derivatives. The flat output of the model is taken to be $y = x_1$. Equation (7.6) is solved with respect to x_4. This gives

$$\begin{aligned} x_4 = \frac{1}{(c_4 y + c_5 x_2 + c_6)}\{(c_4 y\dot{y} + c_5\dot{y}x_2 + c_6\dot{y}) - \\ (-c_1 y - c_1 x_2 - c_1 c_2)(c_4 y + c_5 x_2 + c_6) - c_3 y W_{ca,out}\} \Rightarrow \\ x_4 = f_1(y, \dot{y}, x_2) \end{aligned} \tag{7.11}$$

or equivalently

$$x_4 = f_1(y, \dot{y}, x_2) \tag{7.12}$$

By differentiating in time one obtains also

$$\dot{x}_4 = f_1(y, \dot{y}, \ddot{y}, x_2, \dot{x}_2) \tag{7.13}$$

Substituting Eq. (7.12) into Eq. (7.7) one gets

$$\dot{x}_2 = f_3(y, \dot{y}, x_2) \tag{7.14}$$

By differentiating Eq. (7.6) with respect to time one has

$$\ddot{x}_1 = c_1(\dot{x}_4 - \dot{x}_1 - \dot{x}_2) - \frac{c_3\dot{x}_1 W_{ca,out}(c_4 x_1 + c_5 x_2 + c_6) - c_3 x_1 W_{ca,out}(c_4\dot{x}_1 + c_5\dot{x}_2)}{(c_4 x_1 + c_5 x_2 + c_6)^2} \tag{7.15}$$

Using the flat output's notation $y = x_1$ the previous relation becomes

$$\begin{aligned} \ddot{y} = c_1(f_2(y, \dot{y}, \ddot{y}, x_2, f_3(y, \dot{y}, x_2)) - \dot{y} - f_3(y, \dot{y}, x_2)) - \\ - \frac{c_3\dot{y}W_{ca,out}(c_4 y + c_5 x_2 + c_6) - c_3 y W_{ca,out}(c_4\dot{y} + c_5 f_3(y, \dot{y}, x_2))}{(c_4 y + c_5 x_2 + c_6)^2} \end{aligned} \tag{7.16}$$

The above equation provides a relation between x_2 on the one side and y and its derivatives of the other side. Therefore, one obtains

$$x_2 = f_4(y, \dot{y}, \ddot{y}) \tag{7.17}$$

By substituting Eq. (7.17) into Eq. (7.12) one arrives at

$$x_4 = f_2(y, \dot{y}, f_4(y, \dot{y}, \ddot{y})) \tag{7.18}$$

Next, by differentiating in time Eq. (7.18) one obtains

$$\dot{x}_4 = f_5(y, \dot{y}, \ddot{y}, y^{(3)}) \tag{7.19}$$

Thus, one has that state variables x_1, x_2 and x_3 are differential functions of the flat output y and its derivatives. By substituting Eqs. (7.17), (7.18) and $x_1 = y$ into Eq. (7.9), one obtains that x_3 is also a differential function of the flat output. Thus

$$x_3 = \frac{1}{A}\left[\frac{\dot{x}_4}{c_{14}\{1 + c_{15}[(\frac{x_4}{c_{11}})^{c_{12}} - 1]\}} + c_{16}(x_4 - x_1 - x_2 - c_2)\right] \tag{7.20}$$

From Eq. (7.20) one also obtains that x_3 is a differential function of y, and also

$$\dot{x}_3 = f_6(y, \dot{y}, \ddot{y}, y^{(3)}, y^{(4)}) \tag{7.21}$$

Finally, by solving Eq. (7.8) with respect to the control input u one obtains that

$$u = \frac{1}{c_{13}}\{\dot{x}_3 + c_3 x_3 + c_{10}[(\frac{x_4}{c_{11}})^{c_{12}} - 1]\} \tag{7.22}$$

or equivalently

$$u = f_7(y, \dot{y}, \ddot{y}, y^{(3)}, y^{(4)}) \tag{7.23}$$

7.2.2.2 Transformation of the PEM Fuel Cells Model into a Canonical Form

From Eq. (7.6), and after omitting the disturbance term (unknown stack current) one has

$$\dot{x}_1 = c_1(x_4 - x_1 - x_2 - c_2) - \frac{c_3 x_1 W_{co,out}}{c_4 x_1 + c_5 x_2 + c_6} \tag{7.24}$$

By differentiating with respect to time one gets

$$\ddot{x}_1 = c_1(\dot{x}_4 - \dot{x}_1 - \dot{x}_2) - \frac{(c_3 \dot{x}_1 W_{co,out})(c_4 x_1 + c_5 x_2 + c_6) - (c_3 x_1 W_{ca,out})(c_4 \dot{x}_1 + c_5 \dot{x}_2)}{(c_4 x_1 + c_5 x_2 + c_6)^2} \tag{7.25}$$

By substituting in Eq. (7.25) the derivatives $\dot{x}_1$ from Eq. (1.136), $\dot{x}_2$ from Eq. (1.137) and $\dot{x}_4$ from Eq. (1.139) one gets

$$\begin{aligned}
\ddot{x}_1 = {} & c_1(c_{14}[1 + c_{15}[(\tfrac{x_4}{c_{11}})^{c_{12}} - 1]])[A x_3 - c_{16}(x_4 - x_1 - x_2 - c_2)] - \\
& - c_1(x_4 - x_1 - x_2 - c_2) - \tfrac{(c_3 x_1 W_{ca,out})}{(c_4 x_1 + c_5 x_2 + c_6)} - \\
& - c_8(x_4 - x_1 - x_2 - c_2) - \tfrac{(c_3 x_2 W_{ca,out})}{(c_4 x_1 + c_5 x_2 + c_6)} - \\
& \tfrac{-c_3(c_1(x_4 - x_1 - x_2 - c_2) - \frac{c_3 x_1 W_{ca,out}}{c_4 x_1 + c_5 x_2 + c_6}) W_{ca,out}}{(c_4 x_1 + c_5 x_2 + c_6) +} \\
& + \tfrac{(c_3 x_1 W_{ca,out})}{(c_4 x_1 + c_5 x_2 + c_6)^2}[c_4(c_1(x_4 - x_1 - x_2 - c)) - \tfrac{c_3 x_1 W_{ca,out}}{c_4 x_1 + c_5 x_2 + c_6} \\
& + c_5(c_8(x_4 - x_1 - x_2 - c_2)) - \tfrac{c_3 x_2 W_{ca,out}}{c_4 x_1 + c_5 x_2 + c_6}]
\end{aligned} \tag{7.26}$$

By differentiating the previous relation once more with respect to time one gets

$$\begin{aligned}
x_1^{(3)} &= c_1\{c_{14}[c_{15}[(\tfrac{x_4}{c_{11}}{}^{c_{12}-1)}\dot{x}_3)]][Ax_3 - c_{16}(x_4 - x_1 - x_2 - c_2)]+ \\
&c_{14}[1 + c_{15}[(\tfrac{x_4}{c_{11}}) - 1]][A\dot{x}_3 - c_{16}(\dot{x}_4 - \dot{x}_1 - \dot{x}_2)]- \\
&- c_1(\dot{x}_4 - \dot{x}_1 - \dot{x}_2 - \tfrac{(c_3\dot{x}_1 W_{ca,out})(c_4x_1+c_5x_2+c_6)-(c_3x_1W_{ca,out})(c_4\dot{x}_1+c_5\dot{x}_2)}{(c_4x_1+c_5x_2+c_6)^2} \\
&- c_8(\dot{x}_4 - \dot{x}_1 - \dot{x}_2 - \tfrac{(c_3\dot{x}_2 W_{ca,out})(c_4x_1+c_5x_2+c_6)-(c_3x_2W_{ca,out})(c_4\dot{x}_1+c_5\dot{x}_2)}{(c_4x_1+c_5x_2+c_6)^2}\} \\
&- \tfrac{W_{ca,out}}{(c_4x_1++c_5x_2+c_6)^2}\{-c_3(c_1(\dot{x}_4 - \dot{x}_1 - \dot{x}_2)- \\
&- \tfrac{(c_2\dot{x}_1)W_{ca,out}(c_4x_1+c_5x_2+c_6)-(c_3x_1W_{ca,out})(c_4\dot{x}_1+c_5\dot{x}_2)}{(c_4x_1+c_5x_2+c_6)^2}\} \\
&+ \tfrac{(c_3\dot{x}_1 W_{ca,out})}{(c_4x_1+c_5x_2+c_6)^2}[c_4(c_1(x_4 - x_1 - x_2 - c_2) - \tfrac{c_3x_1W_{ca,out}}{c_4x_1+c_5x_2+c_6}) \\
&- c_5(c_8(x_4 - x_1 - x_2 - c_2) - \tfrac{c_3x_2W_{ca,out}}{c_4x_1+c_5x_2+c_6})]+ \\
&+ \tfrac{(c_3x_1 W_{ca,out})}{(c_4x_1+c_5x_2+c_6)^2}[c_4(c_1(\dot{x}_4 - \dot{x}_1 - \dot{x}_2)- \\
&- \tfrac{(c_3\dot{x}_1 W_{ca,out})(c_4x_1+c_5x_2+c_6)-(c_3x_1W_{ca,out})(c_4\dot{x}_1+c_5\dot{x}_2)}{(c_4x_1+c_5x_2+c_6)^2} \\
&c_5(c_8(\dot{x}_4 - \dot{x}_1 - \dot{x}_2) - \tfrac{(c_3\dot{x}_2 W_{ca,out})(c_4x_1+c_5x_2+c_6)-(c_3x_2W_{ca,out})(c_4\dot{x}_1+c_5\dot{x}_2)}{(c_4x_1+c_5x_2+c_6)^2})]- \\
&- \tfrac{(c_3x_1 W_{ca,out})}{c_4x_1+c_5x_2+c_6{}^2}[c_4(c_1(x_4 - x_1 - x_2 - c_2) - \tfrac{c_3x_1W_{ca,out}}{(c_4x_1+c_5x_2+c_6)}] \\
&+ c_5(c_8(x_4 - x_1 - x_2 - c_2) - \tfrac{c_3x_2W_{ca,out}}{c_4x_1+c_5x_2+c_6})]2(c_4\dot{x}_1 + c_5\dot{x}_2)
\end{aligned} \tag{7.27}$$

By substituting in the previous relation $\dot{x}_3$ from Eq. (7.8) an input-output linearized description of the system's dynamics is obtained in the form

$$x_1^{(3)} = \tilde{f}(x) + \tilde{g}(x)u \tag{7.28}$$

where function $\tilde{f}(x)$ is given by

$$\begin{aligned}
\tilde{f}(x) &= c_1\{c_{14}[c_{15}[(\tfrac{x_4}{c_{11}}{}^{c_{12}-1)}\dot{x}_4)]][Ax_3 - c_{16}(x_4 - x_1 - x_2 - c_2)]+ \\
&c_{14}[1 + c_{15}[(\tfrac{x_4}{c_{11}}) - 1]][A(-c_9x_3 - c_{10}[(\tfrac{x_4}{c_{11}})^{c_{12}} - 1]) - c_{16}(\dot{x}_4 - \dot{x}_1 - \dot{x}_2)]- \\
&- c_1(\dot{x}_4 - \dot{x}_1 - \dot{x}_2 - \tfrac{(c_3\dot{x}_1 W_{ca,out})(c_4x_1+c_5x_2+c_6)-(c_3x_1W_{ca,out})(c_4\dot{x}_1+c_5\dot{x}_2)}{(c_4x_1+c_5x_2+c_6)^2} \\
&- c_8(\dot{x}_4 - \dot{x}_1 - \dot{x}_2 - \tfrac{(c_3\dot{x}_2 W_{ca,out})(c_4x_1+c_5x_2+c_6)-(c_3x_2W_{ca,out})(c_4\dot{x}_1+c_5\dot{x}_2)}{(c_4x_1+c_5x_2+c_6)^2}\} \\
&- \tfrac{W_{ca,out}}{(c_4x_1++c_5x_2+c_6)^2}\{-c_3(c_1(\dot{x}_4 - \dot{x}_1 - \dot{x}_2)- \\
&- \tfrac{(c_2\dot{x}_1)W_{ca,out}(c_4x_1+c_5x_2+c_6)-(c_3x_1W_{ca,out})(c_4\dot{x}_1+c_5\dot{x}_2)}{(c_4x_1+c_5x_2+c_6)^2}\} \\
&+ \tfrac{(c_3\dot{x}_1 W_{ca,out})}{(c_4x_1+c_5x_2+c_6)^2}[c_4(c_1(x_4 - x_1 - x_2 - c_2) - \tfrac{c_3x_1W_{ca,out}}{c_4x_1+c_5x_2+c_6})
\end{aligned}$$

$$
\begin{aligned}
&-c_5(c_8(x_4-x_1-x_2-c_2)-\tfrac{c_3x_2W_{ca,out}}{c_4x_1+c_5x_2+c_6})]+\\
&+\tfrac{(c_3x_1W_{ca,out})}{(c_4x_1+c_5x_2+c_6)^2}[c_4(c_1(\dot{x}_4-\dot{x}_1-\dot{x}_2)-\\
&-\tfrac{(c_3\dot{x}_1W_{ca,out})(c_4x_1+c_5x_2+c_6)-(c_3x_1W_{ca,out})(c_4\dot{x}_1+c_5\dot{x}_2)}{(c_4x_1+c_5x_2+c_6)^2}\\
&c_5(c_8(\dot{x}_4-\dot{x}_1-\dot{x}_2)-\tfrac{(c_3\dot{x}_2W_{ca,out})(c_4x_1+c_5x_2+c_6)-(c_3x_2W_{ca,out})(c_4\dot{x}_1+c_5\dot{x}_2)}{(c_4x_1+c_5x_2+c_6)^2})]-\\
&-\tfrac{(c_3x_1W_{ca,out})}{c_4x_1+c_5x_2+c_6{}^2}[c_4(c_1(x_4-x_1-x_2-c_2)-\tfrac{c_3x_1W_{ca,out}}{(c_4x_1+c_5x_2+c_6)}]\\
&+c_5(c_8(x_4-x_1-x_2-c_2)-\tfrac{c_3x_2W_{ca,out}}{c_4x_1+c_5x_2+c_6})]2(c_4\dot{x}_1+c_5\dot{x}_2)
\end{aligned}
\tag{7.29}
$$

and function $\tilde{g}(x)$ is given by

$$\tilde{g}(x)=c_1(c_{14}[1+c_{15}[(\frac{x_4}{c_{11}})^{c_{12}}-1]]Ac_{13}) \tag{7.30}$$

By proving that the PEM fuel-cells system is differentially flat it can be also assured that it can be transformed into an equivalent linearized form (which is the Brunovsky canonical form).

7.2.3 *Linearization of the Fuel Cells Dynamics Using Lie Algebra*

One can attempt linearization of the PEM fuel cells dynamics using also Lie algebra. The linearizing output $z_1=x_1$ is defined. It holds that

$$
\begin{aligned}
z_2=L_fz_1 \Rightarrow z_2=\tfrac{\partial z_1}{\partial x_1}f_1+\tfrac{\partial z_1}{\partial x_2}f_2+\tfrac{\partial z_1}{\partial x_3}f_3+\tfrac{\partial z_1}{\partial x_4}f_4 \Rightarrow\\
z_2=f_1 \Rightarrow z_2=c_1(x_4-x_1-x_2-c_2)-\tfrac{c_3x_1W_{ca,out}}{c_4x_1+c_5x_2+c_6} \Rightarrow\\
z_2=\dot{z}_1
\end{aligned}
\tag{7.31}
$$

Similarly

$$z_3=L_fz_2 \Rightarrow z_3=\frac{\partial z_2}{\partial x_1}f_1+\frac{\partial z_2}{\partial x_2}f_2+\frac{\partial z_2}{\partial x_3}f_3+\frac{\partial z_2}{\partial x_4}f_4 \tag{7.32}$$

which after intermediate operations gives

$$z_3=L_f^2z_1=\dot{z}_2=\ddot{x}_1 \tag{7.33}$$

Equivalently

$$L_f z_3 = \frac{\partial z_3}{\partial x_1} f_1 + \frac{\partial z_3}{\partial x_2} f_2 + \frac{\partial z_3}{\partial x_3} f_3 + \frac{\partial z_3}{\partial x_4} f_4 \tag{7.34}$$

By performing intermediate operations and by using Eq. (7.29) one finds that

$$L_f z_3 = L_f^2 z_1 = \tilde{f}(x) \tag{7.35}$$

Moreover, one finds that

$$L_g z_1 = \frac{\partial z_1}{\partial x_1} g_1 + \frac{\partial z_1}{\partial x_2} g_2 + \frac{\partial z_1}{\partial x_3} g_3 + \frac{\partial z_1}{\partial x_4} g_4 = 0 \tag{7.36}$$

Similarly

$$L_g L_f z_1 = \frac{\partial z_2}{\partial x_1} g_1 + \frac{\partial z_2}{\partial x_2} g_2 + \frac{\partial z_2}{\partial x_3} g_3 + \frac{\partial z_2}{\partial x_4} g_4 = 0 \tag{7.37}$$

Equivalently

$$\begin{array}{c} L_g L_f^2 z_1 = \frac{\partial z_3}{\partial x_1} g_1 + \frac{\partial z_3}{\partial x_2} g_2 + \frac{\partial z_3}{\partial x_3} g_3 + \frac{\partial z_3}{\partial x_4} g_4 \Rightarrow \\ L_g L_f^2 z_1 = \frac{\partial z_3}{\partial x_3} c_{13} \Rightarrow L_g L_f^2 z_1 = \frac{\partial \ddot{x}_1}{\partial x_3} c_{13} \end{array} \tag{7.38}$$

and using the previously computed relation about $\ddot{x}_1$ that was given in Eq. (7.26) one has the result of Eq. (7.30), that is

$$\begin{array}{c} L_g L_f^2 z_1 = c_1 (c_{14} [1 + c_{15} [(\frac{x_4}{c_{11}})^{c_{12}} - 1]] A c_{13}) \Rightarrow \\ L_g L_f^2 z_1 = \tilde{g}(x) \end{array} \tag{7.39}$$

According to the above, the relative degree of the system is $r - 1 = 2 \Rightarrow r = 3$. Consequently, by applying Lie algebra one arrives again at the input-output linearized description of the system

$$\begin{array}{c} z_1^3 = (L_f^3 z_1) + (L_g L_f^2 z_1) u \\ z_1^3 = \tilde{f}(x) + \tilde{g}(x) u \end{array} \tag{7.40}$$

7.2.4 Flatness-Based Control of the Nonlinear Fuel Cells Dynamics

Using the input-output linearized description of the system, that is

$$x_1^{(3)} = \tilde{f}(x) + \tilde{g}(x) u \tag{7.41}$$

and using also that $x_1 = y$, and by defining $v = \tilde{f}(x) + \tilde{g}(x)u$ one gets

$$y^{(3)} = v \tag{7.42}$$

The control input that is actually applied to the PEM fuel-cells system is

$$u = \frac{1}{\tilde{g}(x)}[v - \tilde{f}(x)] \tag{7.43}$$

Then, a stabilizing feedback controller for the system is defined as

$$v = y_d^{(3)} - k_1(\ddot{y} - \ddot{y}_d) - k_2(\dot{y} - \dot{y}_d) - k_3(y - y_d) \tag{7.44}$$

and one gets the closed-loop dynamics

$$\begin{aligned} (y^{(3)} - y_d^{(3)}) + k_1(\ddot{y} - \ddot{y}_d) + k_2(\dot{y} - \dot{y}_d) + k_3(y - y_d) \Rightarrow \\ e^{(3)} + k_1\ddot{e} + k_2\dot{e} + k_3 e = 0 \end{aligned} \tag{7.45}$$

and by choosing the feedback gains $k_i,\ i = 1, 2, 3$ so as the characteristic polynomial associated with the tracking error's differential equation to be a Hurwitz one, gives

$$\begin{aligned} lim_{t\to\infty}e(t) = 0 \Rightarrow lim_{t\to\infty}y(t) = y_d(t) \Rightarrow \\ lim_{t\to\infty}x_1(t) = x_{1,d}(t) \end{aligned} \tag{7.46}$$

Next, the problem of compensation of model uncertainties and external perturbations (such as the unknown stack current $c_7\zeta$ given in Eq. (7.6)) has to be treated. The linearized dynamics of the system is written as

$$y^{(3)} = v + \tilde{d} \tag{7.47}$$

where $\tilde{d}$ represents the cumulative disturbance terms. Without loss of generality it is assumed that the disturbance is modelled by the associated 3rd order derivative, plus initial conditions. Since estimation is going to be performed with the use of the Kalman Filter and the filter's convergence is not dependent on knowledge of initial conditions, the latter can be omitted from the problem's formulation. The following state variables are defined

$$\begin{aligned} z_1 = y \quad z_2 = \dot{y} \quad z_3 = \ddot{y} \\ z_4 = \tilde{d} \quad z_5 = \dot{\tilde{d}} \quad z_6 = \ddot{\tilde{d}} \end{aligned} \tag{7.48}$$

The extended state vector of the system is $Z = [z_1, z_2, \ldots, z_6]^T$. It holds that

$$\begin{pmatrix} \dot{z}_1 \\ \dot{z}_2 \\ \dot{z}_3 \\ \dot{z}_4 \\ \dot{z}_5 \\ \dot{z}_6 \end{pmatrix} = \begin{pmatrix} 0&1&0&0&0&0 \\ 0&0&1&0&0&0 \\ 0&0&0&0&0&0 \\ 0&0&0&0&1&0 \\ 0&0&0&0&0&1 \\ 0&0&0&0&0&0 \end{pmatrix} \begin{pmatrix} z_1 \\ z_2 \\ z_3 \\ z_4 \\ z_5 \\ z_6 \end{pmatrix} + \begin{pmatrix} 0&0 \\ 0&0 \\ 1&0 \\ 0&0 \\ 0&0 \\ 0&1 \end{pmatrix} \begin{pmatrix} v \\ d^{(3)} \end{pmatrix} \tag{7.49}$$

$$z_1 = \begin{pmatrix} 1&0&0&0&0&0 \end{pmatrix} Z \tag{7.50}$$

Defining the extended control input as $\tilde{v} = [v\, \tilde{d}^{(3)}]^T$, the above state-space description of the extended system is also written as

$$\begin{aligned} \dot{Z} &= AZ + B\tilde{v} \\ z_1 &= CZ \end{aligned} \tag{7.51}$$

Next, a disturbance estimator is designed for the extended state-space model of the system. This has the form

$$\begin{aligned} \dot{\hat{z}} &= A_o\hat{z} + B_o v + K_f(z_1 - \hat{z}_1) \\ \hat{z}_1 &= C_o\hat{Z} \end{aligned} \tag{7.52}$$

where $A_o = A$, $C_o = C$ and

$$B_o = \begin{pmatrix} 0&0&1&0&0&0 \end{pmatrix}^T \tag{7.53}$$

By applying common discretization methods, the discrete-time equivalents of matrices A_o, B_o and C_o are obtained. These are written as A_d, B_d and C_d respectively. In this estimation problem the process and measurement noise covariance matrices are denoted as $Q(k)$ and $R(k)$ respectively, while the estimation error's covariance matrix is denoted as $P(k)$. The disturbance estimator's gain is computed with the use of the Kalman Filter recursion [34, 341, 344].

Measurement update:

$$\begin{aligned} K_f(k) &= P^-(k)C_d^T(k)[C_d(k)P^-(k)C_d^T(k) + R(k)]^{-1} \\ \hat{x}(k) &= \hat{x}^-(k) + K_f(k)[z_1(k) - C_d(k)\hat{Z}(k)] \\ P(k) &= P^-(k) - K_f(k)C_d(k)P^-(k) \end{aligned} \tag{7.54}$$

Time update:

$$\begin{aligned} P^-(k+1) &= A_d(k)P(k)A_d(k)^T + Q(k) \\ \hat{x}^-(k+1) &= A_d\hat{x}(k) + B_d v(k) \end{aligned} \tag{7.55}$$

To compensate for the disturbance's effects, the control input that is actually exerted on the system is

$$v^*(k) = v(k) - \hat{d}(k) \tag{7.56}$$

It is noted that the feedback control input is actually computed with the use of the estimated state vector

$$v^*(k) = y_d(k) - k_1(\hat{\ddot{y}} - \ddot{y}_d) - k_2(\hat{\dot{y}} - \dot{y}_d) - k_3(\hat{y} - y_d) - \hat{d}(k) \tag{7.57}$$

7.2.5 Simulation Tests

The performance of the proposed differential flatness theory-based control scheme has been confirmed through simulation experiments. In the results which are presented in Figs. 7.1, 7.2, 7.3 and 7.4 it can noticed that the developed control scheme

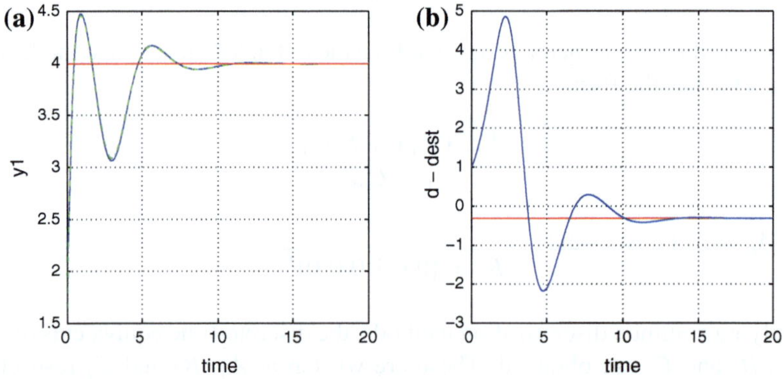

Fig. 7.1 **a** Convergence of state variable $x_1 = P_{O_2}$ (*green line*) to setpoint 1 (*red line*). **b** Kalman Filter-based estimation (*blue line*) of the aggregate disturbance $\tilde{d}$ (*red line*) that affects the PEM fuel cells model

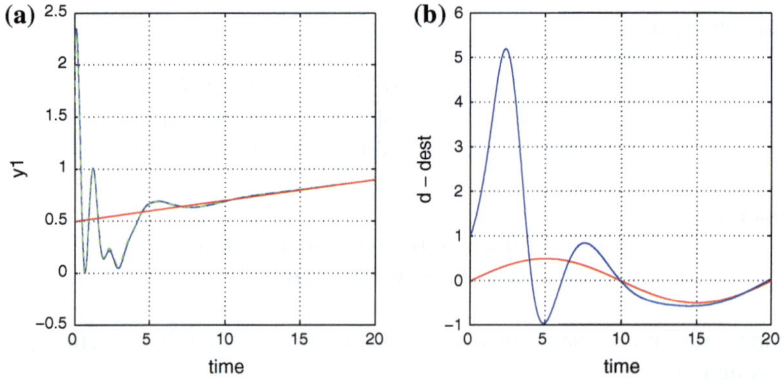

Fig. 7.2 **a** Convergence of state variable $x_1 = P_{O_2}$ (*green line*) to setpoint 2 (*red line*). **b** Kalman Filter-based estimation (*blue line*) of the aggregate disturbance $\tilde{d}$ (*red line*) that affects the PEM fuel cells model

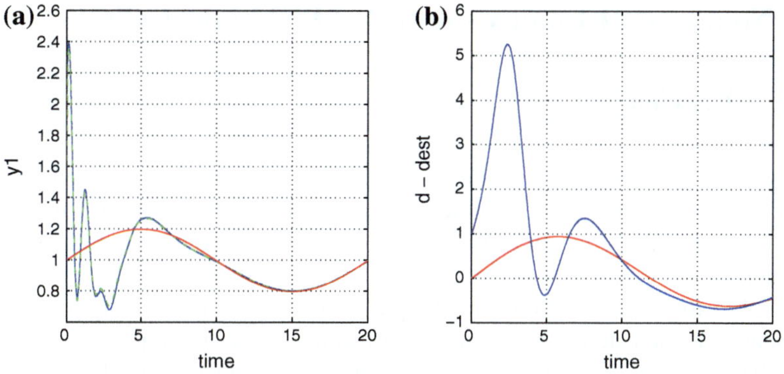

Fig. 7.3 **a** Convergence of state variable $x_1 = P_{O_2}$ (*green line*) to setpoint 3 (*red line*). **b** Kalman Filter-based estimation (*blue line*) of the aggregate disturbance $\tilde{d}$ (*red line*) that affects the PEM fuel cells model

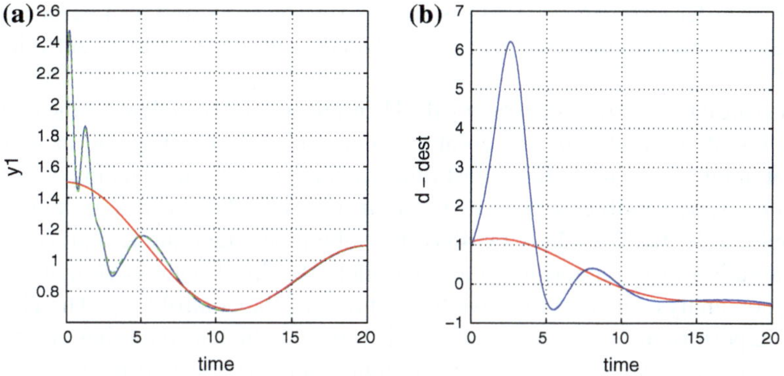

Fig. 7.4 **a** Convergence of state variable $x_1 = P_{O_2}$ (*green line*) to setpoint 4 (*red line*). **b** Kalman Filter-based estimation (*blue line*) of the aggregate disturbance $\tilde{d}$ (*red line*) that affects the PEM fuel cells model

achieves fast and accurate tracking of the reference setpoints. Besides, it can be noticed that the proposed Kalman Filter-based disturbance observer enables to identify fast the aggregate term of model uncertainties and external perturbations that affects the control loop. This permits finally to compensate for the disturbance's effects.

7.3 Nonlinear H-Infinity Control of PEM Fuel Cells

7.3.1 Overview

The problem of nonlinear H-infinity control of PEM fuel cells is solved next. This stands for a different approach to the nonlinear control of such a type of renewable energy systems. The concept of the nonlinear H-infinity control approach has been explained in the case of power generators. First the fuel cells' dynamic model is subjected to local linearization round equilibria. The local operating points, which are updated at each step of the control algorithm, consist of the present value of the fuel cells' state vector and of the last value of the control input that was applied to this system. The local linearization procedure consists also of computation of a Taylor series expansion of the dynamic model of the system and of computation of the associated Jacobian matrices at the local operating points [34, 341, 344]. For the linearized model of the PEM fuel cells an H-infinity nonlinear optimal controller is developed [12, 117, 226, 263, 426]. The modelling error which is due to the truncation of higher order terms in the Taylor series expansion is viewed as a perturbation to the system which is compensated by the robustness of the H-infinity controller [376, 380].

As in the case of power generators, the H-infinity control problem for the linearized dynamics of the fuel cells represents a mini-max differential game in which the control input tries to minimize a cost functional expressing the tracking error for the fuel cells' state variables while the disturbance and model uncertainty input tries to maximize it [110, 327, 328, 443, 484]. By solving an algebraic Riccati equation at each iteration of the algorithm, the controller's feedback gain is computed. Next, Lyapunov analysis is used to prove the stability of the control loop. The H-infinity tracking performance of the control loop is proven, which also implies an improved capability to compensate for model uncertainties and external perturbations affecting the fuel cells' dynamic model. Under moderate conditions, it is also shown that the asymptotic stability of the control loop is assured.

7.3.2 Linearization of the PEM Fuel Cells Model

The PEM fuel cells model (Fig. 7.5) was written in the state-space form

$$\dot{x} = f(x) + g(x){\cdot}u \tag{7.58}$$

where vector fields $f(x)$ and $g(x)$ have been defined as

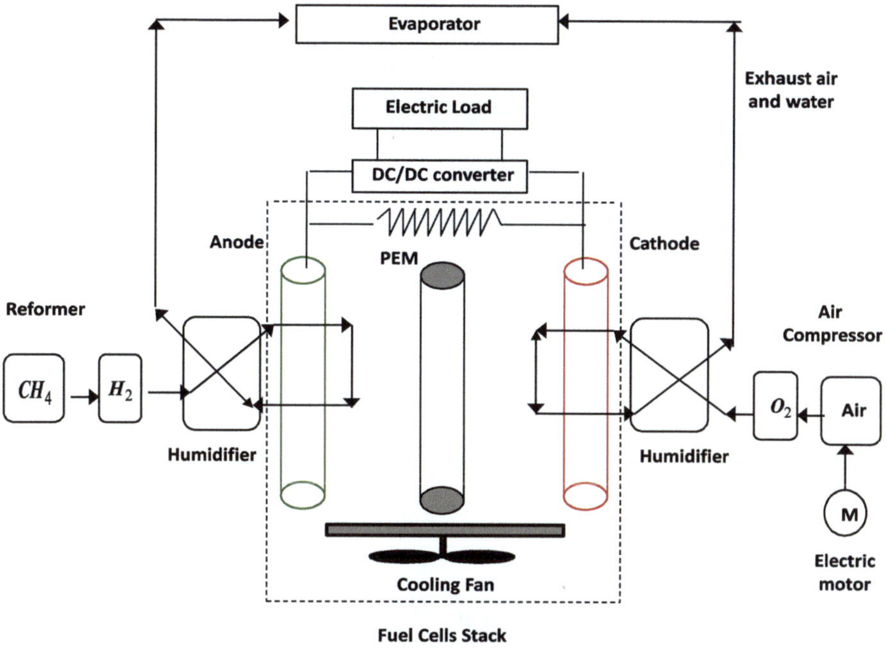

Fig. 7.5 A PEM fuel cells model

$$f(x) = \begin{pmatrix} c_1(x_4 - x_1 - x_2 - c_3) - \frac{c_3 x_1 W_{ca,out}}{c_4 x_1 + c_5 x_2 + c_6} - c_7 \zeta \\ c_8(x_4 - x_1 - x_2 - c_3) - \frac{c_3 x_3 W_{ca,out}}{c_4 x_1 + c_5 x_2 + c_6} \\ -c_3 x_3 - c_{10}[(\frac{x_4}{c_{11}})^{c_{12}} - 1] \\ c_{14}\{1 + c_{15}[(\frac{x_4}{c_{11}})^{c_{12}} - 1]\}[W_{cp} - c_{16}(x_4 - x_1 - x_3 - c_2)] \end{pmatrix} g(x) = \begin{pmatrix} 0 \\ 0 \\ c_{13} \\ 0 \end{pmatrix} \tag{7.59}$$

For the above nonlinear model, the system's Jacobian matrices are computed, through Taylor series expansion, at the operating point (x^*, u^*) where x^* is the present value of the state vector and u^* is the last value of the control input that was exerted on the PEM fuel cells model. Thus, one obtains

$$\begin{aligned} A &= \nabla_x [f(x) + g(x)u] \mid_{(x^*,u^*)} \\ B &= \nabla_u [f(x) + g(x)u] \mid_{(x^*,u^*)} \end{aligned} \tag{7.60}$$

which equivalently gives

$$A = \begin{pmatrix} \frac{\partial f_1}{\partial x_1} & \frac{\partial f_1}{\partial x_2} & \frac{\partial f_1}{\partial x_3} & \frac{\partial f_1}{\partial x_4} \\ \frac{\partial f_2}{\partial x_1} & \frac{\partial f_2}{\partial x_2} & \frac{\partial f_2}{\partial x_3} & \frac{\partial f_2}{\partial x_4} \\ \frac{\partial f_3}{\partial x_1} & \frac{\partial f_3}{\partial x_2} & \frac{\partial f_3}{\partial x_3} & \frac{\partial f_3}{\partial x_4} \\ \frac{\partial f_4}{\partial x_1} & \frac{\partial f_4}{\partial x_2} & \frac{\partial f_4}{\partial x_3} & \frac{\partial f_4}{\partial x_4} \end{pmatrix} \quad B = \begin{pmatrix} 0 \\ 0 \\ c_{13} \\ 0 \end{pmatrix} \tag{7.61}$$

where for the first row of the Jacobian matrix one has

$$\frac{\partial f_1}{\partial x_1} = -c_1 - \frac{(c_3 W_{ca,out})(c_4 x_1 + c_5 x_2 + c_6) - (c_3 x_1 W_{ca,out})c_4}{(c_4 x_1 + c_5 x_2 + c_6)^2}, \ \frac{\partial f_1}{\partial x_2} = -c_1 + \frac{(c_3 x_1 W_{ca,out})c_5}{(c_4 x_1 + c_5 x_2 + c_6)^2} \tag{7.62}$$

$$\frac{\partial f_1}{\partial x_3} = 0, \ \ \frac{\partial f_1}{\partial x_4} = c_1 \tag{7.63}$$

for the second row of the Jacobian matrix one has

$$\frac{\partial f_2}{\partial x_1} = -c_8 + \frac{c_3 x_2 W_{ca,out} c_4}{(c_4 x_1 + c_5 x_2 + c_6)^2}, \ \frac{\partial f_2}{\partial x_2} = -c_8 - \frac{(c_3 W_{ca,out})(c_4 x_1 + c_5 x_2 + c_6) - (c_3 x_2 W_{ca,out})c_5}{(c_4 x_1 + c_5 x_2 + c_6)^2} \tag{7.64}$$

$$\frac{\partial f_2}{\partial x_3} = 0 \ \ \frac{\partial f_2}{\partial x_4} = c_8 \tag{7.65}$$

for the third row of the Jacobian matrix one has

$$\frac{\partial f_3}{\partial x_1} = 0, \ \ \frac{\partial f_3}{\partial x_2} = 0 \tag{7.66}$$

$$\frac{\partial f_3}{\partial x_3} = -c_9, \ \ \frac{\partial f_3}{\partial x_4} = -c_{10} c_{12} (\frac{x_4}{c_{11}})^{c_{12}-1} \tag{7.67}$$

for the fourth row of the Jacobian matrix one has

$$\frac{\partial f_4}{\partial x_1} = c_{14}\{1 + c_{15}[(\frac{x_4}{c_{11}})^{c_{12}} - 1]\}c_{16}, \ \ \frac{\partial f_4}{\partial x_2} = c_{14}\{1 + c_{15}[(\frac{x_4}{c_{11}})^{c_{12}} - 1]\}c_{16} \tag{7.68}$$

$$\frac{\partial f_4}{\partial x_3} = 0, \ \frac{\partial f_4}{\partial x_4} = c_{14}\{c_{15} c_{12} (\frac{x_4}{c_{11}})^{c_{12}-1}\}[W_{cp} - c_{16}(x_4 - x_1 - x_2 - c_2)] - \\ -c_{14}\{1 + c_{15}[(\frac{x_4}{c_{11}}^{c_{12}}) - 1]\}c_{16} \tag{7.69}$$

7.3.3 Design of an H-Infinity Nonlinear Feedback Controller

7.3.3.1 Equivalent Linearized Dynamics of the Fuel Cells

After linearization round its current operating point, the fuel cells' dynamic model is written as

$$\dot{x} = Ax + Bu + d_1 \tag{7.70}$$

Parameter d_1 stands for the linearization error in the fuel cells' dynamic model appearing in Eq. (7.70). The reference setpoints for the fuel cells' state vector are denoted by $\mathbf{x_d} = [x_1^d, \dots, x_4^d]$. Tracking of this trajectory is succeeded after applying the control input u^*. At every time instant the control input u^* is assumed to differ from the control input u appearing in Eq. (7.70) by an amount equal to Δu, that is $u^* = u + \Delta u$

$$\dot{x}_d = Ax_d + Bu^* + d_2 \tag{7.71}$$

The dynamics of the controlled system described in Eq. (7.70) can be also written as

$$\dot{x} = Ax + Bu + Bu^* - Bu^* + d_1 \tag{7.72}$$

and by denoting $d_3 = -Bu^* + d_1$ as an aggregate disturbance term one obtains

$$\dot{x} = Ax + Bu + Bu^* + d_3 \tag{7.73}$$

By subtracting Eq. (7.71) from Eq. (7.73) one has

$$\dot{x} - \dot{x}_d = A(x - x_d) + Bu + d_3 - d_2 \tag{7.74}$$

By denoting the tracking error as $e = x - x_d$ and the aggregate disturbance term as $\tilde{d} = d_3 - d_2$, the tracking error dynamics becomes

$$\dot{e} = Ae + Bu + \tilde{d} \tag{7.75}$$

The above linearized form of the fuel cells' model can be efficiently controlled after applying an H-infinity feedback control scheme.

7.3.3.2 The Nonlinear H-Infinity Control

The initial nonlinear model of the fuel cells is in the form

$$\dot{x} = f(x, u) \;\; x \in R^n, \;\; u \in R^m \tag{7.76}$$

Linearization of the fuel cells' system is performed at each iteration of the control algorithm round its present operating point $(x^*, u^*) = (x(t), u(t - T_s))$. The linearized equivalent of the system is described by

$$\dot{x} = Ax + Bu + L\tilde{d} \;\; x \in R^n, \;\; u \in R^m, \;\; \tilde{d} \in R^q \tag{7.77}$$

where matrices A and B are obtained from the computation of the Jacobians

$$A = \begin{pmatrix} \frac{\partial f_1}{\partial x_1} & \frac{\partial f_1}{\partial x_2} & \cdots & \frac{\partial f_1}{\partial x_n} \\ \frac{\partial f_2}{\partial x_1} & \frac{\partial f_2}{\partial x_2} & \cdots & \frac{\partial f_2}{\partial x_n} \\ \cdots & \cdots & \cdots & \cdots \\ \frac{\partial f_n}{\partial x_1} & \frac{\partial f_n}{\partial x_2} & \cdots & \frac{\partial f_n}{\partial x_n} \end{pmatrix} |_{(x^*,u^*)} B = \begin{pmatrix} \frac{\partial f_1}{\partial u_1} & \frac{\partial f_1}{\partial u_2} & \cdots & \frac{\partial f_1}{\partial u_m} \\ \frac{\partial f_2}{\partial u_1} & \frac{\partial f_2}{\partial u_2} & \cdots & \frac{\partial f_2}{\partial u_m} \\ \cdots & \cdots & \cdots & \cdots \\ \frac{\partial f_n}{\partial u_1} & \frac{\partial f_n}{\partial u_2} & \cdots & \frac{\partial f_n}{\partial u_m} \end{pmatrix} |_{(x^*,u^*)} \tag{7.78}$$

and vector $\tilde{d}$ denotes disturbance terms due to linearization errors. The problem of disturbance rejection for the linearized model that is described by

$$\begin{matrix} \dot{x} = Ax + Bu + L\tilde{d} \\ y = Cx \end{matrix} \tag{7.79}$$

where $x \in R^n$, $u \in R^m$, $\tilde{d} \in R^q$ and $y \in R^p$, cannot be handled efficiently if the classical LQR control scheme is applied. This is because of the existence of the perturbation term $\tilde{d}$. The disturbance term $\tilde{d}$ apart from modeling (parametric) uncertainty and external perturbation terms can also represent noise terms of any distribution.

In the H_∞ control approach, a feedback control scheme is designed for trajectory tracking by the system's state vector and simultaneous disturbance rejection, considering that the disturbance affects the system in the worst possible manner. The disturbances' effects are incorporated in the following quadratic cost function:

$$\begin{matrix} J(t) = \frac{1}{2}\int_0^T [y^T(t)y(t) + \\ +ru^T(t)u(t) - \rho^2\tilde{d}^T(t)\tilde{d}(t)]dt, \;\; r, \rho > 0 \end{matrix} \tag{7.80}$$

As explained in the case of H-infinity control for power generators and converters, the significance of the negative sign in the cost function's term that is associated with the perturbation variable $\tilde{d}(t)$ is that the disturbance tries to maximize the cost function $J(t)$ while the control signal $u(t)$ tries to minimize it. The physical meaning of the relation given above is that the control signal and the disturbances compete to each other within a mini-max differential game. This problem of mini-max optimization can be written as

$$min_u max_{\tilde{d}} J(u, \tilde{d}) \tag{7.81}$$

The objective of the optimization procedure is to compute a control signal $u(t)$ which can compensate for the worst possible disturbance, that is externally imposed to the system. As already analyzed, the solution to the mini-max optimization problem is directly related to the value of the parameter ρ. This means that there is an upper bound in the disturbances' magnitude that can be annihilated by the control signal.

7.3.3.3 Computation of the Feedback Control Gains

For the linearized system given by Eq. (7.79) the cost function of Eq. (7.80) is defined, where the coefficient r determines the penalization of the control input and the weight coefficient ρ determines the reward of the disturbances' effects. Once more it is assumed that: (i) The energy that is transferred from the disturbances signal $\tilde{d}(t)$ is bounded, that is $\int_0^{\infty}\tilde{d}^T(t)\tilde{d}(t)dt < \infty$, (ii) matrices $[A, B]$ and $[A, L]$ are stabilizable, (iii) matrix $[A, C]$ is detectable. Then, the optimal feedback control law is given by

$$u(t) = -Kx(t) \tag{7.82}$$

with

$$K = \frac{1}{r}B^T P \tag{7.83}$$

where P is a positive semi-definite symmetric matrix which is obtained from the solution of the Riccati equation

$$A^T P + PA + Q - P(\frac{1}{r}BB^T - \frac{1}{2\rho^2}LL^T)P = 0 \tag{7.84}$$

where Q is also a positive definite symmetric matrix. The worst case disturbance is given by

$$\tilde{d}(t) = \frac{1}{\rho^2}L^T Px(t) \tag{7.85}$$

The diagram of the considered control loop is depicted in Fig. 7.6.

7.3.4 *Lyapunov Stability Analysis*

Lyapunov stability analysis for fuel cells is as in the case of H-infinity control of power generators and converters. Through Lyapunov stability analysis it will be shown that the proposed nonlinear control scheme assures H_∞ tracking performance for the fuel cells' model, and that in case of bounded disturbance terms asymptotic convergence to the reference setpoints is succeeded.

The tracking error dynamics for the fuel cells' model is written in the form

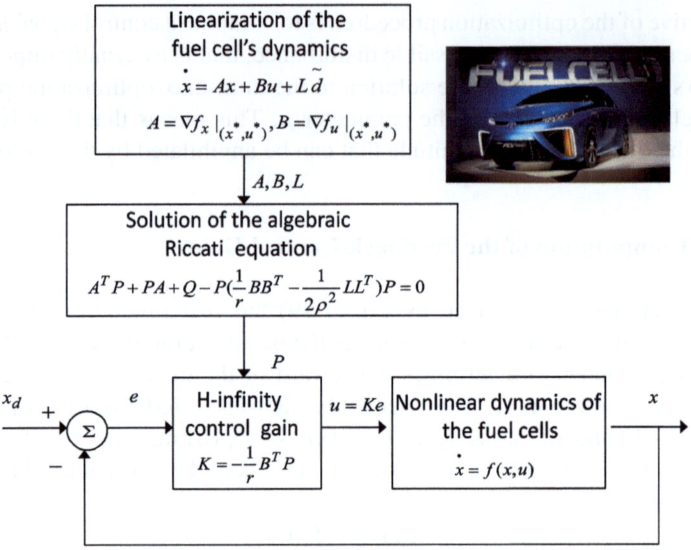

Fig. 7.6 Diagram of the control scheme for the fuel cells

$$\dot{e}=Ae+Bu+L\tilde{d} \tag{7.86}$$

where in the fuel cells' case $L=I\in R^2$ with I being the identity matrix. Variable $\tilde{d}$ denotes model uncertainties and external disturbances of the fuel cells' model. The following Lyapunov equation is considered

$$V=\frac{1}{2}e^TPe \tag{7.87}$$

where $e=x-x_d$ is the tracking error. By differentiating with respect to time one obtains

$$\begin{aligned}\dot{V}&=\tfrac{1}{2}\dot{e}^TPe+\tfrac{1}{2}eP\dot{e}\Rightarrow\\ \dot{V}&=\tfrac{1}{2}[Ae+Bu+L\tilde{d}]^TPe+\tfrac{1}{2}e^TP[Ae+Bu+L\tilde{d}]\Rightarrow\end{aligned} \tag{7.88}$$

$$\begin{aligned}\dot{V}=\tfrac{1}{2}[e^TA^T+u^TB^T+\tilde{d}^TL^T]Pe+\\ +\tfrac{1}{2}e^TP[Ae+Bu+L\tilde{d}]\Rightarrow\end{aligned} \tag{7.89}$$

$$\begin{aligned}\dot{V}=\tfrac{1}{2}e^TA^TPe+\tfrac{1}{2}u^TB^TPe+\tfrac{1}{2}\tilde{d}^TL^TPe+\\ \tfrac{1}{2}e^TPAe+\tfrac{1}{2}e^TPBu+\tfrac{1}{2}e^TPL\tilde{d}\end{aligned} \tag{7.90}$$

The previous equation is rewritten as

$$\dot{V} = \tfrac{1}{2}e^T(A^TP + PA)e + (\tfrac{1}{2}u^TB^TPe + \tfrac{1}{2}e^TPBu) + \\ +(\tfrac{1}{2}\tilde{d}^TL^TPe + \tfrac{1}{2}e^TPL\tilde{d}) \tag{7.91}$$

Assumption: For given positive definite matrix Q and coefficients r and ρ there exists a positive definite matrix P, which is the solution of the following matrix equation

$$A^TP + PA = -Q + P(\frac{2}{r}BB^T - \frac{1}{\rho^2}LL^T)P \tag{7.92}$$

Moreover, the following feedback control law is applied to the system

$$u = -\frac{1}{r}B^TPe \tag{7.93}$$

By substituting Eqs. (7.92) and (7.93) one obtains

$$\dot{V} = \frac{1}{2}e^T[-Q + P(\frac{2}{r}BB^T - \frac{1}{\rho^2}LL^T)P]e + \\ + e^TPB(-\frac{1}{r}B^TPe) + e^TPL\tilde{d} \Rightarrow \tag{7.94}$$

$$\dot{V} = -\frac{1}{2}e^TQe + \frac{1}{r}PBB^TPe - \frac{1}{2\rho^2}e^TPLL^TPe \\ - \frac{1}{r}e^TPBB^TPe) + e^TPL\tilde{d} \tag{7.95}$$

which after intermediate operations gives

$$\dot{V} = -\frac{1}{2}e^TQe - \frac{1}{2\rho^2}e^TPLL^TPe + e^TPL\tilde{d} \tag{7.96}$$

or, equivalently

$$\dot{V} = -\frac{1}{2}e^TQe - \frac{1}{2\rho^2}e^TPLL^TPe + \\ + \frac{1}{2}e^TPL\tilde{d} + \frac{1}{2}\tilde{d}^TL^TPe \tag{7.97}$$

Lemma: The following inequality holds

$$\frac{1}{2}e^TL\tilde{d} + \frac{1}{2}\tilde{d}L^TPe - \frac{1}{2\rho^2}e^TPLL^TPe \le \frac{1}{2}\rho^2\tilde{d}^T\tilde{d} \tag{7.98}$$

Proof The binomial $(\rho\alpha - \frac{1}{\rho}b)^2$ is considered. Expanding the left part of the above inequality one gets

$$\rho^2a^2 + \frac{1}{\rho^2}b^2 - 2ab \geq 0 \Rightarrow \frac{1}{2}\rho^2a^2 + \frac{1}{2\rho^2}b^2 - ab \geq 0 \Rightarrow$$
$$ab - \frac{1}{2\rho^2}b^2 \leq \frac{1}{2}\rho^2a^2 \Rightarrow \frac{1}{2}ab + \frac{1}{2}ab - \frac{1}{2\rho^2}b^2 \leq \frac{1}{2}\rho^2a^2 \tag{7.99}$$

The following substitutions are carried out: $a = \tilde{d}$ and $b = e^TPL$ and the previous relation becomes

$$\frac{1}{2}\tilde{d}^TL^TPe + \frac{1}{2}e^TPL\tilde{d} - \frac{1}{2\rho^2}e^TPLL^TPe \leq \frac{1}{2}\rho^2\tilde{d}^T\tilde{d} \tag{7.100}$$

Eq. (7.100) is substituted in Eq. (7.97) and the inequality is enforced,thus giving

$$\dot{V} \leq -\frac{1}{2}e^TQe + \frac{1}{2}\rho^2\tilde{d}^T\tilde{d} \tag{7.101}$$

Eq. (7.101) shows that the H_∞ tracking performance criterion is satisfied. If the attenuation coefficient ρ is given a sufficiently small value the condition of Eq. (7.101) becomes also an asymptotic stability condition for the closed-loop system. Equivalently, the integration of $\dot{V}$ from 0 to T gives

$$\int_0^T \dot{V}(t)dt \leq -\frac{1}{2}\int_0^T ||e||_Q^2 dt + \frac{1}{2}\rho^2\int_0^T ||\tilde{d}||^2 dt \Rightarrow$$
$$2V(T) + \int_0^T ||e||_Q^2 dt \leq 2V(0) + \rho^2\int_0^T ||\tilde{d}||^2 dt \tag{7.102}$$

Moreover, if there exists a positive constant $M_d > 0$ such that

$$\int_0^\infty ||\tilde{d}||^2 dt \leq M_d \tag{7.103}$$

then one gets

$$\int_0^\infty ||e||_Q^2 dt \leq 2V(0) + \rho^2 M_d \tag{7.104}$$

Thus, the integral $\int_0^\infty ||e||_Q^2 dt$ is bounded. Moreover, $V(T)$ is bounded and from the definition of the Lyapunov function V in Eq. (7.87) it becomes clear that $e(t)$ will be also bounded since $e(t) \in \Omega_e = \{e | e^TPe \leq 2V(0) + \rho^2M_d\}$. According to the above and with the use of Barbalat's Lemma one obtains $lim_{t\to\infty}e(t) = 0$. □

7.3.5 Simulation Tests

The efficiency of the proposed nonlinear H-infinity control method has been tested in the case of tracking of several setpoints. The obtained results have been given in Figs. 7.7, 7.8, 7.9, 7.10, 7.11, 7.12, 7.13, 7.14, 7.15, 7.16, 7.17, 7.18, 7.19, 7.20 and 7.21. It can be observed, that the control scheme succeeds fast and accurate convergence to the reference setpoints while all state variables exhibit smooth transients. Besides, no abrupt variation of the control inputs have been noted. Despite the simplicity of the control concept, the performance of the control loop was indeed satisfactory. The controller was robust to modelling errors caused by the approximate linearization of the fuel cells dynamics and was robust to parametric changes or external disturbances affecting this system. The accuracy of tracking of the reference setpoints by the system's state variables is also outlined in Table 7.1. These

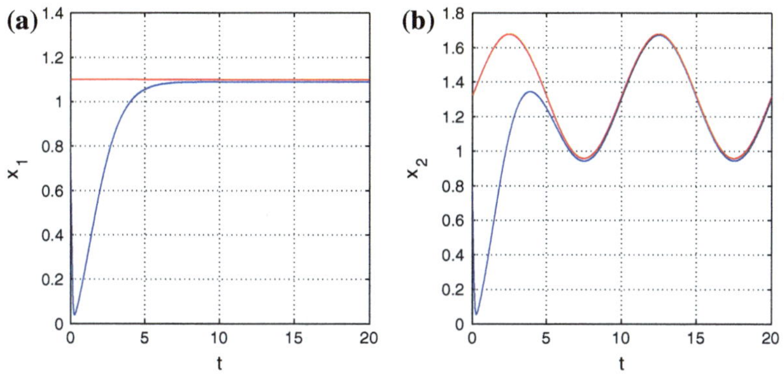

Fig. 7.7 **a** Convergence of state variable $x_1 = P_{O_2}$ (*green line*) to setpoint 1 (*red line*). **b** Convergence of state variable $x_2 = P_{H_2}$ (*green line*) to setpoint 1 (*red line*)

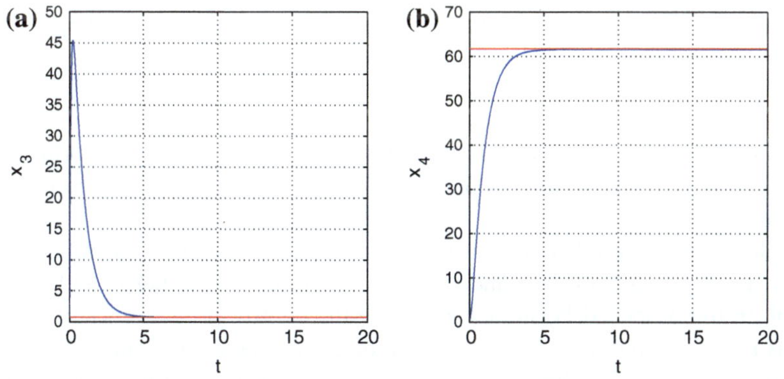

Fig. 7.8 **a** Convergence of state variable $x_3 = \omega_{c_p}$ (*green line*) to setpoint 1 (*red line*). **b** Convergence of state variable $x_4 = p_{s_m}$ (*green line*) to setpoint 1 (*red line*)

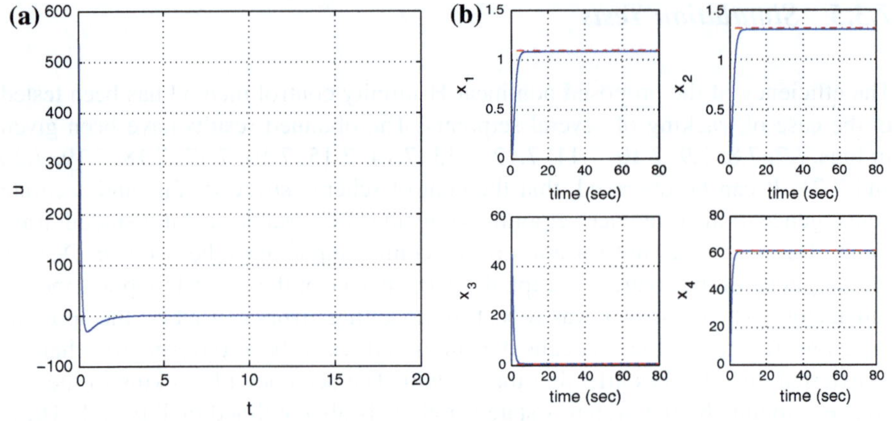

Fig. 7.9 **a** Control input u exerted on the PEM fuel cells model in the case of setpoint 1. **b** Aggregate diagram of state variables control in the case of setpoint 1

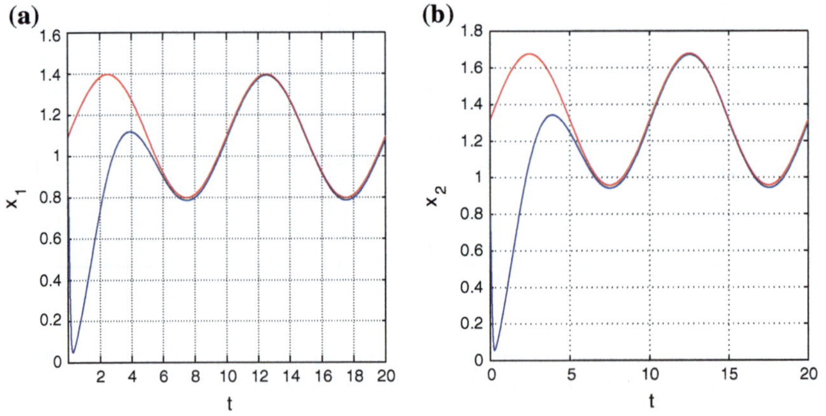

Fig. 7.10 **a** Convergence of state variable $x_1 = P_{O_2}$ (*green line*) to setpoint 2 (*red line*). **b** Convergence of state variable $x_2 = P_{H_2}$ (*green line*) to setpoint 2 (*red line*)

results are only indicative of the capabilities of the proposed control algorithm, while suitable tuning of the controller's parameters can further improve the control loop's performance.

Unlike to model predictive control approaches, the nonlinear H-infinity control method is of proven stability and robustness. Moreover, unlike model predictive control, in the proposed H-infinity approach the control problem is solved in continuous time, without need to use to use a discrete-time equivalent of the system's dynamics. Additional features of the proposed nonlinear H-infinity control method for fuel cells are summarized as follows: (i) the new control approach does not make use of linearizing transformations (diffeomorphisms) of the fuel cells' description

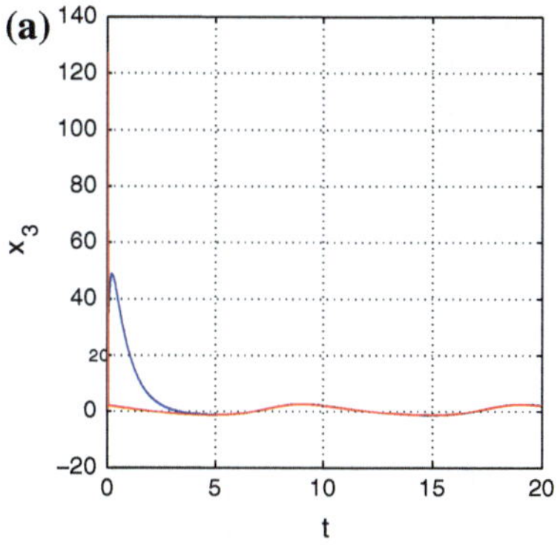

Fig. 7.11 **a** Convergence of state variable $x_3 = \omega_{c_p}$ (*green line*) to setpoint 2 (*red line*). **b** Convergence of state variable $x_4 = p_{s_m}$ (*green line*) to setpoint 2 (*red line*)

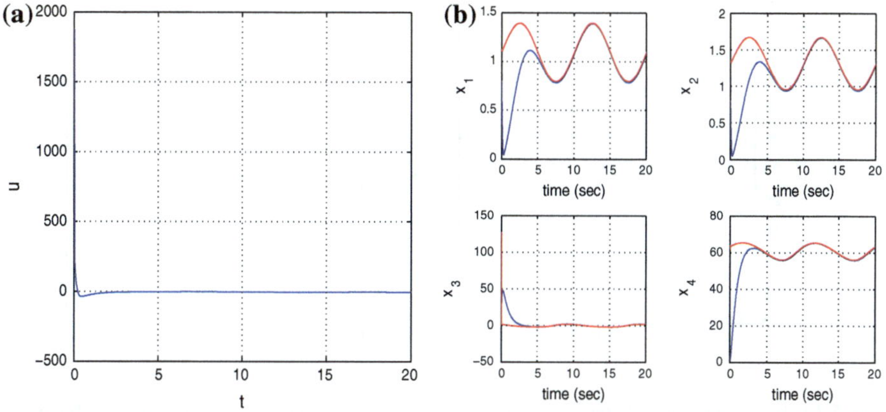

Fig. 7.12 **a** Control input u exerted on the PEM fuel cells model in the case of setpoint 2. **b** Aggregate diagram of state variables control in the case of setpoint 2

but is applied directly on the fuel cells' nonlinear model, (ii) the H-infinity control approach follows optimal control theory for the computation of the control signal (iii) the new control approach is based on local linearization and on the computation of Jacobian matrices in the fuel cells' dynamic model, (iv) the computation of the controller's feedback gains is based on the solution of an algebraic Riccati equation and this is performed at each stage of the control algorithm, (v) yet simple the control method provides good tracking accuracy and exhibits robustness to modelling errors and external perturbations.

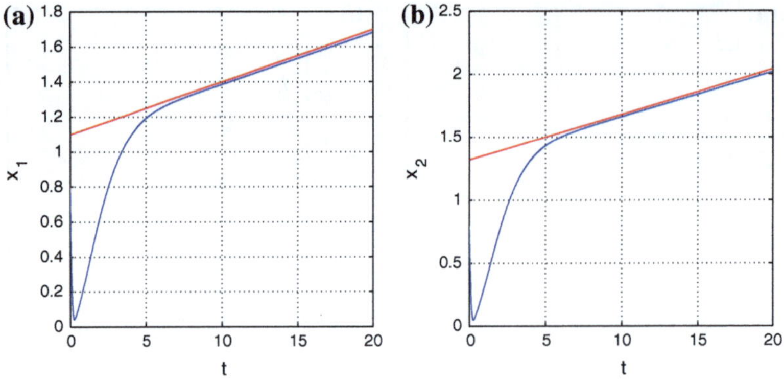

Fig. 7.13 **a** Convergence of state variable $x_1 = P_{O_2}$ (*green line*) to setpoint 2 (*red line*). **b** Convergence of state variable $x_2 = P_{H_2}$ (*green line*) to setpoint 2 (*red line*)

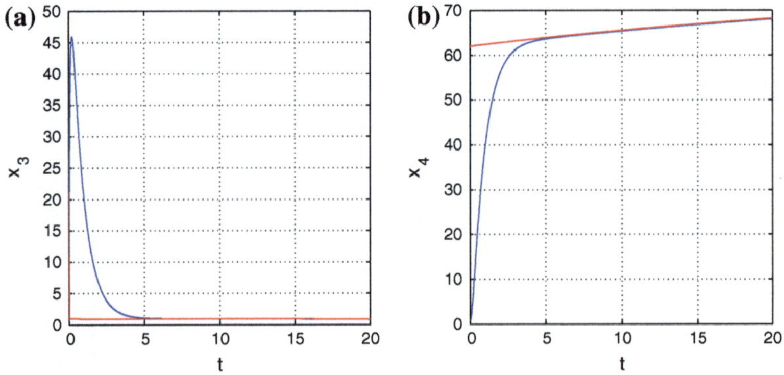

Fig. 7.14 **a** Convergence of state variable $x_3 = \omega_{c_p}$ (*green line*) to setpoint 3 (*red line*). **b** Convergence of state variable $x_4 = p_{s_m}$ (*green line*) to setpoint 3 (*red line*)

7.4 Control of the Diffusion PDE in Li-ion Batteries

Up to now the dynamics of renewable energy systems was taken to be described by nonlinear differential equations. Next, control for renewable energy sources described by PDEs will be analyzed. The need for more efficient power storage has led to the development of new control and estimation methods of the batteries' dynamics which are not based on lumped parameter models but which make use of distributed parameter (PDE) models. Specific interest has been shown in Lithium-ion batteries because these exhibit higher power density than other types of batteries, low self-discharge and prolonged life cycle [97, 218, 416, 417]. The distributed parameter models of Li-ion batteries comprise several coupled PDEs which describe the diffusion of Li ions at the solid electrodes as well as their diffusion in the electrolyte [81, 319, 418, 507]. Control of these diffusion phenomena, by applying external

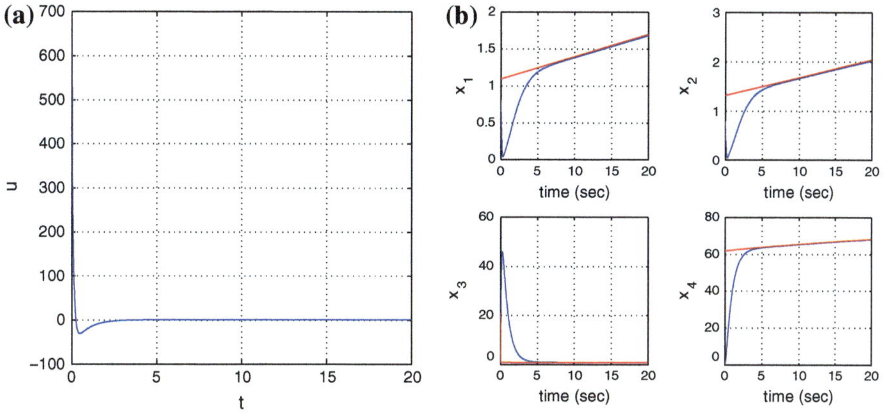

Fig. 7.15 **a** Control input u exerted on the PEM fuel cells model in the case of setpoint 3. **b** Aggregate diagram of state variables control in the case of setpoint 3

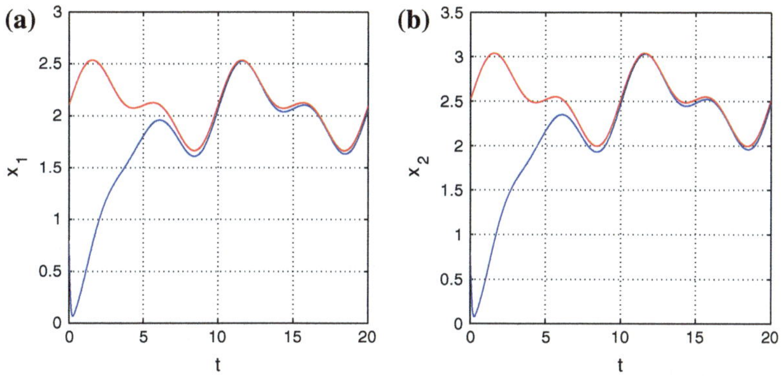

Fig. 7.16 **a** Convergence of state variable $x_1 = P_{O_2}$ (*green line*) to setpoint 4 (*red line*). **b** Convergence of state variable $x_2 = P_{H_2}$ (*green line*) to setpoint 4 (*red line*)

voltage or currents can result in making the battery dynamics follow specific charge or discharge patterns as well to keep its functioning within desirable charge levels. Obviously, this can further improve the performance indexes of Li-ion batteries.

At a first stage the present section develops a solution to the control problem of diffusion dynamics of Li-ion batteries. It is proven that this diffusion PDE dynamics satisfies differential flatness properties and this makes possible to design a stabilizing feedback controller. The present section applied semi-discretization for the numerical representation of the PDE dynamics, which is an approach also implemented in [163, 222, 316, 323, 448]. Methods for feedback stabilization of systems with nonlinear PDE dynamics have been a flourishing research subject in the last years [31, 55, 255, 419]. In particular, feedback control of diffusion-type (parabolic) PDEs has been a subject of extensive research and several remarkable results have been produced

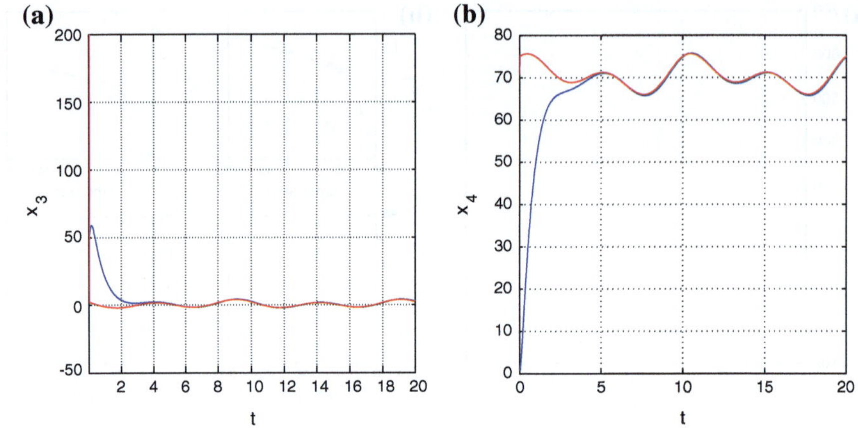

Fig. 7.17 **a** Convergence of state variable $x_3 = \omega_{c_p}$ (*green line*) to setpoint 4 (*red line*). **b** Convergence of state variable $x_4 = p_{s_m}$ (*green line*) to setpoint 4 (*red line*)

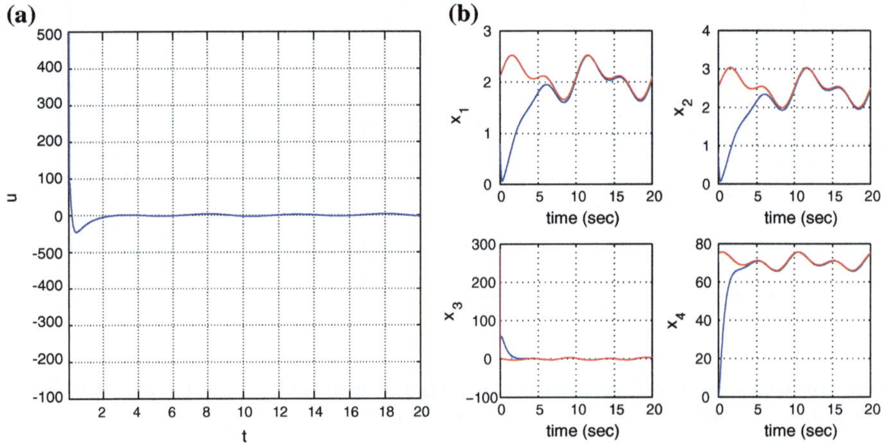

Fig. 7.18 **a** Control input u exerted on the PEM fuel cells model in the case of setpoint 4. **b** Aggregate diagram of state variables control in the case of setpoint 4

[269, 297, 471, 520]. For the control of diffusion PDEs, boundary and distributed control methods have been developed [43, 58, 138, 228, 469, 470].

The control approach followed in this section is as follows. At a first stage, by implementing a finite differences procedure a set of equivalent nonlinear ordinary differential equations is obtained for the diffusion PDE in the Li-ions model. Next, it is shown that the system of the nonlinear ODEs is a differentially flat one. This means that all its state variables and the control inputs can be written as differential functions of one single algebraic variable which is the flat output [57, 240, 241, 375, 388, 411]. By defining specific state variables as virtual control inputs it is shown that each

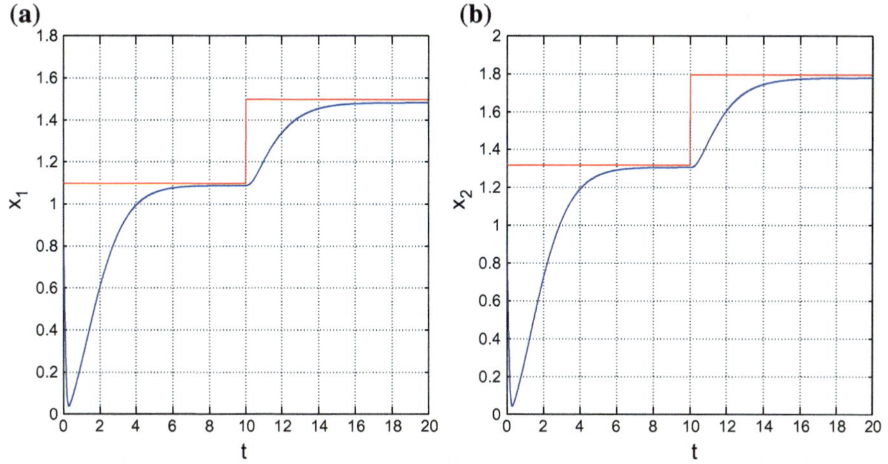

Fig. 7.19 **a** Convergence of state variable $x_1 = P_{O_2}$ (*green line*) to setpoint 5 (*red line*). **b** Convergence of state variable $x_2 = P_{H_2}$ (*green line*) to setpoint 5 (*red line*)

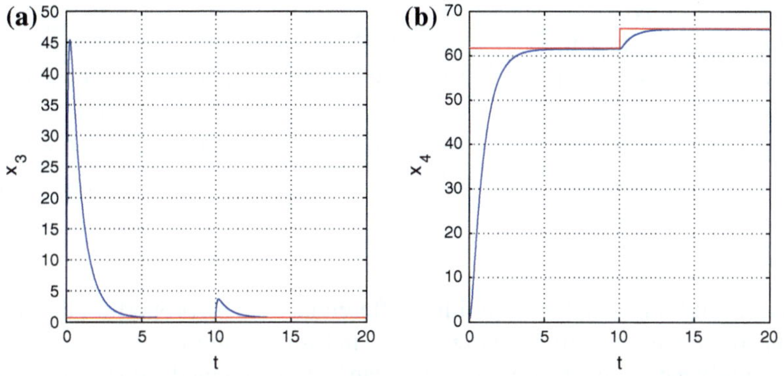

Fig. 7.20 **a** Convergence of state variable $x_3 = \omega_{c_p}$ (*green line*) to setpoint 5 (*red line*). **b** Convergence of state variable $x_4 = p_{s_m}$ (*green line*) to setpoint 5 (*red line*)

row of the state-space model is a differentially flat subsystem for which a feedback control law can be found that achieves elimination of the output's tracking error. From the last row of the state-space model, the control input that is finally exerted on the PDE system is computed and this contains recursively all virtual control inputs of the previous rows of the state-space description. Therefore, by tracing the rows of the state-space model backwards on can calculate the control input that should be applied to the PDE system so as to achieve that all its state vector elements track the reference setpoints.

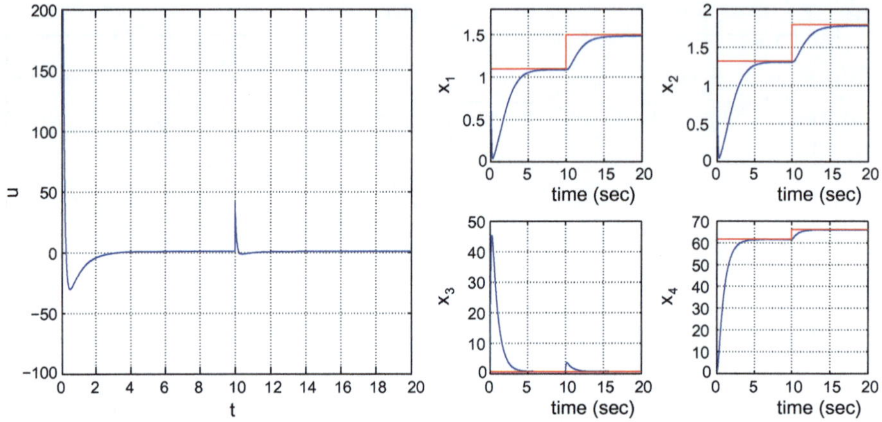

Fig. 7.21 **a** Control input u exerted on the PEM fuel cells model in the case of setpoint 5. **b** Aggregate diagram of state variables control in the case of setpoint 5

Table 7.1 RMSE of the fuel cell's state variables

Parameter	p_{O_2}	p_{H_2}	ω_{cp}	p_{sm}
$RMSE_1$	7.40×10^{-5}	8.88×10^{-5}	1.08×10^{-5}	7.91×10^{-4}
$RMSE_2$	9.82×10^{-5}	1.17×10^{-4}	4.35×10^{-4}	14.0×10^{-4}
$RMSE_3$	7.94×10^{-5}	9.53×10^{-5}	7.52×10^{-6}	6.76×10^{-4}
$RMSE_4$	2.34×10^{-4}	2.81×10^{-4}	9.16×10^{-4}	18.0×10^{-4}
$RMSE_5$	1.86×10^{-4}	2.24×10^{-4}	1.48×10^{-5}	6.64×10^{-4}

At a second stage, and by exploiting differential flatness properties, the state estimation problem for the diffusion PDE model in Li-ions batteries can be solved, which consequently means that feedback control can be implemented with the use of a small number of measurements. It is shown that, being differentially flat, the state-space model of the PDE can be transformed into the canonical form and that state-estimation can be performed with the use of the Derivative-free nonlinear Kalman Filter. As previously noted, this filter consists of the Kalman Filter recursion, applied to the canonical form state-space description of the PDE and of an inverse transformation based on differential flatness theory which provides estimates for the initial state variables of the diffusion model [34, 352, 366, 375]. By performing state estimation it is possible to reconstruct the entire PDE dynamics after taking measurements at a small number of points in the diffusion's spatial grid.

7.4.1 Modeling in State-Space Form of the Li-ions Diffusion PDE

In the previously given diffusion PDE of Eq. (1.141), which describes the concentration of Li ions on the solid electrode one has

$$\frac{\partial c_s}{\partial t} = \frac{D_s}{r^2}\frac{\partial}{\partial r}(r^2\frac{\partial c_s}{\partial r}) \tag{7.105}$$

which in turn is written as

$$\frac{\partial c_s}{\partial t} = \frac{2D_s}{r}\frac{\partial c_s}{\partial r} + D_s\frac{\partial^2 c_s}{\partial r^2} \tag{7.106}$$

which can be also written in the form

$$\frac{\partial c_s}{\partial t} = D_s\nabla^2 c_s + \frac{2D_s}{r}\nabla c_s \tag{7.107}$$

A grid of N points is considered, that is $\{r_1, r_2, \ldots, r_{N-1}, r_N\}$ which are placed at equal distances on the r axis. At the points of spatial discretization it holds

$$\frac{\partial c_s}{\partial t} = D_s\frac{c_{s,i+1} - 2c_{s,i} + c_{s,i-1}}{\Delta r^2} + \frac{2D_s}{r}\frac{c_{s,i+1} - c_{s,i}}{\Delta r} \tag{7.108}$$

where the boundary conditions are $c_s(r,t)|_{r=0} = 0$ and $c_s(r,t)|_{r=R_s}$. The second boundary conditions is taken to act on the system as a control input.
For the i-th ODE one has at sampling point r_i along the r axis one has

$$\frac{\partial c_s}{\partial t} = \frac{Ds}{\Delta r^2}c_{s,i+1} - \frac{2Ds}{\Delta r^2}c_{s,i} + \frac{Ds}{\Delta r^2}c_{s,i-1} + \frac{2D_s}{r}\frac{c_{s,i+1}}{\Delta r} - \frac{2D_s}{r}\frac{c_{s,i}}{\Delta r} \tag{7.109}$$

Eq. (7.109) is also written in the form

$$\frac{\partial c_s}{\partial t} = \frac{Ds}{\Delta r^2}c_{s,i+1} - \frac{2Ds}{\Delta r^2}c_{s,i} + \frac{Ds}{\Delta r^2}c_{s,i-1} + \frac{2D_s}{r}[\frac{c_{s,i+1} - c_{s,i}}{\Delta r}] \tag{7.110}$$

By denoting $K_1 = D_s$, $K_2 = -2D_s$ and $f(c_{s,i}) = \frac{2D_s}{r}[\frac{c_{s,i+1}-c_{s,i}}{\Delta r}]$ or equivalently $f(c_{s,i}) = \frac{\partial c_s}{\partial r}|_{r=r_i}$ one obtains the following description for Eq. (7.110)

$$\frac{\partial c_s}{\partial t} = \frac{K_1}{\Delta r^2}c_{s,i-1} + \frac{K_2}{\Delta r^2}c_{s,i} + \frac{K_1}{\Delta r^2}c_{s,i+1} + f(c_{s,i}) \tag{7.111}$$

7.4.2 Differential Flatness of the Battery's PDE Diffusion Model

Next, the following state vector $\tilde{Y} = [y_{1,1}, y_{1,2}, \ldots, y_{1,i}, \ldots, y_{1,N-1}, y_{1,N}]$ is defined for the PDE model, where $y_{1,1} = c_{s,1}, y_{1,2} = c_{s,2}, \ldots, y_{1,i} = c_{s,i}, \ldots, y_{1,N-1} = c_{s,N-1}$ and $y_{1,N} = c_{s,N}$. It will be shown that the state-space description of the nonlinear PDE dynamics, which has as control input only the boundary condition $c_{s,N+1}$ is differentially flat. One has

$$\dot{y}_{1,1} = \frac{K_1}{\Delta r^2}y_{1,0} + \frac{K_2}{\Delta r^2}y_{1,1} + \frac{K_1}{\Delta r^2}y_{1,2} + f(y_{1,1}) \tag{7.112}$$

$$\dot{y}_{1,2} = \frac{K_1}{\Delta r^2}y_{1,1} + \frac{K_2}{\Delta r^2}y_{1,2} + \frac{K_1}{\Delta r^2}y_{1,3} + f(y_{1,2}) \tag{7.113}$$

$$\dot{y}_{1,3} = \tfrac{K_1}{\Delta r^2}y_{1,2} + \tfrac{K_2}{\Delta r^2}y_{1,3} + \tfrac{K_1}{\Delta r^2}y_{1,4} + f(y_{1,3}) \tag{7.114}$$

$$\cdots\ \cdots$$

$$\dot{y}_{1,i} = \tfrac{K_1}{\Delta r^2}y_{1,i-1} + \tfrac{K_2}{\Delta r^2}y_{1,i} + \tfrac{K_1}{\Delta r^2}y_{1,i+1} + f(y_{1,i}) \tag{7.115}$$

$$\cdots\ \cdots$$

$$\dot{y}_{1,N-1} = \frac{K_1}{\Delta r^2}y_{1,N-2} + \frac{K_2}{\Delta r^2}y_{1,N-1} + \frac{K_1}{\Delta r^2}y_{1,N} + f(y_{1,N-1}) \tag{7.116}$$

$$\dot{y}_{1,N} = \frac{K_1}{\Delta r^2}y_{1,N-1} + \frac{K_2}{\Delta r^2}y_{1,N} + \frac{K_1}{\Delta r^2}y_{1,N+1} + f(y_{1,N}) \tag{7.117}$$

Additionally, it is considered that outside the grid of the N points the value of c_s becomes 0, therefore $y_{1,0} = 0$ while $y_{1,N+1}$ acts as control input. The flat output is considered to be the state variable $y_{1,1}$, which is denoted as $y = y_{1,1}$. Next, it is shown that all state variables which stand also for virtual control inputs of the system $\alpha_i = y_{1,i-1}$, can be written as functions of the flat output $y = y_{1,1}$. From Eq. (7.112) and considering that $y_{1,0} = 0$ one obtains

$$\begin{aligned}\alpha_1 = y_{1,2} = \tfrac{1}{K_1/\Delta r^2}[\dot{y}_{1,1} - \tfrac{K_1}{\Delta r^2}y_{1,0} - \tfrac{K_2}{\Delta r^2}y_{1,1} - f(y_{1,1})] \\ \Rightarrow y_{1,2} = h_1(y, \dot{y}, \ldots)\end{aligned} \tag{7.118}$$

and following a similar procedure, from Eq. (7.113) one gets

$$\begin{aligned}\alpha_2 = y_{1,3} = \tfrac{1}{K_1/\Delta r^2}[\dot{y}_{1,2} - \tfrac{K_1}{\Delta r^2}y_{1,1} - \tfrac{K_2}{\Delta r^2}y_{1,2} - f(y_{1,2})] \\ \Rightarrow \alpha_2 = y_{1,3} = h_2(y, \dot{y}, \ldots)\end{aligned} \tag{7.119}$$

Continuing in a similar manner, from Eq. (7.114) one obtains

$$\alpha_3 = y_{1,4} = \frac{1}{K_1/\Delta r^2}[\dot{y}_{1,3} - \frac{K_1}{\Delta r^2}y_{1,2} - \frac{K_2}{\Delta r^2}y_{1,3} - f(y_{1,3})] \\ \Rightarrow \alpha_3 = y_{1,4} = h_3(y, \dot{y}, \ldots) \tag{7.120}$$

From Eq. (7.115) one obtains

$$\alpha_i = y_{1,i+1} = \frac{1}{K_1/\Delta r^2}[\dot{y}_{1,i} - \frac{K_1}{\Delta r^2}y_{1,i-1} - \frac{K_2}{\Delta r^2}y_{1,i} - f(y_{1,i})] \\ \Rightarrow \alpha_i = y_{1,i+1} = h_i(y, \dot{y}, \ldots) \tag{7.121}$$

From Eq. (7.116) one obtains

$$\alpha_{N-1} = y_{1,N-2} = \frac{1}{K_1/\Delta r^2}[\dot{y}_{1,N-1} - \frac{K_2}{\Delta r^2}y_{1,N-1} - \frac{K_1}{\Delta r^2}y_{1,N} - f(y_{1,N-1})] \\ \Rightarrow \alpha_{N-1} = y_{1,N-2} = h_{N-1}(y, \dot{y}, \ldots) \tag{7.122}$$

Finally, from Eq. (7.117) one obtains

$$\alpha_N = y_{1,N+1} = \frac{1}{K_1/\Delta r^2}[\dot{y}_{1,N} - \frac{K_1}{\Delta r^2}y_{1,N-1} - \frac{K_2}{\Delta r^2}y_{1,N} - f(y_{1,N})] \\ \Rightarrow \alpha_N = y_{1,N+1} = h_N(y, \dot{y}, \ldots) \tag{7.123}$$

The above procedure confirms that all state variables of the model

$$\begin{matrix} y_{1,1} & y_{1,2} & y_{1,3} & \cdots \\ y_{1,i} & \cdots & y_{1,N-1} & y_{1,N} \end{matrix} \tag{7.124}$$

and the control input which is the boundary condition $y_{1,N+1}$ can be written as functions of the flat output $y = y_{1,1}$ and of the flat output's derivatives. Consequently, the state-space model of the PDE dynamics is differentially flat.

Additionally, one can consider decomposition of the PDE state-space equation into submodels, where at each submodel the virtual control input is $\alpha_i = y_{1,i+1}$

$$\dot{y}_{1,i} = \frac{K_1}{\Delta r^2}y_{1,i-1} + \frac{K_2}{\Delta r^2}y_{1,i} + f(y_{1,i}) + \frac{K_1}{\Delta r^2}y_{1,i+1} \tag{7.125}$$

and with local flat output $y_{1,i}$ one can confirm that all such subsystems $i = 1, 2, \ldots, N$ are differentially flat, having as flat output the associated state variable $y_{1,i}$. Next, one can compute the virtual inputs which should be applied to each subsystem.

7.4.3 Computation of a Boundary Conditions-Based Feedback Control Law

To implement boundary feedback control, the feedback control input for Eq. (7.112) is computed first. Following this procedure one arrives to compute the control inputs which are associated with the last row of the state-space model. This uses recursively all virtual control inputs associated with the control of the previous $N - 1$ row of the state-space description of the system.

$$\alpha_1 = y_{1,2}^* = \frac{1}{K_1/\Delta r^2}[\dot{y}_{1,1}^* - \frac{K_1}{\Delta r^2}y_{1,0} - \frac{K_2}{\Delta r^2}y_{1,1} - f(y_{1,1}) - K_{p_1}(y_{1,1} - y_{1,1}^*)] \tag{7.126}$$

$$\begin{aligned}\alpha_2 = y_{1,3}^* &= \frac{1}{K_1/\Delta r^2}[\dot{y}_{1,2}^* - \frac{K_1}{\Delta r^2}y_{1,1} - \frac{K_2}{\Delta r^2}y_{1,2} - f(y_{1,2}) - K_{p_2}(y_{1,2} - y_{1,2}^*)] \Rightarrow \\ \alpha_2 = y_{1,3}^* &= \frac{1}{K_1/\Delta r^2}[\dot{\alpha}_1^* - \frac{K_1}{\Delta r^2}y_{1,1} - \frac{K_2}{\Delta r^2}y_{1,2} - f(y_{1,2}) - K_{p_2}(y_{1,2} - \alpha_1)]\end{aligned} \tag{7.127}$$

$$\begin{aligned}\alpha_3 = y_{1,4}^* &= \frac{1}{K_1/\Delta r^2}[\dot{y}_{1,3}^* - \frac{K_1}{\Delta r^2}y_{1,2} - \frac{K_2}{\Delta r^2}y_{1,3} - f(y_{1,3}) - K_{p_3}(y_{1,3} - y_{1,3}^*)] \Rightarrow \\ \alpha_3 = y_{1,4}^* &= \frac{1}{K_1/\Delta r^2}[\dot{\alpha}_{1,3}^* - \frac{K_1}{\Delta r^2}y_{1,2} - \frac{K_2}{\Delta r^2}y_{1,3} - f(y_{1,3}) - K_{p_3}(y_{1,i} - \alpha_2)]\end{aligned} \tag{7.128}$$

while advancing forward one has

$$\begin{aligned}\alpha_i = y_{1,i+1}^* &= \frac{1}{K_1/\Delta r^2}[\dot{y}_{1,i}^* - \frac{K_1}{\Delta r^2}y_{1,i-1} - \frac{K_2}{\Delta r^2}y_{1,i} - f(y_{1,i}) - K_{p_i}(y_{1,i} - y_{1,i}^*)] \Rightarrow \\ \alpha_i = y_{1,i+1}^* &= \frac{1}{K_1/\Delta r^2}[\dot{\alpha}_{1,i-1}^* - \frac{K_1}{\Delta r^2}y_{1,i-1} - \frac{K_2}{\Delta r^2}y_{1,i} - f(y_{1,i}) - K_{p_i}(y_{1,i} - \alpha_{i-1})]\end{aligned} \tag{7.129}$$

and continuing in a similar manner

$$\begin{aligned}\alpha_{N-2} = y_{1,N-1}^* &= \frac{1}{K_1/\Delta r^2}[\dot{y}_{1,N-2}^* - \frac{K_1}{\Delta r^2}y_{1,N-3} - \\ &- \frac{K_2}{\Delta r^2}y_{1,N-2} - f(y_{1,n-2}) - K_{p_{N-2}}(y_{1,N-2} - y_{1,N-2}^*)] \Rightarrow \\ \alpha_{N-2} = y_{1,N-1}^* &= \frac{1}{K_1/\Delta r^2}[\dot{\alpha}_{1,N-3}^* - \frac{K_1}{\Delta r^2}y_{1,N-3} - \\ &- \frac{K_2}{\Delta r^2}y_{1,N-2} - f(y_{1,n-2}) - K_{p_{N-2}}(y_{1,N-2} - \alpha_{N-3})]\end{aligned} \tag{7.130}$$

$$\begin{aligned}\alpha_{N-1} = y_{1,N}^* &= \frac{1}{K_1/\Delta r^2}[\dot{y}_{1,N-1}^* - \frac{K_1}{\Delta r^2}y_{1,N-2} - \\ &- \frac{K_2}{\Delta r^2}y_{1,N-1} - f(y_{1,N-1}) - K_{p_N}(y_{1,N-1} - y_{1,N-1}^*)] \Rightarrow \\ \alpha_{N-1} = y_{1,N}^* &= \frac{1}{K_1/\Delta r^2}[\dot{\alpha}_{1,N-2}^* - \frac{K_1}{\Delta r^2}y_{1,N} - f(y_{1,N-2}) - \\ &- \frac{K_2}{\Delta r^2}y_{1,N-1} - K_{p_N}(y_{1,N-1} - \alpha_{N-2})]\end{aligned} \tag{7.131}$$

and finally

$$\alpha_N = y_{1,N+1} = \frac{1}{K_1/\Delta r^2}[\dot{y}^*_{1,N} - \frac{K_1}{\Delta r^2}y_{1,i-1} - \frac{K_2}{\Delta r^2}y_{1,N} - K_{p_N}(y_{1,N} - y^*_{1,N})]\Rightarrow$$
$$\alpha_N = y_{1,N+1} = \frac{1}{K_1/\Delta r^2}[\dot{\alpha}_{N-1} - \frac{K_1}{\Delta r^2}y_{1,N-1} - \frac{K_2}{\Delta r^2}y_{1,N} - K_{p_N}(y_{1,N} - \alpha_{N-1})] \tag{7.132}$$

Consequently, the computation of the control input $\alpha_N = y_{1,N+1}$ which is exerted on the PDE model is performed backwards, by substituting recursively into α_N the virtual control inputs $\alpha_{N-1}, \alpha_{N-2}, \ldots, \alpha_i, \ldots, \alpha_2, \alpha_1$.

The control input is the boundary condition $y_{1,N+1} = c_s(r,t)\mid_{r=R_s}$. From Eqs. (1.143) to (1.153) it can be noticed that the current $I(t)$ is the variable that implicitly defines the control input. The measurable output can be obtained from the boundary condition of Eq. (1.143) $\frac{\partial c_s(r,t)}{\partial r}|_{r=R_s} = -\frac{1}{D_s}j^{Li}(x,t)$. Variable c_s is indirectly related to the voltage $V(t)$.

7.4.4 Closed Loop Dynamics

By substituting Eq. (7.132) into Eq. (7.117) of the state-space model of the PDE dynamics, and using the definition $y_{1,N} - y^*_{1,N} = y_{1,N} - \alpha_{N-1} = z_N$ one has

$$\begin{gathered}\dot{y}_{1,N} = \dot{\alpha}_{N-1} - k_{p_N}(y_{1,N} - \alpha_{N-1})\Rightarrow\\ (\dot{y}_{1,N} - \dot{\alpha}_{N-1} - k_{p_N}(y_{1,N} - \alpha_{N-1}) = 0\Rightarrow\\ \dot{z}_N + k_{p_N}z_N = 0\end{gathered} \tag{7.133}$$

Equivalently, by substituting Eq. (7.131) into Eq. (7.116), considering that $(y_{1,N}\rightarrow \alpha_{N-1})$ and by using the definition $y_{1,N-1} - y^*_{1,N-1} = y_{1,N-1} - \alpha_{N-2} = z_{N-1}$ one has

$$\begin{gathered}\dot{y}_{1,N-1} = \dot{\alpha}_{N-2} - k_{p_{N-1}}(y_{1,N-1} - \alpha_{N-2})\Rightarrow\\ (\dot{y}_{1,N-1} - \dot{\alpha}_{N-2} - k_{p_{N-1}}(y_{1,N-1} - \alpha_{N-2}) = 0\Rightarrow\\ \dot{z}_{N-1} + k_{p_{N-1}}z_{N-1} = 0\end{gathered} \tag{7.134}$$

Similarly, continuing with the rest of the equations of the state-space model and by substituting Eq. (7.129) into Eq. (7.115), considering that $(y_{1,i+1}\rightarrow \alpha_i)$ while also using the definition $y_{1,i} - y^*_{1,i} = y_{1,i} - \alpha_{i-1} = z_i$ one has

$$\begin{gathered}\dot{y}_{1,i} = \dot{\alpha}_{i-1} - k_{p_i}(y_{1,i} - \alpha_{i-1})\Rightarrow\\ (\dot{y}_{1,i} - \dot{\alpha}_{i-1} - k_{p_i}(y_{1,i} - \alpha_{i-1}) = 0\Rightarrow\\ \dot{z}_i + k_{p_i}z_i = 0\end{gathered} \tag{7.135}$$

Moving backwards, and by substituting Eq. (7.127) into Eq. (7.113) of the state-space model of the PDE dynamics, considering that $(y_{1,3}\rightarrow \alpha_2)$ while also using the definition $y_{1,2} - y^*_{1,2} = y_{1,2} - \alpha_1 = z_2$ one has

$$\begin{aligned} \dot{y}_{1,2} &= \dot{\alpha}_1 - k_{p_2}(y_{1,2} - \alpha_1) \Rightarrow \\ (\dot{y}_{1,2} - \dot{\alpha}_1 - k_{p_2}(y_{1,2} - \alpha_1) &= 0 \Rightarrow \\ \dot{z}_2 + k_{p_2} z_2 &= 0 \end{aligned} \tag{7.136}$$

Finally, by substituting Eq. (7.126) into Eq. (7.112), and using the definition $y_{1,1} - y^*_{1,1} = z_1$ one obtains

$$\begin{aligned} \dot{y}_{1,N} &= \dot{\alpha}_{N-1} - k_{p,N}(y_{1,N} - \alpha_{N-1}) \Rightarrow \\ (\dot{y}_{1,N} - \dot{\alpha}_{N-1}) + k_{p,N}(y_{1,N} - \alpha_{N-1}) &= 0 \Rightarrow \\ \dot{z}_N + k_{p,N} z_N &= 0 \end{aligned} \tag{7.137}$$

Thus, the dynamics of the closed-loop system asymptotically becomes

$$\begin{aligned} \dot{z}_1 + k_{p,1} z_1 &= 0 \\ \dot{z}_2 + k_{p,2} z_2 &= 0 \\ \dot{z}_3 + k_{p,3} z_3 &= 0 \\ &\cdots \cdots \\ \dot{z}_i + k_{p,i} z_i &= 0 \\ &\cdots \cdots \\ \dot{z}_{N-1} + k_{p,N-1} z_{N-1} &= 0 \\ \dot{z}_N + k_{p,N} z_N &= 0 \end{aligned} \tag{7.138}$$

The dynamics of the closed-loop system can be also written in matrix form

$$\ddot{\tilde{Z}} + K_P \tilde{Z} = 0 \tag{7.139}$$

where

$$\begin{aligned} \tilde{Z} &= [z_1, z_2, z_3, \ldots, z_i, \ldots, z_{N-1}, z_N] \\ K_P &= \mathrm{diag}[k_{p,1}, k_{p,2}, k_{p,3}, \ldots, k_{p,i}, \ldots, k_{p,N-1}, k_{p,N}] \end{aligned} \tag{7.140}$$

After suitable selection of the coefficients $k_{P,i}$, $i = 1, 2, \ldots, N$ such that the monomials $p(s) = s + k_{p,i}$ to have a negative root, it can be assured that $lim_{t \to \infty} z_i(t) = 0$ and that the closed-loop system is asymptotically stable.

Moreover, to prove asymptotic stability for the closed-loop system, the following Lyapunov function can be used

$$V_L = \sum_{i=1}^{N} \frac{1}{2} z_i^2 \tag{7.141}$$

The derivative of this Lyapunov function with respect to time is given by

$$\begin{aligned} &\dot{V}_L = \sum_{i=1}^{N} z_i \dot{z}_i \Rightarrow \dot{V}_L = \sum_{i=1}^{N} z_i(-k_{p_i} z_i) \Rightarrow \\ &\dot{V}_L = -\sum_{i=1}^{N} k_{p,i} z_i^2 < 0 \end{aligned} \tag{7.142}$$

Thus, it is proven again that the closed-loop system is globally asymptotically stable.

7.4.5 State Estimation for the PDE Diffusion Model

The measurable output of the PDE system is considered to be only $y_{1,1} = c_{s,1}$. The rest of the state variables of the model that is $y_{1,2} = c_{s,2}$, $y_{1,3} = c_{s,3}$, …, $y_{1,i} = c_{s,i}$, …, $y_{1,N-1} = c_{s,N-1}$ and $y_{1,N} = c_{s,N}$ have to be reconstructed with the use of a state estimator. As explained before, the system's dynamics consists of the ODEs given in Eqs. (7.112)–(7.117). This is rewritten as

$$\begin{aligned} \dot{y}_{1,1} &= \Delta m_1 + \tfrac{K_1}{\Delta r^2} y_{1,2} \\ \text{where } \Delta m_1 &= \tfrac{K_1}{\Delta r^2} y_{1,0} + \tfrac{K_2}{\Delta r^2} y_{1,1} + f(y_{1,1}) \end{aligned} \tag{7.143}$$

$$\begin{aligned} \dot{y}_{1,2} &= \Delta m_2 + \tfrac{K_1}{\Delta r^2} y_{1,3} \\ \text{where } \Delta m_2 &= \tfrac{K_1}{\Delta r^2} y_{1,1} + \tfrac{K_2}{\Delta r^2} y_{1,2} + f(y_{1,2}) \end{aligned} \tag{7.144}$$

$$\begin{aligned} \dot{y}_{1,3} &= \Delta m_2 + \tfrac{K_1}{\Delta r^2} y_{1,4} \\ \text{where } \Delta m_3 &= \tfrac{K_1}{\Delta r^2} y_{1,2} + \tfrac{K_2}{\Delta r^2} y_{1,3} + f(y_{1,3}) \\ & \cdots \ \cdots \ \cdots \end{aligned} \tag{7.145}$$

$$\begin{aligned} \dot{y}_{1,i} &= \Delta m_i + \tfrac{K_1}{\Delta r^2} y_{1,i+1} \\ \text{where } \Delta m_i &= \tfrac{K_1}{\Delta r^2} y_{1,i-1} + \tfrac{K_2}{\Delta r^2} y_{1,i} + f(y_{1,i}) \\ & \cdots \ \cdots \ \cdots \end{aligned} \tag{7.146}$$

$$\begin{aligned} \dot{y}_{1,N-1} &= \Delta m_{N-1} + \tfrac{K_1}{\Delta r^2} y_{1,N} \\ \text{where } \Delta m_{N-1} &= \tfrac{K_1}{\Delta r^2} y_{1,N-2} + \tfrac{K_2}{\Delta r^2} y_{1,N-1} + f(y_{1,N-1}) \end{aligned} \tag{7.147}$$

$$\begin{aligned} \dot{y}_{1,N} &= \Delta m_N + u \\ \text{where } \Delta m_N &= \tfrac{K_2}{\Delta r^2} y_{1,N} + \tfrac{K_1}{\Delta r^2} y_{1,N+1} + f(y_{1,N}) \text{ and } u = \tfrac{K_1}{\Delta r^2} y_{1,N-1} \end{aligned} \tag{7.148}$$

By deriving the expression about $y_{1,1}$ given in the first row of Eq. (7.143) with respect to time one obtains

$$\begin{aligned} \ddot{y}_{1,1} &= \dot{\Delta m}_1 + \frac{K_1}{\Delta r^2} \dot{y}_{1,2} \Rightarrow \\ \ddot{y}_{1,1} &= \dot{\Delta m}_1 + \frac{K_1}{\Delta r^2} (\Delta m_1 + \frac{K_1}{\Delta r^2} y_{1,3}) \Rightarrow \\ \ddot{y}_{1,1} &= \dot{\Delta m}_1 + \frac{K_1}{\Delta r^2} \Delta m_1 + (\frac{K_1}{\Delta r^2})^2 y_{1,3} \end{aligned} \tag{7.149}$$

By deriving again with respect to time one gets

$$\begin{aligned}\frac{d^3}{dt^3}y_{1,1} &= \ddot{\Delta m}_1 + \frac{K_1}{\Delta r^2}\dot{\Delta m}_1 + (\frac{K_1}{\Delta r^2})^2[\Delta m_3 + \frac{K_1}{\Delta r^2}y_4] \Rightarrow \\ \frac{d^3}{dt^3}y_{1,1} &= \ddot{\Delta m}_1 + \frac{K_1}{\Delta r^2}\dot{\Delta m}_1 + (\frac{K_1}{\Delta r^2})^2\Delta m_3 + (\frac{K_1}{\Delta r^2})^3 y_4\end{aligned} \tag{7.150}$$

Continuing in a similar manner one obtains

$$\frac{d^i}{dt^i}y_{1,1} = \sum_{k=1}^{i}(\frac{K_1}{\Delta r^2})^{k-1}\frac{d^{i-k}}{dt^{i-k}}\Delta m_k + (\frac{K_1}{\Delta r^2})^{i+1}y_{i+1} \tag{7.151}$$

By advancing with the computation of higher-order time derivatives for $y_{1,1}$ one gets

$$\frac{d^{N-1}}{dt^{N-1}}y_{1,1} = \sum_{k=1}^{N-1}(\frac{K_1}{\Delta r^2})^{k-1}\frac{d^{N-1-k}}{dt^{N-1-k}}\Delta m_k + (\frac{K_1}{\Delta r^2})^{N}y_N \tag{7.152}$$

By computing one more derivative of $y_{1,1}$ with respect to time the following relation is obtained

$$\begin{aligned}\frac{d^N}{dt^N}y_{1,1} &= \sum_{k=1}^{N-1}(\frac{K_1}{\Delta r^2})^{k-1}\frac{d^{N-k}}{dt^{N-k}}\Delta m_k + (\frac{K_1}{\Delta r^2})^{N}[\Delta m_N + u] \Rightarrow \\ \frac{d^N}{dt^N}y_{1,1} &= \sum_{k=1}^{N-1}(\frac{K_1}{\Delta r^2})^{k-1}\frac{d^{N-k}}{dt^{N-k}}\Delta m_k + (\frac{K_1}{\Delta r^2})^{N}\Delta m_N + (\frac{K_1}{\Delta r^2})^{N}u\end{aligned} \tag{7.153}$$

This is an input-output linearized description of the system which is a consequence of the differential flatness properties of the diffusion process (since the system is differentially flat it can be transformed to the trivial canonical form). Equation (7.153) can be also written as

$$\begin{aligned}&y_{1,1}^{(N)} = \tilde{f}(y_{1,1}, \dot{y}_{1,1}, \ddot{y}_{1,1}, \dots) + \tilde{g}(y_{1,1}, \dot{y}_{1,1}, \ddot{y}_{1,1}, \dots)u \\ &\text{where } \tilde{f}(y_{1,1}, \dot{y}_{1,1}, \ddot{y}_{1,1}, \dots) = \sum_{k=1}^{N-1}(\frac{K_1}{\Delta r^2})^{k-1}\frac{d^{N-k}}{dt^{N-k}}\Delta m_k + (\frac{K_1}{\Delta r^2})^{N}\Delta m_N \\ &\text{and } \tilde{g}(y_{1,1}, \dot{y}_{1,1}, \ddot{y}_{1,1}, \dots) = (\frac{K_1}{\Delta r^2})^{N}\end{aligned} \tag{7.154}$$

For this description of the system, the estimation problem can be solved by applying the Derivative-free nonlinear Kalman Filter. To this end, the following state variables are defined: $z_1 = y_{1,1}$, $z_2 = \dot{y}_{1,1}$, $z_3 = \ddot{y}_{1,1}$, ..., $z_i = y_{1,1}^{(i-1)}$, ..., $z_{N-1} = y_{1,1}^{(N-2)}$ and $z_N = y_{1,1}^{(N-1)}$.

For the input-output linearized model of the PDE one gets a state-space description of the form

$$\begin{pmatrix} \dot{z}_1 \\ \dot{z}_2 \\ \dot{z}_3 \\ \cdots \\ \dot{z}_i \\ \cdots \\ \dot{z}_{N-1} \\ \dot{z}_N \end{pmatrix} = \begin{pmatrix} 0 & 1 & 0 & 0 & \cdots & 0 & 0 & \cdots & 0 & 0 \\ 0 & 0 & 1 & 0 & \cdots & 0 & 0 & \cdots & 0 & 0 \\ 0 & 0 & 0 & 1 & \cdots & 0 & 0 & \cdots & 0 & 0 \\ \cdots & & & & \cdots & & & \cdots & & \cdots \\ 0 & 0 & 0 & 0 & \cdots & 0 & 1 & \cdots & 0 & 0 \\ \cdots & & & & \cdots & & & \cdots & & \cdots \\ 0 & 0 & 0 & 0 & \cdots & 0 & 0 & \cdots & 0 & 1 \\ 0 & 0 & 0 & 0 & \cdots & 0 & 0 & \cdots & 0 & 0 \end{pmatrix} \cdot \begin{pmatrix} z_1 \\ z_2 \\ z_3 \\ \cdots \\ z_i \\ \cdots \\ z_{N-1} \\ z_N \end{pmatrix} + \begin{pmatrix} 0 \\ 0 \\ 0 \\ \cdots \\ 0 \\ \cdots \\ 0 \\ 1 \end{pmatrix} (\tilde{f} + \tilde{g}u) \quad (7.155)$$

$$z^{meas} = \begin{pmatrix} 1 & 0 & 0 & 0 & \cdots & 0 & 0 & \cdots & 0 & 0 \end{pmatrix} Z \quad (7.156)$$

where $Z = [z_1,\ z_2,\ z_3,\ \ldots\ z_i,\ \ldots\ z_{N-1},\ z_N]^T$ and $v = \tilde{f} + \tilde{g}u$. The state estimation problem for the input-output linearized description of the system is solved with the use of the Derivative-free nonlinear Kalman Filter. The above continuous-time state-space description of the system takes the form

$$\begin{aligned} \dot{z} &= Az + Bv \\ z^{meas} &= CZ \end{aligned} \quad (7.157)$$

The state-space model of Eq. (7.157) is turned into discrete-time form after applying common discretization methods. The discrete-time form of matrices A, B, C is denoted as A_d, B_d and C_d respectively. The Derivative-free nonlinear Kalman Filter is given by

$$\dot{\hat{z}} = A\hat{z} + B\hat{v} + K_f(z^{meas} - \hat{z}) \quad (7.158)$$

where $\hat{v} = \hat{\tilde{f}} + \hat{\tilde{g}}u$. The filter's gain K_f is computed through the following recursion

Measurement update:

$$\begin{aligned} K_f(k) &= P^-(k)C_d^T[C_dP^-(k)C_d^T + R(k)]^{-1} \\ \hat{z}^-(k) &= \hat{z}(k) + K_f[z^{meas} - C\hat{Z}] \\ P(k) &= P^-(k) - K(k)A_dP^-(k) \end{aligned} \quad (7.159)$$

Time update:

$$\begin{aligned} P^-(k+1) &= A_dP(k)A_d^T + Q(k) \\ \hat{z}^-(k+1) &= A_d\hat{z}(k) + B_d\hat{v}(k) \end{aligned} \quad (7.160)$$

Once estimation of $y_{1,1}$ and its derivatives is obtained from the above filtering algorithm it is possible to compute the estimates of all state vector elements $c_{s,i}$ $i = 1, 2, \ldots, N$ of the initial state-space model of the system given in Eq. (7.112) to Eq. (7.117). As explained in Eqs. (7.122) and (7.123) the PDE model is differential flat and all state variables $c_{s,i}$ can be expressed as differential functions of the flat output $y_{1,1}$.

7.4.6 Simulation Tests

Simulation examples about the proposed control method for distributed parameter systems are provided for the case of the Lithium-ions diffusion PDE that was given in Eq. (1.141). The obtained results are depicted in Figs. 7.22, 7.23 and 7.24. The spatial discretization of the PDE model consisted of $N = 15$ points. The sampling period was chosen to be $T_s = 0.01$ s. The c_s axis of the diagrams has been normalized.

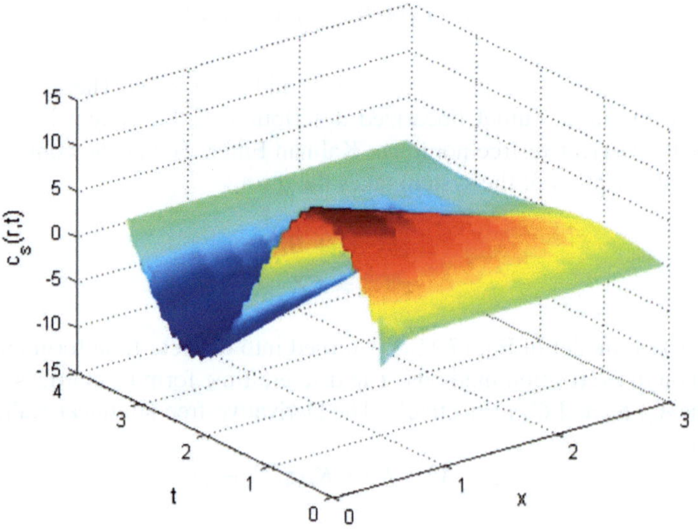

Fig. 7.22 Controlled PDE diffusion in the Li-ion battery

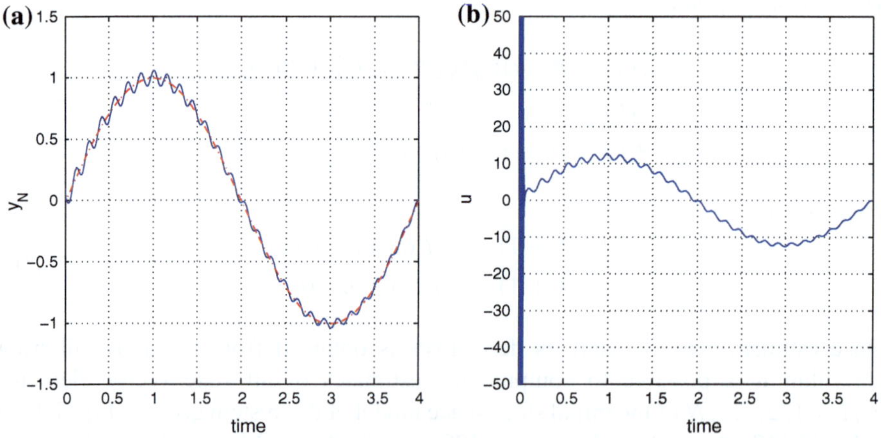

Fig. 7.23 **a** tracking of reference setpoint by the value of the Li-ions PDE at the final grid point x_N. **b** control input

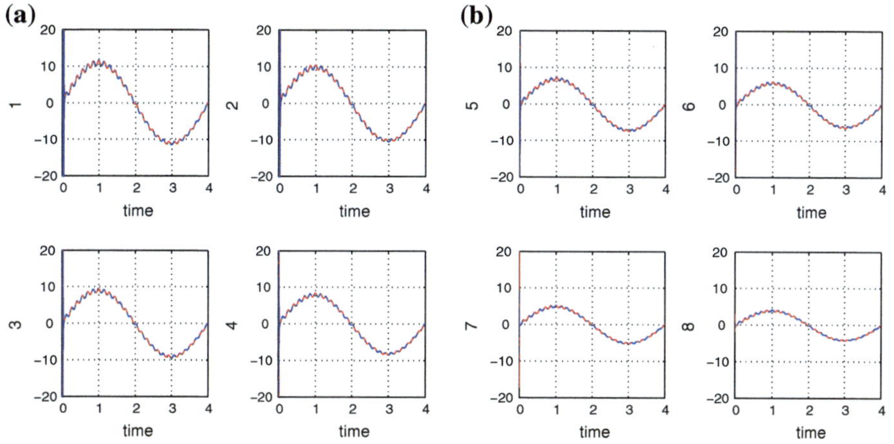

Fig. 7.24 **a** variation in time of the solution of the PDE at grid points x_1 to x_4 (*blue line*) and associated reference setpoint (*red line*). **b** variation in time of the solution of the PDE at grid points x_5 to x_8 (*blue line*) and associated reference setpoint (*red line*)

The boundary conditions of the PDE model were $c_s(r, t) \mid_{r=0}= 0$, while $c_s(r, t) \mid_{r=R_s}$ served as the control input and was computed as each iteration of the control algorithm according to the procedure described in Sect. 7.4.3.

The numerical simulation experiments have confirmed the theoretical findings of this section. It has been shown that by applying the proposed control method, the Li-ions diffusion PDE dynamics can be modified so as to converge to the desirable reference profile. The control input that succeeds this changes smoothly and has a moderate range of variation. The accuracy of tracking of the reference setpoints was quite satisfactory.

Chapter 8
Synchronization and Stabilization of Distributed Power Generation Units

8.1 Outline

The chapter analyzes synchronization and stabilization for distributed power generation units. First, the problem of stabilization and synchronization is solved for a model of distributed interconnected synchronous power generators with the use of flatness-based control. Next, the problem of stabilization and synchronization of the distributed synchronous generators is solved with the use of flatness-based adaptive control. Moreover, the problem of decentralized control and synchronization is solved for a model of distributed interconnected DC power generation units, such as photovoltaics.

First, the chapter develops a control method for distributed interconnected power generation units. The power system comprises Permanent Magnet Synchronous Generators (PMSGs) which are connected to each other through transformers and tie-lines. A derivative-free nonlinear Kalman Filtering approach is introduced aiming at implementing sensorless control of the distributed power generators. In the proposed derivative-free Kalman Filtering method the generator's model is first subjected to a linearization transformation that is based on differential flatness theory and next state estimation is performed by applying the standard Kalman Filter recursion to the linearized model. Unlike Lie algebra-based estimator design methods, the proposed approach provides estimates of the state vector of the permanent magnet synchronous generator without the need for derivatives and Jacobians calculation. Moreover, by redesigning the proposed derivative-free nonlinear Kalman Filter as a disturbance observer it is possible to estimate at the same time the non-measurable elements of each generator's state vector, the unknown input power (torque) and the disturbance terms induced by interarea oscillations. The efficient real-time estimation of the aggregate disturbance that affects each local generator makes possible to introduce a counter-disturbance control term thus maintaining the power system on its nominal operating conditions.

Moreover, the chapter proposes a new nonlinear H-infinity control method for stabilization and synchronization of distributed interconnected synchronous generators.

G. Rigatos, *Intelligent Renewable Energy Systems*,
Green Energy and Technology, DOI 10.1007/978-3-319-39156-4_8

At first stage local linearization of the distributed generators' model is performed round its present operating point. The approximation error that is introduced to the linearized model is due to truncation of higher-order terms in the performed Taylor series expansion and is represented as a disturbance. The control problem is now formulated as a mini-max differential game in which the control input tries to minimize the state vector's tracking error while the disturbances affecting the model try to maximize it. Using the linearized description of the distributed generators' dynamics an H-infinity feedback controller is designed through the solution of a Riccati equation at each step of the control algorithm. The inherent robustness properties of H-infinity control assure that the disturbance effects will be eliminated and the state variables of the individual power generators will converge to the desirable setpoints. The proposed method, stands for a reliable solution to the problem of nonlinear control and stabilization for interconnected synchronous generators. It is also a novel approach, comparing to control of synchronous generators based on global linearization methods. Its efficiency is further confirmed through simulation experiments.

Additionally, the chapter presents an adaptive fuzzy approach to the problem of synchronization and control of distributed power generators, which is based on differential flatness theory and which uses exclusively output feedback. It is shown that the MIMO model of the distributed power generators is a differentially flat one and this permits to transform it to the so-called linear canonical form. For the new description of the system's dynamics the transformed control inputs contain unknown terms which depend on the system's parameters. To identify these terms adaptive fuzzy approximators are used in the control loop. Thus an adaptive fuzzy control scheme is implemented in which the unknown or unmodeled system dynamics is approximated by neurofuzzy networks and next this information is used by a feedback controller that makes the state vector of the distributed power generators converge to the desirable setpoints. This adaptive control scheme is exclusively implemented with the use of output feedback, while the state vector elements which are not directly measured are estimated with the use of a state observer that operates in the control loop.

Finally, the chapter develops decentralized control for parallel inverters connected to the power grid, using differential flatness theory and the Derivative-free nonlinear Kalman Filter. The problem is of elevated difficulty comparing to the control of stand-alone inverters because in this case in the dynamics of each inverter one has also to compensate for interaction terms which are due to the coupling with other inverters. It is proven that the model of the inverters, is a differentially flat one. By exploiting differential flatness properties it is shown that the multiple inverters model can be transformed into a set of local inverter models which are decoupled and linearized. For each local inverter the design of a state feedback controller becomes possible, e.g. using pole placement methods. Such a controller processes measurements not only coming from the individual inverter but also coming from other inverters which are connected to the grid. Moreover, to estimate the non-measurable state variables of each local inverter, the Derivative-free nonlinear Kalman Filter is used. As explained before, this consists of the Kalman Filter recursion applied to the local linearized model of the inverter and of an inverse transformation that is based on differential flatness theory, which enables to compute estimates of the state variables of the initial nonlinear model of the inverter. Furthermore, by redesigning the aforementioned filter as a disturbance observer it becomes also possible to estimate and compensate for disturbance terms that affect each local inverter.

8.2 State Estimation-Based Control of Distributed PMSGs

8.2.1 Outline

In this section, a solution is presented about the problem of distributed control of interconnected generators. Actually, research results are presented on sensorless control of interconnected PMSGs (permanent magnet synchronous generators) with the use of the Derivative-free nonlinear Kalman Filter, in the framework of a multi-area multi-machine model of distributed power generation units. A large-scale power system consists of multiple control areas where local power generation units operate, while these areas are connected through tie lines and power transformers. Due to change in active power loads, the frequencies of the areas variate from their nominal values (synchronous speed) and power exchange through the tie lines will also deviate from the associated setpoints [107, 108, 116, 165, 167, 233, 302, 359, 465, 477, 479, 495]. Therefore, it is necessary to supply power generator units with local controllers capable of maintaining the generators' turn speed at its nominal value. To stabilize the functioning of the distributed power generation system and eliminate frequency deviations several control approaches have been implemented, such as standard PID control, lead-lag compensators where the selection of the controller's parameters is performed with the use of Kharitonov's theorem, H_∞ control or adaptive control [88, 102, 116, 198, 207, 213, 237, 238, 349]. To control efficiently power generators in case that the complete state vector of the electric machine cannot be measured directly, observer-based control schemes have been proposed [121, 267, 290, 291]. Moreover, to control distributed power generators, disturbance rejection control schemes have been developed [60, 262, 265, 315].

A sensorless control scheme is developed for the local power generators. The control approach is based on a diffeomorphism (change of coordinates) that follows differential flatness theory and which enables to transform the initial nonlinear model of the generator into a linear one. The latter model makes possible to design a state feedback controller for the PMSG. Moreover, to estimate the non-measurable elements of the PMSG state vector and to use them in the control loop, the Derivative-free nonlinear Kalman Filter is considered. This filtering approach exploits the previously mentioned transformation of the generator's model in the linear canonical (Brunovsky) form through the application of differential flatness theory, and performs filtering on the linearized equivalent of the generator using the standard Kalman Filter recursion. It will be shown that the sensorless control scheme for the local power generators can be modified so as to provide the desirable robustness to the distributed power generation scheme. By redesigning the Kalman Filter, initially used for estimating the non-measurable elements of the generator's state vector, in the form of a disturbances estimator it becomes possible to estimate in real-time the aggregate disturbance that affects each local generator [98, 99, 169, 292]. Knowing the disturbances' value it is possible to generate a counter-disturbance term that serves as auxiliary control input and which finally compensates for the disturbances' effects. With the previously described control approach each generator keeps on running at its nominal operating conditions and the distributed generation units become robust to phenomena such as mechanical input torque variations and interarea oscillations [15, 439, 474, 483, 504].

The proposed control method is suitable for compensating for both endogenous and exogenous disturbances that affect the generators' model. Endogenous disturbances are associated with parametric uncertainty, e.g. there can be imprecision in the values of reactances, stator and rotor circuit coefficients and the rotor's moment of inertia. Exogenous disturbances are associated with variations in the mechanical input torque, as well as with perturbations in the excitation system of each generator which are due to lack of synchronization with other generators connected to the grid [50]. The efficient compensation of these disturbances enables to maintain synchronism between the individual power generators, thus assuring: (i) robustness improvement of the power grid and maintenance of the synchronization between the distributed electric power generation units even under adverse conditions in the power grid, (ii) improvement of the provided power quality (distortions in power quality due to deviations of the power generation units from their nominal operating conditions will be avoided), (iii) uninterrupted power flow in the grid, ability to incorporate more power generation units without affecting stability, and consequently better response to the increased demand for power supply.

The proposed distributed filtering and control method for the system of the interconnected power generators is a highly data-driven one. Actually, to succeed state estimation and disturbances compensation for a set of n distributed PMSGs it is necessary that at each sampling period each power generator controller receives n variables associated with the turn angles of all generators in the power grid. Thus, the total amount of data that has to be processed consists of n^2 variables. This is a quite intensive computation load which can be managed efficiently thanks to the distributed control structure that has been implemented (each generator computes locally its own control signal). Moreover, taking into account that the exchange of data has to take place in real time one can conclude that communication links of efficient bit rate have to be established between the generators.

8.2.2 Dynamic Model of the Distributed Power Generation Units

8.2.2.1 Dynamics of the Local Power Generator Units

It is considered that the third order model of the PMSG is connected to the power grid thus forming the model of a Single Machine Infinite Bus (SMIB) system. As explained in Chap. 3 PMSG mechanical dynamics can be represented as follows

$$\begin{aligned}\dot{\delta} &= \omega \\ \dot{\omega} &= -\tfrac{D}{2J}(\omega-\omega_0)+\tfrac{\omega_0}{2J}(P_m-P_e)\end{aligned} \tag{8.1}$$

where δ is the turn angle of the generator's rotor, ω is the rotation speed of the rotor with respect to synchronous reference, ω_0 is the synchronous speed of the generator, J is the moment of inertia of the rotor, P_e is the active power of the generator, P_m is the mechanical input torque to the generator which is associated with the mechanical input power, D is the damping constant of the generator and T_e is the electrical torque which is associated to the generated active power. Moreover, the following variables are defined: $\Delta\delta = \delta - \delta_0$ and $\Delta\omega = \omega - \omega_0$ with ω_0 denoting the synchronous speed. The generator's electrical dynamics has already been described in Chap. 1 and is summarized as follows [215, 249, 266]:

$$\dot{E}_q' = \tfrac{1}{T_{d_o}}(E_f - E_q) \tag{8.2}$$

where E_q' is the quadrature-axis transient voltage of the generator, E_q is the quadrature axis voltage of the generator, T_{d_o} is the direct axis open-circuit transient time constant of the generator and E_f is the equivalent voltage in the excitation coil. The algebraic equations of the synchronous generator are given by

$$\begin{aligned}E_q &= \tfrac{x_{d_\Sigma}}{x_{d_\Sigma}'}E_q' - (x_d - x_d')\tfrac{V_s}{x_{d_\Sigma}'}cos(\Delta\delta) \\ I_q &= \tfrac{V_s}{x_{d_\Sigma}'}sin(\Delta\delta) \\ I_d &= \tfrac{E_q'}{x_{d_\Sigma}'} - \tfrac{V_s}{x_{d_\Sigma}'}cos(\Delta\delta) \\ P_e &= \tfrac{V_sE_q'}{x_{d_\Sigma}'}sin(\Delta\delta) \\ Q_e &= \tfrac{V_sE_q'}{x_{d_\Sigma}'}cos(\Delta\delta) - \tfrac{V_s^2}{x_{d_\Sigma}} \\ V_t &= \sqrt{(E_q' - X_d'I_d)^2 + (X_d'I_q)^2}\end{aligned} \tag{8.3}$$

where $x_{d_\Sigma} = x_d + x_T + x_L$, $x_{d_\Sigma}' = x_d' + x_T + x_L$, x_d is the direct-axis synchronous reactance, x_T is the reactance of the transformer, x_d' is the direct axis transient reactance, x_L is the reactance of the transmission line, I_d and I_q are direct and quadrature axis currents of the generator, V_s is the infinite bus voltage, Q_e is the generator reactive power delivered to the infinite bus, and V_t is the terminal voltage of the generator.

Substituting the electrical equations of the PMSG given in Eq. (8.3) into the equations of the electrical and mechanical dynamics of the rotor given in Eqs. (8.1) and (8.2) respectively, the complete model of the Single Machine Infinite Bus model is obtained

$$\begin{array}{c} \dot{\delta} = \omega - \omega_0 \\ \dot{\omega} = -\frac{D}{2J}(\omega - \omega_0) + \omega_0 \frac{P_m}{2J} - \omega_0 \frac{1}{2J} \frac{V_s E_q'}{x_{d\Sigma}'} sin(\Delta\delta) \\ \dot{E}_q' = -\frac{1}{T_d'} E_q' + \frac{1}{T_{do}} \frac{x_d - x_d'}{x_{d\Sigma}'} V_s cos(\Delta\delta) + \frac{1}{T_{do}} E_f \end{array} \tag{8.4}$$

where $T_d' = \frac{x_{d\Sigma}'}{x_{d\Sigma}} T_{d_o}$ is the time constant of the field winding.

The previously analyzed single-machine infinite-bus model of the PMSG is described by a nonlinear state-space model of the form

$$\dot{x} = f(x) + g(x)u \tag{8.5}$$

where the state vector x is defined as $x = \left(\Delta\delta \ \Delta\omega \ E_q'\right)^T$. The vector fields $f(x)$ and $g(x)$ are defined as

$$f(x) = \begin{pmatrix} \omega - \omega_0 \\ -\frac{D}{2J}(\omega - \omega_0) + \omega_0 \frac{P_m}{2J} - \omega_0 \frac{1}{2J} \frac{V_s E_q'}{x_{d\Sigma}'} sin(\Delta\delta) \\ -\frac{1}{T_d'} E_q' + \frac{1}{T_{do}} \frac{x_d - x_d'}{x_{d\Sigma}'} V_s cos(\Delta\delta) \end{pmatrix} \tag{8.6}$$

$$g(x) = \left(0 \ 0 \ \frac{1}{T_{do}}\right)^T \tag{8.7}$$

with control input $u = E_f$ being the field voltage (equivalent voltage in the excitation coil) and measurable output the turn angle of the rotor

$$y = h(x) = \delta - \delta_0 \tag{8.8}$$

8.2.2.2 Dynamics of the Interconnected Power Generators

The interconnection between several local power generation units which are described by the previously analyzed SMIB model results into a multi-area multi-machine distributed power generation system.

A multi-machine power system (MMPS) with n machines, in which the first machine is chosen as the reference machine can be described by the following nonlinear dynamic model

$$\begin{aligned}
\dot{\delta}_i &= \omega_i - \omega_0 \\
\dot{\omega}_i &= -\tfrac{D_i}{2J_i}(\omega_i - \omega_0) + \omega_0 \tfrac{P_{m_i}}{2J_i} - \\
&- \omega_0 \tfrac{1}{2J_i}[G_{ii}{E'_{qi}}^2 + E'_{qi}\textstyle\sum_{j=1,j\neq i}^{n} E'_{qj} G_{ij} sin(\delta_i - \delta_j - \alpha_{ij})] \\
\dot{E}'_{q_i} &= -\tfrac{1}{T'_{d_i}} E'_{q_i} + \tfrac{1}{T_{do_i}} \tfrac{x_{d_i} - x'_{d_i}}{x'_{d_{\Sigma i}}} V_{s_i} cos(\Delta\delta_i) + \tfrac{1}{T_{do_i}} E_{f_i}
\end{aligned} \tag{8.9}$$

where the electric torque P_{e_i} which is associated with the active power at the i-th generator is now given by

$$P_{e_i} = G_{ii}{E'_{qi}}^2 + E'_{qi}\sum_{j=1,j\neq i}^{n} E'_{qj} G_{ij} sin(\delta_i - \delta_j - \alpha_{ij}) \tag{8.10}$$

for $i = 1, 2 \ldots, n$. For a power grid that consists of n generators the aggregate state vector comprises the state vectors of the local machines, i.e. $x = [x^1, x^2, \ldots, x^n]^T$, where $x^i = [x_1^i, x_2^i, x_3^i]^T$, with $x_1^i = \Delta\delta_i$, $x_2^i = \Delta\omega_i$ and $x_3^i = E'_{qi}$ are the state variables for the i-th machine and $i = 1, 2, \ldots, n$.

8.2.3 *Linearization of the Distributed Power Generation System Using Lie Algebra*

As previously analyzed, the dynamic model of the distributed power generation system is defined by Eq. (8.9). The state vector of the i-th local power generator is defined as

$$x_i = [x_1^i, x_2^i, x_3^i]^T = [\Delta\delta_i, \Delta\omega_i, E_{q_i}]^T \tag{8.11}$$

The local i-th power generation system is described by

$$\dot{x}^i = f^i(x) + g^i(x)u^i \tag{8.12}$$

where

$$f^i(x) = \begin{pmatrix} f_1^i \\ f_2^i \\ f_3^i \end{pmatrix} \quad g^i(x) = \begin{pmatrix} 0 \\ 0 \\ \frac{1}{T_{do_i}} \end{pmatrix} \tag{8.13}$$

with

$$\begin{aligned}
f_1^i(x) &= x_2^i \\
f_2^i(x) &= -\tfrac{D}{2J_i} x_2^i + \tfrac{\omega_0}{2J_i} P_{m_i} - \omega_0 \tfrac{1}{2J_i}\{G_{ii}{x_3^i}^2 + x_3^i[G_{ii}{x_3}^{i2} + \\
&+ x_3^i \textstyle\sum_{j=1,j\neq i}^{n} x_3^j G_{ij} sin(x_1^i - x_1^j - a_{ij})]\} \\
f_3^i(x) &= -\tfrac{1}{T'_{d_i}} x_3^i + \tfrac{1}{T_{do_i}} \tfrac{x_{d_i} - x'_{d_i}}{x_{d_{\Sigma_i}}} V_{s_i} cos({x_1}^i).
\end{aligned} \tag{8.14}$$

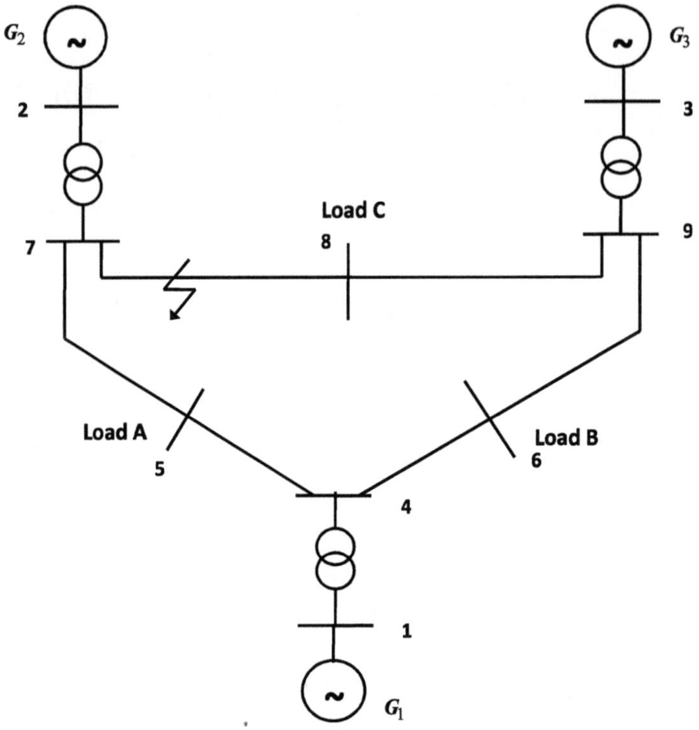

Fig. 8.1 A multi-machine (3-area) distributed power generation model

Generalizing, the distributed power system is described by

$$\dot{x} = f(x) + g(x)u \tag{8.15}$$

where $x = \left(x^1\ x^2\ \ldots\ x^n\right)^T$. Without loss of generality a three-machine distributed power generation model will be considered next ($n = 3$), as shown in Fig. 8.1. The state-space equation of the distributed power generation model is

$$\begin{pmatrix} \dot{x}_1^1 \\ \dot{x}_2^1 \\ \dot{x}_3^1 \\ \dot{x}_1^2 \\ \dot{x}_2^2 \\ \dot{x}_3^2 \\ \dot{x}_1^3 \\ \dot{x}_2^3 \\ \dot{x}_3^3 \end{pmatrix} = \begin{pmatrix} f_1^1(x) \\ f_2^1(x) \\ f_3^1(x) \\ f_1^2(x) \\ f_2^2(x) \\ f_3^2(x) \\ f_1^3(x) \\ f_2^3(x) \\ f_3^3(x) \end{pmatrix} + \begin{pmatrix} 0 & 0 & 0 \\ 0 & 0 & 0 \\ g_1 & 0 & 0 \\ 0 & 0 & 0 \\ 0 & 0 & 0 \\ 0 & g_2 & 0 \\ 0 & 0 & 0 \\ 0 & 0 & 0 \\ 0 & 0 & g_3 \end{pmatrix} \begin{pmatrix} u_1 \\ u_2 \\ u_3 \end{pmatrix} \tag{8.16}$$

The procedure for linearization with the use of Lie algebra is as follows: The following state vector is defined $x = [x_1^1, x_2^1, x_3^1, x_1^2, x_2^2, x_3^2, x_1^3, x_2^3, x_3^3]^T$ while the state vector of the linearized equivalent of the distributed PMSG model is defined as $z \in R^{9\times 1}$. It holds that $z_1^1 = h^1(x) = \Delta\delta^1$, $z_2^1 = L_f h^1(x) = x_2^1$ and $z_3^1 = L_f^2 h^1(x) = f_2^1$. In a similar manner one obtains $z_1^2 = h^2(x) = \Delta\delta^2$, $z_2^2 = x_2^2$ and $z_3^2 = f_2^2$. Moreover, in a similar manner one obtains $z_1^3 = h^3(x) = \Delta\delta^3$, $z_2^2 = L_f h^3(x) = x_2^2$ and $z_3^3 = L_f^2 h^2(x) = f_2^3$.

Furthermore, for the vectors defined as $z_3 = [z_3^1, z_3^2, z_3^3]$, $\tilde{f} = [f_2^1, f_2^2, f_2^3]$ and $\tilde{h} = [h^1, h^2, h^3]$ it holds that

$$\dot{z} = L_f^3 h(x) + L_g L_f^2 h(x) u \tag{8.17}$$

or, equivalently

$$\begin{pmatrix} \dot{z}_3^1 \\ \dot{z}_3^2 \\ \dot{z}_3^3 \end{pmatrix} = \begin{pmatrix} \frac{\partial f_2^1}{\partial x_1} f_1 + \frac{\partial f_2^1}{\partial x_2} f_2 + \cdots + \frac{\partial f_2^1}{\partial x_n} f_n \\ \frac{\partial f_2^2}{\partial x_1} f_1 + \frac{\partial f_2^2}{\partial x_2} f_2 + \cdots + \frac{\partial f_2^2}{\partial x_n} f_n \\ \frac{\partial f_2^3}{\partial x_1} f_1 + \frac{\partial f_2^3}{\partial x_2} f_2 + \cdots + \frac{\partial f_2^3}{\partial x_n} f_n \end{pmatrix} + \begin{pmatrix} \frac{\partial f_2^1}{\partial x_1} g_1 & \frac{\partial f_2^1}{\partial x_2} g_2 & \cdots & \frac{\partial f_2^1}{\partial x_n} g_3 \\ \frac{\partial f_2^2}{\partial x_1} g_1 & \frac{\partial f_2^2}{\partial x_2} g_2 & \cdots & \frac{\partial f_2^2}{\partial x_n} g_3 \\ \frac{\partial f_2^3}{\partial x_1} g_1 & \frac{\partial f_2^3}{\partial x_2} g_2 & \cdots & \frac{\partial f_2^3}{\partial x_n} g_3 \end{pmatrix} \begin{pmatrix} u_1 \\ u_2 \\ u_3 \end{pmatrix} \tag{8.18}$$

Thus, it holds

$$\dot{z}_3 = f_a + Mu = v \tag{8.19}$$

Therefore, setting $\dot{z}_3 = v$, where $v = f_a + Mu$ the dynamics of the distributed power generation system for the i-th power generator $i = 1, 2, 3$ becomes:

$$\begin{pmatrix} \dot{z}_1^i \\ \dot{z}_2^i \\ \dot{z}_3^i \end{pmatrix} \begin{pmatrix} 0 & 1 & 0 \\ 0 & 0 & 1 \\ 0 & 0 & 0 \end{pmatrix} \begin{pmatrix} z_1^i \\ z_2^i \\ z_3^i \end{pmatrix} + \begin{pmatrix} 0 \\ 0 \\ 1 \end{pmatrix} v^i \tag{8.20}$$

In this manner the initial nonlinear power system is transformed into two decoupled linear subsystems which are in the canonical Brunovksy form. For each one of these subsystems $i = 1, 2, 3$ the appropriate control law is

$$v^i = z_d^{(3)^i} - k_3(\ddot{z}^i - \ddot{z}_d^i) - k_2(\dot{z}^i - \dot{z}_d^i) - k_1(z^i - z_d^i) - \tilde{d}^i \tag{8.21}$$

where the feedback control gains k_1, k_2, k_3 are chosen such that the associated characteristic polynomial of the linearized system is a Hurwitz one.

8.2.4 Differential Flatness of the Distributed PMSG Model

8.2.4.1 Properties of Differentially Flat Systems

The main principles of the differential flatness theory have been given in the previous chapters as follows [137, 229, 241, 281, 352, 411, 456]: A finite dimensional system is considered. This can be written in the general form of an ordinary differential equation (ODE), i.e. $S_i(w, \dot{w}, \ddot{w}, \ldots, w^{(i)}),\ \ i = 1, 2, \ldots, q$. The term w denotes the system variables (these variables are for instance the elements of the system's state vector and the control input) while $w^{(i)}$, $i = 1, 2, \ldots, q$ are the associated derivatives. Such a system is said to be differentially flat if there is a collection of m functions $y = (y_1, \ldots, y_m)$ of the system variables and of their time-derivatives, i.e. $y_i = \phi(w, \dot{w}, \ddot{w}, \ldots, w^{(\alpha_i)}),\ i = 1, \ldots, m$ satisfying the following two conditions [137, 229, 241, 281, 352, 411, 456]: (1) There does not exist any differential relation of the form $R(y, \dot{y}, \ldots, y^{(\beta)}) = 0$ which implies that the derivatives of the flat output are not coupled in the sense of an ODE, or equivalently it can be said that the flat output is differentially independent, (2) All system variables (i.e. the elements of the system's state vector w and the control input) can be expressed using only the flat output y and its time derivatives $w_i = \psi_i(y, \dot{y}, \ldots, y^{(\gamma_i)}),\ i = 1, \ldots, s$.

8.2.4.2 Differential Flatness of the Distributed Power Generation Model

Differential flatness for the stand-alone synchronous generator was shown in Chap. 3. It will be proven that the multi-machine power generation system is also a differentially flat one. As flat output of the distributed power generation system, consisting of n PMSGs, the following vector is defined $y = [y_1^1, y_1^2, \ldots, y_1^n]$ or $y = \Delta\delta^1, \Delta\delta^2, \ldots, \Delta\delta^n$. For the n-machines power generation system it holds $x_1^1 = y^1$, $x_1^2 = y^2$, $x_1^3 = y^3$, ..., $x_1^n = y^n$ and $x_2^1 = \Delta\omega^1 = \dot{y}^1$, $x_2^2 = \Delta\omega^2 = \dot{y}^2$, $x_2^3 = \Delta\omega^3 = \dot{y}^3$, ..., $x_2^n = \Delta\omega^n = \dot{y}^n$. Moreover, it holds

$$\begin{aligned} \dot{x}_2^i &= -\tfrac{D_i}{2J_i}x_2^i + \tfrac{\omega_0}{2J_i}P_{mi} - \\ &-\tfrac{\omega_0}{2J_i}[G_{ii}{x_3^i}^2 + x_3^i\textstyle\sum_{j=1,j\neq i}^n[x_3^jG_{ij}sin(x_1^i - x_1^j - \alpha_{ij})] \end{aligned} \tag{8.22}$$

or, using the flat output variables

$$\begin{aligned} \ddot{y}^i &= -\tfrac{D_i}{2J_i}\dot{y}^i + \tfrac{\omega_0}{2J_i}P_{mi} - \\ &-\tfrac{\omega_0}{2J_i}[G_{ii}{x_3^i}^2 + x_3^i\textstyle\sum_{j=1,j\neq i}^n[x_3^jG_{ij}sin(y^i - y^j - \alpha_{ij})] \end{aligned} \tag{8.23}$$

The external mechanical torque P_{mi} is considered to be a piecewise constant variable. For $i = 1, 2, \ldots, n$ one obtains n equations of the form of Eq. (8.23), with unknowns the state variables x_3^i, $i = 1, 2, \ldots, n$. By solving this system of equations with respect to x_3^i, $i = 1, 2, \ldots, n$ one arrives at defining the state variables x_3^i as functions of the elements of the flat outputs vector y^i, $i = 1, 2, \ldots, n$ and of their derivatives. Thus one has $x_3^i = f_{x_3}(y^1, y^2, \ldots, y^n)$. Additionally, from the relation $\dot{E}_{q_i} = -\frac{1}{T_{d_i}}E'_{q_i} + \frac{1}{T_{do_i}}\frac{x_{d_i}-x'_{d_i}}{x_{d\Sigma_i}}V_{s_i}cos(\Delta\delta_i) + \frac{1}{T_{do\,i}}E_{f_i}$ and knowing that the state variables x_1^i, x_2^i, x_3^i, $i = 1, 2, \ldots, n$ can be written as functions of the flat output and its derivatives, one can solve with respect to the control input u_i thus showing that all control inputs u_i, $i = 1, 2, \ldots, n$ can be written as function of the flat output and its derivatives.

By expressing all elements of the multi-machine power generation system and the associated control inputs as functions of the flat output and its derivatives it is proven that the distributed power generation system stands for a differentially flat model.

8.2.4.3 The Case of Unknown Time-Varying Mechanical Input Torque

Functioning of the power generation system under unknown time-varying external mechanical torque P_m^i is the most generic condition, which also comprises the case of piecewise constant external input torque. It will be shown how with the use of the differential flatness theory a controller can be designed for the distributed power system in case of unknown time-varying external mechanical torque. By deriving the expression about $\ddot{y}^i$ once more with respect to time one obtains

$$y^{(3)^i} = a^i(x) + b_1^i(x)g_1u_1 + b_2^i(x)g_2u_2 + b_3^i(x)g_3u_3 \tag{8.24}$$

which means that one arrives at a function of the form $\dot{z}_3^i = a^i(x) + b_1^i(x)g_1u_1 + b_2^i(x)g_2u_2 + b_3^i(x)g_3u_3 + \tilde{d}^i$, where in the case of the distributed power generation that consists of $n = 3$ machines, and considering for instance $i = 1, j = 2, 3$ one has

$$\begin{array}{c} a^i = (\frac{D_i}{2J_i})^2x_2^i + \frac{D_i\omega_0}{(2J_i)^2}[G_{ii}{x_3^i}^2 + x_3^i\sum_{j=1,j\neq i}^n x_3^j G_{ij}sin(x_1^i - x_1^j - \alpha_{ij})] - \\ -\frac{\omega_0}{2J_i}[G_{ii}x_3^i + \sum_{j=1,j\neq i}^n x_3^j G_{ij}sin(x_1^i - x_1^j - 4\alpha_{ij})(-\frac{1}{T'_{d_i}}x_3^i + (\frac{1}{T_{doi}}\frac{x_{d_i}-x'_{d_i}}{x_{d\Sigma_i}'}V_{s_i}cos(x_1^i))] - \\ -\frac{\omega_0}{2J_i}x_3^i\sum_{j=1,j\neq i}^n G_{ij}sin(x_1^i - x_1^j - \alpha_{ij})(-\frac{1}{T'_{d_i}}x_3^i + (\frac{1}{T_{doi}}\frac{x_{d_i}-x'_{d_i}}{x_{d\Sigma_i}'}V_{s_i}cos(x_1^i)) - \\ -\frac{\omega_0}{2J_i}x_3^i\sum_{j=1,j\neq i}^n x_3^j G_{ij}cos(x_1^i - x_1^j - \alpha_{ij})x_2^i\frac{\omega_0}{2J_i}x_3^i\sum_{j=1,j\neq i}^n x_3^j G_{ij}cos(x_1^i - x_1^j - \alpha_{ij})x_2^j \end{array} \tag{8.25}$$

while for functions b_1^i, b_2^i and b_3^i it holds

$$\begin{array}{c} b_1^i = -\frac{\omega_0}{2J_i}[2G_{ii}x_3^i + \sum_{j=1,j\neq i}^n x_3^j G_{ij}sin(x_1^i - x_1^j - \alpha_{ij})]\frac{1}{T_{do\,i}} \\ b_2^i = -\frac{\omega_0}{2J_i}G_{i2}sin(x_1^i - x_1^2 - \alpha_{i2})\frac{1}{T_{do\,2}} \\ b_3^i = -\frac{\omega_0}{2J_i}G_{i3}sin(x_1^i - x_1^3 - \alpha_{i3})\frac{1}{T_{do\,3}} \end{array} \tag{8.26}$$

Finally for the disturbance term one has $\tilde{d}^i = -\frac{D_i\omega_0}{2J_i^2}P_m^i + \frac{\omega_0}{2J_i}\dot{P}_m^i$

Thus, one has the following description of the dynamics of the i-th power generator

$$\begin{aligned}\dot{z}_1^i &= z_2^i \\ \dot{z}_2^i &= z_3^i \\ \dot{z}_3^i &= a^i(x) + {b_1}^i g_1 u_1 + {b_2}^i g_2 u_2 + {b_3}^i g_3 u_3 + \tilde{d}^i\end{aligned} \tag{8.27}$$

For the complete system of the 3 generators one has

$$\begin{aligned}\dot{z}_3^1 &= a^1(x) + {b_1}^1 g_1 u_1 + {b_2}^1 g_2 u_2 + {b_3}^1 g_3 u_3 + \tilde{d}^1 \\ \dot{z}_3^2 &= a^2(x) + {b_1}^2 g_1 u_1 + {b_2}^2 g_2 u_2 + {b_3}^2 g_3 u_3 + \tilde{d}^2 \\ \dot{z}_3^3 &= a^3(x) + {b_1}^3 g_1 u_1 + {b_2}^3 g_2 u_2 + {b_3}^3 g_3 u_3 + \tilde{d}^1\end{aligned} \tag{8.28}$$

or in matrix form

$$\dot{z}_3 = f_a(x) + Mu + \tilde{d} \tag{8.29}$$

where $z_3 = [z_3^1, z_3^2, z_3^3]^T$, $u = [u_1, u_2, u_3]^T$ and $\tilde{d} = [\tilde{d}_1, \tilde{d}_2, \tilde{d}_3]^T$ while

$$f_a(x) = \begin{pmatrix} a^1(x) \\ a^2(x) \\ a^3(x) \end{pmatrix}, \quad M = \begin{pmatrix} {b_1}^1 g_1 & {b_2}^1 g_2 & {b_3}^1 g_3 \\ {b_1}^2 g_1 & {b_2}^2 g_2 & {b_3}^2 g_3 \\ {b_1}^3 g_1 & {b_2}^3 g_2 & {b_3}^3 g_3 \end{pmatrix} \tag{8.30}$$

Setting $v = f_a(x) + Mu + \tilde{d}$, one obtains again the linear canonical form for the i-th generator given by

$$\begin{pmatrix} \dot{z}_1^i \\ \dot{z}_2^i \\ \dot{z}_3^i \end{pmatrix} = \begin{pmatrix} 0 & 1 & 0 \\ 0 & 0 & 1 \\ 0 & 0 & 0 \end{pmatrix} \begin{pmatrix} z_1^i \\ z_2^i \\ z_3^i \end{pmatrix} + \begin{pmatrix} 0 \\ 0 \\ 1 \end{pmatrix} (v^i + \tilde{d}^i) \tag{8.31}$$

In this manner the initial nonlinear power system is transformed into three decoupled linear subsystems which are in the canonical Brunovksy form. For each one of these subsystems the appropriate control law is

$$v^i = z_d^{(3)^i} - k_3(\ddot{z}^i - \ddot{z}_d^i) - k_2(\dot{z}^i - \dot{z}_d^i) - k_1(z^i - z_d^i) - \tilde{d}^i \tag{8.32}$$

Since the disturbance term $\tilde{d}^i$, which is due to the time-varying mechanical input torque at the i-th generator is unknown, in the computation of the control input of Eq. (8.110) it will be substituted by its estimate $\hat{\tilde{d}}^i$, which will be provided by a disturbance observer. For the implementation of the distributed power generation control scheme, the controller at the i-th power generator makes use of not only its own state vector $X^i = [x_1^i, x_2^i, x_3^i]^T$, but also of the state vectors of the rest $n-1$ power generators, i.e. $x^j = [x_1^j, x_2^j, x_3^j]$, $j \neq i$, $j = 1, 2, \ldots, n$. The transformation of the dynamical model of the local power generators into the linear canonical form enables to obtain estimates of the transformed state vector of the system $Y^i = [y^i, \dot{y}^i, \ddot{y}^i]^T$ where $y^i = \Delta\delta^i$, through Kalman Filtering, after processing measurements of only the turn angle difference $\Delta\delta^i$ of the i-th power generator. Therefore one has to compute the estimates $\hat{X}^i = [\hat{x}_1^i, \hat{x}_2^i, \hat{x}_3^i]^T$ after using the estimate provided by the Kalman Filter $\hat{Y}^i = [\hat{y}^i, \hat{\dot{y}}^i, \hat{\ddot{y}}^i]^T$. As already explained, this is a highly data driven application and requires the exchange and processing in real-time of a large amount of data. It holds that

$$\hat{x}_1^i = \hat{y}^i \; \hat{x}_2^i = \hat{\dot{y}}^i \tag{8.33}$$

while for the computation of $\hat{x}_3^i$ for $i = 1, 2, \ldots, n$ one has to solve with respect to $\hat{x}_3^i$ the system of equations

$$\begin{gathered}
\hat{\ddot{y}}^1 = -\frac{D_1}{2J_1}\hat{\dot{y}}^1 + \frac{\omega_0}{2J_1}P_{m1} - \frac{\omega_0}{2J_1}[G_{11}x_3^{1^2} + +\hat{x}_3^1\textstyle\sum_{j=1,j\neq 1}^{n}[\hat{x}_3^j G_{1j} sin(y^1 - y^j - \alpha_{1j})] \\
\hat{\ddot{y}}^2 = -\frac{D_i}{2J_2}\hat{\dot{y}}^2 + \frac{\omega_0}{2J_2}P_{m2} - \frac{\omega_0}{2J_2}[G_{22}x_3^{2^2} + x_3^2\textstyle\sum_{j=1,j\neq 2}^{n}[\hat{x}_3^j G_{2j} sin(y^2 - y^j - \alpha_{2j})] \\
\cdots \\
\hat{\ddot{y}}^n = -\frac{D_n}{2J_n}\hat{\dot{y}}^n + \frac{\omega_0}{2J_n}P_{mn} - \frac{\omega_0}{2J_n}[G_{nn}x_3^{n^2} + \hat{x}_3^n\textstyle\sum_{j=1,j\neq n}^{n}[\hat{x}_3^j G_{nj} sin(y^i - y^j - \alpha_{ij})]
\end{gathered} \tag{8.34}$$

The computation of state estimates for the initial MIMO nonlinear system of the interconnected power generators, out of the estimates obtained for its linearized equivalent, can be also formulated as an optimization problem. To obtain state estimates for the initial system one comes against a set of coupled nonlinear equations described in Eq. (8.34). The latter can be solved with respect to the state vector elements of the initial system using ordinary computation tools for optimization and nonlinear programming [251, 307]. Moreover, by generalizing the results about state estimation in stand-alone synchronous generator to the distributed MIMO case of the interconnected electric power generators, it is possible to compute the state vector estimate for the initial system, through a filter of the form described in Eq. (3.15) that uses the inverse Jacobian matrix of the linearization transformation in its observation gain.

The proposed control scheme, considers as output of the synchronous generator the rotor's difference angle $\Delta\delta$. However, it is possible to avoid the use of encoder readings about the rotor's turn angle, and to indirectly estimate this parameter through the processing of measurement coming from PMUs. Such measurements are the bus voltage magnitude and the associated phase angle, the line current magnitude and the associated angle and the electrical output power at the terminal bus [111, 130, 158, 186].

8.2.5 *Estimation of PMSG Disturbance Input with Kalman Filtering*

8.2.5.1 State Estimation with the Derivative-Free Nonlinear Kalman Filter

As already shown in Chap. 3, with the use of differential flatness theory the nonlinear model of each PMSG can be written in the canonical form of Eqs. (3.11) and (3.54). Thus one has the linear model

$$\begin{aligned}\dot{y}_f &= A_f y_f + B_f v \\ z_f &= C_f y_f\end{aligned} \tag{8.35}$$

where $y_f = [y_1, y_2, y_3]^T$ and matrices A_f, B_f, C_f are in the canonical form

$$A_f = \begin{pmatrix} 0 & 1 & 0 \\ 0 & 0 & 1 \\ 0 & 0 & 0 \end{pmatrix} \quad B_f = \begin{pmatrix} 0 \\ 0 \\ 1 \end{pmatrix} \quad C_f^T = \begin{pmatrix} 1 \\ 0 \\ 0 \end{pmatrix} \tag{8.36}$$

and the measurable variable $y_1 = \delta$ is associated with the turn angle of the rotor. After carrying out discretization of matrices A_f, B_f and C_f with common discretization methods one can perform Kalman Filtering on the linear equivalent model. This is the *Derivative-free nonlinear Kalman filtering* [352, 353, 371].

8.2.5.2 Kalman Filter-Based Estimation of Disturbances

The procedure for identifying additive disturbance inputs that affect the individual power generators have been explained in Chap. 3. In the most generic case it can be assumed that the mechanical input torque P_m varies in time [143, 208]. In that case the aggregate disturbance input exerted on each generator's model is

$$T_m = -\omega_0 \tfrac{D}{(2J)^2} P_m + \tfrac{\omega_0}{2J} \dot{P}_m \tag{8.37}$$

It is also assumed that the dynamics of the disturbance term T_m is defined by its n-th order derivative $T_m^{(n)}$. Considering now that after expressing the system's state variables as functions of the flat outputs and their derivatives the PMSG's dynamics is given by

$$\begin{aligned} y^{(3)} &= f_c(y, \dot{y}, \ddot{y}) + g_c(y, \dot{y}, \ddot{y})u + \tfrac{1}{2H} \dot{P}_m \text{ or} \\ y^{(3)} &= v - \omega_0 \tfrac{D}{(2J)}^2 P_m + \tfrac{\omega_0}{2J} \dot{P}_m \text{ or} \\ y^{(3)} &= v - T_m \end{aligned} \tag{8.38}$$

where

$$f_c(y,\dot{y},\ddot{y}) = (\frac{D}{2J^2})\dot{y} - \omega_0\frac{D}{2J}\frac{P_m}{2J} + \frac{D}{(2J)^2}\omega_0\frac{V_s}{x'_{d\Sigma}}x_3 sin(\dot{y}) + \\ \frac{\omega_0}{2J}\frac{V_s}{x'_{d\Sigma}}\frac{1}{T'_d}x_3 sin(y) - \frac{\omega_0}{2J}\frac{V_s}{x'_{d\Sigma}}\frac{1}{T_{do}}\frac{x_d - x'_d}{x'_{d\Sigma}}V_s cos(y)sin(y) - \\ -\frac{\omega_0}{2H}\frac{V_s}{x'_{d\Sigma}}x_3 cos(y)\dot{y} \tag{8.39}$$

$$g_c(y,\dot{y},\ddot{y}) = -\frac{D}{2M}\frac{\omega_0}{T_{do}}\frac{V_s}{x'_{d\Sigma}}sin(y) \tag{8.40}$$

$$\begin{pmatrix}\dot{y}_1\\ \dot{y}_2\\ \ddot{y}_3\end{pmatrix} = \begin{pmatrix}0&1&0\\0&0&1\\0&0&0\end{pmatrix}\begin{pmatrix}y_1\\y_2\\y_3\end{pmatrix} + \begin{pmatrix}0\\0\\1\end{pmatrix}(v - T_m) \tag{8.41}$$

Next, the state vector of the model of Eq. (8.41) is extended to include as additional state variables the disturbance input T_m [169, 292]. Then, in the new state-space description one has $z_1 = y_1$, $z_2 = y_2$, $z_3 = y_3$, $z_4 = T_m = -\omega_0\frac{D}{(2J)^2}P_m + \frac{\omega_0}{2J}\dot{P}_m$, $z_5 = \dot{T}_m$, and $z_6 = \ddot{T}_m$. Without loss of generality, the disturbance input dynamics is assumed to be described by its third order derivative $\dot{z}_6 = T_m^{(3)}$.

Since the dynamics of the disturbance input are taken to be unknown in the design of the associated disturbances' estimator one has the following dynamics:

$$\dot{z}_o = \tilde{A}_o{\cdot}z + \tilde{B}_o{\cdot}\tilde{v} + K(C_o z - C_o\hat{z}) \tag{8.42}$$

where $\tilde{v} = \left(v\ T_m^{(3)}\right)^T$, $K{\in}R^{6\times 1}$ is the state estimator's gain and

$$\tilde{A}_o = \begin{pmatrix}0&1&0&0&0&0\\0&0&1&0&0&0\\0&0&0&1&0&0\\0&0&0&0&1&0\\0&0&0&0&0&1\\0&0&0&0&0&0\end{pmatrix}\ \tilde{B}_o = \begin{pmatrix}0&0\\0&0\\1&0\\0&0\\0&0\\0&0\end{pmatrix}\ \tilde{C}_o^T = \begin{pmatrix}1\\0\\0\\0\\0\\0\end{pmatrix} \tag{8.43}$$

and the measurable state variable is z_1. It can be confirmed that the disturbance observer model of the PMSG defined in Eq. (8.43) is observable. Defining as $\tilde{A}_d$, $\tilde{B}_d$, and $\tilde{C}_d$, the discrete-time equivalents of matrices $\tilde{A}_o$, $\tilde{B}_o$ and $\tilde{C}_o$ respectively, a Derivative-free nonlinear Kalman Filter can be designed for the aforementioned representation of the system dynamics. The associated Kalman Filter-based disturbance estimator is given by [34, 209, 334, 341, 363]

measurement update:

$$K(k) = P^-(k)\tilde{C}_d^T[\tilde{C}_d{\cdot}P^-(k)\tilde{C}_d^T + R]^{-1} \\ \hat{z}(k) = \hat{z}^-(k) + K(k)[z(k) - \tilde{C}_d\hat{z}^-(k)] \\ P(k) = P^-(k) - K(k)\tilde{C}_d P^-(k) \tag{8.44}$$

time update:

$$P^{-}(k+1) = \tilde{A}_d(k)P(k)\tilde{A}_d^T(k) + Q(k)$$
$$\hat{z}^{-}(k+1) = \tilde{A}_d(k)\hat{z}(k) + \tilde{B}_d(k)\tilde{v}(k) \quad (8.45)$$

To compensate for the effects of the disturbance terms it suffices to use in the control loop the modified control input, which actually compensates for the effects of the external disturbance term T_m. This gives $v^* = v - \hat{T}_m$ or $v^* = v - \hat{z}_4$.

8.2.6 Simulation Experiments

To evaluate the performance of the proposed nonlinear control scheme, that uses Kalman Filtering to estimate the nonmeasurable state vector elements of the distributed PMSGs and the external disturbances, simulation experiments have been carried out. Different rotation speed setpoints had been assumed. Moreover, different input torques (mechanical input power profiles) have been assumed to affect the dynamic model of each local PMSG. Indicative ratings of the PMSG model are as follows: direct axis synchronous reactance x_d = 2.1 p.u., quadrature axis synchronous reactance x_d' = 0.4 p.u., infinite bus voltage V_s = 1.0 p.u., direct axis open circuit time constant T_{d_o} = 5 s, rotor's moment of inertia J = 50 kgm^2. The nonlinear control scheme works efficiently even for distributed synchronous generators with uneven ratings.

The case of distributed PMSG operation under unknown input power (torque) has been examined. The input power at each local generator was considered to be a disturbance input to the PMSG model and it was assumed that its change in time was defined by the third derivative of the associated variable, i.e. $T_m^{(3)}$ where, according to Eq. (8.37), $T_m = -\omega_0 \frac{D}{(2J)^2} P_m + \frac{\omega_0}{2J}\dot{P}_m$. The disturbance dynamics was completely unknown to the controller and its identification was performed in real time by the disturbance estimator. It is shown that the derivative-free nonlinear Kalman Filter redesigned as a disturbance observer is capable of estimating simultaneously the nonmeasurable state variables of the generator (ω and E_q'), as well as the unknown disturbance input T_m. A first set of results is concerned with the tracking performance of the control loop in case of piecewise constant disturbance input. The real value of the state variable of the generator is plotted with blue color, the estimated values is given with green color, while the associated setpoint is printed in red color. This is shown in Figs. 8.2a, 8.3a and 8.4a. The estimation of the piecewise constant disturbance input for the first generator is shown in Fig. 8.2b, for the second generator is given in Fig. 8.3b, while for the third generator it is shown in Fig. 8.4b.

Additionally, results about the tracking performance of the control loop in case of a second piecewise constant disturbance input are shown Figs. 8.3a, 8.5a and 8.7a. The estimation of the piecewise constant disturbance input for the first generator is shown in Fig. 8.5b, for the second generator is given in Fig. 8.6b, while for the third

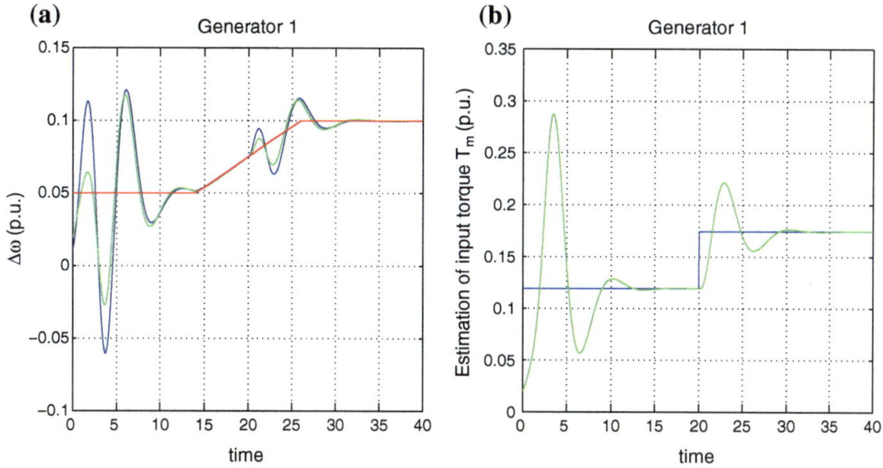

Fig. 8.2 Sensorless control of the PMSG No 1 under non-measurable (piecewise constant) mechanical input torque in case of speed reference setpoint 1: **a** convergence of the real and estimated values of the angular speed difference $\Delta\omega$ **b** estimation of the external mechanical input torque T_m^1

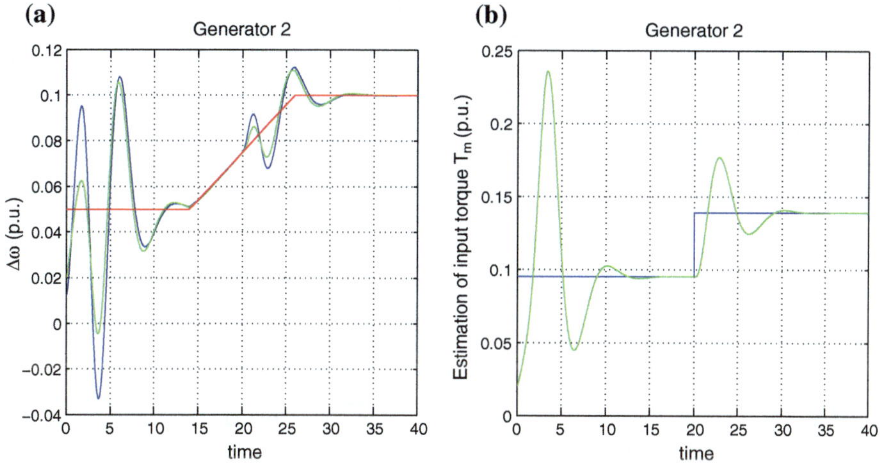

Fig. 8.3 Sensorless control of the PMSG No 2 under non-measurable (piecewise constant) mechanical input torque in case of speed reference setpoint 1: **a** convergence of the real and estimated values of the angular speed difference $\Delta\omega$ **b** estimation of the external mechanical input torque T_m^2

generator it is shown in Fig. 8.7b. The units of the PMSG state variables have been expressed again in p.u. (per unit system).

The simulation experiments have confirmed that the proposed state estimation-based control scheme not only enables implementation of distributed PMSG control through the measurement of a small number of variables (e.g. of only the rotor's turn

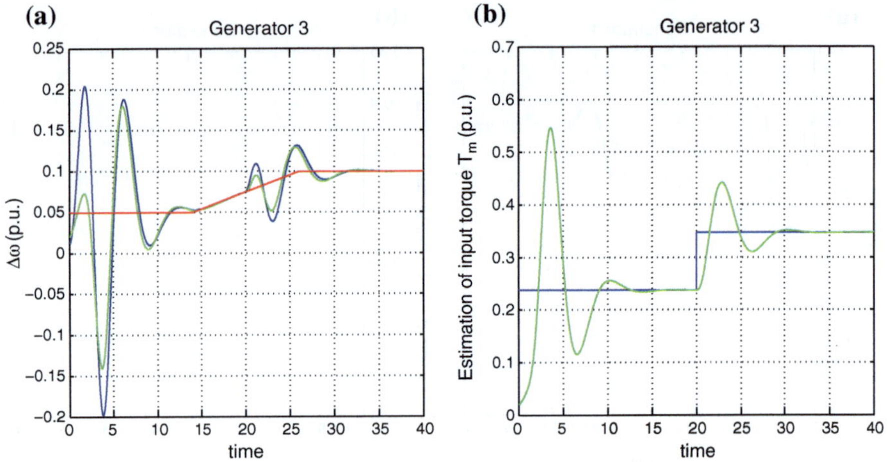

Fig. 8.4 Sensorless control of the PMSG No 3 under non-measurable (piecewise constant) mechanical input torque in case of speed reference setpoint 1: **a** convergence of the real and estimated values of the angular speed difference $\Delta\omega$ **b** estimation of the external mechanical input torque T_m^3

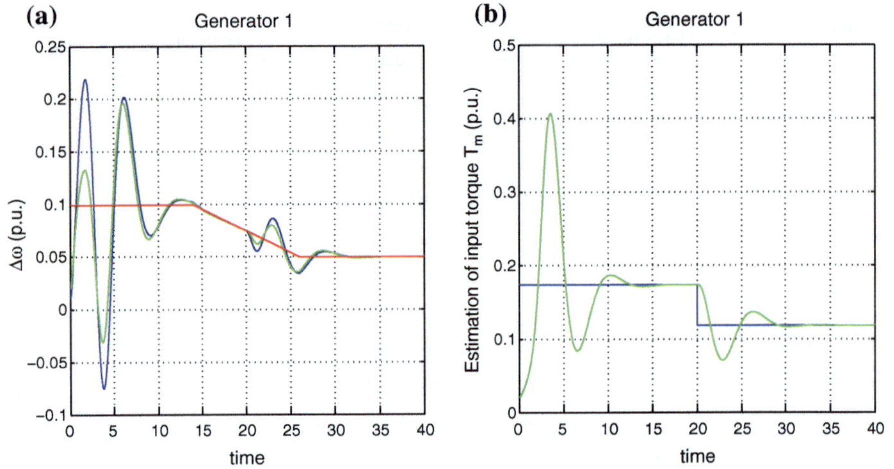

Fig. 8.5 Sensorless control of the PMSG No 1 under non-measurable (piecewise constant) mechanical input torque in case of speed reference setpoint 2: **a** convergence of the real and estimated values of the angular speed difference $\Delta\omega$ **b** estimation of the external mechanical input torque T_m^1

angle) but also improves the robustness of the PMSG control loop in case of varying speed setpoints and varying mechanical input torque.

Comparison to the Extended Kalman Filter, which is the most commonly used nonlinear estimation technique, has shown that the Derivative-free nonlinear Kalman Filter succeeds estimation of smaller variance for both the nonmeasurable state vector elements of the PMSG and the term T_m expressing external torques and disturbances

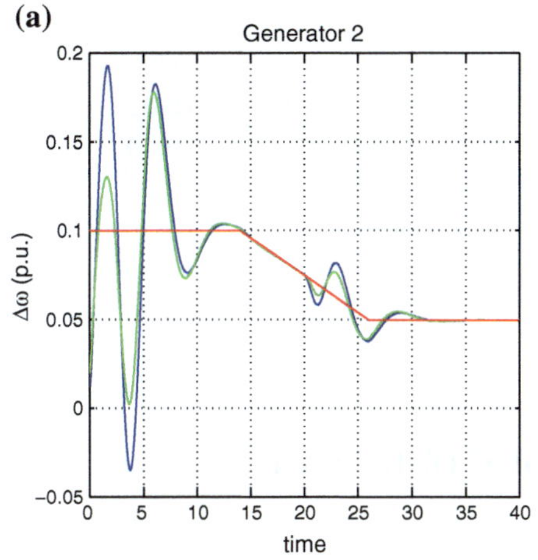

Fig. 8.6 Sensorless control of the PMSG no 2 under non-measurable (piecewise constant) mechanical input torque in case of speed reference setpoint 2: **a** convergence of the real and estimated values of the angular speed difference $\Delta\omega$ **b** estimation of the external mechanical input torque T_m^2

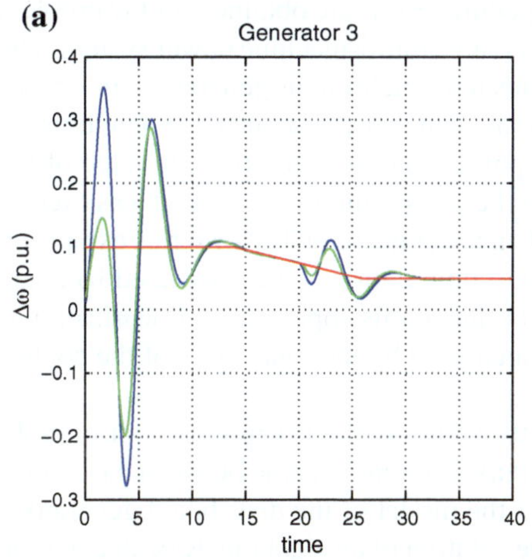

Fig. 8.7 Sensorless control of the PMSG No 3 under non-measurable (piecewise constant) mechanical input torque in case of speed reference setpoint 2: **a** convergence of the real and estimated values of the angular speed difference $\Delta\omega$ **b** estimation of the external mechanical input torque T_m^3

inputs. Estimation with the Extended Kalman Filter was based on local linearization of the PMSG model and on the computation of the Jacobian of Eq. (8.6). The obtained results are summarized in Table 8.1, and correspond to experiments (a) and (b) associated with the two turn speed setpoints given in Figs. 8.2, 8.3, 8.4, 8.5, 8.6 and 8.7 under sinusoidal variations of T_m, and considering different noise levels.

Table 8.1 RMSE for nonlinear filters

	DKF_a	EKF_a	DKF_b	EKF_b
$\Delta\delta$	0.0006	0.0007	0.0011	0.0012
$\Delta\omega$	0.0061	0.0093	0.0069	0.0119
E'_q	0.0167	0.1654	0.0376	0.2253
T_m	0.0172	0.0843	0.0184	0.0810
$\dot{T}_m$	0.0137	0.0374	0.0146	0.0339
$\ddot{T}_m$	0.0054	0.0083	0.0057	0.0080

8.3 Nonlinear H-Infinity Control of Distributed Synchronous Generators

8.3.1 Overview

In this section a new nonlinear H-infinity optimal control method for distributed synchronous generators is developed and specific results are obtained for the problem of synchronization and stabilization of a three-area multi-machine power system. The nonlinear dynamical model of the interconnected synchronous generators undergoes linearization round temporary operating points which are computed at each iteration of the control algorithm. These operating points consist of the present value of the system's state vector and of the last value of the control inputs vector that was exerted on it. This linearization procedure is based on the concept of Taylor series expansion of the system's states-space model and on the computation of the associated Jacobian matrices [34, 341, 344]. The modelling error due to this approximate linearization is considered to be a perturbation that is compensated by the robustness of the control scheme [352, 376, 380].

The feedback control input is based on the solution of an H-infinity control problem [352, 376, 380]. This stands for a mini-max differential game between the control signal and the disturbance terms that affect the model of the distributed generators. The feedback control gain is found from the solution of an algebraic Riccati equation at each iteration of the control algorithm [352, 376, 380]. The stability of the control scheme is demonstrated through Lyapunov analysis. Actually, it is shown that the control loop satisfies an H-infinity tracking performance criterion which signifies elevated robustness against model uncertainty and perturbation inputs. Besides, under moderate conditions the asymptotic stability of the control scheme is proven. Through simulation experiments it is further confirmed that the proposed control method assures fast and accurate tracking of reference setpoints.

8.3.2 Dynamic Model of the Multi-machine Power System

8.3.2.1 State-Space Description of the Distributed SGs Model

The state-space model of the multi-machine power system was analyzed in the previous sections. A multi-machine power system (MMPS) with n machines, in which the first machine is chosen as the reference machine can be described by the following nonlinear dynamic model [352, 375]

$$\begin{aligned}\dot{\delta}_i &= \omega_i - \omega_0 \\ \dot{\omega}_i &= -\frac{D_i}{2J_i}(\omega_i - \omega_0) + \omega_0\frac{P_{m_i}}{2J_i} - \\ -\omega_0\frac{1}{2J_i}[G_{ii}{E'_{qi}}^2 &+ E'_{qi}\sum_{j=1,j\neq i}^{n} E'_{qj}G_{ij}sin(\delta_i - \delta_j - \alpha_{ij})] \\ \dot{E}'_{q_i} &= -\frac{1}{T'_{d_i}}E'_{q_i} + \frac{1}{T_{do_i}}\frac{x_{d_i} - x'_{d_i}}{x'_{d\Sigma i}}V_{s_i}cos(\Delta\delta_i) + \frac{1}{T_{do_i}}E_{fi}\end{aligned} \tag{8.46}$$

where the electric torque P_{e_i} which is associated with the active power at the i-th generator is now given by

$$P_{e_i} = G_{ii}{E'_{qi}}^2 + E'_{qi}\sum_{j=1,j\neq i}^{n} E'_{qj}G_{ij}sin(\delta_i - \delta_j - \alpha_{ij}) \tag{8.47}$$

for $i = 1, 2\ldots, n$. For a power grid that consists of n generators the aggregate state vector comprises the state vectors of the local machines, i.e. $x = [x^1, x^2, \ldots, x^n]^T$, where $x^i = [x_1^i, x_2^i, x_3^i]^T$, with $x_1^i = \Delta\delta_i$, $x_2^i = \Delta\omega_i$ and $x_3^i = E'_{qi}$ are the state variables for the i-th machine and $i = 1, 2, \ldots, n$.

Without loss of generality the three-machine power system of Fig. 8.8 is considered. The associated dynamic model is given by:

$$\dot{x}_1 = x_2 - \omega_0 \tag{8.48}$$

$$\begin{aligned}\dot{x}_2 &= -\frac{D_1}{2J_1}(x_2 - \omega_0) + \omega_0\frac{P_{m_1}}{2J_1} - \\ -\frac{\omega_0}{2J_1}\{G_{11}x_3^2 &+ x_3[x_6G_{12}sin(x_1 - x_4 - a_{12}) + x_9G_{13}sin(x_1 - x_7 - a_{13})]\end{aligned} \tag{8.49}$$

$$\dot{x}_3 = -\frac{1}{T'_{d_1}}x_3 + \frac{1}{T_{do_1}}\frac{x_{d_1} - x'_{d_1}}{x'_{d\Sigma_1}}V_s cos(x_1) + \frac{1}{T_{do_1}}u_1 \tag{8.50}$$

$$\dot{x}_4 = x_5 - \omega_0 \tag{8.51}$$

$$\begin{aligned}\dot{x}_5 &= -\frac{D_2}{2J_2}(x_5 - \omega_0) + \omega_0\frac{P_{m_2}}{2J_2} - \\ -\frac{\omega_0}{2J_2}\{G_{22}x_6^2 &+ x_6[x_3G_{21}sin(x_4 - x_1 - a_{21}) + x_9G_{23}sin(x_4 - x_7 - a_{23})]\end{aligned} \tag{8.52}$$

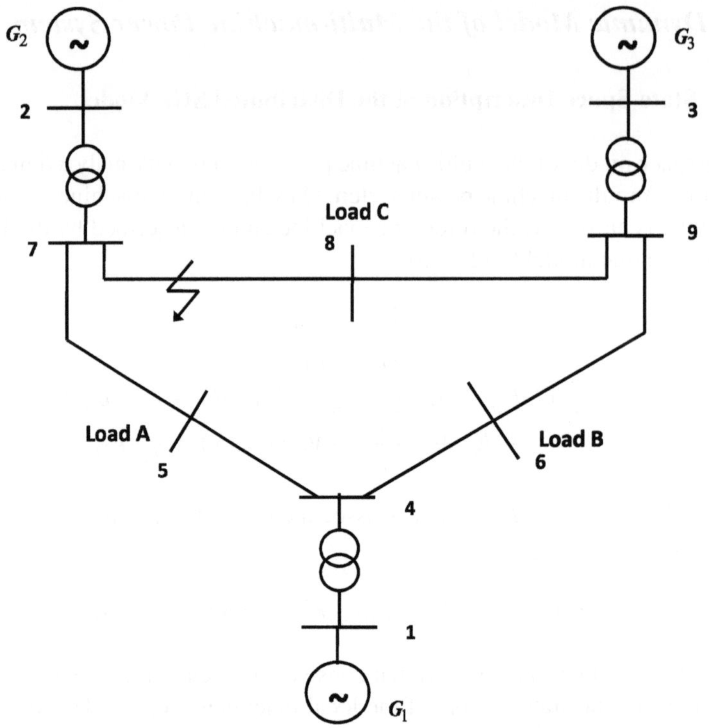

Fig. 8.8 A 3-area distributed power generation model consisting of 3 synchronous machines

$$\dot{x}_6 = -\frac{1}{T'_{d_2}}x_6 + \frac{1}{T_{do_2}}\frac{x_{d_2}-x'_{d_2}}{x'_{d\Sigma_2}}V_s cos(x_4) + \frac{1}{T_{do_2}}u_2 \tag{8.53}$$

$$\dot{x}_7 = x_8 - \omega_0 \tag{8.54}$$

$$\begin{aligned}\dot{x}_8 = -\frac{D_2}{2J_3}(x_8-\omega_0) + \omega_0\frac{P_{m_3}}{2J_3} - \\ -\frac{\omega_0}{2J_3}\{G_{32}x_9^2 + x_9[x_3 G_{31} sin(x_7-x_1-a_{31}) + x_6 G_{32} sin(x_7-x_4-a_{32})]\end{aligned} \tag{8.55}$$

$$\dot{x}_9 = -\frac{1}{T'_{d_3}}x_9 + \frac{1}{T_{do_3}}\frac{x_{d_3}-x'_{d_3}}{x'_{d\Sigma_3}}V_s cos(x_7) + \frac{1}{T_{do_3}}u_3 \tag{8.56}$$

The system is also written in the matrix form:

$$\begin{pmatrix} \dot{x}_1 \\ \dot{x}_2 \\ \dot{x}_3 \\ \dot{x}_4 \\ \dot{x}_5 \\ \dot{x}_6 \\ \dot{x}_6 \\ \dot{x}_7 \\ \dot{x}_8 \\ \dot{x}_9 \end{pmatrix} = \begin{pmatrix} f_1(x) \\ f_2(x) \\ f_3(x) \\ f_4(x) \\ f_5(x) \\ f_6(x) \\ f_7(x) \\ f_8(x) \\ f_9(x) \end{pmatrix} + \begin{pmatrix} 0 & 0 & 0 \\ 0 & 0 & 0 \\ g_1(x) & 0 & 0 \\ 0 & 0 & 0 \\ 0 & 0 & 0 \\ 0 & g_2(x) & 0 \\ 0 & 0 & 0 \\ 0 & 0 & 0 \\ 0 & 0 & g_3(x) \end{pmatrix} \begin{pmatrix} u_1 \\ u_2 \\ u_3 \end{pmatrix} \tag{8.57}$$

where

$$f_1(x) = x_2 - \omega_0 \tag{8.58}$$

$$\begin{array}{c} f_2(x) = -\frac{D_1}{2J_1}(x_2 - \omega_0) + \omega_0 \frac{P_{m_1}}{2J_1} - \\ -\frac{\omega_0}{2J_1}\{G_{11}x_3^2 + x_3[x_6 G_{12} sin(x_1 - x_4 - a_{12}) + x_9 G_{13} sin(x_1 - x_7 - a_{13})] \end{array} \tag{8.59}$$

$$f_3(x) = -\frac{1}{T'_{d_1}}x_3 + \frac{1}{T_{do_1}}\frac{x_{d_1} - x'_{d_1}}{x'_{d\Sigma_1}} V_s cos(x_1) \tag{8.60}$$

$$f_4(x) = x_5 - \omega_0 \tag{8.61}$$

$$\begin{array}{c} f_5(x) = -\frac{D_2}{2J_2}(x_5 - \omega_0) + \omega_0 \frac{P_{m_2}}{2J_2} - \\ -\frac{\omega_0}{2J_2}\{G_{22}x_6^2 + x_6[x_3 G_{21} sin(x_4 - x_1 - a_{21}) + x_9 G_{23} sin(x_4 - x_7 - a_{23})] \end{array} \tag{8.62}$$

$$f_6(x) = -\frac{1}{T'_{d_2}}x_6 + \frac{1}{T_{do_2}}\frac{x_{d_2} - x'_{d_2}}{x'_{d\Sigma_2}} V_s cos(x_4) \tag{8.63}$$

$$f_7(x) = x_8 - \omega_0 \tag{8.64}$$

$$\begin{array}{c} f_8(x) = -\frac{D_2}{2J_3}(x_8 - \omega_0) + \omega_0 \frac{P_{m_3}}{2J_3} - \\ -\frac{\omega_0}{2J_3}\{G_{32}x_9^2 + x_9[x_3 G_{31} sin(x_7 - x_1 - a_{31}) + x_6 G_{32} sin(x_7 - x_4 - a_{32})] \end{array} \tag{8.65}$$

$$f_9(x) = -\frac{1}{T'_{d_3}}x_9 + \frac{1}{T_{do_3}}\frac{x_{d_3} - x'_{d_3}}{x'_{d\Sigma_3}} V_s cos(x_7) \tag{8.66}$$

while one has that

$$g_1 = \frac{1}{T_{do_1}} \; g_2 = \frac{1}{T_{do_2}} \; g_3 = \frac{1}{T_{do_3}} u_3 \tag{8.67}$$

8.3.3 Linearization of the Model of the Distributed Synchronous Generators

Local linearization is performed for the state-space model of the distributed power generators, though Taylor series expansion, round the operating point (x^*, u^*) where x^* is the present value of the system's state vector and u^* is the last value of the control input that was exerted on the machine. Thus, one obtains the linearized description

$$\dot{x} = Ax + Bu + \tilde{d} \tag{8.68}$$

For the previous description of the distributed power generators' model (3-area 3-machine model) by the state-space equation of Eq. (8.57) it holds that $A = \nabla_x \nabla_x [f(x) + g(x)u]$ with

$$A = \nabla_x[f(x) + g(x)u] = \begin{pmatrix} \frac{\partial f_1}{\partial x_1} & \frac{\partial f_1}{\partial x_2} & \frac{\partial f_1}{\partial x_3} & \cdots & \frac{\partial f_1}{\partial x_9} \\ \frac{\partial f_2}{\partial x_1} & \frac{\partial f_2}{\partial x_2} & \frac{\partial f_2}{\partial x_3} & \cdots & \frac{\partial f_2}{\partial x_9} \\ \frac{\partial f_3}{\partial x_1} & \frac{\partial f_3}{\partial x_2} & \frac{\partial f_3}{\partial x_3} & \cdots & \frac{\partial f_3}{\partial x_9} \\ \frac{\partial f_4}{\partial x_1} & \frac{\partial f_4}{\partial x_2} & \frac{\partial f_4}{\partial x_3} & \cdots & \frac{\partial f_4}{\partial x_9} \\ \cdots & \cdots & \cdots & \cdots & \cdots \\ \cdots & \cdots & \cdots & \cdots & \cdots \\ \frac{\partial f_5}{\partial x_1} & \frac{\partial f_5}{\partial x_2} & \frac{\partial f_5}{\partial x_3} & \cdots & \frac{\partial f_5}{\partial x_9} \end{pmatrix} \tag{8.69}$$

For the first row of the Jacobian matrix $A = \nabla_x[f(x) + g(x)u]$ one has: $\frac{\partial f_1}{\partial x_1} = 0$, $\frac{\partial f_1}{\partial x_2} = 1$, $\frac{\partial f_1}{\partial x_3} = 0$, $\frac{\partial f_1}{\partial x_4} = 0$, $\frac{\partial f_1}{\partial x_5} = 0$, $\frac{\partial f_1}{\partial x_6} = 0$, $\frac{\partial f_1}{\partial x_7} = 0$, $\frac{\partial f_1}{\partial x_8} = 0$, $\frac{\partial f_1}{\partial x_9} = 0$.

For the second row of the Jacobian matrix $A = \nabla_x[f(x) + g(x)u]$ one has: $\frac{\partial f_2}{\partial x_1} = -\frac{\omega_0}{2J_1} x_3[x_6 G_{12} cos(x_1 - x_4 - a_{13})] + x_9 G_{13} cos(x_1 - x_7 - a_{13})$, $\frac{\partial f_2}{\partial x_2} = -\frac{D_1}{2J_1}$, $\frac{\partial f_2}{\partial x_3} = \frac{-\omega_0}{2J_1}\{G_{11} 2x_3 + [x_6] G_{12} sin(x_1 - x_4 - a_{12}) + x_9 G_{13} sin(x_1 - x_7 - a_{13})]\}$, $\frac{\partial f_2}{\partial x_4} = \frac{\omega_0}{2J_1} x_3 x_6 G_{12} cos(x_1 - x_4 - a_{12})$, $\frac{\partial f_2}{\partial x_5} = 0$, $\frac{\partial f_2}{\partial x_6} = -\frac{\omega_0}{2J_1} x_3 G_{12} sin(x_1 - x_4 - a_{12})$, $\frac{\partial f_2}{\partial x_7} = \frac{\omega_0}{2J_1} x_3 x_9 G_{13} cos(x_1 - x_7 - a_{13})$, $\frac{\partial f_2}{\partial x_8} = 0$, $\frac{\partial f_2}{\partial x_9} = -\frac{\omega_0}{2J_1} x_3 G_{13} sin(x_1 - x_7 - a_{13})$.

For the third row of the Jacobian matrix $A = \nabla_x[f(x) + g(x)u]$ one has: $\frac{\partial f_3}{\partial x_1} = -\frac{1}{T_{do_1}} \frac{x_{d_1} - x'_{d_1}}{x'_{d\Sigma_1}} V_s sin(x_1)$, $\frac{\partial f_3}{\partial x_2} = 0$, $\frac{\partial f_3}{\partial x_3} = -\frac{1}{T'_{d_1}}$, $\frac{\partial f_3}{\partial x_4} = 0$, $\frac{\partial f_3}{\partial x_5} = 0$, $\frac{\partial f_3}{\partial x_6} = 0$, $\frac{\partial f_3}{\partial x_7} = 0$, $\frac{\partial f_3}{\partial x_8} = 0$, $\frac{\partial f_3}{\partial x_9} = 0$.

For the fourth row of the Jacobian matrix $A = \nabla_x[f(x) + g(x)u]$ one has: $\frac{\partial f_4}{\partial x_1} = 0$, $\frac{\partial f_4}{\partial x_2} = 0$, $\frac{\partial f_4}{\partial x_3} = 0$, $\frac{\partial f_4}{\partial x_4} = 0$, $\frac{\partial f_4}{\partial x_5} = 1$, $\frac{\partial f_4}{\partial x_6} = 0$, $\frac{\partial f_1}{\partial x_7} = 0$, $\frac{\partial f_1}{\partial x_8} = 0$, $\frac{\partial f_1}{\partial x_9} = 0$.

For the fifth row of the Jacobian matrix $A = \nabla_x[f(x) + g(x)u]$ one has: $\frac{\partial f_5}{\partial x_1} = \frac{\omega_0}{2J2} x_6[x_3 G_{21} cos(x_4 - x_1 - a_{21}) + x_3 G_{23} cos(x_4 - x_7 - a_{23})]$, $\frac{\partial f_5}{\partial x_2} = 0$, $\frac{\partial f_5}{\partial x_3} = -\frac{\omega_0}{2J_2} x_6 G_{21} sin(x_4 - x_1 - a_{21})$, $\frac{\partial f_5}{\partial x_4} = -\frac{\omega_0}{2J_2} x_6[x_3 G_{21} cos(x_4 - x_1 - a_{21}) + x_9 G_{23} cos(x_4 -$

$x_7-a_{23})]$, $\frac{\partial f_5}{\partial x_5}=-\frac{D_2}{2J_2}$, $\frac{\partial f_5}{\partial x_6}=-\frac{\omega_0}{2J_2}\{G_{22}2x_6+[x_3G_{21}sin(x_4-x_1-a_{21})+x_9G_{23}sin(x_4-x_7-a_{23})]\}$, $\frac{\partial f_5}{\partial x_7}=\frac{\omega_0}{2J_2}x_6x_9G_{23}cos(x_4-x_7-a_{23})$, $\frac{\partial f_5}{\partial x_8}=0$, $\frac{\partial f_5}{\partial x_9}=-\frac{\omega_0}{2J_2}x_6G_{23}sin(x_4-x_7-a_{23})$.

For the sixth row of the Jacobian matrix $A=\nabla_x[f(x)+g(x)u]$ one has: $\frac{\partial f_6}{\partial x_1}=0$, $\frac{\partial f_6}{\partial x_2}=0$, $\frac{\partial f_6}{\partial x_3}=$, $\frac{\partial f_6}{\partial x_4}=-\frac{1}{T_{do_2}}\frac{x_{d_2}-x'_{d_2}}{x'_{d_{\Sigma_2}}}V_s sin(x_4)$, $\frac{\partial f_6}{\partial x_5}=0$, $\frac{\partial f_6}{\partial x_6}=-\frac{1}{T'_{d_2}}$, $\frac{\partial f_6}{\partial x_7}=0$, $\frac{\partial f_6}{\partial x_8}=0$, $\frac{\partial f_6}{\partial x_9}=0$.

For the seventh row of the Jacobian matrix $A=\nabla_x[f(x)+g(x)u]$ one has: $\frac{\partial f_7}{\partial x_1}=0$, $\frac{\partial f_7}{\partial x_2}=0$, $\frac{\partial f_7}{\partial x_3}=0$, $\frac{\partial f_7}{\partial x_4}=0$, $\frac{\partial f_7}{\partial x_5}=0$, $\frac{\partial f_7}{\partial x_6}=0$, $\frac{\partial f_7}{\partial x_7}=0$, $\frac{\partial f_7}{\partial x_8}=1$, $\frac{\partial f_7}{\partial x_9}=0$.

For the eight row of the Jacobian matrix $A=\nabla_x[f(x)+g(x)u]$ one has: $\frac{\partial f_8}{\partial x_1}=\frac{\omega_0}{2J_3}x_9[x_3G_{31}cos(x_7-x_1-a_{31})+x_6G_{32}cos(x_7-x_4-a_{32})]$, $\frac{\partial f_8}{\partial x_2}=0$, $\frac{\partial f_8}{\partial x_3}=-\frac{\omega_0}{2J_3}x_9G_{31}sin(x_7-x_1-a_{31})$, $\frac{\partial f_8}{\partial x_4}=\frac{\omega_0}{2J_3}x_9x_6G_{32}cos(x_7-x_4-a_{32})$, $\frac{\partial f_8}{\partial x_5}=0$, $\frac{\partial f_8}{\partial x_6}=-\frac{\omega_0}{2J_3}x_9G_{23}sin(x_7-x_4-a_{32})$, $\frac{\partial f_8}{\partial x_7}=-\frac{\omega_0}{2J_3}x_9[x_3G_{31}cos(x_7-x_1-a_{31})+x_6G_{32}cos(x_7-x_4-a_{32})]$, $\frac{\partial f_8}{\partial x_8}=-\frac{D_3}{2J3}$, $\frac{\partial f_8}{\partial x_9}=-\frac{\omega_0}{2J_3}\{G_{33}2x_9+[x_3G_{31}sin(x_7-x_1-a_{31})+x_6G_{32}sin(x_7-x_4-a_{32})]\}$.

For the ninth row of the Jacobian matrix $A=\nabla_x[f(x)+g(x)u]$ one has: $\frac{\partial f_9}{\partial x_1}=0$, $\frac{\partial f_9}{\partial x_2}=0$, $\frac{\partial f_9}{\partial x_3}=0$, $\frac{\partial f_9}{\partial x_4}=0$, $\frac{\partial f_9}{\partial x_6}=0$, $\frac{\partial f_9}{\partial x_7}=-\frac{1}{T_{do_3}}\frac{x_{d_3}-x'_{d_3}}{x'_{d_{\Sigma_3}}}V_s sin(x_7)$, $\frac{\partial f_9}{\partial x_8}=0$, and finally $\frac{\partial f_9}{\partial x_9}=-\frac{1}{T'_{d_3}}$.

Moreover, it holds that $B=\nabla_u\nabla_x[f(x)+g(x)u]$ with

$$B=\nabla_u[f(x)+g(x)u]=\begin{pmatrix} 0 & 0 & 0 \\ 0 & 0 & 0 \\ g_1(x) & 0 & 0 \\ 0 & 0 & 0 \\ 0 & 0 & 0 \\ 0 & g_2(x) & 0 \\ 0 & 0 & 0 \\ 0 & 0 & 0 \\ 0 & 0 & g_3(x) \end{pmatrix} \tag{8.70}$$

8.3.4 *The Nonlinear H-Infinity Control*

8.3.4.1 Mini-Max Control and Disturbance Rejection

The design of an H-infinity controller was clearly analyzed in the case of stand-alone synchronous generators. The initial nonlinear model of the distributed synchronous generators is in the form

$$\dot{x}=f(x,u)\quad x\in R^n,\ u\in R^m \tag{8.71}$$

Linearization of the system (distributed synchronous generators) is performed at each iteration of the control algorithm round its present operating point $(x^*, u^*) = (x(t), u(t - T_s))$. The linearized equivalent of the system is described by

$$\dot{x} = Ax + Bu + L\tilde{d} \quad x \in R^n, \ u \in R^m, \ \tilde{d} \in R^q \tag{8.72}$$

where matrices A and B are obtained from the computation of the Jacobians

$$A = \begin{pmatrix} \frac{\partial f_1}{\partial x_1} & \frac{\partial f_1}{\partial x_2} & \cdots & \frac{\partial f_1}{\partial x_n} \\ \frac{\partial f_2}{\partial x_1} & \frac{\partial f_2}{\partial x_2} & \cdots & \frac{\partial f_2}{\partial x_n} \\ \cdots & \cdots & \cdots & \cdots \\ \frac{\partial f_n}{\partial x_1} & \frac{\partial f_n}{\partial x_2} & \cdots & \frac{\partial f_n}{\partial x_n} \end{pmatrix} |_{(x^*, u^*)} \ B = \begin{pmatrix} \frac{\partial f_1}{\partial u_1} & \frac{\partial f_1}{\partial u_2} & \cdots & \frac{\partial f_1}{\partial u_m} \\ \frac{\partial f_2}{\partial u_1} & \frac{\partial f_2}{\partial u_2} & \cdots & \frac{\partial f_2}{\partial u_m} \\ \cdots & \cdots & \cdots & \cdots \\ \frac{\partial f_n}{\partial u_1} & \frac{\partial f_n}{\partial u_2} & \cdots & \frac{\partial f_n}{\partial u_m} \end{pmatrix} |_{(x^*, u^*)} \tag{8.73}$$

and vector $\tilde{d}$ denotes disturbance terms due to linearization errors. The problem of disturbance rejection for the linearized model that is described by

$$\begin{aligned} \dot{x} &= Ax + Bu + L\tilde{d} \\ y &= Cx \end{aligned} \tag{8.74}$$

where $x \in R^n$, $u \in R^m$, $\tilde{d} \in R^q$ and $y \in R^p$, cannot be handled efficiently if the classical LQR control scheme is applied. This because of the existence of the perturbation term $\tilde{d}$. The disturbance term $\tilde{d}$ apart from modeling (parametric) uncertainty and external perturbation terms can also represent noise terms of any distribution.

In the H_∞ control approach, a feedback control scheme is designed for trajectory tracking by the system's state vector and simultaneous disturbance rejection, considering that the disturbance affects the system in the worst possible manner. The disturbances' effect are incorporated in the following quadratic cost function:

$$J(t) = \frac{1}{2}\int_0^T [y^T(t)y(t) + ru^T(t)u(t) - \rho^2\tilde{d}^T(t)\tilde{d}(t)]dt, \quad r, \rho > 0 \tag{8.75}$$

As in the case of nonlinear H-infinity control for stand-alone generators, in the case of distributed synchronous generators the significance of the negative sign in the cost function's term that is associated with the perturbation variable $\tilde{d}(t)$ is that the disturbance tries to maximize the cost function $J(t)$ while the control signal $u(t)$ tries to minimize it. The physical meaning of the relation given above is that the control signal and the disturbances compete to each other within a mini-max differential game. This problem of mini-max optimization can be written as

$$min_u max_{\tilde{d}} J(u, \tilde{d}) \tag{8.76}$$

Again the objective of the optimization procedure is to compute a control signal $u(t)$ which can compensate for the worst possible disturbance, that is externally imposed to the system. However, the solution to the mini-max optimization problem is directly related to the value of the parameter ρ. This means that there is an upper bound in the disturbances magnitude that can be annihilated by the control signal.

8.3.4.2 H-Infinity Feedback Control

For the linearized system of the distributed generators given by Eq. (8.74) the cost function of Eq. (8.75) is defined, where the coefficient r determines the penalization of the control input and the weight coefficient ρ determines the reward of the disturbances' effects.

It is assumed that (i) The energy that is transferred from the disturbances signal $\tilde{d}(t)$ is bounded, that is $\int_0^\infty \tilde{d}^T(t)\tilde{d}(t)dt < \infty$, (ii) the matrices $[A, B]$ and $[A, L]$ are stabilizable, (iii) the matrix $[A, C]$ is detectable. Then, the optimal feedback control law is given by

$$u(t) = -Kx(t) \tag{8.77}$$

with

$$K = \frac{1}{r}B^T P \tag{8.78}$$

where P is a positive semi-definite symmetric matrix which is obtained from the solution of the Riccati equation

$$A^T P + PA + Q - P(\tfrac{1}{r}BB^T - \tfrac{1}{2\rho^2}LL^T)P = 0 \tag{8.79}$$

where Q is also a positive definite symmetric matrix. The worst case disturbance is given by

$$\tilde{d}(t) = \frac{1}{\rho^2}L^T Px(t) \tag{8.80}$$

The diagram of the considered control loop is depicted in Fig. 8.9.

8.3.5 *Lyapunov Stability Analysis*

The Lyapunov stability analysis for the H-infinity control of the distributed generators follows the procedure that was explained in the case of the stand-alone generators. Through Lyapunov stability analysis it will be shown that the proposed nonlinear control scheme assures H_∞ tracking performance for the distributed synchronous generators, and that in case of bounded disturbance terms asymptotic convergence to the reference setpoints is achieved.

The tracking error dynamics for the distributed synchronous generators is written in the form

$$\dot{e} = Ae + Bu + L\tilde{d} \tag{8.81}$$

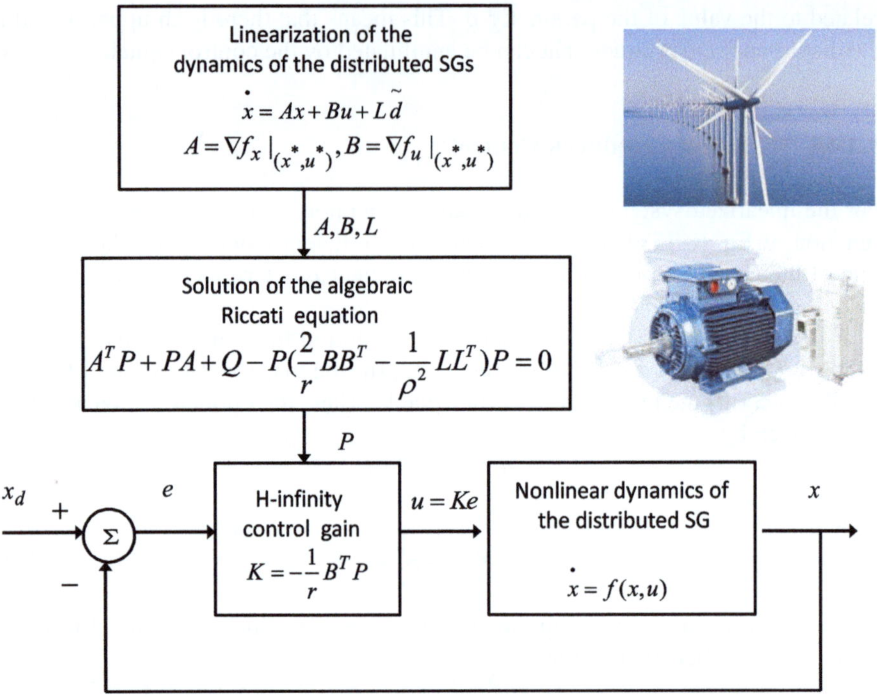

Fig. 8.9 Diagram of the control scheme for the distributed synchronous generators

where in the distributed synchronous generators' case $L = I \in R^{9\times 9}$ with I being the identity matrix. Variable $\tilde{d}$ denotes model uncertainties and external disturbances of the distributed generators' model. The following Lyapunov equation is considered

$$V = \tfrac{1}{2}e^T Pe \tag{8.82}$$

where $e = x - x_d$ is the tracking error. By differentiating with respect to time one obtains

$$\begin{aligned}\dot{V} &= \tfrac{1}{2}\dot{e}^T Pe + \tfrac{1}{2}eP\dot{e} \Rightarrow \\ \dot{V} &= \tfrac{1}{2}[Ae + Bu + L\tilde{d}]^T P + \tfrac{1}{2}e^T P[Ae + Bu + L\tilde{d}] \Rightarrow\end{aligned} \tag{8.83}$$

$$\begin{aligned}\dot{V} = \tfrac{1}{2}[e^T A^T + u^T B^T + \tilde{d}^T L^T]Pe + \\ + \tfrac{1}{2}e^T P[Ae + Bu + L\tilde{d}] \Rightarrow\end{aligned} \tag{8.84}$$

$$\begin{aligned}\dot{V} = \tfrac{1}{2}e^T A^T Pe + \tfrac{1}{2}u^T B^T Pe + \tfrac{1}{2}\tilde{d}^T L^T Pe + \\ \tfrac{1}{2}e^T PAe + \tfrac{1}{2}e^T PBu + \tfrac{1}{2}e^T PL\tilde{d}\end{aligned} \tag{8.85}$$

The previous equation is rewritten as

$$\dot{V}=\tfrac{1}{2}e^T(A^TP+PA)e+(\tfrac{1}{2}u^TB^TPe+\tfrac{1}{2}e^TPBu)+ \\ +(\tfrac{1}{2}\tilde{d}^TL^TPe+\tfrac{1}{2}e^TPL\tilde{d}) \tag{8.86}$$

Assumption: For given positive definite matrix Q and coefficients r and ρ there exists a positive definite matrix P, which is the solution of the following matrix equation

$$A^TP+PA=-Q+P(\tfrac{2}{r}BB^T-\tfrac{1}{\rho^2}LL^T)P \tag{8.87}$$

Moreover, the following feedback control law is applied to the system

$$u=-\tfrac{1}{r}B^TPe \tag{8.88}$$

By substituting Eqs. (8.87) and (8.88) one obtains

$$\dot{V}=\tfrac{1}{2}e^T[-Q+P(\tfrac{2}{r}BB^T-\tfrac{1}{2\rho^2}LL^T)P]e+ \\ +e^TPB(-\tfrac{1}{r}B^TPe)+e^TPL\tilde{d}\Rightarrow \tag{8.89}$$

$$\dot{V}=-\tfrac{1}{2}e^TQe+(\tfrac{2}{r}PBB^TPe-\tfrac{1}{2\rho^2}e^TPLL^T)Pe \\ -\tfrac{1}{r}e^TPBB^TPe)+e^TPL\tilde{d} \tag{8.90}$$

which after intermediate operations gives

$$\dot{V}=-\tfrac{1}{2}e^TQe-\tfrac{1}{2\rho^2}e^TPLL^TPe+e^TPL\tilde{d} \tag{8.91}$$

or, equivalently

$$\dot{V}=-\tfrac{1}{2}e^TQe-\tfrac{1}{2\rho^2}e^TPLL^TPe+ \\ +\tfrac{1}{2}e^TPL\tilde{d}+\tfrac{1}{2}\tilde{d}^TL^TPe \tag{8.92}$$

Lemma: The following inequality holds

$$\tfrac{1}{2}e^TL\tilde{d}+\tfrac{1}{2}\tilde{d}L^TPe-\tfrac{1}{2\rho^2}e^TPLL^TPe\leq\tfrac{1}{2}\rho^2\tilde{d}^T\tilde{d} \tag{8.93}$$

Proof: The binomial $(\rho\alpha-\frac{1}{\rho}b)^2$ is considered. Expanding the left part of the above inequality one gets

$$\rho^2a^2+\tfrac{1}{\rho^2}b^2-2ab\geq 0\Rightarrow\tfrac{1}{2}\rho^2a^2+\tfrac{1}{2\rho^2}b^2-ab\geq 0\Rightarrow \\ ab-\tfrac{1}{2\rho^2}b^2\leq\tfrac{1}{2}\rho^2a^2\Rightarrow\tfrac{1}{2}ab+\tfrac{1}{2}ab-\tfrac{1}{2\rho^2}b^2\leq\tfrac{1}{2}\rho^2a^2 \tag{8.94}$$

The following substitutions are carried out: $a=\tilde{d}$ and $b=e^TPL$ and the previous relation becomes

$$\tfrac{1}{2}\tilde{d}^TL^TPe+\tfrac{1}{2}e^TPL\tilde{d}-\tfrac{1}{2\rho^2}e^TPLL^TPe\leq\tfrac{1}{2}\rho^2\tilde{d}^T\tilde{d} \tag{8.95}$$

Equation (8.95) is substituted in Eq. (8.92) and the inequality is enforced, thus giving

$$\dot{V} \leq -\tfrac{1}{2}e^T Q e + \tfrac{1}{2}\rho^2 \tilde{d}^T \tilde{d} \tag{8.96}$$

Equation (8.96) shows that the H_∞ tracking performance criterion is satisfied. The integration of $\dot{V}$ from 0 to T gives

$$\begin{array}{l} \int_0^T \dot{V}(t)dt \leq -\tfrac{1}{2}\int_0^T ||e||_Q^2 dt + \tfrac{1}{2}\rho^2 \int_0^T ||\tilde{d}||^2 dt \Rightarrow \\ 2V(T) + \int_0^T ||e||_Q^2 dt \leq 2V(0) + \rho^2 \int_0^T ||\tilde{d}||^2 dt \end{array} \tag{8.97}$$

Moreover, if there exists a positive constant $M_d > 0$ such that

$$\int_0^\infty ||\tilde{d}||^2 dt \leq M_d \tag{8.98}$$

then one gets

$$\int_0^\infty ||e||_Q^2 dt \leq 2V(0) + \rho^2 M_d \tag{8.99}$$

Thus, the integral $\int_0^\infty ||e||_Q^2 dt$ is bounded. Additionally, $V(T)$ is bounded and from the definition of the Lyapunov function V in Eq. (8.82) it becomes clear that $e(t)$ will be also bounded since $e(t) \in \Omega_e = \{e | e^T P e \leq 2V(0) + \rho^2 M_d\}$. According to the above and with the use of Barbalat's Lemma one obtains $lim_{t \to \infty} e(t) = 0$.

8.3.6 *Robust State Estimation with the Use of the H_∞ Kalman Filter*

Robust state estimation with the use of the H-infinity Kalman Filter was already analyzed in the case of multi-phase machines. The control loop has to be implemented with the use of information provided by a small number of sensors and by processing only a small number of state variables. To reconstruct the missing information about the state vector of the distributed synchronous generators it is proposed to use a filtering scheme and based on it to apply state estimation-based control [375]. The recursion of the H_∞ Kalman Filter, for the model of the distributed SGs, can be formulated in terms of a *measurement update* and a *time update* part

Measurement update:

$$\begin{array}{l} D(k) = [I - \theta W(k) P^-(k) + C^T(k) R(k)^{-1} C(k) P^-(k)]^{-1} \\ K(k) = P^-(k) D(k) C^T(k) R(k)^{-1} \\ \hat{x}(k) = \hat{x}^-(k) + K(k)[y(k) - C\hat{x}^-(k)] \end{array} \tag{8.100}$$

Time update:

$$\begin{array}{l} \hat{x}^-(k+1) = A(k) x(k) + B(k) u(k) \\ P^-(k+1) = A(k) P^-(k) D(k) A^T(k) + Q(k) \end{array} \tag{8.101}$$

where it is assumed that parameter θ is sufficiently small to assure that the covariance matrix $P^{-}(k) - \theta W(k) + C^{T}(k)R(k)^{-1}C(k)$ will be positive definite. When $\theta = 0$ the H_{∞} Kalman Filter becomes equivalent to the standard Kalman Filter. One can measure only a part of the state vector of the distributed SGs, such as state variables $x_i = \theta_i$, $i = 1, 2, 3$ and can estimate through filtering the rest of the state vector elements.

8.3.7 Simulation Tests

The problem of synchronization and stabilization of the distributed synchronous generators has been solved with the application of a nonlinear H-infinity (optimal) control method. The performance of the control method is evaluated through simulation experiments. It can be observed that the control scheme assures fast and accurate convergence of the rotation speeds of the individual generators to the reference setpoints, as shown in Figs. 8.10, 8.11 and 8.12. As it happens with all optimal control methods the transients of the state variables are affected by values of the controller's gains, that is r, ρ and by values of the elements of matrix Q appearing in Eq. (8.79). The values of these parameters affect also the variation ranges of the control inputs. It can be noted that the H-infinity algorithm exhibited remarkable robustness to uncertainty in the model of the distributed power generators which was due to approximate linearization. It also remained robust to measurement noise and to external perturbations affecting the model of the distributed generators (e.g. variations in the mechanical torque that excites the rotational motion of the machines).

The computational aspects of the proposed control method have been also assessed. For the three-area multi-machine power system the dimension of the state vector was $x \in R^9$, while the dimension of the feedback gain matrix P that was obtained from the solution of the Riccati equation given in Eq. (8.79) was $P \in R^{9\times 9}$. Using an i7 processor at 2.0Ghz the solution of the Riccati equation was performed in msec and definitely in time that was significantly smaller than the sampling period of the control algorithm. For a distributed power system, that comprises more generators and which is of higher dimensionality, the real-time implementation of the control algorithm is not going to be hindered. It can be concluded that the proposed H-infinity control method is computationally very efficient.

8.4 Flatness-Based Adaptive Control of Distributed PMSGs

As previously explained, distributed power generation is becoming progressively an issue of major importance for the functioning of the smart grid [84, 106, 167, 270, 359, 375, 510]. To cover the increasing demand for electric power it has become necessary to interconnect a large number of power generation units, located at different areas. For efficient functioning of such a decentralized power generation scheme stability

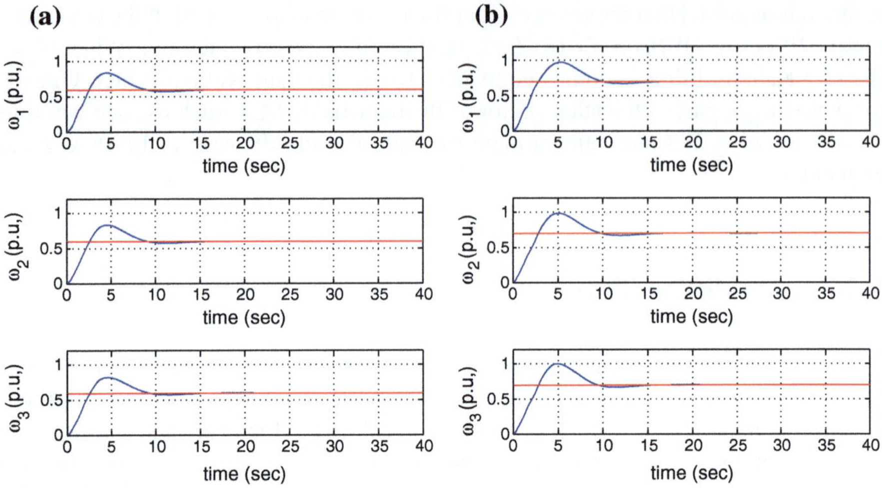

Fig. 8.10 Tracking of reference setpoints by the rotational speeds ω_i $i = 1, 2, 3$ of the distributed synchronous generators **a** setpoint 1 (p.u) **b** setpoint 2 (p.u)

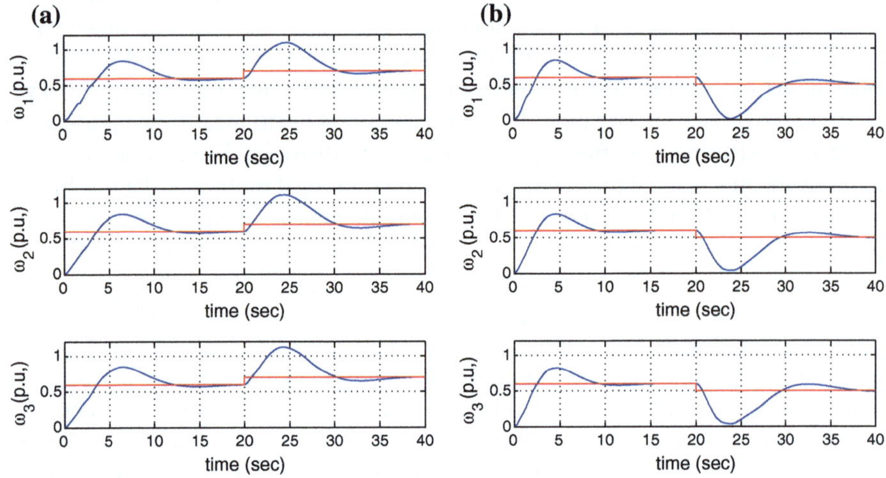

Fig. 8.11 Tracking of reference setpoints by the rotational speeds ω_i $i = 1, 2, 3$ of the distributed synchronous generators **a** setpoint 3 (p.u) **b** setpoint 4 (p.u)

properties should hold (despite modeling uncertainties that may affect the power generators) and synchronization between the individual power generators should be succeeded (despite their exposure to external perturbations) [88, 102, 116, 198, 213, 237, 238, 495]. This section proposes adaptive fuzzy control based on differential flatness theory and using exclusively output feedback, for solving the problem of synchronization between distributed synchronous generators and for succeeding stabilization of a distributed power generation system. The model of the individual

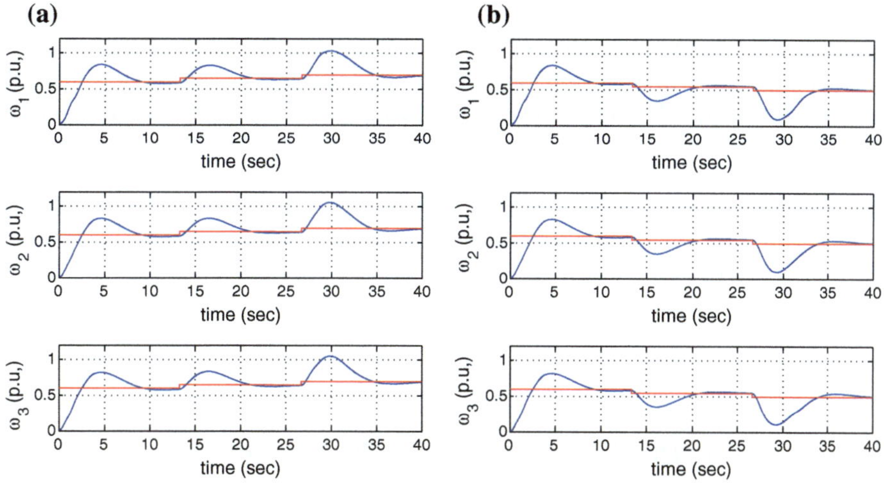

Fig. 8.12 Tracking of reference setpoints by the rotational speeds ω_i $i = 1, 2, 3$ of the distributed synchronous generators **a** setpoint 5 (p.u) **b** setpoint 6 (p.u)

power generator units (synchronous power generators) is considered to be completely unknown while the state vector of these generators is not completely measurable [60, 121, 262, 265, 315]. Thus, finally one has to do with a joint estimation and nonlinear control problem in which the control signal makes use of the estimated state vector of the generators [15, 439, 474].

First, the chapter shows that the dynamic model of the distributed power generation system is a differentially flat one. This means that all its state variables and the control input can be written as functions of one single algebraic variable, which is the flat output, and also as functions of the flat outputs derivatives [20, 137, 229, 241, 281, 375, 411, 456]. This change of variables (differential flatness theory-based diffeomorphism) enables to transform the nonlinear model of the distributed power generators into the linear canonical (Brunovsky) form. In the latter description of the power generators system the transformed control inputs contain elements which are associated with the unknown nonlinear dynamics of the system. These are identified on-line with the use of neurofuzzy approximators and the estimated system dynamics is finally used for the computation of the control signal that will make the generators' state vector track the desirable setpoints. Thus an adaptive fuzzy control scheme is implemented. The learning rate of the neurofuzzy approximators is determined by the requirement that the first derivative of the Lyapunov function of the control loop will always be a negative one [34, 143, 344, 375].

Next, another problem that has to be dealt with is that only output feedback can be used for the implementation of the distributed power generators' control scheme [50, 483]. The nonmeasurable state variables of the power generators have to be reconstructed with the use of a state estimator (observer), which functions again in the control loop [111, 112, 197, 249, 251, 266, 307, 320]. Thus, finally, the Lyapunov function for the proposed control scheme comprises three quadratic terms: (i) a term that describes the tracking error of the generators' state variables, (ii) a term that describes the error in the estimation of the non-measurable state vector elements with respect to the reference setpoints, and (iii) a sum of quadratic terms associated with the distance of the weights of the neurofuzzy approximators from the values that give the best approximation of the unknown dynamics of the generators. It is proven that an adaptive (learning) control law can be found assuring that the first derivative of the Lyapunov function will remain negative, thus assuring that the stability of the control loop will be preserved and that accurate tracking of the setpoints by the systems state variables will be succeeded (H-infinity tracking performance) [117, 215, 226, 263].

8.4.1 An Adaptive Fuzzy Control for the System of the Distributed Synchronous Generators

In the previous sections it was shown that, for the case of the N distributed synchronous generators, the dynamic model of the i-th input-output linearized model of the synchronous machine is given by

$$y^{(3)^i} = a^i(x) + b_1^i(x)g_1u_1 + b_2^i(x)g_2u_2 + b_3^i(x)g_3u_3 \tag{8.102}$$

which means that one arrives at a function of the form $\dot{z}_3^i = a^i(x) + b_1^i(x)g_1u_1 + b_2^i(x)g_2u_2 + b_3^i(x)g_3u_3 + \tilde{d}^i$, where in the case of the distributed power generation that consists of $n = 2$ machines, and considering for instance $i = 1, j = 2$ (Fig. 8.13) one has

$$\begin{aligned} a^i = (\tfrac{D_i}{2J_i})^2 x_2^i + \tfrac{D_i\omega_0}{(2J_i)^2}[G_{ii}{x_3^i}^2 + x_3^i\textstyle\sum_{j=1,j\neq i}^n x_3^j G_{ij}sin(x_1^i - x_1^j - \alpha_{ij})] - \\ -\tfrac{\omega_0}{2J_i}[G_{ii}x_3^i + \textstyle\sum_{j=1,j\neq i}^n x_3^j G_{ij}sin(x_1^i - x_1^j - \alpha_{ij})(-\tfrac{1}{T'_{d_i}}x_3^i + (\tfrac{1}{T_{doi}}\tfrac{x_{d_i}-x'_{d_i}}{x_{d\Sigma_i}{}'}V_{s_i}cos(x_1^i))] - \\ -\tfrac{\omega_0}{2J_i}x_3^i\textstyle\sum_{j=1,j\neq i}^n G_{ij}sin(x_1^i - x_1^j - \alpha_{ij})(-\tfrac{1}{T'_{d_i}}x_3^i + (\tfrac{1}{T_{doi}}\tfrac{x_{d_i}-x'_{d_i}}{x_{d\Sigma_i}{}'}V_{s_i}cos(x_1^i)) - \\ -\tfrac{\omega_0}{2J_i}x_3^i\textstyle\sum_{j=1,j\neq i}^n x_3^j G_{ij}cos(x_1^i - x_1^j - \alpha_{ij})x_2^i\tfrac{\omega_0}{2J_i}x_3^i\textstyle\sum_{j=1,j\neq i}^n x_3^j G_{ij}cos(x_1^i - x_1^j - \alpha_{ij})x_2^j \end{aligned} \tag{8.103}$$

while for functions b_1^i, b_2^i it holds

$$\begin{aligned} b_1^i = -\tfrac{\omega_0}{2J_i}[2G_{ii}x_3^i + \textstyle\sum_{j=1,j\neq i}^n x_3^j G_{ij}sin(x_1^i - x_1^j - \alpha_{ij})]\tfrac{1}{T_{doi}} \\ b_2^i = -\tfrac{\omega_0}{2J_i}G_{i2}sin(x_1^i - x_1^2 - \alpha_{i2})\tfrac{1}{T_{do2}} \end{aligned} \tag{8.104}$$

Finally for the disturbance term one has $\tilde{d}^i = -\frac{D_i\omega_0}{2J_i{}^2}P_m^i + \frac{\omega_0}{2J_i}\dot{P}_m^i$. Thus, one has the following description of the dynamics of the i-th power generator

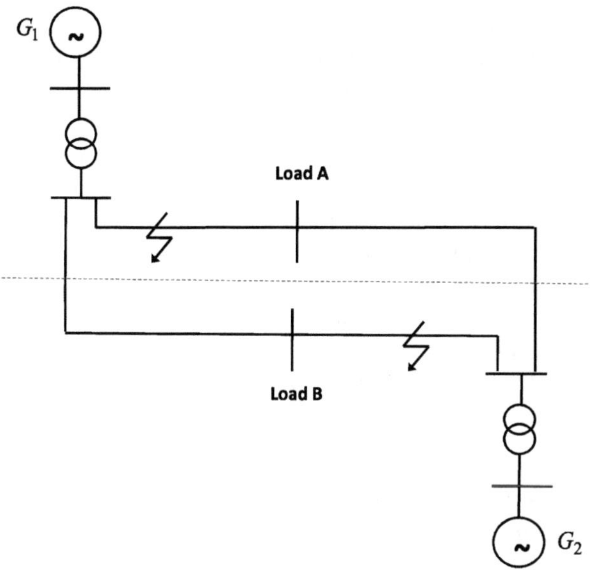

Fig. 8.13 A multi-machine (2-area) distributed power generation model

$$\begin{aligned}\dot{z}_1^i &= z_2^i \\ \dot{z}_2^i &= z_3^i \\ \dot{z}_3^i &= a^i(x) + {b_1}^i g_1 u_1 + {b_2}^i g_2 u_2 + \tilde{d}^i\end{aligned} \tag{8.105}$$

For the complete system of the 2 generators one has

$$\begin{aligned}\dot{z}_3^1 &= a^1(x) + {b_1}^1 g_1 u_1 + {b_2}^1 g_2 u_2 + \tilde{d}^1 \\ \dot{z}_3^2 &= a^2(x) + {b_1}^2 g_1 u_1 + {b_2}^2 g_2 u_2 + \tilde{d}^2\end{aligned} \tag{8.106}$$

or in matrix form

$$\dot{z}_3 = f_a(x) + Mu + \tilde{d} \tag{8.107}$$

where $z_3 = [z_3^1, z_3^2]^T$, $u = [u_1, u_2]^T$ and $\tilde{d} = [\tilde{d}_1, \tilde{d}_2]^T$ while

$$f_a(x) = \begin{pmatrix} a^1(x) \\ a^2(x) \end{pmatrix}, \quad M = \begin{pmatrix} {b_1}^1 g_1 & {b_2}^1 g_2 \\ {b_1}^2 g_1 & {b_2}^2 g_2 \end{pmatrix} \tag{8.108}$$

Setting $v = f_a(x) + Mu + \tilde{d}$, one obtains again the linear canonical form for the i-th generator given by

$$\begin{pmatrix} \dot{z}_1^i \\ \dot{z}_2^i \\ \dot{z}_3^i \end{pmatrix} = \begin{pmatrix} 0 & 1 & 0 \\ 0 & 0 & 1 \\ 0 & 0 & 0 \end{pmatrix} \begin{pmatrix} z_1^i \\ z_2^i \\ z_3^i \end{pmatrix} + \begin{pmatrix} 0 \\ 0 \\ 1 \end{pmatrix} (v^i + \tilde{d}^i) \tag{8.109}$$

In this manner the initial nonlinear power system is transformed into two decoupled linear subsystems which are in the canonical Brunovksy form. For each one of these subsystems the appropriate control law is

$$v^i = z_d^{(3)^i} - k_3(\ddot{z}^i - \ddot{z}_d^i) - k_2(\dot{z}^i - \dot{z}_d^i) - k_1(z^i - z_d^i) - \tilde{d}^i \tag{8.110}$$

The proposed control scheme, considers as output of the synchronous generator the rotor's difference angle $\Delta\delta$. However, it is possible to avoid the use of encoder readings about the rotor's turn angle, and to indirectly estimate this parameter through the processing of measurement coming from PMUs. Such measurements are the bus voltage magnitude and the associated phase angle, the line current magnitude and the associated angle and the electrical output power at the terminal bus [111].

8.4.2 *Flatness-Based Adaptive Fuzzy Control for MIMO Nonlinear Systems*

8.4.2.1 Transformation of MIMO PMSG System into the Brunovsky Form

It is assumed now that after defining the flat outputs of the initial MIMO nonlinear system of the distributed synchronous generators, and after expressing the system state variables and control inputs as functions of the flat output and of the associated derivatives, the system can be transformed in the Brunovsky canonical form [375]:

$$\begin{array}{ll}
\dot{x}_1 = x_2 & \\
\dot{x}_2 = x_3 & \\
\cdots & \\
\dot{x}_{r_1-1} = x_{r_1} & y_1 = x_1 \\
\dot{x}_{r_1} = f_1(x) + \sum_{j=1}^{p} g_{1_j}(x)u_j + d_1 & y_2 = x_{r_1-1} \\
\dot{x}_{r_1+1} = x_{r_1+2} & \cdots \\
\dot{x}_{r_1+2} = x_{r_1+3} & y_p = x_{n-r_p+1} \\
\cdots & \\
\dot{x}_{p-1} = x_p & \\
\dot{x}_p = f_p(x) + \sum_{j=1}^{p} g_{p_j}(x)u_j + d_p &
\end{array} \tag{8.111}$$

where $x = [x_1, \ldots, x_n]^T$ is the state vector of the transformed system (according to the differential flatness formulation), $u = [u_1, \ldots, u_p]^T$ is the set of control inputs, $y = [y_1, \ldots, y_p]^T$ is the output vector, f_i are the drift functions and $g_{i,j}$, $i, j = 1, 2, \ldots, p$ are smooth functions corresponding to the control input gains, while d_j is a variable associated to external disturbances. In holds that $r_1 + r_2 + \cdots + r_p = n$. Having written the initial nonlinear system into the canonical (Brunovsky) form it holds

$$y_i^{(r_i)} = f_i(x) + \sum_{j=1}^{p} g_{ij}(x)u_j + d_j \tag{8.112}$$

Equivalently, in vector form, one has the following description for the system dynamics

$$y^{(r)} = f(x) + g(x)u + d \tag{8.113}$$

where the following vectors and matrices are be defined

$$\begin{array}{l} y^{(r)} = [y_1^{(r_1)}, \dots, y_p^{(r_p)}] \\ f(x) = [f_1(x), \dots, f_p(x)]^T \\ g(x) = [g_1(x), \dots, g_p(x)] \\ with\ g_i(x) = [g_{1i}(x), \dots, g_{pi}(x)]^T \\ A = diag[A_1, \dots, A_p], \ \ B = diag[B_1, \dots, B_p] \\ C^T = diag[C_1, \dots, C_p], \ \ d = [d_1, \dots, d_p]^T \end{array} \tag{8.114}$$

where matrix A has the MIMO canonical form, i.e. with elements

$$\begin{array}{l} A_i = \begin{pmatrix} 0 & 1 & \cdots & 0 \\ 0 & 0 & \cdots & 0 \\ \vdots & \vdots & \cdots & \vdots \\ 0 & 0 & \cdots & 1 \\ 0 & 0 & \cdots & 0 \end{pmatrix}_{r_i \times r_i} \\ B_i^T = \begin{pmatrix} 0 & 0 & \cdots & 0 & 1 \end{pmatrix}_{1 \times r_i} \\ C_i = \begin{pmatrix} 1 & 0 & \cdots & 0 & 0 \end{pmatrix}_{1 \times r_i} \end{array} \tag{8.115}$$

Thus, Eq. (8.112) can be written in state-space form

$$\begin{array}{c} \dot{x} = Ax + B[f(x) + g(x)u + \tilde{d}] \\ y = C^T x \end{array} \tag{8.116}$$

which can be also written in the equivalent form:

$$\begin{array}{c} \dot{x} = Ax + Bv + B\tilde{d} \\ y = C^T x \end{array} \tag{8.117}$$

where $v = f(x) + g(x)u$. The reference setpoints for the system's outputs $y_1, \dots, y_p$ are denoted as $y_{1m}, \dots, y_{pm}$, thus for the associated tracking errors it holds

$$\begin{array}{c} e_1 = y_1 - y_{1m} \\ e_2 = y_2 - y_{2m} \\ \cdots \\ e_p = y_p - y_{pm} \end{array} \tag{8.118}$$

The error vector of the outputs of the transformed MIMO system is denoted as

$$\begin{aligned} E_1 &= [e_1, \dots, e_p]^T \\ y_m &= [y_{1m}, \dots, y_{pm}]^T \\ &\dots \\ y_m^{(r)} &= [y_{1m}^{(r)}, \dots, y_{pm}^{(r)}]^T \end{aligned} \quad (8.119)$$

where $y_{im}^{(r)}$ denotes the r-th order derivative of the i-th reference output of the MIMO dynamical system. Thus, one can also define the following vectors: (i) a vector containing the state variables of the system and the associated derivatives, (ii) a vector containing the reference outputs of the system and the associated derivatives

$$x = [x_1, \dots, x_1^{r_1-1}, \dots, x_p, \dots, x_p^{r_p-1}]^T \quad (8.120)$$

$$Y_m = [y_{1m}, \dots, y_{1m}^{r_1-1}, \dots, y_{pm}, \dots, y_{pm}^{r_p-1}]^T \quad (8.121)$$

while in a similar manner one can define a vector containing the tracking error of the system's outputs and the associated derivatives

$$e = Y_m - x = [e_1, \dots, e_1^{r_1-1}, \dots, e_p, \dots, e_p^{r_p-1}]^T \quad (8.122)$$

It is assumed that matrix $g(x)$ is a nonsingular one, i.e. $g^{-1}(x)$ exists and is bounded for all $x \in U_x$, where $U_x \subset R^n$ is a compact set. In any case, the problem of singularities in matrix $g(x)$ can be handled by appropriately modifying the state feedback-based control input.

The objective of the adaptive fuzzy controller, denoted as $u = u(x, e|\theta)$ is: all the signals involved in the controller's design are bounded and it holds that $lim_{t \to \infty} e = 0$, (ii) the H_∞ tracking performance criterion is succeeded for a prescribed attenuation level.

In the presence of non-gaussian disturbances w_d, successful tracking of the reference signal is denoted by the H_∞ criterion [352, 375]:

$$\int_0^T e^T Q e dt \leq \rho^2 \int_0^T {w_d}^T w_d dt \quad (8.123)$$

where ρ is the attenuation level and corresponds to the maximum singular value of the transfer function $G(s)$ of the linearized model associated to Eqs. (8.116) and (8.117).

8.4.2.2 Control Law

The control signal of the MIMO nonlinear system of the distributed synchronous generators which has been transformed into the Brunovsky form as described by Eq. (8.117) contains the unknown nonlinear functions $f(x)$ and $g(x)$. In case that the complete state vector x is measurable these unknown functions can be approximated by

$$\begin{aligned}\hat{f}(x|\theta_f) &= \Phi_f(x)\theta_f \\ \hat{g}(x|\theta_g) &= \Phi_g(x)\theta_g\end{aligned} \quad (8.124)$$

where

$$\Phi_f(x) = (\xi_f^1(x), \xi_f^2(x), \dots \xi_f^n(x))^T \quad (8.125)$$

with $\xi_f^i(x), \imath = 1, \dots, n$ being the vector of kernel functions (e.g. normalized fuzzy Gaussian membership functions), where

$$\xi_f^i(x) = (\phi_f^{i,1}(x), \phi_f^{i,2}(x), \dots, \phi_f^{i,N}(x)) \quad (8.126)$$

thus giving

$$\Phi_f(x) = \begin{pmatrix} \phi_f^{1,1}(x) & \phi_f^{1,2}(x) & \cdots & \phi_f^{1,N}(x) \\ \phi_f^{2,1}(x) & \phi_f^{2,2}(x) & \cdots & \phi_f^{2,N}(x) \\ \cdots & \cdots & \cdots & \cdots \\ \phi_f^{n,1}(x) & \phi_f^{n,2}(x) & \cdots & \phi_f^{n,N}(x) \end{pmatrix} \quad (8.127)$$

while the weights vector is defined as

$$\theta_f{}^T = \left(\theta_f^1, \theta_f^2, \dots \theta_f^N\right) \quad (8.128)$$

$j = 1, \dots, N$ is the number of basis functions that is used to approximate the components of function f which are denoted as $i = 1, \dots, n$. Thus, one obtains the relation of Eq. (8.124), i.e. $\hat{f}(x|\theta_f) = \Phi_f(x)\theta_f$.

In a similar manner, for the approximation of function g one has

$$\Phi_g(x) = \left(\xi_g^1(x), \xi_g^2(x), \dots \xi_g^N(x)\right)^T \quad (8.129)$$

with $\xi_g^i(x), \imath = 1, \dots, N$ being the vector of kernel functions (e.g. normalized fuzzy Gaussian membership functions), where

$$\xi_g^i(x) = \left(\phi_g^{i,1}(x), \phi_g^{i,2}(x), \dots, \phi_g^{i,N}(x)\right) \quad (8.130)$$

thus giving

$$\Phi_g(x) = \begin{pmatrix} \phi_g^{1,1}(x) & \phi_g^{1,2}(x) & \cdots & \phi_g^{1,N}(x) \\ \phi_g^{2,1}(x) & \phi_g^{2,2}(x) & \cdots & \phi_g^{2,N}(x) \\ \cdots & \cdots & \cdots & \cdots \\ \phi_g^{n,1}(x) & \phi_g^{n,2}(x) & \cdots & \phi_g^{n,N}(x) \end{pmatrix} \quad (8.131)$$

while the weights vector is defined as

$$\theta_g = \left(\theta_g^1, \theta_g^2, \dots, \theta_g^p\right) \quad (8.132)$$

where the components of matrix θ_g are defined as

$$\theta_g^j = \left(\theta_{g_1}^j, \theta_{g_2}^j, \dots \theta_{g_N}^j\right)^T \tag{8.133}$$

$j = 1, \dots, p$ is the number of basis functions that is used to approximate the components of function g which are denoted as $i = 1, \dots, n$. Thus one obtains about matrix $\theta_g \in R^{N \times p}$

$$\theta_g = \begin{pmatrix} \theta_{g_1}^1 & \theta_{g_1}^2 & \cdots & \theta_{g_1}^p \\ \theta_{g_2}^1 & \theta_{g_2}^2 & \cdots & \theta_{g_2}^p \\ \cdots & \cdots & \cdots & \cdots \\ \theta_{g_N}^1 & \theta_{g_N}^2 & \cdots & \theta_{g_N}^p \end{pmatrix} \tag{8.134}$$

It holds that

$$g = \begin{pmatrix} g_1 \\ g_2 \\ \cdots \\ g_n \end{pmatrix} = \begin{pmatrix} g_1^1 & g_1^2 & \cdots & g_1^p \\ g_2^1 & g_2^2 & \cdots & g_2^p \\ \cdots & \cdots & \cdots & \cdots \\ g_n^1 & g_n^2 & \cdots & g_n^p \end{pmatrix} \tag{8.135}$$

Using the above, one finally has the relation of Eq. (8.124), i.e. $\hat{g}(x|\theta_g) = \Phi_g(x)\theta_g$. If the state variables of the system are available for measurement then a state-feedback control law can be formulated as

$$u = \hat{g}^{-1}(x|\theta_g)[-\hat{f}(x|\theta_f) + y_m^{(r)} - K^T e + u_c] \tag{8.136}$$

where $\hat{f}(x|\theta_f)$ and $\hat{g}(x|\theta_g)$ are neurofuzzy models to approximate $f(x)$ and $g(x)$, respectively. u_c is a supervisory control term, e.g. H_∞ control term that is used to compensate for the effects of modelling inaccuracies and external disturbances. Moreover, K^T is the feedback gain matrix that assures that the characteristic polynomial of matrix $A - BK^T$ will be a Hurwitz one.

8.4.2.3 Estimation of the State Vector

The control of the system described by Eq. (8.113) becomes more complicated when the state vector x is not directly measurable and has to be reconstructed through a state observer. The following definitions are used

- error of the state vector $e = x - x_m$
- error of the estimated state vector $\hat{e} = \hat{x} - x_m$
- observation error $\tilde{e} = e - \hat{e} = (x - x_m) - (\hat{x} - x_m)$

When an observer is used to reconstruct the state vector, the control law of Eq. (8.136) is written as

$$u = \hat{g}^{-1}(\hat{x}|\theta_g)[-\hat{f}(\hat{x}|\theta_f) + y_m^{(r)} - K^T \hat{e} + u_c] \tag{8.137}$$

Applying Eq. (8.137) to the nonlinear system described by Eq. (8.113), results into

$$\begin{aligned} y^{(r)} &= f(x) + g(x)\hat{g}^{-1}(\hat{x})[-\hat{f}(\hat{x}) + y_m^{(r)} - K^T\hat{e} + u_c] + d \Rightarrow \\ y^{(r)} &= f(x) + [g(x) - \hat{g}(\hat{x}) + \hat{g}(\hat{x})]\hat{g}^{-1}(\hat{x})[-\hat{f}(\hat{x}) + y_m^{(r)} - K^T\hat{e} + u_c] + d \Rightarrow \\ y^{(r)} &= [f(x) - \hat{f}(\hat{x})] + [g(x) - \hat{g}(\hat{x})]u + y_m^{(r)} - K^T\hat{e} + u_c + d \end{aligned} \tag{8.138}$$

It holds $e = x - x_m \Rightarrow y^{(r)} = e^{(r)} + y_m^{(r)}$. Substituting $y^{(r)}$ in the above equation gives

$$\begin{aligned} e^{(r)} + y_m^{(r)} = y_m^{(r)} - K^T\hat{e} + u_c + [f(x) - \hat{f}(\hat{x})] + \\ + [g(x) - \hat{g}(\hat{x})]u + d \end{aligned} \tag{8.139}$$

and equivalently

$$\begin{aligned} \dot{e} = Ae - BK^T\hat{e} + Bu_c + B\{[f(x) - \hat{f}(\hat{x})] + \\ + [g(x) - \hat{g}(\hat{x})]u + \tilde{d}\} \end{aligned} \tag{8.140}$$

$$e_1 = C^T e \tag{8.141}$$

where $e = [e^1, e^2, \ldots, e^p]^T$ with $e^i = [e_i, \dot{e}_i, \ddot{e}_i, \ldots, e_i^{r_i - 1}]^T$, $i = 1, 2, \ldots, p$ and equivalently $\hat{e} = [\hat{e}^1, \hat{e}^2, \ldots, \hat{e}^p]^T$ with $\hat{e}^i = [\hat{e}_i, \hat{\dot{e}}_i, \hat{\ddot{e}}_i, \ldots, \hat{e}_i^{r_i - 1}]^T$, $i = 1, 2, \ldots, p$. Matrices A,B and C have been defined in Eq. (8.115).

A state observer is designed according to Eqs. (8.140) and (8.141) and is given by [375]:

$$\dot{\hat{e}} = A\hat{e} - BK^T\hat{e} + K_o[e_1 - C^T\hat{e}] \tag{8.142}$$

$$\hat{e}_1 = C^T\hat{e} \tag{8.143}$$

The feedback gain matrix is denoted as $K \in R^{n \times p}$. The observation gain matrix is denoted as $K_o \in R^{n \times p}$ and its elements are selected so as to assure the asymptotic elimination of the observation error.

8.4.3 Application of Flatness-Based Adaptive Fuzzy Control to the Distributed Power Generators' Model

8.4.3.1 Differential Flatness of the Synchronous Power Generator

For the model of the 2-area distributed power generation system it holds that

$$\begin{aligned} x_{1,1}^{(3)} &= f_1(x) + g_1(x)u \\ x_{1,2}^{(3)} &= f_2(x) + g_2(x)u \end{aligned} \tag{8.144}$$

Moreover, it holds that

$$\begin{aligned}\dot{x}_{1,1} &= x_{2,1}\\ \dot{x}_{2,1} &= x_{3,1}\\ \dot{x}_{3,1} &= f_1(x) + g_1(x)u\\ \dot{x}_{1,2} &= x_{2,2}\\ \dot{x}_{2,2} &= x_{2,3}\\ \dot{x}_{3,2} &= f_2(x) + g_2(x)u\end{aligned} \tag{8.145}$$

Moreover, from Eq. (8.145) it holds

$$\begin{aligned}\begin{pmatrix}x_{1,1}^{(3)}\\ x_{1,2}^{(3)}\end{pmatrix} &= \begin{pmatrix}f_1(x)\\ f_2(x)\end{pmatrix} + \begin{pmatrix}g_1(x)\\ g_2(x)\end{pmatrix} u \ i.e.\\ u &= \begin{pmatrix}g_1(x)\\ g_2(x)\end{pmatrix}^{-1} \{\begin{pmatrix}x_{1,1}^{(3)}\\ x_{1,2}^{(3)}\end{pmatrix} - \begin{pmatrix}f_1(x)\\ f_2(x)\end{pmatrix}\}\end{aligned} \tag{8.146}$$

Therefore, the considered power generator system is a differentially flat one. Next, taking into account also the effects of additive disturbances the dynamic model becomes

$$\begin{aligned}x_{1,1}^{(3)} &= f_1(x,t) + g_1(x,t)u + d_1\\ x_{1,2}^{(3)} &= f_2(x,t) + g_2(x,t)u + d_2\end{aligned} \tag{8.147}$$

$$\begin{pmatrix}x_{1,1}^{(3)}\\ x_{1,2}^{(3)}\end{pmatrix} = \begin{pmatrix}f_1(x,t)\\ f_2(x,t)\end{pmatrix} + \begin{pmatrix}g_1(x,t)\\ g_2(x,t)\end{pmatrix} u + \begin{pmatrix}d_1\\ d_2\end{pmatrix} \tag{8.148}$$

The following control input is defined

$$u = \begin{pmatrix}\hat{g}_1(x,t)\\ \hat{g}_2(x,t)\end{pmatrix}^{-1} \{\begin{pmatrix}xd_{1,1}^{(3)}\\ xd_{1,2}^{(3)}\end{pmatrix} - \begin{pmatrix}\hat{f}_1(x,t)\\ \hat{f}_2(x,t)\end{pmatrix} - \begin{pmatrix}K_1^T\\ K_2^T\end{pmatrix} e + \begin{pmatrix}u_{c_1}\\ u_{c_2}\end{pmatrix}\} \tag{8.149}$$

where $[u_{c_1}\ u_{c_2}]^T$ is a robust control term that is used for the compensation of the model's uncertainties and of external disturbances and $K_i^T = [k_1^i, k_2^i, \ldots, k_{n-1}^i, k_n^i]$. Substituting Eq. (8.149) into Eq. (8.148) the closed-loop tracking error dynamics is obtained

$$\begin{aligned}\begin{pmatrix}x_{1,1}^{(3)}\\ x_{1,2}^{(3)}\end{pmatrix} &= \begin{pmatrix}f_1(x,t)\\ f_2(x,t)\end{pmatrix} + \begin{pmatrix}g_1(x,t)\\ g_2(x,t)\end{pmatrix} \begin{pmatrix}\hat{g}_1(x,t)\\ \hat{g}_2(x,t)\end{pmatrix}^{-1} \cdot\\ &\cdot\{\begin{pmatrix}xd_1^{(3)}\\ xd_2^{(3)}\end{pmatrix} - \begin{pmatrix}\hat{f}_1(x,t)\\ \hat{f}_2(x,t)\end{pmatrix} - \begin{pmatrix}K_1^T\\ K_2^T\end{pmatrix} e + \begin{pmatrix}u_{c_1}\\ u_{c_2}\end{pmatrix}\} + \begin{pmatrix}d_1\\ d_2\end{pmatrix}\end{aligned} \tag{8.150}$$

Equation (8.150) can now be written as

$$\begin{pmatrix} x_{1,1}^{(3)} \\ x_{1,2}^{(3)} \end{pmatrix} = \begin{pmatrix} f_1(x,t) \\ f_2(x,t) \end{pmatrix} + \{\begin{pmatrix} g_1(x,t) - \hat{g}_1(x,t) \\ g_2(x,t) - \hat{g}_2(x,t) \end{pmatrix} + \\ + \begin{pmatrix} \hat{g}_1(x,t) \\ \hat{g}_2(x,t) \end{pmatrix}\} \begin{pmatrix} \hat{g}_1(x,t) \\ \hat{g}_2(x,t) \end{pmatrix}^{-1} \cdot \\ \cdot \{\begin{pmatrix} xd_1^{(3)} \\ xd_2^{(3)} \end{pmatrix} - \begin{pmatrix} \hat{f}_1(x,t) \\ \hat{f}_2(x,t) \end{pmatrix} - \begin{pmatrix} K_1^T \\ K_2^T \end{pmatrix} e + \begin{pmatrix} u_{c_1} \\ u_{c_2} \end{pmatrix}\} + \begin{pmatrix} d_1 \\ d_2 \end{pmatrix} \tag{8.151}$$

and using Eq. (8.149) this results into

$$\begin{pmatrix} e_1^{(3)} \\ e_2^{(3)} \end{pmatrix} = \begin{pmatrix} f_1(x,t) - \hat{f}_1(x,t) \\ f_2(x,t) - \hat{f}_2(x,t) \end{pmatrix} + \begin{pmatrix} g_1(x,t) - \hat{g}_1(x,t) \\ g_2(x,t) - \hat{g}_2(x,t) \end{pmatrix} u - \\ - \begin{pmatrix} K_1^T \\ K_2^T \end{pmatrix} e + \begin{pmatrix} u_{c_1} \\ u_{c_2} \end{pmatrix} + \begin{pmatrix} d_1 \\ d_2 \end{pmatrix} \tag{8.152}$$

The following description for the approximation error is defined

$$w = \begin{pmatrix} f_1(x,t) - \hat{f}_1(x,t) \\ f_2(x,t) - \hat{f}_2(x,t) \end{pmatrix} + \begin{pmatrix} g_1(x,t) - \hat{g}_1(x,t) \\ g_2(x,t) - \hat{g}_2(x,t) \end{pmatrix} u \tag{8.153}$$

Moreover, the following matrices are defined

$$A = \begin{pmatrix} 0 & 1 & 0 & 0 & 0 & 0 \\ 0 & 0 & 1 & 0 & 0 & 0 \\ 0 & 0 & 0 & 0 & 0 & 0 \\ 0 & 0 & 0 & 0 & 1 & 0 \\ 0 & 0 & 0 & 0 & 0 & 1 \\ 0 & 0 & 0 & 0 & 0 & 0 \end{pmatrix}, \quad B = \begin{pmatrix} 0 & 0 \\ 0 & 0 \\ 1 & 0 \\ 0 & 0 \\ 0 & 0 \\ 0 & 1 \end{pmatrix} \\ K^T = \begin{pmatrix} K_1^1 & K_2^1 & K_3^1 & K_4^1 & K_5^1 & K_6^1 \\ K_1^2 & K_2^2 & K_3^2 & K_4^2 & K_5^2 & K_6^2 \end{pmatrix} \tag{8.154}$$

Using matrices A, B, K^T, Eq. (8.152) is written in the following form

$$\dot{e} = (A - BK^T)e + Bu_c + B\{\begin{pmatrix} f_1(x,t) - \hat{f}_1(x,t) \\ f_2(x,t) - \hat{f}_2(x,t) \end{pmatrix} + \\ + \begin{pmatrix} g_1(x,t) - \hat{g}_1(x,t) \\ g_2(x,t) - \hat{g}_2(x,t) \end{pmatrix} u + \tilde{d}\} \tag{8.155}$$

When the estimated state vector $\hat{x}$ is used in the feedback control loop, equivalently to Eq. (8.140) one has

$$\dot{e} = Ae - BK^T\hat{e} + Bu_c + B\{\begin{pmatrix} f_1(x,t) - \hat{f}_1(\hat{x},t) \\ f_2(x,t) - \hat{f}_2(\hat{x},t) \end{pmatrix} + \\ + \begin{pmatrix} g_1(x,t) - \hat{g}_1(\hat{x},t) \\ g_2(x,t) - \hat{g}_2(\hat{x},t) \end{pmatrix} u + \tilde{d}\} \tag{8.156}$$

and considering that the approximation error w is now denoted as

$$w = \begin{pmatrix} f_1(x,t) - \hat{f}_1(\hat{x},t) \\ f_2(x,t) - \hat{f}_2(\hat{x},t) \end{pmatrix} + \begin{pmatrix} g_1(x,t) - \hat{g}_1(\hat{x},t) \\ g_2(x,t) - \hat{g}_2(\hat{x},t) \end{pmatrix} u \tag{8.157}$$

Equation (8.156) can be also written as

$$\dot{e} = Ae - BK^T\hat{e} + Bu_c + Bw + B\tilde{d} \tag{8.158}$$

The associated state observer will be described again by Eqs. (8.142) and (8.143).

8.4.3.2 Dynamics of the Observation Error

The observation error is defined as $\tilde{e} = e - \hat{e} = x - \hat{x}$. Subtracting Eq. (8.142) from Eq. (8.140) as well as Eq. (8.143) from Eq. (8.141) one gets

$$\dot{e} - \dot{\hat{e}} = A(e - \hat{e}) + Bu_c + B\{[f(x,t) - \hat{f}(\hat{x},t)] + \\ + [g(x,t) - \hat{g}(\hat{x},t)]u + \tilde{d}\} - K_oC^T(e - \hat{e}) \\ e_1 - \hat{e}_1 = C^T(e - \hat{e})$$

or equivalently

$$\dot{\tilde{e}} = A\tilde{e} + Bu_c + B\{[f(x,t) - \hat{f}(\hat{x},t)] + \\ + [g(x,t) - \hat{g}(\hat{x},t)]u + \tilde{d}\} - K_oC^T\tilde{e} \\ \tilde{e}_1 = C^T\tilde{e}$$

which can be written as

$$\dot{\tilde{e}} = (A - K_oC^T)\tilde{e} + Bu_c + B\{[f(x,t) - \hat{f}(\hat{x},t)] + \\ + [g(x,t) - \hat{g}(\hat{x},t)]u + \tilde{d}\} \tag{8.159}$$

$$\tilde{e}_1 = C^T\tilde{e} \tag{8.160}$$

or equivalently, it can be written as

$$\dot{\tilde{e}} = (A - K_oC^T)\tilde{e} + Bu_c + Bw + \tilde{d}\} \tag{8.161}$$

$$\tilde{e}_1 = C^T\tilde{e} \tag{8.162}$$

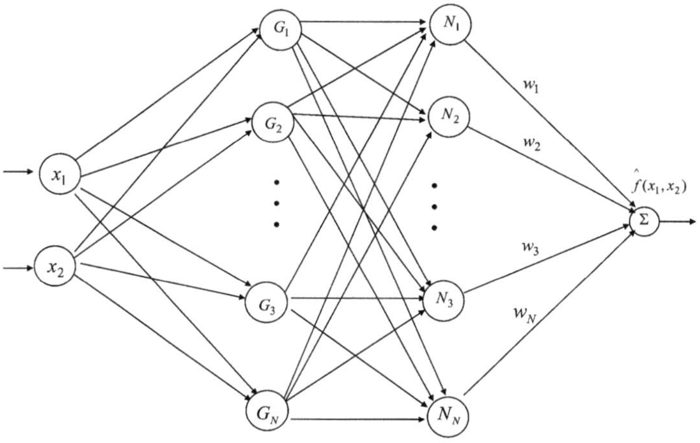

Fig. 8.14 Neurofuzzy approximator used for estimating the unknown dynamics of the distributed generators

8.4.3.3 Approximation of Functions $f(x, t)$ and $g(x, t)$

Next, the following approximators of the unknown dynamics of the distributed power generators are defined

$$\hat{f}(\hat{x}) = \begin{pmatrix} \hat{f}_1(\hat{x}|\theta_f) \ \hat{x} \in R^{6\times 1} \ \hat{f}_1(\hat{x}|\theta_f) \ \in \ R^{1\times 1} \\ \hat{f}_2(\hat{x}|\theta_f) \ \hat{x} \in R^{6\times 1} \ \hat{f}_2(\hat{x}|\theta_f) \ \in \ R^{1\times 1} \end{pmatrix} \tag{8.163}$$

with kernel functions

$$\phi_f^{i,j}(\hat{x}) = \frac{\prod_{j=1}^{n} \mu_{A_j}^{i}(\hat{x}_j)}{\sum_{i=1}^{N} \prod_{j=1}^{n} \mu_{A_j}^{i}(\hat{x}_j)} \tag{8.164}$$

where N is the number of fuzzy rules, n is the number of fuzzy sets in the antecedent part of each rule, $\hat{x}$ is the estimate of the state vector and $\mu_{A_j^i}(\hat{x})$ is the i-th membership function of the antecedent (IF) part of the i-th fuzzy rule. Similarly, the following approximators of the unknown system dynamics are defined (Fig. 8.14)

$$\hat{g}(\hat{x}) = \begin{pmatrix} \hat{g}_1(\hat{x}|\theta_g) \ \hat{x} \in R^{6\times 1} \ \hat{g}_1(\hat{x}|\theta_g) \ \in \ R^{1\times 2} \\ \hat{g}_2(\hat{x}|\theta_g) \ \hat{x} \in R^{6\times 1} \ \hat{g}_2(\hat{x}|\theta_g) \ \in \ R^{1\times 2} \end{pmatrix} \tag{8.165}$$

The values of the weights that result in optimal approximation are

$$\begin{aligned} \theta_f^* &= arg\ min_{\theta_f \in M_{\theta_f}} [sup_{\hat{x} \in U_{\hat{x}}} (f(x) - \hat{f}(\hat{x}|\theta_f))] \\ \theta_g^* &= arg\ min_{\theta_g \in M_{\theta_g}} [sup_{\hat{x} \in U_{\hat{x}}} (g(x) - \hat{g}(\hat{x}|\theta_g))] \end{aligned} \tag{8.166}$$

where the variation ranges for the weights are defined as

$$M_{\theta_f} = \{\theta_f \in R^h : ||\theta_f|| \leq m_{\theta_f}\}$$
$$M_{\theta_g} = \{\theta_g \in R^h : ||\theta_g|| \leq m_{\theta_g}\} \tag{8.167}$$

The value of the approximation error defined in Eq. (8.153) that corresponds to the optimal values of the weights vectors θ_f^* and θ_g^* one has

$$w = \left(f(x,t) - \hat{f}(\hat{x}|\theta_f^*)\right) + \left(g(x,t) - \hat{g}(\hat{x}|\theta_g^*)\right) u \tag{8.168}$$

which is next written as

$$w = \left(f(x,t) - \hat{f}(\hat{x}|\theta_f) + \hat{f}(\hat{x}|\theta_f) - \hat{f}(\hat{x}|\theta_f^*)\right) + \\ + \left(g(x,t) - \hat{g}(\hat{x}|\theta_g) + \hat{g}(\hat{x}|\theta_g) - \hat{g}(\hat{x}|\theta_g^*)\right) u \tag{8.169}$$

which can be also written in the following form

$$w = \left(w_a + w_b\right) \tag{8.170}$$

where

$$w_a = \{[f(x,t) - \hat{f}(\hat{x}|\theta_f)] + [g(x,t) - \hat{g}(\hat{x}|\theta_g)]\}u \tag{8.171}$$

$$w_b = \{[\hat{f}(\hat{x}|\theta_f) - \hat{f}(\hat{x}|\theta_f^*)] + [\hat{g}(\hat{x},\theta_g) - \hat{g}(\hat{x}|\theta_g^*)]\}u \tag{8.172}$$

Moreover, the following weights error vectors are defined

$$\tilde{\theta}_f = \theta_f - \theta_f^*$$
$$\tilde{\theta}_g = \theta_g - \theta_g^* \tag{8.173}$$

8.4.4 Lyapunov Stability Analysis

8.4.4.1 Design of the Lyapunov Function

In case of the distributed synchronous generators, the adaptation law of the neuro-fuzzy approximators weights θ_f and θ_g as well as the equation of the supervisory control term u_c are derived from the requirement for negative definiteness of the Lyapunov function

$$V = \tfrac{1}{2}\hat{e}^T P_1 \hat{e} + \tfrac{1}{2}\tilde{e}^T P_2 \tilde{e} + \tfrac{1}{2\gamma_1}\tilde{\theta}_f^T \tilde{\theta}_f + \tfrac{1}{2\gamma_2} tr[\tilde{\theta}_g^T \tilde{\theta}_g] \tag{8.174}$$

The selection of the Lyapunov function is based on the following principle of indirect adaptive control $\hat{e}$: $\lim_{t\to\infty}\hat{x}(t) = x_d(t)$ and $\tilde{e}$: $\lim_{t\to\infty}\hat{x}(t) = x(t)$. This yields $\lim_{t\to\infty}x(t) = x_d(t)$. Substituting Eqs. (8.142), (8.143) and (8.159), (8.160) into Eq. (8.174) and differentiating results into

$$\begin{array}{c}\dot{V}=\frac{1}{2}\dot{\hat{e}}^TP_1\hat{e}+\frac{1}{2}\hat{e}^TP_1\dot{\hat{e}}+\frac{1}{2}\dot{\tilde{e}}^TP_2\tilde{e}+\frac{1}{2}\tilde{e}^TP_2\dot{\tilde{e}}+\\ +\frac{1}{\gamma_1}\dot{\tilde{\theta}}_f^T\tilde{\theta}_f+\frac{1}{\gamma_2}tr[\dot{\tilde{\theta}}_g^T\tilde{\theta}_g]\Rightarrow\end{array} \tag{8.175}$$

$$\begin{array}{c}\dot{V}=\frac{1}{2}\{(A-BK^T)\hat{e}+K_oC^T\tilde{e}\}^TP_1\hat{e}+\\ +\frac{1}{2}\hat{e}^TP_1\{(A-BK^T)\hat{e}+K_oC^T\tilde{e}\}+\frac{1}{2}\{(A-K_oC^T)\tilde{e}+Bu_c+\\ +B\tilde{d}+Bw\}^TP_2\tilde{e}+\frac{1}{2}\tilde{e}^TP_2\{(A-K_oC^T)\tilde{e}+Bu_c+B\tilde{d}+Bw\}+\\ +\frac{1}{\gamma_1}\dot{\tilde{\theta}}_f^T\tilde{\theta}_f+\frac{1}{\gamma_2}tr[\dot{\tilde{\theta}}_g^T\tilde{\theta}_g]\end{array} \tag{8.176}$$

$$\begin{array}{c}\dot{V}=\frac{1}{2}\{\hat{e}^T(A-BK^T)^T+\tilde{e}^TCK_o^T\}P_1\hat{e}+\frac{1}{2}\hat{e}^TP_1\{(A-BK^T)\hat{e}+K_oC^T\tilde{e}\}+\\ +\frac{1}{2}\{\tilde{e}^T(A-K_oC^T)^T+u_c^TB^T+w^TB^T+\\ +\tilde{d}^TB^T\}P_2\tilde{e}+\frac{1}{2}\tilde{e}^TP_2\{(A-K_oC^T)\tilde{e}+Bu_c+Bw+B\tilde{d}\}+\\ +\frac{1}{\gamma_1}\dot{\tilde{\theta}}_f^T\tilde{\theta}_f+\frac{1}{\gamma_2}tr[\dot{\tilde{\theta}}_g^T\tilde{\theta}_g]\Rightarrow\end{array} \tag{8.177}$$

$$\begin{array}{c}\dot{V}=\frac{1}{2}\hat{e}^T(A-BK^T)^TP_1\hat{e}+\frac{1}{2}\tilde{e}^TCK_o^TP_1\hat{e}+\frac{1}{2}\hat{e}^TP_1(A-BK^T)\hat{e}+\frac{1}{2}\hat{e}^TP_1K_oC^T\tilde{e}+\\ +\frac{1}{2}\tilde{e}^T(A-K_oC^T)^TP_2\tilde{e}+\frac{1}{2}(u_c^T+w^T+\tilde{d}^T)B^TP_2\tilde{e}+\\ +\frac{1}{2}\tilde{e}^TP_2(A-K_oC^T)\tilde{e}+\frac{1}{2}\tilde{e}^TP_2B(u_c+w+\tilde{d})+\\ +\frac{1}{\gamma_1}\dot{\tilde{\theta}}_f^T\tilde{\theta}_f+\frac{1}{\gamma_2}tr[\dot{\tilde{\theta}}_g^T\tilde{\theta}_g]\end{array} \tag{8.178}$$

Assumption 1: For given positive definite matrices Q_1 and Q_2 there exist positive definite matrices P_1 and P_2, which are the solution of the following Riccati equations [375]

$$(A-BK^T)^TP_1+P_1(A-BK^T)+Q_1=0 \tag{8.179}$$

$$\begin{array}{c}(A-K_oC^T)^TP_2+P_2(A-K_oC^T)-\\ -P_2B(\frac{2}{r}-\frac{1}{\rho^2})B^TP_2+Q_2=0\end{array} \tag{8.180}$$

The conditions given in Eqs. (8.179) to (8.180) are related to the requirement that the systems described by Eqs. (8.142), (8.143) and (8.159), (8.160) have stable poles. Substituting Eqs. (8.179) to (8.180) into $\dot{V}$ yields

$$\dot{V}=\frac{1}{2}\hat{e}^T\{(A-BK^T)^TP_1+P_1(A-BK^T)\}\hat{e}+\tilde{e}^TCK_o^TP_1\hat{e}+\frac{1}{2}\tilde{e}^T\{(A-K_oC^T)^TP_2+$$
$$P_2(A-K_oC^T)\}\tilde{e}+\tilde{e}^TP_2B(u_c+w+\tilde{d})+\frac{1}{\gamma_1}\dot{\tilde{\theta}}_f^T\tilde{\theta}_f+\frac{1}{\gamma_2}tr[\dot{\tilde{\theta}}_g^T\tilde{\theta}_g]$$

i.e.

$$\begin{array}{c}\dot{V}=-\frac{1}{2}\hat{e}^TQ_1\hat{e}+\tilde{e}^TCK_o^TP_1\hat{e}-\\ -\frac{1}{2}\tilde{e}^T\{Q_2-P_2B(\frac{2}{r}-\frac{1}{\rho^2})B^TP_2\}\tilde{e}+\tilde{e}^TP_2B(u_c+w+\tilde{d})+\\ +\frac{1}{\gamma_1}\dot{\tilde{\theta}}_f^T\tilde{\theta}_f+\frac{1}{\gamma_2}tr[\dot{\tilde{\theta}}_g^T\tilde{\theta}_g]\end{array} \tag{8.181}$$

The supervisory control u_c is decomposed in two terms, u_a and u_b.

- The control term u_a is given by

$$u_a = -\frac{1}{r}\tilde{e}^T P_2 B + \Delta u_a \tag{8.182}$$

where assuming that the measurable elements of vector $\tilde{e}$ are $\{\tilde{e}_1, \tilde{e}_3, \dots, \tilde{e}_k\}$, the term Δu_a is such that

$$-\tfrac{1}{r}\tilde{e}^T P_2 B + \Delta u_a = -\tfrac{1}{r}\begin{pmatrix} p_{11}\tilde{e}_1 + p_{13}\tilde{e}_3 + \cdots + p_{1k}\tilde{e}_k \\ p_{13}\tilde{e}_1 + p_{33}\tilde{e}_3 + \cdots + p_{3k}\tilde{e}_k \\ \cdots\;\cdots\cdots \\ p_{1k}\tilde{e}_1 + p_{3k}\tilde{e}_3 + \cdots + p_{kk}\tilde{e}_k \end{pmatrix} \tag{8.183}$$

which finally stands for a product between the state vector elements $\{\tilde{e}_1, \tilde{e}_3, \dots, \tilde{e}_k\}$ and the elements of matrix P_2 which is obtained from the solution of the previous Riccati equation.
- The control term u_b is given by

$$u_b = -[(P_2B)^T(P_2B)]^{-1}(P_2B)^T C K_o^T P_1 \hat{e} \tag{8.184}$$

- u_a is an H_∞ control used for the compensation of the approximation error w and the additive disturbance $\tilde{d}$. Its first component $-\frac{1}{r}\tilde{e}^T P_2 B$ has been chosen so as to compensate for the term $\frac{1}{r}\tilde{e}^T P_2 B B^T P_2 \tilde{e}$, which appears in Eq. (8.181). By including also the second component Δu_a, one has that u_a is computed based on the feedback only the measurable variables $\{\tilde{e}_1, \tilde{e}_3, \dots, \tilde{e}_k\}$, out of the complete vector $\tilde{e} = [\tilde{e}_1, \tilde{e}_2, \dots, \tilde{e}_n]$. Equation (8.182) is finally rewritten as $u_a = -\frac{1}{r}\tilde{e}^T P_2 B + \Delta u_a$.
- u_b is a control used for the compensation of the observation error (the control term u_b has been chosen so as to satisfy the condition $\tilde{e}^T P_2 B u_b = -\tilde{e}^T C K_o^T P_1 \hat{e}$.

The control scheme is depicted in Fig. 8.15.

Substituting Eqs. (8.182) and (8.184) in $\dot{V}$ and assuming that Eqs. (8.179) and (8.180) hold, one gets

$$\begin{aligned}\dot{V} = &-\tfrac{1}{2}\hat{e}^T Q_1 \hat{e} + \tilde{e}^T C K_o^T P_1 \hat{e} - \tfrac{1}{2}\tilde{e}^T Q_2 \tilde{e} + \\ &+\tfrac{1}{r}\tilde{e}^T P_2 B B^T P_2 \tilde{e} - \tfrac{1}{2\rho^2}\tilde{e}^T P_2 B B^T P_2 \tilde{e} + \\ &+\tilde{e}^T P_2 B u_a + \tilde{e}^T P_2 B u_b + \tilde{e}^T P_2 B (w + \tilde{d}) + \\ &+\tfrac{1}{\gamma_1}\dot{\tilde{\theta}}_f^T \tilde{\theta}_f + \tfrac{1}{\gamma_2} tr[\dot{\tilde{\theta}}_g^T \tilde{\theta}_g]\end{aligned} \tag{8.185}$$

or equivalently,

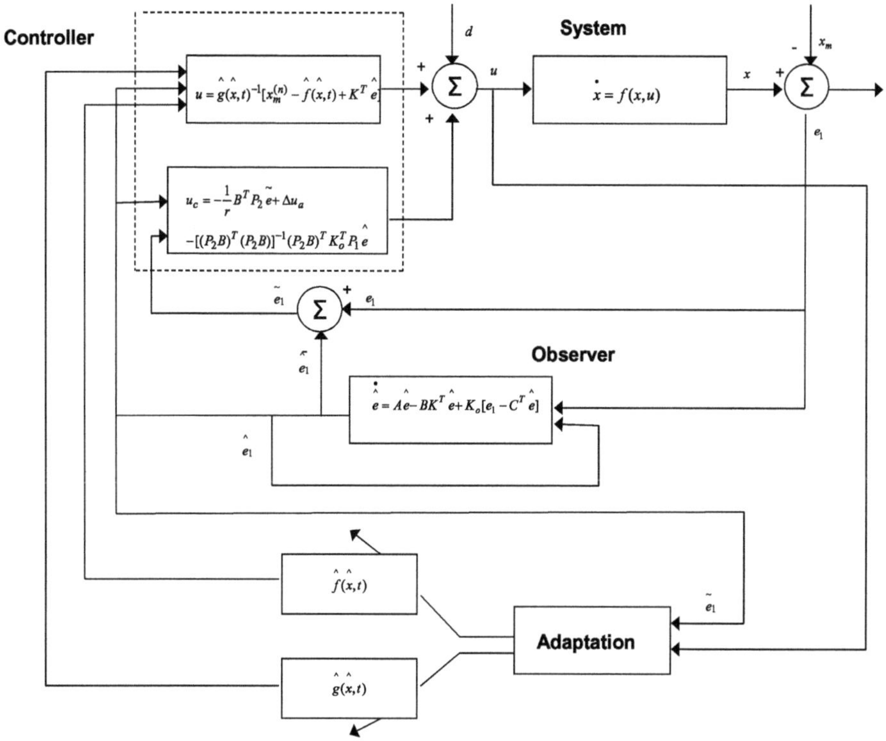

Fig. 8.15 The proposed output feedback-based adaptive fuzzy control scheme for distributed synchronous generators

$$\begin{aligned}\dot{V} = &-\tfrac{1}{2}\hat{e}^T Q_1 \hat{e} - \tfrac{1}{2}\tilde{e}^T Q_2 \tilde{e} - \tfrac{1}{2\rho^2}\tilde{e}^T P_2 B B^T P_2 \tilde{e} + \\ &+ \tilde{e}^T P_2 B(w + \tilde{d} + \Delta u_a) + \tfrac{1}{\gamma_1}\dot{\tilde{\theta}}_f^T \tilde{\theta}_f + \tfrac{1}{\gamma_2} tr[\dot{\tilde{\theta}}_g^T \tilde{\theta}_g]\end{aligned} \tag{8.186}$$

It holds that $\dot{\tilde{\theta}}_f = \dot{\theta}_f - \dot{\theta}_f^* = \dot{\theta}_f$ and $\dot{\tilde{\theta}}_g = \dot{\theta}_g - \dot{\theta}_g^* = \dot{\theta}_g$. The following weights adaptation laws are considered:

$$\begin{aligned}\dot{\theta}_f &= -\gamma_1 \Phi(\hat{x})^T B^T P_2 \tilde{e} \\ \dot{\theta}_g &= -\gamma_2 \Phi(\hat{x})^T B^T P_2 \tilde{e} u^T\end{aligned} \tag{8.187}$$

where assuming N fuzzy rules and associated kernel functions the matrices dimensions are $\theta_f \in R^{N\times 1}$, $\theta_g \in R^{N\times 2}$, $\Phi(x) \in R^{2\times N}$, $B \in R^{6\times 3}$, $P \in R^{6\times 6}$ and $\tilde{e} \in R^{6\times 1}$.

The update of θ_f is a gradient type algorithm. The update of θ_g is also a gradient type algorithm, where u_c implicitly tunes the adaptation gain γ_2 [34, 344]. Substituting Eq. (8.187) in $\dot{V}$ gives

$$\begin{aligned}\dot{V} = -\tfrac{1}{2}\hat{e}^T Q_1\hat{e} - \tfrac{1}{2}\tilde{e}^T Q_2\tilde{e} - \tfrac{1}{2\rho^2}\tilde{e}^T P_2BB^T P_2\tilde{e} + B^T P_2\tilde{e}(w + d + \Delta u_a) + \\ + \tfrac{1}{\gamma_1}(-\gamma_1)\tilde{e}^T P_2B\Phi(\hat{x})(\theta_f - \theta_f^*) + \tfrac{1}{\gamma_2}(-\gamma_2)tr[u\tilde{e}^T P_2B\Phi(\hat{x})(\theta_g - \theta_g^*)]\end{aligned} \tag{8.188}$$

or

$$\begin{aligned}\dot{V} = -\tfrac{1}{2}\hat{e}^T Q_1\hat{e} - \tfrac{1}{2}\tilde{e}^T Q_2\tilde{e} - \tfrac{1}{2\rho^2}\tilde{e}^T P_2BB^T P_2\tilde{e} + B^T P_2\tilde{e}(w + \tilde{d} + \Delta u_a) + \\ + \tfrac{1}{\gamma_1}(-\gamma_1)\tilde{e}^T P_2B\Phi(\hat{x})(\theta_f - \theta_f^*) + \tfrac{1}{\gamma_2}(-\gamma_2)tr[u\tilde{e}^T P_2B(\hat{g}(\hat{x}|\theta_g) - \hat{g}(\hat{x}|\theta_g^*)]\end{aligned} \tag{8.189}$$

Taking into account that $u \in R^{3\times 1}$ and $\tilde{e}^T PB(\hat{g}(x|\theta_g) - \hat{g}(x|\theta_g^*)) \in R^{1\times 3}$ it holds

$$\begin{aligned}\dot{V} = -\tfrac{1}{2}\hat{e}^T Q_1\hat{e} - \tfrac{1}{2}\tilde{e}^T Q_2\tilde{e} - \tfrac{1}{2\rho^2}\tilde{e}^T P_2BB^T P_2\tilde{e} + B^T P_2\tilde{e}(w + \tilde{d} + \Delta u_a) + \\ + \tfrac{1}{\gamma_1}(-\gamma_1)\tilde{e}^T P_2B\Phi(\hat{x})(\theta_f - \theta_f^*) + \tfrac{1}{\gamma_2}(-\gamma_2)tr[\tilde{e}^T P_2B(\hat{g}(\hat{x}|\theta_g) - \hat{g}(\hat{x}|\theta_g^*))u]\end{aligned} \tag{8.190}$$

Since $\tilde{e}^T P_2B(\hat{g}(\hat{x}|\theta_g) - \hat{g}(\hat{x}|\theta_g^*))u \in R^{1\times 1}$ it holds

$$\begin{aligned}tr(\tilde{e}^T P_2B(\hat{g}(x|\theta_g) - \hat{g}(x|\theta_g^*)u) = \\ = \tilde{e}^T P_2B(\hat{g}(x|\theta_g) - \hat{g}(x|\theta_g^*))u\end{aligned} \tag{8.191}$$

Therefore, one finally obtains

$$\begin{aligned}\dot{V} = -\tfrac{1}{2}\hat{e}^T Q_1\hat{e} - \tfrac{1}{2}\tilde{e}^T Q_2\tilde{e} - \tfrac{1}{2\rho^2}\tilde{e}^T P_2BB^T P_2\tilde{e} + B^T P_2\tilde{e}(w + \tilde{d} + \Delta u_a) + \\ + \tfrac{1}{\gamma_1}(-\gamma_1)\tilde{e}^T P_2B\Phi(\hat{x})(\theta_f - \theta_f^*) + \tfrac{1}{\gamma_2}(-\gamma_2)\tilde{e}^T P_2B(\hat{g}(\hat{x}|\theta_g) - \hat{g}(\hat{x}|\theta_g^*))u\end{aligned} \tag{8.192}$$

Next, the following approximation error is defined

$$w_\alpha = [\hat{f}(\hat{x}|\theta_f^*) - \hat{f}(\hat{x}|\theta_f)] + [\hat{g}(\hat{x}|\theta_g^*) - \hat{g}(\hat{x}|\theta_g)]u \tag{8.193}$$

Thus, one obtains

$$\begin{aligned}\dot{V} = -\tfrac{1}{2}\hat{e}^T Q_1\hat{e} - \tfrac{1}{2}\tilde{e}^T Q_2\tilde{e} - \tfrac{1}{2\rho^2}\tilde{e}^T P_2BB^T P_2\tilde{e} + \\ + B^T P_2\tilde{e}(w + \tilde{d}) + \tilde{e}^T P_2Bw_\alpha\end{aligned} \tag{8.194}$$

Denoting the aggregate approximation error and disturbances vector as

$$w_1 = w + \tilde{d} + w_\alpha + \Delta u_a \tag{8.195}$$

the derivative of the Lyapunov function becomes

$$\dot{V} = -\tfrac{1}{2}\hat{e}^T Q_1\hat{e} - \tfrac{1}{2}\tilde{e}^T Q_2\tilde{e} - \tfrac{1}{2\rho^2}\tilde{e}^T P_2BB^T P_2\tilde{e} + \tilde{e}^T P_2Bw_1 \tag{8.196}$$

which in turn is written as

$$\dot{V} = -\tfrac{1}{2}\hat{e}^T Q_1 \hat{e} - \tfrac{1}{2}\tilde{e}^T Q_2 \tilde{e} - \tfrac{1}{2\rho^2}\tilde{e}^T P_2 B B^T P_2 \tilde{e} + \\ + \tfrac{1}{2}\tilde{e}^T P B w_1 + \tfrac{1}{2} w_1^T B^T P_2 \tilde{e} \tag{8.197}$$

Lemma: The following inequality holds

$$\tfrac{1}{2}\tilde{e}^T P_2 B w_1 + \tfrac{1}{2} w_1^T B^T P_2 \tilde{e} - \tfrac{1}{2\rho^2}\tilde{e}^T P_2 B B^T P_2 \tilde{e} \\ \leq \tfrac{1}{2}\rho^2 w_1^T w_1 \tag{8.198}$$

Proof: The binomial $(\rho a - \frac{1}{\rho}b)^2 \geq 0$ is considered. Expanding the left part of the above inequality one gets

$$\rho^2 a^2 + \tfrac{1}{\rho^2}b^2 - 2ab \geq 0 \Rightarrow \tfrac{1}{2}\rho^2 a^2 + \tfrac{1}{2\rho^2}b^2 - ab \geq 0 \Rightarrow \\ ab - \tfrac{1}{2\rho^2}b^2 \leq \tfrac{1}{2}\rho^2 a^2 \Rightarrow \tfrac{1}{2}ab + \tfrac{1}{2}ab - \tfrac{1}{2\rho^2}b^2 \leq \tfrac{1}{2}\rho^2 a^2 \tag{8.199}$$

The following substitutions are carried out: $a = w_1$ and $b = \tilde{e}^T P_2 B$ and the previous relation becomes

$$\tfrac{1}{2} w_1^T B^T P_2 \tilde{e} + \tfrac{1}{2}\tilde{e}^T P_2 B w_1 - \tfrac{1}{2\rho^2}\tilde{e}^T P_2 B B^T P_2 \tilde{e} \\ \leq \tfrac{1}{2}\rho^2 w_1^T w_1 \tag{8.200}$$

The above relation is used in $\dot{V}$, and the right part of the associated inequality is enforced

$$\dot{V} \leq -\frac{1}{2}\hat{e}^T Q_1 \hat{e} - \frac{1}{2}\tilde{e}^T Q_2 \tilde{e} + \frac{1}{2}\rho^2 w_1^T w_1 \tag{8.201}$$

Thus, Eq. (8.201) can be written as

$$\dot{V} \leq -\frac{1}{2}E^T Q E + \frac{1}{2}\rho^2 w_1^T w_1 \tag{8.202}$$

where

$$E = \begin{pmatrix} \hat{e} \\ \tilde{e} \end{pmatrix}, \quad Q = \begin{pmatrix} Q_1 & 0 \\ 0 & Q_2 \end{pmatrix} = diag[Q_1, Q_2] \tag{8.203}$$

Hence, the H_∞ performance criterion is derived. For ρ sufficiently small Eq. (8.201) will be true and the H_∞ tracking criterion will be satisfied. The smallest value of ρ for which the Riccati equation given in Eq. (8.180) admits as solution a positive definite matrix P_2 is the one that provides maximum robustness to the control loop. In that case, the integration of $\dot{V}$ from 0 to T gives

$$\int_0^T \dot{V}(t)dt \leq -\tfrac{1}{2}\int_0^T ||E||^2 dt + \tfrac{1}{2}\rho^2 \int_0^T ||w_1||^2 dt \Rightarrow \\ 2V(T) - 2V(0) \leq -\int_0^T ||E||_Q^2 dt + \rho^2 \int_0^T ||w_1||^2 dt \Rightarrow \\ 2V(T) + \int_0^T ||E||_Q^2 dt \leq 2V(0) + \rho^2 \int_0^T ||w_1||^2 dt \tag{8.204}$$

It is assumed that there exists a positive constant $M_w > 0$ such that $\int_0^{\infty} ||w_1||^2 dt \leq M_w$. Therefore for the integral $\int_0^T ||E||_Q^2 dt$ one gets

$$\int_0^{\infty} ||E||_Q^2 dt \leq 2V(0) + \rho^2 M_w \tag{8.205}$$

Thus, the integral $\int_0^{\infty} ||E||_Q^2 dt$ is bounded and according to Barbalat's Lemma

$$\begin{array}{c} \lim_{t\to\infty} E(t) = 0 \Rightarrow \\ \lim_{t\to\infty} \hat{e}(t) = 0 \\ \lim_{t\to\infty} \tilde{e}(t) = 0 \end{array} \tag{8.206}$$

Therefore $\lim_{t\to\infty} e(t) = 0$.

8.4.5 Simulation Tests

The performance of the proposed adaptive fuzzy control scheme for distributed synchronous generators was tested through simulation experiments. The state feedback gain was $K \in R^{6\times 2}$. The basis functions used in the estimation of $f_i(\hat{x}, t),\ i = 1, \ldots, 3$ and $g_{ij}(\hat{x}, t),\ i = 1, \ldots, 3\ j = 1, \ldots, 3$ were $\mu_{A_j}(\hat{x}) = e^{(\frac{\hat{x}-c_j}{\sigma})^2}, j = 1, \ldots, 2$. Since for each unknown function $f_i(\hat{x}, t)$ and $g_{ij}(\hat{x}, t)$ there are 6 inputs and each one of them consists of 3 fuzzy sets, for the approximation of the functions $f_i(\hat{x}, t)\ i = 1, \ldots, 3$, and $g_{ij}(\hat{x}, t)\ i = 1, \ldots, 3,\ j = 1, \ldots, 3$ there will be 729 fuzzy rules of the form:

$$\begin{array}{l} R^l : IF\ \hat{x}_{1,1}\ is\ A_1^l\ AND\ \hat{x}_{1,2}\ is\ A_2^l\ \cdots \\ AND\ \hat{x}_{3,2}\ is\ THEN\ \hat{f}_i^l\ is\ b^l \end{array} \tag{8.207}$$

The aggregate output of the neuro-fuzzy approximator (rule-base) is

$$\hat{f}_i(\hat{x}, t) = \frac{\sum_{l=1}^{27} \hat{f}_i^l \prod_{i=1}^{3} \mu_{A_i}^l(\hat{x}_i)}{\sum_{l=1}^{27} \prod_{i=1}^{4} \mu_{A_i}^l(\hat{x}_i)} \tag{8.208}$$

The estimation of the control input gain functions $\hat{g}_{ij}(\hat{x}, t)\ i = 1, \ldots, 3$ was derived in a similar way. The sampling period was taken to be 0.01 sec. In the beginning of the training of the neuro-fuzzy approximators their weights were initialized to zero. Moreover, the elements of the system's state vector were also initialized to zero. The positive definite matrices $P_1 \in R^{6\times 6}$ and $P_2 \in R^{6\times 6}$ stem from the solution of the algebraic Riccati equations Eqs. (8.179) and (8.180), for Q_1 and Q_2 also positive definite.

The approximations $\hat{f}$ and $\hat{g}$ were used in the derivation of the control law, given by Eq. (8.137). To show the disturbance rejection capability of the proposed adaptive

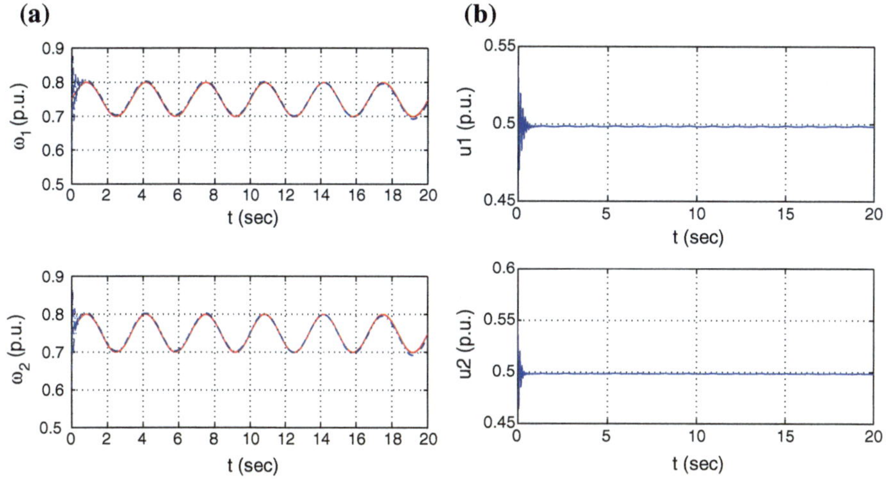

Fig. 8.16 **a** Tracking of reference set-point 1 by the state variables ω_i, $i = 1, \ldots, 2$ of the distributed power generators' model **b** Control inputs u_i, $i = 1, \ldots, 2$ applied to the individual power generators

fuzzy controller, at the beginning of the second half of the simulation time additive sinusoidal disturbances of amplitude equal to 10 % of the mean value of the control inputs were applied to the model of the coupled synchronous generators.

The performance of the differential flatness theory-based adaptive fuzzy control loop was tested in the case of tracking of different reference setpoints. The obtained results are depicted in Figs. 8.16, 8.17, 8.18, 8.19, 8.20 and 8.21. The state variables were measured in the per unit (p.u.) system. It can be observed that the proposed adaptive fuzzy control scheme succeeded fast and accurate tracking of all these setpoints.

The RMSE (root mean square error) of the examined control loop is also calculated (assuming the same parameters of the controller) in the case of tracking of the previous setpoints 1 to 5. The results are summarized in Table 8.2. The tracking accuracy of the control method was remarkable despite the fact that (i) the dynamic model of the power generators was completely unknown, (ii) only output feedback was used in the implementation of the control scheme. It can be also seen that the transient characteristics of the control scheme are quite satisfactory.

8.5 Control and Synchronization of Distributed Inverters

8.5.1 Outline

Next, the problem of decentralized control for stabilization and synchronization of distributed DC power generation units is examined. The integration of distributed DC

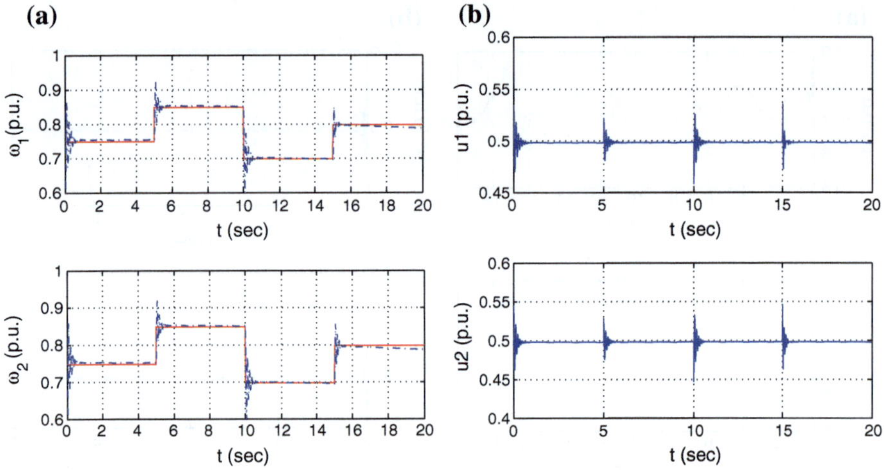

Fig. 8.17 **a** Tracking of reference set-point 2 by the state variables $\omega_i,\ i = 1, \ldots, 2$ of the distributed power generators' model **b** Control inputs $u_i,\ i = 1, \ldots, 2$ applied to the individual power generators

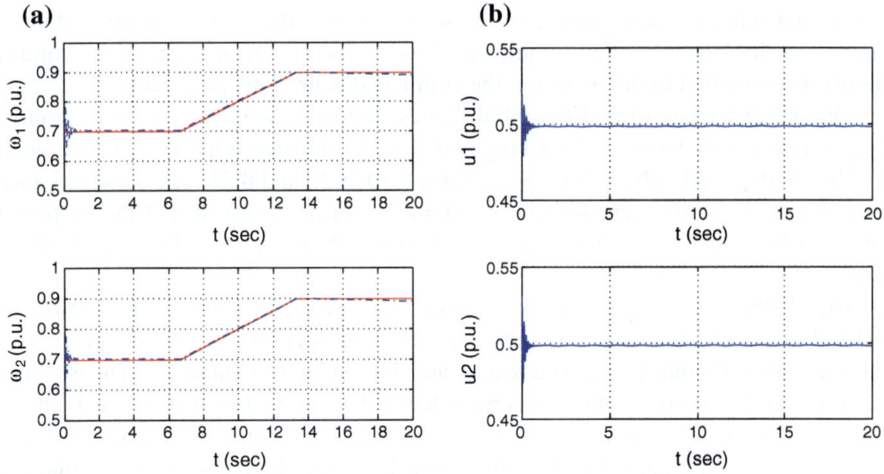

Fig. 8.18 **a** Tracking of reference set-point 3 by the state variables $\omega_i,\ i = 1, \ldots, 2$ of the distributed power generators' model **b** Control inputs $u_i,\ i = 1, \ldots, 2$ applied to the individual power generators

power generation units in microgrids, such as photovoltaics and fuel cells requires the interfacing of such devices with the electricity network through inverters [28, 259, 317, 324, 392, 393]. In this direction, several results on inverters' control have been obtained. In [386], control of photovoltaic inverters is analyzed in fault ride-through situations and voltage sags conditions. In [223], predictive control is proposed for voltage source inverters. A discrete-time model of the inverters is used to compute

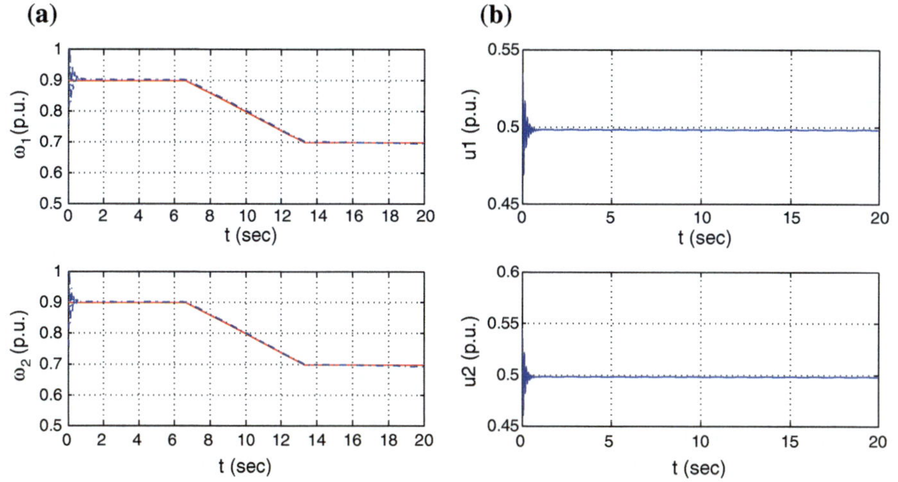

Fig. 8.19 **a** Tracking of reference set-point 4 by the state variables ω_i, $i = 1, \ldots, 2$ of the distributed power generators' model **b** Control inputs u_i, $i = 1, \ldots, 2$ applied to the individual power generators

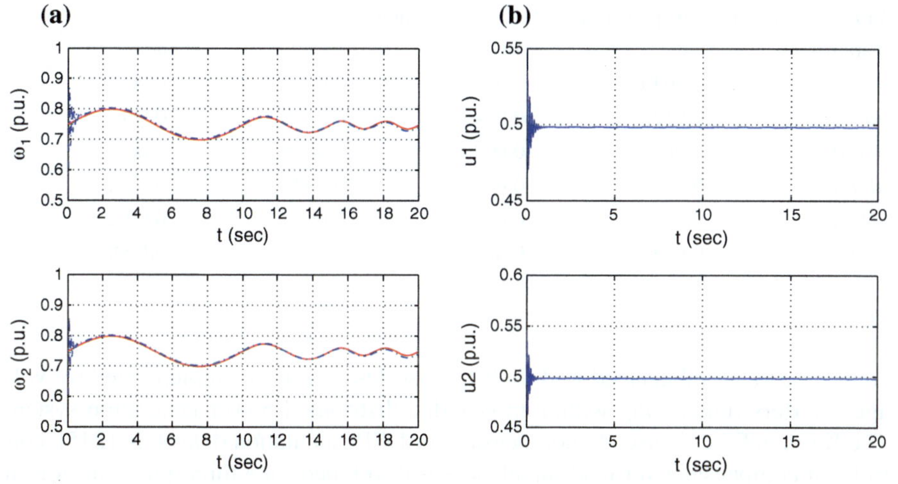

Fig. 8.20 **a** Tracking of reference set-point 5 by the state variables ω_i, $i = 1, \ldots, 2$ of the distributed power generators' model **b** Control inputs u_i, $i = 1, \ldots, 2$ applied to the individual power generators

the control law while Lyapunov stability analysis is also provided. In [164], a control method is proposed for the amplitude and frequency of the output of voltage inverters, in the case of distributed (grid-tied) functioning of the inverters, as well as in the case of stand-alone functioning (islanding). In [413], sliding mode control is proposed for active and reactive power regulation of a three-phase inverter

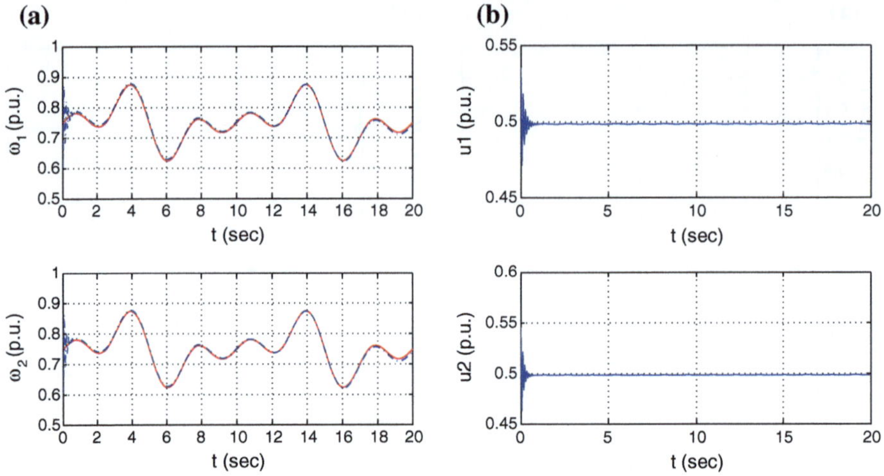

Fig. 8.21 **a** Tracking of reference set-point 6 by the state variables ω_i, $i = 1, \ldots, 2$ of the distributed power generators' model **b** Control inputs u_i, $i = 1, \ldots, 2$ applied to the individual power generators

Table 8.2 RMSE of the power generator's state variables

Parameter	ω_1	$\dot{\omega}_2$	ω_2	$\dot{\omega}_2$
$RMSE_1$	0.0035	0.0002	0.0034	0.0002
$RMSE_2$	0.0123	0.0545	0.0118	0.0602
$RMSE_3$	0.0035	0.0020	0.0035	0.0020
$RMSE_4$	0.0031	0.0020	0.0026	0.0020
$RMSE_5$	0.0034	0.0003	0.0033	0.0002
$RMSE_6$	0.0035	0.0003	0.0033	0.0002

connected to an imbalanced grid. In [9], sensorless control is implemented for voltage inverters functioning within an islanded-distributed power generation system. The Kalman Filter is used to compensate for DSP calculation delays. In [245], control that compensates for load imbalances is developed for a three-phase inverter in an autonomous functioning mode.

Another problem that arises is the synchronization between multiple parallel inverters. In the recent years there has been significant effort in solving control and synchronization problems for voltage inverters, so as to enable connection of a large number of power sources to the grid [512]. In [205] a state-space model is developed for the dynamics of distributed power generation units connected to the grid through inverters. Stability analysis is performed and a simple state feedback control scheme is proposed. In [258] decentralized frequency control of inverters is proposed. Information about the total active power deficiency of the microgrid is shared between the local units. Frequency control is performed to achieve a cooperative frequency recovery. In [105] a Lyapunov theory-based feedback controller is proposed for voltage inverters connected to the grid. In [167] stability of parallel DC sources connected to the grid through inverters is investigated and simple feedback control is proposed using the equivalence of the inverters' model to that of a synchronous generator. In [200] it is considered that each inverter is equivalent to an electronic circuit with oscillatory dynamics and thus the problem of control of parallel inverters becomes a problem of synchronization of coupled electrical oscillators. In [268] Lie algebra is used to transform the nonlinear model of three-phase photovoltaic units into linear systems and based on the linearized description a state feedback controller is designed. In [511] the concept of synchronverters is introduced. This is based on the equivalence between the voltage inverter's model and that of a synchronous generator. It is shown that the controller is in principle a power controller with integrated capability of voltage and frequency regulation. In [49] Lie algebra-based input-output feedback linearization is used to solve the synchronization problem for nonlinear inverters connected in parallel to the grid. It is considered that the power exchange of each local inverter towards the grid is modeled equivalently by the dynamics of a synchronous generator. In [396] a feedback control and synchronization method for distributed DC power generation units is developed using the equivalence between the inverter's model and the model of the synchronous power generator. In [397, 398] stability conditions are provided for distributed DC power generation units using a Lyapunov function approach.

By solving the problem of synchronization between DC distributed power generation units which are connected to the microgrid through voltage inverters a more efficient operation of the power grid can be achieved in terms of power quality and stability [167]. To this end, the present section is proposing a nonlinear feedback control method for parallel inverters which is based on differential flatness theory. Moreover, it is proposing a nonlinear filtering method under the name Derivative-free nonlinear Kalman Filtering which enables to estimate the non-measurable state vector elements of each local inverter, as well as disturbance terms affecting each inverter (the latter can be due to their interaction with other power generation units in the grid or due to external perturbations and modeling uncertainty) [352, 366, 374, 375].

In the following section, the problem of control and synchronization of parallel inverters connected to the grid is primarily examined. As far as the power exchange with the grid is concerned, it is shown that the dynamics of each inverter can be written in a form that is equivalent to the model of the synchronous power generator. Using the latter description one can compute the active power that each local inverter should inject to the network so as to remain synchronized with the rest of the grid. The frequency of this virtual generator model is next used to compute the output voltage setpoints for the electrical part of each local inverter. Thus, finally the synchronization problem of each local inverter is turned into a problem of nonlinear feedback control for the associated inverter's electrical model. It is also shown that in the proposed decentralized control scheme each local controller processes measurements not only coming from the individual inverter but also coming from other inverters which are connected to the grid.

8.5.2 Dynamic Model of the Inverter

The dynamic, model of the three-phase inverter, given in Chap. 1 is reviewed here

$$\frac{d}{dt}\begin{pmatrix} V_{L_d} \\ V_{L_q} \\ i_{I_d} \\ i_{I_q} \end{pmatrix} = \begin{pmatrix} \omega V_{L_q} + \frac{1}{C_f} i_{I_d} - \frac{1}{C_f}\frac{p_f V_{L_d} + q_f V_{L_q}}{V_{L_d}^2 + V_{L_q}^2} + \omega C_f V_{L_q} - \frac{\omega L_f V_{L_q}(i_{I_d}^2 + i_{I_q}^2)}{(V_{L_d}^2 + V_{L_q}^2)} \\ -\omega V_{L_d} + \frac{1}{C_f} i_{I_q} - \frac{1}{C_f}\frac{p_f V_{L_q} - q_f V_{L_d}}{V_{L_d}^2 + V_{L_q}^2} - \omega C_f V_{L_d} + \frac{\omega L_f V_{L_d}(i_{I_d}^2 + i_{I_q}^2)}{(V_{L_d}^2 + V_{L_q}^2)} \\ \omega i_{I_q} - \frac{1}{L_f} V_{L_d} \\ -\omega i_{I_d} - \frac{1}{L_f} V_{L_q} \end{pmatrix} + \begin{pmatrix} 0 & 0 \\ 0 & 0 \\ \frac{1}{L_f} & 0 \\ 0 & \frac{1}{L_f} \end{pmatrix} \begin{pmatrix} V_{I_d} \\ V_{I_q} \end{pmatrix} \tag{8.209}$$

while the measurement equation of the inverter's model is

$$\begin{pmatrix} y_1 \\ y_2 \end{pmatrix} = \begin{pmatrix} V_{L_d} \\ V_{L_q} \end{pmatrix} = \begin{pmatrix} 1 & 0 & 0 & 0 \\ 0 & 1 & 0 & 0 \end{pmatrix} \begin{pmatrix} V_{L_d} \\ V_{L_q} \\ i_{L_d} \\ i_{L_q} \end{pmatrix} \tag{8.210}$$

and by using the state variables notation $x_1 = V_{L_d}$, $x_2 = V_{L_q}$, $x_3 = i_{L_d}$ and $x_4 = i_{L_q}$ one has

$$\frac{d}{dt}\begin{pmatrix} x_1 \\ x_2 \\ x_3 \\ x_4 \end{pmatrix} = \begin{pmatrix} \omega x_2 + \frac{1}{C_f} x_3 - \frac{1}{C_f}\frac{p_f x_1 + q_f x_2}{x_1^2 + x_2^2} + \omega C_f x_2 - \frac{\omega L_f x_2 (x_3^2 + x_4^2)}{(x_1^2 + x_2^2)} \\ -\omega x_1 + \frac{1}{C_f} x_4 - \frac{1}{C_f}\frac{p_f x_2 - q_f x_1}{x_1^2 + x_2^2} - \omega C_f x_1 + \frac{\omega L_f x_1 (x_3^2 + x_4^2)}{(x_1^2 + x_2^2)} \\ \omega x_4 - \frac{1}{L_f} x_1 \\ -\omega x_3 - \frac{1}{L_f} x_2 \end{pmatrix} + \begin{pmatrix} 0 & 0 \\ 0 & 0 \\ \frac{1}{L_f} & 0 \\ 0 & \frac{1}{L_f} \end{pmatrix} \begin{pmatrix} u_1 \\ u_2 \end{pmatrix} \tag{8.211}$$

while the measurement equation of the inverter's model is

$$\begin{pmatrix} y_1 \\ y_2 \end{pmatrix} = \begin{pmatrix} V_{L_d} \\ V_{L_q} \end{pmatrix} = \begin{pmatrix} 1 & 0 & 0 & 0 \\ 0 & 1 & 0 & 0 \end{pmatrix} \begin{pmatrix} x_1 \\ x_2 \\ x_3 \\ x_4 \end{pmatrix} \tag{8.212}$$

thus, the inverter's model is written in the nonlinear state-space form

$$\begin{aligned} \dot{x} &= f(x) + G(x)u \\ y &= h(x) \end{aligned} \tag{8.213}$$

where $f(x) \in R^{4\times 1}$, $G(x) \in R^{4\times 1}$ and $h(x) \in R^{2\times 4}$.

8.5.3 *The Synchronization Problem for Parallel Inverters*

8.5.3.1 Equivalence Between Inverters and Synchronous Generators

By applying appropriate control inputs, inverters can be made to function in a manner equivalent to synchronous generators. Actually, with the regulation of the active power produced by the inverters one can succeed the synchronization between the voltage output of the inverters and the voltage of the grid [513–515]. This synchronization enables better integration of DC renewable energy units to the grid.

The functioning of the i-th inverter is shown to be equivalent to a synchronous generator with turn speed denoted as ω_i [49, 396–398, 511, 512]. In this modelling approach one has

$$\Delta\delta_i = \omega_i - \omega_d = -k_{p_i}(P_i^m - P_i^d) \tag{8.214}$$

where P_i^m is the measured active power of the i-th power generation unit and P_i^d is the desirable active power. Coefficient k_{p_i} is a "droop" gain which is practically computed by dividing the range of variation of the inverter's frequency (ω_{max}–ω_{min}) by the maximum active power $P_{i_{max}}$ that the inverter can produce. Equation (8.214) denotes that when representing the inverter as an equivalent synchronous generator the turn speed should be proportional to the active power that the inverter provides.

Next, a relation is obtained for the measured active power P_i^m and the real active power P_i of the inverter P_i. It holds that $P_i^m(s) = e^{-s\tau_{p_i}}P(s)$ (delay relation in frequency domain) which after intermediate operations gives the differential equation

$$\tau_{p_i}\dot{P}_i^m = -P_i^m + P_i \tag{8.215}$$

Thus the inverter's dynamics is expressed as

$$\begin{aligned} \Delta\dot{\delta}_i &= \Delta\omega_i \\ \tau_{p_i}\dot{P}_i^m &= -P_i^m + P_i \end{aligned} \tag{8.216}$$

By deriving Eq. (8.214) with respect to time one obtains

$$\Delta\dot{\omega}_i = -k_{p_i}\dot{P}_i^m + k_{p_i}\dot{P}_i^d \tag{8.217}$$

Moreover, by solving Eq. (8.215) with respect to $\dot{P}_i^m$ one has

$$\dot{P}_i^m = -\frac{1}{\tau_{p_i}} P_i^m + \frac{1}{\tau_{p_i}} P_i \tag{8.218}$$

By substituting Eq. (8.218) into (8.217) the following relation is obtained

$$\Delta\dot{\omega}_i = -k_{p_i}(-\frac{1}{\tau_{p_i}} P_i^m + \frac{1}{\tau_{p_i}} P_i) + k_{p_i}\dot{P}_i^d \tag{8.219}$$

and by considering that $\dot{P}_i^d = 0$ one gets

$$\Delta\dot{\omega}_i = \frac{k_{p_i}}{\tau_{p_i}} P_i^m - \frac{k_{p_i}}{\tau_{p_i}} P_i \tag{8.220}$$

or equivalently

$$J_i\Delta\dot{\omega}_i = P_i^m - P_i \tag{8.221}$$

where the virtual moment of inertia is defined as $J_i = \tau_{p_i}/k_{p_i}$. Additionally, from Eq. (8.214) one has

$$\begin{array}{c} \omega_i - \omega_d - k_{p_i}P_i^d = -k_{p_i}P_i^m \Rightarrow P_i^m = -\frac{1}{k_{p_i}}\omega_i + \frac{1}{k_{p_i}}\omega_d + P_i^d \Rightarrow \\ P_i^m = -\frac{1}{k_{p_i}}\Delta\omega_i + P_i^d \end{array} \tag{8.222}$$

By substituting Eq. (8.222) into Eq. (8.221) the following relation is obtained

$$J_i\Delta\dot{\omega}_i = -\frac{1}{k_{p_i}}\Delta\omega_i + P_i^d - P_i \tag{8.223}$$

The damping coefficient $D_{p_i} = \frac{1}{k_{p_i}}$ is defined. Using this coefficient in Eq. (8.222) gives

$$J_i\Delta\dot{\omega}_i = -D_{p_i}\Delta\omega_i + P_i^d - P_i \tag{8.224}$$

which is the equation that describes the rotational motion of a synchronous generator. In ideal conditions there is no interaction (power exchange) between distributed power units connected to the same electricity grid. However, frequently such interaction exists and in the latter case Eq. (8.224) should be enhanced by including an interaction term

$$J_i\Delta\dot{\omega}_i = -D_{p_i}\Delta\omega_i + (P_i^d - P_i) + \sum_{j=1, j\neq i}^{n} G_{ij}sin(\delta_i - \delta_j) \tag{8.225}$$

where δ_i is the virtual turn angle that is associated with the i-th power generation unit (inverter). About the coupling coefficients G_{ij} these are functions of the conductance of the grid line which connects the i-th to the j-th power generation unit, as well as of the grid voltage that is measured at points i and j respectively.

Thus, finally the dynamics of the i-th power generation unit is described as

$$\begin{aligned} \Delta\dot{\delta}_i(t) &= \Delta\omega_i(t) \\ J_i\Delta\dot{\omega}_i(t) &= -D_{p_i}\Delta\omega_i(t) + (P_i^d(t) - P_i(t)) + \sum_{j=1}^N G_{ij}sin(\delta_i - \delta_j) \end{aligned} \tag{8.226}$$

In the design of the control and synchronization system for the distributed power generation units, it is considered that the parameters J_i, D_{p_i} and G_{ij} are either known or can be computed from measurements. Moreover, it is considered that the i-th local controller not only processes measurements coming from the associated power generation unit, but also uses measurements coming from the other power units which are connected to the grid (that is the virtual turn angles δ_j).

8.5.3.2 Control for Parallel Inverters Connected to the Grid

In Eq. (8.226), it has be shown that under certain conditions, the dynamics of the inverter becomes equivalent to that of the synchronous power generator. Then one has that the dynamics of the inverter is composed of two parts (i) the rotation part and (ii) the electrical part.

The virtual rotation part is given by:

$$\begin{aligned} \Delta\dot{\delta}_i(t) &= \Delta\omega_i(t) \\ J_i\Delta\dot{\omega}_i(t) &= -D_{p_i}\Delta\omega_i(t) + (P_i^d(t) - P_i(t)) + \sum_{j=1, j\neq i}^N G_{ij}sin(\delta_i - \delta_j) \end{aligned} \tag{8.227}$$

J_i is the virtual inertia, D_{p_i} is the virtual damping coefficient, P_i^d is the desirable active power of the i-th unit and P^i and the measured active power, G_{ij} is a coupling coefficient between the i-th and the j-th power generation unit and $sin(\delta_i - \delta_j)$ is a synchronization index between the i-th and the j-th power generation unit [49, 396–398, 511, 512].

The electrical part has been defined in Chap. 6 and is given by:

$$\frac{d}{dt}\begin{pmatrix} V_{L_d} \\ V_{L_q} \\ i_{I_d} \\ i_{I_q} \end{pmatrix} = \begin{pmatrix} \omega V_{L_q} + \frac{1}{C_f}i_{I_d} - \frac{1}{C_f}\frac{p_f V_{L_d} + q_f V_{L_q}}{V_{L_d}^2 + V_{L_q}^2} + \omega C_f V_{L_q} - \frac{\omega L_f V_{L_q}(i_{I_d}^2 + i_{I_q}^2)}{(V_{L_d}^2 + V_{L_q}^2)} \\ -\omega V_{L_d} + \frac{1}{C_f}i_{I_q} - \frac{1}{C_f}\frac{p_f V_{L_q} - q_f V_{L_d}}{V_{L_d}^2 + V_{L_q}^2} - \omega C_f V_{L_d} + \frac{\omega L_f V_{L_d}(i_{I_d}^2 + i_{I_q}^2)}{(V_{L_d}^2 + V_{L_q}^2)} \\ \omega i_{I_q} - \frac{1}{L_f}V_{L_d} \\ -\omega i_{I_d} - \frac{1}{L_f}V_{L_q} \end{pmatrix} + \begin{pmatrix} 0 & 0 \\ 0 & 0 \\ \frac{1}{L_f} & 0 \\ 0 & \frac{1}{L_f} \end{pmatrix}\begin{pmatrix} V_{I_d} \\ V_{I_q} \end{pmatrix} \tag{8.228}$$

The synchronizing control approach for the i-th inverter makes use of Eq. (8.227) and of the linearized inverter model given in Eq. (8.228). First, the value of P_i, that is the active power that the i-th inverter should inject to the grid, is found from the solution of the control problem of Eq. (8.227). Subsequently P_i is used in the computation of the solution of the control problem of Eq. (8.228). This is depicted in Fig. 8.22.

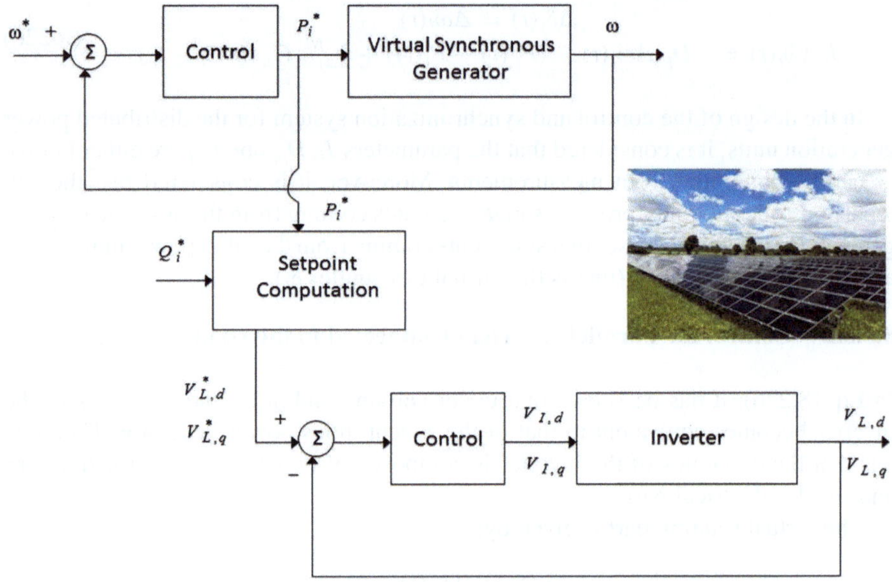

Fig. 8.22 Control loops for the virtual synchronous generator model and for the electrical part of the inverter

The synchronizing control method can be generalized to the case of multiple inverters connected to the grid (see Fig. 8.23). It can be shown again that the model of N distributed power units which are connected to the grid through inverters is also differentially flat. The latter is proven following the procedure described in Sect. 6.2.3 and by defining as flat output a generalization of the vector described in Eq. (6.24), which now becomes $Y = [y_1^1, y_2^1, y_3^1, y_1^2, y_2^2, y_3^2 \ldots, y_1^N, y_2^N, y_3^N]$ or $Y = [\delta^1, V_{L_d}^1, V_{L_q}^1, \delta^2, V_{L_d}^2, V_{L_q}^2, \ldots, \delta^N, V_{L_d}^N, V_{L_q}^N]$. All state variables and control inputs for the model of the N coupled inverters can be expressed as functions of the aforementioned flat output Y and of its derivatives.

Using the previous flat output definition, and the state variables $z_1^i = y_1, z_2^i = \dot{y}_1$, $z_3^i = y_2, z_4^i = \dot{y}_2, z_5^i = y_3, z_6^i = \dot{y}_3$, one has:

$$\begin{pmatrix} \dot{z}_1^i \\ \dot{z}_2^i \\ \dot{z}_3^i \\ \dot{z}_4^i \\ \dot{z}_5^i \\ \dot{z}_6^i \end{pmatrix} = \begin{pmatrix} 0\,1\,0\,0\,0\,0\,0 \\ 0\,0\,0\,0\,0\,0\,0 \\ 0\,0\,0\,1\,0\,0\,0 \\ 0\,0\,0\,0\,0\,0\,0 \\ 0\,0\,0\,0\,0\,1\,0 \\ 0\,0\,0\,0\,0\,0\,0 \end{pmatrix} \begin{pmatrix} z_1^i \\ z_2^i \\ z_3^i \\ z_4^i \\ z_5^i \\ z_6^i \end{pmatrix} + \begin{pmatrix} 0\,0\,0 \\ 1\,0\,0 \\ 0\,0\,0 \\ 0\,1\,0 \\ 0\,0\,0 \\ 0\,0\,1 \end{pmatrix} \begin{pmatrix} v_1^i \\ v_2^i \\ v_3^i \end{pmatrix} \tag{8.229}$$

where the control inputs of this model are defined as

$$v_1^i = \frac{1}{J_i}[-D_{p_i}\Delta\omega_i(t) + (P_i^d(t) - P_i(t)) + \sum_{j=1}^N G_{ij} sin(\delta_i - \delta_j)] \tag{8.230}$$

and using Eq. (6.53) it holds for the i-th inverter's model

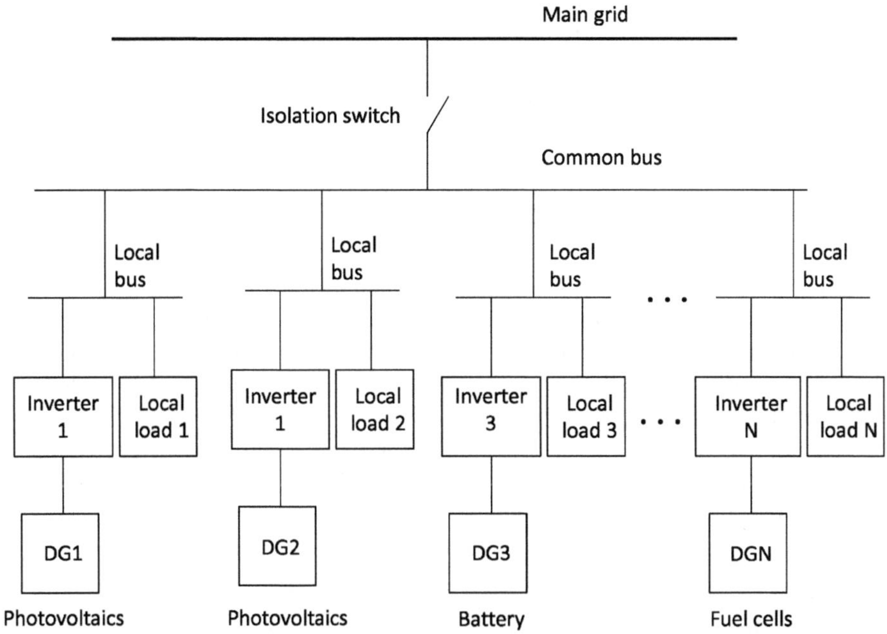

Fig. 8.23 Distributed DC power generation units connected through inverters to the grid

$$\begin{aligned} v_2^i &= L_f^2 {h_1}^i(x) + L_{g_a} L_f h_1^i(x) u_1^i + L_{g_b} L_f h_1^i(x) u_2^i \\ v_3^i &= L_f^2 {h_2}^i(x) + L_{g_a} L_f h_2^i(x) u_1^i + L_{g_b} L_f h_2^i(x) u_2^i \end{aligned} \tag{8.231}$$

If the coupling coefficients $G_{ij},\ i = 1, \ldots, n,\ j = 1, \ldots, n$ are known then the solution of the control problem for Eq. (8.227) can be obtained, by communicating to each power generation unit i the rotation angles of the rest $N-1$ power generation units so as to compute the sum $\sum_{j=1}^{N} G_{ij} sin(\delta_i - \delta_j)$. In such a case, and by denoting $x_{1,i} = \delta_i$, $x_{2,i} = \omega_i$ the control input (power) applied to the i-th power generation unit is

$$\begin{aligned} P_i = -J_i \ddot{x}_{1,i}^d - D_{p_1} x_2 + P_i^d + \textstyle\sum_{j=1}^{N} G_{ij} sin(x_{1,i} - x_{1,j}) + \\ + J_i K_{d_i} (\dot{x}_{1,i} - \dot{x}_{1,i}^d) + J_i K_{p_i} (x_{1,i} - x_{1,i}^d) \end{aligned} \tag{8.232}$$

8.5.4 State and Disturbances Estimation of Parallel Inverters with Nonlinear Kalman Filtering

A state estimator for each local power generation unit can be also designed in the form of a disturbance observer. It is considered that the linearized model of the i-th inverter is affected by additive input disturbances (which are considered to be modelling uncertainties and external perturbations such as load changes, power injected due to the connection of other power generation units to the grid, faults in the grid etc.). With reference to the state-space description given in Eq. (8.229), the disturbance terms can describe both modelling uncertainties and external perturbations.

$$\begin{aligned}\ddot{z}_1^i &= v_1^i + \tilde{d}_1^i \\ \ddot{z}_3^i &= v_2^i + \tilde{d}_2^i \\ \ddot{z}_5^i &= v_3^i + \tilde{d}_3^i\end{aligned} \tag{8.233}$$

The disturbances' dynamics can be represented by the n-th order derivative of the disturbances variables together with the associated initial conditions (however, since the disturbances are going to be estimated by the Kalman Filter, the knowledge of the initial conditions finally becomes obsolete). Thus the additive disturbances are equivalently described in the form $\tilde{d}_1^{(n)} = f_{d_1}$, $\tilde{d}_2^{(n)} = f_{d_2}$ and $\tilde{d}_3^{(n)} = f_{d_3}$.

Without loss of generality it is assumed that in the relation $\tilde{d}_i^{(n)} = f_{d_i}$, the derivative's order is $n = 2$. The state vector is extended by including as additional state variables the disturbances and their derivatives. Thus, one has $z_1^i = x_1$, $z_2^i = \dot{x}_1$, $z_3^i = x_2$, $z_4^i = \dot{x}_2$, $z_5^i = x_3$, $z_6^i = \dot{x}_3$, $z_7^i = \tilde{d}_1$, $z_8^i = \dot{\tilde{d}}_1$, $z_9^i = \tilde{d}_2$, $z_{10}^i = \dot{\tilde{d}}_2$, $z_{11}^i = \tilde{d}_3$ and $z_{12}^i = \dot{\tilde{d}}_3$.

It holds that $\dot{z}_1^i = z_2^i$, $\dot{z}_2^i = z_7 + v_1^i$, $\dot{z}_3^i = z_4^i$, $\dot{z}_4^i = z_9 + v_2^i$, $\dot{z}_5^i = z_6^i$, $\dot{z}_6^i = z_{11}^i + v_3^i$, $\dot{z}_7^i = z_8^i$, $\dot{z}_8^i = f_{d_1}^i$, $\dot{z}_9^i = z_{10}^i$, $\dot{z}_{10} = f_{d_2}^i$, $\dot{z}_{11} = z_{12}$, $\dot{z}_{12}^i = f_{d_3}^i$. Therefore, one has the system's dynamics in the extended spate-space form

$$\dot{z}_e = A_e \hat{z}_e + B_e v_e \tag{8.234}$$

where the extended control inputs vector is $v_e = [v_1^i, v_2^i, v_3^i, f_{d_1}, f_{d_2}, f_{d_3}]^T$ and

$$A_e = \begin{pmatrix} 0&1&0&0&0&0&0&0&0&0&0&0&0 \\ 0&0&0&0&0&0&1&0&0&0&0&0&0 \\ 0&0&0&1&0&0&0&0&0&0&0&0&0 \\ 0&0&0&0&0&0&0&0&0&1&0&0&0 \\ 0&0&0&0&0&1&0&0&0&0&0&0&0 \\ 0&0&0&0&0&0&0&0&0&0&0&1&0 \\ 0&0&0&0&0&0&0&0&1&0&0&0&0 \\ 0&0&0&0&0&0&0&0&0&0&0&0&0 \\ 0&0&0&0&0&0&0&0&0&0&1&0&0 \\ 0&0&0&0&0&0&0&0&0&0&0&0&0 \\ 0&0&0&0&0&0&0&0&0&0&0&0&1 \\ 0&0&0&0&0&0&0&0&0&0&0&0&0 \end{pmatrix} \quad B_e = \begin{pmatrix} 0&0&0&0&0&0 \\ 1&0&0&0&0&0 \\ 0&0&0&0&0&0 \\ 0&1&0&0&0&0 \\ 0&0&0&0&0&0 \\ 0&0&1&0&0&0 \\ 0&0&0&0&0&0 \\ 0&0&0&1&0&0 \\ 0&0&0&0&0&0 \\ 0&0&0&0&1&0 \\ 0&0&0&0&0&0 \\ 0&0&0&0&0&1 \end{pmatrix} \quad C_e^T = \begin{pmatrix} 1&0&0 \\ 0&0&0 \\ 0&1&0 \\ 0&0&0 \\ 0&0&1 \\ 0&0&0 \\ 0&0&0 \\ 0&0&0 \\ 0&0&0 \\ 0&0&0 \\ 0&0&0 \\ 0&0&0 \end{pmatrix} \tag{8.235}$$

For the extended state-space description of the system the state observer becomes

$$\dot{\hat{z}}_e = A_o \hat{z}_e + B_o v_e + K_f (C_o z_e - C_o) \hat{z}_e \tag{8.236}$$

where for matrices A_e and C_e holds $A_o = A_e$ and $C_o = C_e$ while for matrix B_o differs from B_e in the elements of the 10th and 12th row, where all elements are set to zero.

For the extended state-space description of the system it is possible to perform state estimation using the *Derivative-free nonlinear Kalman Filter*. This filter consists of the standard Kalman Filter recursion on the linearized equivalent model of the i-th inverter, and of an inverse transformation based on differential flatness theory (as explained in Sect. 6.2.3) which enables to obtain estimates for the state variables of the initial nonlinear model of the inverter (if the latter is necessary). In the filter's algorithm, the previously defined matrices A_e,B_e and C_e are substituted by their discrete-time equivalents A_{e_d}, B_{e_d} and C_{e_d}. This is done through common discretization methods. The recursion of the Kalman Filter in this case is [34, 341, 344, 367]:

measurement update:

$$\begin{aligned} K_f(k) &= P^-(k)C_d^T[C_{e_d}P^-(k)C_{e_d}^T + R(k)]^{-1} \\ \hat{x}(k) &= \hat{x}^-(k) + K_f(k)[C_{e_d}z(k) - C_{e_d}\hat{z}(k)] \\ P(k) &= P^-(k) - K(k)C_{e_d}P^-(k) \end{aligned} \tag{8.237}$$

time update:

$$\begin{aligned} P^-(k+1) &= A_{e_d}P(k)A_{e_d}^T + Q(k) \\ \hat{x}^-(k+1) &= A_{e_d}\hat{x}(k) + B_{e_d}v(k) \end{aligned} \tag{8.238}$$

After identifying the disturbance terms, the control input of the inverter is modified as follows:

$$\begin{aligned} v_1^i &= \ddot{z}_1 - k_d^1(\dot{z}_1 - \dot{z}_1^d) - k_p^1(z_1 - z_1^d) - \hat{z}_7 \\ v_2^i &= \ddot{z}_3 - k_d^2(\dot{z}_3 - \dot{z}_3^d) - k_p^2(z_3 - z_3^d) - \hat{z}_9 \\ v_3^i &= \ddot{z}_5 - k_d^3(\dot{z}_5 - \dot{z}_5^d) - k_p^3(z_5 - z_5^d) - \hat{z}_{11} \end{aligned} \tag{8.239}$$

The inclusion of the disturbance estimation terms $\hat{z}_7$, $\hat{z}_9$ and $\hat{z}_{11}$ in the feedback control inputs enables to compensate for effects of the perturbations $\tilde{d}_1$, $\tilde{d}_2$ and $\tilde{d}_3$.

8.5.5 Simulation Tests

The performance of the proposed distributed control scheme for the synchronization of parallel inverters was tested through simulation experiments. A model of $N = 3$ distributed DC power generation units was considered, while each one of these units was connected to the grid through an inverter (see Fig. 8.24). The power exchange between each inverter and the grid was considered to be described by the model of a synchronous generator.

The three interconnected inverters, shown in Fig. 8.24, are assumed to have different model parameters which are described in Table 8.3. Moreover, the three inverters are considered to be subjected to different perturbation inputs. Synchronization

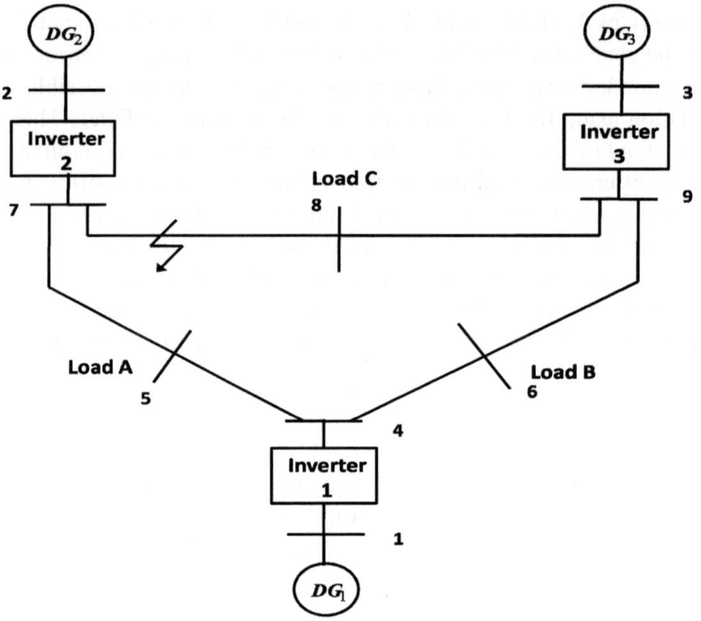

Fig. 8.24 Distributed DC power generation units connected through inverters

Table 8.3 Parameters of the inverters

	Inv_1	Inv_2	Inv_3
L_f (mH)	10.5	10.3	10.1
C_f (mF)	0.04	0.03	0.02
p_f (Kw)	21.1	22.3	23.6

means that given the desirable reference rotation speed ω_i, one can compute the amount of active power P_i contributed to the grid by each local inverter which in turn results into ω_i. Next, knowing P_i one can compute for the inner control loop of Eq. (8.228) the associated voltage reference setpoints V_{L_d} and V_{L_q} and can solve the control problem for the electrical part of the inverter's dynamics. This means that the synchronization problem of each local inverter is finally turned into a problem of nonlinear feedback control for the associated electrical model of the inverter.

The main relation that the section considers for synchronization of the inverters with a reference frequency of the grid is Eq. (8.227). According to this relation the objective is that all inverters finally attain the same frequency ω_i. By knowing the active power P_i which results into virtual rotation speed ω_i, one can compute also the setpoints for the output voltages V_{L_d} and V_{L_q} (see Fig. 8.22) which have to be used in the control problem of the inverter's electrical part described by Eq. (8.228). One can compute these setpoints through Eqs. (1.98) and (1.99). In

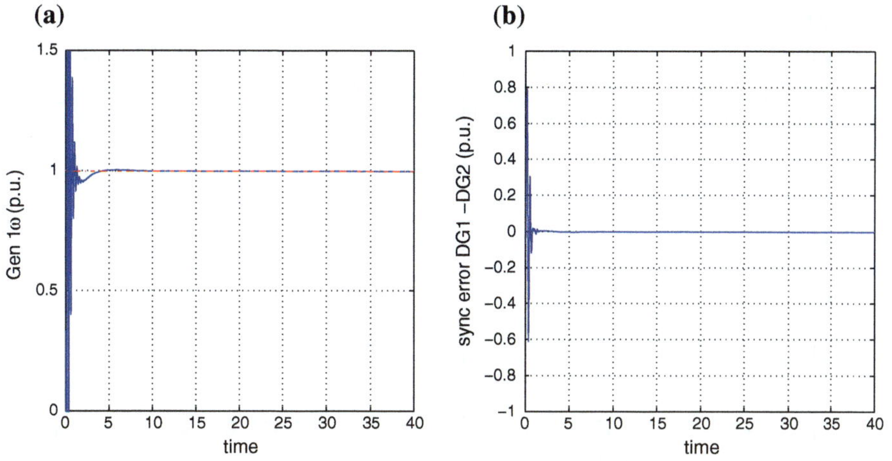

Fig. 8.25 **a** Angular speed of power generation unit 1 (*red line* setpoint, *blue line* real value), **b** synchronization error between power generation units 1 and 2

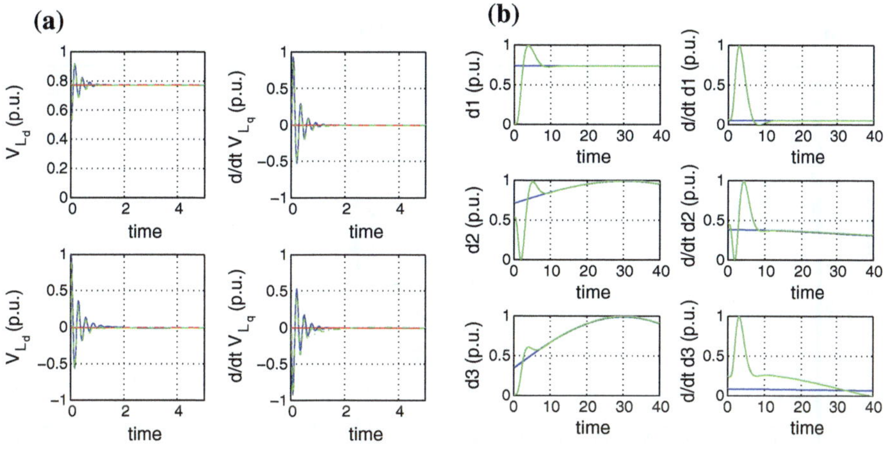

Fig. 8.26 Inverter of power generation unit 1 **a** Voltage vector components V_{L_d} and V_{L_q} and their derivatives, (*red line* setpoint, *blue line* real value, *green line* estimated value) **b** estimation (*green line*) of disturbance inputs (*blue line*) and of their derivatives with the use of the Derivative-free nonlinear Kalman Filter

Eq. (1.98) the active power should be given the value that is computed from the solution of the control problem for the virtual synchronous generator model of Eq. (8.227). Additionally, in Eq. (1.99) a reference value for the reactive power can be used. The currents $i_{L,d}$ and $i_{L,q}$ are considered to be measurable. Thus one finally has a set of two equations with unknowns the output voltages V_{L_d} and V_{L_q}. Solving this system with respect to V_{L_d} and V_{L_q} provides reference setpoints for the output voltages which finally result in the synchronization of the inverters with the grid.

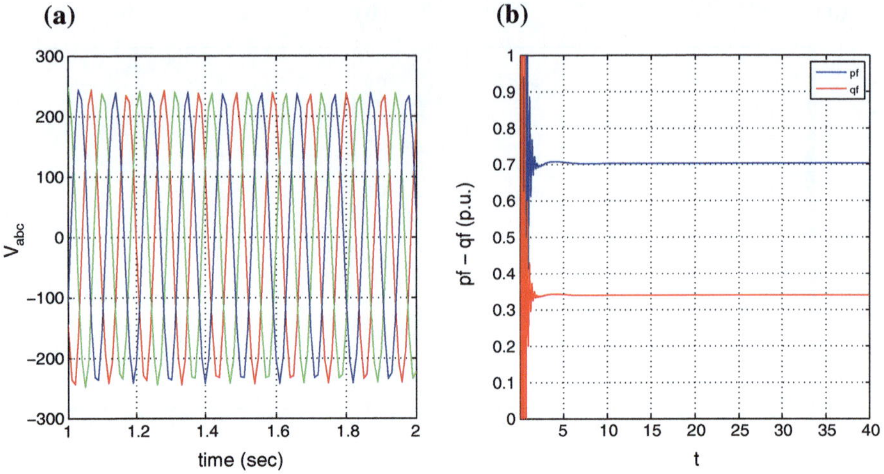

Fig. 8.27 Inverter of power generation unit 1 **a** Three-phase voltage variables V_L, (*red line* V_a, *blue line* V_b, *green line* V_c) **b** active and reactive power of the inverter

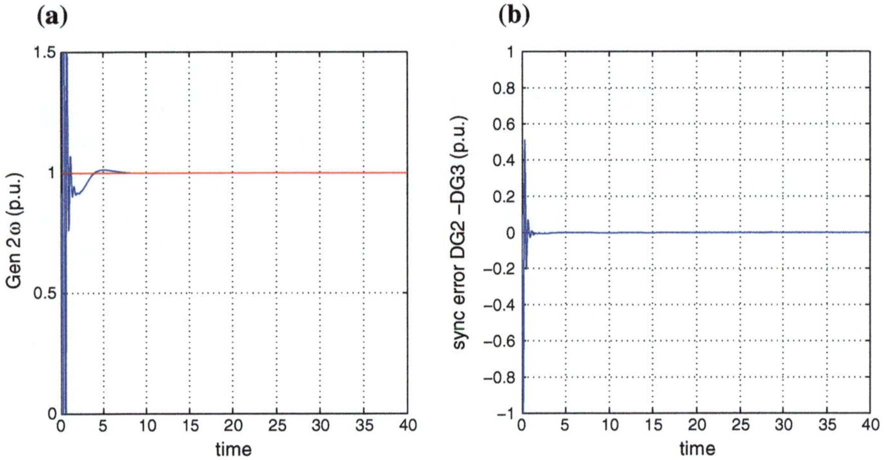

Fig. 8.28 **a** Angular speed of power generation unit 2 (*red line* setpoint, *blue line* real value), **b** synchronization error between power generation units 2 and 3

Moreover, to estimate the nonmeasurable state variables in the inverter models as well as to estimate and compensate for external perturbations affecting the inverter's dynamics the Derivative-free nonlinear Kalman Filter has been used.

The obtained results are depicted in Figs. 8.25, 8.26, 8.27, 8.28, 8.29, 8.30, 8.31, 8.32 and 8.33. The state variables of the distributed power generators are actually measured in SI units. However, in the simulation experiments presented in Figs. 8.25, 8.26, 8.27, 8.28, 8.29, 8.30, 8.31, 8.32 and 8.33, these variables have been measured in

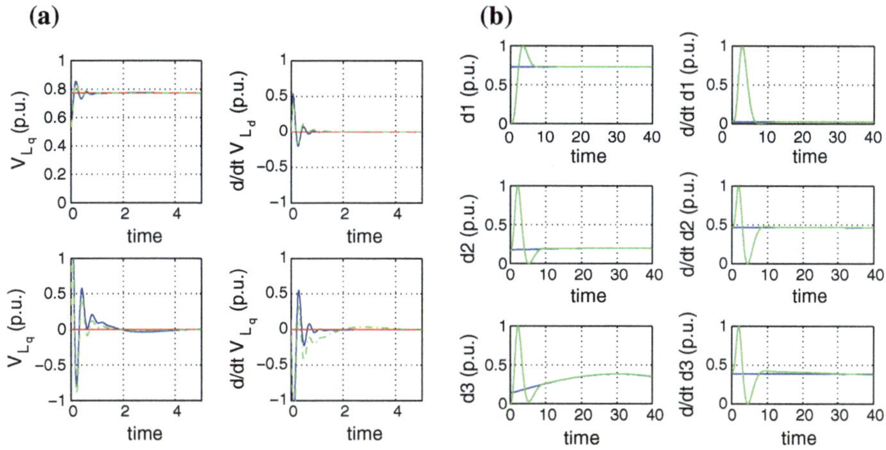

Fig. 8.29 Inverter of power generation unit 2 **a** Voltage vector components V_{L_d} and V_{L_q} and their derivatives, (*red line* setpoint, *blue line* real value, *green line* estimated value) **b** estimation (*green line*) of disturbance inputs (*blue line*) and of their derivatives with the use of the Derivative-free nonlinear Kalman Filter

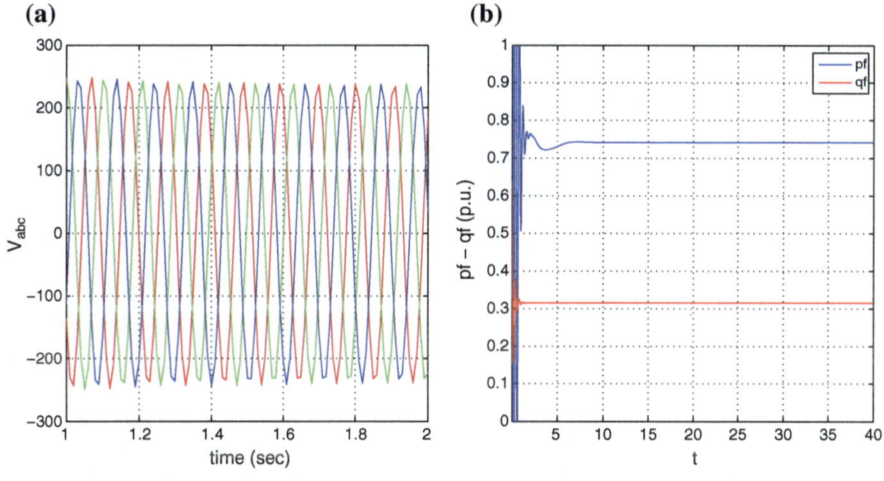

Fig. 8.30 Inverter of power generation unit 2 **a** Three-phase voltage variables V_L, (*red line* V_a, *blue line* V_b, *green line* V_c) **b** active and reactive power of the inverter

the per unit (p.u.) system. It can be noticed that the proposed control and state estimation scheme achieved both satisfactory transients and good tracking performance of the reference setpoints. Moreover, fast synchronization between the distributed power generation units was achieved. It is noted that the proposed control and synchronization approach is scalable and can be applied to a larger number of N interconnected

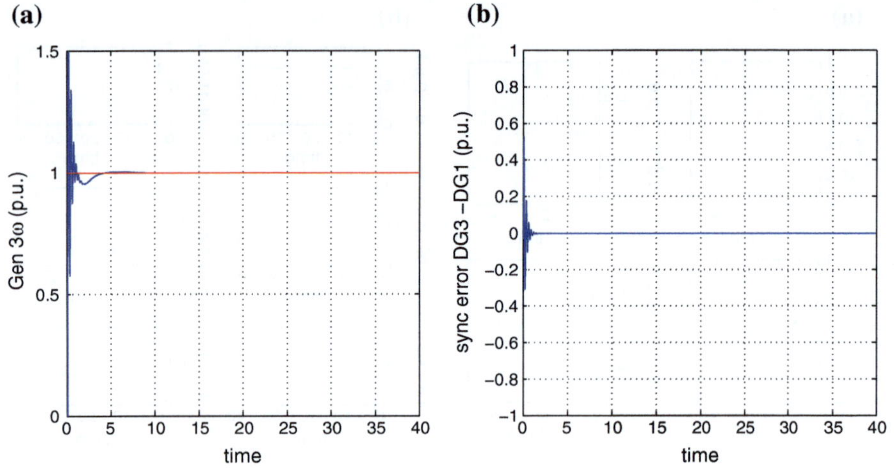

Fig. 8.31 **a** Angular speed of power generation unit 3 (*red line* setpoint, *blue line* real value), **b** synchronization error between power generation units 3 and 1

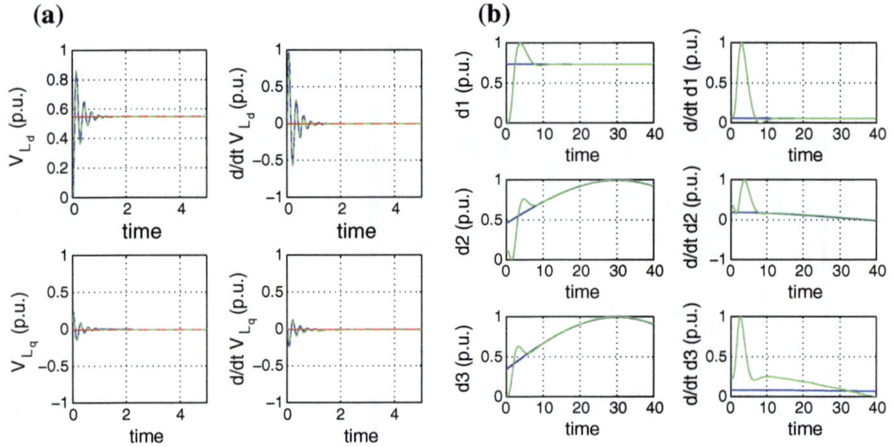

Fig. 8.32 Inverter of power generation unit 3 **a** Voltage vector components V_{L_d} and V_{L_q} and their derivatives, (*red line* setpoint, *blue line* real value, *green line* estimated value) **b** estimation (*green line*) of disturbance inputs (*blue line*) and of their derivatives with the use of the Derivative-free nonlinear Kalman Filter

DC power generation units. The simulation diagrams show convergence of the virtual turn speed ω_i of each inverter, to the associated reference value ω_i^*.

Additionally, it is noted that the section's results confirm the usefulness of the synchronverters concept for control and stabilization of distributed power generation units, e.g. in the case of renewable energy systems [511]. The methodology developed in this section enables a power corporation to control a synchronverter in the same

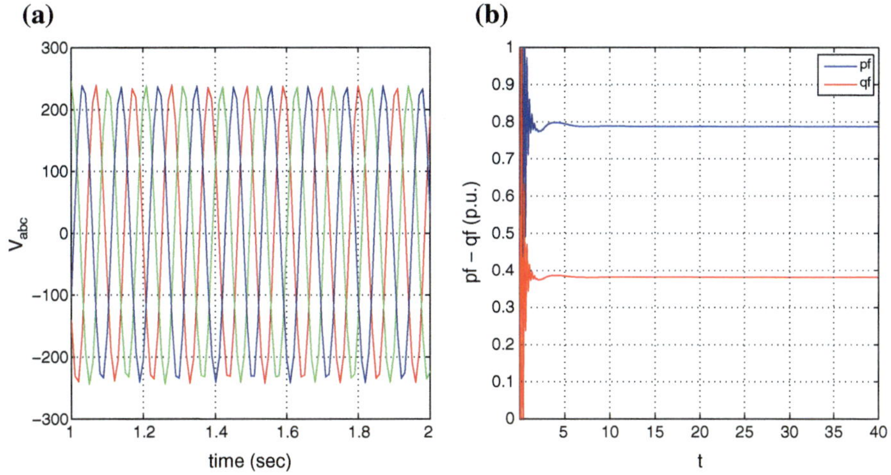

Fig. 8.33 Inverter of power generation unit 3 **a** Three-phase voltage variables V_L, (*red line* V_a, *blue line* V_b, *green line* V_c) **b** active and reactive power of the inverter

Table 8.4 RMSE for the distributed inverters

	$RMSE_1$	$RMSE_2$	$RMSE_3$
ω	0.0225	0.0427	0.0199
V_{L_d}	0.0180	0.0008	0.0003
V_{L_q}	0.0246	0.0020	0.0010

way as controlling synchronous generators, which considerably facilitates the grid connection of renewable energy and smart grid integration.

Finally, in Table 8.4, the RMSE of setpoint tracking by the state variables ω, V_{L_d}, V_{L_q} of the distributed inverters $j = 1, 2, 3$ is presented. It can be noticed that the proposed control and state estimation scheme assures good tracking performance.

The presented simulation experiments demonstrated the efficiency of the control method in tracking rapidly changing reference setpoints while also succeeding good transients. The disturbances appearing in the simulation experiments could be met in adverse operating conditions of the distributed power generation system. Even for the latter case the good performance of the control loop is confirmed. Such disturbances can be due to modelling errors (e.g. parametric changes in the inverters' model) or due to external perturbations (e.g. grid faults or disturbances due to the connection or disconnection from the grid of power generation units).

(a) (b)

Chapter 9
Condition Monitoring and Fault Diagnosis for Electric Power Generators

9.1 Outline

The chapter analyzes condition monitoring and fault diagnosis for electric power generators with the following approaches: (i) fault diagnosis in the time domain for distributed interconnected synchronous generators and with the use of the local statistical approach to fault diagnosis, (ii) fault diagnosis in the time domain for synchronous generators and with the use of nonlinear filtering methods, (iii) fault diagnosis for asynchronous generators, making use of neural networks having activation functions that remain invariant to the Fourier transform.

First the chapter analyzes a Kalman Filtering-based approach for fault diagnosis in distributed and interconnected power generators. Due to the development of distributed interconnected power units, e.g. inland and offshore wind farms, the need for systematic methods for fault diagnosis in such multi-machine power systems has emerged. In this section a new approach to parametric change detection and failure diagnosis for interconnected power units is proposed. The method is based on a nonlinear filtering scheme under the name Derivative-free nonlinear Kalman Filter and on statistical processing of the obtained state estimates, according to the properties of the χ^2 distribution. To apply this fault diagnosis method, first it is shown that the dynamic model of the distributed interconnected power generators is a differentially flat one. Next, by exploiting differential flatness properties, a change of variables (diffeomorphism) is applied to the power system, which enables also to solve the associated state estimation (filtering) problem. The new filtering technique consists of (i) a change of variables (diffeomorphism) which results into a linearized equivalent model for the power system, (ii) application of the Kalman Filter recursion, and (iii) an inverse transformation based again on differential flatness theory which permits to obtain state estimates for the initial nonlinear model. Next, statistical processing is performed for the obtained residuals, that is for the differences between the state vector of the monitored power system and the state vector provided by the aforementioned filter when the latter makes use of a fault-free model. It is shown, that the suitably weighted square of the residuals' vector follows the χ^2 statistical

G. Rigatos, *Intelligent Renewable Energy Systems*,
Green Energy and Technology, DOI 10.1007/978-3-319-39156-4_9

distribution. This property allows to use confidence intervals and to define thresholds that demonstrate whether the distributed power system functions as its fault-free model or whether parametric changes have taken place in it and thus a fault indication should be given. It is also shown that the proposed statistical criterion enables fault isolation to be performed, that is to find out the specific power generators within the distributed power system which have exhibited a failure. The efficiency of the proposed filtering method for condition monitoring in distributed power systems is confirmed through simulation experiments.

Next, the chapter proposes neural modelling and fault detection and isolation (FDI) methods for the early detection of cascading events in electric power systems. A neural-fuzzy network is used to model the dynamics of the power transmission system in fault-free conditions. The output of the neural-fuzzy network is compared to measurements from the power system and the obtained residuals undergo statistical processing according to a fault detection and isolation algorithm. If a fault threshold, defined by the FDI algorithm, is exceeded then deviation from normal operation can be detected at its early stages and an alarm can be launched. In several cases fault isolation can be also performed, i.e. the sources of fault in the power transmission system can be also identified. The performance of the proposed methodology is tested through simulation experiments.

Finally, the chapter proposes the use of a neural network with Gauss-Hermite polynomial activation functions for approximating the nonlinear power system's dynamics out of a set of input-output data. Thus the output of the neural network provides a series expansion that takes the form of a weighted sum of Gauss-Hermite basis functions. Knowing that the Gauss-Hermite basis functions satisfy the orthogonality property and remain unchanged under the Fourier transform, subjected only to a change of scale, one has that the considered neural network provides the spectral analysis of the output of the monitored system. Actually, the squares of the weights of the output layer of the neural network denote the distribution of energy into the associated spectral components for the output signal of the monitored nonlinear system. By observing changes in the amplitude of the aforementioned spectral components one can have also an indication about malfunctioning of the monitored system and can detect the existence of failures. Moreover, since specific faults are associated with amplitude changes of specific spectral components of the system fault isolation can be also performed. As test case, fault diagnosis to power transformers and to doubly fed induction generators is presented.

9.2 Fault Diagnosis for Distributed Power Generators Using Kalman Filtering

9.2.1 Overview

Energy needs grow in relentless manner worldwide. As new power generation units are installed and as distributed power generation sources get interconnected, the dynamics of the electric power generation, transmission and distribution grid becomes more complicated [220, 225, 345, 423, 433]. The monitoring of its condition becomes an elaborated task that can be accomplished only with the use of advanced fault diagnosis tools and methods. To this end, in this chapter a new statistical fault diagnosis method is proposed for detecting and isolating failures in distributed and interconnected power generators [206, 367, 390, 440, 483]. As it has been shown in several studies, by applying fault detection tests based on the χ^2 distribution it can be concluded if the structure remains healthy and if the nominal parameter values for its model still hold. Otherwise, a failure can be detected [34, 37, 191, 344, 497]. Moreover by applying the χ^2 tests in sectors of the monitored power system, the faulty components of it can be also isolated [36, 379, 497, 508].

The proposed fault diagnosis method makes use of the differential flatness properties of the distributed power generation system, that is of the ability to express its dynamics in compact form through a key subset of its state variables, named as flat outputs [240, 241, 388, 411]. Actually, differential flatness theory enables to perform a global linearization on the dynamic model of the monitored system and to transform it to the so-called canonical form [137, 171, 375, 387]. By proving that the model of the distributed power generators is a differentially flat one, the solution of the associated filtering (state estimation problem) becomes possible, using a new nonlinear filtering method known as Derivative-free nonlinear Kalman Filter. The method consists of (i) a nonlinear transformation that enables to rewrite the system's dynamics into the canonical (Brunovsky) form, (ii) application of the Kalman Filter recursion on the linearized equivalent model, (iii) an inverse transformation, based again on differential flatness theory that allows to obtain estimates of the state variables of the initial nonlinear model [352, 359, 366, 375]. Using a dynamic model of the fault-free power system (that is a model that retains the nominal values of the generators' parameters), the filter provides finally estimates of state vector elements of the distributed power units which cannot be directly measured.

The dynamic behavior of the distributed power generators is recorded through suitable sensors (in the form of a sensors network deployed at specific measurement points) and is compared against the response generated by the aforementioned Kalman Filter, the latter functioning under the assumption of a damage-free model. By comparing the two signals, residuals sequences are generated. The processing of the residuals with the use of statistical decision making criteria provides an indication about the existence of parametric changes (damages) in the distributed power system, which otherwise could not have been detected. It is shown, that the suitably weighted square of the residuals' vector follows the χ^2 statistical distribution

[101, 160, 209]. This property allows to use confidence intervals and to define thresholds that demonstrate whether the distributed power system functions as its fault-free model or whether parametric changes have taken place in it and thus a fault indication should be given. It is also shown that the proposed statistical criterion enables fault isolation to be performed, that is to find out the specific power generators within the distributed power system which have exhibited a failure.

9.2.2 Dynamic Model of the Multi-machine Power System

9.2.2.1 Dynamics of the Local Power Generator Units

The dynamic model of the multi-machine power system was previously analyzed in Chap. 8 for control purposes and is now revisited for fault diagnosis purposes. The proposed fault diagnosis method aims at detecting and isolating a faulty generator among a set of n interconnected such generators (Fig. 9.1). The dynamic model of the distributed power generation units is assumed to be that of a synchronous generator. The modeling approach is also applicable to PMSGs (permanent magnet synchronous generators) which are a special case of synchronous electric machines.

As already explained, the dynamic model of the synchronous generator is given by

Fig. 9.1 Fault diagnosis in distributed interconnected power generators

$$\begin{aligned}\dot{\delta} &= \omega - \omega_0 \\ \dot{\omega} &= -\tfrac{D}{2J}(\omega - \omega_0) + \omega_0\tfrac{P_m}{2J} - \omega_0\tfrac{1}{2J}\tfrac{V_s E_q^{'}}{x_{d\Sigma}^{'}}sin(\Delta\delta) \\ \dot{E}_q^{'} &= -\tfrac{1}{T_d^{'}}E_q^{'} + \tfrac{1}{T_{do}}\tfrac{x_d - x_d^{'}}{x_{d\Sigma}^{'}}V_s cos(\Delta\delta) + \tfrac{1}{T_{do}}E_f\end{aligned} \tag{9.1}$$

where $T_d^{'} = \frac{x_{d\Sigma}^{'}}{x_{d\Sigma}}T_{d_o}$ is the time constant of the field winding. The previously analyzed single-machine infinite-bus model of the generator is described by a nonlinear state space model of the form

$$\dot{x} = f(x) + g(x)u \tag{9.2}$$

where the state vector x is defined as $x = \left(\Delta\delta\ \Delta\omega\ E_q^{'}\right)^T$. The vector fields $f(x)$ and $g(x)$ are defined as

$$f(x) = \begin{pmatrix} \omega - \omega_0 \\ -\frac{D}{2J}(\omega - \omega_0) + \omega_0\frac{P_m}{2J} - \omega_0\frac{1}{2J}\frac{V_s E_q^{'}}{x_{d\Sigma}^{'}}sin(\Delta\delta) \\ -\frac{1}{T_d^{'}}E_q^{'} + \frac{1}{T_{do}}\frac{x_d - x_d^{'}}{x_{d\Sigma}^{'}}V_s cos(\Delta\delta)\end{pmatrix} \tag{9.3}$$

$$g(x) = \left(0\ 0\ \tfrac{1}{T_{do}}\right)^T \tag{9.4}$$

with control input $u = E_f$ being the field voltage (equivalent voltage in the excitation coil) and measurable output the turn angle of the rotor

$$y = h(x) = \delta - \delta_0 \tag{9.5}$$

9.2.2.2 Dynamics of the Interconnected Power Generators

The interaction between several local power generation units which are described by the previously analyzed dynamic model results into a multi-area multi-machine distributed power generation system (Fig. 8.1). As explained in Chap. 8, a multi-machine power system (MMPS) with n machines, in which the first machine is chosen as the reference machine can be described by the following nonlinear dynamic model

$$\begin{aligned}\dot{\delta}_i &= \omega_i - \omega_0 \\ \dot{\omega}_i &= -\tfrac{D_i}{2J_i}(\omega_i - \omega_0) + \omega_0\tfrac{P_{m_i}}{2J_i} - \\ &-\omega_0\tfrac{1}{2J_i}[G_{ii}E_{qi}^{'\,2} + E_{qi}^{'}\textstyle\sum_{j=1,j\neq i}^{n}E_{qj}^{'}G_{ij}sin(\delta_i - \delta_j - \alpha_{ij})] \\ \dot{E}_{q_i}^{'} &= -\tfrac{1}{T_{d_i}^{'}}E_{q_i}^{'} + \tfrac{1}{T_{do_i}}\tfrac{x_{d_i} - x_{d_i}^{'}}{x_{d\Sigma i}^{'}}V_{s_i}cos(\Delta\delta_i) + \tfrac{1}{T_{do_i}}E_{fi}\end{aligned} \tag{9.6}$$

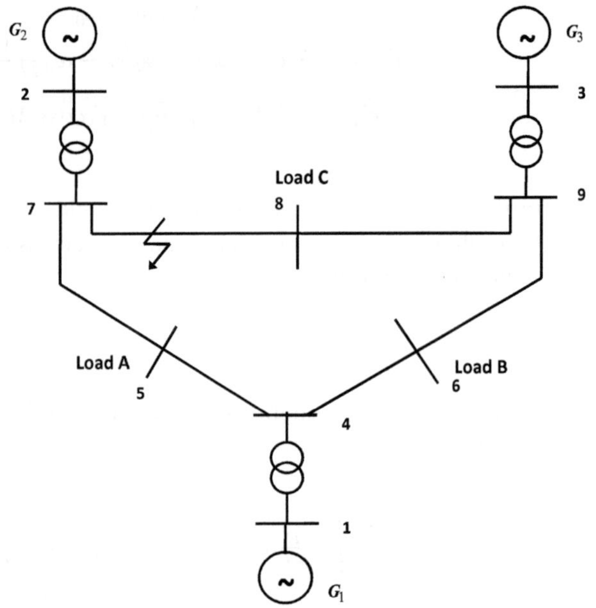

Fig. 9.2 A multi-machine (3-area) distributed power generation model

where the electric torque P_{e_i} which is associated with the active power at the i-th generator is now given by

$$P_{e_i} = G_{ii}{E'_{qi}}^2 + E'_{qi}\sum_{j=1,j\neq i}^{n} E'_{qj}G_{ij}sin(\delta_i - \delta_j - \alpha_{ij}) \tag{9.7}$$

for $i = 1, 2 \ldots, n$. For a power grid that consists of n generators the aggregate state vector comprises the state vectors of the local machines, i.e. $x = [x^1, x^2, \ldots, x^n]^T$, where $x^i = [x_1^i, x_2^i, x_3^i]^T$, with $x_1^i = \Delta\delta_i$, $x_2^i = \Delta\omega_i$ and $x_3^i = E'_{qi}$ are the state variables for the i-th machine and $i = 1, 2, \ldots, n$. Again, the structure of such a multi-area multi-machine power system, for the case of n interconnected generators, is depicted in Fig. 9.2.

9.2.3 *Linearization of the Power Generation System Using Differential Flatness Theory*

9.2.3.1 Linearized Model of the Multi-generator System

Previously, it was proven that the multi-machine power system is differentially flat. The input-output linearized model of the multi-machine power system has been given in Chap. 8. Functioning of the power generation system under time-varying external mechanical torque P_m^i is the most generic condition, which also comprises the case of piecewise constant external input torque. Next, it is overviewed how with the use of the differential flatness theory a controller can be designed for the distributed power system in case of time-varying external mechanical torque. By deriving the expression about $\ddot{y}^i$ once more with respect to time one obtains

$$y^{(3)^i} = a^i(x) + b_1^i(x)g_1u_1 + b_2^i(x)g_2u_2 + b_3^i(x)g_3u_3 \tag{9.8}$$

which means that one arrives at a function of the form $\dot{z}_3^i = a^i(x) + b_1^i(x)g_1u_1 + b_2^i(x)g_2u_2 + b_3^i(x)g_3u_3 + \tilde{d}^i$, where in the case of the distributed power generation that consists of $n = 3$ machines, and considering for instance $i = 1$, $j = 2, 3$ one has

$$\begin{aligned}
a^i = & (\tfrac{D_i}{2J_i})^2 x_2^i + \tfrac{D_i\omega_0}{(2J_i)^2}[G_{ii}{x_3^i}^2 + x_3^i \textstyle\sum_{j=1,j\neq i}^n x_3^j G_{ij}sin(x_1^i - x_1^j - \alpha_{ij})] - \\
& -\tfrac{\omega_0}{2J_i}[G_{ii}x_3^i + \textstyle\sum_{j=1,j\neq i}^n x_3^j G_{ij}sin(x_1^i - x_1^j - \alpha_{ij})(-\tfrac{1}{T_{d_i}'}x_3^i + (\tfrac{1}{T_{doi}}\tfrac{x_{d_i} - x_{d_i}'}{x_{d\Sigma_i}'}V_{s_i}cos(x_1^i))] - \\
& -\tfrac{\omega_0}{2J_i}x_3^i \textstyle\sum_{j=1,j\neq i}^n G_{ij}sin(x_1^i - x_1^j - \alpha_{ij})(-\tfrac{1}{T_{d_i}'}x_3^i + (\tfrac{1}{T_{doi}}\tfrac{x_{d_i} - x_{d_i}'}{x_{d\Sigma_i}'}V_{s_i}cos(x_1^i)) - \\
& -\tfrac{\omega_0}{2J_i}x_3^i \textstyle\sum_{j=1,j\neq i}^n x_3^j G_{ij}cos(x_1^i - x_1^j - \alpha_{ij})x_2^i \tfrac{\omega_0}{2J_i}x_3^i \textstyle\sum_{j=1,j\neq i}^n x_3^j G_{ij}cos(x_1^i - x_1^j - \alpha_{ij})x_2^j
\end{aligned} \tag{9.9}$$

while for functions b_1^i, b_2^i and b_3^i it holds

$$\begin{aligned}
b_1^i &= -\tfrac{\omega_0}{2J_i}[2G_{ii}x_3^i + \textstyle\sum_{j=1,j\neq i}^n x_3^j G_{ij}sin(x_1^i - x_1^j - \alpha_{ij})]\tfrac{1}{T_{do\,i}} \\
b_2^i &= -\tfrac{\omega_0}{2J_i}G_{i2}sin(x_1^i - x_1^2 - \alpha_{i2})\tfrac{1}{T_{do\,2}} \\
b_3^i &= -\tfrac{\omega_0}{2J_i}G_{i3}sin(x_1^i - x_1^3 - \alpha_{i3})\tfrac{1}{T_{do\,3}}
\end{aligned} \tag{9.10}$$

Finally for the additional input term one has $\tilde{d}^i = -\frac{D_i\omega_0}{2J_i^2}P_m^i + \frac{\omega_0}{2J_i}\dot{P}_m^i$. In this term $\tilde{d}^i$ one can one also include additive disturbance inputs which are exerted on each generator's model and which stand for faults.

Thus, one has the following description of the dynamics of the i-th power generator

$$\begin{aligned}
\dot{z}_1^i &= z_2^i \\
\dot{z}_2^i &= z_3^i \\
\dot{z}_3^i &= a^i(x) + {b_1}^i g_1u_1 + {b_2}^i g_2u_2 + {b_3}^i g_3u_3 + \tilde{d}^i
\end{aligned} \tag{9.11}$$

For the complete system of the 3 generators one has

$$\begin{aligned}
\dot{z}_3^1 &= a^1(x) + {b_1}^1 g_1u_1 + {b_2}^1 g_2u_2 + {b_3}^1 g_3u_3 + \tilde{d}^1 \\
\dot{z}_3^2 &= a^2(x) + {b_1}^2 g_1u_1 + {b_2}^2 g_2u_2 + {b_3}^2 g_3u_3 + \tilde{d}^2 \\
\dot{z}_3^3 &= a^3(x) + {b_1}^3 g_1u_1 + {b_2}^3 g_2u_2 + {b_3}^3 g_3u_3 + \tilde{d}^3
\end{aligned} \tag{9.12}$$

or in matrix form

$$\dot{z}_3 = f_a(x) + Mu + \tilde{d} \tag{9.13}$$

where $z_3 = [z_3^1, z_3^2, z_3^3]^T$, $u = [u_1, u_2, u_3]^T$ and $\tilde{d} = [\tilde{d}_1, \tilde{d}_2, \tilde{d}_3]^T$ while

$$f_a(x) = \begin{pmatrix} a^1(x) \\ a^2(x) \\ a^3(x) \end{pmatrix}, \quad M = \begin{pmatrix} {b_1}^1 g_1 & {b_2}^1 g_2 & {b_3}^1 g_3 \\ {b_1}^2 g_1 & {b_2}^2 g_2 & {b_3}^2 g_3 \\ {b_1}^3 g_1 & {b_2}^3 g_2 & {b_3}^3 g_3 \end{pmatrix} \tag{9.14}$$

Setting $v = f_a(x) + Mu + \tilde{d}$, one obtains again the linear canonical form for the i-th generator given by

$$\begin{pmatrix} \dot{z}_1^i \\ \dot{z}_2^i \\ \dot{z}_3^i \end{pmatrix} = \begin{pmatrix} 0 & 1 & 0 \\ 0 & 0 & 1 \\ 0 & 0 & 0 \end{pmatrix} \begin{pmatrix} z_1^i \\ z_2^i \\ z_3^i \end{pmatrix} + \begin{pmatrix} 0 \\ 0 \\ 1 \end{pmatrix} (v^i + \tilde{d}^i) \tag{9.15}$$

In this manner the initial nonlinear power system is transformed into three decoupled linear subsystems which are in the canonical Brunovksy form. For each one of these subsystems the appropriate control law is

$$v^i = {z_d^{(3)}}^i - k_3(\ddot{z}^i - \ddot{z}_d^i) - k_2(\dot{z}^i - \dot{z}_d^i) - k_1(z^i - z_d^i) - \tilde{d}^i \tag{9.16}$$

As in the case of Chap. 8, the implementation of the distributed power generation control scheme, the controller at the i-th power generator makes use of not only its own state vector $X^i = [x_1^i, x_2^i, x_3^i]^T$, but also of the state vectors of the rest $n-1$ power generators, i.e. $x^j = [x_1^j, x_2^j, x_3^j]$, $j \neq i$, $j = 1, 2, \ldots, n$. The transformation of the dynamical model of the local power generators into the linear canonical form enables to obtain estimates of the transformed state vector of the system $Y^i = [y^i, \dot{y}^i, \ddot{y}^i]^T$ where $y^i = \Delta\delta^i$, through Kalman Filtering, after processing measurements of only the turn angle difference $\Delta\delta^i$ of the i-th power generator. Therefore one has to compute the estimates $\hat{X}^i = [\hat{x}_1^i, \hat{x}_2^i, \hat{x}_3^i]^T$ after using the estimate provided by the Kalman Filter $\hat{Y}^i = [\hat{y}^i, \hat{\dot{y}}^i, \hat{\ddot{y}}^i]^T$. As already explained, this requires the exchange and processing in real-time of a large amount of data. It holds that

$$\hat{x}_1^i = \hat{y}^i \quad \hat{x}_2^i = \hat{\dot{y}}^i \tag{9.17}$$

while for the computation of $\hat{x}_3^i$ for $i = 1, 2, \ldots, n$ one has to solve with respect to $\hat{x}_3^i$ the system of equations

$$\begin{aligned}
\hat{\ddot{y}}^1 &= -\tfrac{D_1}{2J_1}\hat{\dot{y}}^1 + \tfrac{\omega_0}{2J_1}P_{m1} - \tfrac{\omega_0}{2J_1}[G_{11}{x_3^1}^2 + \hat{x}_3^1\textstyle\sum_{j=1, j\neq 1}^n[\hat{x}_3^j G_{1j} sin(y^1 - y^j - \alpha_{1j})] \\
\hat{\ddot{y}}^2 &= -\tfrac{D_i}{2J_2}\hat{\dot{y}}^2 + \tfrac{\omega_0}{2J_2}P_{m2} - \tfrac{\omega_0}{2J_2}[G_{22}{x_3^2}^2 + x_3^2\textstyle\sum_{j=1, j\neq 2}^n[\hat{x}_3^j G_{2j} sin(y^2 - y^j - \alpha_{2j})] \\
&\cdots \\
\hat{\ddot{y}}^n &= -\tfrac{D_n}{2J_n}\hat{\dot{y}}^n + \tfrac{\omega_0}{2J_n}P_{mn} - \tfrac{\omega_0}{2J_n}[G_{nn}{x_3^n}^2 + \hat{x}_3^n\textstyle\sum_{j=1, j\neq n}^n[\hat{x}_3^j G_{nj} sin(y^i - y^j - \alpha_{ij})]
\end{aligned} \tag{9.18}$$

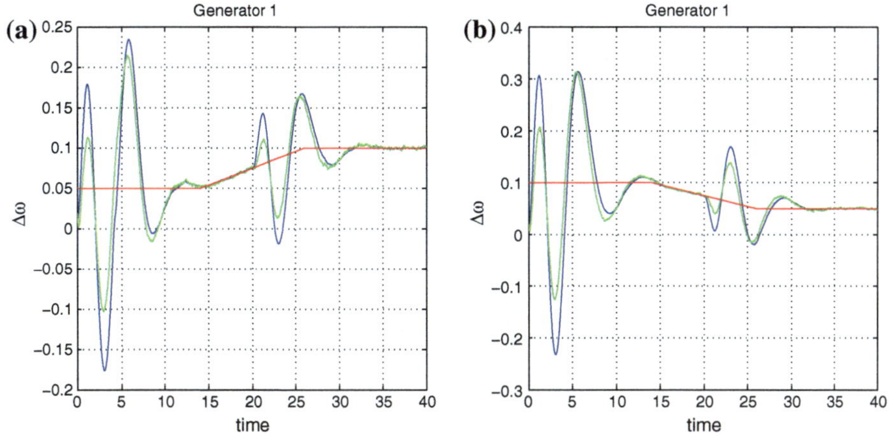

Fig. 9.3 Angular speed of the power generator (*blue line*) and its estimated value (*green line*): **a** tracking of setpoint 1 (*red line*), **b** tracking of setpoint 2 (*red line*)

By applying a stabilizing feedback control scheme the power generators' state vectors can be made to convergence to the reference setpoints and synchronization of their rotor's can be achieved. The associated results are depicted in Fig. 9.3.

9.2.4 Fault Detection with the Use of Statistical Criteria

9.2.4.1 Fault Detection

The residuals' sequence, that is the differences between the real output of the monitored multi-machine power system and the one estimated by the Kalman Filter (Fig. 9.4) is a discrete error process e_k with dimension $m \times 1$ (here $m = N$). Actually, it is a zero-mean Gaussian white-noise process with covariance given by E_k.

A conclusion can be stated based on a measure of certainty that the parameters of the dynamic model of the multi-machine power system remain unchanged. To this end, the following *normalized error square* (NES) is defined [375]

$$\varepsilon_k = e_k^T E_k^{-1} e_k \tag{9.19}$$

The normalized error square follows a χ^2 distribution (Fig. 9.5). An appropriate test for the normalized error sum is to numerically show that the following condition is met within a level of confidence (according to the properties of the χ^2 distribution)

$$E\{\varepsilon_k\} = m \tag{9.20}$$

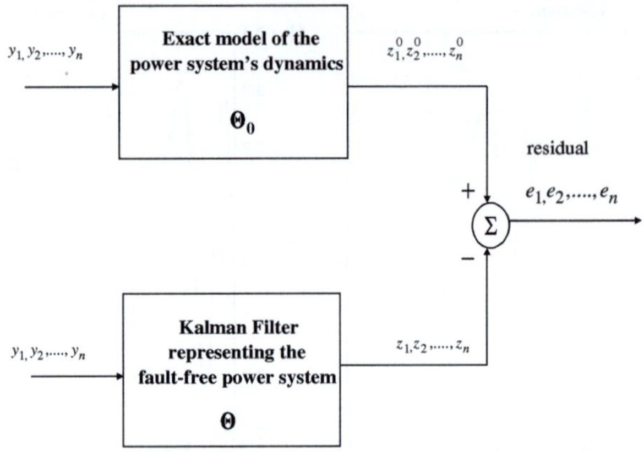

Fig. 9.4 Residuals' generation for the distributed power system, with the use of the Kalman Filtering

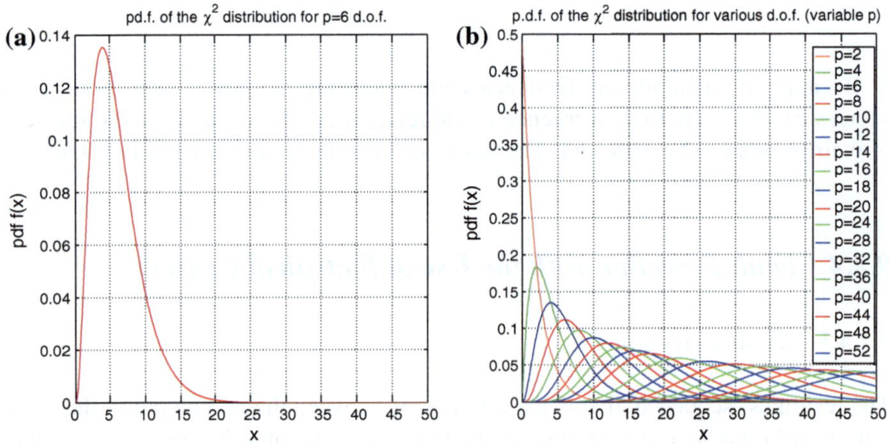

Fig. 9.5 **a** Probability density function of the χ^2 distribution for $p = 6$ degrees of freedom, **b** Probability density function of the distribution for several values of the degrees of freedom (variable p)

This can be achieved using statistical hypothesis testing, which are associated with confidence intervals. A 95 % confidence interval is frequently applied, which is specified using $100(1 - a)$ with $a = 0.05$. Actually, a two-sided probability region is considered cutting-off two end tails of 2.5 % each. For M runs the obtained normalized error square is given by

$$\bar{\varepsilon}_k = \frac{1}{M}\sum_{i=1}^{M}\varepsilon_k(i) = \frac{1}{M}\sum_{i=1}^{M}e_k^T(i)E_k^{-1}(i)e_k(i) \tag{9.21}$$

where ε_i stands for the i-th run at time t_k. Then $M\bar{\varepsilon}_k$ will follow a χ^2 density with Mm degrees of freedom. This condition can be checked using a χ^2 test. The hypothesis holds, if the following condition is satisfied

$$\bar{\varepsilon}_k \in [\zeta_1, \zeta_2] \tag{9.22}$$

where ζ_1 and ζ_2 are derived from the tail probabilities of the χ^2 density. For example, for $m = 20$ and $M = 100$ one has $\chi^2_{Mm}(0.025) = 1878$ and $\chi^2_{Mm}(0.975) = 2126$. Using that $M = 100$ one obtains $\zeta_1 = \chi^2_{Mm}(0.025)/M = 18.78$ and $\zeta_2 = \chi^2_{Mm}(0.975)/M = 21.26$.

9.2.4.2 Fault Isolation

By applying the statistical test into the n individual generators of the multi-machine power system it is also possible to find out the specific generators within the distributed power generation model that has been subjected to a fault [375, 421]. In the case of a single fault one has to carry out n χ^2 statistical change detection tests, where each test is applied to the subset that comprises generators $i-1$, i and $i+1$, $i = 1, 2, \ldots, n$. Actually, out of the n χ^2 statistical change detection tests, the one that exhibit the highest score (or equivalently indicates the largest parameter deviation from the nominal value) are those that identify the generator that has been subjected to failure (the faulty components for this generator are the parameters of its mechanical or electrical model).

In the case of multiple faults one can identify the subset of generators that have been subjected to parametric change by applying the χ^2 statistical change detection test according to a combinatorial sequence. This means that

$$\binom{n}{k} = \frac{n}{k!(n-k)!} \tag{9.23}$$

tests have to take place, for all generators' clusters in the monitored power system, that finally comprise n, $n-1$, $n-2$, ..., 2, 1 generators. Again the χ^2 tests that give the highest scores indicate the generators which are most likely to have been subjected to damage.

9.2.5 Disturbances Estimation with the Derivative-Free Nonlinear Kalman Filter

Kalman Filtering applied on the previously described linearized equivalent model of the distributed power system, is known as Derivative-free nonlinear Kalman Filter [352, 366, 375]. This form of the Kalman Filter is not only a method for performing fault diagnosis in the distributed power system, but is also a tool for estimating the perturbation terms that affect this power system. It is considered that the multi-machine power system's dynamics is affected by additive input disturbances:

$$
\begin{array}{l}
\dot{x}_{1,1} = x_{1,2} \;\; \dot{x}_{1,2} = x_{1,3} \;\; \dot{x}_{1,3} = v_1 + \tilde{d}_1 \;\; \dot{x}_{2,1} = x_{2,2} \\
\dot{x}_{2,2} = x_{2,3} \;\; \dot{x}_{2,3} = v_2 + \tilde{d}_2 \\
\cdots \;\; \cdots \\
\dot{x}_{i,1} = x_{i,2} \;\; \dot{x}_{i,2} = x_{i,3} \;\; \dot{x}_{i,3} = v_i + \tilde{d}_i \;\; \dot{x}_{i+1,1} = x_{i+1,2} \\
\dot{x}_{i+1,2} = x_{i+1,3} \;\; \dot{x}_{i+1,3} = v_{i+1} + \tilde{d}_{i+1} \\
\cdots \;\; \cdots \\
\dot{x}_{n-1,1} = x_{n-1,2} \;\; \dot{x}_{n-1,2} = x_{n-1,3} \;\; \dot{x}_{n-1,3} = v_{n-1} + \tilde{d}_{n-1} \\
\dot{x}_{n,1} = x_{n,2} \;\; \dot{x}_{n,2} = x_{n,3} \;\; \dot{x}_{n,3} = v_n + \tilde{d}_n
\end{array}
\tag{9.24}
$$

It is considered that the dynamics of each perturbation term is described by its n-th order derivative, that is $\tilde{d}^{(n)} = f_d(t)$, and of the associated initial conditions. However, the reconstruction of the signals $\tilde{d}_i$ $i = 1, \ldots, n$ will be performed with the use of Kalman Filtering, and the convergence of the latter estimation method does not depend on initial conditions. Therefore, initial conditions are finally unnecessary for estimating the disturbance terms' evolution in time.

According to the above and without loss of generality it is assumed that $\tilde{d}^{(n)} = f_d(t)$ with $n = 3$. Next, the state vector of the system is extended by introducing as additional state variables the disturbance terms and their derivatives

$$
\begin{array}{l}
z_{1,1} = x_{1,1} \;\; z_{1,2} = x_{1,2} \;\; z_{1,3} = x_{1,3} \;\; z_{2,1} = x_{2,1} \;\; z_{2,2} = x_{2,2} \;\; z_{2,3} = x_{2,3} \\
\cdots \;\; \cdots \;\; \cdots \;\; \cdots \\
z_{i,1} = x_{i,1} \;\; z_{i,2} = x_{i,2} \;\; z_{i,3} = x_{i,2} \;\; z_{i+1,1} = x_{i+1,1} \;\; z_{i+1,2} = x_{i+1,2} \;\; z_{i+1,3} = x_{i+1,3} \\
\cdots \;\; \cdots \;\; \cdots \;\; \cdots \\
z_{n-1,1} = x_{n-1,1} \;\; z_{n-1,2} = x_{n-1,2} \;\; z_{n-1,3} = x_{n-1,3} \;\; z_{n,1} = x_{n,1} \;\; z_{n,2} = x_{n,2} \;\; z_{n,3} = x_{n,3} \\
z_{n+1,1} = \tilde{d}_1 \;\; z_{n+1,2} = \dot{\tilde{d}}_1 \;\; z_{n+1,3} = \ddot{\tilde{d}}_1 \;\; z_{n+2,1} = \tilde{d}_2 \;\; z_{n+2,2} = \dot{\tilde{d}}_2 \;\; z_{n+2,3} = \ddot{\tilde{d}}_2 \\
\cdots \;\; \cdots \;\; \cdots \;\; \cdots \\
z_{n+i+1,1} = \tilde{d}_i \;\; z_{n+i+1,2} = \dot{\tilde{d}}_i \;\; z_{n+i+1,3} = \ddot{\tilde{d}}_i \;\; z_{n+i+2,1} = \tilde{d}_{i+1} \;\; z_{n+i+2,2} = \dot{\tilde{d}}_{i+1} \\
z_{n+i+2,3} = \ddot{\tilde{d}}_{i+1} \\
\cdots \;\; \cdots \;\; \cdots \;\; \cdots \\
z_{2n-1,1} = \tilde{d}_{n-1} \;\; z_{2n-1,2} = \dot{\tilde{d}}_{n-1} \;\; z_{2n-1,3} = \ddot{\tilde{d}}_{n-1} \;\; z_{2n,1} = \tilde{d}_n \;\; z_{2n,2} = \dot{\tilde{d}}_n \;\; z_{2n,3} = \ddot{\tilde{d}}_n
\end{array}
\tag{9.25}
$$

For the extended state-space description of the power system new matrices A, B and C are formulated, comprising a double number of rows comparing to the system's initial description. However, even in this extended state-space form the system remains observable. For example, in the case of a model of three interconnected power generators of Fig. 8.1, the extended state-space description of the system and the system's extended state vector are

$$
\begin{array}{rl}
z_e = & [z_{1,1}, z_{1,2}, z_{1,3}, z_{2,1}, z_{2,2}, z_{2,3}, z_{3,1}, z_{3,2}, z_{3,3}, \\
& [z_{4,1}, z_{4,2}, z_{4,3}, z_{5,1}, z_{5,2}, z_{5,3}, z_{6,1}, z_{6,2}, z_{6,3}]^T
\end{array}
$$

The measurable state variables are $z_{1,1}$, $z_{2,1}$ and $z_{3,1}$. By denoting the extended state vector as $z_e \in R^{18\times 1}$ (and after omitting the disturbance functions $f_{d,i}$, $i = 1, \ldots, 3$ from the control vector) one has the linear state-space equation in the form

$$\begin{aligned} \dot{z}_e &= A_e z_e + B_e v \\ z_e^{meas} &= C_e z_e \end{aligned} \tag{9.26}$$

where

$$A_e = \begin{pmatrix}
0&1&0&0&0&0&0&0&0&0&1&0&0&0&0&0&0&0\\
0&0&1&0&0&0&0&0&0&0&1&0&0&0&0&0&0&0\\
0&0&0&0&0&0&0&0&0&0&0&0&0&0&0&0&0&0\\
0&0&0&1&0&0&0&0&0&0&0&0&1&0&0&0&0&0\\
0&0&0&0&1&0&0&0&0&0&0&0&1&0&0&0&0&0\\
0&0&0&0&0&0&0&0&0&0&0&0&0&0&0&0&0&0\\
0&0&0&0&0&1&0&0&0&0&0&0&0&0&1&0&0&0\\
0&0&0&0&0&0&1&0&0&0&0&0&0&0&1&0&0&0\\
0&0&0&0&0&0&0&0&0&0&0&0&0&0&0&0&0&0\\
0&0&0&0&0&0&0&0&0&1&0&0&0&0&0&0&0&0\\
0&0&0&0&0&0&0&0&0&0&0&0&0&0&0&0&0&0\\
0&0&0&0&0&0&0&0&0&0&0&0&0&0&0&0&0&0\\
0&0&0&0&0&0&0&0&0&0&0&0&1&0&0&0&0&0\\
0&0&0&0&0&0&0&0&0&0&0&0&0&0&0&0&0&0\\
0&0&0&0&0&0&0&0&0&0&0&0&0&0&0&0&0&0\\
0&0&0&0&0&0&0&0&0&0&0&0&0&0&0&1&0&0\\
0&0&0&0&0&0&0&0&0&0&0&0&0&0&0&0&0&0\\
0&0&0&0&0&0&0&0&0&0&0&0&0&0&0&0&0&0
\end{pmatrix} \quad B_e = \begin{pmatrix}
0&0&0\\
0&0&0\\
1&0&0\\
0&0&0\\
0&0&0\\
0&1&0\\
0&0&0\\
0&0&0\\
0&0&1\\
0&0&0\\
0&0&0\\
0&0&0\\
0&0&0\\
0&0&0\\
0&0&0\\
0&0&0\\
0&0&0\\
0&0&0
\end{pmatrix} \tag{9.27}$$

$$C_e = \begin{pmatrix}
1&0&0&0&0&0&0&0&0&0&0&0&0&0&0&0&0&0\\
0&0&0&1&0&0&0&0&0&0&0&0&0&0&0&0&0&0\\
0&0&0&0&0&0&1&0&0&0&0&0&0&0&0&0&0&0
\end{pmatrix} \tag{9.28}$$

A state estimator for the system of Eq. (9.26) has the form

$$\begin{aligned} \dot{\hat{z}} &= A_e \hat{z} + B_e v + K(z^{meas} - \hat{z}^{meas}) \\ \hat{z}^{meas} &= C_e \hat{z} \end{aligned} \tag{9.29}$$

The computation of the estimator's gain K can be performed with the use of the Kalman Filter recursion, which consists of two stages: the *measurement update* and the *time update*.

measurement update:

$$\begin{aligned} K &= P^- C_{e,d}^T [C_{e,d} P^- C_{e,d}^T + R]^{-1} \\ \hat{z}_e &= \hat{z}_e^- + K_f (z^{meas} - \hat{z}^{meas}) \\ P(k) &= P^-(k) - K(k) C_{e,d} P^-(k) \end{aligned} \tag{9.30}$$

time update:

$$\begin{aligned} P^{-}(k+1) &= A_{e,d}P(k)A_{e,d}^{T} + Q \\ \hat{z}(k+1) &= A_{e,d}\hat{z}_e(k) + B_{e,d}v(k) \end{aligned} \tag{9.31}$$

9.2.6 Simulation Tests

The Kalman Filter-based fault diagnosis algorithm was applied to each individual power generator included in the 3-area power system that is described by Fig. 8.1. The fault thresholds are determined by the confidence intervals of the χ^2 distribution. The χ^2 distribution has $d = 3$ degrees of freedom. The number of iterations was $M = 2000$. Thus, for a 98 % confidence interval the associated upper and lower fault thresholds are $U = 2.8886$ and $L = 3.1136$. The model of the distributed power generators was brought to the form described by Eq. (9.12). The first test case was concerned with the functioning of the distributed power generation system under no fault. In such a case it was considered that the value of the additive external disturbance $\tilde{d}_i$ that was exerted on the power generators was of the order of 10^{-4}. In the next three test cases $i = 1, \ldots, 3$ an additive input disturbance (fault) $\tilde{d}_i$ with value of the order of 10^{-1} was assumed to be exerted on the i-th power generator. Such a disturbance can be due for instance to a change in the damping coefficient D_i that defines the turn speed of the rotor of the i-th generator. It can be also due to a variation in the electromagnetic torque that is exerted on the rotor and this can be caused by a change in the reactance parameter x'_{d_Σ} of the generator, as given in Eq. (8.3).

The performance of the proposed fault diagnosis scheme for distributed power generators was tested through simulation experiments. For the fault-free case the obtained results are depicted in Figs. 9.6, 9.7 and 9.8. For the case of faults appearing in the individual power generators the obtained results are depicted in Figs. 9.9, 9.10, 9.11, 9.12, 9.13, 9.14, 9.15, 9.16 and 9.17. It can be noticed that the proposed diagnosis test that is based on Kalman Filtering and on the statistical properties of the χ^2 distribution, achieved detection and isolation of the failures taking place at the individual generators. It is noted that comparing to the RMSE (root mean square error) index, the proposed fault diagnosis method which is based on the statistical properties of the χ^2 distribution is much more efficient. Actually, for the additive faults described above, the RMSE index was of the order of 10^{-5}. This RMSE indication misleads to the conclusion that the generators function properly, although faults have affected them.

The proposed fault diagnosis method is characterized by its simplicity and can be easily implemented to several types of distributed power generation systems. One should also consider the problem of incipient fault detection, that is of the ability of the fault diagnosis method to identify parametric changes of very small magnitude. This is particularly important for detecting failures in the power generators at their early stages, and for getting enough time for restoration actions. To this end, optimal fault

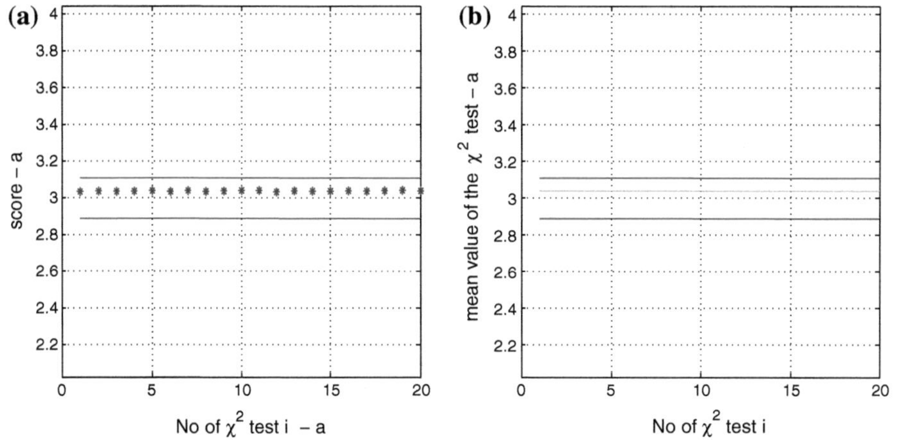

Fig. 9.6 Fault-free case: **a** consecutive χ^2 tests performed at Generator 1, **b** the mean value (*green line*) of the χ^2 tests performed at Generator 1 remains within the thresholds (*red lines*) indicating healthy condition

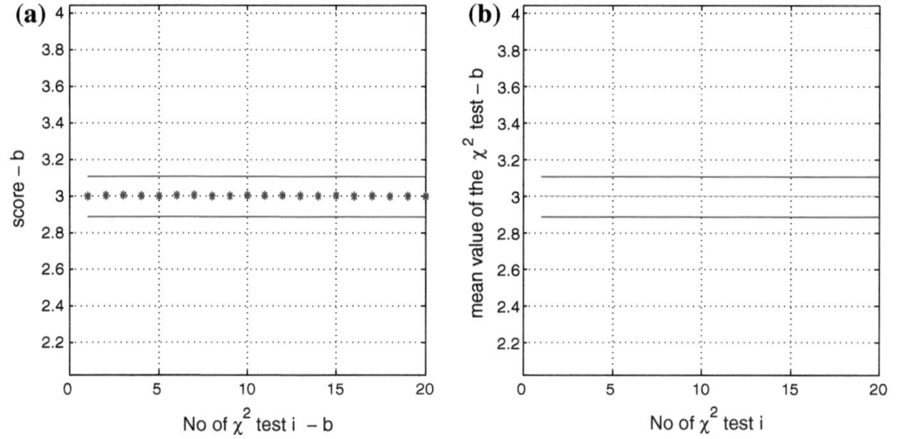

Fig. 9.7 Fault-free case: **a** consecutive χ^2 tests performed at Generator 2, **b** the mean value (*green line*) of the χ^2 tests performed at Generator 2 remains within the thresholds (*red lines*) indicating healthy condition

diagnosis methods based on the generalized likelihood ratio and on the χ^2 distribution of a transformed residuals's sequence have been developed. Such methods constitute the *local statistical approach to fault diagnosis* [34, 36, 37, 344, 345, 379, 497, 508].

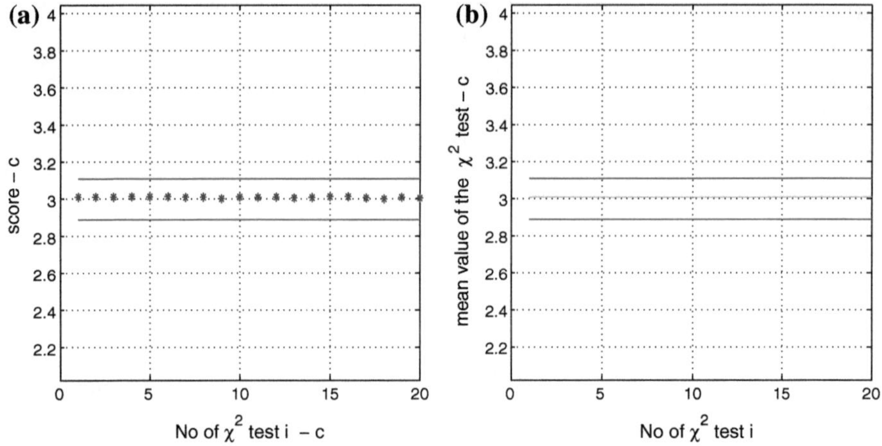

Fig. 9.8 Fault-free case: **a** consecutive χ^2 tests performed at Generator 3, **b** the mean value (*green line*) of the χ^2 tests performed at Generator 3 remains within the thresholds (*red lines*) indicating healthy condition

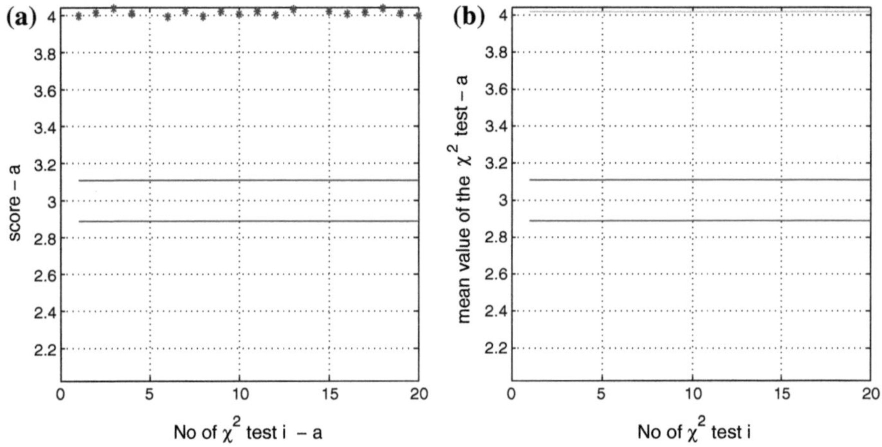

Fig. 9.9 Fault at power Generator 1: **a** consecutive χ^2 tests performed at Generator 1, **b** the mean value (*green line*) of the χ^2 tests performed at Generator 1 exceeds the thresholds (*red lines*) indicating healthy condition

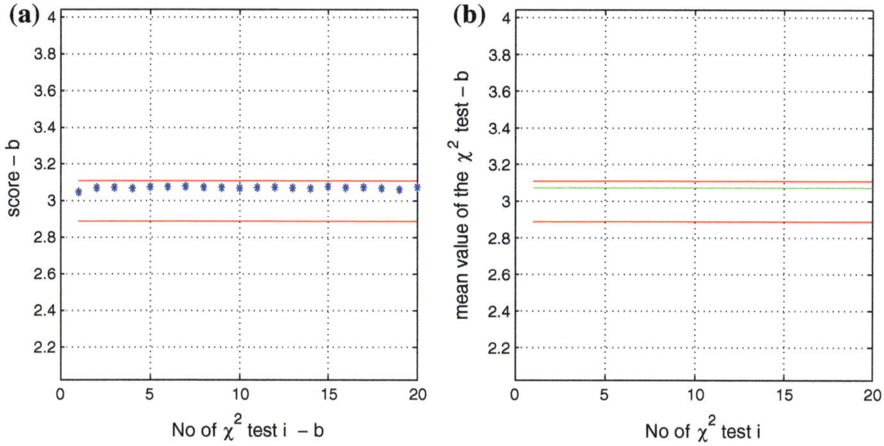

Fig. 9.10 Fault at power Generator 1: **a** consecutive χ^2 tests performed at Generator 2, **b** the mean value (*green line*) of the χ^2 tests performed at Generator 2 remains within the thresholds (*red lines*) indicating healthy condition

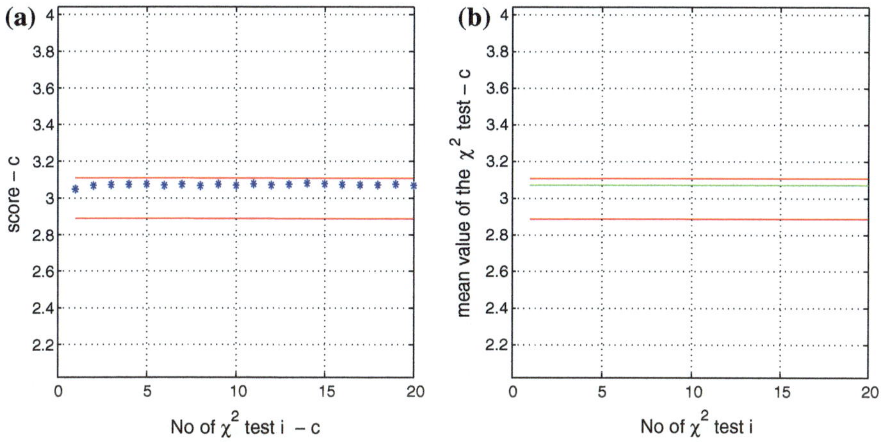

Fig. 9.11 Fault at power Generator 1: **a** consecutive χ^2 tests performed at Generator 3, **b** the mean value (*green line*) of the χ^2 tests performed at Generator 3 remains within the thresholds (*red lines*) indicating healthy condition

9.3 Neural Network-Based Fault Diagnosis in Distributed Power Generators

9.3.1 Outline

As noted, the problem of fault diagnosis comes to complement that of control in the design and development of renewable energy systems. Large-scale power systems

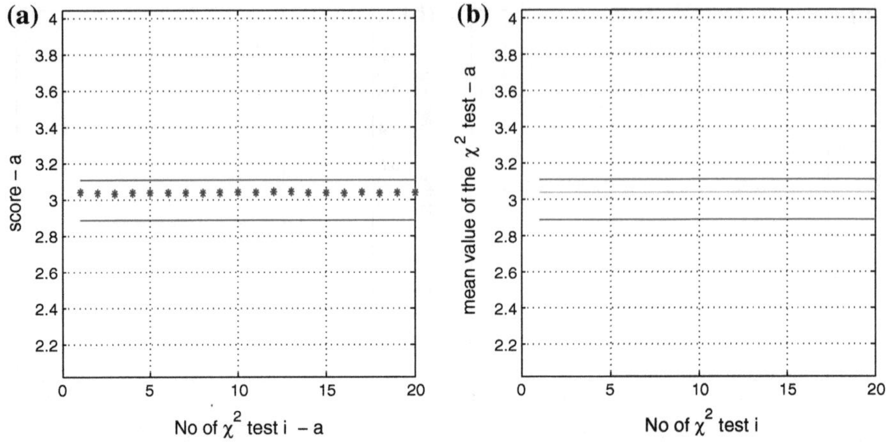

Fig. 9.12 Fault at power Generator 2: **a** consecutive χ^2 tests performed at Generator 1, **b** the mean value (*green line*) of the χ^2 tests performed at Generator 1 remains within the thresholds (*red lines*) indicating healthy condition

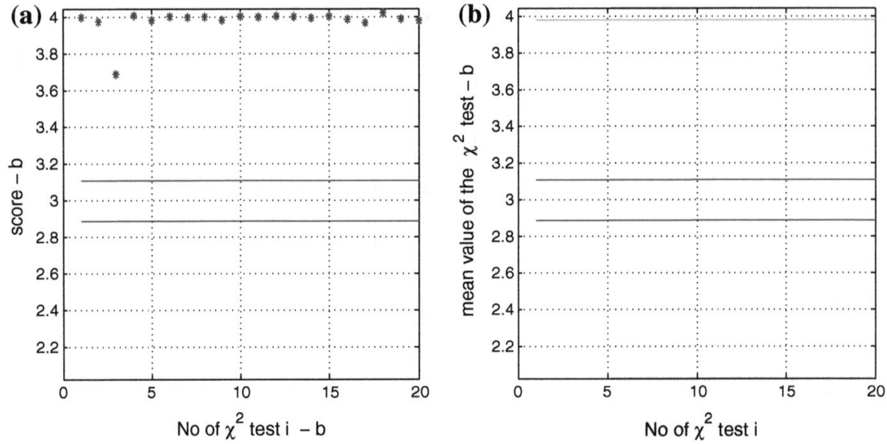

Fig. 9.13 Fault at power Generator 2: **a** consecutive χ^2 tests performed at Generator 2, **b** the mean value (*green line*) of the χ^2 tests performed at Generator 2 exceeds the thresholds (*red lines*) indicating healthy condition

have commonly experienced adverse impacts on the system operation and cascading events due to the under-damped low-frequency oscillation. Low-frequency oscillations is a main cause for instability in power systems. Such oscillations are classified into two modes: (i) local modes representing oscillations between one generator and the rest of the power grid or oscillations among several adjacent synchronous power generators, and (ii) inter-area modes representing swings among different power grids interconnected through tie-lines [220, 225]. Monitoring of these oscillations

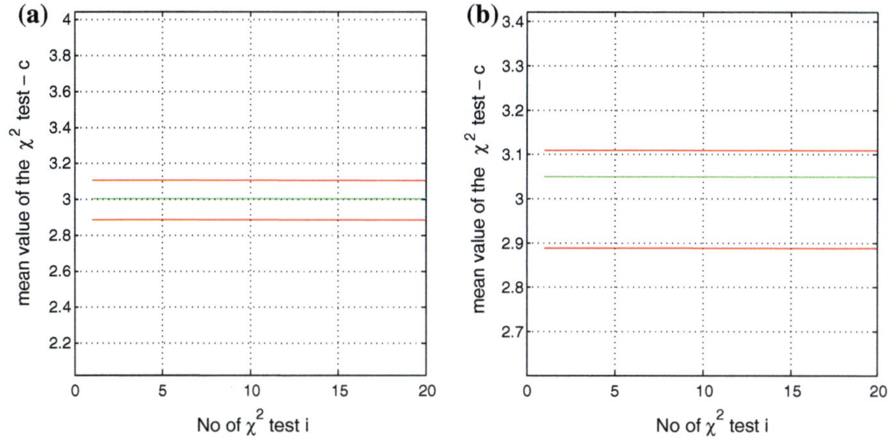

Fig. 9.14 Fault at power Generator 2: **a** consecutive χ^2 tests performed at Generator 3, **b** the mean value (*green line*) of the χ^2 tests performed at Generator 3 remains within the thresholds (*red lines*) indicating healthy condition

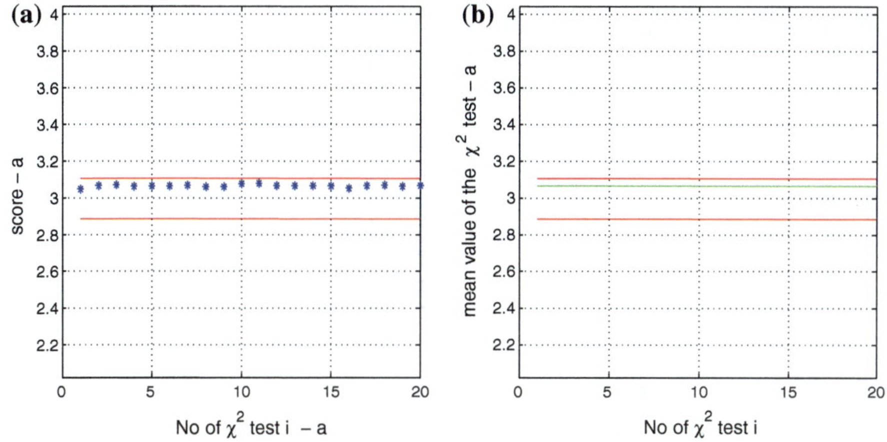

Fig. 9.15 Fault at power Generator 3: **a** consecutive χ^2 tests performed at Generator 1, **b** the mean value (*green line*) of the χ^2 tests performed at Generator 1 remains within the thresholds (*red lines*) indicating healthy condition

can help to diagnose at its early stages the risk of power system destabilization [206, 390, 433]. To succeed failure diagnosis and to enable restoration action, models of the power system have been developed.

The behavior of the electric power transmission system is usually described by a multi-area multi-machine power system model. Slow and fast modes can be distinguished. For instance slow modes correspond to the weak connection between areas. Early detection of the system's tendency to become unstable permits the

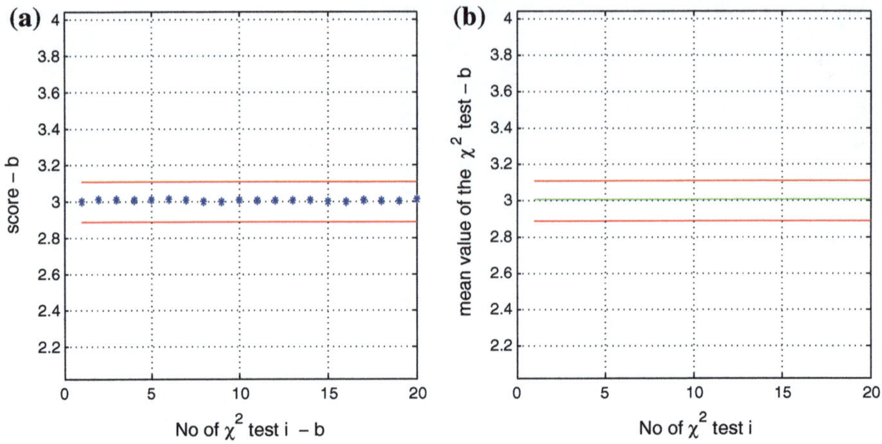

Fig. 9.16 Fault at power Generator 3: **a** consecutive χ^2 tests performed at Generator 3, **b** the mean value (*green line*) of the χ^2 tests performed at Generator 3 exceeds the thresholds (*red lines*) indicating healthy condition

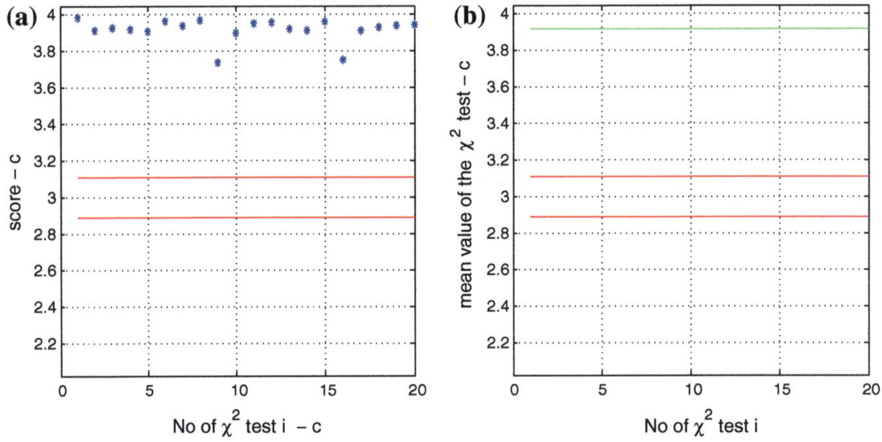

Fig. 9.17 Fault at power Generator 3: **a** consecutive χ^2 tests performed at Generator 3, **b** the mean value (*green line*) of the χ^2 tests performed at Generator 3 exceeds the thresholds (*red lines*) indicating healthy condition

activation of automatic voltage regulators which can damp-out inter-area oscillations. Thus cascading events (black-outs) can be prevented. The objective of this research work is to model the dynamics of power transmission system, and to detect faults at their early stages, thus enabling measures for fault restoration and actions for power system stabilization.

Linear models or NARX representations (nonlinear autoregressive models with exogenous inputs) and NARMAX representations (nonlinear autoregressive moving average models with exogenous inputs) have been proven in several cases inadequate to model the complete dynamics of power systems. On the other hand, neural-fuzzy networks have demonstrated a good performance in problems of modeling the nonlinear dynamics of electric power systems [30, 63, 119, 146, 189, 275, 341, 435, 436, 445, 447]. In this section modelling of the multi-machine multi-area power system dynamics (in fault-free operating conditions) will be carried out with non-parametric estimators, such as a neuro-fuzzy network (alternatively, one can use neural networks that are able to capture the multi-frequency characteristics of the power system dynamics, such as wavelet networks, or feed-forward neural networks with Hermite polynomial activation functions [154, 339, 498]. Comparing the measurements from the power system with the output of the neural model, the early detection of cascading events (black-outs) is possible. To this end, the statistical fault detection and isolation (FDI) algorithm analyzed in [34] is used. If the deviation exceeds a threshold, which is defined by the aforementioned FDI algorithm, then an alarm can be launched [36, 37, 508]. In several cases fault isolation can be also performed, i.e. the sources of fault in the power transmission system can be also identified. The use of pattern recognition methods for FDI in the electric power system is also possible [191, 336].

9.3.2 Power System Faults and Cascading Events

Many research efforts aim at understanding and finding ways to prevent or mitigate cascading events: study of the cascade model, dynamic decision event tree analysis, wide area backup protection, relay hidden failure analysis, special protection scheme, self-healing system with the aid of multi-agent technology, etc. [68, 87, 120, 254, 433]. The above mentioned techniques are still far from being an established practice in solving the cascading event problem. In general cascading events can be prevented by in-time operator actions.

Various methods have tried to analyze the oscillations in real-time, such as peak-value detecting [177], oscillation decay time detecting [467], FFT analysis and wavelet transform [178, 180]. However, enough detailed information about the oscillations from real-time measurements cannot be acquired by using these methods. The detailed oscillatory modes can be instead obtained with the stochastic subspace identification method [159] and the Kalman filtering method [221]. However none of them is suitable for online applications as the former is only appropriate for small disturbance responses and the latter requires a high computation time. The Prony algorithm, provided with fairly good performance [390], has been widely used in power systems [24, 168, 184, 492, 494, 501] as it can be used for online monitoring of system behavior and prediction of system instability. Nevertheless, it cannot ensure accurate mode identification and deal with the distorted real-time signals.

Normally there are two stages of a cascading event. First, there is a stage of slowly evolving consecutive events that can be approximated with steady state analysis. Several new disturbances following one another can make the system operating conditions worse. Then, as a consequence of a succession of several major disturbances a fast transient process takes place that results in cascading events, and finally the system collapses. When the total system collapse starts, normally it is too late to stop it. Nevertheless, much can be done during the slow steady state successions at the first stage and early proper control actions at the steady state stage can prevent the possible cascading event [423]. Analysis of blackouts of power systems has shown that these outages are caused by a cascading sequence of events involving line outages, overloading of certain lines in the transmission system, malfunctions of protection systems, power oscillations and voltage problems and finally system separation and collapse. A dynamical system model of the failure phenomena in the power transmission network has been presented in [113] while a probabilistic model of the propagation of failures in a power system has been given in [333].

Apart from low frequency oscillations, some examples of basic patterns of cascading events are: (i) line tripping due to overloading (a line fault can cause power flows to be rerouted, leading to overload of other lines. As a result the overloaded lines may be tripped due to a depressed voltage and /or high current resulting in a low apparent impedance seen by the impedance relay), (ii) generator tripping due to over-excitation (line outages may cause low voltages and high reactive power demand on generators nearby, leading to tripping of these generator units by over-excitation protection equipment), (iii) line tripping due to loss of synchronism (system disturbances such as faults, line outages and generator tripping can cause oscillation in machine rotor angles and system bus voltages leading to a power flow swing. Furthermore, when the angle difference between two buses is sufficient large due to a power swing, the impedance seen by a relay may trigger line tripping), (iv) generator tripping due to abnormal voltage and frequency system condition (a loss of a generator may affect the system voltage and frequency and sustained low voltage may cause other generator tripping. Furthermore, high voltage on some part of the system can also contribute to generator tripping. Additionally, under-frequency or over-frequency can trigger relays thus provoking generator tripping), (v) under-frequency / voltage load shedding (automatic under-frequency load shedding or under-voltage load-shedding may be triggered when the system voltage or frequency in some areas falls below a pre-specified value) [7, 17, 61, 274, 309, 485].

The relation between the major network problems and the time available to control actions is shown in Table 9.1.

Cascading events can be stopped by a number of complementary and corrective actions, including load rejection, controlled separation, load shedding and low-frequency isolation scheme.

Table 9.1 Relationship between network faults and the time available to control actions

Major network problems	Time available to control action	Blackout prevention by operation actions	Blackout prevention by emergency controls
Overloads	Second to minutes	Partially possible	Possible
Low transmission voltage	Second to minutes	Partially possible	Possible
Low frequency	0.1 seconds to seconds	Impossible	Possible
Loss of synchronism	Milliseconds to seconds	Impossible	Possible

9.3.3 Neural Networks for Power System Identification

9.3.3.1 Neuro-Fuzzy Models for Power System Identification

Because of linear models or NARX representations and NARMAX representations (nonlinear autoregressive moving average models with exogenous inputs) have been proved in several cases inadequate to model the complete dynamics of power systems, neural-fuzzy network models have been considered in this section. Artificial neural networks (ANNs) proved to be suitable at identifying and controlling nonlinear systems with complex dynamic and transient processes and they can easily identify the interactions between the system's inputs and outputs for multivariable applications [155, 188, 301, 453, 454]. It has been shown that the complex and nonlinear dynamics of a single machine infinite bus configuration can be identified with sufficient accuracy by a Multilayer Perceptron neural network using deviation signals as inputs [454].

ANNs have also been used for dynamic security assessment [30, 59, 119, 275]. An ANN is used to predict system stability in [59] and to estimate maximum generators swing angles in [119]. The transient energy margins are used for contingencies screening and ranking in [275]. In [30], recurrent radial basis function, and multilayer perceptron ANN schemes are used for dynamic system modeling and generators' angles and angular velocities prediction for multi-machine power systems. Transient stability is assessed based on monitoring generators' angles and angular velocities with time, and checking whether they exceed the specified limits for system stability or not. In [435, 436] a new method based on neural networks for eigenvalue predictions of critical stability modes of power systems is proposed, while the interest is focused on inter-area oscillations in the interconnected power system.

Neural-fuzzy modeling using the data of the power system is described as follows: a collection of N data in a $n+1$-dimensional space is considered, and a generic neuro-fuzzy model is presented as a set of fuzzy rules in the following form:

$$R^l \ : \ IF\, x_1 \ is\ A_1^l \ \ AND \ \ x_2\, is \ \ A_2^l\, AND \cdots AND\, x_n \ is \ \ A_n^l \ THEN\, \bar{y}^l = z^l(x) \tag{9.32}$$

where $x = [x_1, x_2, \ldots, x_n] \in U$ are input variables, $\bar{y}^l \in V$ is the output variable, A_i^l are the fuzzy sets in which $U^i \in R$ is divided and $z^l(x)$ is a function of the input variables.

The modeling procedure begins with the model initialization stage that includes data processing, prior knowledge utilization, and initial rule-base generation. The optimal number of fuzzy rules is determined and the appropriate type of fuzzy rules is selected. Model optimization follows which involves parameter learning and rule-base simplification. Finally, the acquired neurofuzzy model is validated under certain performance indexes. If the model performance is not satisfactory, further modification including structure and parameter optimization is required [79, 91, 118, 174, 175, 325, 402, 414, 415, 431, 462].

In the sequel fuzzy rules of the Takagi-Sugeno type will be considered. These have the form:

$$\begin{array}{l} R_l \ : \ IF\, x_1\ is\ A_1^l\, AND\, x_2\ is\ A_2^l\, AND\ \cdots AND\, x_n\ is\ A_n^l \\ \quad\ \ THEN\, \bar{y}^l = \sum_{i=1}^n w_i^l x_i + b^l \ l = 1, 2, \ldots, L \end{array} \tag{9.33}$$

where R^l is the l-th rule, $x = [x_1, x_2, \ldots, x_n]^T$ is the input (antecedent) variable, $\bar{y}^l$ is the output (consequent) variable, and w_i^l, b^l are the parameters of the local linear models. The above model is a Takagi-Sugeno model of order 1. Setting $w_i^l = 0$ results in the zero order Takagi-Sugeno model [152, 192, 305]. The output of the Takagi-Sugeno model is given by the weighted average of the rules consequents (Fig. 9.18):

$$\hat{y} = \frac{\sum_{l=1}^L \bar{y}^l \prod_{i=1}^n \mu_{A_i^l}(x_i)}{\sum_{l=1}^L \prod_{i=1}^n \mu_{A_i^l}(x_i)} \tag{9.34}$$

where $\mu_{A_i^l}(x_i) \ : \ R \rightarrow [0, 1]$ is the membership function of the fuzzy set A_i^l in the antecedent part of the rule R^l. In the case of a zero order TS system the output of the l-th local model is $\bar{y}^l = b^l$, while in the case of a first order TS system the output of the l-th local model is given by $\bar{y}^l = \sum_{l=1}^L w_i^l x_i + b^l$.

The individual steps of data-driven fuzzy modelling are discussed in [193–195] and are summarized in Fig. 9.19. Methods for optimization of neuro-fuzzy models parameters has been extensively discussed in [193, 195, 401, 403].

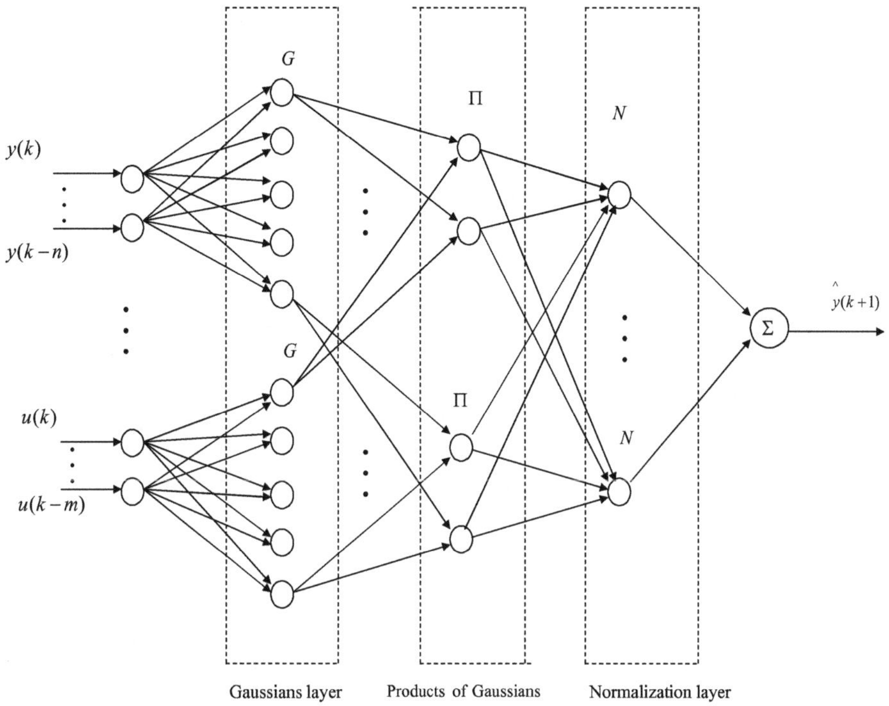

Fig. 9.18 Neural-fuzzy approximator

Fig. 9.19 General scheme of data-driven neural-fuzzy modelling

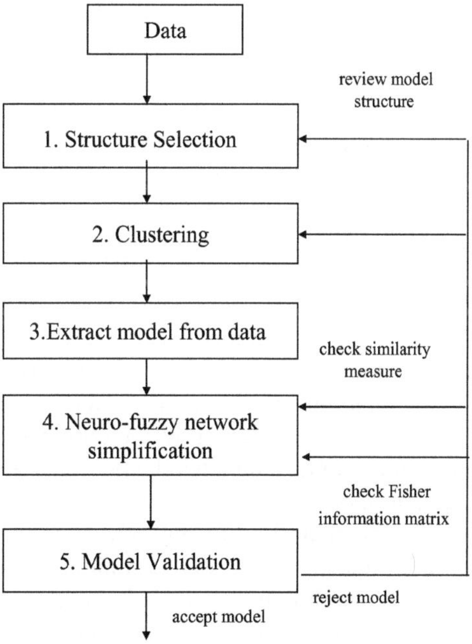

9.3.4 *Fault Diagnosis for Electric Power Transmission Systems*

9.3.4.1 Fault Detection with Statistical Methods

A statistical FDI method which can be used for the early detection of cascading events in electric power systems is the so called *Local Statistical Approach to change detection* [34]. The local statistical approach has been successfully applied to several FDI problems, such as [35, 38, 39, 40, 282, 497, 509]: (i) Vibration monitoring in linear dynamical systems: this application concerns power plants (rotating machines, nuclear power plants installations), civil engineering (large buildings subject to earthquakes, bridges, dams and offshore structures), aeronautics (wings and other structures subject to strength), automobile and train transportation, (ii) Nonlinear static systems, such as the gas turbine: this example refers to the detection of faults in the combustion chamber of gas turbines, which are widely used as electric power generators, and aircraft engines, (iii) Nonlinear dynamical systems, such as the catalyst used in vehicles: this application concerns the detection of faults in the catalytic converter and the oxygen sensors of the automobiles.

Based on a small parametric disturbance assumption, the proposed FDI method aims at transforming complex detection problems concerning a parameterized stochastic process into the problem of monitoring the mean of a Gaussian vector. The local statistical approach consists of two stages: (i) the global χ^2 test which indicates the existence of a change in some parameters of the fuzzy model, (ii) the diagnostics tests (sensitivity or min-max) which isolate the parameter affected by the change. The local statistical approach is suitable for fault diagnosis in renewable energy systems. The method's stages are analyzed first, following closely the method presented in [34, 37]:

As shown in Fig. 9.20 the proposed method is based on the definition of the residual e_i described as the difference between the output from the real power system y_i^0 and the output from a neural model y_i. The neural model is used to simulate the physical power system in a stable undistorted state and has been extracted from input/output data when the power system is in a stable state. To verify the proposed method performance, the real power system has been simulated by using the so-called *exact model*. In order to have the neural model and the exact model with the same number of parameters, the exact model can be also represented by a neural model extracted from input/output data of the power system. Therefore, when the power system is affected by slight parameters variations which can lead to system instability, the output of the exact model will differ from the output of the neural model. In other words, while the neural model simulates the power system or a part of it in an undistorted stable state, the exact model simulates the real power system or a part of it in all conditions and is extracted from the real power system data.

The concept of this FDI technique is as follows: the dynamics of the fault-free power transmission system is learned by the neural-fuzzy network. At each time instant the neural network's output is compared to the real condition of the system.

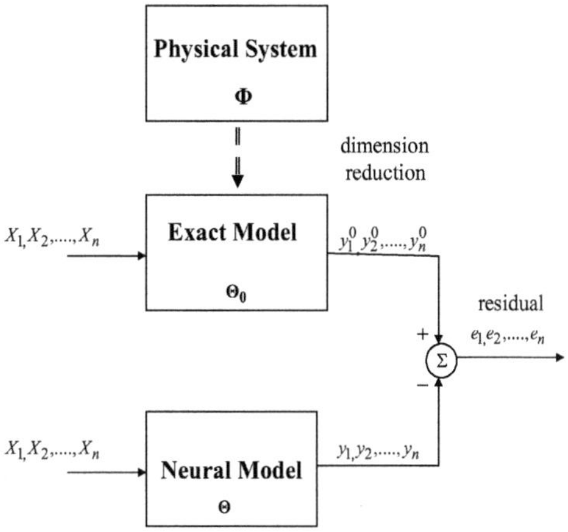

Fig. 9.20 Residual between the exact model and the neuro-fuzzy model

The difference between the real condition of the power system and the output of the neural network is called residual. The statistical processing of a sufficiently large number of residuals through the aforementioned FDI method provides an index-variable that is compared against a fault threshold and which can give early indication about deviation of the power system from the normal operating conditions. Therefore alarm launching can be activated at the early stages of power system destabilization, and restoration measures can be taken. Under certain conditions (detectability of changes) the proposed FDI method enables also fault isolation, i.e. to identify the source of fault within the power transmission system.

The partial derivative of the residual square is:

$$H(\theta, y_i) = \frac{\partial e_i^2}{\partial \theta} = e_i \frac{\partial y_i}{\partial \theta} \tag{9.35}$$

where θ is the vector of model's parameters. The vector H having as elements the above $H(\theta, y_i)$ is called primary residual. In the case of neuro-fuzzy models of Eq. (9.34) the gradient of the output with respect to the consequent parameters w_i^l is given by

$$\frac{\partial \hat{y}}{\partial w_i^l} = \frac{x_i \mu_{R^l}(x)}{\sum_{l=1}^{L} \mu_{R^l}(x)} \tag{9.36}$$

The gradient with respect to the center c_i^l is

$$\frac{\partial \hat{y}}{\partial c_i^l} = \sum_{l=1}^{L} \frac{\overline{y}^l \frac{2(x_i - c_i^l)}{v_i^l} \mu_{R^l}(x_i)[\sum_{j=1}^{L} \mu_{R^j}(x_i) - \mu_{R^l}(x_i)]}{[\sum_{l=1}^{L} \mu_{R^l}(x_i)]^2} \tag{9.37}$$

The gradient with respect to the spread v_i^l is

$$\frac{\partial \hat{y}}{\partial v_i^l} = \sum_{l=1}^{L} \frac{\bar{y}^l \frac{2(x_i - c_i^l)^2}{v_i^{l^3}} \mu_{R^l}(x_i)[\sum_{j=1}^{L} \mu_{R^j}(x_i) - \mu_{R^l}(x_i)]}{[\sum_{l=1}^{L} \mu_{R^l}(x_i)]^2} \tag{9.38}$$

Next, having calculated the partial derivatives of Eqs. (9.36), (9.37) and (9.38), the rows of the Jacobian matrix J are found by

$$J(\theta_0, y_k) = \left. \frac{\partial \hat{y}_k(\theta)}{\partial \theta} \right|_{\theta=\theta_0} \tag{9.39}$$

where θ_0 represents the nominal value of the parameters. The problem of change detection with the χ^2 test consists of monitoring a change in the mean of the Gaussian variable which for the one-dimensional parameter vector θ is formulated as

$$X = \frac{1}{\sqrt{N}} \sum_{i=1}^{N} e_k \frac{\partial \hat{y}_k}{\partial \theta} \sim N(\mu, \sigma^2) \tag{9.40}$$

where $\hat{y}_k$ is the output of the neural model generated by the input pattern x_k, e_k is the associated residual and θ is the vector of the model's parameters. It is noted that X is the monitored parameter for the FDI test, which means that when the mean value of X is 0 the system is in the fault-free condition, while when the mean value of X has moved away from 0 the system is in a faulty condition. For a multivariable parameter vector θ should hold $X \sim N(M\delta\theta, S)$, where S denotes the covariance matrix of X. In order to decide if the power system is in fault-free operating conditions, given a set of data of N measurements, let θ_* be the value of the parameters vector μ minimizing the RMSE. The notation is introduced only for the convenience of problem formulation, and its actual value does not need to be known. Then the model validation problem amounts to make a decision between the two hypotheses:

$$\begin{aligned} H_0 &: \theta_* = \theta_0 \\ H_1 &: \theta_* = \theta_0 + \tfrac{1}{\sqrt{N}}\delta\theta \end{aligned} \tag{9.41}$$

where $\delta\theta \neq 0$. It is known from the central limit theorem that for a large data sample, the normalized residual given by Eq. (9.40) asymptotically follows a Gaussian distribution when $N \rightarrow \infty$ [34, 37]. More specifically, the hypothesis that has to be tested is:

$$\begin{aligned} H_0 &: X \sim N(0, S) \\ H_1 &: X \sim N(M\delta\theta, S) \end{aligned}$$

where M is the sensitivity matrix (see Eq. (9.42)), $\delta\theta$ is the change in the parameters' vector and S is the convariance matrix (see Eq. (9.43)). The product $M\delta\theta$ denotes the new center of the monitored Gaussian variable X, after a change on the system's parameter θ. The sensitivity matrix M of $\frac{1}{\sqrt{N}}X$ is defined as the mean value of the partial derivative with respect to θ of the primary residual defined in Eq. (9.35), i.e. $E\{\frac{\partial}{\partial\theta}H(\theta, y_k)\}$ and is approximated by [8]:

$$M(\theta_0)\simeq\frac{\partial}{\partial\theta}\frac{1}{N}\sum_{k=1}^{N}H(\theta_0, y_k)\simeq\frac{1}{N}J^TJ \tag{9.42}$$

The covariance matrix S is defined as $E\{H(\theta, y_k)H^T(\theta, y_{k+m})\}$, $m = 0, \pm 1, \ldots$ and is approximated by [37]:

$$\begin{aligned} S = \simeq\sum_{k=1}^{N}[H(\theta_0, y_k)H^T(\theta_0, y_k)]+ \\ +\sum_{m=1}^{I}\frac{1}{N-m}\sum_{k=1}^{N-m}[H(\theta_0, y_k)H^T(\theta_0, y_{k+m}) + H(\theta_0, y_{k+m})H^T(\theta_0, y_k)] \end{aligned} \tag{9.43}$$

where an acceptable value for I is 3. The decision tool is the likelihood ratio $s(X) = ln\frac{p_{\theta_1(x)}}{p_{\theta_0(x)}}$, where $p_{\theta_1}(X) = e^{[X-\mu(X)]^T S^{-1}[X-\mu(X)]}$ and $p_{\theta_0}(X) = e^{X^T S^{-1} X}$. The center of the Gaussian distribution of the changed system is denoted as $\mu(X) = M\delta\theta$ where $\delta\theta$ is the change in the parameters vector. The *Generalized Likelihood Ratio* (GLR) is calculated by maximizing the likelihood ratio with respect to $\delta\theta$ [37]. This means that the most likely case of parameter change is taken into account. This gives the global χ^2 test t:

$$t = X^T S^{-1}M(M^T S^{-1}M)^{-1}M^T S^{-1}X \tag{9.44}$$

Since X asymptotically follows a Gaussian distribution, the statistics defined in Eq. (9.44) follows a χ^2 distribution with n degrees of freedom. Mapping the change detection problem to this χ^2 distribution enables the choice of the change threshold. Assume that the desired probability of false alarm is α then the change threshold λ should be chosen from the relation

$$\int_{\lambda}^{\infty}\chi_n^2(s)ds = \alpha, \tag{9.45}$$

where $\chi_n^2(s)$ is the probability density function (p.d.f.) of a variable that follows the χ^2 distribution with n degrees of freedom.

9.3.4.2 Statistical Fault Isolation with the Sensitivity Test

Fault isolation is needed to identify the source of faults in the electric power transmission system. A first approach to change isolation is to focus only on a subset of the parameters while considering that the rest of the parameters remain unchanged [37]. The parameters vector η can be written as $\eta = [\phi, \psi]^T$, where ϕ contains those parameters to be subject to the isolation test, while ψ contains those parameters to be excluded from the isolation test. M_ϕ contains the columns of the sensitivity matrix M which are associated with the parameters subject to the isolation test. Similarly M_ψ

contains the columns of M that are associated with the parameters to be excluded from the sensitivity test.

Assume that among the parameters η, it is only the subset ϕ that is suspected to have undergone a change. Thus η is restricted to $\eta = [\phi, 0]^T$. The associated columns of the sensitivity matrix are given by M_ϕ and the mean of the Gaussian to be monitored is $\mu = M_\phi \phi$, i.e.

$$\mu = MA\phi, \quad A = [0, I]^T \tag{9.46}$$

Matrix A is used to select the parameters that will be subject to the fault isolation test. The rows of A correspond to the total set of parameters while the columns of A correspond only to the parameters selected for the test. Thus the fault diagnosis (χ^2) test of Eq. (9.44) can be restated as:

$$t_\phi = X^T S^{-1} M_\phi (M_\phi^T S^{-1} M_\phi)^{-1} M_\phi^T S^{-1} X \tag{9.47}$$

9.3.4.3 Statistical Fault Isolation with the Min-Max Test

In this approach the aim is to find a statistic that will be able to detect a change on the part ϕ of the parameters vector η and which will be robust to a change in the non observed part ψ [37]. Assume the vector partition $\eta = [\phi, \psi]^T$. The following notation is used:

$$M^T S^{-1} M = \begin{pmatrix} I_{\varphi\varphi} & I_{\varphi\psi} \\ I_{\psi\varphi} & I_{\psi\psi} \end{pmatrix} \tag{9.48}$$

$$\gamma = \begin{pmatrix} \varphi \\ \psi \end{pmatrix}^T \cdot \begin{pmatrix} I_{\varphi\varphi} & I_{\varphi\psi} \\ I_{\psi\varphi} & I_{\psi\psi} \end{pmatrix} \cdot \begin{pmatrix} \varphi \\ \psi \end{pmatrix} \tag{9.49}$$

where S is the previously defined covariance matrix. The min-max test aims to minimize the non-centrality parameter γ with respect to the parameters that are not suspected for change. The minimum of γ with respect to ψ is given for:

$$\psi^* = \arg\min_{\psi} \gamma = \varphi^T (I_{\varphi\varphi} - I_{\varphi\psi} I_{\psi\psi}^{-1} I_{\psi\varphi}) \varphi \tag{9.50}$$

and is found to be

$$\begin{aligned} \gamma^* &= \min_{\psi} \gamma = \varphi^T (I_{\varphi\varphi} - I_{\varphi\psi} I_{\psi\psi}^{-1} I_{\psi\varphi}) \varphi = \\ &= \begin{pmatrix} \varphi \\ -I_{\psi\psi}^{-1} I_{\psi\varphi} \varphi \end{pmatrix}^T \begin{pmatrix} I_{\varphi\varphi} & I_{\varphi\psi} \\ I_{\psi\varphi} & I_{\psi\psi} \end{pmatrix} \begin{pmatrix} \varphi \\ -I_{\psi\psi}^{-1} I_{\psi\varphi} \varphi \end{pmatrix} \end{aligned} \tag{9.51}$$

which results in

$$\gamma^* = \varphi^T\{[I, -I_{\varphi\psi}I_{\psi\psi}^{-1}]M^T\Sigma^{-1}\}\,\Sigma^{-1}\{\Sigma^{-1}M[I, -I_{\varphi\psi}I_{\psi\psi}^{-1}]\}\varphi \tag{9.52}$$

The following linear transformation of the observations is considered:

$$X_\phi^* = [I, -I_{\varphi\psi}I_{\psi\psi}^{-1}]M^T\Sigma^{-1}X \tag{9.53}$$

The transformed variable X_ϕ^* follows a Gaussian distribution $N(\mu_\phi^*, I_\phi^*)$ with mean:

$$\mu_\varphi^* = I_\varphi^*\varphi \tag{9.54}$$

and with covariance:

$$I_\varphi^* = I_{\varphi\varphi} - I_{\varphi\psi}I_{\psi\psi}^{-1}I_{\psi\varphi} \tag{9.55}$$

The max-min test decides between the hypotheses:

$$H_0^* : \mu^* = 0$$
$$H_1^* : \mu^* = I_\varphi^*\varphi$$

and is described by:

$$\tau_\varphi^* = X_\varphi^{*T} I_\varphi^{*-1} X_\varphi^* \tag{9.56}$$

The stages of fault detection and isolation (FDI) with the use of the local statistical approach are summarized in the following Table 9.2:

Table 9.2 Stages of the local statistical approach for FDI

1.	Generate the residuals partial derivative given by Eq. (9.35)
2.	Calculate the Jacobian matrix J given by Eq. (9.39)
3.	Calculate the sensitivity matrix M given by Eq. (9.42)
4.	Calculate the covariance matrix S given by Eq. (9.43)
5.	Apply the χ^2 test for change detection of Eq. (9.44)
6.	Apply the change isolation tests of Eq. (9.47) or Eq. (9.56)

9.3.5 Simulation Tests

9.3.5.1 Test System

In order to evaluate the effectiveness of the proposed fault diagnosis method, Kundur's multi-area multi-machine power system is considered here [225]. Figure 9.21 shows Kundur's 2-area 4-machine power system model consisting of two symmetrical areas interconnected by a 220 km two-looped weak tie-line. G1-G4 are synchronous generators with rating of 900 MVA and 20 kV which consist two power areas-Area 1 and Area 2 respectively. The transmission system nominal voltage is 230 kV. The local loads of two areas and the active power transmitted on tie-line are also shown in Fig. 9.21.

A parametric representation of the linearized dynamics of the single-machine power system is given in the form [225]:

$$\begin{pmatrix} \Delta\dot{\omega}_r \\ \Delta\dot{\delta} \\ \Delta\dot{\psi}_{fd} \\ \Delta\dot{\psi}_{1d} \\ \Delta\dot{\psi}_{1q} \\ \Delta\dot{\psi}_{2q} \end{pmatrix} = \begin{pmatrix} \alpha_{11} & \alpha_{12} & \alpha_{13} & \alpha_{14} & \alpha_{15} & \alpha_{16} \\ \alpha_{21} & 0 & 0 & 0 & 0 & 0 \\ 0 & \alpha_{32} & \alpha_{33} & \alpha_{34} & \alpha_{35} & \alpha_{36} \\ 0 & \alpha_{42} & \alpha_{43} & \alpha_{44} & \alpha_{45} & \alpha_{46} \\ 0 & \alpha_{52} & \alpha_{53} & \alpha_{54} & \alpha_{55} & \alpha_{56} \\ 0 & \alpha_{62} & \alpha_{63} & 0 & \alpha_{65} & \alpha_{66} \end{pmatrix} \cdot \begin{pmatrix} \Delta\omega_r \\ \Delta\delta \\ \Delta\psi_{fd} \\ \Delta\psi_{1d} \\ \Delta\psi_{1q} \\ \Delta\psi_{2q} \end{pmatrix} + \begin{pmatrix} b_{11} & 0 \\ 0 & 0 \\ 0 & b_{32} \\ 0 & 0 \\ 0 & 0 \end{pmatrix} \cdot \begin{pmatrix} \Delta T_m \\ \Delta E_{fd} \end{pmatrix} \tag{9.57}$$

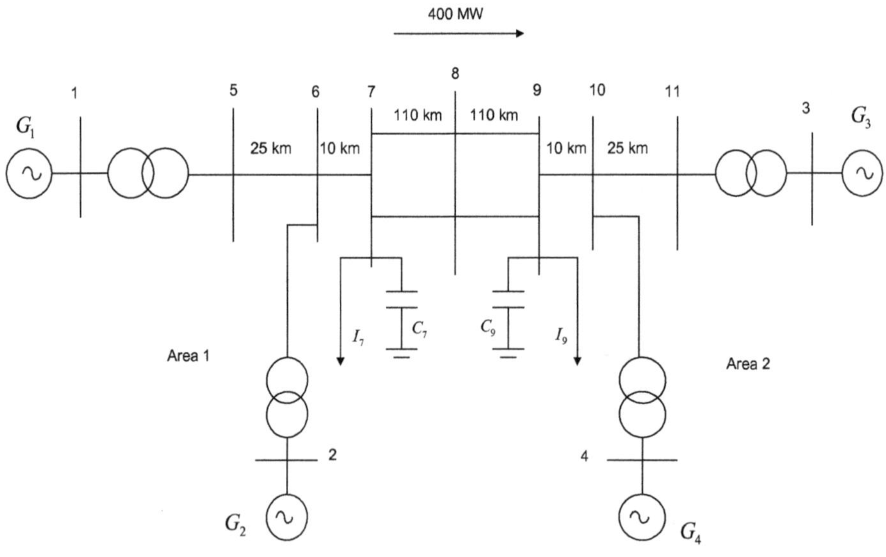

Fig. 9.21 Four-machine two-area model

Using the $d-q$ reference frame to express the magnetic flux and assuming that the model includes one d-axis amortisseur and two q-axis amortisseurs, the components of the state vector are defined as follows: $\Delta\omega_r$ is the rotor's speed difference, and $\Delta\delta$ is the rotor's angle difference. The rest of the parameters of the state vector are defined as $\Delta\psi_{fd}$ is the difference of the field (stator) flux along axis d, $\Delta\psi_{1d}$, is the difference of the rotor's flux of the d-axis amortisseur, and $\Delta\psi_{1q}$ is the difference of the rotor's flux of the first amortisseur of the q-axis, $\Delta\psi_{2d}$ is the difference of the rotor's flux of the second amortisseur of the q-axis. Moreover, the components of the control input vector are defined as follows: ΔT_m is the difference of the rotor's input torque and ΔE_{fd} is the difference of the excitation (stator's) field along axis d. The parameters a_{ij}, $i = 1,\dots,6$ and $j = 1,\dots,6$ and b_{ij} $i = 1,\dots,6$ and $j = 1,\dots,2$ are described in [225].

For the 2-area four-machine model depicted in Fig. 9.21 the state vector of the dynamic model is $x \in R^{24\times 1}$. The state vector can be written as $x = [x^1, x^2, x^3, x^4]$, where x^i, $i = 1,\dots,4$ is the state vector of the i-th generator given by $x^i = [\Delta\dot{\omega}_r, \Delta\dot{\delta}, \Delta\dot{\psi}_{fd}, \Delta\dot{\psi}_{1d}, \Delta\dot{\psi}_{1q}, \Delta\dot{\psi}_{2q}]^T$. Low frequency oscillation modes in Fig. 9.21 include two local modes and one inter-area mode. The two local oscillation modes are the swing between G1 and G2, and the swing between G3 and G4, respectively. The inter-area mode is the oscillation between Area 1 and Area 2 connected by transmission tie line. Indicative values of the system modes, when a linear model of the power system is considered, are given in Table 9.3 [220, 225].

To stabilize the multi-area, multi-machine power system the distributed generators G_i are equipped with Automatic Voltage Regulators (AVR). For instance in Fig. 9.21 G1 and G2 are equipped with local Power System Stabilizer (PSS) in order to restrain the poorly damped local oscillation mode between G1 and G2. In order to eliminate poorly damped inter-area oscillation between Area 1 and Area 2, the following variables can be measured and used by algorithm for control of supplementary excitation: (1) Remote/global signals of area speed difference $\Delta\omega_1 + \Delta\omega_2 - \Delta\omega_3 - \Delta\omega_4$, (2) Local signals at a generator G_i, such as shaft speed ω_i, terminal voltage V_{t_i}, excitation voltage E_{fd_i} and accelerating power P_{α_i}.

9.3.5.2 Simulation Results

Neuro-fuzzy modelling of the distributed power system has been carried out. The power system dynamics can be modelled using a neural network for each generator, having as output the rotor's speed difference $\Delta\omega_r(k)$ and as inputs $\Delta\omega_r(k-1)$, $\Delta\omega_r(k-2)$ and $\Delta\delta(k-1)$. Neural models with a different output, such as $\Delta\delta$ and a larger number of inputs could be also considered. The input space was segmented using the input dimension (grid) partition (additionally, a neural network with Hermite basis functions was used to model the dynamics of the power system. In the latter case the Hermite basis functions can capture with increased accuracy the multi-frequency characteristics of the oscillatory behavior of the power transmission system).

Table 9.3 Modes (eigenvalues) for the two-area four-machines power system

No.	Eig. real	Eig. imag.	Frequency	Damping ratio	Dominant states
1,2	$0.76E-3$	$\pm 0.22E-2$	0.0003	0.331	$\Delta\omega$, $\Delta\delta$ of G_1, G_2, G_3, G_4
3	$-0.96E-1$	–	–	–	"
4,5	–0.111	±3.43	0.545	0.032	"
6	–0.117	–	–	–	"
7	–0.265	–	–	–	$\Delta\psi_{fd}$ of G_3 and G_4
8	–0.276	–	–	–	$\Delta\psi_{fd}$ of G_1 and G_2
9, 10	–0.492	±6.82	1.087	0.072	$\Delta\omega$, $\Delta\delta$ of G_1, G_2
11, 12	–0.506	±7.02	1.117	0.072	$\Delta\omega$, $\Delta\delta$ of G_3, G_4
13	-3.428	–	–	–	d, q amortisseur flux linkages
14	-4.139	–	–	–	"
15	-5.287	–	–	–	"
16	-5.303	–	–	–	"
17	-31.03	–	–	–	"
18	-32.45	–	–	–	"
19	-34.07	–	–	–	"
20	-35.53	–	–	–	"
21,22	-37.89	±0.142	0.023	≃1.0	"
23,24	-38.01	$\pm 0.38E-1$	0.006	≃1.0	"

For a fault-free condition, a comparison between the neural model output and the exact model output is shown in Fig. 9.22, for generator G_1. The exact model has been obtained by using Kundur's power system model simulated in a power systems software library [293]. The neural model has been identified considering both a neural network with Hermite polynomial basis functions and a neuro-fuzzy network of the Takagi-Sugeno type.

As pointed out in Sect. 9.3.4, the neural model is used to simulate the physical power system in a stable undistorted state and the real power system is simulated by the exact model (a neural model extracted from input/output data of the power system). Due to parameters variation in the exact model, its output differs from the output of the neural model, and thus the residuals described in Sect. 9.3.4 have been calculated.

Regarding the general description of the training set used to identify both the neural model and the general description of the exact model, this is given by:

$$\begin{pmatrix} u(k-m) & \cdots & u(k-1) & y(k-n) & \cdots & y(k-1) & \rightarrow y(k) \\ u(k-m-1) & \cdots & u(k-2) & y(k-n-1) & \cdots & y(k-2) & \rightarrow y(k-1) \\ \vdots & \vdots & \vdots & \vdots & \vdots & \vdots & \vdots \\ u(k) & \cdots & u(m-1) & y(1) & \cdots & y(n-1) & \rightarrow y(n) \end{pmatrix} \tag{9.58}$$

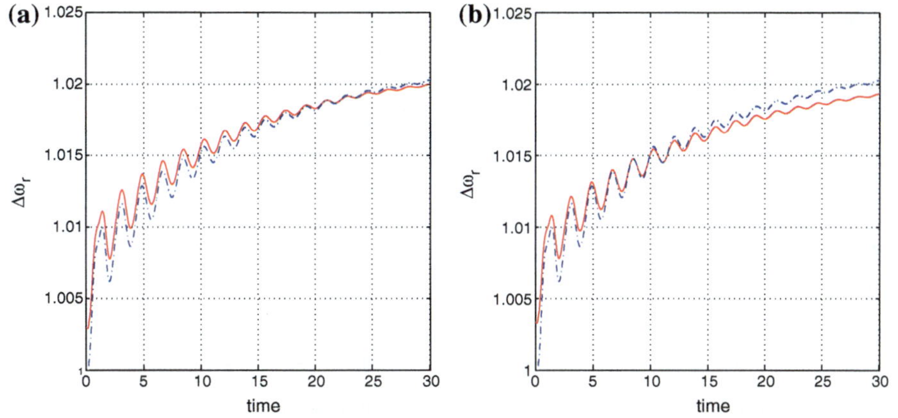

Fig. 9.22 **a** Approximation of the oscillatory behavior of the electric power system (*dashed line*) by a neural network with Hermite polynomial basis functions (*continuous line*), **b** Approximation of the oscillatory behavior of the electric power system (*dashed line*) by a neuro-fuzzy network of the Takagi-Sugeno type (*continuous line*)

An indicative neurofuzzy model of the power system consists of rules of the form:

$$R^l : IF\ y(k-3)\ is\ A_i\ AND\ y(k-2)\ is\ B_j\ AND\ y(k-1)\ is\ C_k\ \ THEN\ y(k)\ is\ D_m \tag{9.59}$$

with $i, j, k, m = 1, 2, 3, 4$. The size of the training set was 3000. The LMS (Least Mean Square) algorithm was used for the adaptation of the linear weights $w_1^{(l)}$. The rule base consists of 64 rules (3 input variables partitioned in 4 fuzzy subsets each). All fuzzy sets where assumed to have the same spread. Inactive rules have been removed from the fuzzy rule base.

The reduced-size rule base contained 22 rules. Therefore, the dimension of the parameters vector is 34, (22 linear weights in the antecedent part of the rules, and 12 nonlinear centers). Using Eq. (9.45) and setting the false alarm rate $\alpha = 0.5$, from the table of the χ^2 distribution one obtains that the value of the change threshold λ should be set to $\eta = 34$. In the case of absence of change to the parameters of the rule base the global χ^2 test was (averaging over 10 trials) $t = 38.6216$. The condition of the Fisher Information Matrix was $cond(M^T \Sigma^{-1} M) = 4.4314 \times 10^9$.

Next, faults were imposed to the linear and the nonlinear parameters of the exact model that describes the power system dynamics. Detailed diagrams on the performance of the change detection and isolation tests, in the case of a change in parameter $w_1^{(10)} = 1.195923$ of the exact model, are given in Fig. 9.23. The success rate of the two tests is depicted in Fig. 9.23a. The mean value of the global χ^2 is depicted in Fig. 9.23b.

Detailed diagrams on the performance of the change detection and isolation tests, in the case of a change in parameter $c_1^{(1)} = 1.954000$ of the exact model, are given

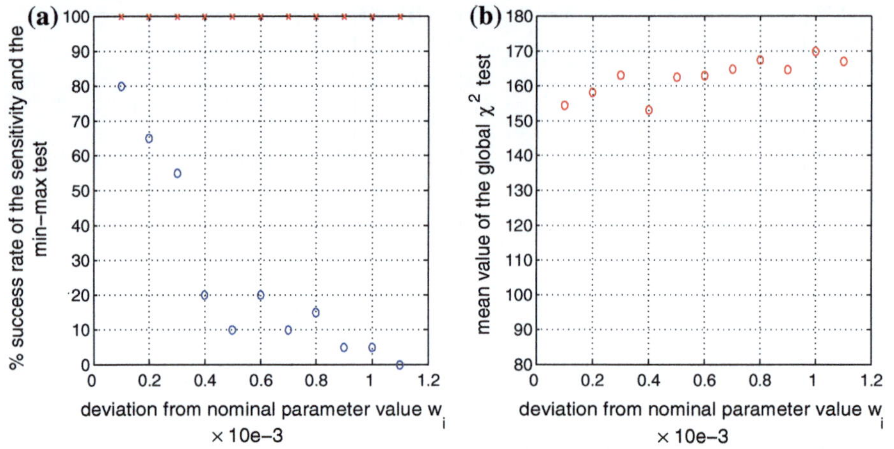

Fig. 9.23 Success rate of the sensitivity ($\times$) and the min-max (o) tests in case of a change (fault) in parameter $w_1^{(10)}$ of the exact model of the electric power transmission system. **b** Mean value of the global χ^2 test

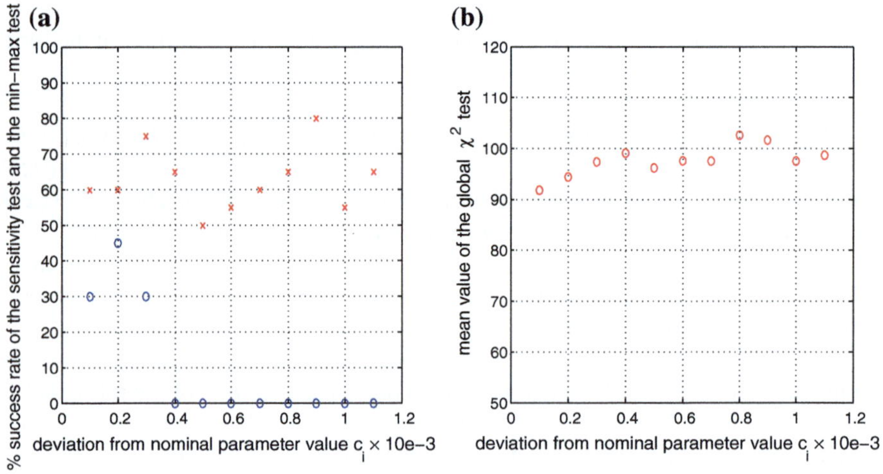

Fig. 9.24 Success rate of the sensitivity ($\times$) and the min-max (o) tests in case of a change (fault) in parameter $c_1^{(1)}$ of the exact model of the electric power transmission system. **b** Mean value of the global χ^2 test

in Fig. 9.24. The parameters vector contained only the centers of the fuzzy model. The success rate of the two tests is depicted in Fig. 9.24a.

The mean value of the global χ^2 is depicted in Fig. 9.24b. It is clear that when $c_1^{(1)}$ is close to its nominal value the χ^2 test takes a value close to the change threshold $\eta = 34$. On the other hand when the deviation of $c_1^{(1)}$ from its nominal value increases the value of the χ^2 test grows significantly.

Table 9.4 % parameter change and % success rate of fault isolation tests

$w_1^{(10)} = 1.195931$	Sensitivity	Min-max	$c_1^{(1)} = 1.954000$	Sensitivity	Min-max
0.0083	100	80	0.0083	60	30
0.0166	100	65	0.0166	60	45
0.0249	100	55	0.0249	75	30
0.0332	100	20	0.0332	65	0
0.0415	100	10	0.0415	50	0
0.0498	100	20	0.0498	55	0
0.0581	100	10	0.0581	60	0
0.0664	100	15	0.0664	65	0
0.0747	100	5	0.0747	80	0
0.0830	100	5	0.0830	55	0

To increase the success rate of the change isolation tests one had better perform the tests separately to the linear and the nonlinear parameters of the fuzzy model [344, 335]. This is summarized as follows: (i) perform the diagnostic test to the linear parameters. If it finds a fault then stop, otherwise (ii) perform the test to the nonlinear parameters.

The success rate of the change isolation tests is also affected by the following factors [508]: (a) the magnitude of the parameter change: a parameter change close to the nominal value results in increased success of the sensitivity test, (b) the size of the data set used for the statistical tests: if the number of data used in the statistical tests is large then the success rate is high, (c) the signal-to-noise ratio.

The changes on the parameters of the power system model that were used in the simulation experiments are summarized in Table 9.4.

9.4 Fault Diagnosis for Power Generators Using Spectral Analysis Methods

9.4.1 Outline

A neural network with basis functions that remain invariant under the Fourier transform is used next for fault diagnosis of nonlinear power systems. The proposed neural model follows the concept of wavelet networks [4, 211, 498]. By employing Gauss-Hermite activation functions which are localized both in space and frequency, the neural network allows better approximation of the multi-frequency characteristics of the monitored power system [42, 65, 212, 246, 427]. Gauss-Hermite basis functions have some interesting properties [326, 339]: (i) they remain almost unchanged by the Fourier transform and satisfy an orthogonality property, which means that the weights of the associated neural network demonstrate the energy which is distributed to the various eigenmodes of the nonlinear system's dynamics, (ii) unlike wavelet basis functions the Gauss-Hermite basis functions have a clear physical meaning since they represent the solutions of differential equations describing stochastic oscillators and each neuron can be regarded as the frequency filter of the respective eigenfrequency [338, 339].

The concept of the proposed Fault Detection and Isolation (FDI) method is as follows: The neural network with Gauss-Hermite polynomial activation functions is used for approximating the nonlinear system's dynamics out of a set of input-output data. Thus the output of the neural network provides a series expansion that takes the form of a weighted sum of Gauss-Hermite basis function. Knowing that the Gauss-Hermite basis functions are orthogonal and remain unchanged under the Fourier transform, subjected only to a change of scale, one has that the considered neural network provides the spectral analysis of the output of the monitored system. Actually, the squares of the weights of the output layer of the neural network denote the distribution of the energy of the monitored signal from the nonlinear system into the associated spectral components. Moreover, since Gauss-Hermite basis functions satisfy an orthogonality property the sum of the square of the output layer weights of these neural networks stands for a measure of the energy contained in the output of the monitored system. By observing changes in the amplitude of the aforementioned spectral components of the nonlinear system's dynamics one can have also an indication about malfunctioning of the monitored system and can detect the existence of failures. Moreover, since specific faults are associated with amplitude changes of specific spectral components of the system, fault isolation can be also performed.

As a case study the problem of fault diagnosis of the doubly-fed induction generator has been examined [143, 144, 322, 345, 357, 363, 365]. The dynamics of the rotor current has been modeled with the use of a Gauss-Hermite neural network and the associated spectral components have been obtained. Variation in the energy spectrum of the rotor's current provided again information about the existence of failures and about the association of faults with specific components of the turbine-generator system.

9.4.2 *Feed-Forward Neural Networks for Nonlinear Systems Modelling*

The proposed fault diagnosis approach for renewable energy systems, can be implemented with the use of feed-forward neural networks. The idea of function approximation with the use of feed-forward neural networks (FNN) comes from generalized Fourier series. It is known that any function $\psi(x)$ in a L^2 space can be expanded, using generalized Fourier, series in a given orthonormal basis, i.e.

$$\psi(x) = \sum_{k=1}^{\infty} c_k \psi_k(x), \ a \le x \le b \tag{9.60}$$

Truncation of the series yields in the sum

$$S_M(x) = \sum_{k=1}^{M} a_k \psi_k(x) \tag{9.61}$$

If the coefficients a_k are taken to be equal to the generalized Fourier coefficients, i.e. when $a_k = c_k = \int_a^b \psi(x)\psi_k(x)dx$, then Eq. (9.61) is a mean square optimal approximation of $\psi(x)$.

Unlike generalized Fourier series, in FNN the basis functions are not necessarily orthogonal. The hidden units in a FNN usually have the same activation functions and are often selected as sigmoidal functions or Gaussians. A typical feed-forward neural network consists of n inputs x_i, $i = 1, 2, \ldots, n$, a hidden layer of m neurons with activation function $h : R \rightarrow R$ and a single output unit (see Fig. 9.25a). The FNN's output is given by

$$\psi(x) = \sum_{j=1}^{n} c_j h(\sum_{i=1}^{n} w_{ji} x_i + b_j) \tag{9.62}$$

The root mean square error in the approximation of function $\psi(x)$ by the FNN is given by

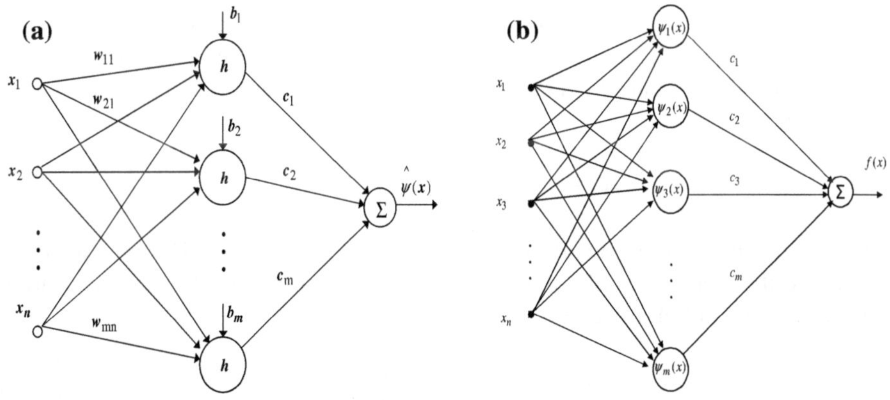

Fig. 9.25 **a** Feed-forward neural network. **b** Neural network with Gauss-Hermite basis functions

$$E_{RMS} = \sqrt{\frac{1}{N}\sum_{k=1}^{N}(\psi(x^k) - \hat{\psi}(x^k))^2} \tag{9.63}$$

where $x^k = [x_1^k, x_2^k, \ldots, x_n^k]$ is the k-th input vector of the neural network. The activation function is usually a sigmoidal function $h(x) = \frac{1}{1+e^{-x}}$ while in the case of radial basis functions networks it is a Gaussian [175]. Several learning algorithms for neural networks have been studied. The objective of all these algorithms is to find numerical values for the network's weights so as to minimize the mean square error E_{RMS} of Eq. (9.63). The algorithms are usually based on first and second order gradient techniques. These algorithms belong to: (i) batch-mode learning, where to perform parameters update the outputs of a large training set are accumulated and the mean square error is calculated (back-propagation algorithm, Gauss-Newton method, Levenberg-Marquardt method, etc.), gradient learning in batch mode (ii) pattern-mode learning, in which training examples are run in cycles and the parameters update is carried out each time a new datum appears (Extended Kalman Filter algorithm) 344.

Unlike conventional FNN with sigmoidal or Gaussian basis functions, Hermite polynomial-based FNN remain closer to Fourier series expansions by employing activation functions which satisfy the property of orthogonality [518]. Other basis functions with the property of orthogonality are Hermite, Legendre, Chebyshev, and Volterra polynomials [326, 338, 339, 487].

9.4.3 Neural Networks Using Hermite Activation Functions

9.4.3.1 The Gauss-Hermite Series Expansion

Next, as orthogonal basis functions of the feed-forward neural network Gauss-Hermite activation functions are considered. These are the spatial components $X_k(x)$ of the solution of Schrödinger's differential equation and describe a stochastic oscillation:

$$X_k(x) = H_k(x)e^{\frac{-x^2}{2}}, \ k = 0, 1, 2, \ldots \tag{9.64}$$

where $H_k(x)$ are the Hermite orthogonal functions (Fig. 9.26). The Hermite functions $H_k(x)$ are the eigenstates of the quantum harmonic oscillator. The general relation for the Hermite polynomials is

$$H_k(x) = (-1)^k e^{x^2} \frac{d^{(k)}}{dx^{(k)}} e^{-x^2} \tag{9.65}$$

According to Eq. (9.65) the first five Hermite polynomials are:

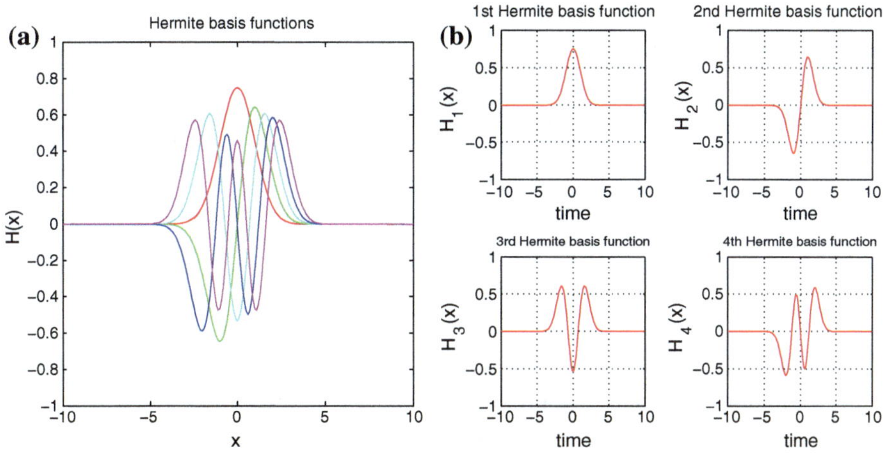

Fig. 9.26 **a** First five one-dimensional Hermite basis functions. **b** Analytical representation of the 1D Hermite basis function

$H_0(x) = 1$, $H_1(x) = 2x$, $H_2(x) = 4x^2 - 2$, $H_3(x) = 8x^3 - 12x$, $H_4(x) = 16x^4 - 48x^2 + 12$.

It is known that Hermite polynomials are orthogonal, i.e. it holds

$$\int_{-\infty}^{+\infty} e^{-x^2} H_m(x) H_k(x) dx = \begin{cases} 2^k k! \sqrt{\pi} \; if \; m = k \\ 0 \; if \; m \neq k \end{cases} \tag{9.66}$$

Using now, Eq. (9.66), the following basis functions can be defined 326:

$$\psi_k(x) = [2^k \pi^{\frac{1}{2}} k!]^{-\frac{1}{2}} H_k(x) e^{-\frac{x^2}{2}} \tag{9.67}$$

where $H_k(x)$ is the associated Hermite polynomial. From Eq. (9.66), the orthogonality of basis functions of Eq. (9.67) can be concluded, which means

$$\int_{-\infty}^{+\infty} \psi_m(x) \psi_k(x) dx = \begin{cases} 1 \; if \; m = k \\ 0 \; if \; m \neq k \end{cases} \tag{9.68}$$

Moreover, to succeed multi-resolution analysis Gauss-Hermite basis functions of Eq. (9.67) are multiplied with the scale coefficient α. Thus the following basis functions are derived [326]

$$\beta_k(x, \alpha) = \alpha^{-\frac{1}{2}} \psi_k(\alpha^{-1} x) \tag{9.69}$$

which also satisfy orthogonality condition

$$\int_{-\infty}^{+\infty} \beta_m(x, \alpha) \beta_k(x, \alpha) dx = \begin{cases} 1 \; if \; m = k \\ 0 \; if \; m \neq k \end{cases} \tag{9.70}$$

Any function $f(x),\ x \in R$ can be written as a weighted sum of the above orthogonal basis functions, i.e.

$$f(x) = \sum_{k=0}^{\infty} c_k \beta_k(x, \alpha) \tag{9.71}$$

where coefficients c_k are calculated using the orthogonality condition

$$c_k = \int_{-\infty}^{+\infty} f(x)\beta_k(x, \alpha)dx \tag{9.72}$$

Assuming now that instead of infinite terms in the expansion of Eq. (9.71), M terms are maintained, then an approximation of $f(x)$ is succeeded. The expansion of $f(x)$ using Eq. (9.71) is a Gauss-Hermite series. Equation (9.71) is a form of Fourier expansion for $f(x)$. Equation (9.71) can be considered as the Fourier transform of $f(x)$ subject only to a scale change. Indeed, the Fourier transform of $f(x)$ is given by

$$F(s) = \frac{1}{2\pi}\int_{-\infty}^{+\infty} f(x)e^{-jsx}dx \Rightarrow f(x) = \frac{1}{2\pi}\int_{-\infty}^{+\infty} F(s)e^{jsx}ds \tag{9.73}$$

The Fourier transform of the basis function $\psi_k(x)$ of Eq. (9.67) satisfies [326]

$$\Psi_k(s) = j^k \psi_k(s) \tag{9.74}$$

while for the basis functions $\beta_k(x, \alpha)$ using scale coefficient α it holds that

$$B_k(s, \alpha) = j^k \beta_k(s, \alpha^{-1}) \tag{9.75}$$

Therefore, it holds

$$f(x) = \sum_{k=0}^{\infty} c_k \beta_k(x, \alpha) \overset{F}{\rightarrow} F(s) = \sum_{k=0}^{\infty} c_k j^n \beta_k(s, \alpha^{-1}) \tag{9.76}$$

which means that the Fourier transform of Eq. (9.71) is the same as the initial function, subject only to a change of scale. The structure of a a feed-forward neural network with Hermite basis functions is depicted in Fig. 9.25b.

9.4.3.2 Neural Networks Using 2D Hermite Activation Functions

Two-dimensional Hermite polynomial-based neural networks can be constructed by taking products of the one dimensional basis functions $B_k(x,\alpha)$ [326]. Thus, setting $x=[x_1,x_2]^T$ one can define the following basis functions [326]

$$B_k(x,\alpha)=\frac{1}{\alpha}B_{k_1}(x_1,\alpha)B_{k_2}(x_2,\alpha) \tag{9.77}$$

These two dimensional basis functions are again orthonormal, i.e. it holds

$$\int d^2x B_n(x,\alpha)B_m(x,\alpha)=\delta_{n_1m_1}\delta_{n_2m_2} \tag{9.78}$$

The basis functions $B_k(x)$ are the eigenstates of the two dimensional harmonic oscillator and form a complete basis for integrable functions of two variables. A two dimensional function $f(x)$ can thus be written in the series expansion:

$$f(x)=\sum_{k_1,k_2}^{\infty}c_k B_k(x,\alpha) \tag{9.79}$$

The choice of an appropriate scale coefficient α and maximum order k_{max} is of practical interest. The coefficients c_k are given by

$$c_k=\int dx^2 f(x)B_k(x,\alpha) \tag{9.80}$$

Indicative basis functions $B_2(x,\alpha)$, $B_6(x,\alpha)$, $B_9(x,\alpha)$, $B_{11}(x,\alpha)$ and $B_{13}(x,\alpha)$, $B_{15}(x,\alpha)$ of a 2D feed-forward neural network with Hermite basis functions are depicted in Figs. 9.27, 9.28, and 9.29. Following, the same method N-dimensional Hermite polynomial-based neural networks ($N>2$) can be constructed. The associated high-dimensional Gauss-Hermite activation functions preserve the properties of orthogonality and invariance to Fourier transform.

9.4.4 *Signals Power Spectrum and the Fourier Transform*

9.4.4.1 Parseval's Theorem

To find the spectral density of a signal $\psi(t)$ with the use of its Fourier transform $\Psi(j\omega)$, the following definition is used:

$$\begin{matrix} E_{\psi}=\int_{-\infty}^{+\infty}(\psi(t))^2dt=\frac{1}{2\pi}\int_{-\infty}^{+\infty}\psi(t)(\int_{-\infty}^{+\infty}\Psi(j\omega)e^{j\omega}d\omega)dt \ \text{i.e.} \\ E=\frac{1}{2\pi}\int_{-\infty}^{+\infty}\Psi(j\omega)\Psi(-j\omega)d\omega \end{matrix} \tag{9.81}$$

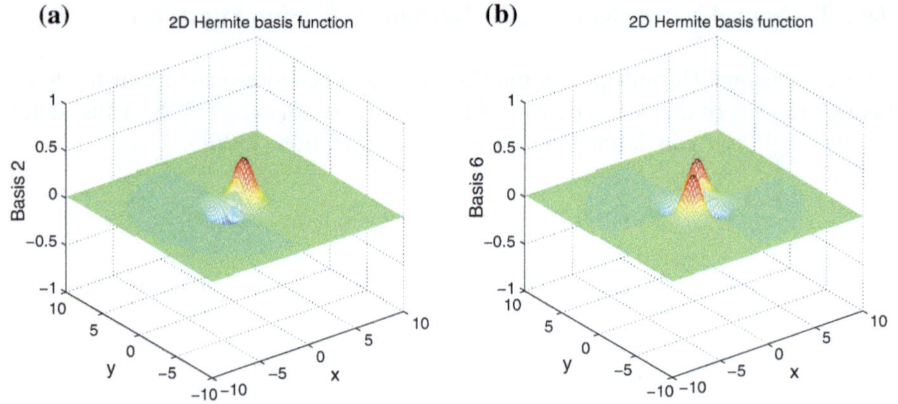

Fig. 9.27 2D Hermite polynomial activation functions: **a** basis function $B_2(x, \alpha)$, **b** basis function $B_6(x, \alpha)$

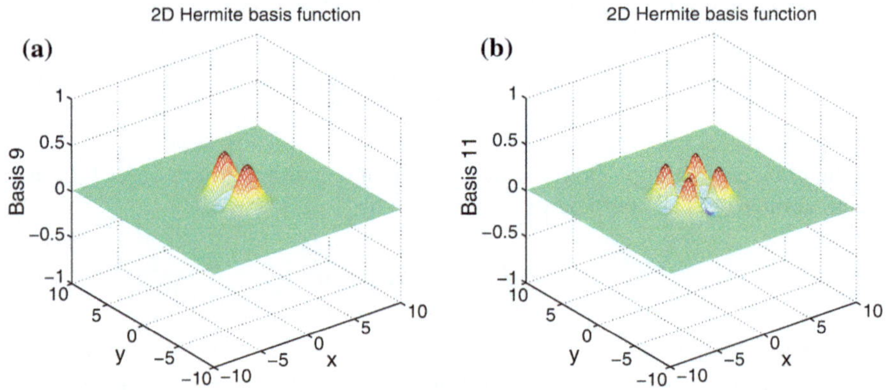

Fig. 9.28 2D Hermite polynomial activation functions: **a** basis function $B_9(x, \alpha)$, **b** basis function $B_{11}(x, \alpha)$

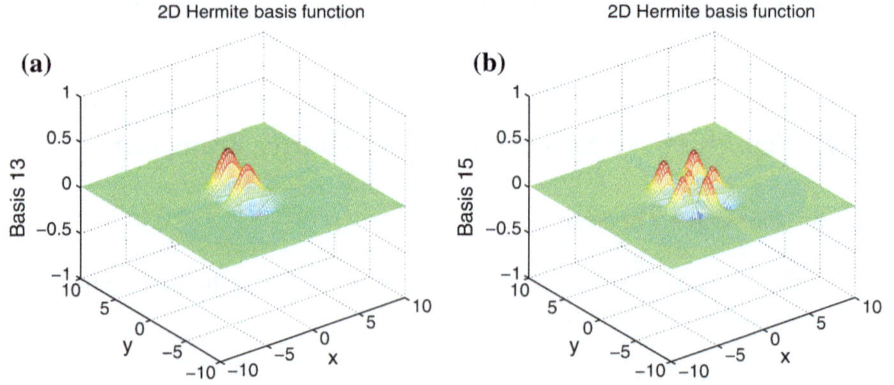

Fig. 9.29 2D Hermite polynomial activation functions: **a** basis function $B_{13}(x, \alpha)$, **b** basis function $B_{15}(x, \alpha)$

Taking that $\psi(t)$ is a real signal it holds that $\Psi(-j\omega) = \Psi^*(j\omega)$ which is the signal's complex conjugate. Using this in Eq. (9.81) one obtains

$$\begin{aligned} E_\psi &= \tfrac{1}{2\pi}\int_{-\infty}^{+\infty}\Psi(j\omega)\Psi^*(j\omega)d\omega \;\text{ or} \\ E_\psi &= \tfrac{1}{2\pi}\int_{-\infty}^{+\infty}|\Psi(j\omega)|^2 d\omega \end{aligned} \tag{9.82}$$

This means that the energy of the signal is equal to $\frac{1}{2\pi}$ times the integral over frequency of the square of the magnitude of the signal's Fourier transform. This is *Parseval's theorem*. The integrated term $|\Psi(j\omega)|^2$ is the energy density per unit of frequency and has units of magnitude squared per Hertz.

9.4.4.2 Power Spectrum of the Signal Using the Gauss-Hermite Expansion

As shown in Eqs. (9.66) and (9.78) the Gauss-Hermite basis functions satisfy the orthogonality property, i.e. for these functions it holds

$$\int_{-\infty}^{+\infty}\psi_m(x)\psi_k(x)dx = \begin{cases} 1 \; if \; m=k \\ 0 \; if \; m \neq k \end{cases}$$

Therefore, using the definition of the signal's energy one has

$$E = \int_{-\infty}^{+\infty}(\psi(t))^2 dt = \int_{-\infty}^{+\infty}[\textstyle\sum_{k=1}^{N} c_k\psi_k(t)]^2 \tag{9.83}$$

and exploiting the orthogonality property one obtains

$$E = \textstyle\sum_{k=1}^{N} c_k^2 \tag{9.84}$$

Therefore the square of the coefficients c_k provides an indication of the distribution of the signal's energy to the associated basis functions. One could arrive at the same results using the Fourier transformed description of the signal and Parseval's theorem. It has been shown that the Gauss-Hermite basis functions remain invariant under the Fourier transform subject only to a change of scale. Denoting by $\Psi(j\omega)$ the Fourier transformed signal of $\psi(t)$ and by $\Psi_k(j\omega)$ the Fourier transform of the k-th Gauss-Hermite basis function one obtains

$$\Psi(j\omega) = \textstyle\sum_{k=1}^{N} c_k\Psi_k(j\omega) \tag{9.85}$$

and the energy of the signal is computed as

$$E_\psi = \tfrac{1}{2\pi}\int_{-\infty}^{+\infty}|\Psi(j\omega)|^2 d\omega \tag{9.86}$$

Substituting Eqs. (9.85) into (9.86) one obtains

$$E_{\psi} = \frac{1}{2\pi}\int_{-\infty}^{+\infty}|\sum_{k=1}^{N}c_k\Psi_k(j\omega)|^2 d\omega \tag{9.87}$$

and using the invariance of the Gauss-Hermite basis functions under Fourier transform one gets

$$E_{\psi} = \frac{1}{2\pi}\int_{-\infty}^{+\infty}|\sum_{k=1}^{N}c_k\alpha^{-\frac{1}{2}}\psi_k(\alpha^{-1}j\omega)|^2 d\omega \tag{9.88}$$

while performing the change of variable $\omega_1 = \alpha^{-1}\omega$ it holds that

$$E_{\psi} = \frac{1}{2\pi}\int_{-\infty}^{+\infty}|\sum_{k=1}^{N}c_k\alpha^{\frac{1}{2}}\psi_k(j\omega_1)|^2 d\omega_1 \tag{9.89}$$

Next, by exploiting the orthogonality property of the Gauss-Hermite basis functions one gets that the signal's energy is proportional to the sum of the squares of the coefficients c_k which are associated with the Gauss-Hermite basis functions, i.e. a relation of the form

$$E_{\psi} = \sum_{k=1}^{N}c_k^2 \tag{9.90}$$

9.4.5 Gauss-Hermite Modeling of Electric Power Generators

9.4.5.1 Model of the Doubly-Fed Induction Generator

As explained in Chap. 2, the doubly-fed induction generator (DFIG) is not only the most widely used technology in wind turbines due to its good performance, but it is also used in many other fields such as hydro-power generation, pumped storage plants and flywheel energy storage systems. The DFIG model is derived from the voltage equations of the stator and rotor. It is assumed that the stator and rotor windings are symmetrical and symmetrically fed. The saturation of the inductances, iron losses, skin effect, and bearing friction is neglected. The winding resistance is considered to be constant.

This type of wound-rotor machine is connected to the grid by both the rotor and stator side. The DFIG stator can be directly connected to the electric power grid while the rotor is interfaced through back-to-back converters (see Fig. 1.2). By decoupling the power system electrical frequency and the rotor mechanical frequency the converter allows a variable speed operation of the wind turbine.

In a compact form the doubly-fed induction generator can be described by the following set of equations in the $d-q$ reference frame that rotates at an arbitrary speed denoted as ω_{dq} [141]

$$\frac{d\psi_{s_q}}{dt} = -\frac{1}{\tau_s}\psi_{s_q} - \omega_{dq}\psi_{s_d} + \frac{M}{\tau_s}i_{r_q} + v_{s_q} \tag{9.91}$$

$$\frac{d\psi_{s_d}}{dt} = \omega_{dq}\psi_{s_q} - \frac{1}{\tau_s}\psi_{s_d} + \frac{M}{\tau_s}i_{r_d} + v_{s_d} \tag{9.92}$$

$$\begin{aligned} \frac{di_{r_q}}{dt} &= \frac{\beta}{\tau_s}\psi_{s_q} + \beta\omega_r\psi_{s_d} - \\ -\gamma_2 i_{r_q} &- (\omega_{dq} - \omega_r)i_{r_d} - \beta v_{s_q} + \frac{1}{\sigma L_r}v_{r_q} \end{aligned} \tag{9.93}$$

$$\begin{aligned} \frac{di_{r_d}}{dt} &= -\beta\omega_r\psi_{s_q} + \frac{\beta}{\tau_s}\psi_{s_d} + (\omega_{dq} - \omega_r)i_{r_q} - \\ &-\gamma_2 i_{r_d} - \beta v_{s_d} + \frac{1}{\sigma L_r}v_{r_d} \end{aligned} \tag{9.94}$$

where $\lambda_{s_q}, \lambda_{s_d}, i_{r_q}, i_{r_d}$ are the stator flux and the rotor currents, $v_{s_q}, v_{s_d}, v_{r_q}, v_{r_d}$ are the stator and rotor voltages, L_s and L_r are the stator and rotor inductances, ω_r is the rotor's angular velocity, M is the magnetizing inductance. Moreover, denoting as R_s and R_r the stator and rotor resistances the following parameters are defined

$$\begin{array}{ccc} \sigma = 1 - \frac{M^2}{L_r L_s} & \beta = \frac{1-\sigma}{M\sigma} & \tau_s = \frac{L_s}{R_s} \\ \tau_r = \frac{L_r}{R_r} & \gamma_2 = (\frac{1-\sigma}{\sigma\tau_s}) & \end{array} \tag{9.95}$$

The dynamic model of the doubly-fed induction generator has been also written in state space equations form by defining the following state variables: $x_1 = \theta$, $x_2 = \omega_r$, $x_3 = \psi_{s_d}$, $x_4 = \psi_{s_q}$, $x_5 = i_{r_d}$ and $x_6 = i_{r_q}$. It holds that [56, 357]

$$\dot{x} = f(x) + g_a(x)v_{r_d} + g_b(x)v_{r_q} \tag{9.96}$$

where $x = [x_1, x_2, x_3, x_4, x_5, x_6]^T$ and

$$f(x) = \begin{pmatrix} x_2 \\ -\frac{K_m}{J}x_2 - \frac{T_m}{J} + \frac{n}{J}(i_{s_q}x_3 - i_{s_d}x_4) \\ -\frac{1}{\tau_s}x_3 + \omega_{dq}x_4 + \frac{M}{\tau_s}x_5 + v_{s_d} \\ -\omega_{dq}x_3 - \frac{1}{\tau_s}x_4 + \frac{M}{\tau_s}x_6 + v_{s_q} \\ -\beta x_2 x_4 + \frac{\beta}{\tau_s}x_3 + (\omega_{dq} - x_2)x_6 - \gamma_2 x_5 - \beta v_{s_d} \\ \frac{\beta}{\tau_s}x_4 + \beta x_2 x_3 + (\omega_{dq} - x_2)x_5 - \gamma_2 x_6 \end{pmatrix} \tag{9.97}$$

$$g_a(x) = \left(0\;0\;0\;0\;\frac{1}{\sigma L_r}\;0\right) \tag{9.98}$$

$$g_b(x) = \left(0\;0\;0\;0\;0\;\frac{1}{\sigma L_r}\right) \tag{9.99}$$

Indicative numerical values for the parameters of the considered doubly-fed induction generator model have been given in Chap. 2, Table I.

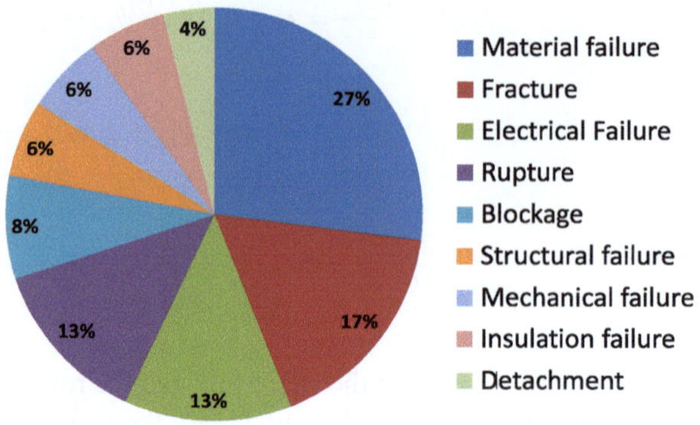

Fig. 9.30 Frequency of faults in components of the turbine-power generator system

9.4.6 Fault Diagnosis for Doubly-Fed Induction Generators

The components of the turbine-power generator system are exposed to harsh operating conditions and exhibit failures (see Fig. 9.30). The generator's dynamic model, i.e. variations of the rotor's current on the d-axis, has been identified considering the previously analyzed neural network with Gauss-Hermite basis functions. The inputs/outputs configuration of the neural model of the power generator dynamics is shown in Fig. 9.30. Real-time measurements of the rotor current in the $a-b$ reference frame are available which after the application of a rotation transformation can provide the associated rotor current measurements in the $d-q$ reference frame. To approximate the variations of the rotor current i_{r_d} described in a data set consisting of 2000 quadruplets of the form $[i_{rd}(k-1), i_{rq}(k-2), \omega(k-1)|i_{rd}(k)]$ a feed-forward neural network with 3-D Gauss-Hermite basis functions has been used, containing 64 nodes in its hidden layer. Neural models with the same output, such as $i_{rd}(k)$ and a larger number of inputs, i.e. including more past values of the rotor's current and of the rotation speed, could be also considered (Fig. 9.31).

Fault cases 1 to 2 were associated with a change in the value of the stator's resistance R_s under two different set-points for the rotor's angular speed. Fault cases 3 to 4 were associated with a change in the value of the rotor's inductance L_r under two different set-points for the rotor's angular speed. As shown in Figs. 9.32 and 9.33, thanks to the multi-frequency characteristics of the Gauss-Hermite basis functions, such a neural model can capture with increased accuracy spikes and abrupt changes in the rotor's current [344, 339, 498]. The RMSE of training the Gauss-Hermite neural model was of the order of 4×10^{-3}.

The update of the output layer weights of the neural network is given by a gradient equation of the LMS-type (Least Mean Squares) given by

$$w^i(k+1) = w^i(k) - \eta e(k)\phi^T(k) \tag{9.100}$$

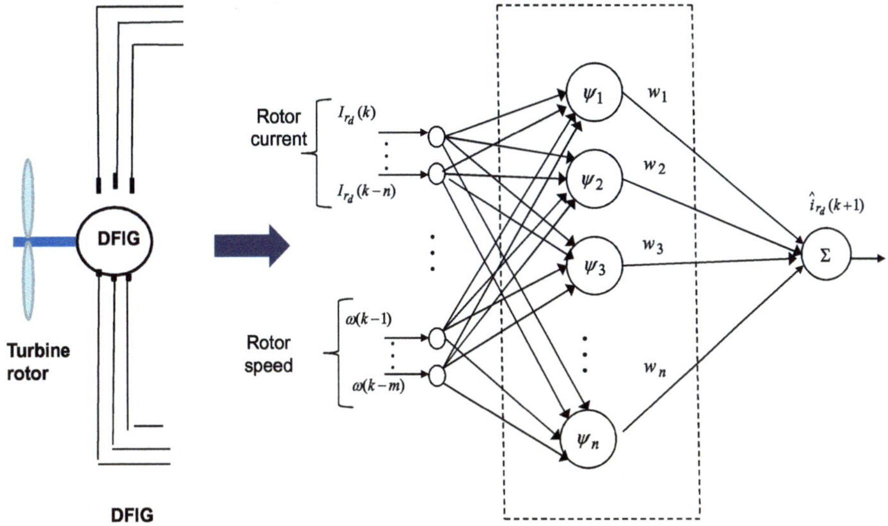

Fig. 9.31 Inputs/outputs configuration of the neural model of the power generator dynamics

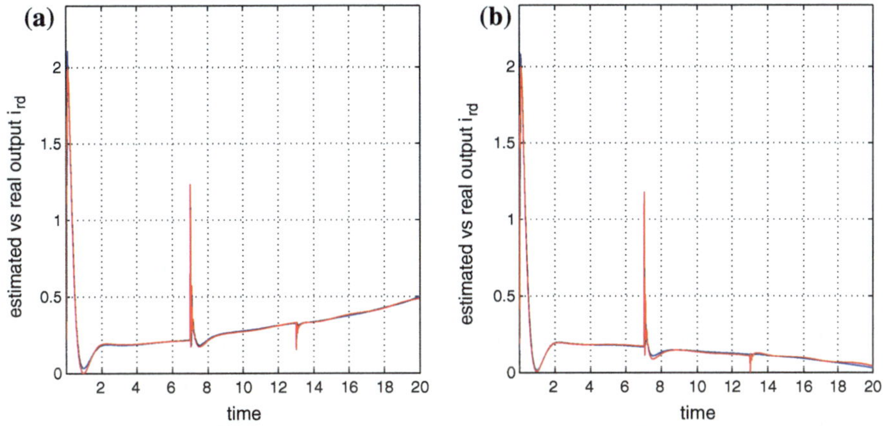

Fig. 9.32 Approximation of the rotor's current i_{r_d} of the electric power generator (*red line*) by a neural network with Hermite polynomial basis functions (*blue-line*), **a** d-axis rotor's current under fault in stator's resistance-case 1 **b** d-axis rotor's current under fault in stator's resistance-case 2

where $e(k) = y(k) - y_d(k)$ is the output estimation error at time instant k and $\phi^T(k)$ is the regressor vector having as elements the values $\phi(x(k))$ of the Gauss-Hermite basis functions for input vector $x(k)$.

To approximate the rotor current variations described in a data set consisting of 2000 quadruplets of the form $i_{rd}(k-1), i_{rd}(k-2), \omega(k-1)|i_{rd}(k)$ a feed-forward neural network with 3-D Gauss-Hermite basis functions has been used, containing 64 nodes in its hidden layer.

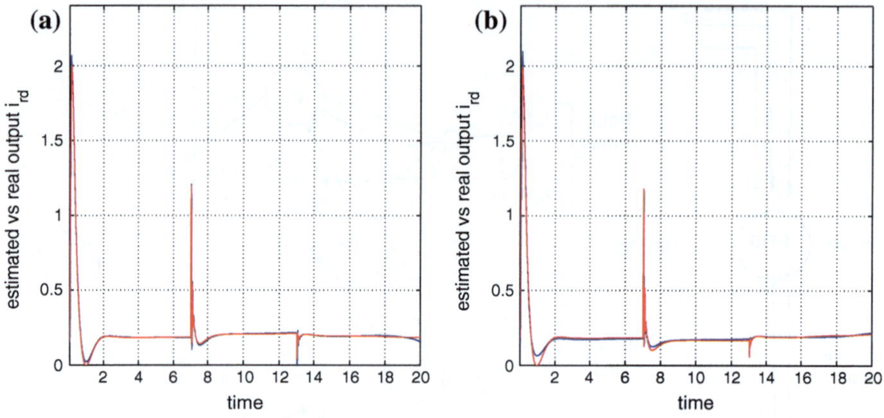

Fig. 9.33 Approximation of the rotor's current i_{r_d} of the electric power generator (*red line*) by a neural network with Hermite polynomial basis functions (*blue-line*), **a** d-axis rotor's current under fault in rotor's inductance-case 1, **b** d-axis rotor's current under fault in rotor's inductance-case 2

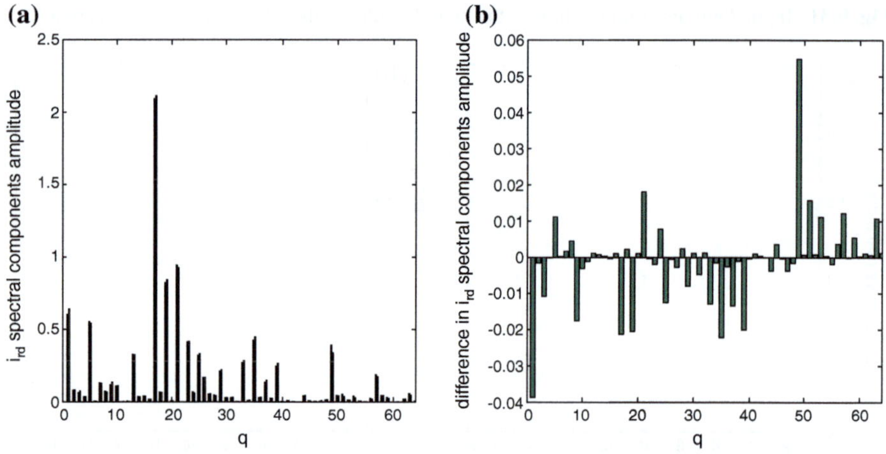

Fig. 9.34 d-axis rotor current i_{r_d} under fault in stator's resistance-case 1: **a** Amplitude of the spectral components of the rotor's current i_{r_d} measured from the electric power generator in the fault free case (*red bar line*) and when a fault had taken place (*yellow bar line*). **b** Differences in the amplitudes of the spectral components between the fault-free and the faulty case (*green bar line*)

The spectral components of the i_{r_d} signal for both the fault-free and the under-fault operation of the power generator have been shown in Figs. 9.34, 9.35, 9.36 and 9.37. It can be noticed that after a fault has occurred, the amplitude of the aforementioned spectral components changes and this can be a clear indication about failure of the monitored transformer.

Again, the proposed spectral decomposition of the monitored signal, with series expansion in Gauss-Hermite basis functions can be used for fault detection tasks.

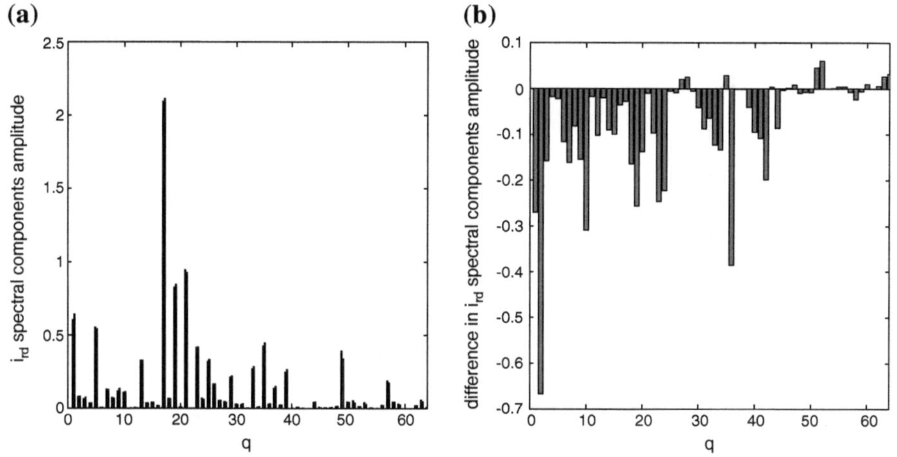

Fig. 9.35 d-axis rotor current i_{r_d} under fault in stator's resistance-case 2: **a** Amplitude of the spectral components of the rotor's current i_{r_d} measured from the electric power generator in the fault free case (*red bar line*) and when a fault had taken place (*yellow bar line*). **b** Differences in the amplitudes of the spectral components between the fault-free and the faulty case (*green bar line*)

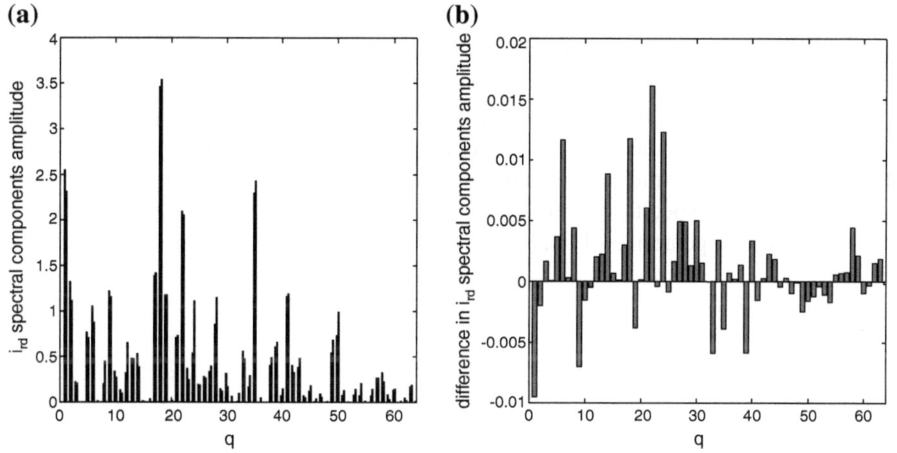

Fig. 9.36 d-axis rotor current i_{r_d} under fault in rotor's inductance-case 1: **a** Amplitude of the spectral components of the rotor's current i_{r_d} measured from the electric power generator in the fault free case (*red bar line*) and when a fault had taken place (*yellow bar line*). **b** Differences in the amplitudes of the spectral components between the fault-free and the faulty case (*green bar line*)

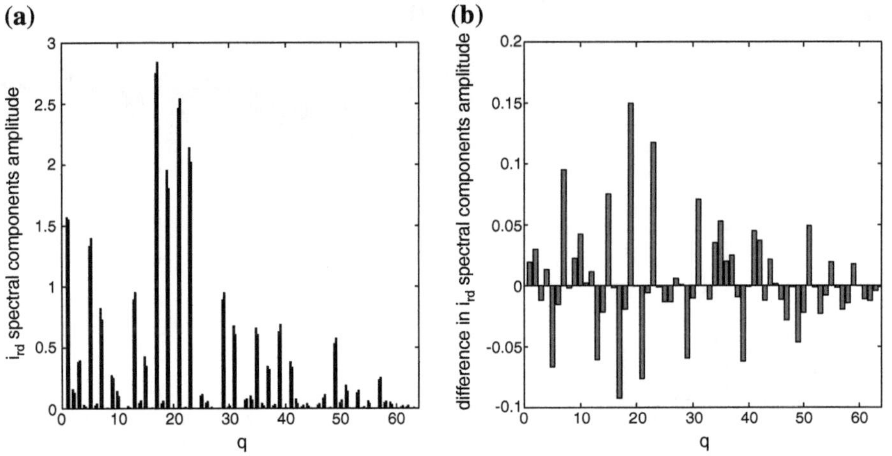

Fig. 9.37 d-axis rotor current i_{r_d} under fault in rotor's inductance-case 2: **a** Amplitude of the spectral components of the rotor's current i_{r_d} measured from the electric power generator in the fault free case (*red bar line*) and when a fault had taken place (*yellow bar line*). **b** Differences in the amplitudes of the spectral components between the fault-free and the faulty case (*green bar line*)

As it can be seen in Figs. 9.34, 9.35, 9.36 and 9.37, in case of failure, the spectral components of the monitored signal differ from the ones which are obtained when the system is free of fault. Moreover, the fact that certain spectral components exhibit greater sensitivity to the fault and change value in a more abrupt manner is a feature which can be exploited for fault isolation. Specific failures can be associated with variations of specific spectral components. Therefore, they can provide indication about the appearance of specific types of failures and specific malfunctioning components.

Chapter 10
Condition Monitoring of the Electric Power Transmission and Distribution System

10.1 Outline

This chapter is concerned with condition monitoring and fault diagnosis for components of the electric power transmission and distribution system. First, neural modelling and the local statistical approach to fault diagnosis is used for incipient fault detection and isolation in power transformers. Next, the chapter proposes distributed nonlinear filtering for condition monitoring of the power distribution system and for the detection of changes in the quality of the provided electric power (voltage sags, change of harmonics, etc).

Initially, the chapter analyzes on-line monitoring of electric power transformers. This can provide a clear indication of their status and ageing behavior. This chapter proposes neural modeling and the local statistical approach to fault diagnosis for the detection of incipient faults in power transformers. The method can detect transformer failures at their early stages and consequently can deter critical conditions for the power grid. A neural-fuzzy network is used to model the thermal condition of the power transformer in fault-free operation (the thermal condition is associated to a temperature variable known as hot-spot temperature). The output of the neural-fuzzy network is compared to measurements from the power transformer and the obtained residuals undergo statistical processing according to a fault detection and isolation algorithm. If a fault threshold (that is optimally defined according to detection theory) is exceeded, then deviation from normal operation can be detected at its early stages and an alarm can be launched. In several cases fault isolation can be also performed, that is the sources of fault in the power transformer model can be also identified. The performance of the proposed methodology is tested through simulation experiments.

Next, the chapter analyzes distributed state estimation methods for condition monitoring of electric power transmission and distribution systems. When a fault occurs in such large-scale systems, it is usually difficult to detect it and to determine its exact position. Moreover, due to the cost of installation and maintenance of measurement devices and due to the excessive size of the electric power grid, the complete monitoring of the associated infrastructure is impractical. Therefore, to monitor the condition

G. Rigatos, *Intelligent Renewable Energy Systems*,
Green Energy and Technology, DOI 10.1007/978-3-319-39156-4_10

of the power grid, some form of estimation is required. As suitable approaches for distributed state estimation the chapter proposes the Extended Information Filter (EIF) and the Unscented Information Filter (UIF). The Extended Information Filter is actually an implementation of distributed Extended Kalman Filtering while the Unscented Information Filter is an implementation of distributed Unscented Kalman Filtering. With the use of the aforementioned filtering algorithms on processing units located at different parts of a distributed renewable energy system, one can produce local estimates of the system's state vector which in turn can be fused into an aggregate state estimation. The produced global state estimate enables continuous monitoring of the condition of the electric power system and early fault diagnosis if used by a suitable fault detection and isolation algorithm.

10.2 Fault Diagnosis in Power Transformers Using Statistical Signal Processing

10.2.1 Outline

Fault diagnosis methods are developed for the components of the power grid, enabling the uninterrupted transmission and distribution of the power that is generated by renewable sources. Power transformers are among the most expensive equipment of the electric power transmission and distribution system and their condition monitoring is important for the uninterrupted and reliable functioning of the power grid. Transformer life management has been a topic of intensive research during the last years because of the need for operating the electric power grid under more harsh conditions and because of the increased demand for electric energy. According to an IEEE survey, oil immersed transformer failure rate per year is 0.00625. Therefore, in a set of 100 transformers, ten will have problem in the next 16 years. Typical failure rates for power transformers are in the range of 1–2 % per year for the large power transformers (operating voltages up to 300 kV). Load growth has contributed to an increase of the transformer's Hot Spot Temperature (HST), i.e. of a parameter that is directly associated to the ageing of the transformer and to the probability of failures of the transformer's components. The average HST a few decades ago was 50 °C, while under present operating conditions it is around 73 °C [27, 289].

Transformers operating beyond their ratings exhibit the following symptoms: (i) increase in temperature of windings, insulation and oil, (ii) increase in leakage flux density outside the core, causing additional eddy current heating in the metallic parts, (iii) moisture and gas content increases with the increase in temperature, (iv) bushings, tap-changers, and cables are exposed to higher stresses, (v) deterioration of the windings insulation appears due to higher thermal stresses [3, 196, 247, 452]. Obviously, there is a significant safety and environmental risk of operating aged transformer units close to their loading limits without surveillance and assessment. On the other hand, on-line monitoring of power transformers can provide a clear indication of their status and ageing behavior. Analysis of critical parameters collected from power transformers allows avoidance of irreversible failures and permits preventive maintenance.

During the last years research efforts have been carried out to develop thermal models of improved accuracy for power transformers [145, 428, 429, 430]. The load-current profile, the top-oil temperature profile and the weather conditions (ambient temperature, solar heating, wind speed, rain conditions, etc.) are among the parameters that influence the transformer's thermal behavior. As mentioned, significant indications about the thermal condition of a power transformer and the associated failure risks can be obtained through monitoring the transformer's Hot Spot Temperature (HST). A deviation of HST from the anticipated temperature profile is probably an indication of ageing of the transformer or in some cases of pre-failure situations. Analytical, as well numerical (neural/fuzzy) models for HST have been developed [34, 189, 322]. These models associate the hot spot temperature to inputs such as: the ambient temperature, the top oil temperature rise over the ambient temperature, and the maximum winding hot-spot rise over the top-oil temperature.

The development of a model of the power transformer's thermal behavior in the fault-free condition and the comparison of the outputs of such a model with online measurements of the real transformer output enables to implement fault detection and isolation (FDI). A statistical FDI method that can be used to find incipient failures in the transformer's components is the so-called Local Statistical Approach to change detection [44, 37, 174, 497, 508]. The proposed statistical fault diagnosis method can conclude the existence of a fault through the processing of the residuals sequence, where the residuals are defined as the differences between the measured and the estimated HST values at every sampling instant. The proposed FDI method transforms the complex detection problem into the problem of monitoring the mean of a Gaussian vector. The local statistical approach consists of two stages: (i) the global χ^2 test which indicates the existence of a change in some parameters of the transformer's model, (ii) the diagnostics tests (sensitivity or min-max) which isolate the parameter affected by the change [44, 37, 174, 497, 508]. The local statistical approach is suitable for detecting incipient faults in the power transformer, thus enabling preventing maintenance.

The concept of the proposed FDI technique is as follows: the thermal profile of the fault-free power transformer system is learned by a neural-fuzzy network. At each time instant the neural network's output is compared to the real Hot-Spot Temperature of the power transformer. The difference between the real condition of the power transformer and the output of the neural network forms a residual. The statistical processing of a sufficiently large number of residuals through the aforementioned FDI method provides an index-variable that is compared against a fault threshold and which can give early indication about deviation of the transformer from the normal operating conditions. Therefore alarm launching can be activated at the early stages of power transformer failure, and repair measures can be taken. Under certain conditions (detectability of changes) the proposed FDI method enables also fault isolation, i.e. it makes possible to identify the source of fault within the power transformer model [344].

10.2.1.1 Transformers in the Electric Power Grid

As mentioned, electric power transformers are the most expensive single asset in a transmission and distribution network, and the minimization of their risk of failures is important for improving the operation of the power grid. Fault detection and isolation (FDI) for power transformers aims at continuously assessing the transformer's condition through the monitoring of associated critical parameters and at determining if the transformer is on the verge of a failure (this can be due to an internal fault or due to aging). To implement FDI it is necessary to develop a model of the transformer's functioning that associates its internal state to environmental conditions thus (i) prohibiting the erroneous interpretation of the monitored critical parameters and (ii) avoiding the launch of false alarms (for example if the transformer is operating in a heat wave, its oil temperature could be expected to be unusually high, but the transformer's FDI system should ascribe the temperature rise to the environmental conditions rather than a transformer problem) [3, 27, 71, 72, 247, 289, 452].

A final objective is to implement condition monitoring for power transformers within a smart-grid architecture which means that (i) the functioning of the power transformer should not depend exclusively on local observations and on the outcome of the local FDI system but should also be determined by interpreting the information coming from other parts of the power grid (e.g. the function mode of the transformer should be modified if information about a remote short-circuit or line tripping is sent to the distribution transformers substation). Considering that the power transformer's FDI system is a local software agent, its functioning should be viewed in the framework of a multi-agent system which means that there should be exchange of information and coordination with other local FDI modules of the grid (ii) in turn, the local FDI system of the power transformers should be in position to initiate a procedure for modifying the status of the power grid when a failure is detected (e.g. if it is known that a transformer is experiencing problems, power flow management can favor the load reduction on the transformer to a certain percentage of its rating until maintenance is possible. Additionally, if the FDI system provides indications that the transformer might fail, an alternate network configuration can be implemented to assure the uninterrupted operation of the grid until the faulty transformer is restored) [300, 345].

10.2.2 Reasons for Failures in Electric Power Transformers

Common failures in power transformers are (see Fig. 10.1):

(i) *Insulation breakdown in windings*. As the transformer ages the windings insulation is weakened to the point that it can no longer sustain the mechanical stresses due to a fault (e.g. in case of a short circuit). Turn-to-turn insulation suffers a dielectric failure. Winding failures cause in turn degradation of the overall insulation system, including the transformer's oil. Ageing of transformer oil is characterized by partial discharge (gas evolution starts from the oil) and thermal degradation (the rise of

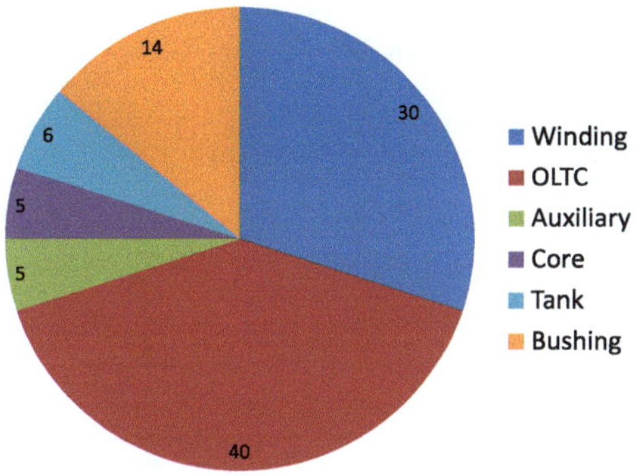

Fig. 10.1 Frequency of faults in components of electric power transformers

the transformer's temperature and the associated thermal stresses accelerate the oil's decomposition).

(ii) *The on-load tap changers (OLTC) failures*. On-load tap changers are used to change the tapping connection of the transformer's windings while the transformer is energized. The tap changers suffer from ageing. The insulating oil inside the tap changer becomes dirty due to switching arcs, which leads to weakened insulation properties. Another effect of the switching arcs is the wear of the arcing contacts. An additional ageing mechanism is the so-called long-term effect on the changeover selector. This effect starts with the formation of a thin layer of oil. The increased contact resistance due to the oil film layer can cause creation of hard and porous carbon material at places where the load current flows.

(iii) *Bushings failures*. An electrical bushing is an insulating structure including a through conductor or providing a central passage for such a conductor aiming at transmitting electric power in or out of the transformer. Oil-paper insulation is widely used in power transformer bushings. However, prolonged exposure to extreme electrical, thermal, mechanical and environmental stresses can deteriorate the insulation's condition and can break the cellulose bonds of the paper. This can lead to the formation of gas byproducts and bubbles and in turn can result in partial discharge and conducting tracts ending at shorting out one or more layers of the bushings.

10.2.3 Condition Monitoring Methods for Power Transformers

The cost savings from performing preventive maintenance for power transformers (e.g. due to deterring critical conditions in the power grid and subsequent cascading events) can be significantly more important than the monitoring cost itself. Condition monitoring for power transformers is helpful in many aspects such as planning of maintenance schedules, obtaining knowledge of the health of equipment, estimating the remaining life of the equipment, finding areas of further improvement, refining

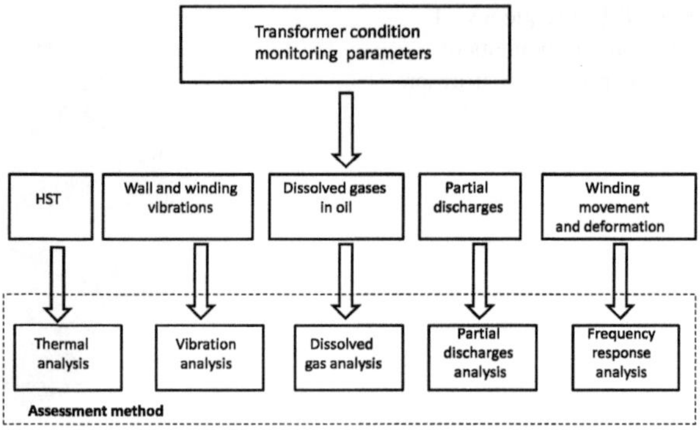

Fig. 10.2 Condition monitoring methods for power transformers and the associated processed parameters

product specifications, etc. Established methods for preventive maintenance of power transformers are (see Fig. 10.2):

(a) *the dissolved gas analysis method.* Transformers generate different gases at normal operating temperatures. However, the concentration of these gases increases in the presence of an abnormality (fault), such as thermal, partial discharge and arcing faults, usually associated with insulation degradation. Thermal faults such as long-term overloads and high values of the Hot-Spot Temperature produce also gases. The dissolved gas analysis method performs off-line estimation of the concentration of specific gases in the transformer's oil and compares the associated values to empirically defined alarm thresholds.

(b) *partial discharge measurement.* Partial discharges occur when the electric field strength exceeds the dielectric breakdown strength of a certain localized area, in which an electrical discharge partially bridges the insulation between the conductors. The dielectric properties of the insulator may be severely affected if subjected to continuous partial discharge activity over a long period of time. If the partial discharge remains untreated, it may lead to complete failure of the transformer. Partial discharge can be detected and measured using piezo-electric sensors, optical-fiber sensors and ultra high frequency sensors. On-site partial discharge measurement is often affected by strong coupled electromagnetic interference that increases the difficulty of extracting partial discharge signals without noise.

(c) *frequency response analysis.* The method aims at measuring the impedance of offline transformer's windings over a wide range of frequencies and at comparing these results with a reference set taken either at installation or at a later point of time. Winding damage detection can be accomplished by comparing the frequency response of a healthy winding (or the calculated response using an equivalent circuit) with the frequency response of the damaged winding. Changes in frequency response can be used to estimate the extent of damage in the windings.

(d) *acoustic detection and vibration analysis of partial discharge.* The acoustic wave induced by partial discharge can be measured and used for monitoring, diagnosing and locating potential failures in the transformers. These generated vibrations propagate through the transformer oil until they reach the transformer walls where they can be collected by vibration sensors. The health condition of the core and of the windings of the transformer can be assessed through the analysis of the vibrations signature, i.e. of the pattern of the signal generated by the vibrations and its evolution in time.

(e) *thermal analysis*. The method aims at monitoring of the evolution in time of the transformer's Hot-Spot Temperature which can be an indication of the ageing and degradation of the windings or of operating the transformer under overload conditions. Most of the faults cause change in the thermal behavior of the transformer. Such abnormal conditions can be detected by analysing the Hot-Spot Temperature (HST). The most common abnormal condition of the transformer that can be detected with the use of thermal analysis is the overload. Transformer life is severely affected if the Hot Spot Temperature remains for long time intervals more than 110 °C. In this section a method for preventive maintenance of power transformers based on Hot-Spot Temperature monitoring (thermal analysis) will be presented.

10.2.4 Fault Management Practices for Power Transformers

To avoid exposure of power transformers to hazardous conditions (which can be also due to faults), protective devices are mounted external to the transformer. Typically, a fuse cutout which is a combination of a fuse and switch, is installed adjacent to the transformer, to disconnect it from the primary underload conditions or an internal transformer failure. Similarly, a surge arrester is installed adjacent to the conventional transformer primary aiming at protecting it from transient over-voltages due to switching and lightning surges. There are also completely self-protected (CSP) transformers, where a primary fuse is located within the transformer tank. The surge arrester is mounted outside the tank but connected to the primary bushing. Circuit breakers on the secondary side of the CSP transformers provide protection from overloads and are coordinated with primary fuses.

In the case of power grid substations, if a transformer has an outage then the rest of the transformers are used to carry out the complete of the substation's load. Typically, there are two emergency loading criteria for distribution substation transformers: (i) a 2 h emergency rating, which gives time to perform switching operations and to reduce loads, (ii) a longer-duration emergency rating (10–30 days) which gives time to replace a failed transformer with a spare that is in stock.

10.2.5 Analytical Thermal Model of Electric Power Transformers

10.2.5.1 Thermal Modeling of Power Transformers

The stages for obtaining an analytical model of the power transformer's thermal behavior are as follows [189]:

• Calculate at each time step the ultimate top oil temperature rise in the transformer from the load current at that instant, using:

$$\Delta\Theta_{TO,U} = \Delta\Theta_{TO,R}[\tfrac{I_L^2 R+1}{R+1}]^q \quad (10.1)$$

where $\Delta\Theta_{TO,U}$ is ultimate top oil temperature (TOT) rise, [°C], $\Delta\Theta_{TO,R}$ is the rated TOT rise over ambient, [°C], I_L is the load current normalised to rated current, [p.u.], q is an empirically derived exponent to approximately account for effects of change of resistance with change in load, R is the ratio of rated-load loss to no-load loss at applicable tap position.

• Calculate the increment in the TOT from the ultimate top oil rise and the ambient temperature at each time step using the differential equation:

$$\tau_{TO}\tfrac{d\Theta_{TO}}{dt} = [\Delta\Theta_{TO,U} + \Theta_A] - \Theta_{TO} \quad (10.2)$$

where Θ_{TO} is the TOT, [°C], τ_{TO} is the top oil rise time constant, and Θ_A is the ambient temperature, [°C].

• Calculate the ultimate hot spot temperature rise using:

$$\Delta\Theta_{HS,U} = \Delta\Theta_{HS,R} I_L^{2\beta} \quad (10.3)$$

where β is an empirically derived exponent, dependent on the cooling method, $\Delta\Theta_{HS,U}$ is the ultimate HST rise over top oil (for a given load current), [°C], $\Delta\Theta_{HS,R}$ is the rated HST rise over top oil (for rated load current), [°C].

• Calculate the increment in the HST rise, using the differential equation:

$$\tau_{HS}\{\tfrac{d\Delta\Theta_{HS}}{dt}\} = \Delta\Theta_{HS,U} - \Delta\Theta_{HS} \quad (10.4)$$

where Θ_{HS} is the hot spot winding temperature, [°C], is the HST rise above top oil, [°C], is the hot spot rise time constant, [h].

• Finally, add the TOT to the hot spot temperature rise to get the HST, using:

$$\Theta_{HS} = \Theta_{TO} + \Delta\Theta_{HS} \quad (10.5)$$

The model of Eqs. (10.1)–(10.5), named top-oil rise model, is based on some simplifying assumptions and its accuracy can deteriorate due to parameter variations. As a result, in order to protect power transformers, conservative safety factors have been introduced that prevent the transformer's overheating. Consequently, the calculated maximum power transfer may be 20–30 % less or worse than the real transformer capability.

10.2.5.2 HST as an Ageing and Failure Indication for Power Transformers

There are two types of aging in power transformers. The first one is the so-called intransitive aging and indicates the degradation of the transformers' components and particularly the degradation of the insulating material to withstand the designed stresses, such as electrical, mechanical, thermal and physical. The second one is the so-called transitive aging and denotes the rapid degradation of the transformer's components and particularly of the windings insulation due to abnormal operating conditions. The sustainable high Hot-Spot Temperature results in transitive transformer ageing (see Fig. 10.3).

Methods for assessing the intransitive ageing of electric power transformers are usually based on sampling and off-line analysis of electrochemical properties of the transformer's insulation (e.g. degree of polymerization, furanic compounds, dissolved gas analysis and recovery voltage measurement) or on sampling and offline testing of mechanical strength properties of the insulating material (retained tensile strength) [3, 27, 289, 452].

On the other hand the transitive ageing of electric power transformers can be detected by monitoring the transformer's Hot-Spot Temperature. The increase of the transformer HST accelerates the end of the transformer lifetime and vice-versa. The relationship between the HST and the transformer life consumption is governed by the Arhennius reaction rate theory (IEEE Standard C57.91-1995) which states that

$$\begin{aligned} \text{remaining life} &= Ae^{B/T}, \text{ or} \\ \text{per unit life} &= Ae^{B/\Theta_{HS}+273} \end{aligned} \tag{10.6}$$

where A and B are empirical constants. The A and B constants are based on materials characteristics of the insulation and they are determined such that per unit life is unity at HST of 110 °C. Indicative values of A and B are 9.8×10^{-18} and 15×10^3. According to the previous formula one can calculate the lost lifetime of the transformer as a function of the HST. Thus, one sees that transitive ageing causes acceleration of the transformer's end of life which is mainly due to the increase of the HST.

10.2.6 Neuro-Fuzzy Modelling of Power Transformers' Thermal Condition

The approach followed in this section, for extracting a neuro-fuzzy model of HST variations, results in improved modeling of the transformer's thermal behavior. This is in accordance to the stages of neurofuzzy modelling presented in Sect. 9.3. Neurofuzzy models have been also used in identification and fault diagnosis for nonlinear systems. In the sequel, fuzzy rules of the Takagi-Sugeno type will be considered. These have the form:

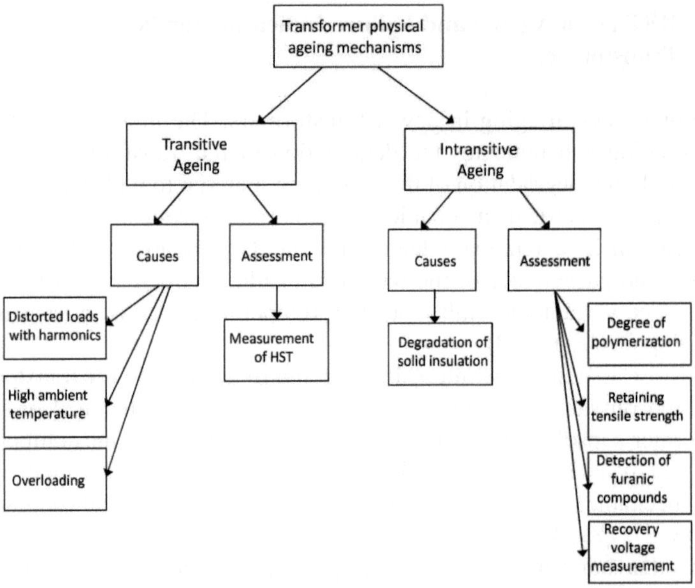

Fig. 10.3 Transitive and intransitive ageing mechanisms of power transformers

$$\begin{aligned} R_l \ : \ & IF\ x_1\ is\ A_1^l\ AND\ x_2\ is\ A_2^l\ AND\ \cdots AND\ x_n\ is\ A_n^l \\ & THEN\ \bar{y}^l = \textstyle\sum_{i=1}^{n} w_i^l x_i + b^l\ l = 1, 2, \ldots, L \end{aligned} \tag{10.7}$$

where R^l is the l-th rule, $x = [x_1, x_2, \ldots, x_n]^T$ is the input (antecedent) variable, $\bar{y}^l$ is the output (consequent) variable, and w_i^l, b^l are the parameters of the local linear models. As explained in Chap. 9, the above model is a Takagi-Sugeno model of order 1. Setting $w_i^l = 0$ results in the zero order Takagi-Sugeno model [344]. The output of the Takagi-Sugeno model is given by the weighted average of the rules consequents (Fig. 10.4):

$$\hat{y} = \frac{\sum_{l=1}^{L} \bar{y}^l \prod_{i=1}^{n} \mu_{A_i^l}(x_i)}{\sum_{l=1}^{L} \prod_{i=1}^{n} \mu_{A_i^l}(x_i)} \tag{10.8}$$

where $\mu_{A_i^l}(x_i) \ : \ R \rightarrow [0, 1]$ is the membership function of the fuzzy set A_i^l in the antecedent part of the rule R^l. In the case of a zero order TS system the output of the l-th local model is $\bar{y}^l = b^l$, while in the case of a first order TS system the output of the l-th local model is given by $\bar{y}^l = \sum_{l=1}^{L} w_i^l x_i + b^l$.

If the numerically extracted neural-fuzzy model does not approximate efficiently the monitored physical system then a refinement of the partitioning of the patterns space may be required, and improved placement of the centers of the Gaussian activation functions can be attempted [174, 462]. The individual steps of data-driven fuzzy modelling for nonlinear function approximation are discussed in [344, 355]. These stages have been previously demonstrated in Fig. 9.19.

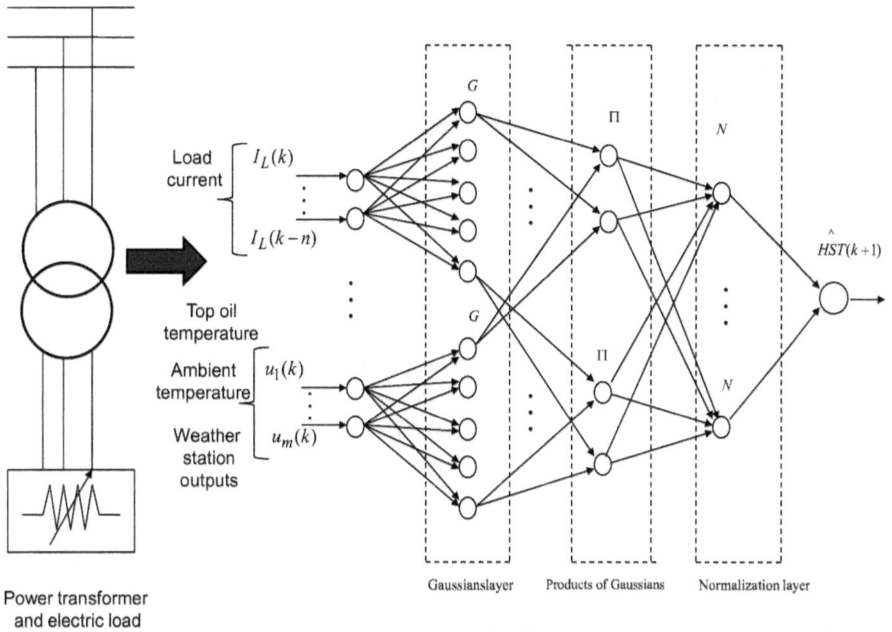

Fig. 10.4 Inputs/outputs configuration of the neural model of the power transformer thermal dynamics

10.2.7 Simulation Tests

The previously described methodology was employed to identify a neuro-fuzzy model able to estimate with high accuracy the winding HST of a laboratory prototype mineral-oil-immersed power transformer. The transformer main characteristics are resumed in Table 10.1. A measurement station has been set up consisting of thermocouples that were monitoring (a) the Hot Spot Temperature of the medium and voltage windings and (b) the Top Oil Temperature. The Hot Spot Temperature could have been also measured with optical fiber sensors. The manufacturer's specifications give, the most probable hot-spot position. A hall effect current transducer has been used in order to measure the load current.

Neuro-fuzzy modelling of the power transformer has been carried out. The electric power transformer was modeled with the use of a neuro-fuzzy network, having as output the estimated hot-spot temperature $HST(k)$ and as inputs past values of the top-oil temperature, e.g. $\Theta_{TO}(k-1)$, $\Theta_{TO}(k-2)$ and past values of the load current. e.g. $I_L(k-1)$. Neural models with the same output, such as $HST(k)$ and a larger number of inputs, i.e. including more past values of the top-oil temperature and of the load current, could be also considered. The transformer's model has been identified considering both a neural network with Hermite polynomial basis functions and a neuro-fuzzy network of the Takagi-Sugeno type. In the first case,

Table 10.1 Ratings of the modeled power transformer

Nameplate rating	25 kVA
Vprimary/Vsecondary	10 kV/400 V
Iron losses	195 Watt
Copper losses (full load)	776 Watt
Top oil temperature rise at full load	73.1 °C
Weight of core and coil assembly	136 Kg
Weight of oil	62 Kg
Density	757 Kg/m^3
Total weight	310 Kg
Length, width and height of tank	$64 \times 16 \times 80$ cm
Type of cooling	ONAN
Factory/year	MACE/87

a neural network with Hermite basis functions was used to model the variations of the power transformer HST. As shown in Fig. 10.5a thanks to the inherent multi-frequency characteristics of the Hermite polynomial basis functions, such a neural model can capture with increased accuracy spikes and abrupt changes in the HST profile [498, 339]. In the second case, the neural/fuzzy TSK model of Fig. 10.4 was used to estimate the variations of HST. The obtained approximation is shown in Fig. 10.5b. To decide where the basis functions should be placed, the input space was segmented using the input dimension (grid) partition [344]. The RMSE of training the Hermite and the TSK model was of the order of 4×10^{-3}.

The size of the training set was 300. The LMS (Least Mean Square) algorithm was used for the adaptation of the linear weights $w_1^{(l)}$ [174, 462]. This neurofuzzy model actually stands for a rule base that consists of 64 rules (3 input variables partitioned in 4 fuzzy subsets each). All fuzzy sets (Gaussian activation functions) were assumed to have the same spread.

Fault diagnosis for the transformer's model was performed using the *Local statistical approach to fault diagnosis* and the χ^2 change detection test which where described in Sect. 9.3. The TSK fuzzy model of the transformer that was extracted from real power transformer HST, TOT and load current data consisted initially of 64 rules. It comprised 64 linear parameters (weights) and 12 nonlinear parameters (centers of fuzzy sets). The size of the model was reduced to 27 rules after omitting those rules which received small activation from the existing data set. The parameters set in the new TSK fuzzy model consisted of 39 parameters (27 linear parameters which were the output layer weights and 12 nonlinear parameters which were the centers of the fuzzy sets). This means that by applying the local statistical approach to FDI and the χ^2 change detection test to the considered electric power transformer model, the fault threshold should be equal to 39.

The numerical tests confirmed theory. In case that no fault was assumed for the power transformer the mean value of the χ^2 test over a number of trials was found to

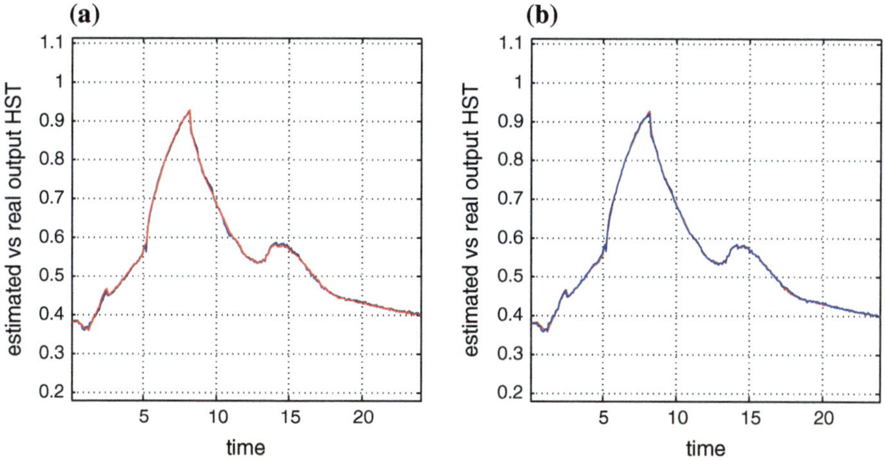

Fig. 10.5 Approximation of the Hot Spot Temperature of the electric power transformer (*red line*) **a** by a neural network with Hermite polynomial basis functions (*blue-line*), **b** by a fuzzy TSK network (*blue-line*)

be equal to 38.713. Such a value was anticipated according to the theoretical analysis of the χ^2 test. For slight deviations of the parameters of the power transformer from their nominal (fault-free) values, the global χ^2 test was capable of giving a clear indication about the existence of a fault. Thus for changes which varied between 0.01 and 1 % of the nominal parameter's value (either for a linear or a nonlinear parameter) the score of the χ^2 test deviated significantly from the fault threshold (which as mentioned before was set equal to 39). As shown in Fig. 10.6, a small fault (deviation from the nominal parameter's value) suffices to generate an output of the χ^2 test that exceeds several times the value of the fault threshold.

As far as fault isolation is concerned, the numerical results showed that the sensitivity method for fault isolation was very efficient in distinguishing the parameter subject to fault among all 39 parameters in the power transformer's parameter set. The sensitivity fault isolation test was performed for both a linear parameter (weight w_1) and for a nonlinear parameter (center $c_1^{(1)}$) of the transformer's model. In Fig. 10.7 it can be observed that the success rate for the aforementioned fault magnitudes attained the value of 100 %.

Finally, fault isolation tests for detecting changes in both linear (weight w_3) and nonlinear parameters (center $c_1^{(23)}$) of the power transformer model were also performed with the use of the max-min method. The associated results are depicted in Fig. 10.8. It can be observed that the max-min fault isolation method succeeded also a high success rate in finding the faulty parameter in the transformer's thermal model.

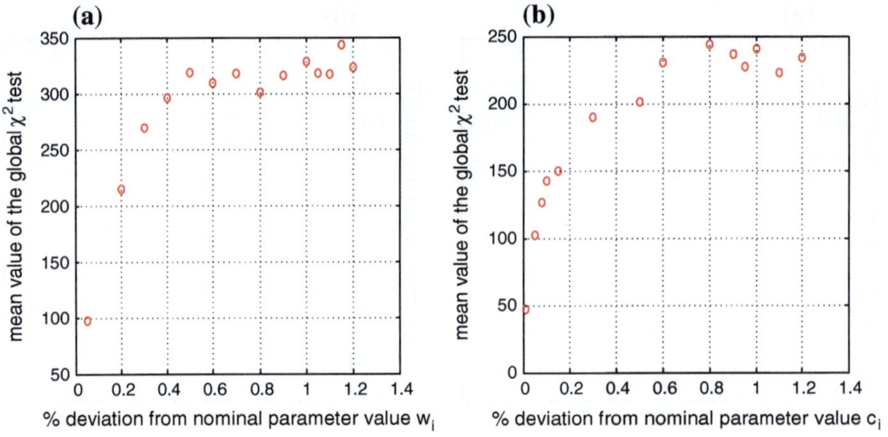

Fig. 10.6 **a** Score of the global χ^2 for changes in a linear parameter of the electric power transformer model, ranging between 0.01 and 1.0 % of the nominal value. **b** Score of the global χ^2 for changes in a nonlinear parameter of the electric power transformer model, ranging between 0.01 and 1.0 % of the nominal value

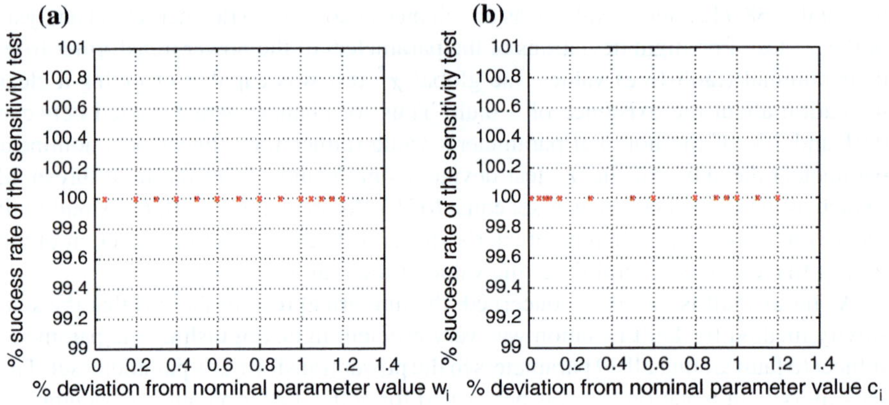

Fig. 10.7 Success rate of the fault isolation test (sensitivity method) **a** for changes in a linear parameter of the electric power transformer model, ranging between 0.01 and 1.0 % of the nominal value, **b** for changes in the a nonlinear parameter of the electric power transformer model, ranging between 0.01 and 1.0 % of the nominal value

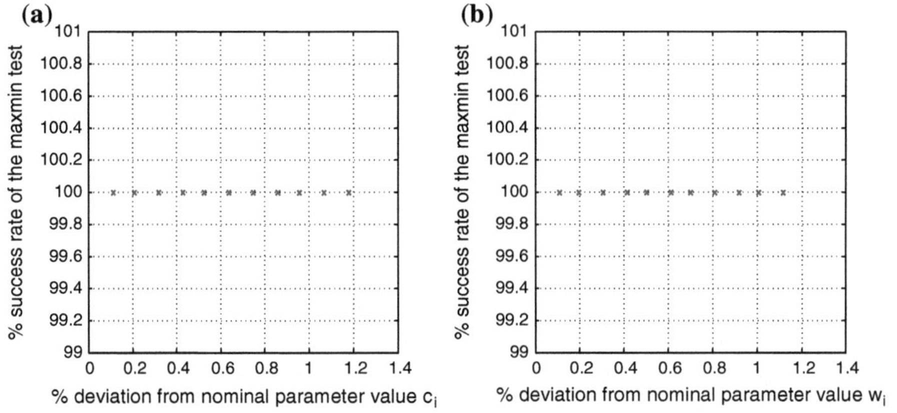

Fig. 10.8 Success rate of the fault isolation test (max-min method) **a** for changes in a linear parameter of the electric power transformer model, ranging between 0.01 and 1.0 % of the nominal value, **b** for changes in the a nonlinear parameter of the electric power transformer model, ranging between 0.01 and 1.0 % of the nominal value

10.3 Distributed Filtering for Condition Monitoring of the Electric Power Grid

10.3.1 Outline

This section aims at developing new solutions to the problem of distributed estimation for condition monitoring of distributed nonlinear power systems, such as the electric power transmission and distribution grid, so as to enable early detection of faults and the take up of efficient restoration measures. Fault detection in distributed nonlinear dynamical systems exhibits elevated difficulty. Moreover, when a fault occurs in such large-scale systems, it is usually difficult to determine its exact position. On the other hand, it is important to identify quickly the existence and position of a fault so as to carry out appropriate actions and to preserve the secure operation of the system. However, due to the measurement cost and the excessive size of the system, the complete monitoring of the associated infrastructure is impractical. Therefore, to monitor the condition and to detect faults in distributed nonlinear dynamical systems, some form of estimation is required.

Most of the existing state estimation and condition monitoring methods for nonlinear electric power systems are centralized, which means that all sensing data are collected and processed at one unit. This not only causes excessive communication and computational burdens, but also creates a single point of failure [41, 103, 345, 348, 350]. To overcome these problems, the section proposes distributed state estimation [41, 103, 350]. Necessary stages for performing distributed state estimation are the transmission of measurements to local processing units over a communication network and the fusion of local state estimates into an aggregate state vector

[312, 391]. Obtaining estimates of the state variables of a large-scale system, such as the electric power transmission and distribution system, enables the continuous monitoring of its condition.

During the last years there have been several attempts towards the development of distributed estimation methodologies for the electric power system. In [294], a distributed implementation of the particle filter is proposed for state estimation in large-scale dynamical systems such as wireless sensor networks and power distribution systems. In [296] it is concluded that most state estimation programs in practical use are formulated and solved as weighted least squares (WLS) problems. In [156] weighted least squares and Extended Kalman Filter are proposed for state estimation in the electric power distribution grid. The fast convergence properties of the Kalman Filter estimator are shown. In [466] state estimation techniques are proposed to reduce the cost for installing a large number of sensors and monitoring devices and the cost of the associated telecommunication equipment. To implement localized state estimation, least squares methods and Kalman Filtering approaches are considered. In [468] Kalman Filtering is applied to the problem of detecting model changes in large interconnected power systems. Short-term alarming procedures are developed based on the statistics of the power spectral density of the Kalman Filter innovation. The new technique is tested on both simulated data and real data obtained from power systems in normal operation. More papers treating the problem of distributed state estimation for the power grid are [29, 94, 96, 308, 311, 384, 404, 457].

The current section proposes the Extended Information Filter (EIF) and the Unscented Information Filter (UIF) as suitable approaches for fusing the state estimates provided by local processing units over the electric power grid, making the assumption of Gaussian noises. The Extended Information Filter is a generalization of the Information Filter in which the local filters do not exchange raw measurements but send to an aggregation filter their local information matrices (local inverse covariance matrices) and their associated local information state vectors (products of the local information matrices with the local state vectors) [230, 231, 306]. In the case of the Unscented Information Filter there is no linearization of the model of the electric power distribution system [202, 203]. However the application of the Information Filter algorithm is possible through an implicit linearization which is performed by approximating the Jacobian matrix of the system's output equation by the product of the inverse of the state vector's covariance matrix with the cross-covariance matrix between the system's state vector and the system's output [230, 455]. Again, the local information matrices and the local information state vectors are transferred to an aggregation filter which produces the global estimation of the system's state vector. At a later stage, the EIF and the UIF can be also used by a Fault Detection and Isolation (FDI) algorithm for residuals generation [34, 344, 508].

10.3.2 *State of the Art in State Estimation and Fault Diagnosis for the Power Grid*

10.3.2.1 Causes of Faults in the Power Transmission and Distribution System

During the last years, there have been some remarkable blackouts and incidents related to electric power quality in national and transnational power grids. In each of these cases, official investigations mostly associate these events to faults of critical equipment of the power grid. A fault in a power system is an abnormal condition that involves the failure of power system equipment. Generally two types of failure can occur. The first is an insulation failure that results in a short circuit fault and can occur as a result of overstressing and degradation of the insulation over time or due to a sudden overvoltage condition. The second is a failure that results in a cessation of current flow, also known as open circuit fault [457].

Short-circuit faults can occur between phases, or between phases and earth, or both. Short circuits may be one-phase to earth, phase to phase, two-phase to earth, three-phase clear of earth and three-phase to earth. The three-phase fault is a fault that symmetrically affects all phases of the three-phase circuit and is called "balanced fault", whereas all other types of faults are unbalanced. Simultaneous faults are a combination of two or more faults that occur at the same time. They may be of the same or different types and may occur at the same or different locations. A broken overhead line conductor that falls to earth is a simultaneous one-phase open-circuit and one-phase short-circuit fault at one location. A short-circuit fault occurring at the same time on each circuit of a double-circuit overhead line, where the two circuits are strung on the same tower, is a simultaneous fault condition. A one-phase to earth short-circuit fault in a high impedance earthed distribution system may cause a sufficient voltage rise on a healthy phase elsewhere in the system that a flashover and short-circuit fault occurs. This is known as a cross-country fault. Most faults do not change in type during the fault period but some faults do change and evolve a one-phase to earth short-circuit to include a second phase where it changes to a two-phase earth short circuit fault. This can occur on overhead lines or in substations where the flashover arc of the faulted phase spreads to other healthy phases. Internal short circuits to earth and open-circuit faults can also occur on windings of transformers, power generators as well as faults between a number of winding turns of the shame phase.

Open-circuit faults may be caused by the failure of joints on cables or overhead lines or the failure of all the three phases of a circuit-breaker to open or close. For example, two phases of a circuit-breaker may close and latch but not the third phase or two phases may properly open but the third remains stuck on the closed position. Except of mainly underground systems, the vast majority of short-circuit faults are weather related followed by equipment failure. The weather factors that usually cause short-circuit faults are: lighting strikes, accumulation of snow or ice, heavy rain, strong winds or gales, pollution depositing on insulators on overhead lines and in substations, floods and fires adjacent to electrical equipment, e.g. beneath overhead lines.

10.3.2.2 Effects of Faults in the Power Transmission and Distribution System

The effects of short and open circuit faults in power transmission and distribution systems are summarized as follows: (i) changes of frequency, (ii) changes in steady-state voltage magnitude (under or over voltage), (iii) variation in the peak of the RMS voltage (sags or dips), (iv) variations of harmonics (and inter-harmonics), (v) variations of transients (spikes, impulses or surges). In order to minimize the disruption caused by faults the disturbance source must be quickly identified and rapid remedial actions must be taken. It is important that efficient tools and analysis methods are available to meet this requirement.

It is noted that voltage dips and harmonics distortion are common voltage disturbances affecting the power grid [147, 194, 493]. An efficient condition monitoring system for the power grid should be capable to detect disturbances like voltage dips and harmonics distortion: (i) Voltage dips are reductions of the voltage magnitude at a point in the electrical system below a specified threshold which is followed by voltage recovery after a short period of time (dip duration), lasting from half a cycle to a few seconds. The source of voltage dips is usually associated with a fault somewhere in the distribution system. Voltage dips duration and magnitude are factors that can directly affect electric equipment and can cause shutdown or malfunction of customers' load. (ii) Harmonics variation results into a distorted voltage waveform which in turn can be represented as summation of various higher frequency sinusoidal components known as harmonics (harmonics are integer multiple of a fundamental frequency). These harmonics have large impact on operational efficiency and reliability of power system, and on loads or protective relaying. The presence of harmonics sometimes leads to failure of devices like compensating capacitors. Communication interference is another serious side effect of harmonics.

10.3.2.3 Localized State Estimation in the Power Transmission and Distribution System

Nonlinear least squares: Parameter estimation in static nonlinear power systems can be performed with the use of least squares methods. The recursive solution to the nonlinear least squares problems can be performed (i) with the use of second-order gradient algorithms such as the Gauss-Newton or the Levenberg-Marquardt algorithm that process data in batch mode, i.e. the state estimate is updated once the complete set of training data has been received, (ii) with the use of the Extended Kalman Filter algorithm that processes data in sequential mode, i.e. the parameters estimate is updated once a new measurement is received.

Nonlinear Kalman Filtering: Two main approaches for state estimation in nonlinear dynamical systems are the Extended Kalman Filter and the Unscented Kalman Filter. The Extended Kalman Filter is the linearization of the standard Kalman Filter state estimation algorithm to the nonlinear case. The standard Kalman Filter is an optimal state observer in the sense that it can compensate in optimal way for the effect that process and measurement noises have on the estimation of the system's state vector. The Extended Kalman Filter is based on a local linearization of the nonlinear dynamical model around the current state estimation and uses Jacobian matrices in the place of the state transition matrix and measurement matrix of the standard Kalman Filter. This first-order local linearization procedure introduces approximation errors due to the truncation of higher order term that appear in the associated Taylor series expansion [341].

The Unscented Kalman Filter overcomes the flaws of the Extended Kalman Filter. Unlike EKF no analytical Jacobians of the system equations need to be calculated. This makes the UKF approach suitable for application in black-box models where analytical expressions of the system dynamics are either not available or not in a form that allows easy linearization. This is achieved through a different approach for calculating the posterior 1st and 2nd order statistics of a random variable that undergoes a nonlinear transformation. The state distribution is represented again by a Gaussian Random Variable but is now specified using a minimal set of deterministically chosen weighted sample points [202, 203].

Particle Filtering: Particle filtering has improved performance over the established nonlinear filtering approaches (e.g. the EKF), since it can provide optimal estimation in nonlinear non-Gaussian state-space models, as well as in the estimation of nonlinear models. Particle filters can estimate the system states sufficiently when the number of particles (estimations of the state vectors which evolve in parallel) is large. Particle filtering has been mainly applied to state estimation for nonlinear systems fault diagnosis. The particle filtering algorithm is reminiscent of evolutionary algorithms where a number of particles is subject to a mutation mechanism which corresponds to the prediction stage, and to selection mechanism which corresponds to the correction stage [242, 347, 352].

10.3.2.4 Distributed State Estimation in the Power Transmission and Distribution System

Distributed state estimation is a rather new topic in the area of modelling and condition monitoring of electric power grid. Usually the term distributed state estimation is associated to static models of the electric power system, i.e. models that consist of algebraic equations and ignore transient phenomena. Under the assumption of a static power system model, the *Weighted Least Squares* form an established approach for distributed state estimation, which in this case means identification of the parameters of the power system model. More recently other approaches to distributed state estimation have been developed, this time taking into account transient phenomena in the system dynamics and assuming that the power grid is described by differential equations. Considering a linear power system model the Information Filter has been implemented for distributed estimation of the electric power grid [41, 103]. Moreover, the distributed particle filter has been proposed in [294] as a suitable methodology for decentralized state estimation of the electric power distribution system. On the other hand distributed state estimation methodologies for nonlinear dynamical systems, such as the Extended Information Filter and the Unscented Information Filter have not been tested in the problem of decentralized state estimation for large scale systems, such as the electric power grid. The current section, offers new results on the use of the aforementioned distributed state estimation methods, i.e. Extended Information Filter (EIF) and Unscented Information Filter, in the case of large-scale systems such as the electric power transmission and distribution system.

Weighted least squares: The Weighted Least Squares (WLS) method tries to determine the most probable state of the system given the quantities that are measured, using measurements and residuals that come from distributed measurement devices [296]. The assumptions that are made by the WLS method are that: (i) The measurements are normally distributed. (ii) The mean value of the measurement errors is zero. (iii) The measurement errors are independent. To determine the most probable state, the WLS method minimizes the sum of the weighted squares of the residuals. The residuals are provided by distributed information processing units where the measurements from local sensors are collected. In case of the distribution grid, the measurement consists of voltage, or current. The weights coefficients that appear in the residuals cost function represent the certainty on the measurements. The certainty of a measurement is given by the inverse of its error variance. The optimal solution is found by computing the state variable which minimizes the weighted squares of the residuals. The solution of the weighted least squares problem is found iteratively with the use of gradient algorithms.

Distributed Kalman Filtering: The Extended Information Filter (EIF) and the Unscented Information Filter (UIF) can be proposed as possible approaches for fusing the state estimates provided by the local monitoring stations, under the assumption of Gaussian noises [348, 351]. The Extended Information Filter is a generalization of the Information Filter in which the local filters do not exchange raw measurements but send to an aggregation filter their local information matrices (local inverse covariance matrices) and their associated local information state vectors (products of the local information matrices with the local state vectors). In the case of the Unscented Information Filter there is no linearization of the power system observation equation. However the application of the Information Filter algorithm is possible through an implicit linearization which is performed by approximating the Jacobian matrix of the system's output equation with the product of the inverse of the state vector's covariance matrix (information matrix) with the cross-correlation covariance matrix between the system's state vector and the system's output. Again, the local information matrices and the local information state vectors are transferred to an aggregation filter which produces the global estimation of the system's state vector (Fig. 10.9).

Distributed Particle Filtering: The Distributed Particle Filter (DPF) is also proposed for fusing the state estimates provided by the local monitoring stations (local filters). The reason for using DPF is that it is well-suited to accommodate

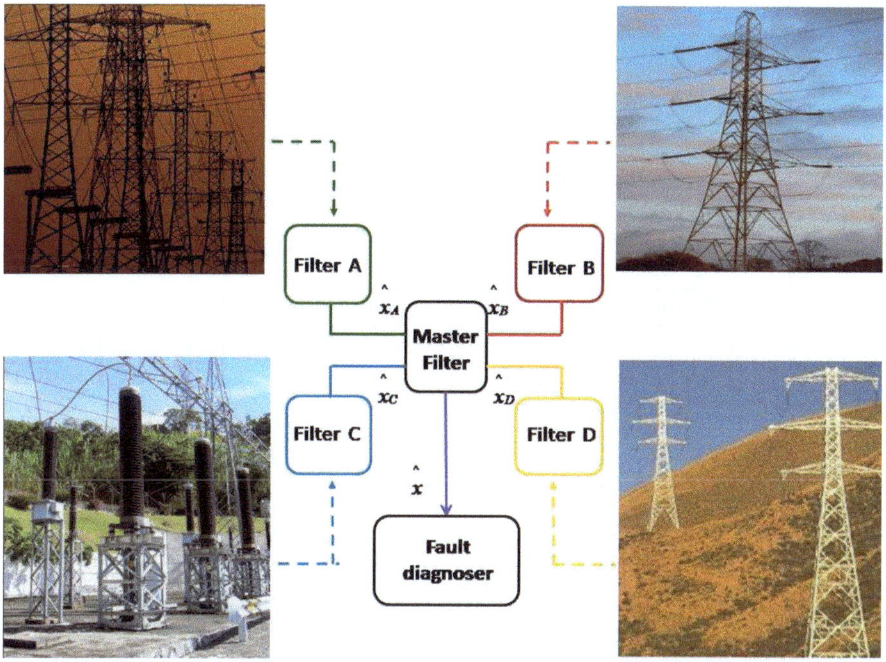

Fig. 10.9 Distributed state estimation and distributed fault diagnosis for the power grid

non-Gaussian measurements [294, 348]. A difficulty in implementing distributed particle filtering is that particles from one particle set (which correspond to a local particle filter) do not have the same support (do not cover the same area and points on the samples space) as particles from another particle set (which are associated with another particle filter). This can be resolved by transforming the particles sets into Gaussian mixtures, and defining the global probability distribution on the common support set of the probability density functions associated with the local filters.

10.3.3 Fault Diagnosis in the Power Transmission and Distribution System

Model-based methods of fault detection use the relation between several measured variables to extract information on possible changes caused by faults. The relations between the input and output signals are represented by a mathematical model. Such models can be obtained through nonlinear identification techniques or through nonlinear state estimation and filtering methods. Decisions on the existence and cause of a fault are usually based on the comparison between the observed system outputs

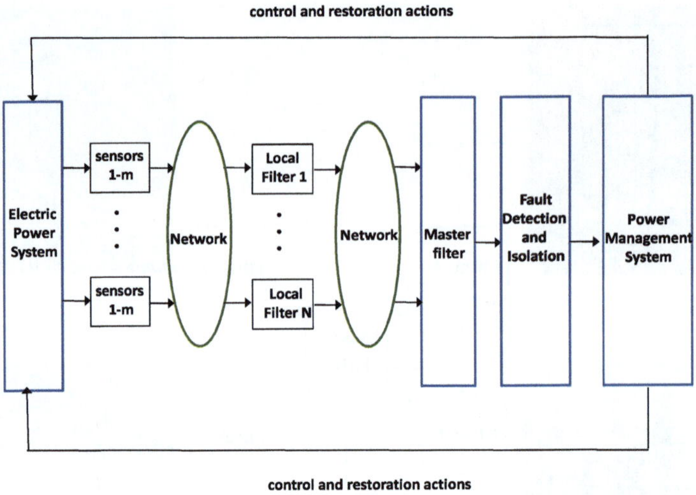

Fig. 10.10 Structure of distributed filtering for condition monitoring of the electric power distribution system

with their nominal values (see Fig. 10.10). The differences between the nominal and the observed system features are also called "residuals".

Apart from residuals generation through modeling of the monitored system's function in the fault-free condition, another significant problem for fault detection and isolation is the optimal selection of the fault threshold. The value of the fault threshold should be such that small faults can be detected at their early stages and also the rate of false alarms is minimized. A solution to the problem of fault threshold selection in fault diagnosis systems has been provided by the *Local Statistical Approach* to change detection. As already explained, the local Statistical Approach has been successfully applied to several nonlinear FDI problems. Based on a small parametric disturbance assumption, the Local Statistical Approach to fault diagnosis aims at transforming complex detection problems concerning a parameterized stochastic process into the problem of monitoring the mean of a Gaussian vector. The local statistical approach consists of two stages: (i) the global test which indicates the existence of a change in some parameters of the system's model, (ii) the diagnostics tests (sensitivity or min-max) which isolate the parameter affected by the change [34, 344, 365, 508, 509].

10.3.4 State Estimation with the Extended Information Filter

10.3.4.1 Extended Kalman Filtering at Local Processing Units

The distributed Extended Kalman Filter, also known as Extended Information Filter, performs fusion of the state estimates which are provided by local Extended Kalman

Filters. Thus, the functioning of the local Extended Kalman Filters should be analyzed first. The following discrete-time state-space model is considered [352]:

$$\begin{aligned} x(k+1) &= \phi(x(k)) + L(k)u(k) + w(k) \\ z(k) &= \gamma(x(k)) + v(k) \end{aligned} \tag{10.9}$$

where $x \in R^{m\times 1}$ is the system's state vector and $z \in R^{p\times 1}$ is the system's output, while $w(k)$ and $v(k)$ are uncorrelated, Gaussian zero-mean noise processes with covariance matrices $Q(k)$ and $R(k)$ respectively. The operators $\phi(x)$ and $\gamma(x)$ are $\phi(x) = [\phi_1(x), \phi_2(x), \ldots, \phi_m(x)]^T$, and $\gamma(x) = [\gamma_1(x), \gamma_2(x), \ldots, \gamma_p(x)]^T$, respectively. It is assumed that ϕ and γ are sufficiently smooth in x so that each one has a valid series Taylor expansion. Following a linearization procedure, ϕ is expanded into Taylor series about $\hat{x}$, i.e.

$$\phi(x(k)) = \phi(\hat{x}(k)) + J_\phi(\hat{x}(k))[x(k) - \hat{x}(k)] + \cdots \tag{10.10}$$

where $J_\phi(x)$ is the Jacobian of ϕ calculated at $\hat{x}(k)$. Likewise, γ is expanded about $\hat{x}^-(k)$, i.e.

$$\gamma(x(k)) = \gamma(\hat{x}^-(k)) + J_\gamma[x(k) - \hat{x}^-(k)] + \cdots \tag{10.11}$$

where $\hat{x}^-(k)$ is the estimation of the state vector $x(k)$ before measurement at the k-th instant to be received and $\hat{x}(k)$ is the updated estimation of the state vector after measurement at the k-th instant has been received. $J_\gamma(x)$ is the Jacobian of γ calculated at $\hat{x}(k)$. The resulting expressions create first order approximations of ϕ and γ. Thus the linearized version of the system is obtained:

$$\begin{aligned} x(k+1) &= \phi(\hat{x}(k)) + J_\phi(\hat{x}(k))[x(k) - \hat{x}(k)] + w(k) \\ z(k) &= \gamma(\hat{x}^-(k)) + J_\gamma(\hat{x}^-(k))[x(k) - \hat{x}^-(k)] + v(k). \end{aligned} \tag{10.12}$$

Now, the EKF recursion is as follows: First the time update is considered: by $\hat{x}(k)$ the estimation of the state vector at instant k is denoted. Given initial conditions $\hat{x}^-(0)$ and $P^-(0)$ the recursion proceeds as:

Measurement update. Acquire $z(k)$ and compute:

$$\begin{aligned} K(k) &= P^-(k)J_\gamma^T(\hat{x}^-(k)) \cdot \\ &\cdot [J_\gamma(\hat{x}^-(k))P^-(k)J_\gamma^T(\hat{x}^-(k)) + R(k)]^{-1} \\ \hat{x}(k) &= \hat{x}^-(k) + K(k)[z(k) - \gamma(\hat{x}^-(k))] \\ P(k) &= P^-(k) - K(k)J_\gamma(\hat{x}^-(k))P^-(k) \end{aligned} \tag{10.13}$$

Time update. Compute:

$$\begin{aligned} P^-(k+1) &= J_\phi(\hat{x}(k))P(k)J_\phi^T(\hat{x}(k)) + Q(k) \\ \hat{x}^-(k+1) &= \phi(\hat{x}(k)) + L(k)u(k) \end{aligned} \tag{10.14}$$

10.3.4.2 Calculation of Local Estimations in EIF Terms

Again the discrete-time nonlinear system of Eq. (10.9) is considered. The Extended Information Filter (EIF) performs fusion of the local state vector estimates which are provided by the local Extended Kalman Filters, using the Information matrix and the Information state vector [230, 276, 455]. The Information Matrix is the inverse of the state vector covariance matrix, and can be also associated to the Fisher Information matrix [344]. The Information state vector is the product between the Information matrix and the local state vector estimate

$$\begin{aligned} Y(k) &= P^{-1}(k) = I(k) \\ \hat{y}(k) &= P^{-}(k)^{-1}\hat{x}(k) = Y(k)\hat{x}(k) \end{aligned} \tag{10.15}$$

The update equation for the Information Matrix and the Information state vector are given by

$$\begin{aligned} Y(k) &= P^{-}(k)^{-1} + J_{\gamma}^{T}(k)R^{-1}(k)J_{\gamma}(k) \\ &= Y^{-}(k) + I(k) \\ \hat{y}(k) &= \hat{y}^{-}(k) + J_{\gamma}^{T}R(k)^{-1}[z(k) - \gamma(x(k)) + J_{\gamma}\hat{x}^{-}(k)] \\ &= \hat{y}^{-}(k) + i(k) \end{aligned} \tag{10.16}$$

where $I(k) = J_{\gamma}^{T}(k)R(k)^{-1}J_{\gamma}(k)$ is the associated information matrix while, the information state contribution is $i(k) = J_{\gamma}^{T}R^{(}k)^{-1}[(z(k) - \gamma(x(k))) + J_{\gamma}\hat{x}^{-}(k)]$. The predicted information state vector and Information matrix are obtained from

$$\begin{aligned} \hat{y}^{-}(k) &= P^{-}(k)^{-1}\hat{x}^{-}(k) \\ Y^{-}(k) = P^{-}(k)^{-1} &= [J_{\phi}(k)P^{-}(k)J_{\phi}(k)^{T} + Q(k)]^{-1} \end{aligned} \tag{10.17}$$

It is assumed that an observation vector $z^{i}(k)$ is available for the N different sensor sites (processing units) $i = 1, 2, \ldots, N$ and each processing unit observes the power system according to the local observation model, expressed by $z^{i}(k) = \gamma(x(k)) + v^{i}(k)$, $i = 1, 2, \ldots, N$, where the local noise vector $v^{i}(k) \sim N(0, R^{i})$ is assumed to be white Gaussian and uncorrelated between sensors. The variance of a composite observation noise vector v_k is expressed in terms of the block diagonal matrix $R(k) = diag[R^{1}(k), \ldots, R^{N}(k)]^{T}$. The information contribution can be expressed by a linear combination of each local information state contribution i^{i} and the associated information matrix I^{i} at the i-th sensor site

$$\begin{aligned} i(k) &= \textstyle\sum_{i=1}^{N} {J_{\gamma}^{i}}^{T}(k)R^{i}(k)^{-1} \cdot \\ &\cdot [z^{i}(k) - \gamma^{i}(x(k)) + J_{\gamma}^{i}(k)\hat{x}^{-}(k)] \\ I(k) &= \textstyle\sum_{i=1}^{N} {J_{\gamma}^{i}}^{T}(k)R^{i}(k)^{-1}J_{\gamma}^{i}(k) \end{aligned} \tag{10.18}$$

Thus, the update equations for fusing the local state estimates become

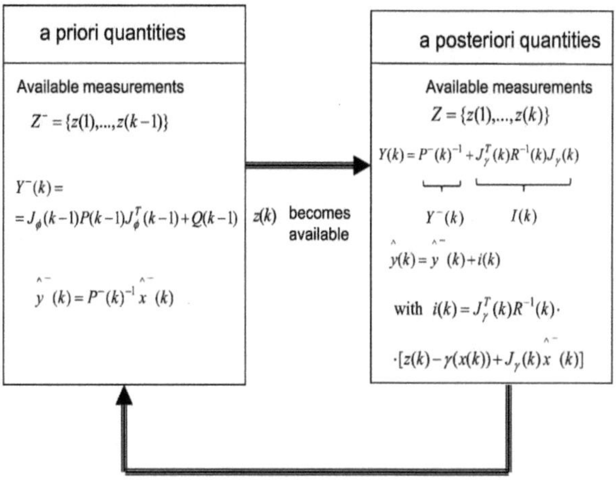

Fig. 10.11 Schematic diagram of the Extended Information Filter loop

$$\begin{array}{c}\hat{y}(k) = \hat{y}^-(k) + \sum_{i=1}^{N} {J_\gamma^i}^T(k)R^i(k)^{-1}\cdot \\ \cdot[z^i(k) - \gamma^i(x(k)) + J_\gamma^i(k)\hat{x}^-(k)] \\ Y(k) = Y^-(k) + \sum_{i=1}^{N} {J_\gamma^i}^T(k)R^i(k)^{-1}J_\gamma^i(k)\end{array} \tag{10.19}$$

It is noted that in the Extended Information Filter an aggregation (master) fusion filter produces a global estimate by using the local sensor information provided by each local filter. As in the case of the Extended Kalman Filter the local filters which constitute the Extended information Filter can be written in terms of *time update* and a *measurement update* equation (see Fig. 10.11).

Measurement update: Acquire $z(k)$ and compute

$$\begin{array}{c}Y(k) = P^-(k)^{-1} + J_\gamma^T(k)R(k)^{-1}J_\gamma(k) \Rightarrow \\ Y(k) = Y^-(k) + I(k), \;\; I(k) = J_\gamma^T(k)R^{-1}(k)J_\gamma(k) \\ \hat{y}(k) = \hat{y}^-(k) + J_\gamma^T(k)R(k)^{-1}\cdot \\ \cdot[z(k) - \gamma(\hat{x}(k)) + J_\gamma\hat{x}^-(k)] \\ \text{or } \hat{y}(k) = \hat{y}^-(k) + i(k)\end{array} \tag{10.20}$$

Time update: Compute

$$\begin{array}{c}Y^-(k+1) = P^-(k+1)^{-1} = \\ = [J_\phi(k)P(k)J_\phi(k)^T + Q(k)]^{-1} \\ \text{and also } y^-(k+1) = P^-(k+1)^{-1}\hat{x}^-(k+1)\end{array} \tag{10.21}$$

10.3.4.3 Extended Information Filtering for State Estimates Fusion

The outputs of the local filters are treated as measurements which are fed into the aggregation fusion filter [230, 455]. Then each local filter is expressed by its respective error covariance and estimate in terms of information contributions and is described by

$$\begin{aligned} P_i^{-1}(k) &= {P_i^-(k)}^{-1} + J_\gamma^T R(k)^{-1} J_\gamma(k) \\ \hat{x}_i(k) &= P_i(k)(P_i^-(k)^{-1}\hat{x}_i^-(k)) + \\ &+ J_\gamma^T R(k)^{-1}[z^i(k) - \gamma^i(x(k)) + J_\gamma^i(k)\hat{x}_i^-(k)] \end{aligned} \tag{10.22}$$

It is noted that the local estimates are suboptimal and also conditionally independent given their own measurements. The global estimate and the associated error covariance for the aggregate fusion filter can be rewritten in terms of the computed estimates and covariances from the local filters using the relations

$$\begin{aligned} J_\gamma^T(k)R(k)^{-1}J_\gamma(k) &= P_i(k)^{-1} - P_i^-(k)^{-1} \\ J_\gamma^T(k)R(k)^{-1}[z^i(k) - \gamma^i(x(k)) + J_\gamma^i(k)\hat{x}^-(k)] &= \\ = P_i(k)^{-1}\hat{x}_i(k) - P_i(k)^{-1}\hat{x}_i(k-1) \end{aligned} \tag{10.23}$$

For the general case of N local filters $i = 1, \ldots, N$, the distributed filtering architecture is described by the following equations

$$\begin{aligned} P(k)^{-1} &= P^-(k)^{-1} + \sum_{i=1}^N [P_i(k)^{-1} - P_i^-(k)^{-1}] \\ \hat{x}(k) &= P(k)[P^-(k)^{-1}\hat{x}^-(k) + \\ &+ \sum_{i=1}^N (P_i(k)^{-1}\hat{x}_i(k) - P_i^-(k)^{-1}\hat{x}_i^-(k))] \end{aligned} \tag{10.24}$$

The global state update equation in the above distributed filter can be written in terms of the information state vector and of the information matrix, i.e.

$$\begin{aligned} \hat{y}(k) &= \hat{y}^-(k) + \sum_{i=1}^N (\hat{y}_i(k) - \hat{y}_i^-(k)) \\ \hat{Y}(k) &= \hat{Y}^-(k) + \sum_{i=1}^N (\hat{Y}_i(k) - \hat{Y}_i^-(k)) \end{aligned} \tag{10.25}$$

The local filters provide their own local estimates and repeat the cycle at step $k+1$. In turn the global filter can predict its global estimate and repeat the cycle at the next time step $k+1$ when the new state $\hat{x}(k+1)$ and the new global covariance matrix $P(k+1)$ are calculated. From Eq. (10.24) it can be seen that if a local filter (processing station) fails, then the local covariance matrices and the local state estimates provided by the rest of the filters will enable an accurate computation of the power system's state vector.

10.3.5 State Estimation with the Unscented Information Filter

10.3.5.1 Unscented Information Filtering

Again the nonlinear state model of Eq. (10.9) is considered [352]. For such a non-linear model the Unscented Information Filter (UIF) can also perform fusion of the state vector estimates which are obtained from local Unscented Kalman Filters, by weighting these estimates with local Information matrices (inverse of the local state vector covariance matrices), also known as Fisher information matrices [344]. First, an augmented state vector ${x_\alpha}^-(k)$ is considered, along with the process noise vector, and the associated covariance matrix is introduced:

$$\begin{aligned} \hat{x}_\alpha^-(k) &= [\hat{x}^-(k), \hat{w}^-(k)]^T \\ {P_\alpha}^-(k) &= diag\{P^-(k), Q^-(k\} \end{aligned} \tag{10.26}$$

As in the case of local (lumped) Unscented Kalman Filters, a set of weighted sigma points ${X_\alpha^i}^-(k)$ is generated as

$$\begin{aligned} &X_{\alpha,0}^-(k) = \hat{x}_\alpha^-(k) \\ &X_{\alpha,i}^-(k) = \hat{x}_\alpha^-(k) + [\sqrt{(n_\alpha+\lambda)P_\alpha^-(k-1)}]_i \\ &i = 1, \ldots, n \\ &X_{\alpha,i}^-(k) = \hat{x}_\alpha^-(k) - [\sqrt{(n_\alpha+\lambda)P_\alpha^-(k-1)}]_i \\ &i = n+1, \ldots, 2n, \end{aligned} \tag{10.27}$$

where $\lambda = \alpha^2(n_\alpha+\kappa)-n_\alpha$ is a scaling, while $0 \leq \alpha \leq 1$ and κ are constant parameters. The corresponding weights for the mean and covariance are defined as in the case of the lumped Unscented Kalman Filter

$$\begin{aligned} W_0^{(m)} &= \tfrac{\lambda}{n_\alpha+\lambda} \quad W_0^{(c)} = \tfrac{\lambda}{(n_\alpha+\lambda)+(1-\alpha^2+\beta)} \\ W_i^{(m)} &= \tfrac{1}{2(n_\alpha+\lambda)} \quad W_i^{(C)} = \tfrac{1}{2(n_\alpha+\lambda)} \end{aligned} \tag{10.28}$$

where $i = 1, \ldots, 2n_\alpha$ and β is again a constant parameter [203]. The equations of the prediction stage (measurement update) of the information filter, i.e. the calculation of the information matrix and the information state vector are [230, 455]:

$$\begin{aligned} \hat{y}^-(k) &= Y^-(k)\textstyle\sum_{i=0}^{2n_\alpha} W_i^{(m)} X_i^x(k) \\ Y^-(k) &= P^-(k)^{-1} \end{aligned} \tag{10.29}$$

where X_i^x are the predicted state vectors when using the sigma point vectors X_i^w in the state equation

$$X_i^x(k+1) = \phi(X_i^{w-}(k)) + L(k)U(k) \tag{10.30}$$

The predicted state covariance matrix is computed as

$$P^-(k) = \sum_{i=0}^{2n_\alpha} W_i^{(c)}[X_i^x(k) - \hat{x}^-(k)][X_i^x(k) - \hat{x}^-(k)]^T \tag{10.31}$$

In the equations of the Unscented Kalman Filter (UKF) there is no linearization of the system dynamics, thus the UKF cannot be included directly into the Extended Information Filter (EIF) equations [230]. Instead, it is assumed that the nonlinear measurement equation of the system given in Eq. (10.9) can be mapped into a linear function of its statistical mean and covariance, which makes possible to use the information update equations of the EIF. Denoting $Y_i(k) = \gamma(X_i^x(k))$ (i.e. the output of the system calculated through the propagation of the i-th sigma point X^i through the system's nonlinear equation) the observation covariance and its cross-covariance are approximated by

$$\begin{aligned} P_{YY}^-(k) &= E[(z(k) - \hat{z}(k)^-)(z(k) - \hat{z}(k)^-)^T] \\ &\simeq J_\gamma(k)P^-(k)J_\gamma(k)^T \\ P_{XY}^-(k) &= E[(x_k - \hat{x}(k)^-)(z_k - \hat{z}(k)^-)^T] \\ &\simeq P(k)J_\gamma(k)^T \end{aligned} \tag{10.32}$$

where $z(k) = \gamma(x(k))$ and $J_\gamma(k)$ is the Jacobian of the output equation $\gamma(x(k))$. Next, multiplying the predicted covariance and its inverse term on the right side of the information matrix and replacing $P(k)J_\gamma(k)^T$ with $P_{XY}^-(k)$ gives the following representation of the information matrix [230, 455]:

$$\begin{aligned} I(k) &= J_\gamma(k)^T R(k)^{-1} J_\gamma(k) = \\ P^-(k)^{-1}&P^-(k)J_\gamma(k)^T R(k)^{-1} J_\gamma^-(k)P^-(k)^T(P^-(k)^{-1})^T \\ &= P^-(k)^{-1}P_{XY}(k)R(k)^{-1}P_{XY}(k)^T(P^-(k)^{-1})^T \end{aligned} \tag{10.33}$$

where $P^-(k)^{-1}$ is calculated according to Eq. (10.31) and the cross-correlation matrix $P_{XY}(k)$ is calculated from

$$P_{XY}^-(k) = \sum_{i=0}^{2n_\alpha} W_i^{(c)}[X_i^x(k) - \hat{x}^-(k)][Y_i(k) - \hat{z}^-(k)]^T \tag{10.34}$$

where $Y_i(k) = \gamma(X_i^x(k))$ and the predicted measurement vector $\hat{z}^-(k)$ is obtained by $\hat{z}^-(k) = \sum_{i=0}^{2n_\alpha} W_i^{(m)} Y_i(k)$. Similarly, the information state vector $i(k)$ can be rewritten as

$$\begin{aligned} i(k) &= J_\gamma(k)^T R(k)^{-1}[z(k) - \gamma(x(k)) + J_\gamma(k)^T \hat{x}^-(k)] \\ &\Rightarrow i(k) = P(k)^{-1}P_{XY}^-(k)R(k)^{-1} \cdot \\ &\cdot [z(k) - \gamma(x(k)) + P_{XY}^-(k)(P^-(k)^{-1})^T \hat{x}^-(k)] \end{aligned} \tag{10.35}$$

A "measurement" matrix $H^T(k)$ is defined as

$$H(k)^T = P^-(k)^{-1}P_{XY}^-(k) \tag{10.36}$$

while the information contributions equations are written as

$$\begin{aligned} i(k) &= H^T(k)R(k)^{-1}[z(k) - \gamma(x(k)) + H(k)\hat{x}^-(k)] \\ I(k) &= H^T(k)R(k)^{-1}H(k) \end{aligned} \tag{10.37}$$

The above procedure leads to an implicit linearization in which the nonlinear measurement equation of the system given in Eq. (10.9) is approximated by the statistical error variance and its mean

$$\begin{aligned} z(k) &= \gamma(x(k)) \simeq H(k)x(k) + \bar{u}(k) \\ \text{where } \bar{u}(k) &= \gamma(\hat{x}^-(k)) - H(k)\hat{x}^-(k) \end{aligned} \tag{10.38}$$

is a measurement residual term.

10.3.5.2 Local Estimations in UIF Terms

It is assumed that the observation vector $\bar{z}_i(k+1)$ is available from N different sensors, and that each sensor observes a common state according to the local observation model, expressed by

$$\bar{z}_i(k) = H_i(k)x(k) + \bar{u}_i(k) + v_i(k) \tag{10.39}$$

where the noise vector $v_i(k)$ is taken to be white Gaussian and uncorrelated between sensors. Then one can define the local information matrix $I_i(k)$ and the local information state vector $i_i(k)$ at the i-th sensor, as follows

$$\begin{aligned} i_i(k) &= H_i^T(k)R_i(k)^{-1}[z_i(k) - \gamma_i(x(k)) + H_i(k)\hat{x}^-(k)] \\ I_i(k) &= H_i^T(k)R_i(k)^{-1}H_i(k) \end{aligned} \tag{10.40}$$

The update equations for the multiple state estimation and data fusion are

$$\begin{aligned} \hat{y}(k) &= \hat{y}^-(k) + \textstyle\sum_{i=1}^N i_i(k) \\ Y(k) &= Y^-(k) + \textstyle\sum_{i=1}^N I_i(k) \end{aligned} \tag{10.41}$$

The Unscented Information Filter running at the i-th measurement processing unit can be written in terms of *measurement update* and *time update* equations:

Measurement update: Acquire measurement $z(k)$ and compute

$$\begin{aligned} &Y(k) = P^-(k)^{-1} + H^T(k)R(k)^{-1}H(k) \ \Rightarrow \\ &Y(k) = Y^-(k) + I(k), \ \ I(k) = H^T(k)R(k)^{-1}H(k) \\ &\hat{y}(k) = \hat{y}^-(k) + H^T(k)R(k)^{-1} \cdot \\ &\cdot [z(k) - \gamma(\hat{x}(k)) + H(k)\hat{x}^-(k)] \\ &\text{or } \hat{y}(k) = \hat{y}^-(k) + i(k) \end{aligned} \tag{10.42}$$

Time update: Compute

$$
\begin{aligned}
&Y^{-}(k+1) = (P^{-}(k+1))^{-1} \\
&P^{-}(k+1) = \sum_{i=0}^{2n_a} W_i^{(c)} \cdot \\
&\cdot [X_i^x(k+1) - \hat{x}^{-}(k+1)][X_i^x(k+1) - \hat{x}^{-}(k+1)]^T \\
&\hat{y}^{-}(k+1) = Y^{-}(k+1) \sum_{i=0}^{2n_a} W_i^{(m)} X_i^x(k+1) \\
&\text{where } X_i^x(k+1) = \phi(X_i^w(k)) + L(k)U(k)
\end{aligned}
\tag{10.43}
$$

10.3.5.3 Distributed UIF for State Estimates Fusion

It is assumed that the local Unscented Kalman Filters do not have access to each other row measurements and are allowed to communicate only their information matrices and their local information state vectors. Then Eq. (10.29) gives

$$
\begin{aligned}
&P_i(k)^{-1} = P_i^{-}(k)^{-1} + H_i^T(k) R_i(k)^{-1} H_i(k) \\
&\hat{x}_i = P_i(k)(P_i^{-}(k)\hat{x}_i^{-}(k) + \\
&+ H_i^T(k) R_i(k)^{-1}[z_i(k) - \gamma_i(x(k)) + H_i(k)\hat{x}^{-}(k)])
\end{aligned}
\tag{10.44}
$$

Using Eq. (10.44), each local information state contribution i_i and its associated information matrix I_i at the i-th filter are rewritten in terms of the computed estimates and covariances of the local filters, i.e.

$$
\begin{aligned}
&H_i^T(k) R_i(k)^{-1} H_i(k) = P_i^{-1}(k) - P_i^{--1}(k) \\
&H_i^T(k) R_i(k)^{-1}[z_i(k) - \gamma_i(x(k)) + H_i(k)\hat{x}^{-}(k)]) = \\
&= P_i(k)^{-1}\hat{x}_i - (P_i^{-}(k)^{-1})\hat{x}_i^{-}(k)
\end{aligned}
\tag{10.45}
$$

where according to Eq. (10.36) it holds $H_i(k) = P_i^{-}(k)^{-1} P_{XY,i}^{-}(k)$. Next, the aggregate estimates of the UIF, can be written in terms of covariances [230, 455]:

$$
\begin{aligned}
&P(k)^{-1} = P^{-}(k)^{-1} + \sum_{i=1}^{N}[P_i(k)^{-1} - P_i^{-}(k)^{-1}] \\
&\hat{x}(k) = P(k)[P^{-}(k)^{-1}\hat{x}^{-}(k) + \\
&+ \sum_{i=1}^{N}(P_i(k)^{-1}\hat{x}_i(k) - P_i^{-}(k)^{-1}\hat{x}_i^{-}(k))]
\end{aligned}
\tag{10.46}
$$

and also in terms of the information state vector and of the information matrix, i.e.

$$
\begin{aligned}
&\hat{y}(k) = \hat{y}^{-}(k) + \sum_{i=1}^{N}(\hat{y}_i(k) - \hat{y}_i^{-}(k)) \\
&Y(k) = Y^{-}(k) + \sum_{i=1}^{N}[Y_i(k) - Y_i^{-}(k)].
\end{aligned}
\tag{10.47}
$$

Since the proposed distributed filtering schemes (EIF or UIF) are scalable, there is no limitation in the number of the local state estimates that can be processed by the aggregation filter. Therefore, the distributed filtering scheme is suitable for state estimation in large scale complex systems, such as the electric power transmission and distribution system. There is no restriction about the dimensionality of the state vector that can be estimated by the distributed filter. Moreover, there is no requirement

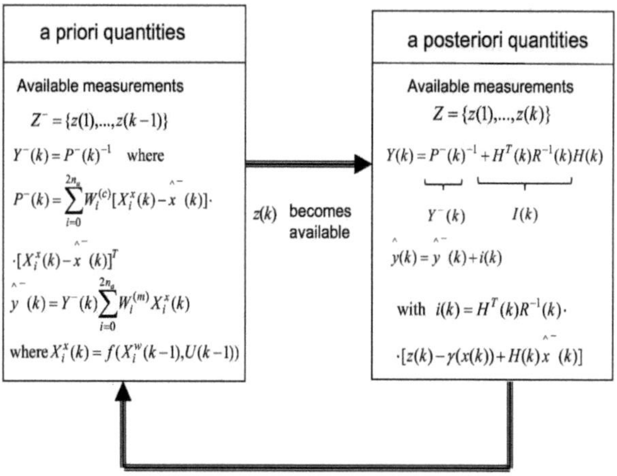

Fig. 10.12 Schematic diagram of the Unscented Information Filter loop

for using a significant number of measurement sensors. Measurements can be taken at specific locations of the power grid and thus there is no need for installing and maintaining costly measurements equipment (Fig. 10.12).

Moreover, it is noted that there is no need to assume a specific network topology since the only information about the significance of a local state estimate and its contribution to the aggregate state estimate comes from the associated covariance matrix. The separation of the local information processing units into clusters could be useful in case that one wanted to investigate the effects of delays in the transmission of the local state estimates towards the aggregate filter. However, even in the latter case this clustering may be obsolete since time delays and packet loses are characterized by randomness and are not directly associated with specific zones where the local information processing units are found.

Additionally, it is pointed out that optimal positioning of the local information processing units can be computed according to some minimization procedure for the trace of the aggregate estimation error covariance matrix or through minimization of indexes such as the Cramer-Rao lower bound which in turn is associated to the Fisher Information matrix [344].

10.3.6 State Estimates Fusion with the Covariance Intersection Method

The previous formulation of EIF and UIF-based distributed filtering given in Sects. 10.3.4 and 10.3.5 respectively assumes that the local state estimates are independent and there is no cross-covariance between them. However, this assumption may not hold and omitting cross-covariance of the local estimates from the aggregation filter results in sub-optimal estimation. To resolve this the Covariance Intersection (CI) method is applied which results in modified equations of the recursion

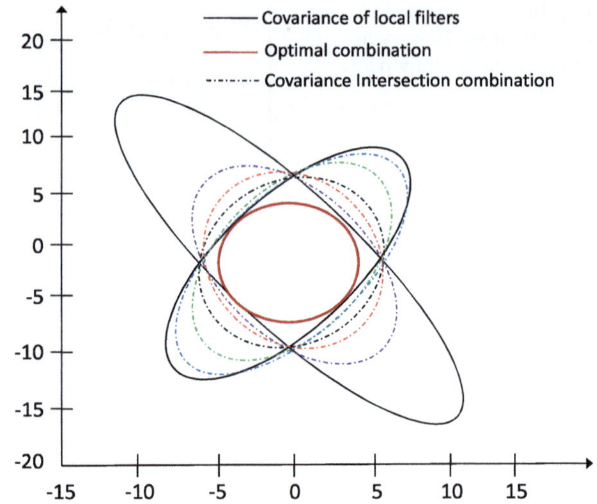

Fig. 10.13 Diagrams of covariance intersection

for distributed state estimation. The Covariance intersection (CI) method enables to combine distributed state estimates and covariances and to maintain consistency [101, 204].

An example of the Covariance Intersection method can be given, in the case of fusion of 2D state vectors provided by local Kalman filters. When the cross covariance is known exactly, then the covariance of the fused estimation lies always within the intersection of the individual covariances. The form of the estimate and the covariance is identical to the standard Kalman Filter when independence is given. When the cross-covariance is unknown, a consistent estimate still exists when its covariance encloses the intersection region (see Fig. 10.13). When the cross-covariance information is known, methods exist that provide optimal fusion and a centralized estimate can be obtained taking into account this information.

A centralized state estimate can be obtained even in the case of unknown cross-covariance. In Fig. 10.13 an example of the covariance intersection process is shown. The local filter estimates correspond to covariance ellipses shown with a continuous black line. The centralized solution, which is the optimal solution produces an ellipse that is within the intersection of the individual ones. The Covariance Intersection (CI) solution produces a solution that always passes through the intersection of the local estimations covariance ellipses. Actually, the Covariance Intersection method provides an estimate and a covariance matrix whose covariance ellipsoid encloses the intersection region. A family of solutions is also possible, and one solution can be chosen by minimizing the trace of the determinant of the combined covariance matrix. In the Covariance intersection approach a scalar weighted average of the covariance matrices is used. The Covariance intersection method is however conservative in the sense that its error covariance ellipsoid is larger than the true one.

Consider two local filters and consequently two local state estimates with covariance pairs $\{a, P_{aa}\}$ and $\{b, P_{bb}\}$. The true values of the covariances matrices are denoted by an overbar, i.e. $\bar{P}_{aa} = E\{\tilde{a}\tilde{a}^T\}$ and $\bar{P}_{bb} = E\{\tilde{b}\tilde{b}^T\}$, where $\tilde{a} = a - \bar{a}$ and $\tilde{b} = b - \bar{b}$. It is assumed that the estimates for a and b are consistent, so that $P_{aa} - \bar{P}_{aa} \geq 0$ and $P_{bb} - \bar{P}_{bb} \geq 0$. The optimal filter incorporates $\bar{P}_{ab}$ in its state covariance computation. However, this information is lost, i.e. unknown in a decentralized system. A consistent estimate when fusing state estimates a and b is given by

$$\begin{aligned} P_{cc}^{-1} &= \omega P_{\alpha\alpha}^{-1} + (1-\omega)P_{bb}^{-1} \\ c &= P_{cc}[\omega P_{\alpha\alpha}^{-1}\alpha + (1-\omega)P_{bb}^{-1}b] \end{aligned} \tag{10.48}$$

where $\omega \in [0,\ 1]$ is a scalar weight. The requirement for ω ensures that the covariance $P_{cc} \geq 0$, $P_{aa} \geq P_{cc}$ and $P_{bb} \geq P_{cc}$. It has been proven that the estimate c is consistent for all P_{ab} and ω [204]. That is $P_{cc} - \bar{P}_{cc} \geq 0$, where $\bar{P}_{cc} = E\{\tilde{c}\tilde{c}^T\}$ with $\tilde{c} = c - \bar{c}$.

The optimal weights can be found using an optimization scheme that minimizes the trace of the determinant of P_{cc} at every step. The trace and the determinant of P_{cc} characterize the size of the Gaussian uncertainty ellipsoid associated with P_{cc}. Minimizing the trace of the determinant of P_{cc} is a convex optimization problem. This means that the cost function has only one local optimum of ω in the range of [0, 1], which is also the global optimum.

One can apply the covariance intersection approach to fuse multiple estimates. Given a set of M estimates $\hat{x}_1, \hat{x}_2, \ldots, \hat{x}_M$ and associated covariances $P_1, P_2, \ldots, P_M$ a consistent estimate is given by

$$\begin{aligned} P^{-1} &= \textstyle\sum_{i=1}^{M} \omega_i P_i^{-1} \\ \hat{x} &= P \textstyle\sum_{i=1}^{M} \omega_i P_i^{-1} \hat{x}_i \end{aligned} \tag{10.49}$$

where the weights satisfy $\sum_{i=1}^{M} \omega_i = 1$ and $\omega_i \in [0,\ 1]$. The weights ω_i can be found by minimizing the trace of the determinant of P subject to the aforementioned constraints.

In the case of the local state estimators, optimality of estimation can be shown only for the linear Kalman Filter, in terms of minimization of the trace of the covariance matrix of the estimation error vector. There is no proof of convergence of the Extended Kalman Filter in the stochastic case and the truncation of higher order terms in the associated Taylor series expansion introduces cumulative error which deteriorates the accuracy of estimation or can even risk the convergence of the local filter. On the other hand, the Unscented Kalman Filter is a derivative-free state estimator that does not feed back in its recursion any linearization or approximation error and in that sense is more robust and accurate than the Extended Kalman Filter. In the case of the aggregate filter, as already explained, a sub-optimal state estimate is obtained if the cross-covariance between the local state estimates is neglected. However this can be avoided with the covariance intersection method and thus the accuracy of the global state estimate can be also improved. There is no analytical proof about the effect that feedback from the aggregate state estimators to the local state estimators, could have on the convergence and the accuracy of both the local filters and the of the aggregate filter.

It is also noted that comparing to centralized state estimation the proposed distributed state estimators have significant advantages: (i) they are fault tolerant: if a local processing unit is subject to a fault then state estimation is still possible and accurate, (ii) the computation load is distributed between local processing units and since there is no need to exchange a large amount of information, the associated communication bandwidth is low, (iii) the communication bandwidth remains low since in the EIF and UIF case there is no need to send raw data from the local processing units to the aggregate filter, but it is only needed to transmit information covariance matrices and information state vectors.

10.3.7 Simulation Tests

10.3.7.1 Distributed State Estimation for the Electric Power Distribution System

For testing the performance of the proposed distributed filtering approaches, first a power grid with five states and three observation nodes is simulated [294]. At each node a local state estimator runs. The overall state estimation is given by

$$x(k+1) = \phi(x(k)) + v(k) \tag{10.50}$$

where the system's state vector $x(k) \in R^{5\times 1}$ and noise vector $v(k) \in R^{5\times 1}$ are defined as

$$\begin{aligned} x(k) &= [x_1(k), x_2(k), x_3(k), x_4(k), x_5(k)]^T \\ v(k) &= [v_1(k), v_2(k), v_3(k), v_4(k), v_5(k)]^T \end{aligned} \tag{10.51}$$

while vector $\phi(x(k)) \in R^{5\times 1}$ is defined as

$$\phi(x(k)) = \begin{pmatrix} 1 + sin(0.04\pi x_1(k)) + x_2(k) \\ 1 + sin(0.04\pi x_2(k)) + x_4(k) \\ 1 + sin(0.04\pi x_1(k)) + x_3(k) \\ 1 + sin(0.04\pi x_4(k)) + x_5(k) \\ 1 + sin(0.04\pi x_5(k)) + x_4(k) \end{pmatrix} \tag{10.52}$$

The process noise terms $v_i(k),\ i = 1, \ldots, 5$ are assumed to follow a Gaussian distribution. The observation model is composed of 3 measurement processing nodes and is given by

$$y^i(k) = C^i x^i(k) + w^i(k),\ i = 1, 2, 3 \tag{10.53}$$

where $y(k) \in R^{3\times 1}$, $C \in R^{3\times 5}$ and $x(k) \in R^{5\times 1}$. Moreover, $w(k) \in R^{3\times 1}$ is the measurements' Gaussian noise vector. A possible form for matrix C is given by [294]

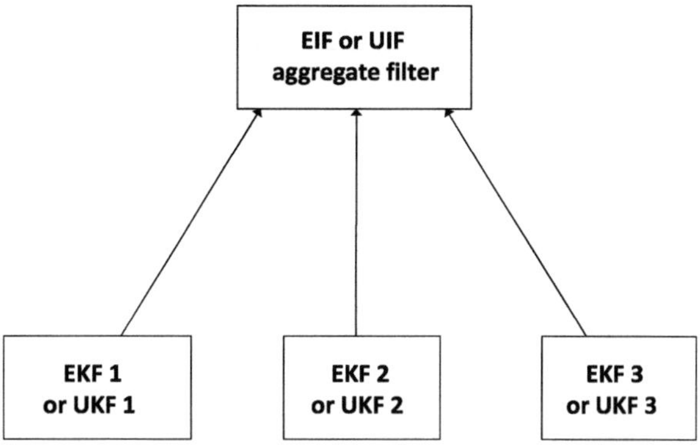

Fig. 10.14 Structure of the Extended Information Filter or the unscented information filter for condition monitoring of the electric power distribution system

$$C = \begin{pmatrix} 0 & 1 & 1 & 0 & 1 \\ 1 & 0 & 1 & 0 & 1 \\ 1 & 1 & 0 & 1 & 0 \end{pmatrix} \tag{10.54}$$

The Jacobian $J_\phi(x)$ of the $\phi(x)$ is calculated as follows:

$$J_\phi(x) = \begin{pmatrix} \frac{\partial \phi_1}{\partial x_1} & \frac{\partial \phi_1}{\partial x_2} & \cdots & \frac{\partial \phi_1}{\partial x_5} \\ \frac{\partial \phi_2}{\partial x_1} & \frac{\partial \phi_2}{\partial x_2} & \cdots & \frac{\partial \phi_2}{\partial x_5} \\ \cdots & \cdots & \cdots & \cdots \\ \frac{\partial \phi_5}{\partial x_1} & \frac{\partial \phi_5}{\partial x_2} & \cdots & \frac{\partial \phi_5}{\partial x_5} \end{pmatrix} \tag{10.55}$$

The numerical values of the Jacobian's elements are: $J_\phi^{11} = J_\phi^{22} = J_\phi^{31} = J_\phi^{43} = J_\phi^{55} = 0.04\pi cos(0.04\pi x_1(k))$, $J_\phi^{12} = J_\phi^{14} = J_\phi^{33} = J_\phi^{44} = J_\phi^{45} = 1$, while the rest of the elements are 0. The distributed filtering architecture which is used for condition monitoring of the electric power distribution system is shown in Fig. 10.14.

On the other hand, the Unscented Information Filter needs neither the linearization of the electric power system model, nor the calculation of Jacobian matrices. The implementation of the Unscented Information Filter in the simulation experiments was performed according to the stages described in Sect. 10.3.5. In terms of accuracy of the provided state estimation, both the EIF and UIF give state estimates of the nonlinear electric power system within acceptable accuracy levels, which can be used for condition monitoring tasks. It is noted however, that the UIF does not introduce errors due to linearization approximations, thus in that sense it results in more robust distributed state estimation.

As it can be noticed from the simulation experiments it is possible to obtain a global estimate of the state vector of the power grid, through the fusion of the local state estimates, provided by the local Extended Kalman Filters or the local Unscented Kalman Filters. In case that the power distribution system is in the fault-free condition, then the state estimates provided by the Extended Information Filter or the Unscented Information Filter converge asympotically to the real state variables

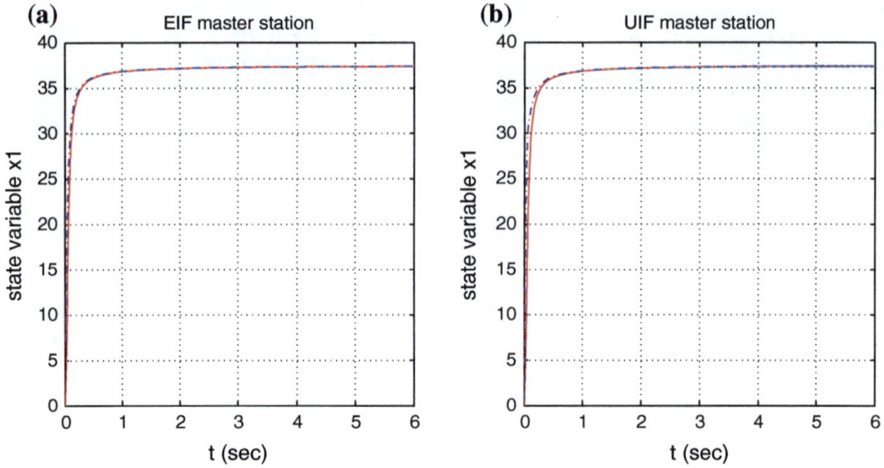

Fig. 10.15 State variable x_1 (*continuous line*) and distributed filtering-based estimation of state variable $\hat{x}_1$ (*dashed line*) of the nonlinear electric power system in the fault-free condition: **a** obtained by the EIF, **b** obtained by the UIF

(see Fig. 10.15). When a fault exists a discrepancy between a state variable and the associated estimated value is observed.

The previously explained EIF and UIF-based distributed state estimation was tested in the case of an additive fault affecting the sensors used to monitor the outputs of the power distribution system. As it can be observed in Figs. 10.16 and 10.20, the existence of a fault causes discrepancy between the real and the estimated state variables of the power distribution system. Further processing of the residuals (differences between the real and the estimated state variables) is possible, with the use of fault diagnosis algorithms [242, 347]. A statistical FDI method that can be used to find incipient failures in the electric power system is the so-called *Local Statistical Approach* to change detection. Äs previously noted, the local statistical approach has been successfully applied to several FDI problems, such as vibration monitoring in dynamic systems, while electric power systems and renewable energy is also a promising application field [34, 344, 508] (Figs. 10.17, 10.18 and 10.19).

10.3.8 Distributed State Estimation for Detection of Voltage Dips and Harmonics Variation

Voltage dips are among the most important power quality disturbances because they occur frequently and can affect electrical and electronic equipment that is sensitive to such voltage variation. The use of efficient methods for detecting voltage dips is important for avoiding the negative economic impact that abrupt voltage variations

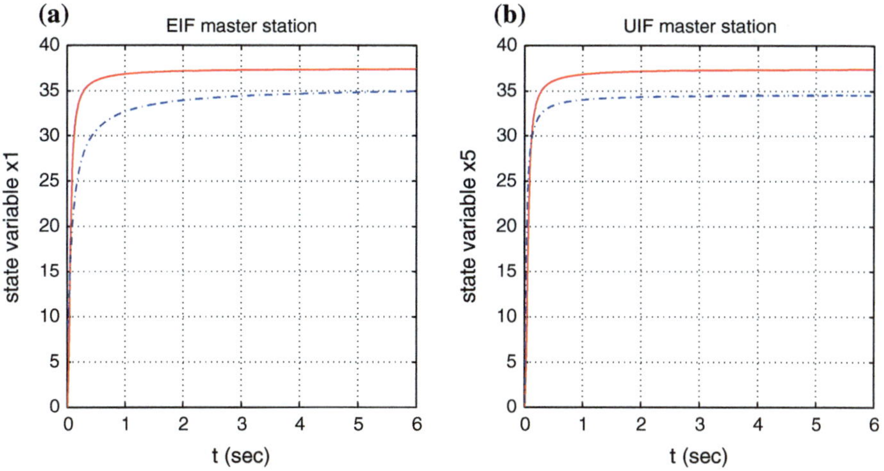

Fig. 10.16 State variable x_1 (*continuous line*) and distributed filtering-based estimation of state variable $\hat{x}_1$ (*dashed line*) of the nonlinear electric power system in a faulty condition: **a** obtained by the EIF, **b** obtained by the UIF

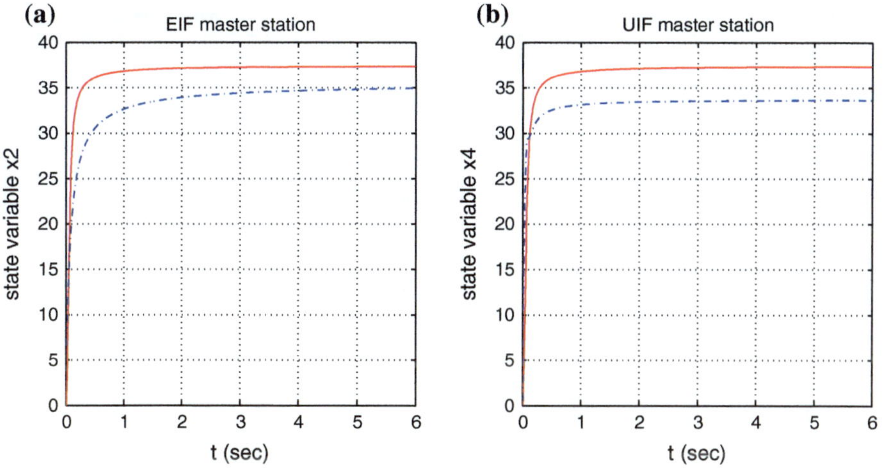

Fig. 10.17 State variable x_2 (*continuous line*) and distributed filtering-based estimation of state variable $\hat{x}_2$ (*dashed line*) of the nonlinear electric power system in the faulty condition: **a** obtained by the EIF, **b** obtained by the UIF

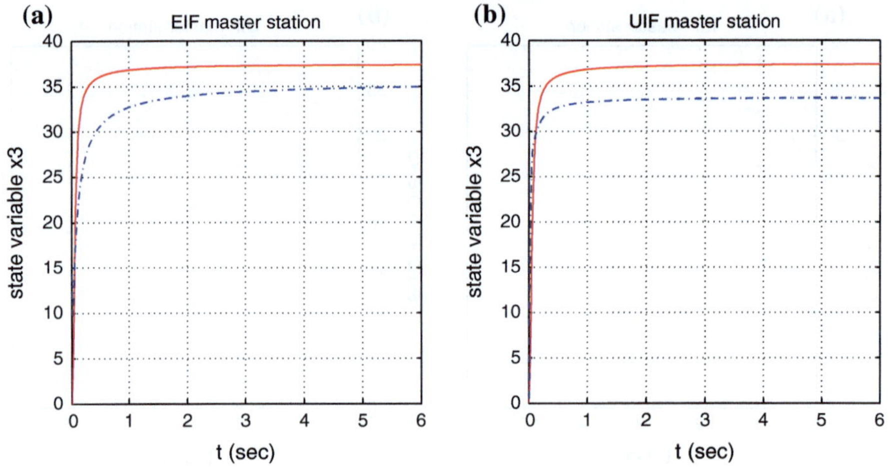

Fig. 10.18 State variable x_3 (*continuous line*) and distributed filtering-based estimation of state variable $\hat{x}_3$ (*dashed line*) of the nonlinear electric power system in a faulty condition: **a** obtained by the EIF, **b** obtained by the UIF

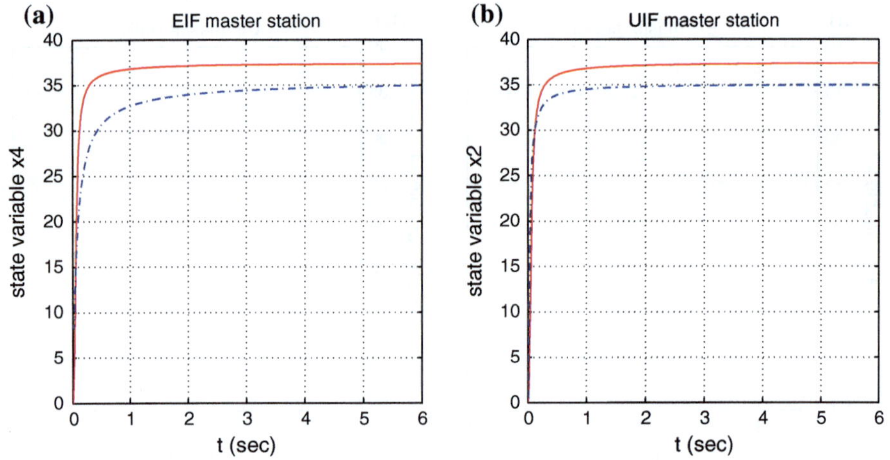

Fig. 10.19 State variable x_4 (*continuous line*) and distributed filtering-based estimation of state variable $\hat{x}_4$ (*dashed line*) of the nonlinear electric power system in a faulty condition: **a** obtained by the EIF, **b** obtained by the UIF

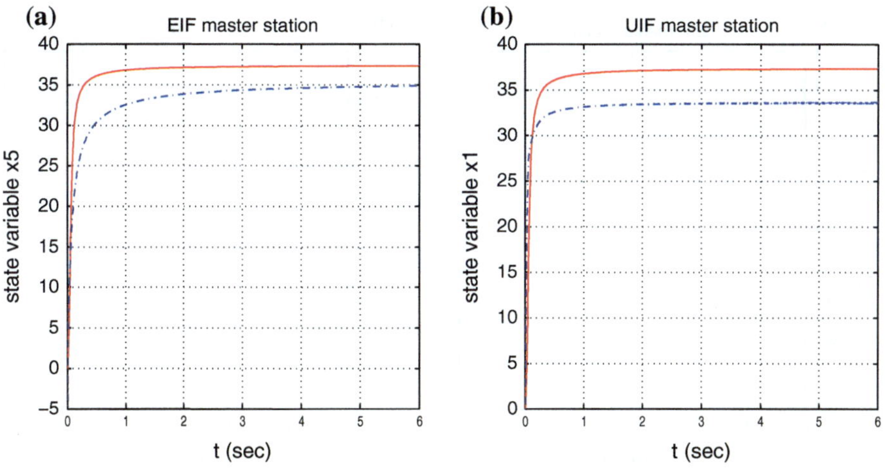

Fig. 10.20 State variable x_5 (*continuous line*) and distributed filtering-based estimation of state variable $\hat{x}_5$ (*dashed line*) of the nonlinear electric power system in a faulty condition: **a** obtained by the EIF, **b** obtained by the UIF

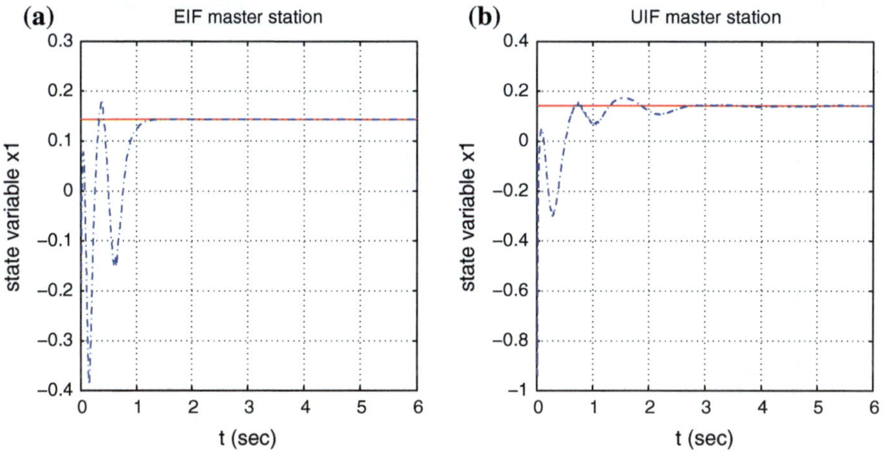

Fig. 10.21 Estimation of the state variable x_1 resulting from decomposition of the voltage signal into harmonics **a** using the extended information filtering, **b** using the unscented information filter

may have to industrial and commercial distribution systems. Previous results on state estimation methods for the detection of abrupt voltage variations such as voltage dips, sags and flickers are based on localized Kalman Filtering approaches [122, 321]. Here, the Extended Information Filter and the Unscented Information Filter will be used, in a distributed state estimation scheme, for the detection of changes in the voltage signal profile. It is assumed that the locally measured voltage signal can

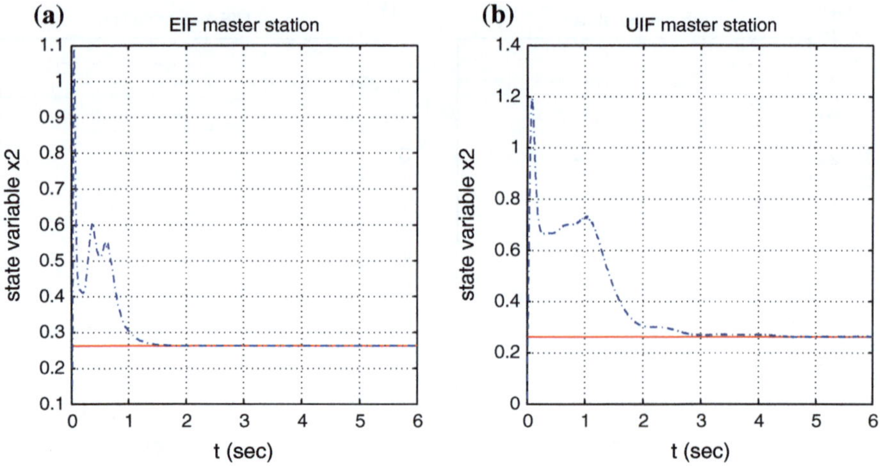

Fig. 10.22 Estimation of the state variable x_2 resulting from decomposition of the voltage signal into harmonics **a** using the extended information filtering, **b** using the unscented information filter

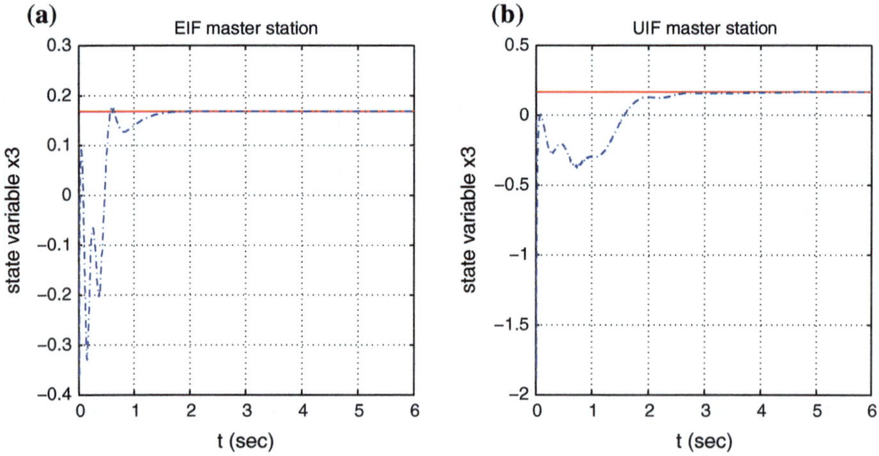

Fig. 10.23 Estimation of the state variable x_3 resulting from decomposition of the voltage signal into harmonics **a** using the extended information filtering, **b** using the unscented information filter

be decomposed in n harmonic components, which in discrete-time form are written as

$$z(k) = \sum_{i=1}^{N} A_i sin(i\omega k T_s + \theta_i) \tag{10.56}$$

where i is the harmonic order, A_i and θ_i are the amplitude and phase angle of each harmonic component, T_s is the sampling period and $\omega = 2\pi f$ is the power system frequency. Denoting the state variables

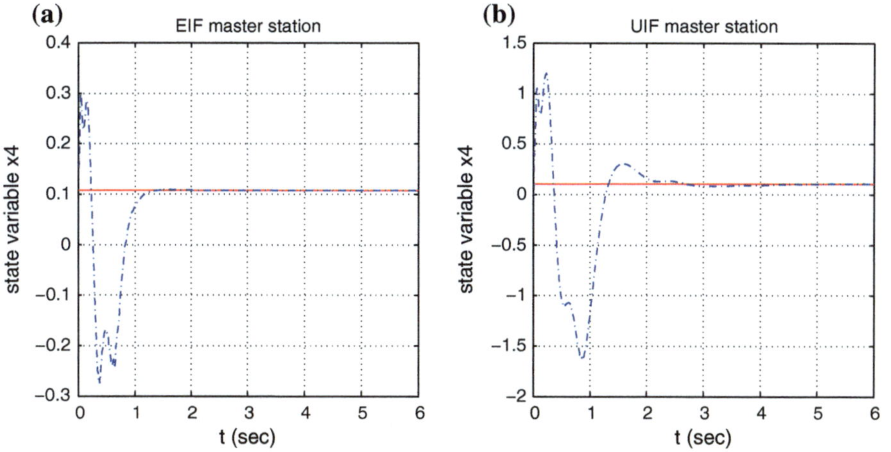

Fig. 10.24 Estimation of the state variable x_4 resulting from decomposition of the voltage signal into harmonics **a** using the extended information filtering, **b** using the unscented information filter

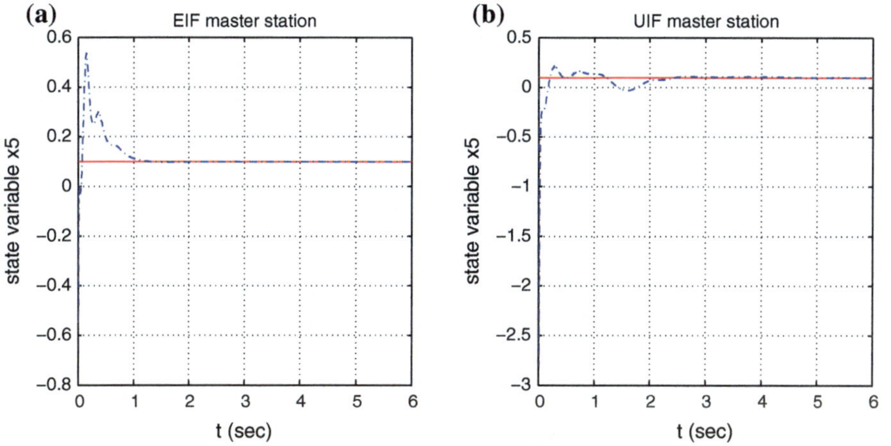

Fig. 10.25 Estimation of the state variable x_5 resulting from decomposition of the voltage signal into harmonics **a** using the extended information filtering, **b** using the unscented information filter

$$
\begin{array}{ll}
x_1 = A_1 cos(\theta_1) & x_2 = A_1 sin(\theta_1) \\
x_3 = A_2 cos(\theta_2) & x_4 = A_2 sin(\theta_2) \\
x_5 = A_3 cos(\theta_3) & x_6 = A_3 sin(\theta_3) \\
\cdots & \cdots \\
x_{2N-1} = A_N cos(\theta_N) & x_{2N} = A_N sin(\theta_N)
\end{array}
\quad (10.57)
$$

one obtains the following description for the voltage dynamics

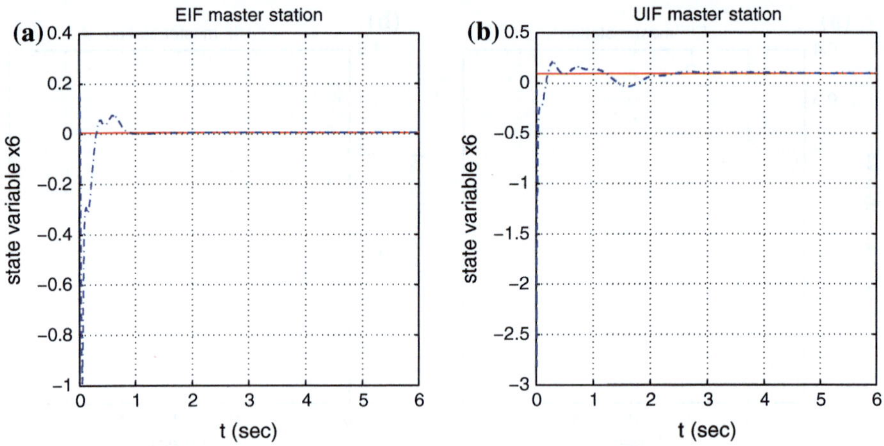

Fig. 10.26 Estimation of the state variable x_6 resulting from decomposition of the voltage signal into harmonics **a** using the extended information filtering, **b** using the unscented information filter

$$\begin{aligned} x(k+1) &= \phi(x(k)) + w(k) \\ z(k) &= \gamma(x(k)) + v(k) \end{aligned} \tag{10.58}$$

with

$$\phi(x(k)) = \Phi \cdot x(k) \text{ where} \\ \Phi = \begin{pmatrix} 1 & 0 & 0 & \cdots 0 \\ 0 & 1 & 0 & \cdots 0 \\ 0 & 0 & 1 & \cdots 0 \\ \cdots & \cdots & \cdots & \cdots \\ 0 & 0 & 0 & \cdots 1 \end{pmatrix} \tag{10.59}$$

$$\begin{aligned} \gamma(x(k)) = & \, sin(\omega k T_s)x_1 + cos(\omega k T_s)x_2 + \\ & + sin(2\omega k T_s)x_3 + cos(2\omega k T_s)x_4 + \\ & + sin(3\omega k T_s)x_5 + cos(3\omega k T_s)x_6 + \\ & \cdots\cdots \\ & + sin(N\omega k T_s)x_{2N-1} + cos(N\omega k T_s)x_{2N} \end{aligned} \tag{10.60}$$

The associated Jacobian matrices are calculated as follows:

$$J_\phi = \frac{\partial \phi(x)}{\partial x}|_{x(k)=\hat{x}(k)} = \Phi \tag{10.61}$$

$$J_\gamma = \frac{\partial \gamma(x)}{\partial x}|_{x(k)=\hat{x}(k)}, \ i.e. \tag{10.62}$$

$J_\gamma = [sin(\omega k T_s), cos(\omega k T_s), sin(2\omega k T_s), cos(2\omega k T_s),$
$\ldots, sin(N\omega k T_s), cos(N\omega k T_s)].$

The simulation experiments that follow have been produced assuming that the number of harmonics that constitute the voltage signal was $N = 3$. The obtained results are depicted in Figs. 10.21, 10.22, 10.23, 10.24, 10.25 and 10.26. The estimated state variables are denoted with a dashed blue line, while the associated reference setpoints are marked as continuous red lines. As in the case of Sect. 10.3.7.1 it was assumed that voltage measurements were collected at three different observation nodes where local state estimates where produced.

It can be observed that using distributed state estimation based either on the Extended Information Filter or on the Unscented Information Filter, the local state estimates produced at local measurement stations could be fused into an aggregate state estimate. This global state estimate provided an indication about changes in the voltage signal. For example, voltage dips and harmonics distortions are associated to changes of the values of state variables x_1 to x_6 of the state-space model of Eqs. (10.58) and (10.59). This means also that, voltage dips and harmonics distortions are associated to variations of the parameters appearing in Eq. (10.57), i.e. the amplitude $A_i, i = 1, \ldots, 3$ and the phase angle $\theta_i, i = 1, \ldots, 3$. As long as, no deviation of x_1 to x_6 is observed with respect to the setpoints appearing in Figs. 10.21, 10.22, 10.23, 10.24, 10.25 and 10.26, no voltage dips or harmonics distortions are detected in the line's voltage.

It is worth mentioning that the local Kalman Filters recursion involved in the previously analyzed distributed state estimation scheme can be reformulated so as to compensate for delayed measurements and packet drops in measurements transmissions [461, 478]. Two cases are distinguished: (i) there are time delays and packet drops in the transmission of information between the distributed local filters and the master filter, (ii) there are time delays and packet drops in the transmission of information from distributed sensors to each one of the local filters. In the first case, the structure and calculations of the master filter for estimating the aggregate state vector remain unchanged. In the second case, the effect of the random delays and packets drops has to be taken into account in the redesign of the local Kalman Filters, which implies a modified Riccati equation for the computation of the covariance matrix of the state vector estimation error, as well as the use of a correction (smoothing) term in the update of the state vectors estimate so as to compensate for delayed measurements arriving at the local Kalman Filters.

Glossary

AC to DC voltage source converters: Voltage source converters (VSC), are three-phase filtered rectifiers, that are widely used in the electric power grid (mainly for power flow control). Voltage source converters are the main building blocks of power flow controllers in transmission lines. For example, VSCs are contained in Unified Power Flow Controllers (UPFCs), or distribution-static synchronous compensators (D-STATCOMs). VSCs enable control of the amplitude and phase angle of the AC terminal voltages. Moreover, their bidirectional power flow capabilities allow VSCs to perform real and/or reactive power flow control in AC transmission lines. The dynamic model of voltage source converters is a nonlinear one and requires the development of nonlinear control methods.

Boundary Control: An approach to the control of partial differential equations in which the control action is exerted to the PDE through its boundary conditions. This is different to distributed or pointwise control of PDEs, in which the control action is exerted at several points of the systems state space.

Control of renewable energy systems with global linearization methods: these are methods based on the transformation of the systems' dynamics to equivalent state-space descriptions for which one can design stabilizing feedback controllers and can solve the filtering (state estimation problem). In this area one can classify methods based on differential flatness theory or Lie algebra, which finally permit the design of controllers and filters of high precision and robustness. In this field belongs also the new nonlinear filtering method under the name Derivative-free nonlinear Kalman Filter.

Control of renewable energy systems with approximate linearization methods: in this area one can distingusih robust and adaptive control methods which use the decomposition of the system's dynamics into local linear models. There are methods of nonlinear feedback control making use of linearized description of the system round local operating points. The parameters of the local controllers is performed in a manner that assures the stability and robustness of the closed loop. The controllers achieve asymptotically (as time advances) the compensation of the nonlinear dynamics of the system. There are results about control with the solution of Linear

G. Rigatos, *Intelligent Renewable Energy Systems*,
Green Energy and Technology, DOI 10.1007/978-3-319-39156-4

Matrix Inequalities (LMIs) and solution of multiple Riccati equations. There are also results on linearization of the system's dynamics round temporary equilibria, with the use of Taylor series expansion and the computation of Jacobian matrices.

Control of renewable energy systems with Lyapunov methods: In this area one can distinguish nonlinear control methods based on the minimization of Lyapunov functions. For the development of Lyapunov controllers one can either use a model of the system's dynamics or can avoid completely the dependence on a dynamic model, as for instance in the case of adaptive control. In this latter case, the system's dynamics is completely unknown and is approximated online with the use of adaptive algorithms. Lyapunov methods assure the stability and robustness of the control loop.

DC to DC voltage converters: DC-DC converters are used for the exploitation of the power produced by photovoltaic units or fuel cells, since the DC-DC converter control the level of the produced output voltage. In turn the DC output voltage can be fed into DC motors, as for instance in the case of actuators, mechatronic devices and pumps. Pulse Width Modulation (PWM) is applied for such converters' control. The amplitude of the output voltage is determined by the duty cycle of the PWM.

DC to AC converters (inverters): The dynamic model of the inverter is a nonlinear multi-variable one. The control inputs are the voltages generated by an H-bridge circuit while the state vector comprises as elements the voltages at the load's side and the currents at the H-bridge circuit output. Efficient control of inverters is important for the reliable operation of the electric power grid and the connection to it of distributed DC power generation units (such as photovoltaic arrays providing power to the main grid or local photovoltaic cells connected to a microgrid infrastructure). Moreover, inverters find extensive use in HVDCs (High Voltage DC) lines and in their connection with the rest of the AC electric power network.

Derivative-free nonlinear Kalman Filter: A conditionally optimal state estimator for nonlinear dynamical systems. The constituents of the filter are: (i) a transformation that is based on differential flatness theory and which enables to express the system's dynamics into an equivalent linear canonical (Brunovsky) form, (ii) solution of the state estimation problem for the lineaarized equivalent model of the system, using the Kalman Filter recursion, (iii) an inverse transformation that is based again on differential flatness theory which enables to obtain estimates for the state variables of the initial nonlinear model.

Differential Flatness Theory: This is a primary research direction in the area of nonlinear dynamical systems control. It considers that instead of describing the systems dynamics through its entire state vector, one can use for this purpose specific algebraic variables which are called flat outputs and which are dependent only on certain elements of the state vector. Differential flatness theory enables to succeed global linearization for complicated nonlinear dynamics and in this manner to solve the associated control and state estimation problems.

Distributed parameter system: A dynamical system that evolves not only in time but also in space. Otherwise stated the system exhibits spatiotemporal dynamics along the time axis and along one or more spatial axes. Systems described by partial differential equations are distributed parameter ones.

Distributed power generation systems: A large-scale power system consists of multiple control areas where local power generation units operate, while these areas are connected through tie lines and power transformers. Due to change in active power loads, the frequencies of the areas variate from their nominal values (synchronous speed) and power exchange through the tie lines will also deviate from the associated setpoints. To stabilize the functioning of the distributed power generation system and eliminate frequency deviations elaborated nonlinear control approaches have to be implemented. Moreover, the integration of distributed DC power generation units in microgrids, such as photovoltaics and fuel cells requires the interfacing of such devices with the electricity network through inverters. Another problem that arises is the synchronization between multiple parallel inverters.

Doubly-fed induction generators: In DFIGs the rotor's magnetic field is considered to be rotating at an asynchronous speed that is not dependent on grid's reference frequency. DFIGs have been widely used in variable-speed fixed frequency hydro-power generation systems, wind-power generation systems and turbine engine power generation systems. Doubly-fed induction generators have proven to be more efficient the synchronous generator systems in terms of cost and losses of the associated power electronics converters. DFIG systems can operate either in grid-connected mode or in stand-alone mode.

Fault diagnosis for distributed power generation units: Apart from control and stabilization another problem that has to be dealt with in the functioning of renewable power generation units is that of condition monitoring and fault diagnosis. There are several fault detection and isolation problems associated with power generators, either in stand alone or in grid-connected functioning. Furthermore, there are several fault detection and isolation problems associated with the power electronics which are used for connecting the renewable energy sources to the grid, such as converters, inverters and power transformers. Finally, there are fault detection and isolation issues related to the transmission and distribution grid. The development of elaborated fault diagnosis tools can result if efficient preventive maintenance of renewable power generation units, early detection of their components' failure, safer operation of the power grid and uninterrupted power supply.

Fuel cells: Fuel cells are electrochemical energy devices that convert the chemical energy of the reaction between hydrogen and oxygen into electricity and heat giving also water as by product of the reaction. Fuel cells are a renewable power generation source and their use gets widely deployed in the smart grid. Fuel Cell of the Proton Exchange Membrane type consist of a polymer electrolyte membrane which is placed between the electrodes (anode and cathode). Ions can be diffused through the membrane. If an electrical circuit is established between the anode and the cathode, there will be also a flow of electrons and a potential will appear between the electrodes. Gas supply at the fuel cells' cathode and anode controls the produced electric power. Fuel cells generate DC voltage therefore, at levels which can be varied with the use of a DC to DC converter.

Kalman Filtering: An optimal state estimation method, which enables to reconstruct the entire state vector of a dynamical system out of indirect measurement. For linear dynamical systems and under white Gaussian noise the method exhibits optimality in the sense that it minimizes the trace of the covariance matrix of the estimation error.

Li-ion batteries: Lithium-ion batteries exhibit higher power density than other types of batteries, low self-discharge and prolonged life cycle. The distributed parameter models of Li-ion batteries comprise several coupled PDEs which describe the diffusion of Li ions at the solid electrodes as well as their diffusion in the electrolyte. Control of these diffusion phenomena, by applying external voltage or currents can result in making the battery dynamics follow specific charge or discharge patterns as well to keep its functioning within desirable charge levels.

Lyapunov function: This is an energy function of the dynamical system which depends on quadratic terms of the systems state vector error. It takes positive values apart from the equilibrium where it becomes zero. A system is stabilized when the associated Lyapunov function becomes zero.

Multi-phase machines: These power generators have multiple phases (windings) at the stator. Thus one can use multi-phase synchronous generators or doubly-fed reluctance generators. Multi-phase synchronous machines are a generalization of the case of synchronous machines after including additional windings at the stator. On the other side, in Doubly-fed reluctance machines there is no winding and no permanent magnet at the rotor while an additional winding is included in the stator through which the power of the machine is controlled.

Nonlinear H-infinity control: An approach to nonlinear optimal control under model uncertainty and external perturbations. The method provides optimal solution to a differential game taking place between the control input (trying to minimize deviation of the state vector from the reference setpoint) and the disturbance input (trying to maximize the aforementioned deviation).

Renewable Energy Systems: Renewable energy systems produce electricity out of renewable energy sources, such as wind power, solar power, hydro-power and tidal power. The generated electric power in AC or DC form has to be transmitted and distributed through the electricity grid, therefore renewable energy generation is also concerned with connection to the grid through the use of power electronics. Basic components of renewable energy systems are: (i) power generators (PMSG, DFIG, multi-phase machines, and Doubly-Fed Reluctance machines), (ii) power electronics for integrating generators to the grid (DC to AC inverters, AC to DC and DC to DC converters, active power filters, power transformers).

Stability: A property of a dynamical system denoting that the state vector of the system converges to a specific point in the state space, which is called equilibrium or to a bounded region in the state-space which is called domain of attraction and remains there.

Synchronous generators: In synchronous generators the rotor's magnetic field is considered to be rotating at the synchronous speed which is determined by grid's reference frequency. Comparing to DFIGs the associated dynamic model is of smaller dimension.

Power electronics: The power generated by renewable energy sources has to be transmitted and distributed through the electricity grid and to this end the renewable power generation units have to be connected to the grid with the use of power electronics. These can be three-phase AC to DC converters, inverters (DC to AC converters), DC to DC converters and power transformers.

References

1. Abdeddaim S, Betka A, Drid S, Becherif M (2014) Implementation of MRAC controller of a DFIG based variable speed grid. Energy Convers Manag Elsevier 79:281–288 (connected wind turbine)
2. Abjadi NR (2014) Sliding-mode control of a six-phase series/parallel connected two induction motors drive. ISA Trans Elsevier 53:1847–1856
3. Abu-Elanien A, Salama M (2010) Asset management techniques for transformers. Electr Power Syst Res 80:465–466
4. Addison PS (2002) The illustrared wavelet transform handbook. Institute of Physics Publishing
5. Ademi S, Jovanovic M (2015) Vector control methods for brushless doubly fed reluctance machines. IEEE Trans Indus Electron 62(1):96–104
6. Ademi S, Jovanovic M (2016) Control of doubly-fed reluctance generators for wind power applications. Renew Energy Elsevier 85:171–180
7. Adibi MM, Kafka RJ, Maram S, Mili LM (2006) On power system controlled separation. IEEE Trans Power Syst 21(4):1894–1902
8. Aguglia P, Voiarouge P, Wamkeue R, Cros J (2010) Determination of fault operation dynamical constraints for the design of wind turbine DFIG drives. Math Comput Simul Elsevier 81: 252–262
9. Ahmed KH, Massoud AM, Finney SJ, Williams BW (2009) Autonomous adaptive sensorless controller of inverter-based islanded-distributed generation system. IET Power Electron 2(3): 256–266
10. Al-Baiyat SA (2005) Power system transient stability enhancement by STATCOM with nonlinear H_∞ stabilizer. Electr Power Syst Res Elsevier 73:45–52
11. Alden MJ, Wang X (2015) Robust H-infinity control of time delayed power systems. Syst Sci Control Eng Taylor and Francis 3:253261
12. Ali HS, Boutat-Baddas L, Becis-Aubry Y, Darouach M (2006) H_∞ control of a SCARA robot using polytopic LPV approach. In: IEEE MED 2006, 14th Mediterranean Conference on Control and Automation, Ancina, Italy
13. Allag A, Hammoudi M, Mimoune SM, Ayad MY, Becherif M, Miraoui A (2007) Tracking control via adaptive backstepping approach for a three-phase PWM AC-Dc converter. In: IEEE ISIE 2007 International symposium on industrial electronics
14. Alonge F, D'Ippolito A (2010) Robustness analysis of an extended Kalman filter for sensorless control of induction motors. In: 2010 IEEE International Symposium on Industrial Electronics, ISIE 2010. Bari, Italy

G. Rigatos, *Intelligent Renewable Energy Systems*,
Green Energy and Technology, DOI 10.1007/978-3-319-39156-4

15. Altifai MT, Hassan MF, Zribi M (2011) Decentralized load frequency controller for a multi-area interconnected power system. Electr Power Energy Syst Elsevier 33:198–209
16. An L, Dah-Chuan D (2015) Lu, design of a single-switch DC/DC converter for a PV-battery-powered pump system with PFM+PWM Control. IEEE Trans Indus Electron 62(2):910–921
17. Anderson G, Donalek P, Farmer R, Hatziargyriou N, Kamwa I, Kundur P, Martins N, Paserba J, Pourbeik P, Sanchez-Gasca J, Schulz R, Stankovic A, Taylor C, Vittal V (2005) Causes of the 2003 major grid blackouts in North America and Europe, and recommended means to improve system dynamic performance. IEEE Trans Power Syst 20(4):1922–1928
18. Andoulssi R, Draou A, Jerbi H, Alghonamy A, Khiari B (2013) Nonlinear control of a photovoltaic water pumping system. Energy Procedia Elsevier 42:328–336
19. Angulo M, Ruiz-Caballero DA, Lago J, Heldwein ML, Musso SA (2013) Active power filter control strategy with implicit closed-loop current control and resonant controller. IEEE Trans Ind Electron 60(7):2721–2730
20. Anene EC, Aliya UO, Lévine J, Vemayagamoorthy GK (2007) Flatness-based feedback linearization of a synchronous machine model with static excitation and fast turbine valving. In: IEEE 2007 power engineering. General Meeting, Tampa, Florida USA
21. Angulo-Nunez MI, Sira-Ramirez H (1998) Flatness in the passivity based control of DC-to-DC power converters. In: Proceedings of the 37th IEEE conference on decision and control, Tampa, Florida, USA
22. Antritter F, Maurer P, Reger J (2006) Flatness-based control of a buck-converter driven DC motor. IFAC Mechatronic Systems, vol 4, no 1, Ruprecht-Karls-University, Germany
23. Arahal MR, Barrero F, Toral S, Duran M, Gregor R (2009) Multi-phase current control using finite-state model-predictive control. Control Eng Pract Elsevier 17:579–587
24. Archer BA, Annakkage UD (2005) Monitoring and predicting power system behavior by tracking dominant modes of oscillation. In: Proceedings of the IEEE power engineering society general meeting, vol 2, pp 1475–1482
25. Arioua L, Marinescu B, Monmasson E (2014) Control of high voltage direct current links with overall large-scale grid objectives. IET Gen Transm Distrib 8(5):945–956
26. Arioua L, Marinescu B (2014) Large-scale implementation and validation of multivariable control with grid objectives of an HVDC link. In: 18th Power systems corporations conference, Wroclau, Poland
27. Arshad M, Islam SM (2006) A novel fuzzy logic technique for power transformer asset management. In: IEEE industry applications conference, conference record of the 2006 41st IAS annual meeting, pp 276–286, Tampa, Florida, USA
28. Arul P, Ramachandaramuthy V, Rajkumar R (2015) Control strategies for hybrid renewable energy systems: a review. Renew Sustain Energy Rev Elsevier 42:597–608
29. Baalbergen F, Gibescu M, Van Sluis LD (2009) Modern state estimation methods in power systems. In: 2009 IEEE/PES Power systems conference and exposition, PSCE 2009, Seattle 2009
30. Bahbah AG, Girgis AA (2004) New method for generators angles and angular velocities prediction for transient stability assessment of multimachine power systems using recurrent artificial neural network'. IEEE Trans Power Syst 19:1015–1022
31. Balogh A, Kristic M (2002) Infinite dimensional backstepping style feedback transformations for a heat equation with an arbitrary level of instability. Eur J Control 8:165–175
32. Bao X, Zhuo F, Tian Y, Tan P (2013) Simplified feedback linearization control of three-phase photovoltaic inverter with an LCL filter. IEEE Trans Power Electron 28(6):2739–2752
33. Barrera-Cardiel E, Pastor-Gomez N (1999) Microcontroller-based power-angle instrument for a power system laboratory. In: IEEE powering engineering society summer meeting, pp 1008–1012
34. Basseville M, Nikiforov I (1993) Detection of abrupt changes: theory and applications. Prentice-Hall
35. Basseville M (1997) Information criteria for residual generation and fault detection and isolation. Automatica Elsevier 33(5):783–803

36. Basseville M (1998) On-board component fault detection and isolation using the statistical local approach. Automatica Elsevier 34(11):1391–1415
37. Basseville M, Benveniste A, Zhang Q (1996) Surveilliance d' installations industrielles: démarche générale et conception de l' algorithmique. IRISA Publication Interne No 1010
38. Basseville M, Abdelghani M, Benveniste A (1997) Sub-space-based fault detection and isolation methods - application to vibration monitoring'. Publication Interne IRISA, no 1143
39. Basseville M, Benveniste A, Mathis G, Zhang Q (1994) Monitoring the combustion set of a gas turbine. In: Proceedings of the Safeprocess 1994, Helsinki, Finland
40. Basseville M, Benveniste A, Gach-Devauchelle B, Goursat M, Bonnecase D, Dorey P, Prevosto M, Olagnon M (1993) Damage monitoring in vibration mechanics: issues in diagnostics and predictive maintenance. Mech Syst Signal Process Elsevier 7(5):401–423
41. Benigni A, D' Antona G, Ghisla U, Monti A, Ponci F (2010) A decentralized observer for ship power system applications: implementation and experimental validation. IEEE Trans Instrum Meas 59(2):440–449
42. Bernard C, Slotine J (1997) Adaptive control with multiresolution bases. In: IEEE CDC 97, 36th IEEE International conference on decision and control, San Diego, California, USA
43. Bensoussan A, Prato GD, Delfour MC, Mitter SK (2006) Representation and control of infinite dimensional systems. Birkahaüser, Boston
44. Benveniste A, Basseville M, Moustakides G (1987) The asymptotic local approach to change detection and model validation. IEEE Trans Autom Control 32(7):583–592
45. Bertsekas DP (1996) Incremental least squares methods and the extended Kalman filter. SIAM J. Optim 6(3):807–822
46. Betz R, Jovanovic M (2002) Theoretical analysis of control properties for the brushless doubly-fed reluctance machine. IEEE Trans Energy Convers 17(3):332–339
47. Besançon G, Zhang Q, Hammouri H (2004) High-gain observer based state and parameter estimation in nonlinear systems. In: IFAC Symposium on nonlinear control systems (NOLCOS), Stuttgart, Germany
48. Bhattacharyya SP, Chapelat H, Keel LH (1997) Robust control: the parametric approach. Prentice Hall
49. Bidram A, Davoudi A, Lewis FL, Guerrero JM (2013) Distributed cooperative secondary control of microgrids using feedback linearization. IEEE Trans Power Syst 28(3):3462–3470
50. Blaabjerg F, Teodorescu R, Liserre M, Timbus AV (2006) Overview of control and grid synchronization for distributed power generation systems. IEEE Trans Ind Electron 53(5):1398–1409
51. Blasco V, Kaura V (1997) A new mathematical model and control of a three-phase AC-DC voltage source converter. IEEE Trans Power Electron 12(1):116–123
52. Bodson M, Chiasson J, Novotnak R (1994) High-performance induction motor control via input-output linearization. IEEE Control Syst Mag:24–33
53. Boizot N, Busvelle E, Gauthier J-P (2010) An adaptive high-gain observer for nonlinear systems. Automatica Elsevier 46:1483–1488
54. Bojoi R, Cavagnino A, Tenconi A, Vaschetto S (2015) Control of shaft-line-embedded multiphase starter/generator for aero-engine. IEEE Trans Ind Electron
55. Boskovic DM, Krstic M, Liu WJ (2002) Boundary control of an unstable heat equation via measurement of domain averaged temperature. IEEE Trans Autom Control 46:2022–2028
56. Boukhezzar B, Siguerdidjane H (2009) Nonlinear control with wind estimation of a DFIG variable speed wind turbine for power capture optimization. Energy Convers Manag Elsevier 50:885–892
57. Bououden S, Boutat D, Zheng G, Barbot JP, Kratz F (2011) A triangular canonical form for a class of 0-flat nonlinear systems. Int J Control Taylor and Francis 84(2):261–269
58. Boussaada I, Cela A, Mounier H, Niculescu SI (2013) Control of drilling vibrations: a time-delay system-based approach. In: 11th Workshop on time delay systems, France
59. Brandwajn V, Kumar ABR, Ipakchi A, Bose A, Kuo SD (1997) Severity indices for contingency screening in dynamic security assessment. IEEE Trans Power Syst 12:1136–1142

60. Brahmi J, Krichen L, Ouali A (2009) A comparative study between three sensorless control strategies for PMSG in wind energy conversion systems. Appl Energy, Elsevier 86:1565–1573
61. Bruno S, de Benedictis M, Delfanti M, LaScala M (2005) Preventing blackouts through reactive rescheduling under dynamical and protection system constraints. In: Proceedings of the IEEE PowerTech, St. Petersburg, Russia, pp 1–8
62. Calderaro V, Galdi V, Piccolo A, Siano P (2008) A fuzzy controller for maximum energy extraction from variable speed wind power generation systems. Electr Power Syst Res, Elsevier 78:11091118
63. Calderaro V, Galdi V, Piccolo A, Siano P (2007) Adaptive relays for overhead line protection. Electr Power Syst Res Elsevier 77:1552–1559
64. Campos-Gaona D, Moreno-Goytia EL, Anaya-Lara O (2013) Fault ride-through improvement of DFIG-NT by integrating a two-degrees-of-freedom internal model control. IEEE Trans Ind Electron 60(3):1133–1145
65. Cannon M, Slotine J (1995) Space-frequency localized basis function networks for nonlinear system estimation and control. Neurocomputing 9:293–342
66. Capece SL, Cecati C, Rotondale N (2003) A new three-phase active rectifier based on power matching modulation. In: IEEE IECON 2003, 29th Annual conference of the industrial electronics society, vol 1, pp 202–207
67. Capecce SL, Cecati C, Rotondale N (2003) A sensorless control technique for low cost AC/DC converters. In: IEEE industry applications conference 2003, 38th IAS Annual meeting, vol 3, pp 1546–1551
68. Carreras BA, Lynch VE, Dobson I (2004) Dynamical and probabilistic approaches to the study of blackout vulnerability of the power transmission grid. In: Proceedings of the 37th Annual Hawaii International conference on system sciences, pp 55–61
69. Castelli-Dezza F, Iocchetti MF, Perini P (2013) An observer for sensorless DFIM drives based on the natural fifth harmonic of the line voltage, without stator current measurements. IEEE Trans Ind Electron 60(10):4301–4309
70. Castilla M, Miret J, Camacho A, Matas J, Garca de Vicua L (2013) Reduction of current harmonic distortion in three-phase grid-connected photovoltaic inverters via resonant current control. IEEE Trans Ind Electron 60(4):1464–1472
71. Catterson V, Davidson E, McArthur S (2010) Agents for active network management and condition monitoring in the smart grid. In: Proceedings of the first international workshop on agent technologies for energy systems, ATES, 2010, Toronto, Canada
72. Catterson V, Davidson E, McArthur S (2011) Embedded intelligence for electrical network operation and control. IEEE Intell Syst 26:38–45
73. Cecati C, Dell' Aquila A, Lecci A, Liserre M (2005) Implementation issues of a fuzzy-logic-based three-phase active rectifier employing only voltage sensors. IEEE Trans Ind Electron 52(2):378–385
74. Cecati C, Ciancetta F, Siano P (2010) A multilevel inverter for photovoltaic systems with fuzzy logic control. IEEE Trans Ind Electron 57(12):4115–4125
75. Cecati C, Ciancetta F, Siano P (2009) A FPGA/fuzzy logic - based multilevel inverter. In: Proceedings of the ISIE 2009, 18th IEEE International symposium on industrial electronics, pp 706–711
76. Cecati C, Dell' Aquila A, Liserre M, Ometto A (2003) A fuzzy-logic-based controller for active rectifier. IEEE Trans Ind Appl 39(1):105–112
77. Chaal H, Jovanovic M (2012) Practical implementation of sensorless torque and reactive power control of doubly fed machines. IEEE Trans Ind Electron 59(6):2645–2653
78. Chaer TA, Rambault L, Gaumbert GP, Najjur M (2008) Static output feedback control with H_∞ performance for a three-phase shunt active filter. In: American control conference, Seattle Washington, USA
79. Chak CK, Feng G, Mia J (1998) An adaptive fuzzy neural network for MIMO system model approximation in high dimensional spaces, IEEE Trans. on Syst. Man and Cybern. - Part B. Cybernetics 28(3):436–446

80. Chang CH, Lin YH, Chen YM, Chang YR (2014) Simplified reactive power control for single-phase grid-connected photovoltaic inverters. IEEE Trans Ind Electron 61(5): 2286–2296
81. Chaturvedi NA, Klein R, Christensen J, Ahmed J, Kojic A (2010) Algorithms for advanced battery-management systems. IEEE Control Syst Mag:49–68
82. Chavarra J, Biel D, Guinjoan F, Meza C, Negroni JJ (2013) Energy-balance control of PV cascaded multilevel grid-connected inverters under level-shifted and phase-shifted PWMs. IEEE Trans Ind Electron 60(1):98–111
83. Chen G, Zhang L, Cai X, Zhang W, Yin C (2011) Nonlinear control of the doubly fed induction generator by input-output linearizing strategy. In: Hu W (ed) Electronics and signal processing. Lecture notes in electrical engineering, vol. 97. Springer pp. 601–608
84. Chen H, Yu R, Wang X, Lu R (2015) Cooperative control of power system load and frequency by using differential games. IEEE Trans Control Syst Technol
85. Chen WH, Ballance DJ, Gawthrop PJ, Reilly JO (2000) A nonlinear disturbance observer for robotic manipulators. IEEE Trans Ind Electron 47(4):932–938
86. Chen S, Malik O (1996) H_∞ optimization-based power system stabilizer design. In: IEEE Proceedings on generation transmission and distribution, vol 193, pp 253–257
87. Chen Q, Zhu K, McCalley JD (2001) Dynamic decision-event trees for rapid response to unfolding events in bulk transmission systems. In: Proceedings of the IEEE 2001 Power technology conference
88. Chen XB, Siljiak DD, Stankovic SS (2002) Decentralized H_∞ design of automatic generation control. In: IFAC 15th World congress, Barcelona, Spain
89. Chen Y, Dai J, Damm G, Lamnabhi-Lagarrigue F (2013) Nonlinear control design of a multi-terminal VSC-HVDC system. In: ECC 2013 European control conference, Zurich
90. Chen Y, Damm G, Benchaib A, Lamnabhi-Lagarrigue F (2014) Feedback linearization for the DC voltage control of a VSC-HVDC terminal. In: European control conference, Strasbourg, France
91. Chen MY, Linkens DA (2001) A systematic neuro-fuzzy modeling framework with application to material property prediction. IEEE Trans Syst Man Cybern—Part B: Cybern 31(5):781–790
92. Chiasson J (1998) A new approach to dynamic feedback linearization control of an induction motor. IEEE Trans Autom Control 43(3):391–397
93. Chihab A, Ouadi H, Giri F, Ahmed-Ali T (2014) Adaptive nonlinear control of three-phase four-wire shunt active power filters for unbalanced and nonlinear loads. In: 19th IFAC World congress. Cape Town, South Africa
94. Chilard O, Grenard S, Devaux O, de Alavaro Garcia L (2009) Distribution state estimation based on voltage state variables: assessment of results and limitations. In; CIRED 2009, 20th International conference on electricity distribution, Prague
95. Cirrinzione M, Pucci M, Vitale G, Miraoui A (2007) Correct harmonic compensation by a single-phase shunt active power filter controlled by adaptive neural filtering. IEEE Trans Ind Electron 56(8):3128–3143
96. Cobelo I, Shafiu A, Jenkins N, Strbac G (2007) State estimation of networks with distributed generation. Eur Trans Electr Power 17(1):21–36
97. Corno M, Bhatt N, Savaresi SM, Verhaegen M (2015) Electrochemical model-based state of charge estimation in Li-ion cells. IEEE Trans Control Syst Technol 23(1):112–127
98. Cortesao R, Park J, Khatib O (2005) Real-time adaptive control for haptic telemanipulation with Kalman active observers. IEEE Trans Robot 22(5):987–999
99. Cortesao R (2006) On Kalman active observers. J Intell Robot Syst, Springer 48(2):131–155
100. da Costa Lopes F, Watanabe EH, Rolim L (2015) A control-oriented model of a PEM fuel cell stack based on NARX and NOE neural networks. IEEE Trans Ind Electron 62(8):5155–5163
101. Crassidis JL,Junkins JL (2012) Optimal estimation of dynamic systems, 2nd edn. CRC Press, Boca Raton
102. Damm G, Marino R, Lamnabhi-Lagarrigue R (2004) Adaptive nonlinear output feedback for stabilization and voltage regulation of power generators with unknown parameters, Intl. J Robust Nonlinear Control 14:833–855

103. D' Antona G, Monti A, Ponci F (2007) A decentralized state estimator for non-linear electric power systems. In: 1st Annual IEEE systems conference, Hawaii, USA
104. Dash PK, Rout PK, Moharana AK (2010) Nonlinear control of parallel AC voltage source converter high-voltage DC system using a hybrid input-output linearization evolutionary fuzzy controller. Electr Power Compon Syst, Taylor and Francis 28:881–899
105. Dasgupta S, Mohan SN, Sahoo SK, Panda SK (2013) Lyapunov function-based current controller to control active and reactive power flow from a renewable energy source to a generalized three-phase microgrid system. IEEE Trans Ind Electron 60(2):799–813
106. Delfino F, Pampararo F, Procopio R, Rossi M (2012) A feedback linearization control scheme for the integration of wind energy conversion systems into distributed grids. IEEE Syst J 6(1): 85–93
107. Dib W, Ortega R, Barabanov A, Lamnabhi-Lagarrigue F (2009) A globally convergent controller for multi-machine power systems using structure-preserving models. IEEE Trans Autom Control 54(9):2179–2185
108. Dib W, Ortega R, Astolfi A, Hill D (2011) Improving transient stability of multimachine power systems: synchronization via immersion of a pendular system. In: American control conference, San Francisco, CA, USA
109. Dong L, Zhang Y, Gao Z (2012) A robust decentralized load frequency controlller for interconnected power systems. ISA Trans, Elsevier 51:410–419
110. Du H, Zhang N (2007) H_∞ control of active vehicle suspensions with actuator time delay. J Sound Vib, Elsevier 301:236–252
111. Del Angel A, Geurts P, Ernst D, Glavic M, Wehenkel L (2007) Estimation of rotor angles of synchronous machines using artificial neural networks and local PMU-based quantities. Neurocomputing, Elsevier 70:26682678
112. De Leon Morales J, Busawon K, Acha-Duza S (2001) A robust observer-based controller for synchronous generators. Electr Power Energy Syst, Elsevier 23:195–211
113. DeMarco CL (2001) A phase transition model for cascading network failure. IEEE Control Syst Mag:40–51
114. Dendouga A, Abdessemed R, Bendaas ML, Chaiba A (2007) Decoupled active and reactive power control of a doubly-fed induction generator (DFIG). In: Proceedings of the 15th mediterranean conference on control and automation, Athens, Greece
115. Do TD, Leu VQ, Choi YS, Choi HH, Jung JW (2013) An adaptive voltage control strategy of three-phase inverter for stand-alone distributed generation systems. IEEE Trans Ind Electron 60(12):5660–5672
116. Dong L, Zhang Y, Gao Z (2012) A robust decentralized load frequency controlller for interconnected power systems. ISA Trans, Elsevier 51:410–419
117. Doyle JC, Glover K, Khargonekar PP, Francis BA (1989) State-space solutions to standard H_2 and H_∞ control problems. IEEE Trans Autom Control 34:831–847
118. Dubois D, Prade H, Ughetto L (1997) Checking the coherence and redundancy of fuzzy knowledge bases. IEEE Trans Fuzzy Syst 5(3):398–417
119. Edwards AR, Chan RW, Dunn RW, Daniels AR (1996) Transient stability screening using artificial neural networks within a dynamic security assessment system. Proceedings of the Instrumentation Electrical Engineering, Generation, Transmission Distribution, March 1996, 143, pp 129–134
120. Elizondo DC, de La Ree J, Phadke AG, Horowitz S (2001) Hidden failures in protection systems and their impact on wide-area disturbances. In: Proceedings of the IEEE 2001 PES Winter meeting, vol 2, pp 710–714
121. El Magri A, Giri F, Bensançon G, El Fadili A, Dagard L, Chaoui FZ (2013) Sensorless adaoptive output feedback control of wind energy systems with PMS generators. Control Eng Pract, Elsevier 21:530–543
122. Elnady A, Salama MMA (2005) Unified approach for mitigating voltage sag and voltage flicker using the DSTATCOM. IEEE Trans Power Deliv 20(2):992–1000
123. Eren S, Pahlevaninezhad M, Bakhshai A, Jain P (2010) Grid-connected voltage source inverter for renewable energy conversion system with sensorless current control. In: APEC 2010,

Twenty-Fifth Annual IEEE applied power electronics conference and exposition (APEC), Palm Springs, California USA
124. Elkington K, Knazkins V, Ghanhari M (2008) On the stability of power systems containing doubly-fed induction generator-based generation. Electr Power Syst Res, Elsevier 78: 1477–1484
125. Elshafei AL, Metwally KA, Shaltout AA (2005) A variable-structure adaptive fuzzy-logic stabilizer for single and multi-machine power systems. Control Eng Pract, Elsevier 13: 413–423
126. Eriksson R, Söder L (2012) On the coordinated control of multiple HVDC links using input-output exact linearization in large power systems. Electr Power Energy Syst, Elsevier 43: 118–125
127. Eriksson R, Knazkins V, Söder L (2010) Coordinated control of multiple HVDC links using input-output exact linearization. Electr Power Syst Res 80:1406–1412
128. Escobar G, Chevreau D, Ortega R, Mendes E (2001) An adaptive passivity-based controller for a unity power factor rectifier. IEEE Trans Control Syst Technol 9(4):637–644
129. Fadil HE, Giri F, Haloua M, Ouadi H (2003) Nonlinear and adaptive control of buck power converters. In: Proceedings of the 42nd IEEE conference on decision and control, Maui, Hawaii USA
130. Fan L, Wehbe Y (2013) Extended Kalman filtering based real-time dynamic state and parameter estimation using PMU data. Electr Power Syst Res, Elsevier 103:168, 177
131. Fang C, Xu L, Li J, Ouyang M (2015) Feedback linearizatio-based air pressure and mass flow regulation for PEMFC. In: IEEE SICE Annual conference 2015, Hagzhou, China
132. Fang Y, Wang Z, Fei J (2014) Adaptive neural sliding mode control of active power filter using feedback linearization. In: ICCAS 2014 14th International conference on control, automation and systems, Oct 2014, Gyeonggi-do, Korea
133. Fei J, Ma K, Zhang S, Yan W, Yuan Z (2013) Adaptive current control with PI-fuzzy compound controller for shunt active power filter. Math Probl Eng, Hindawi Publications 2013:article ID 546842
134. Fei J, Wang Z (2012) Adaptive RBF neural network control for three-phase active power filter. Int J Adv Robot Syst, InTech Publications
135. Fei J, Hou S, Yang Y (2013) Robust adaptive fuzzy control based on feedback linearization for three-phase active power filters. In: Proceedings of the 32nd Chinese control conference, Xi'an, China
136. Fei J, Hou S (2012) Adaptive fuzzy control with supervisory compensator for three-phase active power filter. J Appl Math, Hindawi Publications 2012:article ID 654937
137. Fliess M, Mounier H (1999) Tracking control and π-freeness of infinite dimensional linear systems, In: Picci G, Gilliam DS (eds) Dynamical systems, control, coding and computer vision, vol 258. Birkhaüser, pp 41–68
138. Fliess M, Mounier H (2002) An algebraic framework for infinite-dimensional linear systems. In: Proceedings of international school on automatic control of Lille, Control of distributed parameter systems: theory and applications, Grenoble, France
139. Fnaiech MA, Betin F, Capolino GA, Fnaiech F (2010) Fuzzy logic and sliding-mode controls applied to six-phase induction machine with open phases. IEEE Trans Ind Electron 57(1): 354–364
140. Forchetti DG, Solsona JA, Garcia GO, Valla MI (2007) A control strategy for stand-alone wound rotor induction machine. Electr Power Syst Res, Elsevier 77:163–169
141. Forchetti DG, Garcia GO, Valla MI (2009) Adaptive observer for sensorless control of stand-alone doubly-fed induction generator. IEEE Trans Ind Electron 56(10):4174–4180
142. Freijedo FD, Doval-Gondoy J, Lopez O, Fernandez-Comesanu P, Martinez-Penolver C (2009) A signal processing adaptive algorithm for selective curernt harmonic cancellation in active power filters. IEEE Trans Ind Electron 56(8):2829–2840
143. Galdi V, Piccolo A, Siano P (2008) Designing an adaptive fuzzy controller for maximum wind energy extraction. IEEE Trans Energy Convers 23(2):559569

144. Galdi V, Piccolo A, Siano P (2009) Exploiting maximum energy from variable speed wind power generation systems by using an adaptive Takagi-Sugeno-Kang fuzzy model. Energy Convers Manag 50:413–421
145. Galdi V, Ippolito L, Piccolo A, Vaccaro A (2000) Neural diagnostic system for transformer thermal overload protection. IEE Proc Power Appl 147:415–421
146. Galdi V, Piccolo A, Siano P (2008) Designing an adaptive fuzzy controller for maximum wind energy extraction. IEEE Trans Energy Convers 28:559–569
147. Gallo D, Landi C, Luiso M (2010) Severity assessment issues for short voltage dips. Measurement, Elsevier 43:10401048
148. Gensior A, Nguyen TMP, Rudolph J, Guldner H Flatness-based loss optimization and control of a doubly-fed induction generator system. IEEE Trans Control Syst Technol
149. Gensior A, Rudolph J, Güldner H (2005) Flatness-based control of three-phase boost rectifiers, In: 11th European conference on power electronics and applications, EPE 2005, Dresden, Germany
150. Gensior A, Sira-Ramirez H, Rudolph J, Guldner H (2009) On some nonlinear current controllers for three-phase boost rectifiers. IEEE Trans Ind Electron 56(2):360–370
151. Georges D, De Wit C, Ramirez J (1999) Nonlinear H_2 and H_∞ optimal controllers for current-fed induction motors. IEEE Trans Autom Control 44(7):1430–1435
152. Glorennec PY (1999) Algorithmes d' apprentissage pour systèmes d' inférence floue. Hermes Science Publications
153. Grelle C, Ippolito L, Loia V, Siano P (2006) Agent-based Architecture for Designing Hybrid Control Systems. Inf Sci 176:1103–1130
154. Gaing ZL (2004) Wavelet-based neural network for power disturbance recognition and classification. IEEE Trans Power Deliv 19(4):1560–1568
155. Ge SS, Hang CC, Zhang T (1999) Adaptive neural network control of nonlinear systems by state and output feedback. IEEE Trans Syst Man Cybern—Part B Cybern 29:818–828
156. Gelagaev R, Vermeyen P, Driesen J (2008) State estimation in distribution grids. In: IEEE ICHQP 2008, 13th International conference on harmonics and quality of power, New South Wales, Australia
157. Duncan Glover J, Sarma M, Overbye T (2012) Power system: analysis and design. Cengage Publishing
158. Ghahremani E, Kamwa I (2011) Dynamic state estimation in power system by applying the extended Kalman filter with unknown inputs to phasor measurements. IEEE Trans Power Syst 26(4):2556–2566
159. Ghasemi H, Canizares C, Moshref A (2006) Oscillatory stability limit prediction using stochastic subspace identification. IEEE Trans Power Syst 21(2):736–745
160. Gibbs BP (2011) Advanced Kalman filtering. A practical handbook. Least Squares and Modelling. Wiley
161. Gros M (2014) A distributed algorithm for NMPC-based wind farm control. In: 53rd IEEE conference on decision and control, Los Angeles, California, USA
162. Guerrero JM, Chandorkar M, Lee TL, Loh PC (2013) Advanced control architectures for intelligent MicrogridsPart I: decentralized and hierarchical control. IEEE Trans Ind Electron 60(4):1260–1262
163. Guo L, Billings SA (2007) State-space reconstruction and spatio-temporal prediction of lattice dynamical systems. IEEE Trans Autom Control 52(4):622–632
164. Guo Z, Sha D, Liao X (2014) Voltage magnitude and frequency control of three-phase voltage source inverter for seamless transfer. IET Power Electron 7(1):200–208
165. Guo Y, Hill DJ, Wang Y (2000) Nonlinear decentralized control of large-scale power systems. Automatica, Elsevier 36:1275–1289
166. Gopal C, Mohanraj M, Chandramohan P, Chandrasekar P (2013) Renewable energy source water pumping systems A literature review. Renew Sustain Energy Rev, Elsevier 25:351–370
167. Guerrero JM, Chandorkar M, Lee TL, Loh PC (2013) Advanced control architectures for intelligent microgrids Part I: decentralized and hierarchical control. IEEE Trans Ind Electron 60(3):1254–1262

168. Guoqiang H, Renmu H, Huachun Y, Peng W, Rui M (2005) Iterative Prony method based power system low frequency oscillation mode analysis and PSS design. In: Proceedings of the transmission and distribution conference and exhibition, vol 1, pp 1–6
169. Gupta A, Malley MKO (2011) Disturbance-observer-based force estimation for haptic feedback. ASME J Dyn Syst Meas Control 133(1):Article number 014505
170. Guzman H, Duran MJ, Barrero F, Bogado B, Toral S (2014) Speed control of five-phase induction motors with integrated open-phase fault operation using model-based predictive current control techniques. IEEE Trans Ind Electron 69(9):4474–4484
171. Hagenmeyer V, Delaleau E (2003) Robustness analysis of exact feedforward linearization based on differential flatness. Automatica, Elsevier 39:1941–1946
172. Hagenmeyer V, Delaleau E (2010) Robustness analysis with respect to exogenous perturbations for flatness-based exact feedforward linearization. IEEE Trans Autom Control 55(3): 727–731
173. Han Y, Xu L, Khan M, Chen C, Yao G, Zhou LD (2011) Robust deadbeat control scheme for a hybrid APF with resetting filter and ADALINE-based harmonic estimation algorithm. IEEE Trans Ind Electron 58(9):3893–3904
174. Harris C, Hong X, Gan Q (2002) Adaptive modelling, estimation and fusion from data: a neurofuzzy approach. Springer
175. Haykin S (1994) Neural networks: a comprehensive foundation. McMillan
176. Hardiansyah A, Furuya S, Irisawa J (2006) A robust H_∞ power system stabilizer design using reduced-order models. Electr Power Energy Syst, Elsevier 28:21–28
177. Hiyama T, Suzuki N, Kita T, Funakoshi T (1999) Development of real time stability monitoring system for power system operation. In: Proceedings of the IEEE power engineering society winter meeting, vol 1, pp 525–530
178. Hiyama T, Suzuki N, Funakoshi T (2000) On-line coherency monitoring of power system oscillations. In: Proceedings of the IEEE power engineering society summer meeting, vol 3, pp 839–1844
179. Hofmann H, Sanders S, El-Antably A (2004) Stator-flux-oriented vector control of synchronous reluctance machines with maximized efficiency. IEEE Trans Ind Electron 51(5):1066–1072
180. Hojo M, Ohnishi T, Mitani Y, Saeki O, Ukai H (2003) Observation of frequency oscillation in western Japan 60 Hz power system based on multiple synchronized phasor measurements. In: Proceedings of the IEEE power technology conference, vol 2, pp 1–6
181. Holtz J (2002) Sensorless control of induction motor drives. Proc IEEE 90(8):1359–1394
182. Hossain MJ, Saha TK, Mithulananthan N, Pota HR (2013) Control strategies for augmenting LVRT capability of DFIGs in interconnected power systems. IEEE Trans Ind Electron 60(6): 2510–2522
183. Houari A, Renaudineau H, Martin JP, Pierfederici S, Meibody-Tabar F (2012) Flatness-based control of three-phase inverter with output LC filter. IEEE Trans Ind Electron 59(7): 2890–2897
184. Hua Y, Yutian L, Xinsheng N (2006) Low frequency oscillation analysis and damping based on Prony method and sparse eigenvalue technique. In: Proceedings of the IEEE International conference on networking, sensing and control, vol 1, pp 1006–1010
185. Huang YJ, Wang YJ (2000) Robust PID tuning strategy for uncertain plants based on the Kharitonov theorem. ISA Trans, Elsevier 39:419–431
186. Huang Z, Schneider K, Nieplocha J (2007) Feasibility studies of applying Kalman filter techniques to power system dynamic state estimation. In: IPEC 2007 Power engineering international conference, pp 376–382
187. Huerta F, Pizzaro D, Cobreces S, Rodriguez F, Giron C, Rodriguez A (2012) LQG servo controller for the current control of LCL grid-connected voltage-source converters. IEEE Trans Ind Electron 59(11):4272–4284
188. Hunt KJ, Sbarbaro D, Zbikowski R, Gawthrop PJ (1992) Neural networks for control systems-A survey. Automatica 28(6):1083–1112

189. Ippolito L, Siano P (2004) Identification of Tagaki-Sugeno-Kang fuzzy model for power transformers predictive overload system. In: IEE Proceedings—Generation, transmission and distribution, vol 151, issue no 5, pp 582–589
190. Ippolito L, Loia V, Siano P (2003) Extended fuzzy C-Means and genetic algorithms to optimize power flow management in hybrid electric vehicles. Fuzzy Optim Decis Making 2:359–374
191. Isermann R (2006) Fault-diagnosis systems: an introduction from fault detection to fault tolerance. Springer
192. Isermann R, Schreiber A (2006) Identification methods for experimental modelling of nonlinear combustion processes. In: Proceedings of the Collection of Technical Papers—4th International energy conversion. Engineering conference, vol 2, pp 1468–1485
193. Isermann R (1996) Local basis function networks for identification of a turbocharger. Proc IEE Conf 427(1):7–12
194. Jain SK, Singh SN (2011) Harmonics estimation in emerging power system: Key issues and challenges. Electr Power Syst Res, Elsevier 81:1754–1766
195. Jang JSR, Sun CT, Mizutani E (1997) Neurofuzzy and soft-computing. A computational approach to learning and machine intelligence. Prentice-Hall
196. Jardini JA, Brittes JL, Magrini LC, Bini MA, Yasuoka J (2005) Power transformer temperature evaluation for overloading conditions. IEEE Trans Power Deliv 20(1):179–184
197. Jiang L, Wu QH, Zhang C, Zhou XX (2001) Observer-based nonlinear control of synchronous generators with pertubation estimation. Electr Power Energy Syst, Elsevier 23:359–367
198. Jiang L, Wu QH, Wen JY (2004) Decentralized nonlinear adaptive control for multimachine power systems via high-gain perturbation observer. IEEE Trans Circuits Syst-I 51(10): 2052–2059
199. Jiang X, Huang W, Cao R, Hao Z, Jiang W (2015) Electric-drive system of dual-winding fault-tolerant pearmanent magnet motor for aerospace applications. IEEE Trans Ind Electron
200. Johnson BB, Dhople SV, Hamadeh AC, Krein PT (2014) Synchronization of nonlinear oscillators in an LTI electrical power network. IEEE Trans Circuits Syst—Part I 61(3):834–844
201. Jovanovic M, Betz R, Yu J (2002) The use of doubly fed reluctance machines for large pumps and wind turbines. IEEE Trans Ind Appl 38(6):1508–1516
202. Julier S, Uhlmann J, Durrant-Whyte HF (2000) A new method for the nonlinear transformations of means and covariances in filters and estimators. IEEE Trans Autom Control 45(3):477–482
203. Julier SJ, Uhlmann JK (2004) Unscented filtering and nonlinear estimation. Proc IEEE 92:401–422
204. Julier SJ, Uhlmann JK (1997) A non-divergent estimation algorithm in the presence of unknown correlations. In: Proceedings of the American control conference, vol 4, Albuquerque, NM, pp 2369–2373
205. Kahrobaeian A, Abdel-Rady A, Mohamed I (2014) Analysis and mitigation of low-frequency instabilities in autonomous medium-voltage converter-based microgrids with dynamic loads. IEEE Trans Ind Electron 61(4):1643–1657
206. Kakimoto N, Sugumi M, Makino T, Tomiyama K (2006) Monitoring of inter-area oscillation mode by synchronized phasor measurement. IEEE Trans Power Syst 21:260–268
207. Kalsi K, Lian J, Zak SH (2009) Decentralized control of multi-machine power systems. In: American control conference, St, Louis, MO, USA
208. Kamal E, Aitouche A (2013) Robust fault tolerant control of DFIG wind energy systems with unknown inputs. Renew Energy, Elsevier 56:2–15
209. Kamen EW, Su JK (1999) Introduction to optimal estimation. Springer
210. Kanieski JM, Cardoso R, Pinheiro H, Groundling HA (2013) Kalman Filter-based control system for power quality conditioning devices. IEEE Trans Ind Electron 60(11):5214–5227
211. Karimi H, Lohmann B (2006) A computational method to robust vibration control of vehicle engine-body system using haar wavelets. In: Proceedings of the 2006 IEEE International conference on control applications, pp 169-174, Munich, Germany
212. Krzyzak A, Sasiadek J (1991) Flexible robot identification using nonparametric techniques. In: Proceedings of the 30th IEEE conference on decision and control, Brighton UK

213. Kenné G, Gome R, Nkwawo H, Lamnabhi-Lagarrique F, Arzandé A, Vannier JC (2010) An improved direct feedback linearization techique for transient stability enhancement and voltage regulation of power generators. Electr Power Energy Syst 32:809–816
214. Kennel RM (2006) Why do incremental encoders do a reasonably good job in electrical drives with digital control? In: Industry applications conference, IEEE 41st IAS Annual meeting. Conference Record
215. Khalil HK (1996) Nonlinear systems, 2nd edn. Prentice Hall
216. Kim DE, Lee DC (2010) Feedback Linearization Control of Three-Phase UPS Inverter Systems. IEEE Trans Ind Electron 57(3):963–968
217. Kim K, Jeung Y, Lee D, Kim H (2012) LVRT scheme of PMSG wind power systems based on feedback linearization. IEEE Trans Power Electron 27(5):2376–2384
218. Klein R, Chaturvedi NA, Christensen J, Ahmed J, Findeisen R, Kojic A (2013) Electrochemical model-based observer design for a Lithium-Ion battery. IEEE Trans Control Syst Technol 21(2):289–301
219. Klein R, Chaturvedi NA, Christensen J, Ahmed J, Findeisen R, Kojic A (2010) State estimation of a reduced electrochemical model of a lithium-ion battery. In: 2010 American control conference, Baltimore MD, USA
220. Klein M, Rogers GJ, Murty S, Kundur P (1992) Analytical investigation of factors influencing power systems stabilizers performance. IEEE Trans Energy Convers 7(3):382–390
221. Korba P, Larsson M, Rehtanz C (2003) Detection of oscillations in power systems using Kalman filtering techniques. In: Proceedings of the IEEE conference on control applications, vol 1, pp 183–188
222. Kröner A (2011) Adaptive finite element methods for optimal control second order hyberbolic equations. Comput Methods Appl Math 11(2):214–240
223. Kwak S, Yoo S, Park J (2014) Finite control set predictive control based on Lyapunov function for three-phase voltage source inverters. IET Power Electron 7(11):2726–2732
224. Kwon S, Chung WK (2003) Combined synthesis of state estimator and perturbation observer. J Dyn Syst Meas Control, ASME 125:19–26
225. Kundur P (1994) Power system stability and control. McGraw-Hill, New York
226. Kurylowicz A, Jaworska I, Tzafestas SG (1993) Robust stabilizing control: an overview. In: Tzafestas SG (ed) Applied control—current trends and modern methodologies. Marcel Dekker, pp 289–324
227. Laghrouche S, Liu J, Ahmed F, Harmouche M, Wack M (2015) Adaptive second-order sliding mode observer-based fault reconstruction for PEM fuel cell air-feed system. IEEE Trans Control Syst Technol 23(3):1098–1109
228. Laroche B, Martin P, Rouchon P (2000) Motion planning of the heat equation. Int J Robust Nonlinear Control, Wiley 40(8):629–643
229. Laroche B, Martin P, Petit N (2007) Commande par platitude: Equations différentielles ordinaires et aux derivées partielles. Ecole Nationale Supérieure des Techniques Avancées, Paris
230. Lee DJ (2008) Nonlinear estimation and multiple sensor fusion using unscented information filtering. IEEE Signal Process Lett 15:861–864
231. Lee DJ (2008) Unscented information filtering for distributed estimation and multiple sensor fusion. In: AIAA Guidance, navigation and control conference and exhibit, Hawai, USA
232. Lee HD, Kang SJ, Sul SK (1999) Efficiency-optimized direct torque control of synchronous reluctance motor using feedback linearization. IEEE Trans Ind Electron 46(1):192–198
233. Lee SJ, Ahn HS (2012) Output synchronization of multi-machine power system using dissipative hamiltonian realization. In: IEEE International symposium on intelligent control (ISIC) Part of 2012 IEEE Multi-conference on systems and control, Dubrovnik, Croatia, 3–5 Oct 2012
234. Lee K, Jahns TJ, Lipo TA, Blasko V (2010) New control method including state observer of voltage unbalance for grid voltage-source converters. IEEE Trans Ind Electron 57(6): 2054–2065
235. Lee TS (2004) Lagrangian modeling and passivity-based control of three-phase AC/DC voltage-source converters. IEEE Trans Ind Electron 51(4):892–901

236. Leon AE, Solsona JS, Busada C, Chiacchiarini H, Valla MI (2009) High-performance control of a three-phase voltage-source converter including feedforward compensation of the estimated load current. Energy Convers Manag, Elsevier 50:2000–2008
237. Leon AE, Solsona JA, Valla MI (2012) Comparison among nonlinear excitation control strategies used for damping power system oscillations. Energy Convers Manag 53:55–67
238. Leon AE, Mauricio JM, Solsona JA (2012) Multi-machine power system stability improvement. using an observer-based nonlinear controller. Electr Power Syst Res, Elsevier 89: 202–214
239. Levi E (2009) Multiphase electric machines for variable-speed applications. IEEE Trans Ind Electron 55(5):1893–1909
240. Lévine J (2009) Analysis and control of nonlinear systems: a flatness-based approach. Springer
241. Lévine J (2011) On necessary and sufficient conditions for differential flatness. Appl Algebra Eng Commun Comput, Springer 22(1):47–90
242. Li P, Kadirkamanathan V (2001) Particle filtering based likelihood ratio approach to fault diagnosis in nonlinear stochastic systems. IEEE Trans Syst Man Cybern—Part C 31:337–343
243. Li G, Lie TT, Soh CB, Yang GH (2002) Design of state-feedback decentralized H_∞ controllers in power systems. Electr Power Energy Syst, Elsevier 24:601–610
244. Li Q, Chen W, Wang Y (2009) Nonlinear robust control of proton exhange membrane fuel cell by state feedback exact linerization. J Power Sources 124(1):338–348
245. Lidozzi A, Calzo GL, Salevo L, Crescimbini F (2014) Integral-resonant control for standalone voltage source inverters. IET Power Electron 7(2):271–278
246. Lin C (2006) Nonsingular terminal sliding mode control of robot manipulators using fuzzy wavelet networks. IEEE Trans Fuzzy Syst 14:849–859
247. Lin CH, Wub CH, Huang PZ (2009) Grey clustering analysis for incipient fault diagnosis in oil-immersed transformers. Expert Syst Appl, Elsevier 36:1371–1379
248. Lim JS, Park C, Han J, Lee JL (2014) Robust tracking of a three-phase DC-AC inverter for UPS applications. IEEE Trans Ind Electron 61(8):4142–4151
249. Lin WM, Hong CM, Cheng FS (2010) On-line designed hybrid controller with adaptive observer for variable-speed wind generation system. Energy, Elsevier 35:3022–3030
250. Lin FJ, Hung YC, Tsai MT (2013) Fault-tolerant control for six-phase PMSM drive system via intelligent complementary sliding-mode control using TSKFNN-AMF. IEEE Trans Ind Electron 60(12):5747–5762
251. Lin SS, Horng SC (2011) A more general parallel dual-type method and application to state estimation. Electric Power Energy Syst, Elsevier 33:799–804
252. Linares-Flores J, Sira-Ramrez H (2004) DC motor velocity control through a DC-to-DC power converter. In: 43rd IEEE conference on decision and control, Bahamas
253. Linares-Flores J, Hernndez A, Mndez C, Garca-Rodrguez, Sira-Ramrez (2014) Robust nonlinear adaptive control of a boost converter via algebraic parameter identification. IEEE Trans Ind Electron 61(8):4105–4114
254. Liu CC, Jung J, Heydt GT, Vittal V, Phadke AG (2000) The strategic power infrastructure defense (SPID) system: a conceptual design. IEEE Control Syst Mag 20(4):40–52
255. Liu WJ (2003) Boundary stabilization of an unstable heat equation. SIAM J Control Optim 42:1033–1043
256. Liu G, Sun J, Chen Z, Qian P, Liao Z (2013) Suppression of parameters perturbation in active power filter based on sliding-mode control. In: Proceedings of the 32nd Chinese control conference, Xi'an, China
257. Liu J, Laghrouche S, Wack M (2013) Differential flatness-based observer design for a PEM fuel cell using adaptive-gain sliding mode differentiators. In: 2013 European control conference (ECC), Zurich, Switzerland
258. Liu W, Gu W, Sheng W, Meng X, Wu Z, Chen W (2014) Decentralized multi-agent system-based cooperative frequency control for autonomous microgrids with communication constraints. IEEE Trans Sustain Energy 5(2):446–456
259. Liserre M, Teodorescu B, Blaabjerg F (2006) Stability of photovoltaic and wind turbine cross-connected inverters for a large set of grid impedance values. IEEE Trans Power Electron 2(1):263–271

260. Lopez-Luque R, Reca J, Martnez J (2015) Optimal design of a standalone direct pumping photovoltaic system for deficit irrigation of olive orchards. Appl Energy, Elsevier 149:13–23
261. Low KS, Cao R (2008) Model predictive control of parallel-connected inverters for uninterruptible power supplies. IEEE Trans Ind Electron 55(8):2884–2893
262. Lu Q, Mei S, Hu W, Wu FF, Ni Y, Shen T (2001) Nonlinear decentralized disturbance attenuation excitation control via new recursive design for multi-machine power systems. IEEE Trans Power Syst 16(4):729–736
263. Lublin L, Athans M (1995) An experimental comparison of and designs for interferometer testbed. In: Francis B, Tannenbaum A (eds) Lectures notes in control and information sciences: feedback control , nonlinear systems and complexity. Springer, pp 150–172
264. McNamara P, Negenborn R, de Schutter B, Lightbody G (2013) Optimal coordination of a multiple HVDC link system using centralized and distributed control. IEEE Trans Control Syst Technol 21(2):302–314
265. Machowski J, Robak S, Bialek JW, Bumby JR, Abi-Samra N (2000) Decentralized stability-enhancing control of synchronous generator. IEEE Trans Power Syst 15(4):1336–1344
266. Mahmoud MA, Rota HK, Hussain MJ (2012) Full-order nonlinear observer-based excitation controller design for interconnected power systems via exact linearization approach. Electric Power Energy Syst, Elsevier 41:54–62
267. Mahmud MA, Hossain MJ, Pota HR, Roy NK (2014) Robust nonlinear excitation controller design for multimachine power systems. In: IEEE PES general meeting—Conference and exposition, National Harbor, MD, USA
268. Mahmud MA, Pota HR, Hossain MJ (2012) Design nonlinear controller, for single-phase grid-connected photovoltaic systems using partial feedback linearization. In: Australian Control Conference 2012, Sydney, Australia
269. Maidi A, Corriou JP (2014) Distributed control of nonlinear diffusion systems by input-output linearization. Int J Robust Nonlinear Control 26:389–405
270. Majica-Nava E, Macana CA, Quijano N (2014) Dynamic population games for optimal dispatch on hierarchical microgrid control. IEEE Trans Syst Man Cybern 44(1):306–317
271. Malinkowski M, Kazimierkowski MP, Hansen S, Blaabjerg F, Marques GD (2001) Virtual-flux-based direct power control of three-phase PWM rectifiers. IEEE Trans Ind Appl 37(4): 1019–1027
272. Malinkowski M, Jasinksi M, Kazmeirkowski MP (2004) Simple direct power control of three-phase PWM rectifier using space-vector modulation (DPC-SVM). IEEE Trans Ind Electron 51(2):447–454
273. Majumber R, Chaudhuri B, Pal BC, Zhong QC (2005) A unified Smith predictor approach for power system damping control design using remote signals. IEEE Trans Control Syst Technol 13:1063–1068
274. Makarov YV, Voropai NI, Efimov DN (2008) Complex emergency control system against blackout in Russia. In: Proceedings of the IEEE power and energy society general meeting—Conversion and delivery of electrical energy in the 21st Century, Pittsburg, USA, pp 1–8
275. Mansour Y, Vaahedi E, EL-Sharkawi SA (1997) Large scale dynamic security screening and ranking using neural networks. IEEE Trans Power Syst 12:954–960
276. Manyika J, Durrant-Whyte H (1994) Data fusion and sensor management: a decentralized information theoretic approach. Prentice Hall, Englewood Cliffs, NJ
277. Marino R, Peresada S, Valigi P (1993) Adaptive input-output linearizing control of induction motors. IEEE Trans Autom Control 38(2):208–221
278. Marino R, Peresada S, Tomei P (1999) Global adaptive output feedback control of induction motors with uncertain rotor resistance. IEEE Trans Autom Control 44(5):967–983
279. Marques GD, Sousa DM (2012) Stator flux active damping methods for field-oriented doubly fed induction generator. IEEE Trans Energy Convers 27(3):799–806
280. Martin P, Rouchon P (1996) Two remarks on induction motors. In: CESA '96 IMACS multiconference, vol 1, pp 76–79, Lille, France

281. Martin P, Rouchon P (1999) Systèmes plats: planification et suivi des trajectoires. Journées X-UPS, École des Mines de Paris, Centre Automatique et Systèmes, Mai
282. Mathis G (1994) Outils de détection des ruptures et de Diagnostic - Application à la surveillance de Turbines à Gaz'. Université de Rennes I, Thèse de Doctorat
283. Matraji I, Laghrouche S, Jemei S, Wack M (2013) Robust control of the PEM fuel cell air-feed system via sub-optimal second order sliding mode. Appl Energy, Elsevier 104:945–957
284. Mazumder SK, Nayfeh AH, Borojevic D (2002) Robust control of parallel DC-DC buck converters by combining integral-variable-structure and multiple-sliding-surface control schemes. IEEE Trans Power Electron 17(3):428–437
285. Mekri F, Benelghali S, Benbouzid M, Charpentier J-F (2011) A fault-tolerant multiphase permanent magnet generator for marine current turbine applications. In: 2011 IEEE International symposium on industrial electronics, Gdansk, Poland, pp 2079–2084
286. Mekri F, Benelghali S, Benbouzid M (2013) Fault-tolerant control performance comparison of three- and five-phase PMSG for marine current turbine applications. IEEE Trans Sustain Energy 4(2):425–433
287. Mendalek N, Al-Haddad K, Fnaiech F, Dessaint LA (2003) Nonlinear control techique to enhance dynamic performance of a shunt active power filter. IEE Proc Electr Power Appl 150(4):373–379
288. Menhour L, d'Andréa-Novel B, Fliess M, Mounier H (2014) Coupled nonlinear vehicle control: flatness-based setting with algebraic estimation techniques. Control Eng Pract 2:135146
289. Metwally I (2011) Failures monitoring and new trends of power transformers. IEEE Potentials 30:36–43
290. Mielczarski W (1987) Observing the state of a synchronous generator. Part 1. Theory. Int J Control, Taylor and Francis 45(3):987–1000
291. Mielczarski W (1987) Observing the state of a synchronous generator. Part 2. Applications. Int J Control, Taylor and Francis 45(3):1001–1021
292. Miklosovic R, Radke A, Gao Z (2006) Discrete implementation and generalization of the extended state observer. In: Proceedings of the 2006 Americal control conference, Minneapolis, Minnesota, USA
293. Milano F (2006) An open-source power system analysis toolbox'. IEEE Trans Power Syst 20(3):1199–1206
294. Mohammadi A, Asif A (2009) Distributed particle filtering for large scale dynamical systems. In: 13th IEEE INCIM International conference, Islamabad, Pakistan
295. Mohseni M, Islam SM (2012) Transient control of DFIG-based wind power plants in compliance with the Australian grid code. IEEE Trans Power Electron 27(6):2813–2824
296. Monticelli A (2000) Electric power system state estimation. Proc IEEE 88(2):261–282
297. Mounier H, Rudolph J, Wouttenneck F (2010) Boundary value problems and convolutional systems over rings of ultradistributions. Advances in the theory of control, signal and systems with physical modelling. Lecture notes in control an information sciences. Springer, pp 179–188
298. Na W, Gou B, Diong B (2007) Nonlinear control of PEM fuel cells by exact linearization. IEEE Trans Ind Appl 43(6):1426–1433
299. Na W, Gou B (2008) Feedback-linearization-based nonlinear control for PEM fuel cells. IEEE Trans Energy Convers 21(1):179–190
300. Namburu SM, Azam MS, Luo J, Choi K, Pattipati KR (2007) Data-driven modeling, fault diagnosis and optimal sensor selection for HVAC chillers. IEEE Trans Autom Sci Eng 4(3): 469–473
301. Narendra KS, Parthasarathy K (1990) Identification and control of dynamical systems using neural networks. IEEE Trans Neural Netw 1(1):4–27
302. Nazaripouya H, Mehraeen Shahab (2016) Modeling and nonlinear optimal control of weak/islanded grids using FACTS device in a game theoretic approach. IEEE Trans Control Syst Technol 24(1):158–171
303. Nayak N, Routray SK, Rout PK (2012) State feedback robust H_∞ controller for transient stability enhancement of VSC-HVDC transmission systems. Procedia Technol, Elsevier 4: 652–660

304. Negrea AC, Imecs M, Incze II, Pop A, Szabo C (2012) Error compensation methods in speed identification using incremental encoder. In: International conference and exposition on electrical and power engineering (EPE), Iasi, Romania
305. Nelles O, Fischer M (1996) Local linear model trees (LOLIMOT) for linear system identification of a cooling blast. In: Proceedings of the EUFIT 96 conference, Aachen, Germany
306. Nettleton E, Durrant-Whyte H, Sukkarieh S (2003) A robust architecture for decentralized data fusion. In: ICAR03, 11th International conference on advanced robotics, Coimbra, Portugal
307. Niknam T, Firouzi B (2009) A practical algorithm for distribution state estimation including renewable energy sources. Renew Energy, Elsevier 34:23092316
308. Nordman MM, Lehtonen M (2005) Distributed agent-based state estimation for electrical distribution networks. IEEE Trans Power Syst 20(2):652–658
309. Novosel D, Begovic MM, Madani Y (2006) Shedding light on blackouts. IEEE Power Energy Mag 2(1):32–43
310. Ng GW (2003) Intelligent systems—Fusion. Tracking and control. Research Studies Press
311. Nguyen NV, Shin V, Shevlyakov G (2010) Power system state estimation with fusion method. In: 2nd International conference on computer and automation engineering, ICCAE 2010, vol 5, Singapore, 2010, pp 71–76
312. Okatan A, Hajiyev C, Hajiyeva U (2001) Fault detection in sensor information fusion Kalman filter. Int J Electron Commun, Elsevier 47(4):1657–1665
313. Ouachani I, Rabhi A, Tidhaf B, Zouggar S, Elhajjaji A (2013) Optimization and control for a photovoltaic pumping system. In; 2013 International conference on renewable energy research and applications, Madrid, Spain
314. Okou FA, Gauthier S (2010) A novel robust nonlinear control of a three-phase NPC inverter-based active power filter. In: 2010 American control conference, Baltimore, USA
315. Okou F, Dessaint L-A, Akhrif O (2005) Power systems stability enhancement using a wide-area signals based hierarchical controller. IEEE Trans Power Syst 20(3):1465–1477
316. Olivier F, Sedoglavic A (2001) A generalization of flatness to nonlinear systems of partial differential equations: application to the control of a flexible rod. In: Proceedings of the 5th IFAC symposium on nonlinear control systems, Saint-Petersbourg
317. Ovalle A, Ramos G, Bacha S, Hably A, Rameau A (2015) Decentralized control of voltage source converters in microgrids based on the application of instantaneous power theory. IEEE Trans Ind Electron 62(2):1152–1162
318. Park J, Kalev C, Hofmann H (2008) Control of high-speed solid-rotor synchronous reluctance motor/generator for flywheel-based uninterruptible power supplies. IEEE Trans Ind Electron 55(8):3038–3046
319. Pena R, Cardenas R, Reyes E, Clare J, Wheeler P (2011) Control of a doubly fed induction generator via an indirect matrix converter with changing DC voltage. IEEE Trans Ind Electron 58(10):4664–4674
320. Peresada S, Tilli A, Tonielli A (2004) Power control of a doubly fed induction machine via output feedback. Control Eng Pract, Elsevier 12:4157
321. Perez E, Barros J (2008) An extended Kalman filtering approach for detection and analysis of voltage dips in power systems. Electr Power Syst Res, Elsevier 78:618–625
322. Piccolo A, Siano P, Rigatos G (2010) An adaptive framework for power components dynamic loadability. In: Rigatos GG (ed) Modelling and automation of intelligent industrial systems. IGI Publications
323. Pinsky M (1991) Partial differential equations and boundary value problems. Prentice-Hall
324. Planas E, Andrea J, Garate J, Martinez de Alegria I, Iborra E (2015) AC and DC technology in microgrids: a review. Renew Sustain Energy Rev, Elsevier 43:726–749
325. Psichogios DC, Ungar LH (1994) SVD-NET: an algorithm that automatically selects network structure. IEEE Trans Neural Netw 5(3):513–515
326. Refregier A (2003) Shapelets—I. A method for image analysis. Mon Not R Astron Soc 338:35–47
327. Raffo GV, Ortega MG, Rubio FR (2011) Nonlinear H-infinity controller for the quad-rotor helicopter with input coupling. In: 18th World congress of the IFAC, 2011, Milan, Italy

328. Raffo GV, Ortega MG, Rubio FR (2011) Path tracking of a UAV via an underactuated H-infinity control strategy. Eur J Control, Elsevier 17:194–213
329. Rahmani S, Hamadi A, A-Haddad K (2012) A Lyapunov function-based control for a three-phase shunt active filter. IEEE Trans Ind Electron 59(3):1418–1429
330. Rahman S, Hamadi A, Al-Haddad K, Dessaint LA (2014) A combination of shunt hybrid power filter and thyristor-controlled reactor for power quality. IEEE Trans Ind Electron 61(5):2152–2164
331. Ramadan HS, Siguerdidjane H, Petit M (2009) A robust stabilizing nonlinear control design for VSC-HVDC systems: a comparative study. In: IEEE International conference of industrial technology, ICIT09, Australia, p 16
332. Ramos-Paja C, Giral R, Martinez-Salamero L, Romano J, Romero A, Spagnuolo G (2010) A PEM fuel-cell model featuring oxygen-excess-ratio estimation and power-electronics interaction. IEEE Trans Ind Electron 57(6):1914–1924
333. Ren H, Dobson I (2008) Using transmission line outage data to estimate cascading failure propagation in an electric power system. IEEE Trans Circuits Syst-II 55(9):927–931
334. Rigatos G, Zhang Q (2001) Fuzzy model validation using the local statistical approach. Publication Interne IRISA No 1417, Rennes, France
335. Rigatos GG, Zhang Q (2002) Fuzzy model validation using the local statistical approach. In: IEEE SMC 02 conference, Hammamet, Tunisia
336. Rigatos GG, Tzafestas SG (2004) Fuzzy automata for fault diagnosis: a syntactic analysis approach. Lecture notes in artificial intelligence, vol 3025. Springer
337. Rigatos GG, Tzafestas SG (2006) Adaptive fuzzy control for the ship steering problem. J Mechatron, Elsevier 16(6):479–489
338. Rigatos G, Tzafestas S (2006) Feed-forward neural networks using Hermite polynomial activation functions. Lecture notes in artificial intelligence, vol 3925. Springer, pp 323–333
339. Rigatos GG, Tzafestas SG (2006) Neural structures using the eigenstates of the quantum harmonic oscillator. Open Syst Inf Dyn, Springer 13(1):27–41
340. Rigatos GG, Tzafestas SG (2007) H_∞ tracking of uncertain SISO nonlinear systems: an observer-based adaptive fuzzy approach, Intl. J Syst Sci, Taylor and Francis 38:459–472
341. Rigatos GG, Tzafestas SG (2007) Extended kalman filtering for fuzzy modelling and multi-sensor fusion. Math Comput Modell Dyn Syst, Taylor & Francis 13:251–266
342. Rigatos GG (2008) Particle filtering for state estimation in industrial robotic systems. IMeche J Syst Control Eng 222(6):437–455
343. Rigatos GG (2009) Particle filtering for state estimation in nonlinear industrial systems. IEEE Trans Instrum Meas 58(11):3885–3901
344. Rigatos G, Zhang Q (2009) Fuzzy model validation using the local statistical approach. Fuzzy Sets Syst, Elsevier 60(7):882–904
345. Rigatos GG, Siano P, Piccolo A (2009) A neural network-based approach for early detection of cascading events in electric power systems. IET J Gener Transm Distrib 3(7):650–665
346. Rigatos GG (2009) Adaptive fuzzy control of DC motors using state and output feedback. Electr Power Syst Res, Elsevier 79(11):1579–1592
347. Rigatos GG (2009) Particle and Kalman filtering for fault diagnosis in DC motors. In: IEEE VPPC 2009, IEEE VPPC 2009, IEEE 5th Vehicle power propulsion conference, Michigan, USA
348. Rigatos GG (2010) A derivative-free Kalman Filtering approach for sensorless control of nonlinear systems. In: IEEE ISIE 2010. IEEE International symposium on industrial electronics, Bari, Italy
349. Rigatos G, Siano P (2011) Design of robust electric power system stabilizer using Kharitonov's theorem. Math Comput Simul, Elsevier 82(1):181–191
350. Rigatos GG, Siano P (2011) Distributed state estimation for condition monitoring of nonlinear electric power systems. In: IEEE ISIE 2011. International symposium on industrial electronics, Gdansk, Poland
351. Rigatos GG (2011) A derivative-free distributed filtering approach for sensorless control of nonlinear systems. Int J Syst Sci, Taylor and Francis

352. Rigatos GG (2011) Modelling and control for intelligent industrial systems. Adaptive algorithms in robotics and industrial engineering. Springer
353. Rigatos GG (2011) Adaptive fuzzy control for field-oriented induction motor drives. Neural computing and applications. Springer
354. Rigatos GG (2011) Flatness-based adaptive fuzzy control for nonlinear dynamical systems. In: AIM 2011, IEEE/ASME, International conference on advanced intelligent mechatronics, Budapest, Hungary
355. Rigatos GG (2012) Nonlinear Kalman filters and particle filters for integrated navigation of unmanned aerial vehicles. Robot Auton Syst, Elsevier
356. Rigatos GG (2012) Distributed filtering over sensor networks for autonomous navigation of UAVs. Intell Serv Robot, Springer 5(3):179–198
357. Rigatos G, Siano P (2012) DFIG control using Differential flatness theory and extended Kalman filtering. In: IFAC INCOM 2012, 14th IFAC International conference on information control problems in manufacturing, Bucharest, Romania
358. Rigatos GG (2012) A derivative-free Kalman filtering approach to state estimation-based control of nonlinear systems. IEEE Trans Ind Electron 59(10):3987–3997
359. Rigatos G, Siano P, Zervos N (2012) Derivative-free nonlinear Kalman filtering for PMSG sensorless control. In: Habib(ed) Mechatronics engineering: research development and education. Wiley
360. Rigatos G, Al-Khazraji A (2011) Flatness-based adaptive fuzzy control for MIMO nonlinear dynamical systems. In: Nonlinear estimation and applications to industrial systems control. Nova Publications
361. Rigatos GG (2012) Derivative-free nonlinear Kalman Filtering for MIMO dynamical systems: application to multi-DOF robotic manipulators. J Adv Robot Syst (special issue on Robot Manipulators)
362. Rigatos G, Siano P (2012) Sensorless control of electric motors with Kalman Filters: applications to robotic and industrial systems. Int J Adv Robot Syst (Special Issue Robot Manipulators)8:62–80
363. Rigatos G, Siano P, Zervos N (2012) A distributed state estimation approach to condition monitoring of nonlinear electric power systems. Asian J Control, Wiley
364. Rigatos G, Siano P (2012) Sensorless nonlinear control of induction motors using Unscented Kalman filtering. In: Proceedings of the IEEE IECON, 38th International conference of the industrial electronics society, Montreal, Canada
365. Rigatos G, Siano P, Piccolo A (2012) Incipient fault detection for electric power transformers using neural modeling and the local statistical approach to fault diagnosis. In: IEEE SAS 2012, 2012 IEEE Sensors applications symposium. University of Brescia, Italy
366. Rigatos G (2013) Advanced models of neural networks: nonlinear dynamics and stochasticity in biological neurons. Springer
367. Rigatos G, Siano P, Zervos N (2013) PMSG sensorless control with the Derivative-free nonlinear Kalman filter for distributed generation units. In: 11th IFAC International workshop on adaptation and learning in control and signal processing (ALCOSP'2013), Caen, France
368. Rigatos G, Siano P, Zervos N (2013) PMSG sensorless control with the use of the derivative-free nonlinear Kalman filter. In: IEEE ICCEP 2013, IEEE International conference on clean electrical power, Alghero, Sardinia Italy
369. Rigatos G, Siano P, Zervos N (2013) Control and disturbances compensation for doubly-fed induction generators using the derivative-free nonlinear Kalman filter. In: IEEE IECON 2013, 39th IEEE Conference on industrial electronics, Vienna, Austria
370. Rigatos G, Siano P, Zervos N, Cecati C (2013) Derivative-free nonlinear Kalman Filtering for control of three-phase voltage source converters. In: IEEE IECON 2013, 39th IEEE Industrial electronics conference, Austria, Vienna
371. Rigatos G (2013) Derivative-free distributed filtering for MIMO robotic systems under delays and packet drops. Int J Adv Robot Syst, Wang Y-T (ed), Intech Publications, doi:10.5772/54186

372. Rigatos GG (2014) A differential flatness theory approach to observer-based adaptive fuzzy control of MIMO nonlinear dynamical systems. Nonlinear Dyn, Springer 76(2):1335–1354
373. Rigatos G, Siano P (2014) An H-infinity feedback control approach to autonomous robot navigation. In: IEEE IECON 2014, Dallas, Texas
374. Rigatos G, Siano P, Zervos N (2014) Sensorless control of distributed power generators with the derivative-free nonlinear Kalman filter. IEEE Trans Ind Electron
375. Rigatos G (2015) Nonlinear control and filtering using differential flatness theory approaches: applications to electromechanical systems. Springer
376. Rigatos G, Siano P, Wira P, Profumo F (2015) Nonlinear H-infinity feedback control for asynchronous motors of electric trains. J Intell Ind Syst, Springer 1(3)
377. Rigatos G, Siano P, Zervos N, Cecati C (2015) Control and disturbances compensation for doubly-fed induction generators using the derivative-free nonlinear Kalman filter. IEEE Trans Power Electron
378. Rigatos G, Siano P, Zervos N, Cecati C (2015) Decentralized control of parallel inverters connected to microgrid using the Derivative-free nonlinear Kalman filter. IET Power Electron
379. Rigatos G, Siano P (2015) A nonlinear H-infinity feedback control approach for asynchronous generators. In: IEEE ICCEP 2015, 5th International conference on clean electrical power, Taormina, Sicily, Italy
380. Rigatos G, Siano P, Cecati C (2015) A new nonlinear H-infinity feedback control approach for three-phase voltage source converters. Electr Power Compoments Syst, Taylor and Francis
381. Rigatos G, Siano P, Zervos N (2015) A new concept on flatness-based control of nonlinear dynamical systems. In: IEEE INDIN 2015, 4th International conference on industrial informatics, Cambridge UK
382. Rigatos G, Siano P (2016) Power transformers condition monitoring using neural modelling and the local statistical approach to fault diagnosis. Electr Power Energy Syst, Elsevier 80: 150–159
383. Robak S (2009) Robust SVC controller design and analysis for uncertain power systems. Control Eng Pract, Elsevier 17:1280–1290
384. Rodriguez-Cortes H, Stankovic A, Hadjicostis CN (2006) Model-based fault detection in power systems. In: International symposium on industrial electronics, ISIE 2006, Montreal 2006, pp 1922–1927
385. Rodutz S, Paganelli G, Guzzella L (2003) Optimizing air supply control of a PEM fuel cells system. In: Proceedings of the IEEE American control conference, vol 3, pp 2043–2048
386. Roldon-Pérez J, Garcia-Cerrada A, Zamara-Macho JL, Ochea-Giménez M (2014) Helping all generations of photovoltaic inverters ride-through voltage sags. IET Power Electron 7(10): 2555–2563
387. Rouchon P (2005) Flatness-based control of oscillators. J Appl Math Mech, Wiley 85(6): 411–421
388. Rudolph J (2003) Flatness Based Control of Distributed Parameter Systems. Aachen, Steuerungs- und Regelungstechnik. Shaker Verlag
389. Salehifar M, Moreno-Eguilaz JM, Salehi R, Sala V, Romeral M, Jose L (2014) Observer-based open transistor fault diagnosis and fault-tolerant control of five-phase PM motor drive for application in electric vehicles. IET Power Electron
390. Sanchez-Gasca JJ, Chow JH (1999) Performance comparison of three identification methods for the analysis of electromechanical oscillation. IEEE Trans Power Syst 14:995–1002
391. Salahshoor K, Mosallaei M, Bayat M (2008) Centralized and decentralized process and sensor fault monitoring using data fusion based on adaptive Extended Kalman Filter algorithm. Measurement, Elsevier 41:1059–1076
392. Shafiee Q, Stefanovic C, Draginevic T, Popovskii P, Vasquez J, Guerrero JM (2014) Robust networked control scheme for distributed secondary control of islanded microgrids. IEEE Trans Ind Electron 61(10):5363–5374
393. Shafiee Q, Vasquez J, Guerrero JM (2012) Distributed secondary control for islanded microgrids: a networked control systems approach. In: IEEE IECON, 2012 38th Annual conference of the IEEE industrial electronics society, Montréal, Canada

394. Schiffer J, Ortega R, Astolfi A, Raisch J, Sezi T (2014) Conditions for stability of droop-controlled inverter-based microgrids. Automatica, Elsevier 50(10):2457–2469
395. Schmuck C, Woittennek F, Gension A, Rudolph J (2014) Feedforward control of the HVDC power transmission network. IEEE Trans Control Syst Technol 22(2):597–605
396. Schiffer J, Goldin D, Raisch J, Sezi T (2013) Synchronization of droop-controlled microgrids with distributed rotational and electronic generation. In: IEEE Conference on decision and control CDC 2013, Firenze, Italy, pp 2334–2339
397. Schiffer J, Ortega R, Astolfi A, Raisch J, Sezi T (2014) Conditions for stability of droop-controlled inverter-based microgrids. Automatica, Elsevier 50(10):2457–2469
398. Schiffer J, Ortega R, Astolfi A, Raisch A, Sezi T (2014) Stability of synchronized motions of inverter-based microgrids under droop control. In: 19th IFAC World congress, Cape Town, South Africa
399. Schultz M, Horn J (2012) PEM fuel cell system power control based on a feedback-linearization approach. In: IEEE MED 2012, Meditteranean conference on control and automation, Barcelona, Spain
400. Scuiller F, Charpentier JF, Semail E (2010) Multi-star multi-phase winding for a high power naval propulsion machine with low ripple torques and high fault tolerant ability. In: IEEE VPPC'10, IEEE International vehicular power propulsion conference, France, pp 1–5
401. Setnes M, Babuska R (2000) Fuzzy modeling for predictive control. In: Farinwata F, Langari (eds) Fuzzy control—Synthesis and analysis. Wiley, pp 23–46
402. Setnes M, Babuska R, Kaymak U, van Nauta HR (1998) Lemke, Similarity measures in fuzzy rule base simplification. IEEE Trans Syst, Man Cybern-Part B Cybern 28(3):376–386
403. Setnes M, Kaymak U (1998) Extended fuzzy c-means with volume prototypes and cluster merging. In: Proceedings of the EUFIT'98, Aachen, Germany
404. Shafiu A, Jenkins N, Strbac G (2005) Measurement location for state estimation of distribution networks with generation. IEE Proc: Gener, Transm Distrib 152(2):240–246
405. Shya KK, Yang MJ, Chen YM, Lin YF (2008) Model reference adaptive control design for a shunt active-power-filter design. IEEE Trans Ind Electron 55(1):97–106
406. Silva-Ortigoza R, Hernandez-Guzman VM, Antonio-Cruz M, Munoz-Carrillo D (2015) DC-DC buck power converter as a smooth starter for a DC motor based on a hierarchical control. IEEE Trans Power Electron 30(2):1076–1084
407. Singh B, Niwas R (2016) Performance of synchronous reluctance generator for DG set based stand alone supply system. Electr Power Syst Res, Elsevier 133:93103
408. Simon D (2006) A game theory approach to constrained minimax state estimation. IEEE Trans Signal Process 54(2):405–412
409. Singh B, Niwas R (2016) Performance of synchronous reluctance generator for DG set based stand alone supply system. Electr Power Syst Res, Elsevier 133:93103
410. Sira-Ramirez H (1999) Flatness and trajectory tracking in sliding mode based regulation of DC-to-AC conversion schemes. In: Proceedings of the 38th conference on decision & control phoenix, Arizona USA
411. Sira-Ramirez H, Agrawal S (2004) Differentially flat systems. Marcel Dekker, New York
412. Sira-Ramirez H, Fliess M (2004) On the output feedback control of a synchronous generator. In: 43rd IEEE Conference decision and control, Bahamas
413. Shang L, Sun D, Hu J (2011) Sliding mode-based direct power control of grid-connected voltage-sourced inverters under unbalanced conditions. IET Power Electron 4(5):570–579
414. Simon D (2002) Training fuzzy systems with the extended Kalman filter. Fuzzy Sets Syst, Elsevier 132:189–199
415. Simon D (2002) Sum normal optimization of fuzzy membership functions. Int J Uncertain, Fuzziness Knowl-Based Syst 10:363–384
416. Smith K, Rahn C, Wang CY (2010) Model-based electrochemical estimation and constraint management for pulse operation of Litium-ion batteris. IEEE Trans Control Syst Technol 10(18):654–665
417. Smith KA, Rahn CD, Wang CY (2007) Control oriented 1D electrochemical model of lithium-ion battery. Energy Convers Manag, Elsevier 48:2565–2578

418. Smith K (2010) Electrochemical control of lithium-ion batteries. IEEE Control Syst Mag: 18–25
419. Smyshlyaev A, Krstic M (2010) Adaptive control of parabolic PDEs. Princeton University Press
420. Sobczuk D (2006) Feedback linearization control of inverter fed induction motor—with sliding mode speed and flux observers. In: IECON 2006, 32nd Annual conference on IEEE industrial electronics, Paris, France
421. Rigatos G, Soldatos E (2014) Structural condition monitoring with the use of the derivative-free nonlinear Kalman filter, In: Habib M (ed) Handbook on research in robotics and mechatronics. IGI Publications
422. Soliman HM, Elshafei AL, Shaltout AA, Morsi MF (2000) Robust power system stabilizer. IEE Proc Electr Power Appl 147:285–291
423. Song S, Kezunovic M (2007) A new analysis method for early detection and prevention of cascading events. Electr Power Syst Res 77:1132–1142
424. Song E, Lynch AF, Dinavahi V (2007) Experimental validation of a flatness-based control for a voltage source converter. In: Proceedings of the 2007 American control conference, New York, USA
425. Staudt T (2015) Brushless doubly-fed reluctance machine modeling, design and optimization. Thèse de Doctorat, Universit Grenoble Alpes et lUniversit Fdrale de Santa Catarina, Grenoble
426. Stout WL, Sawan ME (1992) Application of H-infinity theory to robot manipulator control. In: First IEEE Conference on control applications, Dayton, Ohio, USA
427. Sureshbabu N, Farell J (1999) Wavelet-based system identification for nonlinear control. IEEE Trans Autom Control 44:412–417
428. Susa D, Nordman H (2005) Dynamic thermal modelling of power transformers. IEEE Trans Power Deliv 20(1):197–204
429. Susa D, Nordman H (2006) Dynamic thermal modelling of power transformers—Further development: Part I. IEEE Trans Power Deliv 21(4):1961–1970
430. Susa D, Nordman H (2006) Dynamic thermal modelling of power transformers—Further development: part II. IEEE Trans Power Deliv 21(4):1971–1980
431. Takagi T, Sugeno M (1985) Fuzzy identification of systems and its applications to modeling and control. IEEE Trans Syst, Man Cybern 15:116–132
432. Talj R, Hissel D, Ortega R, Becherif M, Hilairet M (2010) Experimental validation of a PEM fuel-cell reduced-order model and a moto-compressor higher order sliding-mode control. IEEE Trans Ind Electron 57(6):1906–1913
433. Tan JC, Crossley PA, McLaren PG (2002) Application of a wide area backup protection expert system to prevent cascading outages. IEEE Trans Power Deliv 17(2):375–380
434. Tan N, Kaya I, Yeroglu C, Atherton DP (2006) Computation of stabilizing PI and PID controllers using the stability boundary locus. Energy Convers Manag, Elsevier 47:3045–3058
435. Teeuwsen SP, Erlich I, Fischer A, El-Sharkawi NA (2001) Assessment of the small signal stability of the European interconnected electric power system using neural networks. In: Proceedings of the large engineering systems conference on power engineering, pp 158–161
436. Teeuwsen SP, Erlich I, El-Sharkawi NA (2003) Small-signal stability assessment based on advanced neural network methods. In: IEEE power engineering society general meeting, vol 4, pp 13–17
437. Thao NGM, Uchida K (2013) Control the active and reactive powers of three-phase grid-connected photovoltaic inverters using feedback linearization and fuzzy logic. In: 2013 Australian control conference, Perth Australia
438. Thounthong P, Pierfederici S (2010) A new control law based on the differential flatness principle for multiphase interleaved DC-DC converter. IEEE Trans Circuits Syst II: Express Briefs 57(11):903–907
439. Tian G Gao Z (2007) Frequency response analysis of active disturbance rejection based control system. In: 16th IEEE International conference on control applications, Singapore
440. Tleis N (2008) Power systems modelling and fault analysis: theory and Practice. Elsevier, Oxford, UK

441. Tong S, Li H-X, Chen G (2004) Adaptive fuzzy decentralized control for a class of large-scale nonlinear systems. IEEE Trans Syst, Man Cybern-Part B: Cybern 34(1):770–775
442. Toscano R, Lyonnet P (2010) Robust static output feedback controller synthesis using Kharitonov's theorem and evolutionary algorithms. Inf Sci, Elsevier 180:2023–2028
443. Toussaint G, Başar T, Bullo F (2000) H_∞ optimal tracking control techniques for nonlinear underactuated systems. In: Proceedings of the 39th IEEE conference on decision and control, Sydney, Australia
444. Tsai MT, Tsai WI (1999) Analysis and design of three-phase AC-to-DC converters with high power factor and near-optimum feedforward control. IEEE Trans Ind Electron 46(3):535–543
445. Tzafestas SG, Rigatos GG (2000) Self-tuning multivariable fuzzy and neural control using genetic algorithms. J Inf Optim Sci, IOS Press 21(2):257–287
446. Tzafestas SG, Vagelatos GA, Rigatos GG (2001) Design of fuzzy-gain scheduled robust controllers using Kharitonov's extremal gain margin theorem. J Intell Fuzzy Syst 10:39–56
447. Tzafestas SG, Skoundrianos EN, Rigatos GG (2000) Nonlinear neural control of discrete—Time systems using local model networks. Knowl-Based Intell Eng Syst, IOS Press 4
448. Utz T, Meurer T, Kugi A (2011) Trajectory planning for two-dimensional quasi-linear parabolic PDE based on finite diffference semi-discretization. In: 18th IFAC World congress, Milano, Italy
449. Valensiaga F (2010) Second order sliding power control for a variable speed-constant frequency energy conversion system. Energy Convers Manag, Elsevier 51:3000–3008
450. Valverde L, Bordons C, Rosa F (2015) Integration of fuel cell technologies in renewable energy-based microgrids optimizing operational cost and durability. IEEE Trans Ind Electron
451. Variani M, Tomsovic K (2015) Two-level control of doubly-fed induction generator using flatness-based approach. IEEE Trans Power Syst 31(1):518–525
452. Velasquez-Contreras J, Sanz-Bobi M, Arellano S (2011) General asset management model in the context of an electric utility: application to power transformers. Electr Power Syst Res, Elsevier 81:2015–2037
453. Venayagamoorthy GK, Harley RG (1999) Implementation of an adaptive neural network identifier for effective control of turbogenerators. In: Proceedings of the IEEE Budapest powertech conference, paper BPT99-431-23
454. Venayagamoorthy GK, Harley RG (2001) A continually online trained neurocontroller for excitation and turbine control of a turbogenerator. IEEE Trans Energy Convers 16(3):261–269
455. Vercauteren T, Wang X (2005) Decentralized sigma-point information filters for target tracking in collaborative sensor networks. IEEE Trans Signal Process 53(8):2997–3009
456. Villagra J, d'Andrea-Novel B, Mounier H, Pengov M (2007) Flatness-based vehicle steering control strategy with SDRE feedback gains tuned via a sensitivity approach. IEEE Trans Control Syst Technol 15:554–565
457. Wai RJ, Wang WH, Lin CY (2008) High-performance stand-alone photovoltaic generation system. IEEE Trans Ind Electron 55(1):240–250
458. Wang W, Beddard A, Barnes M, Marjanovic O (2014) Analysis of active power control for VSC-HVDC. IEEE Trans Power Deliv 29(4):1978–1988
459. Wei MY, Liu TH (2013) Design and implementation of an online tuning adaptive controller for synchronous reluctance motor drives. IEEE Trans Ind Electron 60(9):3644–3657
460. Wan X, Fang H (2011) Fault detection for a class of discrete-time nonlinear systems subject to network-induced uncertainties. Asian J Control Wiley. doi:10.1002/asjc.445
461. Wang LX (1998) A course in fuzzy systems and control. Prentice-Hall
462. Wang L, Truong D (2013) Stability enhancement of a power system with a PMSG-Based and a DFIG-based offshore wind farm using a SVC with an adaptive-network-based fuzzy inference system. IEEE Trans Ind Electron 60(7):2799–2807
463. Wang X, Blaabjerg F, Wu W (2014) Modelling and analysis of harmonic stability in an AC power-electronics-based system. IEEE Trans Power Electron 29(12):6421–6432
464. Wang Y, Cheng D, Li C, Ge Y (2003) Dissipative hamiltonian realization and energy-based-disturbance attenuation control of multimachine power systems. IEEE Trans Autom Control 48(8):1428–1433

465. Watson NR (2010) Power quality state estimation. Eur Trans Electr Power, Wiley 20:19–33
466. Wilson DH (2004) Managing oscillatory stability using online dynamics measurements. In: Proceedings of the IEEE Power systems conference and exposition, vol 2, pp 1101–1108
467. Wiltshire RA, Ledwich G, OShea P (2007) A Kalman Filterign approach to rapidly detecting modal changes in power systems. IEEE Trans Power Syst 22(4):1698–1706
468. Winkler FJ, Lohmann B (2010) Design of a decoupling controller structure for first order hyperbolic PDEs with distributed control action. In: 2010 American control conference, Baltimore, MD USA
469. Winkler F, Krause I, Lohmann B (2009) Flatness-based control of a continuous furnace. In: 18th International conference on control applications, Part of, IEEE Multi-conference on systems and control, Saint Petersburg, Russia
470. Woitteneck F, Mounier H (2010) Controllability of networks of spatially one-dimensional second order PDEs—an algebraic approach. SIAM J Control Optim 48(6):3882–3902
471. Wu B, Lang Y, Zargari N, Kouro S (2011) Power conversion and control of wind energy systems. Wiley
472. Wu F, Zhang XP, Ju P, Sterling MJH (2008) Decentralized nonlinear control of wind turbine with doubly-fed induction generator. IEEE Trans Power Syst 23(2):613–621
473. Wu D, Chen K (2009) Design and analysis of precision active disturbance rejection control for nonsingular turning process. IEEE Trans Ind Electron 56(7):2746–2753
474. Wu W, He Y, Blaabjerg F (2013) A new design method for the passive damped LCL and LLCL filter-based single-phase grid-tied inverters. IEEE Trans Ind Electron 60(10):4339–4350
475. Xi Z, Feng G, Cheng D, Lu Q (2003) Nonlinear decentralized saturated controller design for power systems. IEEE Trans Control Syst Technol 11(4):539–547
476. Xia Z, Chen D, Lu Q, Mei S (2002) Nonlinear decentralized controller design for multimachine power systems using Hamiltonian function method. Automatica, Elsevier 38:527–534
477. Xia Y, Shang J, Chen J, Liu GP (2009) Networked data fusion with packet losses and variable delays. IEEE Trans Syst, Man Cybern-Part B: Cybern 39(5):1107–1119
478. Xie X, Cui W, Han Y (2005) Simultaneously tuning decentralized nonlinear optimal excitation controllers in multimachine power systems. Electr Power Syst Res, Elsevier 74:371378
479. Xiong J (2008) An introduction to stochastic filtering theory. Oxford University Press
480. Xu H, Hu J, He Y (2012) Operation of wind-turbine-driven DFIG systems under distorted grid voltage conditions: analysis and experimental validations. IEEE Trans Power Electron 27(5):2354–2366
481. Xue Y, Tai N (2012) System frequency regulation in doubly fed induction generators. Electr Power Syst Res 43:977983
482. Yadaiah N, Venkata Ramara N (2007) Linearization of multi-machine power system: modelling and control—A survey. Electr Power Syst Res 29:297–311
483. Yamashita M, Fujimori K, Hakayawa K, Kimura H (1994) Application of H_∞ control to active suspension systems. Automatica, Elsevier 30(11):1717–1729
484. Yamashita K, Joo SK, Li J, Zhang P, Liu CC (2008) Analysis, control and economic impact assessment of major blackout events. Eur Trans Electr Power, Wiley 18:854–871
485. Yang S, Ajjarapu V (2010) A speed-adaptive reduced-order observer for sensorless vector control of doubly-fed induction generator-based variable-speed wind turbines. IEEE Trans Energy Convers 25(3):891–900
486. Yang S, Cheng C (1996) An orthogonal neural network for function approximation. IEEE Trans Syst, Man Cybern-Part B: Cybern 26:779–784
487. Yang S, Lei Q, Peng FZ, Qian Z (2011) A robust control scheme for grid-connected voltage-source inverters. IEEE Trans Ind Electron 58(1):202–212
488. Yao J, Li H, Chen Z, Xia X, Li XQ, Liao Y (2013) Enhanced control of a DFIG-based wind-power generation system with series grid-side converter under unbalanced grid voltage conditions. IEEE Trans Power Electron 28(7):3167–3181
489. Yao J, Li H, Chen Z, Xia X (2013) Enhanced control of a DFIG-Based wind-power generation system with series grid-side converter under unbalanced grid voltage conditions. IEEE Trans Power Electron 28(7):3167–3181

490. Yen J, Wang L (1999) Simplifying fuzzy rule-based models using orthogonal transformation methods. IEEE Trans Syst, Man Cybern-Part B: Cybern 29:13–24
491. Ying H, Zheng X, Wulue P (2005) A practical analysis method of low frequency oscillation for large power systems. In: Proceedings of the Power engineering society general meeting, vol 2, pp 1623–1629
492. Yun SY, Kim JC (2003) An evaluation method of voltage sag using a risk assessment model in power distribution systems. Electr Power Energy Syst 25:829839
493. Yutian L, Chi X, Yuanyuan S (2005) Real-time phasor measurement for low frequency oscillation in power system. In: Proceedings of the IEEE International symposium on circuits and systems, vol 4, pp 3889–3893
494. Zamani MA, Sidha TS, Yazdani A (2014) Investigations into the control and protection of existing distribution network to operate as a microgrid: A case study. IEEE Trans Ind Electron 61(4):1904–1915
495. Zanetti D, Ortega R, Benchaib A (2014) Modelling and control of high-voltage direct current transmission systems: From theory to practice and back. Elsevier, Automatica
496. Zhang Q (1997) Fault detection and isolation with nonlinear black-box models. In: Proceedings of the SYSID 1997, Kitakyushu, Japan
497. Zhang Q, Benveniste A (1993) Wavelet networks. IEEE Trans Neural Netw 3(6):869–898
498. Zhang Q, Basseville M, Benveniste A (1994) Early warning of slight changes in systems. Automatica, Elsevier (Special Issue on Statistical Methods in Signal Processing and Control) 30(1):95–113
499. Zhang Q, Basseville M, Benveniste A (1993) Fault detection and isolation in nonlinear dynamic systems: a combined input-output and local approach. Automatica, Elsevier 34(11):1359–1373
500. Zhang S, Xie X, Wu J (2008) WAMS-based detection and early-warning of low-frequency oscillations in large-scale power systems. Electr Power Energy Syst, Elsevier 78(5):897–906
501. Zhang X, Foo G, Vilathgamuwa D, Maskell D (2015) An improved robust field-weakeaning algorithm for direct-torque-controlled synchronous-reluctance-motor drives. IEEE Trans Ind Electron 62(5):3255–3264
502. Zhang Y, Ma H (2012) Theoretical and experimental investigation of networked control for parallel operation of inverters. IEEE Trans Ind Electron 59(4):1961–1970
503. Zhao C, Low S (2014) Control optimal decentralized primary frequency, in power networks. In: 53rd IEEE Conference on decision and control, 15–17 Dec 2014, Los Angeles, California, USA
504. Zhong QC (2013) Robust droop controller for accurate proportional load sharing among inverters operated in parallel. IEEE Trans Ind Electron 60(4):1281–1290
505. Zhou Z, Wang C, Liu Y, Holland PM, Igic P (2012) Load current observer based feed-forward DC bus voltage control for active rectifiers. Electr Power Syst Res, Elsevier 84:165–0173
506. Zou C, Mansic C, Anway S (2014) Control-oriented modelling of a lithium-ion battery for fast charging. In: 19th IFAC World Congress, pp 3912–3917, Cape Town, South Africa
507. Zhang Q, Basseville M, Benveniste A (1998) Fault detection and isolation in nonlinear dynamic systems: a combined input-output and local approach. Automatica, Elsevier 34(1):1359–1373
508. Zhang Q, Basseville M, Benveniste A (1994) Early warning of slight changes in systems. Automatica (special issue on Statistical Methods in Signal Processing and Control)30:95113
509. Zhao C, Topcu U, Li N, Low S (2014) Design and stability for load-side primary frequency control in power ystems. IEEE Trans Autom Control 59(5):1177–1189
510. Zhong QC, Nguyen FL, Ma Z, Sheng W (2014) Self-synchronized synchronverters: inverters without a dedicated synchronization unit. IEEE Trans Power Electron 29(2):617–629
511. Zhong QC (2013) Robust droop controller for accurate proportional load sharing among inverters operated in parallel. IEEE Trans Ind Electron 60(4):1281–1290
512. Zhong QC, Hornik J (2013) Control of power inverters in renewable energy and smart grid integration. Wiley, New York

513. Zhong QC, Weiss G (2009) Static synchronous generators for distributed generation and renewable energy. In: Proceedings of the IEEE PES systems control conference exhibition, pp 1–6
514. Zhong QC, Weiss G (2011) Synchronverters: inverters that mimic synchronous generators. IEEE Trans Ind Electron 58(4):1259–1267
515. Zhu G, Lévine J, Praly L, Peter YA (2006) Flatness-based control of electrostatically actuated MEMS with application to adaptive optics: a simulation study. IEEE J Microelectromech Syst 15(5):1165–1174
516. Zribi M, Mahmoud MS, Karkoub M, Lie TT (2000) H-infinity controllers for linearised time-delay power systems. IEE Proc Gen Transm Distrib 147(3):401–408
517. Zuo W, Zhu Y, Cai L (2009) Fourier-neural-network-based learning control for a class of nonlinear systems with flexible components. IEEE Trans Neural Netw 20:139–151
518. Zurita-Bustamante EW, Linares-Flores J, Guzmn-Ramrez E, Sira-Ramrez H (2011) A comparison between the GPI and PID controllersfor the stabilization of a DCDC buck converter: a field programmable gate array implementation. IEEE Trans Ind Electron 58(11):5251–5263
519. Zwart H, Le Gorrec Y, Maschke B (2011) Linking hyperbolic and parabolic PDEs. In: 2011 50th IEEE Conference on decision and control and European control conference, CDC-ECC, Orlando, Florida, USA

Index

G. Rigatos, *Intelligent Renewable Energy Systems*,
Green Energy and Technology, DOI 10.1007/978-3-319-39156-4

MIX
Papier aus verantwortungsvollen Quellen
Paper from responsible sources
FSC® C105338

If you have any concerns about our products,
you can contact us on
ProductSafety@springernature.com

In case Publisher is established outside the EU,
the EU authorized representative is:
Springer Nature Customer Service Center GmbH
Europaplatz 3, 69115 Heidelberg, Germany

Printed by Libri Plureos GmbH
in Hamburg, Germany